STRUCTURAL ANALYSIS OF HISTORIC BUILDINGS

STRUCTURAL ANALYSIS OF HISTORIC BUILDINGS

Restoration, Preservation, and Adaptive Reuse Applications for Architects and Engineers

J. Stanley Rabun

John Wiley & Sons, Inc.
New York, Chichester, Weinheim, Brisbane, Singapore, Toronto

> A NOTE TO THE READER
> This book has been electronically reproduced from
> digital information stored at John Wiley & Sons, Inc.
> We are pleased that the use of this new technology
> will enable us to keep works of enduring scholarly
> value in print as long as there is a reasonable demand
> for them. The content of this book is identical to
> previous printings.

This book is printed on acid-free paper. ⊗

Copyright © 2000 by John Wiley & Sons. All rights reserved.

Published sumultaneously in Canada.

No part of this publication may be reproduced, stored in a retrieval system or transmitted in any form or by any means, electronic, mechanical, photocopying, recording, scanning or otherwise, except as permitted under Sections 107 or 108 of the 1976 United States Copyright Act, without either the prior written permission of the Publisher, or authorization through payment of the appropriate per-copy fee to the Copyright Clearance Center, 222 Rosewood Drive, Danvers, MA 01923, (978) 750-8400, fax (978) 750-4470. Requests to the Publisher for permission should be addressed to the Permissions Department, John Wiley & Sons, Inc., 111 River Street, Hoboken, NJ 07030, (201) 748-6011, fax (201) 748-6008.

This publication is designed to provide accurate and authoritative information in regard to the subject matter covered. It is sold with the understanding that the publisher is not engaged in rendering professional services. If professional advice or other expert assistance is required, the services of a competent professional person should be sought.

Library of Congress Cataloging-in-Publication Data:
Rabun, J. Stanley.
    Structural analysis of historic buildings : restoration,
preservation, and adaptive reuse applications for achitects and
engineers / J. Stanley Rabun.
       p.   cm.
    "Published simultaneously in Canada."
    Revision of the author's thesis (doctoral)—Institute of Advanced
Architectural Studies, York, England.
    Includes index.
    ISBN 0-471-31545-1 (alk. paper)
    1. Structural analysis (Engineering) 2. Historic buildings—
Conservation and restoration. 3. Historic buildings—Remodeling
for other use. 4. Structural analysis (Engineering) 5. Historic
buildings—Conservation and restoration. I. Title.
TA645.R32 2000
690'.24—dc21
                                                                                                        99-36214

Printed in the United States of America.

10 9 8 7 6 5

Contents

Preface vii

1 Assessment Methodology: Material Chronology, Early Building Laws, and Loads 1

2 Foundation Systems of American Historic Buildings 53

3 Historic American Building Systems: Walls and Columns 117

4 Historic American Floor Systems—Beams 257

5 Historic American Roof Systems: Lateral Bracing of Buildings 387

6 The Historic Material Assessment 453

Bibliography 493

Index 497

PREFACE

The need for a reference of this type became apparent over 25 years ago when I was involved in a situation where an evaluation of in situ structural components was needed in order to "certify" a historic building for a new occupant use. The existing fabric was in excellent condition and the wood beams and joists were fairly easy to evaluate, as were the masonry walls. We had to make a lot of assumptions concerning allowable stresses as the data on the period materials were not known. However, very conservative approaches provided the necessary load capacity and we felt secure. The cast iron columns along the center of the building proved to be another matter. We literally spent weeks searching for data on cast iron columns, load tables, or another method of determining the capacity of these columns, with no results. Finally, after we had contacted every library and university we felt might contain technical information, and many engineers and architects who might have had need for the same data, it become apparent that this was an area where no resources could be found. In fact, everyone in the field was in about the same situation, guessing allowable stresses and approaching everything with extreme conservatism. Liabilities even caused many designers to refuse work in this arena.

We finally made some measurements of size, thickness, area, length, and so on and utilized modern formulae along with the ultimate strength of cast iron and a large factor of safety before making a very careful certification. The project was completed and the columns carried the "actual loads" that the floors produced. About a year later, I found a Carnegie Steel Manual in a flea market, and inside the manual was a page of cast iron column loads. To my surprise, the column I had worked so hard on and worried so much about was in the table, and the capacity was almost exactly the number that I had calculated. The discovery of that volume was the beginning of a very interesting 25 years of hunting and collecting period books on construction, materials, architecture, and structural engineering in America and England. The work of locating and investigating these volumes, now totaling approximately 200, has evolved beyond a Doctorate of Philosophy on the subject to this reference book, which is a modified version of my dissertation.

Several publications now exist that outline methods of assessment of existing buildings. These publications deal with diagnostic methods of investigation of material deterioration from exterior influences. Each publication is quite good in setting procedures and methods of determining material conditions, deterioration of structural members and in situ testing methods. *Structural Assessment,* published by the Association for Preservation Technology in 1986 as a reprint of Guideline No. 9, *The Guideline for Structural Assessment,* by the National Institute of Building Science, originally released in 1982, is an excellent source for the establishment of a procedure for assessment and inspection policies. It outlines methods of visual as well as technical inspection of buildings, documents procedures for testing of materials and components, and, when the need exists, outlines methods for an in-depth structural analysis. The in-depth structural analysis is a pathological analysis of every component and structural system within a building. This analysis requires data on the period components and members, allowable stresses, and analytical methods at the time of construction. This analysis is the subject of the current work. Once one verifies the original design of a member, the member data and period allowable stresses can be combined with the modern analytical formulae and the member can be "certified" as capable (or not capable) of carrying the required loads of the new use.

The data provided by this work will enable architects, engineers, and practitioners in other disciplines to analyze historic building components such as cast iron columns, wrought-iron beams and columns, early steel beams and columns, patented shapes, load-bearing masonry, and wood structural systems with a predictable degree of accuracy. Thus, historic building analysis will become a reality and material components and systems can be analyzed by the same methods as was the original design. The original design methods can be checked by methods of modern analysis, and in many instances the structural capacity of an existing building may be adequate for the proposed use and expensive modifications prove not to be needed. In addition, with these methods of analysis and with greater knowledge, many buildings can be modified and/or have components strengthened in an economical manner. In many instances, existing buildings have been modified unnecessarily or even demolished due to the lack of a rational method of determining the capacities of the structural systems. Visual inspection methodology, analysis criteria, and data on structural shapes and the allowable stresses for dated or historic building materials are needed on a daily basis by professionals who work with historic and outdated buildings. This work is limited to American commercial building applications from the period 1820–1940, which is the period to which over 75 percent of American building inventory is attributed. It provides data, tables of loads and allowable stresses, and design methodology obtained from reference books and journals of the period.

My original hypothesis was that the historic American building was designed, through empirical or analytical methods, to meet or exceed the load requirements of the original intended use. Further, I wanted to prove that these original designs were not only adequate for the period requirements, but often have reserve capacity resulting from the conservative nature of the design methodology of these periods. Many older buildings are revised in good faith today; however, in many cases the actual need for the revision is not proven. The system or member capacity is not known and therefore conservative practices and designer liability require that the system or member be reinforced or modified via modern materials that have known characteristics. Expensive consequences dictate that we must be sure of the capacity of systems and members rather than rely too much on intuition. It is not imperative that every building be treated in exactly the same manner, that is, the total reconstruction of the original analysis. The designer in charge of the work must exercise professional judgment in deciding how much analysis is to be done. The recommended approach is to use the modern methods of analysis, the historic member section properties (its geometric properties), and the allowable stresses of the period. This method is still conservative and may require the designer to modify allowable stresses in cases where good engineering judgment permits. For instance, early structural steel allowable stresses were specified at 50 percent of the yield stress of the material. We have been utilizing two-thirds of the yield value since the 1950s, with remarkable success. Specific design dimensions, character of end conditions, bearing, among other factors, require that certain modifications of allowable stresses be made, and engineering judgment is again required.

I am grateful to Dr. Derek Linstrum, RIBA, the Major Professor for my dissertation at the Institute of Advanced Architectural Studies, York, England, Professor John Worthington, Director of the Institute of Advanced Architectural Studies, and Mr. Poul Beckmann, ISE, Hon. RIBA, of Ove Arp and Partners, and considered by most to be the world's premier structural engineer. Dr. Linstrum put up with me for a year-plus in residence at the Institute doing coursework and for the seven years during which I returned to England two to three times annually for meetings on the work. John Worthington and Poul Beckmann very graciously agreed to be the external readers, which obligated them to administer the oral examination prior to the awarding of the degree. I am also very grateful to my wife, Dr. Josette H. Rabun, ASID, for her daily support and encouragement.

1

Assessment Methodology: Material Chronology, Early Building Laws, and Loads

INTRODUCTION

The existing buildings inventory of the United States and of most other countries shares many common characteristics of style, type, use, and structural system or systems. Another parallel, unfortunately, is the lack of original data on the structural analysis as performed by the architect/engineer designers of the buildings. In the United States, original drawings exist for less than 5 percent of the existing building stock, while records of the structural analysis exist for less than 1 per cent. The method or thought process of the original designer, an important analysis that would greatly assist architectural conservators and designers of contemporary uses, is not only unavailable, it is extremely difficult to reconstruct. The allowable stresses on hand-made brick, cast iron, wrought iron, and early steel, and the dimensional data and section properties needed by the architects and engineers responsible for certification of historic buildings, have been lost as current texts and academic courses concentrate on contemporary building materials. The problem is more complex than it appears because the allowable stresses on the early metal components changed regularly as grades of the materials evolved over time. Producers did not utilize any type of marking system identifying, for example, the type of steel, and rolling mills manufactured the same shapes in the new grade as it became available. Dating methods, testing of extracted samples, and in situ testing methods exist that can be utilized in the certification process. This work will enable architects and engineers involved in this important area of design to understand the building in question, make preliminary appraisals as they may apply, and if necessary reconstruct the original analysis and the basis of the design and identify original factors of safety. It will further describe methods of verifying capacities of structural components and the overall building system through use of the period allowable stresses, the design methods that were in effect at that period, and contemporary assessment strategies.

In evaluating an existing building, whether historic or simply a utilitarian structure, for reuse as commercial, housing, or mixed-use redevelopment, most of today's professional designers, architects, and engineers are reluctant to include in their analyses the full original capacity of the existing structural fabric to carry loads and to maintain lateral stability regardless of its condition. In some instances, the existing components, such as bearing walls and floor joists, are utilized to some extent in a very conservative manner, or they are used in conjunction with some form of contemporary "strengthening system." Many buildings were originally designed for heavier loads than a proposed new use requires and yet are still modified due to lack of a rational analysis. In some cases, buildings have been demolished because of lack of knowledge about the ability of the historic material to withstand loads. In the same manner, historic buildings that are being restored for continued same uses or contemporary new uses are often unnecessarily modified as a result of a lack of knowledge in this specialized area of design. Generally, the structural engineer consultant takes a very conservative approach when working on historic buildings, in response to the lack of quantitative information to rely upon in making important professional judgments. Pressures of professional liability and the lack of original design data also tend to affect decisions and encourage conservative reaction. The structural design is within the professional realm of the licensed engineer, and law requires that the seal of a professional engineer appear on all plans for modifications and reuse projects.

THE CERTIFICATION PROCESS

Several publications now exist that outline methods of assessment of existing buildings. Most of these publications deal with diagnostic methods of identification of material deterioration from exterior influences. Each is quite good in setting procedures and providing methods of determining structural conditions, deterioration of structural members, and in situ testing methods. *Structural Assessment* is an excellent source for the establishment of a procedure for assessment and inspection policies. It outlines methods of visual as well as technical inspection of buildings, documents procedures for testing of materials and components, and, when the need exists, outlines methods for an in-depth structural analysis. The in-depth structural analysis is a pathological analysis of every component and structural system within a building. This volume was inspired by an excellent document produced by the Institution of Structural Engineers in England, *Appraisal of Existing Structures,* which is now in its second edition (1996).

The *"Standards for Rehabilitation* initiated by the Secretary of the Interior are used as one of the available parameters or guidelines for the restoration and adaptive reuse of historic buildings. They require sensitive treatment of interior features and structural systems as well as the total restoration of the exterior facade as a part of the Investment Tax Act qualification process. The Standards encourage designers to maintain the structural integrity of the existing building, but do not require total utilization of existing structural systems. Restoring significant features to their original appearances while removing nonsignificant or later alterations is a basic requirement. Most structural modifications are very damaging to a building. The Standards and Investment Tax Act process, including Section 106 review, required when federal funds are involved in rehabilitation work, will not be discussed further in this work, as any professional involved in this area will be adequately informed on this process.

Code officials also need information on historic materials to use in reviewing and evaluating plans and specifications for the restoration and adaptive reuse of older and historic structures. The official who reviews and approves proposed restoration or rehabilitation projects must be willing to participate in the process of working out solutions to code considerations that will least impact the building. Such solutions should be complementary and sensitive to the historic buildings and in keeping with the requirement of protecting the public from unnecessary risk. Building laws require that life safety standards and building code criteria be satisfied before a project is undertaken.

Designs are reviewed and approved by code officials and local fire marshals concerned only with code and life safety measures. Very few local ordinances require any esthetic considerations, and even those that do quickly dissolve when code requirements conflict.

The difficulty professional designers and code officials encounter in knowing how to approach the evaluation of historic buildings arises from a combination of circumstances:

1. Textbooks, reference books, and periodicals of today are devoted primarily to contemporary materials. Current articles on conservation focus extensively on cosmetic approaches, light material technology, and conservation of ma-

terial finishes. Preservation, restoration, and conservation are rarely seen as serious contemporary sciences in the mainstream of the design and construction industry.
2. In the period since World War II, structural engineers have been taught conservative approaches toward the use of contemporary materials and have acquired a distrust of outdated materials. The lack of data provides a designer with no analytical basis for use in evaluating the historic material, nor are there any guidelines for code officials to use in the process of approving proposed projects. Evaluation of the in situ building fabric is an expensive concept in the modern approach and thus compounds the problem.

The Investment Tax Act process for private work and the Section 106 review process required for federally funded projects utilize the words "certification" and "certified" in more than one very meaningful way. The Investment Tax Act allows certain Investment Tax Credits for "Certified Rehabilitations" on "Certified Historic Structures." The "Certified Rehabilitation" is a completed project ready for occupancy that followed all of the rules of the Investment Tax Act and certain IRS requirements and was inspected by the State Historic Preservation Officer or his designee. The project is truly certified only after a letter is issued to the owner(s) by the National Park Service. In addition, the project must have been performed on a "Certified Historic Structure," which is an individually listed building on the National Register of Historic Places, or be certified as a Contributing Building within a National Register Historic District.

The structural certification is a different legal certification process that requires a licensed engineer or architect to certify that the structure is sound and will carry all imposed loads as generated by the new use and that the structure is within code compliance. This certification carries the legal implications of professional liability.

This work is limited to American building applications from the period 1840–1960, which is the period to which over 85 percent of American building inventory is attributed. This work will provide data, tables of loads and allowable stresses, and design methodology obtained from reference books and journals of the period. Case studies and sample computations for most design situations will be included. The material provided will be sufficient for use in analyzing and certifying the structural capacity of historic buildings.

PHASE ONE ASSESSMENT

Any project begins with the owner, architect, and engineer assembling all of the readily available information on the building. Usually these informational data consist of zoning and planning requirements, square footages per floor, fire-stair placement, and the new program requirements for the adaptive reuse of the building. The architect and the structural engineer must carry this data-gathering process one step further by making the assessment of the building's structural system or systems. This may be the first or earliest expenditure of real monies on the project. If the project requires the purchase of a building, however, this process may be needed even earlier because the information acquired may be pertinent even before a building is purchased. If there are structural problems that may be expensive to repair or restructure, this information may allow the owner to compile a pro forma indicating that the building is not suited for the project because it may not provide sufficient return on the investment or may actually lose money.

The structural assessment requires expertise, judgment, and interpretation in making determinations as to the structural integrity of the building. The process is a systematic appraisal of existing data and visual observations and should be carried out in a manner that provides the necessary information for a reliable conclusion in an expedient time frame. The preliminary assessment is intended to identify structural deficiencies or damages and ascertain the need, if any, for further in-depth investigations, more detailed mathematical analysis, scientific testing, or no further investigation. In a few instances, the Phase One Assessment may verify the capacity of the building as adequate for the use proposed. The building may have been in continuous use, have been properly maintained by a discriminating owner, and be in excellent condition, while the proposed use may be sympathetic to the size and physical arrangement of the building.

The personnel involved in the preliminary investigation should plan ahead and wear proper protective clothing and aspirators. The minimum equipment necessary for the Phase One Assessment is as follows:

adequate flashlights, spare batteries	brushes, cloths
area lighting	scale, tapes
air quality monitor	sample bags
camera with flash	pencils, paper, clipboard
video tape recorder	audio tape recorder
cellular phone	mirror
moisturemeter	string and string level
ladders, stepstools	monoscope
ropes, hammers, handsaw	48″ level, torpedo level
screwdrivers, pliers, ice pick	

Care should be taken around pigeon droppings and the possible presence of friable asbestos. The air quality monitor should be utilized at all times. If the building is in the ownership of the developer, an asbestos survey and mitigation undertaking should be performed as necessary to remove all asbestos prior to further work. Tests of any pigeon droppings should be performed, and if the results show "toxic," all pigeon droppings should also be removed prior to larger work crews being moved into the building.

The Phase One Assessment should proceed in an orderly, organized manner. A minimum outline of the visual observations and the items to be investigated is as follows:

1. Study documentation available (data immediately available).
2. Identify type of construction, load paths.
   (a) Determine original use of building.
   (b) Determine structural mechanism or mechanisms of building.
   (c) Determine adjacent buildings.
   (d) Lateral stability—use of adjacent buildings.
3. Inspection process—identify defects visually.
   (a) Defective members, trusses, joists, beams, columns.
   (b) Roof decks, floor decks, trussed walls.
   (c) Loadbearing walls, lintels
   (d) Connections, beam pockets, bearing points.
   (e) Foundation condition (visual observations).
      (i) Evidence of settlement exists.
      (ii) Areas of foundation failure.
      (iii) Underpinning (is the building adequately underpinned?).
4. Failure considerations.
   (a) Are any parts, members, or mechanisms in danger of failure (cracks, buckling, deflected, separated, etc.)?
   (b) Local failure types, conditions of failure.
   (c) Consequences of local failures; does another mechanism take the load?
   (d) Global structural failure possibilities (can structural collapse occur, how and when?).
5. Estimate actual loads on the building.
   (a) Determine actual dead loads, required live loads.
   (b) Roof loads, floor loads, wall loads, foundation loads.
      Lateral loads, uplift, diaphragm transfers.
   (c) Existing factors of safety.
      Factor of safety at time of inspection.

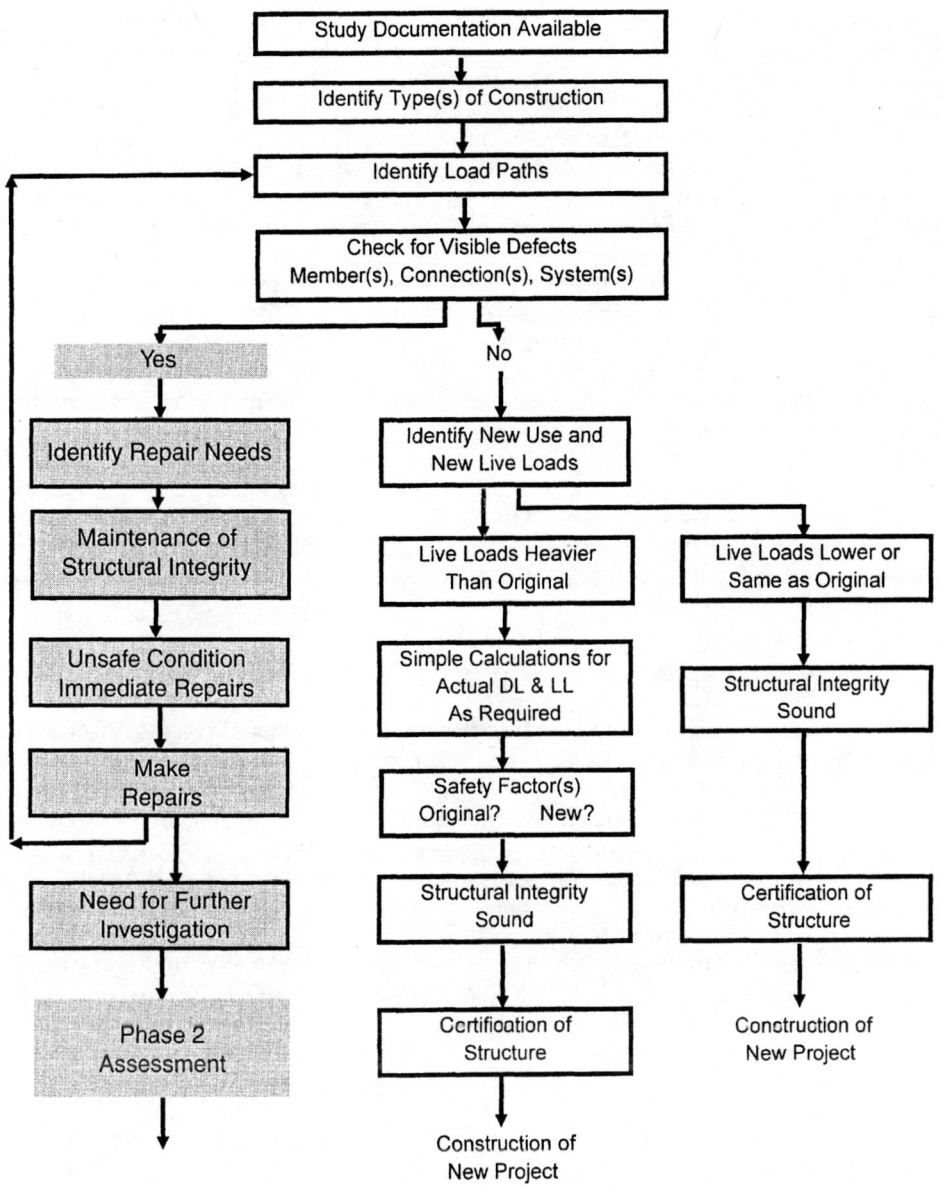

Figure 1-1. The Phase One Assessment (visual, on-site).

Factor of safety with loads associated with new use.
6. Structure adequate as exists for unchanged use.
   (a) Adequate without any repairs.
   (b) Adequate with minor proposed repairs.
7. Determine the need for a more detailed assessment. (If visual observations suggest inadequacy, a more detailed second-level assessment is warranted.)

Some buildings are relatively simple to analyze, and visual inspection of the Initial Assessment Process can adequately verify the structural capacity. Experience and professional judgment are required. In some instances, the structural system loses only a portion of its original factor of safety when the live load of a proposed new use is only marginally larger than the design live load. In many cases, the design live loads are not easily defined. However, experience allows the professional to reconstruct the design loads through an intuitive feel for member sizes, wall thicknesses, and so on, which enables one to determine the existing live load capacity by subtracting the dead load from the total load. The total load capacity is determined by defining the capacity of a floor system through the intuitive process as described above. The lateral load capacity of the building may be very difficult to define. If the building stands alone (without adjacent structures) or more importantly, if adjacent buildings have been removed, there may be a severe lack of lateral resistance. Many groups of adjacent buildings in historic downtown blocks are dependent upon each other for lateral stability. This was not by design; it was an inherited quality. The one inherent weakness in nineteenth-century one- to four-story commercial buildings is their inability to withstand wind loads (calculated wind pressures). Modern calculations often prove that a building will not resist design wind loads, yet it has been standing for over 100 years—proof positive that these buildings actually have more capacity than that calculated or, more probably, that the factor of safety of the original design was larger than the original assumption. At close inspection, however many of these buildings do show signs of small lateral displacement.

The capacity of the system of subassemblies may be greater than the calculated capacity of the assembled parts, but more likely than not, the allowable stresses were and still are extremely conservative. On occasion, the engineer or architrect will exercise professional judgment and determine that the global structural integrity may also be determined within the accuracies as needed for an overall certification through the initial assessment process.

PHASE TWO ASSESSMENT

If evidence of deterioration from building envelope failure exists, or a loss of structural capacity from man-inflicted damage has occurred as determined by the initial assessment process, a more detailed analysis may be needed in certain areas or for the overall structural system. Also, an overall change in use requiring new heavier live loads may require that the architect or engineer make a detailed analysis to determine the capacity of the structural components and overall system to carry the loads. From this analysis, the professional must determine the need for repairs, reinforcement, structural modifications, or redefinition of the structural system to bring the structure into modern code compliance. Building codes have detailed requirements concerning the cost or proportion of the building rehabilitation and the need for bringing the total building "up to code." These requirements normally pertain to utilities systems, life safety measures, and so on. It is always understood that the building must be certified structurally sound by a professional. There is no room for compromise where the structural liability is concerned.

The responsibility to determine the need for a further in-depth analysis is the professional realm of the engineer or architect, who must determine whether the Phase Two Assessment must be performed and in what manner is the work to be done. This phase requires some semidestructive investigative methods, detailed mathematical analysis, possibly scientific tests of materials, and usually a series of in situ load tests. Members and member connections that may be hidden behind finish systems must be accessed through removal of areas of the finish system to allow the designers to see all aspects of the structural components that are necessary for certification. Often it is possible to acquire the needed data with minimal intervention, and careful removal of certain areas will allow a relatively simple "patch" to restore the original appearance. Once all of the member data and connection data are available, the mathematical model can be redefined to produce a more accurate analysis. Actual dead loads can be more accurately determined in an effort to identify excesses, etc. The critical member or connection can be identified and the possibility of rerouting the load paths considered. Any members that are of an unidentified material must be identified and checked as a part of the certification process.

If the computational analysis using modern allowable stresses determines that the structural system or a part of the structural components is only marginally deficit in capacity, a more refined mathematical model that takes into account a comparison between the analytical methods of the period of construction and those of today should be performed. The allowable stresses of the period of construction should be identified and checked against the results of scientific testing to ascertain whether additional capacity can be discovered.

If the initial visual assessment determines that the structure will be severely overloaded as a result of the new use, or visual structural damage or overstressed members are present or a preliminary mathematical analysis verifies that the new loads overstress the existing structural components, the second phase of the assessment process and some form of structural modification or structural intrusion are required.

The Phase Two Assessment must be comprehensive and should include at a minimum the following visual observations, testing, verification of hidden elements, and investigative forensic methods:

House to Small Commercial Building

1. *Determination of the Phase Two Assessment strategy.*
   (a) Where is the detailed analysis needed?
       Roof deck, joists, trusses, purlins

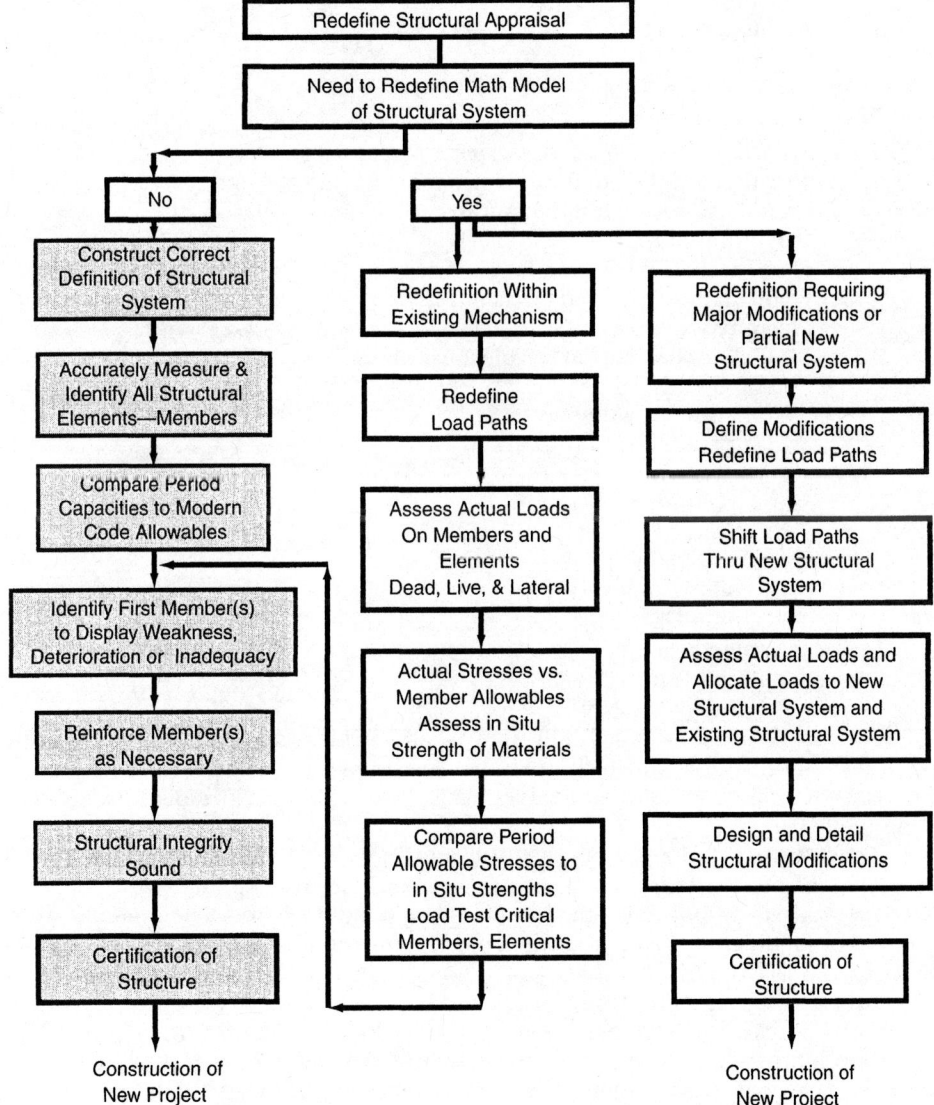

Figure 1-2. The Phase Two Assessment (visual, on-site, testing and office analysis).

Longitudinal bracing system
Floor deck, floor joists, floor trusses
Floor diaphragm, load transfers
Masonry wall system
   Thickness of masonry, thickness changes
   Length-to-height ratios
   Transverse bracing systems
   Front facade system
   Openings, headers, lintels
Wood wall system
   Braced frame system
   Stud, size and spacing
   Balloon or Platform
Metal, column and beam
   Cast iron, wrought iron, steel
   Curtain wall system
Foundation system
   Wall footings, type
   Column footings, type
   Piling, piling caps

Medium Commercial Building to 19th Century Skyscraper

2. *Determination of the Phase Two Assessment strategy.*
   (a) Where is the detailed analysis needed?
   Roof deck, joists, trusses, purlins
   Longitudinal bracing system
   Floor deck, floor joists, floor trusses
   Floor diaphragm, load transfers
   Masonry wall system
      Thickness of masonry, thickness changes
      Length-to-height ratios
      Transverse bracing systems
      Front facade system
      Openings, headers, lintels
   Metal, column and beam
      Cast iron, wrought iron, steel
      Curtain wall system
      Lateral bracing system
        Trussed girder
        Knee bracing
        Portal frame
        Cantilever system

3. *In situ material testing.* Determine where verifications are required and the type of testing or samples for testing are needed. Many types of in situ tests are very expensive and will generate only specific data. Chapter 6 provides data on available testing methods.

4. *Load testing.* Probably the cheapest and most definitive test available is the load testing of the structure. Load testing can be done to advantage on a member-by-member basis without damaging much of the building fabric. The method can be as simple as loading the joists or girders individually by placing weights at points on a floor or roof. Hydraulic tensioning of rods that have been anchored below foundation strata can give very accurate results at relatively moderate expense. Either method can provide data on load capacities as well as deflection verification.

5. Development of the mathematical model. The mathematical model of the structure, including modifications and new systems with proper allocation of loads, can be achieved in most cases on historic buildings with relative ease and not too much expense, in some instances by the use of the computer or the computer in

addition to manual methods. Load Resistance Factored Design methods can be used on steel; however, it is recommended that Allowable Stress Design (ASD) be used for the initial verification process because it will be easier to work it into the historical methods, which were all Allowable Stress methods. If, for instance, a member is just under or just shy of verification, LRFD methods may enable a designer to push the member analysis above the verification line. The consensus of designers' opinions is that LRFD produces a member about one size smaller than that produced by ASD design procedures.

MATERIAL CHRONOLOGY

The evolution of building construction in the United States is a complicated issue. Communication over the vast distances of the country contributed in very distinct ways. The major metropolitan areas of the East, Boston, New York, Philadelphia, and Charleston, would adopt architectural styles before the remote areas between them and the Midwest. Pittsburgh and Cleveland would be exceptions; they would get the latest developments within a few weeks of the eastern cities. The Midwest, Chicago, Detroit, Minneapolis, Denver, and Memphis, were each centers of progressive movements and would also have the styles within a few weeks of the eastern cities. The West, Seattle, San Francisco, and Los Angeles, would get the latest styles and technology by means of materials received via overland express railroads and oceangoing vessels. The more remote areas and rural areas would often lag months or even years behind in architectural styles and technology.

During the period 1820–1940, many changes in technology occurred, materials invented or developed, and transport methods, the railroad networks, developed throughout the country. However, during all of this progress, small towns and remote areas did not pick up the new technologies or materials as rapidly as the urban areas. Obviously, the need for larger buildings, longer spans, and greater volumes was not the same in the small towns and rural areas. Many areas were also geographically isolated and thus were left out when it came to many advancements. This phenomenon still exists today, even though the instant communication provided by the mass media does introduce new advancements to remote areas. It is still a matter of need versus utilization when it comes to technical advances.

The designer who is responsible for the certification of existing structures and the safety of the occupants needs former design methodology and data. In addition to the design methods and member sizes developed in this work, the designer needs dates that identify the first time that a material or technological advance was used in a building. The Chronology of Availability in Table 1-1 gives the dates for the first time of use of certain materials and the type of building (or building) that utilized these materials. In some instances, a material was used once initially and there was a small time delay before the material began to be utilized simultaneously in several buildings or locations. It seems that in some cases there was a waiting period to see if the material or technology was dependable or proven.

The dates provided in the Chronology are either an exact or an approximate date as indicated in several references. It is difficult to document certain beginnings, such as the first all-steel frame. The evidence conflicts on this topic: the Rand McNally Building, Chicago, 1889–90 (Corydon Purdy, Eng.) may have been the first, or perhaps the Old Colony Building, Chicago, 1893–94 (Holabird and Roche), or the American Surety, New York, 1894–95 (Bruce Price). Each claims certain attributes that make it the "first." The title of first skeletal frame building is generally given to the Home Insurance Building, 1884, designed by William LeBaron Jenny. It was a hybrid frame mixing steel, wrought iron, and cast iron in a manner that was considered most efficient at the time.

Designers can utilize the Chronology to assist in the analysis of historic buildings by correlating the structural framing in a building with the first date of the materials' availability. Many framing materials are obvious and can be identified by visual inspection. Cast iron is one example, requiring at most a magnifying glass and a pen knife. The allowable stresses for cast iron columns and beams are found in this work.

Table 1-1. Chronology of the Availability of Materials and Technology in America

	Approx. Start of Usuage in Urban Areas	Building Type and/or Typ. Size
Foundation and Foundation Systems:		
Brick, Hand-Made, Reverse Corbeling, Bearing on Ground	1632+	Houses
Brick, Machine-Made, Reverse Corbeling, Bearing on Beton Footing	1790s	Houses & Commercial buildings
Stone, Reverse Corbeling, Bearing on Ground	1632+	Houses
Stone, Reverse Corbelling, Bearing on Beton Footing	1800s	Houses & Commercial buildings
Inverted Arches, Brick and/or Stone	1860s	Commercial buildings
Wood Grillage, Bearing on Ground	1820s	Houses & Commercial buildings
Timber Footing and Grillage, Bearing on Ground	1820s	Houses & Commercial buildings
Wood Piles, Wood Platform, Wood Grillage	1830s	Commerical buildings
Wood Piles, Wood Platform, Stone or Beton Bed Top of Piles	1830s	Commercial buildings
Sand Piles	1850s	Houses & Commercial buildings
Cast Iron Screw Piles	1850s	Houses & Commercial buildings
Wrought Iron Screw Piles	1860s	Houses & Commercial buildings
Rail or Beam Grillage Footings, Beton Encasement	1860s	Commercial buildings
Plain Concrete Footings	1880s	Commercial buildings
Reinforced Concrete Footings	1880s	Commerical buildings
Caissons, Pneumatic	1893	Manhattan Life, NYC
Cylinders, Pneumatic, 4' to 10' Diameter	1890s	Commercial buildings
Building Structural Systems:		
Wood Log Houses	1630s	Houses, Wood 2 story
Wood Braced Frame	1636+	Houses, Wood 2 story
Wood Baloon Frame	1832	Houses, Wood 2 story
Wood Platform Frame (Western Frame)	1850s	Houses, Wood 2 story
Masonry Load Bearing, Hand-Made Brick in LIme Sand Mortar	1632	House & Buildings 2 to 4 story
Masonry Load Bearing, Stone in Lime-Sand Mortar	1630s	House & Buildings 2 to 4 story
Masonry Load Bearing, Hard-Pressed Brick in Lime-Sand Mortar	1850s	House & Buildings 2 to 4 story
Masonry Load Brg., High Str. Machine-Made Brick in Lime-Sand Mortar	1870s	House & Buildings 4 to 8 story
Masonry Load Brg., High Str. Machine-Made Brick in Rosendale Mortar	1870s	House & Buildings 4 to 8 story
Masonry Load Brg., High Str. Machine-Made Brick in P. Cem. Mortar	1880s	House & Buildings 4 to 8 story
Cast Iron, Columns, Column Pentals, Beams, Trussed Girter Plynths	1826	Navy Home, Phil. Pa.
Cast Iron Building Fronts, Framing, Columns, Etc.	1857	Haughwout building, NYC
Wrought Iron, Shapes, Plates, Rods, Etc.	1859+	Commercial Buildings
Riveted Section	1880	Commercial Buildings
Concrete Floor, Nonreinforced on Metal Deck on W.I. Joists	1871	Nixon Building, Chicago
Steel Structural, Hot Rolled Shapes, Columns, Beams, Etc.	1880	Commerical Buildings
Steel, ASTM A9, Buildings, Medium Grade	1900	Commercial Buildings
Steel, ASTM A9, Buildings, Structural Steel	1909	Commercial Buildings
Steel, ASTM A140-32T, Buildings, Structural Steel	1932	Commercial Buildings
Steel, ASTM A9 (Revised), Buildings, Structural Steel	1933	Commercial Buildings
Steel, ASTM A9-33T, Buildings, Structural Steel	1933	Commercial Buildings
Steel, ASTM A9-34, Buildings, Structural Steel	1934	Commercial Buildings
Steel, ASTM A7-39, Buildings, Structural Steel	1939	Commercial Buildings
Steel, ASTM A7-49, Buildings, Structural Steel	1949	Commercial Buildings
Steel, Welding	1925+	
Steel, W. I. & C. I. Fireproofing, Terra Cotta	1870+	Commercial Buildings
Steel, W. I. & C. I. Fireproofing, Metal Lathe & Plaster	1880+	Commercial Buildings
Steel, W. I. & C. I. Fireproofing, Sprayed Insulation on Material	1890+	Commercial Buildings
7" Rail Beams, Brick Floor Arches	1854	Copper Union, NYC

Table 1-1. (Continued)

	Approx. Start of Usuage in Urban Areas	Building Type and/or Typ. Size
Terra Cotta Floor Arches	1872	Equitable Building, Chicago
Skeletal Frame, Original Skyscrapers, Curtain Wall Systems	1883	Home Insurance Building, Chicago
Plain Concrete Columns, Walls Footings	1880+	
Plain Concrete Encasement of Steel Shapes (Fireproofing & Structural)	1885+	
Reinforced Concrete, Columns, Beams	1880	
Reinforced Concrete Building Structure, Columns, Beams, Slabs	1903	United Shoe Mach., Mass.
Reinforced Concrete Building Structure, Columns, Beams, Slabs, Trusses	1904	Terminal Station, Atlanta
Reinforced Concrete Building Structure, Office Building, 16 Stories	1904	Ingals Building, Cincinnatti

All dates Approximate. Time lag can be months or years in remote rural areas.

Wrought iron is also fairly easy to identify, and the designer can correlate the date of the building with the date of the allowable stresses. Steel is the most difficult, as there are no identifying marks on it. Once a beam or column has been identified as steel, there is no means of identifying the grade and thus the allowable stress without some form of testing. If one can accurately date the building and verify that there appears to have been no remodeling or additions, the maximum allowable stress used would be the stress that fits the grade of steel that was on the market about six months to a year before the building was completed. A conservative method would be to use no greater stress for steel than that which was available one year before the completion of the building.

The structural analysis is complex, and the loads and allowable stresses are not the only variables. One very important additional variable is the connections, and another is the condition of the material used as for beams, columns, and connections, including the bolts or rivets. The inspection and analysis must include all variables involved in the existing structural system.

EARLY BUILDING LAWS

The dead loads on the building are the actual weights of the material used in the construction, the structural system, floors, fixed partitions, and other permanent systems. These dead loads are fixed as a permanent part of the building and are now determined to a high degree of precision. The live loads are the occupant loads, movable loads, and basically the in-and-out user loads, which could be called "temporary loads." These live loads can also be of an external nature, such as the effect of wind and snow upon the structural system. Today the live loads are obtained from a "mandated system" adopted by the local governing body. Modern regulation is through the use of Building Codes, which were generally written and adopted immediately after the end of World War II as America began its post-War building boom. The National Building Code was first published in 1905 by the American Insurance Association. It was far from national in scope or acceptance by American cities; rather, it was used as a set of guidelines by clients of the insurance company. Special premium rates were probably offered to owners of buildings that complied with the requirements by constructing in the prescribed manner, or coverage itself may have been denied unless certain degrees of compliance were met. Danger of fires in buildings during the period 1820–1940 was ever-present. The prime reasons for the codes were fire prevention and life safety.

Building codes address many other building concerns besides the setting out of a consistent and reasonable set of rules for determining the required loadings that a building must withstand not only in daily use but in predictable "special circumstances" such as earthquakes and hurricanes. Without this specification for the live loading parameters according to buildings' occupancy classification as required by the building codes, building owners and users would have no assurance of a building's stability under load. All of the requirements of the building codes center around the concern for life safety, interior movement, and egress procedures (Comer 1942, 8). Material finishes and wall systems designed to prevent spread of fire are dictated by building codes, along with many other provisions such as the needs for tenant separation through wall and floor systems and fire-rated stair enclosures.

The code research component of this work is concerned only with the building loads and the capability of the building and its component structural systems to perform their required function. The codes are intended not only to make buildings more habitable and safer and provide the least risk to the general public, the user," but also to ensure the uniformity of performance expectations for owners and users during the life expectancy of the structures (ibid., 11). Codes do not dictate esthetic or durable quality, material finishes, comfort control, or functional performance. As has been stated, the main function of the building codes can be reduced to two words: "life safety."

Specifications and building codes are the backbone of the design and construction process:

> To the Architect and Engineer, it is a guide to safe and accepted design procedures, a convenience in selecting structural members and outlining construction methods. To the contractor and building code official, it is a document setting forth rules of safe construction that must be strictly followed. And to the owner, it is a guarantee that the resulting structure will comply with basic standards that ensure safety, utility, and economy (Beedle 1964, 25–26).

The earliest forms of building regulations were laws enacted by individual cities in response to a specific need. These laws began in the early nineteenth century in the larger cities, where living conditions were very crowded and the potential for disastrous fires was most prevalent. New York was one of the first cities to regulate building construction, with the Building Law of 1813 (Comer 1942, 8). In the years that followed, revisions to the New York building laws occurred numerous times and other cities followed suit, using the New York law as their model. *They naturally adapted specific clauses to their own needs, eliminated some clauses, and wrote new requirements where necessary.* By the middle of the fourth quarter of the nineteenth century, most larger cities in America had a building law. While the building laws and codes addressed many aspects of design and construction, in this work only those portions that deal with dead loads and occupant and externally applied live loads will be investigated. It will be necessary as a building is analyzed to have at least conceptual knowledge of which loads the original architect or engineer was designing for. Although the laws of the various cities were generally similar, they differed significantly in their live load requirements:

> Each city has its own building code to which the buildings of that city must conform. There is great lack of uniformity in these codes even where there is no reason for variation. For identical conditions, the floor loads, the allowable stresses, the wall thickness and many other items vary through a wide range. This results either in a waste of material or a sacrifice of safety and leads to confusion among architects and engineers whose practice is not confined to one city. (Huntington 1929, 4–5)

In the 1920s, many agencies were attempting to correct this situation. The U.S. Department of Commerce, the American Society for Testing Materials, and the National Board of Fire Underwriters were researching the problems and issuing recommendations in the form of reports. The Pacific Coast Building Officials Conference published the Uniform Building Code in the 1920s, which was specific to the requirements of the California area, with its high risk of earthquakes. Building laws are one of the responsibilities that come under the provisions of the "police powers" of a city. The "police power" is that inherent power of government that protects the people against harmful acts of individuals, so far as matters of safety, health, morals, or the like are concerned, and unless a code requirement can be shown to be necessary for such protection, it will not be supported by the courts (ibid., 5).

The early settlers of the Virginia Colony brought English building traditions of the late medieval period to America and initially used the building materials that were readily available and that they were familiar with, notably wood. The earliest permanent buildings were houses, and their builders utilized the "braced frame" building tradition that the same generation had been using in England. In remote areas, these traditions continued for long periods of time. However, in coastal areas, the inhabitants were in closer touch with their homeland and with new arrivals from England and the continent of Europe, and technical progress in the home lands was thus transmitted to the New World with the arrival of new immigrants. In addition, with time, wealth was accumulating in many of the original colonies and larger, more permanent houses were being constructed of brick molded and kiln baked on the site. In the Virginia Colony in the Tidewater area, two buildings, a house and a church less than thirty miles apart, were constructed of brick made at the site in 1632.

The continuing reference to the architectural styles of England is further evidenced in the houses in Tidewater Virginia in the eighteenth century, when a simple form of architecture based upon colonists' visual intrepretations of the English Georgian was being constructed at and around Williamsburg. By this time, native-born American craftsmen, and immigrant craftsmen from England who brought over the latest styles (often handbooks on architecture were now available), were combining their efforts to construct a regional style that responded to climate, material availability, and the functional needs of a predominantly agricultural society (Peterson 1976, 55, 56). Interestingly, a simple form of braced frame was used on wood houses and farm buildings, while brick structures utilizing load-bearing masonry with wood joists and flooring were emerging as the building type of the more wealthy owners.

The builders were sometimes traveling craftsmen who had either apprenticed under a master in their home country, if he was an emigrant, or, if he was first-generation American, had been trained in the old school tradition, with several years of apprenticeship. The craftsmen followed the rules of thumb that were handed down through the trades. Within these rules, some basis of empirical analysis for the limitations of wood girders and beams appears to have been worked out through experience and actual in-place construction. There appeared also to have been an appreciation for the load-carrying ability of the various different species of wood. It must be emphasized that American building followed tradition rather than developing new technology until well into the nineteenth century.

With the introduction of the "balloon frame" in 1832, Americans invented a new method of construction. Necessity, through the lack of native timber in the Great Plains of the Midwest, brought about new technology through framing houses and small commercial buildings with minimum-sized studs for vertical walls and deeper sections for floor joists. An equally important feature of this new construction method was that for the first time mills were set up to produce material in standardized sizes in central locations and the product of the mill was shipped to the construction site, ready for use. The concept of the material supplier, the lumber yard, began with this development. An intermediate vendor sold to the builder rather than the builder being required to go directly to the forest or mill site.

English architects and engineers were years ahead in the field of analytical analysis and had developed methodologies for designing through stress analysis by the 1830s (Timoshenco 1983, 222, 223, 260–270). The development and growth of the railroad in England necessitated that engineers be able to design masonry arches, trusses, girders, and columns for the extremely heavy loads of locomotives. Empirical methods of proportioning cast iron columns and compression members in arch bridges and roof structures began in England in the 1790s and in America in the 1820s. (Beedle 1964, 11). Mathematicians, scientists, and engineers were continuing to develop analytical methods for calculation of bending stresses in beams and formulas for buckling of columns under loads. These analytical solutions were based upon mathematical theory and checked or verified by testing apparatus of the period. At the same time, the iron producers were improving their products and the invention of new production techniques and processes resulted in new materials such as wrought iron and early forms of structural steels. The science of producing these new forms of iron and the technical expertise required for their implementation into the structural forms required by bridges and buildings were slow to develop. Acceptance of new concepts by builders was not the only limitation within the emerging field; often consumers were slow to break away from traditional systems until new methods or materials were proven. Initially, without a convenient shape to use for floor joists, innovative builders began using railroad rails for floor joists in some early forms of "fireproof floor systems." In 1848, the Iron Works of Cooper & Hewitt, Trenton, New Jersey, manufactured a 92-pound rail that was 7 inches deep for the Camden and Amboy Railroad. This rail proved to be very acceptable for short spans as joist members for brick arches, which were being used in early forms of fireproof floor systems (Jandl 1983, 75–77).

Traditional commercial buildings in the 1840s and 1850s in the East were constructed of masonry load-bearing walls with wood joists and wood floor decking. Most of these buildings were constructed in a manner in which wall placement was dictated by the allowable span of the wood joists, usually a maximum of 18–24 feet. The same limits of joist or beam spans were seen in timber construction, with the walls constructed that same distance apart. As construction technology advanced, interior walls were omitted and columns and girders were placed in the line where some interior walls previously would have existed. This ability to make a two-bay building allowed spatial enclosures to be twice as wide as the earlier form, with only a row of columns penetrating the floor plan. This opening up of spaces within buildings began the emergence of the system of bays that is still in use today as a structural system and order of space. By 1880, buildings of this type were constructed several bays wide, often with as many as four or five rows of columns opening up very wide spaces. In areas where the larger joists were available, wall or bay spacing could be as wide as 30 feet. Lateral stability was the only limitation on the number of bays that could be opened in each direction. Designers of this period were not sure of the predictability of long exterior brick, masonry, or wood walls withstanding the lateral wind forces and remaining stable without intermediate supports such as shear walls, pilasters, and trussed partitions to brace the walls.

In the 1840s and 1850s, there was also an emerging need for "specialized" types of buildings such as state houses, county courthouses, monumental churches, opera houses, banks, and other large commercial buildings. Each of these types of buildings would require special architectural and structural characteristics that would utilize unique design solutions. The simple utilitarian building would no longer serve every need. At this time, designers, architects, and engineers needed a more refined approach to the art of building design and construction. Building codes played an integral part in the development of these new types of buildings and building uses. The building codes developed with the concept of protection of the building user, while building contents were of secondary value. Life safety was and is the primary responsibility of the building codes. The public, the user of a commercial building, by law has the right to experience a building in safety. The larger the building becomes programmatically, the larger it becomes in volume and square footage, and that projects to a larger number of people that can be found in the building at a specific time. Codes are necessary to protect the building user, and the period 1840–1920 saw the development of extensive building regulations in America. The following chapters will expand on the criteria for design of the various components of the building structure.

BUILDING LAWS AND STRENGTH REQUIREMENTS

Building laws evolved slowly in America, with larger cities being the first to dictate environmental requirements, such as those designed to prevent slum landlords from renting single rooms to immigrant families. Fire safety and emergency escape routes parallel environmental provisions, while loads were not dictated by local ordinances until later as further refinements came with experience. The earliest mandates for material sizes or system expectations came in the form of thickness requirements for brick and masonry walls. The thickness of the masonry walls was specified according to height, beginning at the top of the building, increasing as floors were stacked beneath, and ending with the thickness at foundation level. The London Building Laws contained this type of graphic presentation of wall thickness as early as 1774. The

London Building Law of 1877 addressed buildings as high as 100 feet and was also one of the first to consider wall length (see Figure 1-3).

These two acts and the various acts that were enacted in between were responses to the need for thickness of non combustible material and to the load-carrying capacity of the walls. The wall thickness provided a natural means of reducing the heat transmission through the wall and thus could be relied upon to contain a fire on one

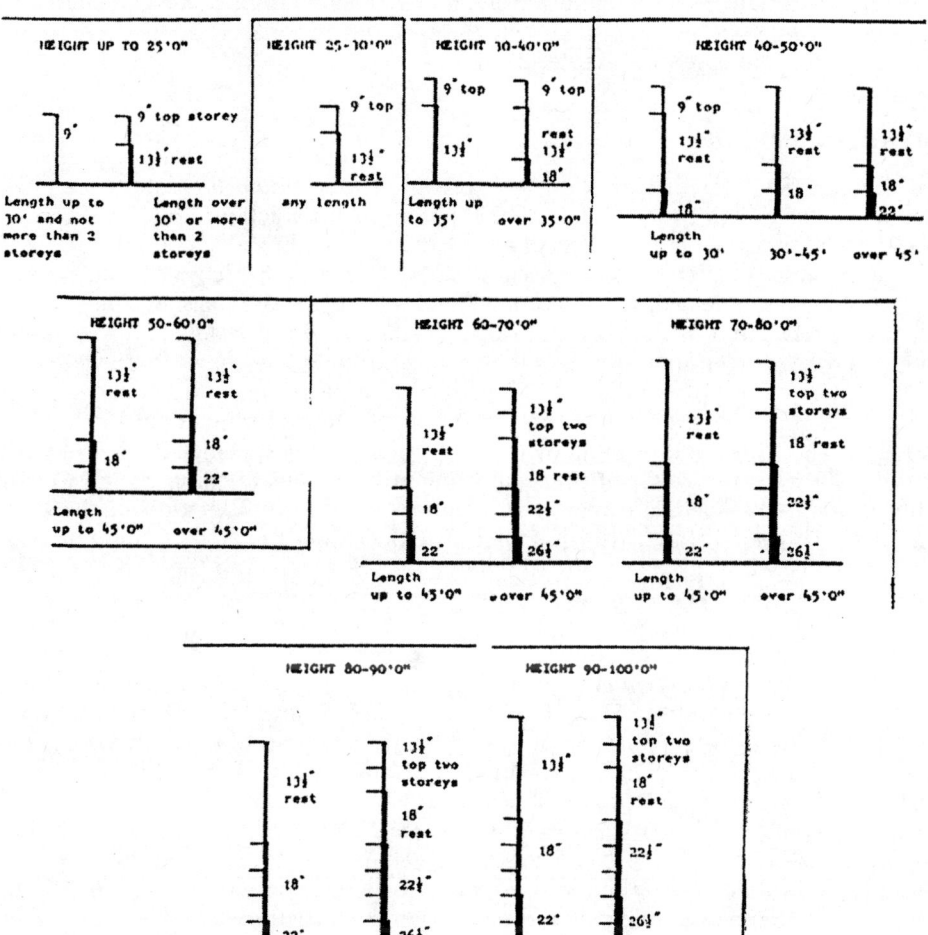

Figure 1-3. London Building Law, 1877. Reprinted, by permission, from R. H. Harper, *The Evolution of the English Building Regulations*, Table 8, Sheets 11–14. © 1985 Mansell Publications Ltd.

side of such a wall. The loads on the wall that support one-half of the floor system, through bearing of the floor joists upon the wall, are transmitted down the wall to the foundation system. The wall has a certain compressive strength that allows it to carry load without buckling or failing due to crushing. The required thickness of the wall is a function of the allowable stresses the wall can carry and the load on the wall. Each of these factors is a part of the overall design considerations that must be investigated.

As may be seen from the notes that accompany the London Building Law of 1877, the British engineers of the period were very advanced in their analysis. The criteria for adding piers to the design of the wall system are indicative of the beginnings of a combined stress analysis on the wall. This method of analyzing brick walls through combining axial and flexural stress analysis to determine whether any zone of the wall is in tension was an important concept, since lime-sand mortar, used in many areas until the 1920s, had practically no tensile ability. This combined stress analysis was not used in America until approximately the turn of the century. It will be expanded upon and fully explained in Chapter 3.

AMERICAN BUILDING LAWS

The Building Law of 1882 in New York provided a schedule for determining wall thicknesses for buildings to be constructed within its jurisdiction. The New York Law also required that the bearing walls of all buildings exceeding 105 feet in length without a cross-wall, or piers or buttresses, shall be increased 4 inches in thickness for each additional 105 feet in length or part thereof. In the same manner, the Law provided for a reduction in the required wall thickness when the walls of any buildings were less than 20 feet apart and less than 40 feet in length or when there were cross walls or piers or buttresses that served to strengthen the walls. The thickness of interior walls could be reduced at the discretion of the Superintendent of Buildings (Kidder 1902, 156, 157). It is not known whether the criteria specified by the New York Law of 1882 were the result of practical experience, technical computations, or scientific testing. We speculate that it was the result of a combination of the three, combined with sound analytical judgment. At this time, material testing and engineering theory was only beginning to emerge on the scene, and the vast network of testing organizations and code and insurance underwriting systems in use today was in its infancy.

Boston had a similar schedule for use of designers and builders in its Building Law of 1892. Boston used much the same system of categories in its law as did New York, utilizing the headings of Dwelling Houses, and Buildings of the First and Second Class (other than dwellings). Boston also provided for granite or brick foundations and exterior load-bearing walls as for brick for both uses. This is an indicator of the local availability of a regional material.

Kidder (1902) publishes excerpts from the 1882 New York Law and the 1892 Boston Law, as shown in Table 1-2. The Kidder volume, in its many editions, was the authority on building construction in America for a long period of time, 1885 to the early 1940s, when all commercial and residential construction halted due to the war needs. The *Pocketbook* contained information on all phases of architectural building construction and was used by designers in all regions of the country. Despite its name, it could hardly be considered a "pocketbook," since it contained nearly 2000 pages. The *Pocketbook* and other Kidder volumes were in high demand and were published in many editions over the years. Some volumes are still in use today, and the information contained in Kidder's books is invaluable to designers in the field of historic preservation. Early editions are extremely difficult to find.

In the United States, the same graphic presentation used in England was used to dictate the thickness requirements to builders in the reference books of the 1890s. Brick sizes in the United States were slightly smaller as a rule than in England, but in both countries brick sizes varied within regions. Specified wall thicknesses are given in inches or wythes (number of bricks in the thickness dimension of the wall). One of the earliest representations of wall thickness requirements for American buldings

Table 1-2. Thickness of Walls Required in New York (1882) and Boston (1892)[a]

Thickness of Walls Required in the City of New York (Laws of 1882)
Dwelling houses, Apartment houses, Hotels, and Schools

Height of Walls, Measured from the Curb opposite Center of Building.	Foundations Stone	Foundations Brick	Outside and Bearing Walls	Remarks
Not exceeding 35 feet.	20″	16″	12″ in Basement, 8″ above.	8″ partition walls, not
35 feet—not > 50 ft.	20″	16″	12″ above foundation	over 50 feet in
50 feet—not > 60 ft.	24″	20″	16″ in first story, 12″ above.	height. [The height
60 feet—not > 75 ft.	24″	20″	16″ for 25 feet, 12″ above.	in all cases to be
75 feet—not > 85 ft.	28″	24″	20″ for 20′, 16″ to 60′, 12″ above.	taken to the nearest
85 feet—not > 100 ft.	32″	28″	24″ for 35′, 20″ to 75′, 16′ above.	tier of floor beams.]
100 feet—not > 115 ft.	36″	32″	28″ for 25′, 24″ to 50′, 20″ to 90′, and 16″ above.	

Walls exceeding 115 feet in height to be increased at the bottom 4″ for every additional 25 feet in height or part thereof, the upper 115 feet remaining the same as specified for walls of that height.

Walls for Warehouses

Not exceeding 40 feet.	20″	16″	12″	[If there is a clear span
40 feet—not > 60 ft.	24″	20″	16″ for 40 ft, 12″ above.	of over 25′ betw'n
60 feet—not > 75 ft.	28″	24″	20″ for 25 ft, 16″ above.	walls, the bearing
75 feet—not > 85 ft.	32″	28″	24″ for 25′, 20″ to 60′, 16″ above.	walls shall be 4″
85 feet—not > 100 ft.	36″	32″	28″ for 25′, 24″ to 50′, 20″ to 75′, 16″ above.	more in tk. than here spec. for each 12.5′ or fraction thr'of., that said walls are more than 25′ apart.]

Walls exceeding 100 feet in height to be increased at the bottom 4″ for every additional 24 feet in height or part thereof, the upper 100 feet remaining the same as specified for walls of that height.

33′ high or < 20′ wide, & 40′ deep.	16″	12″	8 inches
33′ high and ≤ 60′	20″	16″	12 inches.
60′ high and ≤ 70′	24″	20″	16″ to top 2nd floor, 12″ above.
70′ high and ≤ 80′	28″	24″	20″ to top 2nd floor, 16″ to top of upper floor & to within 15′ of the roof, and 12 ″ above.

Walls exceeding 80 feet in height shall have, for the upper 80 feet, the thickness required for buildings between 70 and 80 feet in height, and every section of 25 feet or part thereof, below buildings between 70 and 80 feet in height, and every section of 25 feet or part thereof, below such upper 80 feet, shall have a thickness of 4″ or more than is required for the section above.

Buildings of the First and Second Class, Other than Dwellilngs

40 feet or less	24″	20″	16″ to top of 2nd floor, 12″ for remaining height.
40′ & < 60′	28″	24″	20″ to top of 2nd floor, 16″ for remaining height.
60′ & < 80′	32″	28″	24″ to top of 1st floor, 20″ to top of upper floor and to within 15′ of the roof, 16″ above.

Walls exceeding 80 feet in height shall have, for the upper 80 feet, the thickness required for buildings between 60′ and 80′ in height, and every section of 25′ or part thereof, below such upper 80′ shall have a thickness of 4″ more than is required for the section next above it.

[a] From Kidder 1905, 157.

was found by this author in *A Treatise on Architecture and Building Construction* (International Correspondence Schools 1899) (see Figures 1-4 and 1-5).

A provision in the New York Building Law of 1882 required additional wall thickness when parallel walls were to be separated by more than 25 feet: "If there is to be a clear span of over 25 feet between walls, the bearing walls shall be 4 inches more in thickness than is heretofore specified, for every 12.5 ft. or fraction thereof, that said walls are more than 25 feet apart" (ibid., vol. 2, sec. 7, 101).

18 ◇ Structural Analysis of Historic Buildings

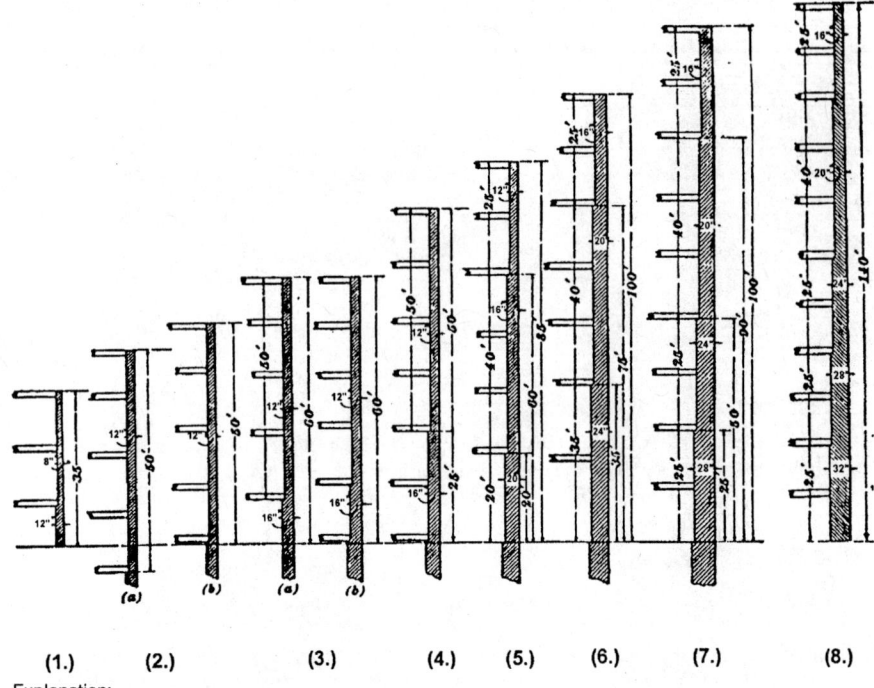

(1.)   (2.)   (3.)   (4.)   (5.)   (6.)   (7.)   (8.)

Explanation:

(1.) The basement walls of dwelling houses not over 35 feet in height and not over 20 feet in width shall not be less than 12 inches thick, if of brick; the outer walls shall not be less than 8 inches thick; but no party wall in any such building shall be less than 12 inches thick.

(2.) (a & b) The walls of all dwelling houses, whether called tenement houses, apartment houses, flats, hotels or other buildings, which are to be used for residence purposes, and are 26 feet or less in width between walls, and also the walls of school-houses over 35 feet in height and not over 50 feet in height, shall not be less than 12 inches thick above the foundation wall; but no wall shall be built having a 12 inch portion measuring vertically more than 50 feet.

(3.) (a & b) If between 50 and 60 feet in height, the walls shall not be less than 12 inches thick above the basement, if a high stoop house, and not less than 16 inches in the first story if not a high stoop house.

(4.) If between 60 and 75 feet in height, the walls shall not be less than 16 inches thick to the height of 25 feet, or to the nearest tier of beams to that height, above that 12 inches thick.

(5.) If between 75 and 85 feet in height, the walls shall not be less than 20 inches thick to the height of 20 feet, or the nearest tier of beams, above that 16 inches thick to the height of 60 feet, above that not less than 12 inches thick to the top.

(6.) If between 85 and 100 feet in height, the walls shall not be less than 24 inches thick to the height of 35 feet, or the nearest tier of beams, above that not less than 20 inches thick to the height of 75 feet, above that not less than 16 inches thick to the top.

(7.) If between 100 and 115 feet in height, the walls shall not be less than 28 inches thick to the height of 25 feet, or the nearest tier of beams, above that not less than 24 inches thick to the height of 50 feet, or the nearest tier of beams, above that not less than 20 inches thick to the height of 90 feet, or the nearest tier of beams, above that not less than 16 inches thick to the top.

(8.) If over 115 feet in height each additional 0 to 25 feet shall be increased 4 inches in thickness beyond that which is beneath while the upper 115 feet shall remain as specified.

Partition walls of 8 inches may be built to support joists provided that clear spans of joists do not exceed 33 feet, and the clear vertical height shall not exceed 26 feet. If any partition wall exceeds 50 feet in height, thickness shall be increased 4 inches for each 25 feet of height.

Figure 1-4. Thickness of brick walls for "dwelling houses" (from the New York Building Law of 1882). (From International Correspondence Schools 1899, vol. 2, sec. 7, 97–98.)

Figures 1-4 and 1-5 and the written explanations above are indicative of the type of design information provided in the building laws of the period. New York, as the largest city, was the leader, and most cities of sufficient size to need building regulations followed in format, if not in total compliance. The designer of the period may have been determining the actual stress on the wall that was the result of the dead and live loads from the materials and the occupants; however, he was required to utilize wall thicknesses at least as thick as those required by the diagrams and/or the tabular lists. The requirements were of the minimal criteria, which meant that as long

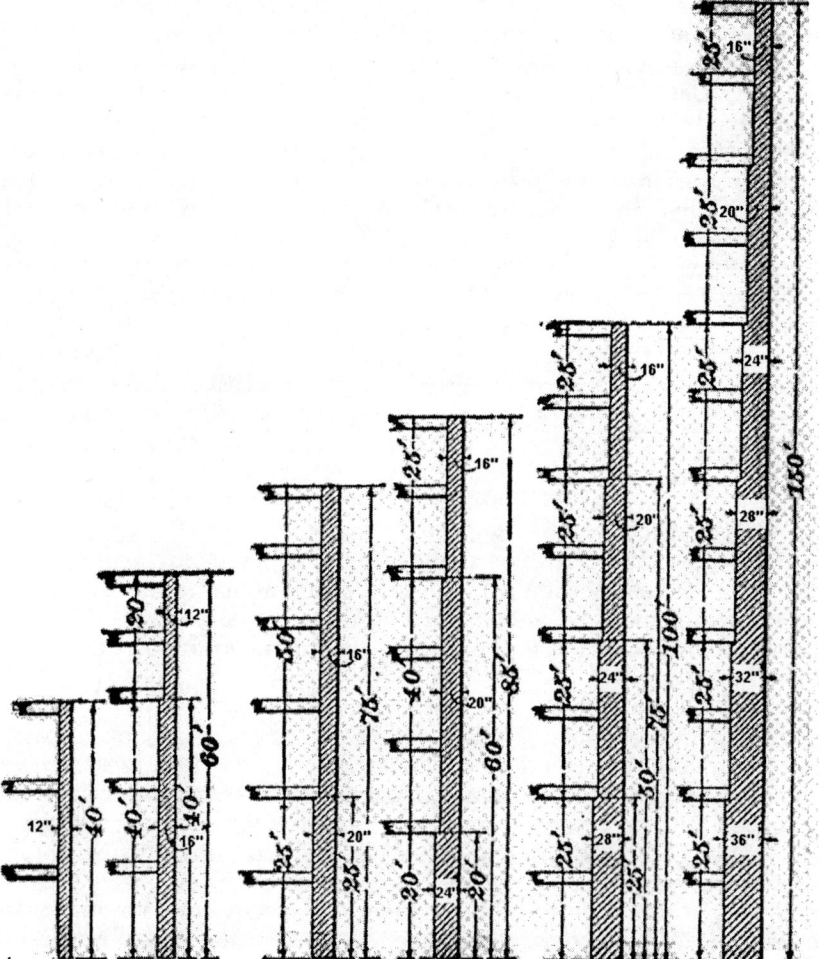

Explanation:
(1.) The walls of all Warehouses, Stores, Factories, and Stables 25 feet or less in width between walls shall not be less than 12 inches thick to the height of 40 feet.
(2.) If over 40 feet, up to 60 feet in height, thickness shall be 16 inches thick up to 40 feet or the nearest tier of joists, above 12 inches thick minimum.
(3.) If over 60 feet, up to 75 feet in height, thickness shall be 20 inches thick up to 25 feet or the nearest tier of joists, above 16 inches thick to top.
(4.) If over 75 feet, up to 85 feet in height, thickness shall be 24 inches up to 20 feet or nearest tier of joists, above 20 inches up to the height of 60 feet or nearest tier of joists, above 16 inches thick to the top.
(5.) If over 85 feet, up to 100 feet in height, thickness shall be 28 inches up to 25 feet or nearest tier of joists, above 24 inches up to the height of 50 feet or nearest tier of joists, above 20 inches up to the height of 75 feet or nearest tier of joists, above 16 inches thick to the top.
(6.) If over 100 feet in height, each additional 25 feet in height above ground, wall thickness is to be increased by 4 inches. The upper 100 feet shall be the same as that specified for a 100 foot building.

Figure 1-5. Thickness of brick walls for warehouses, stores, factories, and stables (from the New York Building Law of 1882). (*From International Correspondence Schools 1899, vol. 2, sec. 7, 98.*)

as the wall thickness provided was equal to or greater than that specified by law, it was in compliance. Evidently no provision for design data "to prove a thinner wall was adequate" existed.

INNOVATIONS IN EDUCATION AND MANUALS

The International Correspondence School was one of the leaders in the field of technical education with its series of publications presented for home study. At this time in the United States, few colleges and universities existed in the more remote areas of the country. Learning at the higher level was concentrated in the larger cities in the East, and therefore learning at home through correspondence courses was very popular. The quality of education received through a correspondence course from the International Correspondence School was excellent. This system produced many engineers and architects. The International Correspondence School (1899) also includes, in its wall thickness discussion, a "Table of the Thickness of Walls in Inches for Warehouses, Etc." for six major cities in America. The previous figures indicate the minimum wall thickness by "height of wall in feet." This table indicates thickness requirements by building height in "stories." The explanation further states that the tops of second-floor beams are taken at 19 feet above the sidewalk, while the heights of the other stories are 13 feet 4 inches from floor to floor. The sidewalk is to be considered as having been within inches of the level of the first floor. There is no provision for basement wall thicknesses in the table, and therefore it may have been the interpretation of the designers to use the same thickness as that tabulated for the first floor. The table is further complicated by the fact that while New York and Boston Laws give the wall heights in feet, the Chicago laws indicate that the maximum heights of stories for the given wall thickness are 18 feet for the first story, 15 feet for the second, 13 feet 6 inches for the third, and 12 feet from the fourth upward (ibid., vol. 2, sec. 7, 102) (see Table 1-3).

The discussions, tables, and charts above are all based upon the use of a standard handmade brick, fired locally if not on the jobsite, and upon the use of a jobsite mixed limesand mortar of the standard proportional mix of the period. Strengths of the individual components and size of brick for the various locations will be expanded upon in later chapters.

As can be seen from the above tables and discussions, there was a wide variance in the requirement of the building laws. Obviously, at least a regional standardized requirement would have been desirable. A look at the six cities presented in Table 1-3 shows representative figures from regional geographic locations. Smaller cities in an area looked to the larger metropolitan area closest to them for guidance for cultural as well as technical progress. A number of obvious areas are missing from the six, Atlanta probably being the most notable.) Since each of the six was the leader in its geographic area, it was logical for builders in the respective areas to follow the literature in print in reference texts and the requirements of the larger city if there were no local laws.

During the same period, the last two decades of the nineteenth century, loads on buildings were apparently taken from building laws where applicable, or from reference texts for designers and builders constructing in areas not governed by local laws. Bulding laws began in about 1880 to dictate live loads to be utilized by designers in America. Designers working in areas not under the jurisdiction of building laws were obviously making conscious decisions with each design as to the magnitude of the live loads for the required uses. It can only be assumed that these designers were following the design criteria that were published in the textbooks and periodicals of the time. Most of the volumes of the period used the laws of the major cities as their models. A part of a modern analysis would be to attempt to ascertain the original design loads as a part of the criteria for determining the capacity of the components of the structure to carry load.

LIVE LOADS/OCCUPANT LOADS

Live loads on buildings are the temporary loads that come from the occupants on floor systems and wind and snow on the roofs. The existence of these loads was known to the medieval builders, but they lacked the skill required to analyze buildings floor systems etc., and to compute sizes of individual members. Selection of beams and columns was purely empirical, as was the determination of masonry wall thickness, in the United States until the 1880s, when methods of analysis were introduced and reference books on the technology of building construction published. These reference books, published in England and the United States simultaneously, began recommending live loads for buildings and presented weights of building materials in a

Table 1-3. Thickness of Walls in Inches for Warehouses, Etc.[a]

Height of Building	City	1st	2nd	3rd	4th	5th	6th	7th	8th	9th	10th
Two Stories	Boston	16"	12"								
	New York	12"	12"								
	Chicago	12"	12"								
	Minneapolis	12"	12"								
	Memphis	18"	13"								
	Denver	13"	13"								
Three Stories	Boston	20"	16"	16"							
	New York	16"	16"	12"							
	Chicago	16"	12"	12"							
	Minneapolis	16"	12"	12"							
	Memphis	22.5"	18"	13"							
	Denver	17"	13"	13"							
Four Stories	Boston	20"	16"	16"	16"						
	New York	16"	16"	16"	12"						
	Chicago	20"	16"	16"	12"						
	Minneapolis	16"	16"	12"	12"						
	Memphis	27"	22.5"	18"	13"						
	Denver	21"	17"	13"	13"						
Five Stories	Boston	20"	20"	20"	20"	16"					
	New York	20"	16"	16"	16"	16"					
	Chicago	20"	20"	16"	16"	16"					
	Minneapolis	20"	16"	16"	12"	12"					
	Memphis	31.5"	27"	22.5"	18"	13"					
	Denver	21"	21"	17"	17"	13"					
Six Stories	Boston	24"	20"	20"	20"	20"	16"				
	New York	24"	20"	20"	20"	16"	16"				
	Chicago	20"	20"	20"	16"	16"	16"				
	Minneapolis	20"	20"	16"	16"	16"	12"				
	Memphis	36"	31.5"	27"	22.5"	18"	13"				
	Denver	26"	21"	21"	17"	17"	13"				
Seven Stories	Boston	24"	20"	20"	20"	20"	20"	16"			
	New York	28"	24"	24"	20"	20"	16"	16"			
	Chicago	20"	20"	20"	20"	16"	16"	16"			
	Minneapolis	20"	20"	20"	16"	16"	16"	12"			
	Memphis	40.5"	36"	31.5"	27"	22.5"	18"	13"			
	Denver	26"	21"	21"	17"	17"	17"	13"			
Eight Stories	Boston	28"	24"	20"	20"	20"	20"	20"	16"		
	New York	32"	28"	24"	24"	20"	20"	16"	16"		
	Chicago	24"	24"	20"	20"	20"	16"	16"	16"		
	Minneapolis	24"	20"	20"	20"	16"	16"	16"	12"		
	Memphis	45"	40.5"	36"	31.5"	27"	22.5"	18"	13"		
	Denver	30"	26"	21"	21"	21"	17"	17"	17"		
Nine Stories	Boston	28"	24"	20"	20"	20"	20"	20"	20"	20"	
	New York	32"	32"	28"	24"	24"	20"	20"	16"	16"	
	Chicago	24"	24"	24"	20"	20"	20"	16"	16"	16"	
	Minneapolis	24"	24"	20"	20"	20"	16"	16"	16"	12"	
	Memphis	49.5"	45"	40.5"	36"	31.5"	27"	22.5"	18"	13"	
	Denver	30"	26"	26"	21"	21"	21"	17"	17"	17"	
Ten Stories	Boston	28"	28"	24"	24"	20"	20"	20"	20"	20"	16"
	New York	36"	32"	32"	28"	24"	24"	20"	20"	16"	16"
	Chicago	28"	28"	24"	24"	24"	20"	20"	20"	16"	16"
	Minneapolis	24"	24"	24"	20"	20"	20"	16"	16"	15"	12"
	Memphis	54"	49.5"	45"	40.5"	36"	31.5"	27"	22.5"	18"	13"
	Denver	30"	30"	26"	26"	21"	21"	21"	17"	17"	17"

[a] From International Correspondence Schools 1899, vol. 2, sec. 7, 100–101.

format that enabled a more accurate calculation of the dead loads of the structure. Kidder (1902, 127) describes live load as a load that is applied suddenly or accompanied by vibrations, while a dead load is a load that is applied by imperceptible degrees and remains steady, such as the weight of the structure itself. Kidder further states that the effect of a live load on a beam or other piece of material is twice as severe as that of a dead load of the same weight, and that when designing for a live load, one should use a factor of safety twice as large as required for a dead load. This procedure was the state of the art at that time and reflected the concerns for design of bridges and other structures in which sudden and/or moving loads were recognized. Present methods of analysis use an impact factor for that type of live load. Moving loads and sudden vibratory loads rarely apply to the design of buildings in modern usage, with two possible exceptions: parking garages and heavy materials warehouses. A specialized building may also include a factory or fabricating plant that employs moving overhead cranes that lift and move heavy loads and are subject to starting and stopping action. Since the 1880s, live loads by building usage have varied in magnitude, beginning conservatively and progressing gradually toward the current values. Cities began dictating the live loads by building type at approximately the same time.

International Correspondence Schools (1899) presents two tables and a discussion of the live loads for buildings. It states that the live load is the people in the building, furniture, movable stocks of goods, small safes, and varying weights of any character. In stating that large safes and heavy machinery require special provision, the ICS introduces the criterion of sound engineering judgment, which is required (see Table 1-4).

The discussion of the table states that the load of 70 psf for dwellings will probably never be realized, but recommends that it be used because a house in a city may at some time be used for other purposes. For a "country house," hotel, or building of "like character" where economy is necessary, 40 psf is ample for rooms not used for public assembly. For public assembly, a live load of 80 psf is sufficient, since it has been demonstrated that "a floor cannot be crowded to more." Retail stores should have live loads of 100 psf or more. Printing houses with heavy stock should have 250 psf, while special equipment such as heavy presses should be proportioned for twice its static load (dead loads). The indication here is that the author recognizes the impact load of the movement of or within the machinery and the resultant effect upon the structure.

It should be noted that architects and engineers of the period recognized that the loads published in reference books were very conservative. We do not know what actual loads these designers used during this peroid, but we can use these figures as the basis of analysis and for verification from the sizes of structural members provided. Major cities dictated the live load, and New York was so concerned with structural stability that it required its building department to check structural computations. These structural computations were required to be detailed and to be submitted in such a form that they could be checked by a professional employed in the City's Department of Buildings.

In addition, the submission was to have its "correctness sworn to" by the person making the submission (ibid., sec. 5, 13). New York further required that commercial

Table 1-4. Live Loads per Square Foot of Floor Surface[a]

Character of Building	Pounds/SF
Dwellings	70
Offices	70
Hotels and apartment houses	70
Theatres	120
Churches	120
Ballrooms and drill halls	120
Factories from	150 up
Warehouses from	150 to 250

[a] From International Correspondence Schools 1899, vol. 7, sec. 5, 4.

buildings owners place a certificate of the approved loading in the building within public view. This provision was designed to give future owners notice of the capability of the building when new uses were contemplated. The first item, the computations, sworn to by the designer (an engineer or an architect), is a predecessor of the "seal or license" that is required today. The second is a very sensible measure in this society where uses are permitted to change without any form of structural investigation unless major modifications are required by the new user.

In engineers' handbooks of the period, warehouses were considered to be special structures due to the widely varying conditions of loading resulting from the types and weights of material stored and the geographic location, as can be seen from a brief analysis of Table 1-3. Memphis was obviously concerned about buildings being used as cotton warehouses and made special requirements in the wall thickness schedules for the weight of cotton bales. Other cities also responded to anticipated special loadings for all buildings, rather than allowing the owner of a building to state the building's proposed use and design on that basis.

In 1900, the Boston Manufacturer's Mutual Fire Insurance Company prepared extensive data for architects and engineers detailing live loads for approximately 100 types of materials. A portion of this list, which contained solid materials, bales, barrels, and cartons, is reproduced in Table 1-5.

The allowable live loads on buildings continued to be modified by individual cities as knowledge of more accurate live load criteria for a given condition evolved. Cities revised their positions on loads through revisions of the laws, and in the normal manner smaller cities followed the lead of the major ones. Table 1-6 reproduces a table of the "Allowable Live Loads" for four of the largest U.S. cities at the turn of the century (International Correspondence Schools, 1905). The table gives us our first look at Philadelphia law.

Clearly, the need for uniformity through national or at least regional codes was becoming obvious. Not only were the cities changing their requirements, but values for live loads continued to vary widely from city to city. Table 1-9 tabulates the laws for the cities shown for ca. 1910. For the first time we see a live load listed for snow, and there is a differentiation for varying roof pitches. Earlier volumes mention snow loads in discussions of roofs, trusses, etc. and state that the designer must make ample provision for the weight of the snow to be expected for the location.

Kidder (1905) *The Architect's and Builder's Pocket-Book* has two tables of live loads in its section on floor systems. One duplicates the live loads dictated by law in six cities during the 1890s, while the second presents what Kidder considers "reasonable live loads" for use by designers when working in areas not covered by laws. This

Table 1-5. Weights of Merchandise for Calculating Live Loads[a]

Materials Floor Area	Measurements		Approximate Weight		
	Contents Sq. Ft	Total Cu. Ft	Pounds	Pounds per Sq. Ft	Pounds per Cu. Ft
Cotton, etc. Bale	8.1	44.2	515	64	12
Bale of compressed	4.1	21.6	550	134	25
Bale of American Cotton Co.	4.0	11.0	263	66	24
Bale of Planters Comp. Co.	2.3	7.2	254	110	35
Bale of Jute	2.4	9.9	300	125	30
Bale of Jute lashings	2.6	10.5	450	172	43
Cotton Goods, Bale of unbleached jeans	4.0	12.5	300	72	24
Piece of duck	1.1	2.3	75	68	33
Case of prints	4.5	13.4	420	93	31
Bale of tickings	3.3	8.8	325	99	37
Grain, Wheat in bags	4.2	4.2	165	39	39
Wheat in bulk					44
Flour in barrels on side	4.1	5.4	218	53	40
Corn in bags	3.6	3.6	112	31	31
Wool, Bale, South American	7.0	34	1000	143	29
Misc., Crate of crockery	9.9	39.6	1600	162	40

[a] From International Correspondence Schools 1905, sec. 5, 11–12.

Table 1-6. The Allowable Live Loads on Floors in Different Cities[a]

Character of Building	Pounds per Square Foot (ca. 1900)			
	New York	Philadelphia	Chicago	Boston
Buildings for Public Assembly	90	100	120	150
Buildings for ordinary stores, light manufacturing, and light storage	120	100	120	
Dwellings, apartment houses, hotels, tenement houses, or lodging houses	60	40	70	50
Office Buildings, first floor	150	100	100	100
Office Buildings, above 1st floor	75	100	100	100
Public Buildings, except schools				150
Roofs, pitch < than 20 degrees	50	25	30	25
Roofs, pitch > than 20 degrees	30	25	30	25
Schools or places of instruction	75			80
Stables or carriage houses less than 500 sq. ft in area	75	40		
Stables or carriage houses more than 500 sq. ft in area	75	100		
Stores for heavy materials, warehouses, and factories	150		150	250
Sidewalks	300			

The roof loads shown above do not include the wind load. Boston Law required that an additional 30 psf (vertical surface) be provided when designing roofs.
[a] From International Correspondence Schools 1905, sec. 5, 19.

volume contains an interesting discussion of the "live load" question and its quantitative value. Kidder presents live load values that were reasonable, economical, and safe. He reports the results of tests by Professor L. J. Johnson of Harvard University (Johnson 1904) to ascertain the weight of crowds (of men), in which Johnson obtained weights of 134.2, 143.9, 148.1, and 156.9 psf. The live load of 156.9 psf was obtained by packing 67 men in a room about 11′ × 6′ in size. Johnson also found that with 50 men in the room, giving a weight of 122 psf, the crowd was compacted "so that a man could elbow his way through it only with perseverance and determined effort" (Kidder 1902, 653). Kidder also presents a table of the various live loads required by six cities at about the turn of the century (see Table 1-7).

Kidder was one of the leading authorities in the United States on building design, detailing, and construction. His volumes were the recognized authority of the period and were most probably followed by anyone who worked outside the jurisdiction of the larger cities. As previously stated, Kidder (1905) has a table of live loads that Kidder

Table 1-7. Minimum Safe Superimposed Loads for Floors Required by Various Building Laws[a]

Class of Buildings	Minimum Live Load Per Square Foot of Floor					
	Buffalo 1896	Boston 1895	Chicago 1895	Denver 1898	New York 1899	St. Louis 1897
Dwellings	40	50	70	40	60	70
Hotels, tenements, and lodging houses	70	50	70	50[b]	60	70
Offices	70	100	70	70	75[c]	70[c]
Buildings for public assembly	100	150	70	80[d]	90	120[b]
Stores, warehouses, and manufacturing bldgs.	120	250	150[e]	150[e]	120[e]	150[e]

The author further states that in his opinion there is a more reasonable set of live loads that when "taken in connection with the values given for the safe strength of beams, will provide absolute safety with proper allowance for economy."
[a] From Kidder 1902, 653.
[b] Also includes schoolhouses.
[c] First floor = 150 psf.
[d] With fixed desks.
[e] And upwards.

suggests for use in the design of buildings. Kidder's suggested live loads are shown in Table 1-8.

Kidder further explains that the live loads found by investigation (the 33.3 psf) may not be the desirable value for use in designing the floor beams, etc., due to the fact that portions of the floor are possibly loaded above that amount; thus, 40 psf over the entire floor area is suggested as an adequate, safe value. Interestingly, Kidder does suggest that the lower value should be adequate for calculating the loads for the sizing of the columns. We see here an early consideration of what we know today as "live load reductions for floor areas to be used in the proportioning of columns." This consideration will be discussed later in this chapter.

The live load requirements were constantly researched and discussed by all of the authorities of the period in an effort to reach industry standard. However, the size of and geographical differences in this country made this task most difficult. Cities continued to respond to the results of researchers and scholars of the period by changing load requirements in their laws, but differences continued to exist. Ketchum (1918), one of the leading structural designers of the early twentieth century, includes a very detailed table of live load requirements by cities in this chapter on design of buildings. This table represents the live load values in force in nine U.S. cities in 1913 (see Table 1-9). As one can see, use categories were increasing in number as the need for building types in this new industrial world constantly expanded. It appears that many types of buildings were under consideration at this time. Uses were coming into existence which historically were not identified as separate or individual concepts. Designers were realizing that in planning a building for a specific purpose, it was proper to design for loads that were adequate for the planned use only. This concept, while it was expedient for the purpose at hand, made no provision for possible changes in use. Further needs beyond the present owners' requirements, or any other anticipated changes in spatial allowances, were not provided for.

Without realizing it, the designers and owners were bringing about the phenomenon of "building life cycle" or "Building Obsolescence" and the twentieth-century concept of "adaptive use of historic buildings." It was inevitable that this segregation by building type would become a reality in the construction of modern buildings. There have always been buildings for specific purposes, but, as we have seen occur throughout this century, buildings are becoming totally specialized. Economy is and must be a factor, and it was as impractical in earlier periods as it is today to construct more structure for a building than is necessary. Earlier materials had more inherited capacity through a more limited selection of products and sizes than the materials of today. The early designer was required to produce a building in which there might be several categories of loadings. Occupant types might have varied from floor to floor in the same manner as today, but required loads were heavier. At the end of the nineteenth century, many new materials for exterior finishes were available, as well as new structural systems such as skeletal steel and reinforced concrete frames, that offered many advantages. In 1900, the architectural engineer was emerging as the

Table 1-8. Kidder's Suggested Live Loads[a]

For dwellings, sleeping and lodging rooms	40 psf
For schoolrooms	50 psf
For offices (upper stories)	60 psf
For offices (first story)	80 psf
For stables and carriage houses	65 psf
For banking-rooms, churches, and theatres	80 psf
For assembly halls, dancing halls, and the corridors of all public buildings, including hotels	120 psf
For drill rooms	150 psf

Further data included in the discussion following the table explained:
"Floors for ordinary stores, light manufacturing and light storage should be computed for not less than 120 pounds per square foot, and to sustain a concentrated load at any point of 4000 pounds.
Office loads: Blackall & Everett of Boston, found that the average live load in 210 offices, in three prominent office buildings in Boston was between 16 and 17 psf, while the average load for the 10 heaviest offices was 33.3 psf. He further states that the loads found were not evenly distributed, with some portions of the floor loaded heavier than other. Therefore the suggested live load of 40 psf should be fully adequate for offices."
[a] From Kidder 1902, 654.

Table 1-9. Floors and Roofs: Minimum Live Loads, Pounds per Square Foot, by Building Laws of Various Cities[a]

Kind of Building	Boston 1912	New York 1906	Philadelphia 1913	Baltimore 1908	Pittsburgh 1913	Cleveland 1911	Chicago 1911	St. Louis 1910	San Francisco 1910
Apartments	50	60	70	60	50	50	40	60	60
Public rooms[b] and halls	100					100			
Assembly halls	125	90	120		125	100		100	125
Fixed seat auditoriums				75	125	80	100		75
Movable seat auditoriums				125		100	100		125
Churches		90		75	125		100		125
Dance halls	200			150	150	100			
Drill rooms	200				150				
Riding schools	200				150				
Dwellings	50	60	70	60	50	40	40	60	60
Public rooms[b]	100								
Hotels	50	60	70	60	70	50	50	60	60
First floors								100	
Corridors						80			
Office floors	100					80			
Public rooms[b]	100								
Manufacturing	125	120	120	125	125		100		125
Light factories		120	150		125			150	
Mercantile									
Heavy storehouses		150	150	250	200	200			250
Retail stores	125		120	125	125	125	100	150	125
Warehouses	250	150	150		200		100	150	250
Offices	100	75	100	75	70	60	50	70	60
First floor	100	150		150				150	150
Corridors						100			
Schools (class rooms)	60	75		75	70	60	40	100	75
Assembly rooms, halls	125	90		75	70	60	40	100	125
Sidewalks		300		200		200			150
Stables, carriage houses		75		100		80	100		75
Area less than 500 sq. ft							40		
Stairways and landings	70					80	100		
Fire escapes	70					80			
Roofs, flat[c]	40	50		40	50[d]	40	25	40	30
Horizontal projection									
Steeps roofs		30		20	50[d]		25		20
Superficial surface			30		50[d]	40			
Wind pressure		30	30[e]	30	25	30[e]	20	30	20

[a] From Ketchum 1918, 71.
[b] Area greater than 500 sq. ft.
[c] Slopes less than 20°.
[d] Dead and Live, except for one-story steel frame buildings, corrugated iron roofs, 35 lbs.
[e] High buildings, built-up districts, 35 lbs; 14 stories or over, 25 lbs at 10th story, 2½ pounds less each story below.

individual responsible for the actual computation and design of the structural elements.

Still, the situation was not as complicated as it may seem. Many uses were compatible, that is, their live loads were the same. When a new use is contemplated for an existing building that requires heavier live loads than originally utilized in the design, the building may have to be modified. In many cases, with older buildings even the heavier live loads can be accommodated with minimal interference. It becomes obvious that a designer of today would be much better informed and able to perform to more advantage if the original design concept were known or were easily determined by observation and analysis.

The Chicago Building Ordinance of 1911 contained a table that designated the thickness requirements for brick, stone, and solid concrete walls for construction within the jurisdiction of the City of Chicago (see Table 1-10).

It is interesting to see that while the concept of designing from specified live loads seemed to be in accepted use in 1911, cities were still designating the thickness of masonry walls. This is an indication that "load-bearing masonry" buildings were still

Table 1-10. Thickness of Walls (Inches): Chicago Building Ordinance (1911)[a]

	Basement	Stories											
		1	2	3	4	5	6	7	8	9	10	11	12
One-story	12	12											
Two-story	16	12	12										
Three-story	16	16	12	12									
Four-story	20	20	16	16	12								
Five-story	24	20	20	16	16	16							
Six-story	24	20	20	20	16	16	16						
Seven-story	24	20	20	20	20	16	16	16					
Eight-story	24	24	24	20	20	20	16	16	16				
Nine-story	28	24	24	24	20	20	20	16	16	16			
Ten-story	28	28	28	24	24	24	20	20	20	16	16		
Eleven-story	28	28	24	24	24	24	20	20	20	16	16	16	
Twelve-story	32	28	28	28	24	24	24	20	20	20	16	16	16

[a] From Ketchum 1918, 76.

being constructed. A requirement in the Chicago Building Ordinance of 1911 specified the minimum thickness of curtain walls in steel skeletal construction to be 12 inches for brick or concrete and 8 inches for reinforced concrete (ibid., 75). In addition, city building officials were considering the fact that the design live load for columns could be reduced because the probability of a condition of full live load on every floor simultaneously was highly improbable. As stated, this live load reduction criterion will be developed later in this chapter.

In the 1890s, William Birkmire published four very informative books on skeletal construction for "high office buildings." The first building with an all-metal frame (skeletal construction) was the Home Insurance Company Building, 1884, designed by W. L. Jenny in Chicago. The first all-steel frame was probably the Rand McNally Building, Chicago, 1889–90, designed by the engineer Corydon Purdy (see Table 1-1 and the comments thereon). The skeletal frame has a row of columns along each exterior wall, with beams framing between the columns. The beams are located at each floor, and in addition to carrying the floors along the perimeter they carry the section of wall for that floor. Each successive floor and wall for that floor is carried independently in the same manner. This system passes all of the floor and wall loads into the columns and thus accumulates the loads toward the lower levels of the structure, terminating in the column footings. In this manner, exterior walls can be constructed or assembled not necessarily from the ground up; it was often more advantageous to the builders to leave the lower facades for later completion. Skeletal construction and the steel frame from whence its name was derived liberated the height of buildings in a way that the designers of the period did not fully realize. Suddenly the height of buildings was doubled and the skyscraper era of the 1890s began. The height of buildings was now limited only by the ability or the reluctance of designers to attempt greater heights. They could actually have gone higher with the technology they were using. The 200–250-foot height of the new building type was so controversial that it was probably the most highly discussed architectural issue in our history. People spread rumors that these high buildings with offices 200 feet above the ground were not healthy, that the environment at those heights would be physiologically detrimental, and some critics argued that buildings of that height would not stand up, that wind or thunderstorms would turn them over.

At least the latter fear, that of safety, has been proven wrong by the actual performance of the buildings over the years; the former, regarding the environment within and in the immediate vicinity of the skyscraper, still persists. The term *skyscraper* is now used for the modern buildings of 1950 to the present, and the high office buildings of the 1880s to the 1930s are now called by that term only by knowledgeable architects and historians. Today's heights have again more than doubled over those of the 1880s. In this study it will be emphasized that the skyscrapers of the 1880s were the predecessors of the current buildings through the kinship of utilizing the same structural concepts in their construction. The technical developments that occurred, first to increase the heights from 250 feet to 500 feet, and then to the 1000-plus feet of the modern buildings, have been progressive.

The introduction of structural steels in the 1870s brought about the technical innovations that produced the skeletal-frame concept by the 1880s. Engineers and architects were quick to realize the potential for replacing the very thick, heavy load-bearing walls of the masonry system with the thinner, lighter construction that came with the skeletal frame. Developments in modern structural steels were occurring parallel to and literally in conjunction with the skyscrapers. Architects and engineers were experimenting with shape combinations that produced the most efficient columns, girders, beams, etc. Rivets were used to create the combination sections, composed mainly of angles, plates, and channel sections. Wide flange shapes in common use today were not hot rolled in the United States until 1907, six years after their invention by Gray in Luxembourg in 1901. Riveting continued to be the method of connecting components in the steel frame until the modern shear connection was introduced in the 1930s. The shear connection was a less intense connection, with the web of the beam being connected to the web of the girder, etc. Top and bottom angles attaching flanges to the girder or column were no longer included. The new shear connection did not gain total acceptance until after World War II with the introduction of bolting rather than riveting. The increases in height meant a more efficient use of land with the larger number of stories that could be constructed on a site. Prime land in the center of cities was already priced at a premium by this time, transportation was not what it is today, and center cities were the center of commercial activity. The potential for decreasing the thickness of walls at street level and above (in the lower floor regions) meant a large increase in the rentable space, the space that was worth the most to an owner, since ground floors were then, as now, prime commercial spaces. Traditional masonry load-bearing construction at its highest limits at the time, 15 to 17 stories, would require at least 7-foot-thick walls at ground level for a building of a height of 250–300 feet in some regions. The new, skeletal-frame structures might have required only 12–16-inch-thick masonry curtain walls as compared to load-bearing, walls which would be several feet thick at the ground level.

A comparison between load-bearing masonry wall and the steel skeletal frame systems for a building of the same external dimensions, 60 × 80 feet, 300 feet tall, could be striking, as shown in Table 1-11. Naturally, if the gross dimensions of the building were somewhat larger, the percentage of increase in the net usable space would be less. However, the dramatic increase in the net usable interior space makes the immediate acceptance of the steel skeletal frame more understandable.

The increase in rentable square footage is obvious. When several floors are considered, the financial returns of these large increases in rentable space would easily justify the use of the skeletal frame on any building. Few data are recorded on any rate of return calculations for economic feasibility analysis in the late nineteenth or early twentieth century, but the gain in usable space is obvious. Evidence indicates that there was little if any increase in cost in the new system. The immediate adoption of the new technique supports this conclusion.

Birkmire (1898, 98) states that the changes that had taken place in the construction of high office buildings had come as a result of the employment of the steel skeletal frame, in which the loads are carried directly to the foundation by the vertical columns, and that when the skeletal frame is brought to its fullest development, ma-

Table 1-11. Usable Ground Floor Square Footage: Comparison between Load-Bearing Walls and Curtain Wall Systems

Building Type	Gross Size	Wall Tk.	Net Size	Clear Area	% Increase
Load-bearing masonry	60' × 80'	84'[a]	46' × 66'	3036 sq. ft	Basis
Skeletal-frame	60' × 80'	16"	57.3' × 77.3'	4435 sq. ft	46%
Skeletal-frame	60' × 80'	12"	58' × 78'	4524 sq. ft	49%

The nonrental lobby, elevator closets, stairwells, bathrooms, etc. would be approximately the same size for either type of construction, so the increase in net rentable space comparison is valid.
[a] The walls of the Monadnock Building, Chicago, 1891 were seven feet thick at sidewalk level.
The Monadnock was one of the last load bearing masonry buildings of its size in Chicago.

sonry walls as supports count for absolutely nothing. He further states that masonry is to be utilized as protective covering (fireproofing) for the frame and as exterior facade treatment. The inability of steels to withstand the temperatures of a fire in a building was evidently known by designers at the time of the introduction of the skeletal frame.

Building laws were revised to include provision for the allowable thickness for masonry curtain walls as the skeletal-frame form of construction became popular among designers and owners. A section of the New York Building Law of 1892 applies to curtain walls for steel skeletal frame construction:

> Curtain walls of brick built in between iron or steel columns, and supported wholly or in part on iron and steel girders, shall not be less than twelve inches thick for fifty feet of the uppermost height thereof, or to the nearest tiers of beams to that measurement in any building so constructed; and every lower section of fifty feet or to the nearest tier of beams to such vertical measurement, or part thereof, shall have a thickness of four inches more than is required for the section next above it down to the tier of beams closest to the curb level; and thence downwardly the thickness of walls shall increase in the ratio prescribed in section 474 of this title for the thickness of foundation-walls. (Ibid., 101)

Birkmire presents figures comparing the curtain wall thicknesses allowed by the cities of New York and Chicago with that proposed by Birkmire as sufficient. See Figure 1-6 for a diagram of the thickness of curtain walls in New York according to the 1892 Law.

Birkmire recommends a still greater reduction in these exterior curtain walls. He proposes a 12″ thickness for the 12 upper stories and 16″ or 20″ thick walls for the lower stories to protect the building from any extreme exposure from fire from adjoining lower nonfireproof buildings. Therefore, the lower floors would be the same height as adjoining nonfireproof buildings. The curtain wall section as recommended by Birkmire in 1894 is shown in Figure 1-7.

The Chicago Law was revised that same year, 1894, to include curtain wall provisions that were considered more liberal than those in the New York Law. Birkmire quotes the Chicago Law: "The four upper stories of a twelve story building may be 12 inches, the next six lower stories 16 inches, then one story at 20 inches, and the basement 24 inches" (see Figure 1-8).

Birkmire also states that it is customary for designers of skeletal frame buildings in New York to include thin walls facing courts or other internal openings and that the New York Board of Examiners frequently grants an amendment to the law for that provision. He further quotes from the Boston Law of the same era (early 1890s): "External walls may be built in part of iron or steel, and when so built may be of less thickness than is required for external walls, providing that all construction parts are wholly protected from heat by brick or terra-cotta, or by plastering three quarters of an inch thick with iron furring and wiring" (ibid., 103).

As one can see from the discussions and varying rules dictated by different cities through their building laws, the skeletal frame and its inherent curtain wall enclosure system were both widely received and very controversial in the building industry. Every designer had his own theories as to the required curtain wall thickness and attempted to influence the various building authorities.

It must be pointed out that this was a totally new type of construction, a concept totally foreign to designers and builders of the time. It was not exactly the same as the curtain wall system of today; rather, it was the forerunner of today's very light weight glass or thin-skin type of curtain walls. As with any new concept, there were some initial transitional forms, such as the freestanding exterior masonry walls, which were constructed around a skeletal frame that carried the floors. These walls would be semiattached to the steel frame, but carry their own weight and bearing on their own foundations.

The Central Bank Building, 1893, was constructed in accordance with the New York Building Law of 1892, and Birkmire shows two curtain wall sections from the building along with a section that Birkmire proposes as the thickness that should be adequate for the building height. See Figure 1-9 for a diagram of the curtain wall sections of the Central Bank Building and Birkmire's recommended wall.

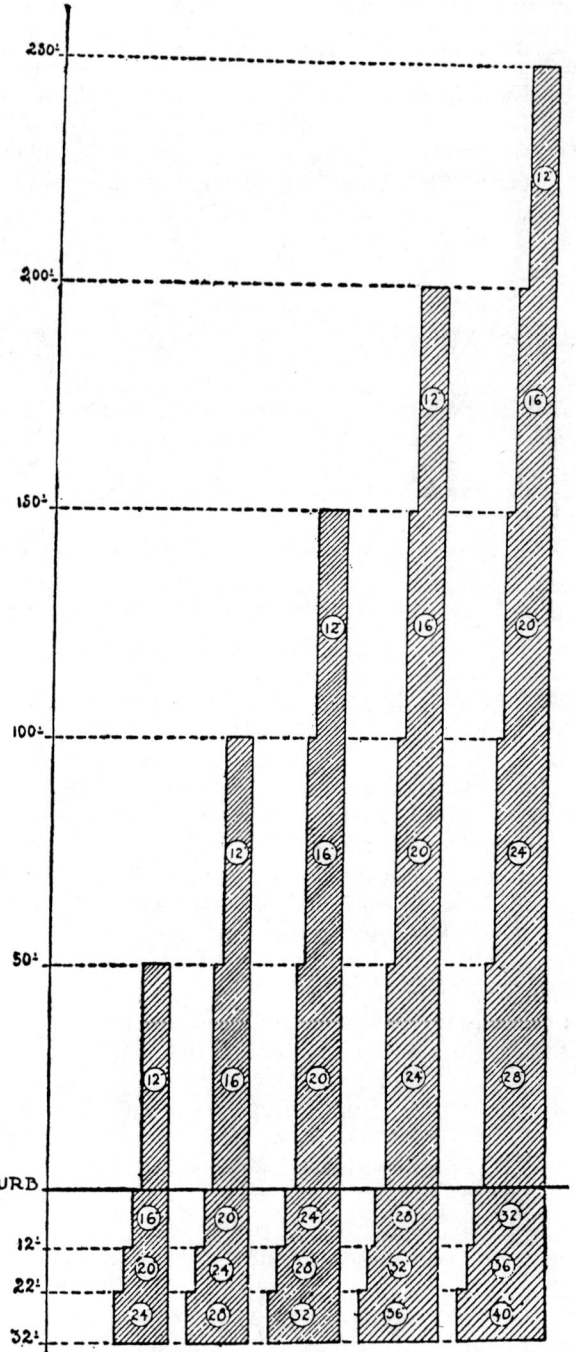

Figure 1-6. New York Building Law, 1892: Curtain wall construction. (*From Birkmire 1898, 100.*)

When the curtain wall system was introduced, the prevailing commercial styles of architecture still dictated the use of masonry and stone facades with heavily detailed and often rusticated finishes. Deep cornices, banding, and arched windows were still in fashion, and the new concept of construction was required to adopt the architectural style of the day. As the true nonload-bearing type of curtain wall evolved, each level of the framing carried the portion of the wall at that level, and thus all of the building loads were transferred to the foundations through the columns. Many advantages would evolve, such as the curtain walls being able to begin at the second

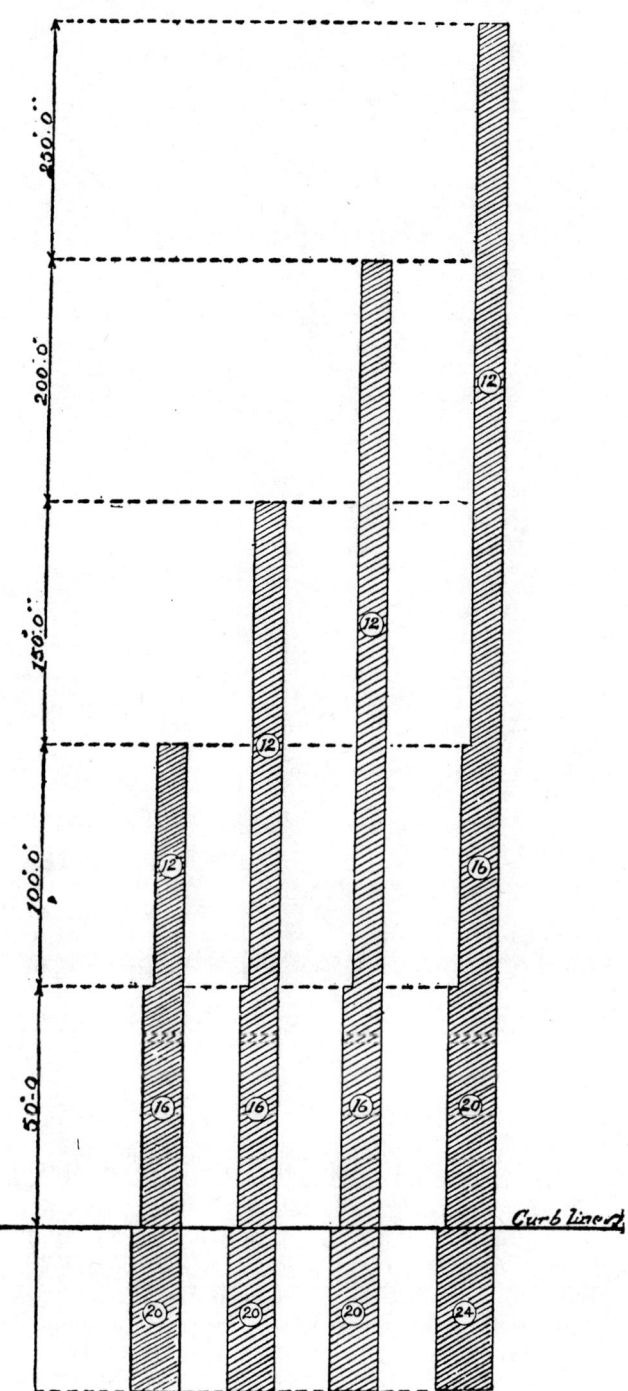

Figure 1-7. Birkmire's recommended curtain walls for use in steel skeletal frame construction. (*From Birkmire 1898, 102.*)

or third story, allowing the contractor the flexibility of having the ground floors open during construction. This would allow for large pieces of equipment to arrive at the site at almost any time rather than before major enclosure walls were built. Masonry curtain walls could also be partially built from the interior of the building, and therefore the extensive scaffolding system around the perimeter of the building could be attached at each level instead of from the ground up.

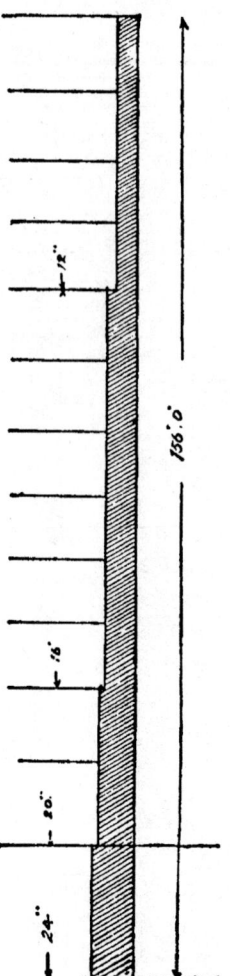

Figure 1-8. Curtain Wall Section: Chicago Building Law, 1894. (*From Birkmire 1898, p. 98.*)

The curtain wall sections shown in Figure 1-9 illustrate the manner in which the beams are framed between the columns on the exterior facades of the building. The beams shown in section are spaced apart to provide a shelf for the brick or terra cotta of the wall above that level (see Figure 1-10).

In the early twentieth century, the steel producers were promoting their product through issuing design manuals for professionals in the field. Such design manuals had appeared in the last quarter of the nineteenth century but were few in number and followed a catalogue format that gave the sizes, thickness, and properties of the sections produced by the company. The twentieth-century versions were more refined in their approach to the subject of design. It was apparent that if companies expected designers to use the product, they had to provide design criteria and examples of the analysis and design of building components such as beams and columns. Structural steels were relatively new to the market, and the steel skeletal frame was a new concept that was being utilized more with each year's experience. The skeletal frame was to become the standard structural system of the twentieth century because it freed the limits on heights of buildings and made for a more flexible plan layout in the sense that the designer did not have to follow load-bearing walls through the structure from the uppermost story down to the foundation. The new materials were also rapidly gaining on traditional methods because of their economy. Not only did the new frame systems require less material, they also were much less labor-intensive. At this time, reference books and product manuals were beginning to publish data on the

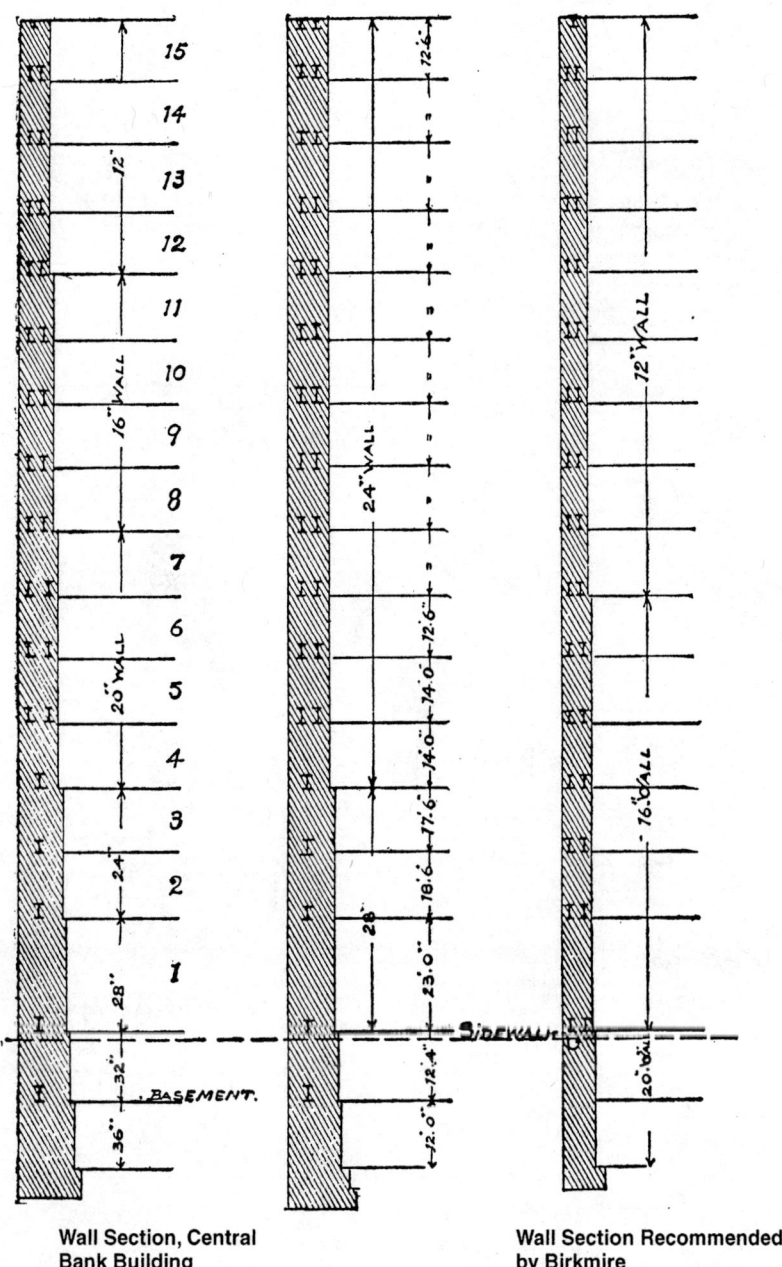

Figure 1-9. Central Bank Building, New York, 1893: Curtain Wall sections. (*From Birkmire 1894, 110.*)

relative weights of buildings per cubic foot of enclosure. Economy was beginning to take a prominent part in the decision-making process of building design, a role that has evolved to the very height of prominence today. These manuals also provided detailed design methodology and current criteria for minimum live loads as required by various cities. The Carnegie Steel Company of Pittsburgh, Pennsylvania, issued its *Pocket Companion* annually for several years, providing information for designers, including a table of suggested live loads (see Table 1-12).

As can be seen from Table 1-12 and the notes that accompany it, there are now many categories of loading, as a large number of building uses have now been identified. A large regional variance still existed in the live loads as dictated by building laws of the various cities at this time; however, World War I probably slowed research

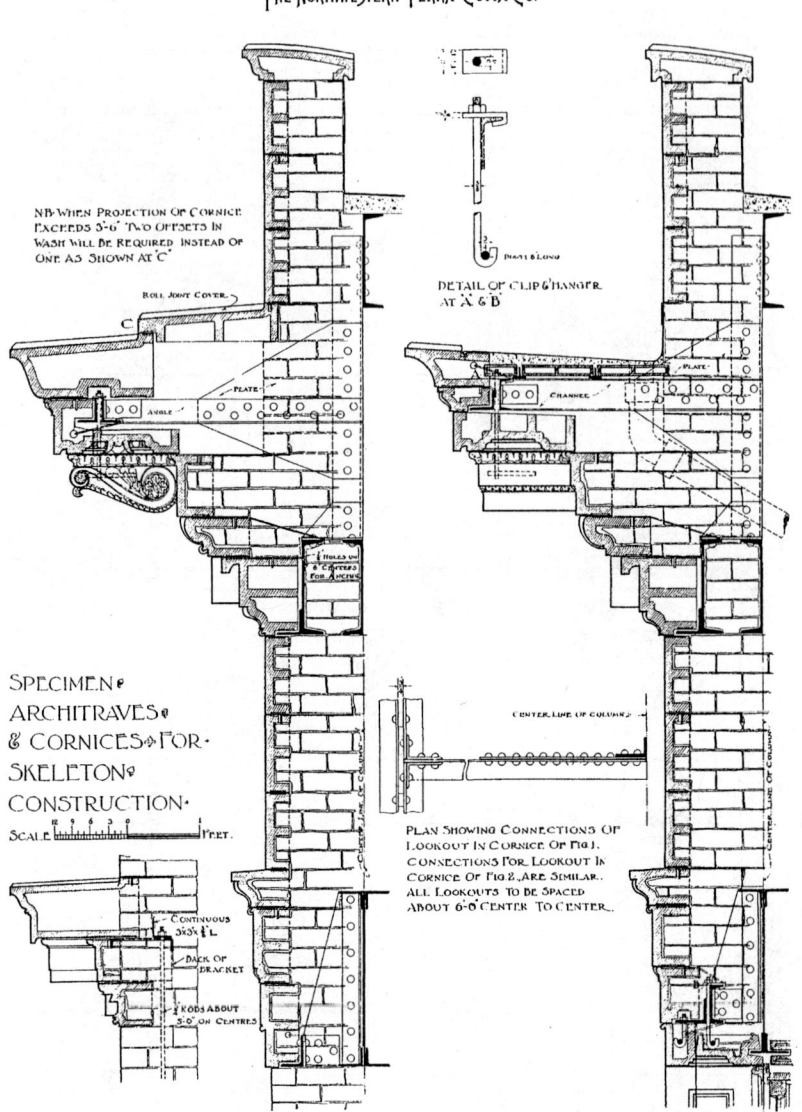

Figure 1-10. Section at Intermediate Wall Support: Skeletal frame construction. (*From Northwestern Terra-Cotta Co. catalog, p. 59.*)

and progress on codes, and cities were presumably continuing with the laws of preceding years or merely remaining at the status quo.

In 1921, the Department of Commerce of the U.S. Government organized a Building Code Committee to study the problem of the wide range of live load values required by the various jurisdictions throughout the country. The Committee's report, issued in 1925, made recommendations for a set of uniform live loads. The report stated: "It was found that live loads assumed in designing many types of buildings were largely matters of tradition and had scant scientific basis. The result was that accuracy in stress computations was defeated because of ignorance of the loads causing stress. The building professions for years have busied themselves with tests of materials, but have given little attention to the complementary factor of loads." See Table 1-13 for the Building Code Committee's live load recommendations.

The loads recommended by the Building Code Committee of the Department of Commerce reflect the state of the art of the period. The loads represent a very reasonable value for most of the categories, with the possible exception of the 250 psf for general storage. Many variables must be considered in identifying a live load for a storage warehouse; however, a realistic figure may have been 150 psf, with a special

Table 1-12. Floors and Roofs: Minimum Live Loads, Pounds per Square Foot, by Building Laws of Various Cities[a]

Description of Building	New York 1917	Chicago 1919	Philadelphia 1919	St. Louis 1917	Boston 1919	Cleveland 1920	Baltimore 1908	Pittsburgh 1914	Cincinnati 1917
Floors for rooms									
Apartments and dwellings	40	40	70	50	50	70[b]	60	50	40
Asylums, Hospitals, etc.	100	50	70	50	50[c]			70	40
Detention buildings etc.	100	50			50[c]	80			60
Factories:									
Light manufacture	120[e]	100[d]	120[d]	100[e]	125[e]		125[e]	125[e]	100[e]
Heavier manufacture			150[d]	150[e]	250[e]		125[e]		150[e]
Hotels, Lodgingg houses	40	50	70	50	50[d]	70	60	70	40[b]
Office buildings, etc	60	50	100	60[b]	75[c]	70[c]	75[c]	70	50[b]
Public buildings:									
Municipal buildings	100				75[d]	100			100
Churches	100	100	120	75	100	80	75	125	100
Libraries, museums	100				100	125		200	
Theatres	100	100	120	100	100	80	75	125	100
Schools, Colleges, etc.	75	75		75	50	70	75	70	60
Stores, light goods	120	100	120	100	125	100[c]	125	125	100
Heavier goods			150	150	250		175		150
Warehouses			150	150	250		250	200	150
Floors for assembly halls, etc.									
Auditoriums, fixed seats	100	100	120	100	100	80	75	125	100
Moveable seats	100	100	120	100	100	125	125	125	100
Arrmories, dance halls, etc.	100	100			100	150		150	150
Miscellaneous									
Garages, stables	120	100[e]		100	150[f]	150[f]	100		75
Corridors, hallways	100	100		100	75[g]	70[h]			80[h]
Stairways, fire escapes	100	100		100	75[g]	100[i]			80[h]
Sidewalks	300				250	200	200		300
Roofs:									
Flat, slope ≤ 20°	40	25	30[j]	30	40	35[j]	40	50[l]	25
Steep, slope ≥ 20°	30	25	30[j]		25[k]	30[j]	20	50[l]	25
Wind pressure	30[m]	20	30[n]	10–20[o]	20[p]	25	20[q]		

[a] From Parker 1949, 831.
[b] Dwellings, Cleveland = 60.
[c] First Floors: St. Louis = 100, Boston = 125, Cleveland = 125, Baltimore = 150, Cincinnati = 100.
[d] Public floors of hospitals, hotels, public buildings, etc. Boston = 100.
[e] Floor loads do not include the weight or the impact load of machinery.
[f] Garages, private: Chicago = 40, Boston = 75, Garages, public, upper floors, Cleveland = 100, Stables: Cleveland = 80.
[g] Corridors, stairways, etc., for assembly halls, armories, etc.: Boston = 100.
[h] Except in dwellings where floor loads are less.
[i] Stairways, etc., for apartment houses = 80; dwellings = 60.
[j] Loads per sq. ft of superficial roof area; other roof loads are for the projected area.
[k] Loads include wind pressure: 10 lbs up to ⅔ slope, 15 up to ½ slope, 20 over ½ slope.
[l] Dead and live load; snow load 25 lbs reduced 1 lb each degree between 20 and 45 degrees.
[m] For buildings over 150 ft high, or where height is over four times least horizontal dimension.
[n] Wind pressure for high buildings in built-up districts: 25 lbs at 10th story, 2½ lbs less for each story below, 2½ lbs more for each story above, up to 35 pounds per sq. ft.
[o] For buildings 40 ft high, 10 lbs; up to 80 ft, 15 lbs; over 80 ft, 20 lbs.
[p] Wind pressure on curtain walls, 30 lbs.
[q] For buildings over 100 ft high, or where height is over 3 times average width of base.

asterisk that required the designer to provide a proper load for a specific use that was heavier than average. Further, we see the category of stables still remaining in the list of building types.

Stables as a building type would no longer be listed by 1929, with the automobile replacing horse-drawn vehicles. Also, in the years since this report was presented, we have seen the weight of the automobile more than double; however, modern codes (1987 and later) require only 100 psf for parking garages, with a provision for shear on the deck for wheel load of 2000 lbs and a lateral impact factor usually of 25 percent.

The Depression years curtailed construction in America, while government work programs provided funds for research that enabled building officials, architects, and engineers to combine resources in research and documentation of building regula-

Table 1-13. U.S. Department Commerce, Minimum Live Loads Allowable for Use in Design of Buildings (1925)[a]

	psf
Floor Loads	
Human occupancy:	
Rooms of private dwellings, hospitals, hotels	40
Offices, rooms with fixed seats such as in churches, school class rooms and theatres	50
Aisles, corridors, lobbies, banquet rooms, assembly halls without fixed seats, gymnasiums.	100
Industrial and commercial occupancy:	
Storage (general)	250
Storage (special)	100
Manufacturing (light)	75
Printing plants	100
Wholesale stores (light merchandise)	100
Retail salesrooms (light merchandise)	75
Stables	75
Garages for all types of vehicles	100
Garages for passenger cars only	80
Sidewalks, 800 lb concentrated load or	250

The live loads for which each floor or differing parts thereof of a commercial or industrial building is designed shall be certified by the building official and shall be conspicuously posted in that part of each story where they apply, using durable metal signs.

The occupant of the building shall be responsible for keeping the actual loads below the certified limits. Adequate measures shall also be taken by the building official to insure that these loadings are not exceeded.

Roof loads (in lbs per sq. ft of horizontal projection)

For slopes of 4 in. or less per ft	30
For slopes of from 4 in. to 12 in. per ft	20

For slopes of over 12 in. per ft no vertical load need be assumed but provision shall be made for a wind force acting normal to the roof surface of 20 lb per sq. ft of such surface.

[a] From Huntington 1929, 13; Parker 1949, 831–834; American Institute of Steel Construction 1970, 52–53.

tions and to promote the adoption of regional codes. World War II ended the Depression but brought an already unstable construction industry to a complete standstill. Special war effort construction did occur with the building of defense factories and the retrofitting of other peacetime operations to war production units.

The design of buildings in America was reserved for the professional during the period 1900–1940. Even though professional registration was not yet a reality in all states, only educated architects and engineers were employed on projects of size and importance. We can assume that sizable buildings in metropolitan areas were designed by qualified individuals. Research continued on the live loads for buildings, with the American Standards Association (ASA) producing the authoritative load tables of the period. They became the model for all codes and building regulators. The A58 Committee was formed in the early 1940s and issued a report (A58 Committee 1945) that was much more inclusive than the 1925 Report of the Building Code Committee of the U.S. Department of Commerce. The ASA picked up on the work of the Building Code Committee, which had produced a summary of live loads based upon available sources and the judgment of experienced architects and engineers, and issued reports in 1945 and 1955. The 1955 Report of the ASA Sectional Committee on Building Code Requirements, called A58.1-1955, was issued in 1955 and contained the most complete set of live load criteria introduced to date. It was very comprehensive in its listing of building uses by categories and subcategories where appropiate for institutional, commercial, industrial, and residential uses. This live load table was presented as the standard for adoption by the various building codes in the country. Many code bodies saw it as an adoptable standard because of the authenticity and legitimacy of the American Standards Association. See Table 1-14 for a reproduction of the ASA 1955 recommendations on live loads for buildings.

Table 1-14. Minimum Design Loads in Buildings and Other Structures[a]

Occupancy or Use	Live Load psf	Occupancy or Use	Live Load psf
Apartments (see Residential)		Residential:	
Armories and drill rooms	150	Multifamily houses:	
Assembly halls, other places of assembly:		Private apartments	40
Fixed seats	60	Public rooms	100
Movable seats	100	Corridors	60
Balcony (exterior)	100	Dwellings:	
Bowling alleys, poolrooms, and similar recreational areas	75	First floor	40
		Second floor and habitable attics	30
Corridors:		Uninhabitable attics	20
First floor	100	Hotels:	
Other floors, same as occupancy served except as indicated		Guest rooms	40
		Public rooms	100
Dance halls	100	Corridors serving public	
Dwellings (see Residential)		Private corridors	40
Garages (passenger cars)	100	Reviewing stands and bleachers[b]	
Floors shall be designed to carry 150 percent of the maximum wheel load anywhere on the floor.		Schools:	
		Classrooms	40
Grandstands (See Reviewing stands)		Corridors	100
Gymnasiums, main floors and balconies	100	Sidewalks, vehicular driveways, and yards subject to trucks	250
Hospitals:			
Operating rooms	60	Skating rinks	100
Private rooms	40	Stairs, fire escapes, and exitways	100
Wards	40	Storage warehouse, light	125
Hotels (see Residential)		Storage warehouse, heavy	250
Libraries:		Stores:	
Reading rooms	60	Retail:	
Stack rooms	150	First floor, rooms	100
Manufacturing	125	Upper floors	75
Marquees	75	Wholesale	125
Office buildings:		Theatres:	
Offices	80	Aisles, corridors, and lobbies	100
Lobbies	100	Orchestra floors	60
Penal institutions:		Balconies	60
Cell blocks	40	Stage floors	150
Corridors	100	Yards and terraces, pedestrians	100

[a] Reprinted, by permission, from National Bureau of Standards, Minimum Design Loads in Buildings and Other Structures, ASA A58.1-1955, 7. © 1955 National Bureau of Standards.
[b] For detailed recommendations see American Standard Places of Outdoor Assembly, Grandstands and tents, Z20.3-1950 or latest revision by ASA.

The ASA Building Code also allows for an alternative concentrated load that can be used instead of the normal uniform live loads. It states that floors shall be designed to support safely the inherent dead loads of the structure and the recommended live loads or the concentrated loads (live load alternative) as suggested by Table 1-15, whichever produces the greater stress. It further states that the concentrated load shall be assumed to occupy an area of 2.5 feet square and shall be so located as to produce the maximum stress conditions in the structural members.

The concept of the concentrated load as an alternative to the minimum uniform live load for stress analysis of flexural members has not been picked up by all of the codes in the United States. It has, however, been endorsed by the Building Officials and Code Administrators International, Inc., Basic/National Building Code, (BOCA), and both alternatives have been refined and improved through testing and experience to their present form. Sections 1100–1115 of the 1987 BOCA National Building Code contain the required live loads for buildings as defined by occupancy categories (see Table 1-16).

Following are the notes that accompany the BOCA National Building Code, 1987:

Note (a) Minimum live loads shall be in accordance with lane loads of AASHTO HB-12 listed in Appendix A, but shall not be less than 50 psf.

Note (b) Live load need be applied to joists or to bottom chords of trusses or trussed rafters only in those portions of attic space having a clear height of 42 inches

Table 1-15. Alternative Concentrated Loads[a]

Location	Load, lbs
Elevator machine room grating (on area of 4 sq.in.)	300
Finish light floor plate construction (on area of 1 sq. in.)	200
Office floors	2000
Scuttles, skylight ribs, and accessible ceilings	200
Sidewalks	8000
Stair treads (on center of tread)	300

[a] Reprinted, by permission, from National Bureau of Standards, Minimum Design loads in Buildings and Other Strutures, ASA A58.1-1955, 8. @ 1955 National Bureau of Standards.
[b] Provision shall be made in the structural design for uses and loads which involve unusual
vibration and/or impact forces.
[c] All moving elevator loads shall be increased 100 percent for impact.
[d] For the purposes of design, the weight of heavy machinery and moving loads shall be increased not less than 25 percent for impact.

or more between joist and rafter in conventional rafter construction; and between bottom chord and any other member in trusses or trussed rafter construction. However, joists or the bottom chords of trusses or trussed rafters shall be designed to sustain the imposed dead load or 10 psf, whichever is greater, uniformly distributed over the entire span.

Note (c) Loads of 120 pounds per lineal foot on footboards and seatboards shall be used. Lateral sway bracing loads of 24 pounds per foot parallel to and 10 pounds per lineal foot perpendicular to seats and footboards shall be used.

Note (d) One inch = 25.4 mm; one psf = 4.882 kg per sq. meter.

IMPACT LOADS

1. The live loads shown above shall be assumed to include adequate allowance for ordinary impact conditions. Provisions shall be made in the structural design for special uses and loads which involve vibration and impact forces.

2. All moving elevator loads shall be increased 100% for impact and the structural supports shall be designed within the limits of deflection prescribed by ANSI A17.1 listed in Appendix A.

3. For purpose of design, the weight of machinery and moving loads shall be increased as follows to allow for impact as follows:

    Elevator machinery 100 percent
    Light machinery, shaft or motor driven 20 percent
    Reciprocating machinery or power driven units 50 percent
    Hangers for floor balconies 33 percent

The BOCA National Building Code specifies that the floors of buildings must be designed for the minimum uniformly distributed live loads as shown above or the concentrated loads shown in Table 1-16b. The provisions for impact indicated above also pertain to the alternative minimum concentrated loads. In either case, uniform load plus impact or concentrated load plus impact, the larger combination of load, impact, and consideration of the maximum deflection will be the basis of the design. The impact of such requirements is minimal when compared to the cost of providing safe working and living environments for the general public. Other tangible costs, such as insurance on the building, components, and furnishings and liability for damages to others, are much harder to define accurately. They are variable, and such considerations as building construction and type of business are all a part of the exposure and have direct bearing on costs of coverage. Often a change in construction material may be added first costs, but the overall result may be a substantial savings in annual insurance costs and enhanced overall feasibility of continued ownership.

A designer must certify that a building has the required strength in floor members, roof systems, columns, and lateral capacity to withstand the loads that must be borne by the structure to comply with the code in the local jurisdiction at the present date.

Table 1-16.(a) Minimum Uniformly Distrubited Live Loads[a]

Occupancy or Use	(psf)	Occupancy or Use	(psf)
Apartments (see Residential)		Manufacturing:	
Armories and drill rooms	150	Light	100
Assembly areas;		Heavy	150
Fixed seats	50	Marquees:	75
Moveable seats	100	Office buildings:	
Platforms (assembly)	100	Offices	50
Stage floors	100	Lobbies	100
Balcony (exterior)	100	Corridors, above 1st floor	80
One, two family dwellings only	60	File and computer rooms require heavier loads	
Bowling alleys, poolrooms and billiard rooms	75	Penal Institutions:	
Cornices	60	Cell blocks	40
Corridors, except as indicated	100	Residential:	
Dwellings (see Residential)		Attics	20[c]
Fire escapes	100	Multifamily dwellings:	
Single-family residential buildings only	40	Dwelling units	40
Garages:		Public rooms	100
Passenger cars	50	Corridors	80
Trucks and buses[b]		One- and two-family dwellings	40
Grandstands (see Reviewing stands)		Sleeping rooms	30
Gymnasiums, main floors and balconies	100	Hotels:	
Hospitals:		Guest rooms	40
Operating rooms, laboratories	60	Public rooms	100
Private rooms	40	Corridors to public rooms	100
Wards	40	Corridors	80
Corridors, above first floor	80	Reviewing stands, grandstands and bleachers	100[d]
Hotels (see Residential)		Schools:	
Institutional-residential care (see Residential)		Classrooms	40
Libraries:		Corridors	80
Reading rooms	60	Sidewalks, vehicular driveways, subject to trucking	250
Stack rooms	150	Skating rinks	100
Stairs and exits	100	Stores:	
Storage areas:		Retail	75
Light	125	Wholesale	100
Heavy	250	Yards and terraces, pedestrians	100

Table 1-16.(b) Minimum Concentrated Loads[a]

Location	Pounds[b]
Elevator machine room grating, (on area of 4 sq. in.)	300
Finish light floor plate construction (on area of 1 sq. in.)	200
Garages[c]	
Greenhouse roof bars, purlins, and rafters	100
Hospitals and ward rooms	1000
Libraries	1000
Manufacturing and storage buildings	2000
Mercantile areas	2000
Offices	2000
Schools	1000
Scuttles, skylight ribs, and accessible ceilings	200
Sidewalks or vehicular driveways subject to trucking	8000
Stair treads (on area of 4 sq. in. of tread)	300

[a] From BOCA 1987.
[b] Applied to an area of 2.5 ft × 2.5 ft (762 mm × 762 mm) unless otherwise indicated. These are the minimum concentrated loads; if anticipated loads are higher, the actual loads are used.
[c] Minimum concentrated loads for garages or portions of buildings used for parking motor vehicles shall be:

1. For passenger cars accommodating not more than nine passengers, 2000 lbs acting on an area of 20 sq. in.
2. Mechanical parking structures without slab, passenger cars only, 1500 lbs per wheel; and
3. For trucks or buses, on slabs, maximum axle load on an area of 20 square inches.

[d] One lb = 4448N; One sq. in. = 645.16 sq. mm.

Since the 1930s, states have required architects and engineers to be licensed (by experience or by education and examination) and to prepare plans and specifications for the construction, of buildings that meet code requirements. The designer uses the building codes in addition to the owner's program requirements to provide the basis for a safe, functional, and esthetic product.

WIND LOADS AND SNOW LOADS

Wind loads are considered lateral live loads. Cities have made special provisions to prescribe wind load requirements since slightly after 1900. Kidder (1902) gives a very detailed tabulation of the wind pressure per square foot normal to surfaces inclined at different angles to the horizon. The table is for a horizontal wind pressure of 40 psf, apparently the recommended pressure according to Kidder (see Table 1-17).

Kidder further states, "Until of late years it has been the general custom to add the wind pressure in with the weight of snow and roof; and, although this is evidently not the proper way to do, yet for wooden trusses it gives results which are perhaps sufficiently accurate for all practical purposes; and, if caution is taken to put in extra bracing wherever any four sided figure occurs, this method will answer very well for wooden trusses" (ibid., 523).

Kidder also advises that the preferred method of analysis is to compute the stresses from the live and dead loads separately and combine them, as applicable, to obtain the highest overall stress in each individual member. This is the accepted method of analyzing stresses in members today. The object of combining the analysis of dead load, occupant live load, and wind and snow loads is to determine the combination of dead and live loads that produces the maximum stress in the member. Member selection is then based upon utilizing the member that has an actual stress that approaches but does not exceed the allowable stress while still satisfying the code. Code allowable stresses will be discussed in the chapters on the various materials. The allowable stresses for structural materials have a history similar to that of the allowable loads; it will be necessary to understand fully the dates of availability of each material.

The 40 psf wind pressure utilized by Kidder in his table of normal wind pressures (see Table 1-17) represents a wind velocity of approximately 90 mph. Kidder credits this figure as being from *The Builder's Guide and Price Book* (see Table 1-18). Trautwine (1888) has a section on wind that credits Smeaton with the development of the table reproduced in Table 1-18, which Trautwine states was based upon the following equation (ibid., 216):

$$F_{psf} = \frac{(Vel_{psf})^2}{200}$$

where:
F = pressure
Vel = velocity in miles per hour

Table 1-17. Normal Wind Pressure (Horizontal Wind Pressure of 40 psf)[a]

Angle of Roof		Normal Pressure	Angle of Roof		Normal Pressure
Degrees	Rise in./ft.		Degrees	Rise in./ft.	
5°	1.0 in.	5.2 psf	35°	8.4 in.	31.1 psf
10°	2.13 in.	9.6 psf	40°	10.0 in.	33.4 psf
15°	3.20 in.	14.0 psf	45°	12.0 in.	36.1 psf
20°	4.38 in.	18.3 psf	50°	14.3 in.	38.1 psf
25°	5.5 in.	22.5 psf	55°	17.13 in.	39.6 psf
30°	6.9 in.	26.5 psf	60°	20.75 in.	40.0 psf

[a] From Kidder 1902, 523.

Table 1-18. Force of the Wind[a]

Miles per Hour	Force (psf)	Description
1	0.005	Hardly perceptible
2	0.020	
3	0.044	Just perceptible
4	0.079	
5	0.123	Gentle breeze
10	0.492	
15	1.107	Pleasant breeze
20	1.970	
25	3.067	Brisk gale
30	4.429	
35	6.027	High wind
40	7.870	
45	9.900	Very high wind
50	12.304	Storm
60	17.733	
70	24.153	Great storm
80	31.490	
100	49.200	Hurricane

[a]From Kiddler 1902, 725.

This formula as credited to Smeaton is considered probably defective by Trautwine. However, he offers no other solution. Trautwine does present theories on the effect of wind pressures on flat surfaces versus curved surfaces which are similar in concept to todays methods of analysis. And he states that the effect of the force of the wind on a large surface is much greater than that on a smaller surface. Trautwine also states that "Tredgold recommends to allow 40 psf. for roofs for the pressure of wind against them" (ibid.).

Trautwine states that he does not believe that roofs will receive the full force of 40 psf due to the fact that roofs slope and the inclination of the roof will result in a smaller force than a wind produces against a vertical surface. He states that probably the force in such cases varies approximately as the "sines of the angle" of the slopes. He further states that the proximity of the building to others and the shape of the building will have an effect on the amount of force that results from the wind pressure. Trautwine cites examples of wind velocities and the resultant pressures:

> Liverpool, England, 1860, a wind of 38 mph produced a pressure of 14 psf on a perpendicular object.
> Liverpool, England, 1860, a wind of 70 mph (a record) produced a pressure of 42 psf.

Trautwine states that these figures are the equivalent of the Smeaton formula with a divisor of 100 instead of 200. This observation is fairly accurate in that the denominator that produces these observed pressures is approximately 100, but the formula itself is not the same as that in use today. He further states that the Smeaton rule is used by the U.S. Signal Service. In a footnote, Trautwine states that 8 psf for ordinary double-sloping roofs, or 16 psf for shed roofs, is sufficient for design. In a later chapter, in an analysis of trusses, a suggested combined snow and wind load of 20 psf is utilized. This truss example utilizes 12 psf for snow, which allows 8 psf for wind. The loads were considered to be perpendicular (normal) to the slope of the roof—a precept that was beginning to evolve in the late 1880s and is still in use today. International Correspondence Schools (1905) gives a very detailed method for analyzing wind pressures on roofs and vertical walls of buildings. The notes to Table 1-12 indicate that Boston in 1919 required a pressure from wind of 30 psf on vertical surfaces for "roof design." There is no indication of what is considered the appropiate method of proportioning the wind load for inclined surfaces. It may be that designers used the "sine of the angle" of inclination, as may have been standard practice. See Figure 1-11 for a diagram by this author that depicts the geometry of the roof and the typical application of the wind force. The load on the roof was considered to be applied to

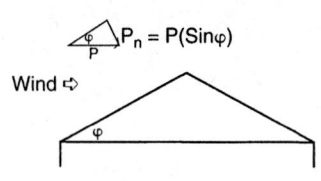

Figure 1-11.

the windward side of the roof only. There was no recognition of the fact that the leeward side of the roof would sustain an actual uplift force from a positive wind force on the windward side.

Patton (1893) refers to the formula published by Trautwine (1888). Mahan (1885) devotes a paragraph to the force of the wind on roofs. Mahan states that Smeaton has observed a force on a flat surface of 32 psf resulting from a hurricane. Mahan also states that Tredgold recommends an allowance of 40 psf (Mahan 1885, 394). He also quotes the record wind at Liverpool that produced 42 psf and states that during a violent gale in Scotland a wind gauge once indicated 45 psf (an indication that a gauge was calibrated to read wind force directly). He makes no mention of the formula that makes this conversion or the method that he proposes to apply the force of the wind to the roof—a clear indication of the lack of experiments or the development of theories by the profession in this period. International Correspondence Schools (1912) states that the wind is considered as blowing in a horizontal direction, but the resulting pressure upon the roof is always taken normal (at right angles) to the slope. This is similar to taking the sine of the angle of inclination, (see Figure 1-11). However, there is no mention of the use of the sine, and the values given for the normal pressures are not the result of that application. This ICS volume also states that the U.S. Signal Service experiment station at Mt. Washington, New Hampshire, has developed pressure forces corresponding to the wind velocities that may be encountered by buildings (see Table 1-19).

This ICS volume states that the wind pressure on a circular cylindrical surface is one-half that upon a flat surface of the same height and width (ibid., 67). While no proof in theory is given or a source quoted, it is evident that results of experimental work are now beginning to appear. This ICS volume also presents a table of wind pressures on roofs. The table gives the designer the wind pressure normal to a roof of varying slopes (see Table 1-20). Kidder (1905) has some very comprehensive paragraphs on the effect of wind on tall buildings. The new methods of design and bracing of skeletal frame buildings will be discussed in detail below in the sections that deal with the portal and cantilever methods of wind analysis.

In addition, in an appendix, Kidder (1905) gives the designer the latest information on the forces that result from various velocities of wind. In the section "Force of the Wind," Kidder recognizes experiments made by Professor Marvin of the U.S. Signal Service in 1890 as giving the most accurate data available at this time. He states Marvin's formula for the relationship between wind velocity and force pressure as follows (ibid., 1510):

$$P = 0.004V^2$$

In Marvin's formula the constant of 0.004 is different from Smeaton's constant of 0.005 for the same formula. This formula will be revised again before reaching its present form. Refinements and further research will produce the theory that wind velocities increase in height zones vertically, and therefore, design pressures on the vertical facade of a building increase in horizontal zones as building height increases. In another section, Kidder (1905) has an explanation and tables on the use of wind force on roofs of varying inclinations. See Tables 1-21 and 1-22 which reproduce three of Kidder's tables in which he presents new theories on wind loads.

Table 1-19. Wind Velocities and Pressures (U.S. Signal Service)[a]

Velocity (mph)	Pressure (psf)	
10	0.4	Fresh breeze
20	1.6	Stiff breeze
30	3.6	Strong wind
40	6.4	High wind
50	10.0	Storm
60	14.4	Violent storm
80	25.6	Hurricane
100	40.0	Violent hurricane

[a] From International Correspondence Schools 1912, 67.

Table 1-20. Wind Pressure on Roofs (psf)[a]

Rise in. per Foot of Run	Angle with Horizontal	Pitch Proportion of Rise to Span	Wind Pressure Normal to Slope
4	18°—25'	1/6	16.8 psf
6	26°—33'	1/4	23.7 psf
8	33°—41'	1/3	29.1 psf
12	45°—00'	1/2	36.1 psf
16	53°—07'	2/3	38.7 psf
18	56°—20'	3/4	39.3 psf
24	63°—27'	1	40.0 psf

[a] From International Correspondence Schools 1912, 68.

Kidder (1905) introduces several new thoughts on the design for wind forces:

1. Regional values for snow and wind combined according to geographical location
2. Introduction of a horizontal component of wind pressure on an inclined roof surface in addition to the normal pressure component that had been the standard to date
3. The introduction of a concept of wind bracing for tall buildings, both masonry and skeletal frame

Table 1-21.(a) Force of the Wind[a]

Miles per Hour	Feet per Minute	Feet per Second	Force, in Pounds, per Square Foot	Description
1	88	1.47	0.004	Hardly perceptible
2	176	2.93	0.014	
3	264	4.40	0.036	Just perceptible
4	352	5.87	0.064	
5	440	7.33	0.100	Gentle breeze
10	880	14.67	0.400	
15	1320	22.0	0.900	Pleasant breeze
20	1760	29.30	1.600	
25	2200	36.67	2.500	Brisk gale
30	2640	44.00	3.600	
35	3080	51.30	4.900	High wind
40	3520	58.60	6.400	
45	3960	66.00	8.100	Very high wind
50	4400	73.00	10.000	Storm
60	5280	88.00	14.400	
70	6160	102.70	19.600	Great storm
80	7040	117.30	25.600	
100	8800	146.60	40.000	Hurricane

Table 1-21.(b) Allowance for Wind and Snow Combined in Pounds per Square Foot of Roof Surface[b]

| Location | Pitch of Roof | | | | | |
	60°	45°	1/3	1/4	1/5	1/6
Northwest states	30	30	25	30	37	45
New England states	30	30	25	25	35	40
Rocky Mountain states	30	30	25	25	27	35
Central states	30	30	25	25	22	30
Southern & Pacific states	30	30	25	25	22	20

[a] From Kidder 1905, 1510. Kidder notes: "The above table is based on Prof. Marvin's formula and is quoted by Profs. Turneaure and Ketchum and is utilized by Trautwine's Pocket-Book."
[b] From ibid., 950. Kidder notes: "No roof truss should be proportioned for a total load of less than 40 psf except flat roofs in warm climates."

Table 1-22. Normal and Horizontal Wind Pressure on Roofs for 30 psf: Horizontal Pressure Against a Vertical Surface[a]

Inclination	Normal psf	Horizontal psf
5°	3.9	0.3
10°	7.2	1.2
15°	10.5	
18°–26′ (1/6 pitch)	13.0	4.0
20°	13.7	4.5
21°–48′ (1/5 pitch)	15.0	6.0
25°	16.9	
26°–34′ (1/4 pitch)	18.0	8.0
30°	19.9	10.0
30°–41′ (1/3 pitch)	22.0	12.0
35°	22.6	
40°	25.1	15.9
45° (1/2 pitch)	27.1	19.0
50°	28.6	21.9
55°	29.7	
60°	30.0	25.5

For horizontal wind pressure of 40 psf the pressure forces given above should be increased by one-third.
[a] From Kidder 1905, 951.

Ketchum (1918) publishes live loads from building laws of 1913 and earlier that contained wind load requirements (see Table 1-9). The third edition of the same work (Ketchum 1924) has many references and example design solutions for wind loads on buildings, roofs, mill buildings, towers, and chimneys. These are by far the most comprehensive data in any of the early volumes researched. Ketchum states the wind pressure, P, in pounds per square foot on a flat surface normal to the direction of the wind for any given velocity, V, in miles per hour as (ibid., 5):

$$P = 0.004 \, V^2$$

From the note above, as with all innovations and refinements on values such as loads, there is always the tendency by liberal designers to bring values below safe tolerances. Ketchum also states that the pressure on other than flat surfaces may be taken as a percentage of the value of P above:

1. 80 percent for a rectangular building
2. 67 percent on the convex side of cylinders
3. 115–130 percent on the concave side of cylinders, channels, and flat cups
4. 130–170 percent on the concave sides of spheres and deep cups

The recommended wind pressures for buildings are stated as 30 psf on the side and the normal component of a horizontal pressure of 30 psf on the roof, except for exposed locations. Judgment is recommended for exposed locations. The pressure on inclined surfaces is calculated by several methods in this volume. See Figure 1-11 for a graphical presentation of an inclined roof and the force vectors involved in the resolution of horizontal wind pressures. The state of the art in computation of wind forces on buildings was in a period of unusual transition during the second and third decades of the twentieth century. Many theorists were working on experiments with wind on buildings and mathematical equations for design solutions.

Computation of the normal pressure component of the horizontal wind pressure, P_n, by each method is as follows:

Hutton's formula: $P_n = P (\text{Sin } A)^{(1.842 \text{Cos} A - 1)}$

Duchemin's formula: $P_n = \dfrac{P(2 \text{ Sin } A)}{(1 + \text{Sin}^2 A)}$

$P_h = \dfrac{P(2 \text{ Sin}^2 A)}{(1 + \text{Sin}^2 A)}$

$P_l = \dfrac{P(2 \text{ Sin } A)(\text{Cos } A)}{(1 + \text{Sin}^2 A)}$

Straight line method: $P_n = P \left(\dfrac{A}{45°} \right)$

Historical method: $P_n = P (\text{Sin } A)$

where (for all formulas):
- P_n = pressure component normal to the roof
- P_h = pressure component in the horizontal direction (Not used today; the normal component is considered adequate since it has a horizontal vector)
- P_l = pressure at right angles to the wind; it may be inward pressure or outward pressure
- A = angle of inclination of the roof, measured from the horizontal

Ketchum states that the Hutton formula is based upon very crude and probably erroneous experiments and is therefore not recommended. He states that the Duchemin formula was based upon very careful experiments and should be considered the most accurate of those to choose from. He also states that the straight line formula agrees with experiments quite closely and that it was preferred by many engineers because of its simplicity (Ketchum 1924, 5). See Figure 1-12 for a graphical comparison of the formulae for normal wind pressures based upon horizontal wind pressures of 20 psf, 30 psf, and 40 psf from Ketchum. The graph shows three of the methods in a plot of normal pressures for angles of the inclination of a roof.

The 1987 BOCA National Building Code provides extensive live load requirements for roofs for snow and wind and makes extensive recommendations based upon occupancy categories. The required loads from snow are increased in intensity for buildings of the type that has a higher human occupancy while also being a variable that is dependent upon geographical location. The 1987 Code has a table of minimum roof loads that are required for design.

WIND AND THE MODERN BUILDING CODE

Contemporary building codes require that all exposed structures be designed to resist wind in any direction. In most codes, the designer must first determine the maximum wind velocity in miles per hour for the site from a map of the United States and then obtain the actual design pressures in accordance with the proper exposure coefficient from the tables that indicate pressures for the various wind speeds and building heights. In addition, buildings must be designed to resist the overturning moment due to the wind load. The overturning moment shall not exceed two-thirds of the dead load stabilizing moment unless the building is anchored to resist the excess moment. Horizontal shear at the base must also be considered. Wind force tends to push a building. If the shear capacity at the base is insufficient to prevent sliding, proper anchorage must also be provided for sliding. Building skin or cladding must be capable of assuming the wind force and distributing the force to the building's primary structural system, which is responsible for the dissipation of the overall wind force. Wind must also be considered during construction and erection of the building. In some instances, this is most important, but because this work is basically oriented to restoration and rehabilitation of historic structures, design requirements relating to new construction will not be considered.

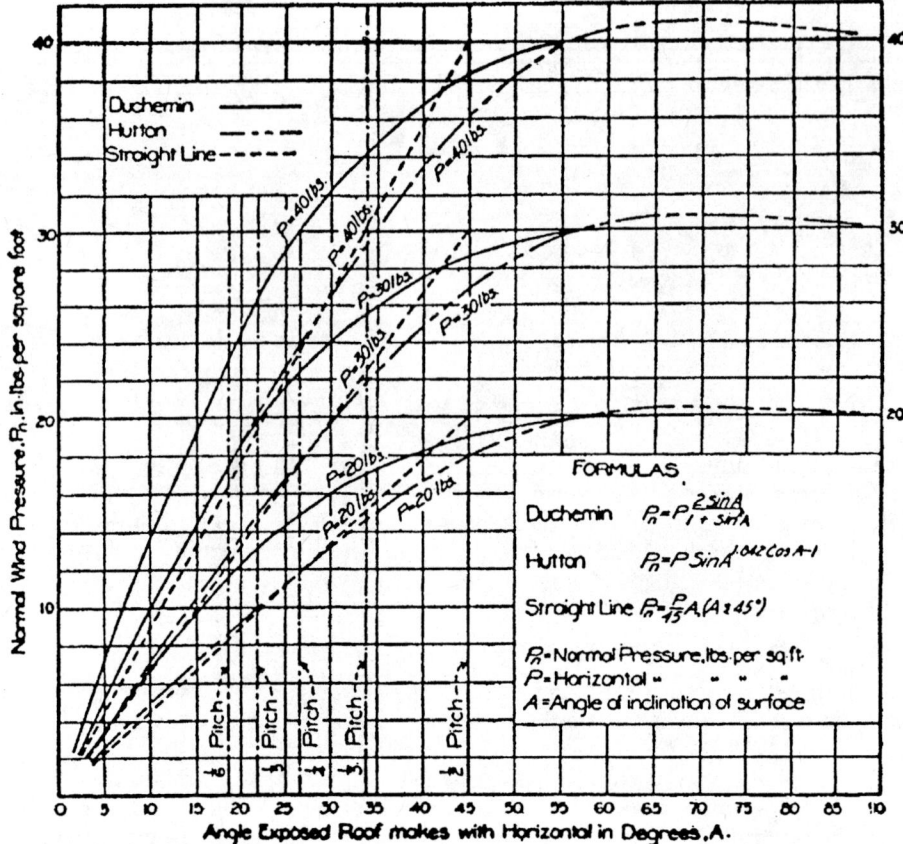

Figure 1-12. Normal wind load on roof according to different formulas. (*From Ketchum 1924, p. 6.*)

Design wind pressures for the main structural system should be determined by the following formula (BOCA 1987, 238):

$$P_d = P_e(I^2)(C_p)$$

$$P_d = P_z + P_h$$

where:
- P_d = design pressure for the main wind resisting system, in psf
- P_z = windward design pressure at height z above grade, in psf
- P_h = leeward or sidewall pressure at height z above grade, in psf
- P_e = effective velocity pressure including gust effect for exposure type B or C
- I = importance factor of the building
- C_p = external pressure coefficient

In earlier times, designers did not know the full effect of wind as we know it today from near-full-scale experiments. The existence of positive and negative pressures that result from the direction and velocity of the wind has been the subject of much deliberation for many years. Wind analysis of today is more a product of the combination of experience and experimentation than any other area of structural design. For years, wind analysis has been conducted by experts on the subject, and they have had many theories. As has been shown in this study, the existence of negative pressures on leeward sides of gable roofs has been known only since the second quarter of the twentieth century. Negative pressures on the windward side of gable roofs with very low pitches have been known for only a few years, and a part of code requirements as a positive/negative combination for less than 10 years. At lower pitches the pressure on the windward side of a roof is negative, and it is positive from a pitch of 35° (8 in 12) and larger.

The reason we are still experiencing changes in methods of wind analysis is that our knowledge is still expanding on the subject. The results of a windstorm are obvious to the observer, but the direction and magnitude of the damaging gusts of wind are not totally predictable. We do not have gauges and/or monitoring equipment on every building or structure, nor are we funded to the extent that continuous tests or monitoring over a broad geographical area can be performed. Wind analysis is still mainly a laboratory science. We are able to produce large-magnitude winds in a laboratory situation, but they are mostly unidirectional, and we are not reproducing gusts that change left/right and up/down at any instance as in nature. The effects of adjacent buildings, wind corridors, and other factors are not always duplicated in a laboratory setting.

The basic wind speed, in miles per hour, that must be used is dependent upon the geographical location of the building. The wind velocity is given on the isoline map of the country; however, the designer is on notice to consider that in some areas the local prevailing conditions require that a higher value must be used. The BOCA (1987) map of the wind speeds used in the design of buildings and other structures indicates those areas that require special consideration. For purposes of designing for wind loads, buildings have been classified by BOCA officials in accordance with their relative importance in terms of human occupancy or their value for civil defense. The importance is chosen from a table of classifications by building usage and is placed in formulae as a factor to be applied to the effective velocity pressure as a part of the process of determining the design pressure. Shape or form coefficients shall be integrated into the design for wind pressures as factors that are used in the equations for the total design pressure, P_f, as required for the specific structure, size, or shape.

The 1987 BOCA National Building Code also incorporates new and comprehensive earthquake requirements into the design of every structure that meets the category of need or the combination of use, frequency zone, or importance factor that necessitates this protection. These requirements are general and inclusive, and it is notable that the code assumes that earthquake design is a requirement in all but the cases that it specifically excepts. Because it is only recently that the 1987 version of the Code has introduced this requirement and most jurisdictions have adopted it for use, there is no record of its application to any historic buildings. Excerpts of the earthquake requirements are not included in this work due to the fact that most designers of adaptive reuse or rehabilitation projects on historic buildings will undoubtedly apply for and be granted exceptions from this portion of the code. One of the bases for the exception will be the age and proof of structural integrity of the buildings over the years. Owners will rightfully argue that a hundred-year-old building has proven itself capable of withstanding the forces of nature by virtue of its own longevity. Exceptions would be the prerogative of the local jurisdiction.

Normally, when a building is to be rehabilitated, if the proposed expenditures exceed 50 percent of the current market value of the building, the total structure must be brought up to code. This means that live load considerations must be analyzed in light of the new use, and the dead plus live load combinations must be within the modern code requirement. In most instances, historic buildings are capable of withstanding an earthquake of small magnitude with minimal damage, but major earthquakes can inflict a high degree of damage. The probability of major earthquakes is known for most areas, and designers must work with local code officials to determine the necessity of preparing a historic building for earthquake conditions in any area in which they are working.

LIVE LOAD REDUCTIONS

The principle of live load reductions is based upon the fact that, at any given hour, a building with multiple floors will not actually experience the design live load simultaneously on the entire area of every floor. Early designers contended that column and bearing walls should be designed for a reduced live load because, they felt the probability of all floors being fully loaded with 100 percent of their live loads at the same time was practically zero. As has been pointed out in this chapter, there was an ongoing controversy over the magnitude of the live loads for different occupancy categories.

Kidder (1905) states that in a tall building it is customary to reduce the column loads somewhat from the loads used in calculating floor beams, based on the theory that it is quite impossible for the entire floor area in every story to be loaded to the maximum live load limit at the same time. Kidder recommends that for all buildings except warehouses it would seem good practice to design the columns to carry all the dead loads and 75 percent of the assumed live load (ibid., 459). He also quotes The Building Law of Greater New York, which specifies that for buildings exceeding five stories in height the column loads shall be made up as follows:

> For the roof and top floor the full live loads shall be used;
> For each succeeding lower floor it shall be permissable to reduce the live load by 5 percent, until 50 percent of the live load is reached, when such reduced loads shall be used for all remaining floors (Ibid., 460).

International Correspondence Schools (1905), states that in warehouses built especially for the storage of heavy merchandise, where floors are likely at any time to be fully loaded, the beams, girders, columns, and foundations are always proportioned for the entire live and dead loads on all floors. Further, where the building exceeds four or five stories in height and is used for any purpose other than storage—for instance, a modern office building—it is customary to assume that certain members, while proportioned for the entire dead load, carry only a certain percentage of the live loads (ibid., 30). For an office building or similar structure it is permissible to calculate only 90 percent of the live load on girders, and on floor joists it is necessary to design for the full live load. The "5 percent rule" allowable for New York is generally followed by most principal American cities. Where permanent partitions exist, they should always be figured in the dead load; and where they are directly above a beam or girder, the member should be proportioned to sustain the additional weight without appreciable deflection. And where movable partitions occur or where there is a probability of the locations of permanent partitions being changed, it is usual to add 20 psf to the overall dead load to provide for such contingencies. The author does state that the foundations for an office building should be proportioned for the entire dead load and none of the live load, the latter being provided for by making the unit pressure on the footings and piers well within the safe unit bearing value of the soil (ibid., 32).

This volume has a table of the live load reductions, based upon the New York Law, that presents a case study of the live load reductions for columns for a 60-psf live load with a comparison of the same building for full live load consideration and a tabulation showing the percentage of savings. Note that the example does not provide for any live load reduction for the roof or for the floor immediately below the roof. It was considered wise to require the full live load on the uppermost floor level and on the roof. Roof loads were fairly well known by this time, and it was common for uppermost floors to be used for social gatherings. The designer of the building, then as now, was responsible for provision of structure to carry the loads required by the programmed use of the space. The table allows the designer to follow its layout for his own project needs (see Table 1-23).

International Correspondence Schools (1923) describes the method used by the City of New York as the "Live Load Reduction Factors for Vertical Supports," which is almost identical to the 5 percent rule. The reductions applied to commercial and office buildings not used as warehouses; warehouses were excluded from reductions. Columns were sized to withstand the total dead loads plus the full live load on the roof and topmost story and a reduced live load of 5 percent per floor below until 50 percent of the live load was reached. At the point of 50 percent, live loads on each successive floor had to remain at 50 percent until footings were encountered. To calculate the pressure under the footings, the full dead loads and the figured live loads on the lowest tier of columns, piers, or walls were taken (ibid., ch. 5, 13). This volume

Table 1-23. Live Load Reductions for Tall Buildings (60 psf Live Load on Floor, Column Loads Roof to Foundation)

Floors	Live Load psf on Each Floor	Live Load psf on Column from Floor above at 5% Reduction $(a_1) = (a-0.05a)$	Live Load psf on Column from All Floors above if No Reduction Made sum of (a)'s	Live Load psf on Column from All Floors above Increment of Reduction sum of (a_1)'s	Theoretical Percentage of Saving Instituted by Reduction of 5% at Each Floor (sum a's−sum a_1's) sum (a)'s
Roof	20	20.00	20	20.00	0
18	60	60.00	80	80.00	0
17	60	57.00	140	137.00	2.1
16	60	54.15	200	191.15	4.4
15	60	51.44	260	252.59	6.7
14	60	48.87	320	291.46	8.9
13	60	46.43	380	337.89	11.1
12	60	44.11	440	382.00	13.2
11	60	41.90	500	423.90	15.2
10	60	39.80	560	463.70	17.2
9	60	37.81	620	501.51	19.1
8	60	35.92	680	537.43	21.0
7	60	34.12	740	571.55	22.8
6	60	32.41	800	603.96	24.5
5	60	30.79	860	634.75	26.2
4	60	30.00	920	664.75	27.7
3	60	30.00	980	694.75	29.1
2	60	30.00	1040	724.75	30.3
1	60	30.00	1100	754.75	31.4

[a] From International Correspondence Schools 1905, 31.

includes a table of the live load reductions as an example for designers (see Table 1-24).

Ketchum (1924) presents criteria from the 1911 Chicago Building Ordinance pertaining to live load reductions. This ordinance notably states that columns must be designed for full dead and live loads or a minimum of 20,000 pounds, whichever is

Table 1-24. Reduced Live Loads for a 16-Story Office Building According to the New York City Building Code[a]

Floor	Maximum Live Loads as per Building Law (psf)	Allowable Percentage of Max. Live Load	Net Reduced Live Load on Columns and Walls (psf)
Roof	40	100	40
16	60	100	60
15	60	95	57
14	60	90	54
13	60	85	51
12	60	80	48
11	60	75	45
10	50	70	42
9	60	65	39
8	60	60	36
7	60	55	33
6	60	50	30
5	60	50	30
4	60	50	30
3	60	50	30
2	60	50	30
1	120[b]	50	60
TOTAL	1060		715

[a] From International Correspondence Schools 1923, ch. 5, 14.
[b] First floor, mercantile type loading—a requirement of the times.

larger. Further, the full live load must be used on roofs and all floors of buildings of five floors and less, and on buildings higher than five floors, the walls, piers, and columns of all buildings shall be designed to carry the full dead loads and not less than the proportions of the live loads given in Table 1-25.

The Steel Construction Manual (American Institute of Steel Construction 1930), the forerunner of the authorities on today's methodology, suggests reductions in live loads for buildings (except storage warehouses). The reduction criteria, attributed to the recommendations of the Building Code Committee of the Bureau of Standards of the U.S. Department of Commerce, allow for reduced live loads in the design of columns, piers or walls, foundations, trusses, and girders, as shown in Table 1-26.

The BOCA Code of 1987 allows a live load reduction for the design computations for beams and columns, with certain qualifying requirements and limitations (BOCA 1987, 275):

Members having an influence area of 400 square feet or more may be designed for a reduced live load determined by applying the following equations:

$$LL_{DESIGN} = L_o(0.25 + 15/(A_i)^{.5})$$

LL_{DESIGN} = Reduced design live load in psf.

L_o = Unreduced design live load from BOCA Code.

A_i = Influence Area in square feet.

(4 times the tributary area for a column.)

(2 times the tributary area for a beam.)

(1 times the panel area for two-way slab.)

Limitations:

The reduced live load shall not be less than 50% of the unreduced live load for members supporting one floor and not less than 40% of the unreduced live load for members supporting more than one floor.

Table 1-25. Percentage of Live Loads for Columns, Walls, and Piers: Chicago Building Ordinance (1911)[a]

Floor	17	16	15	14	13	12	11	10	9	8	7	6	5	4	3	2	1
17	85 percent																
16	80	85															
15	75	80	85														
14	70	75	80	85													
13	65	70	75	80	85												
12	60	65	70	75	80	85											
11	55	60	65	70	75	80	85										
10	50	55	60	65	70	75	80	85									
9	50	50	55	60	65	70	75	80	85								
8	50	50	50	55	60	65	70	75	80	85							
7	50	50	50	50	55	60	65	70	75	80	85						
6	50	50	50	50	50	55	60	65	70	75	80	85					
5	50	50	50	50	50	50	55	60	65	70	75	80	85				
4	50	50	50	50	50	50	50	55	60	65	70	75	80	85			
3	50	50	50	50	50	50	50	50	55	60	65	70	75	80	85		
2	50	50	50	50	50	50	50	50	50	55	60	65	70	75	80	85	
1	50	50	50	50	50	50	50	50	50	50	55	60	65	70	75	80	85

The entire Dead Loads and the percentage of the Live Loads shown above for walls, columns and piers for buildings above grade and the same shall apply to basement walls, columns and piers in determining the stress in foundations.
[a] From Ketchum 1924, 98.

Table 1-26. Reductions of Total Live Loads Carried

	Percent
Carrying one floor	0
Carrying two floors	10
Carrying three floors	20
Carrying four floors	30
Carrying five floors	40
Carrying six floors	45
Carrying seven or more floors	50

This criterion states that the reductions can be increased above this amount in determining the area of footings for buildings for human occupancy (private dwellings, hospitals, hotel guest rooms, offices, schools, reading rooms, etc.) a further reduction of one-half of the remaining live load from the table above is permitted.

[a] Reprinted, by permission, from American Institute of Steel Construction, *The Steel Construction Manual*, 1st ed., 53. © 1930 The American Institute of Steel Construction, Inc.

Live load reductions are an accepted part of the process that is followed today in the design of buildings. Reductions are allowed for the same reasons as in earlier times; the methods are different, but the result is similar. Girders and columns in buildings subject to normal loads are sized for a reduced live load determined by a method derived from statistical and theoretical data.

2
Foundation Systems of American Historic Buildings

EARLY AMERICAN BUILDINGS (TO 1860)

By the 1820s (the beginning point of this reference), there existed a complete network of building trades in the metropolitan areas of the country and a smaller, less organized system of traveling builders that served rural areas, smaller towns, and in some areas the frontier zones. As with architectural styles in the postcolonial period, the technology of building moved slowly from the East and metropolitan areas such as Chicago and Minneapolis toward the West. The era of the railroad shortened this time lag from what had been years in earlier periods to only a matter of weeks, and in some cases days, by the 1850s. In addition, building materials began to be transported by rail, and the centralized mill and/or production facilities were gradually beginning to evolve.

In the cities, architects were designing in the styles of the period, and innovation within the rules or guidelines was beginning to occur. Limitations of factors such as height and span were products of the capacities of the materials available to the designers. The limitations on buildings as imposed by architectural styles were generally not a factor. Only the larger metropolitan areas benefited much from building science or knowledgeable designers. Technology was advanced in major metropolitan areas; stress analysis methods used by knowledgeable designers from 1840–1940 were the same methods in use today. Smaller cities and towns, especially in remote areas, were utilizing empirical methods until well beyond the middle of the twentieth century. In many ways this technology gap still exists today. In the 1990s we know a lot more about internal stresses, residual stresses, and other stresses than was known 100 years ago, and we have advanced very far in calculation methods. Indeterminate structural analysis and methods of moment distribution were in general use by the late 1930s. Virtual work and matrix analysis were certainly not available to the designers of the period of the building of the first skyscrapers, nor was the digital computer; however, many of these new methods of calculating are applied to the same theories of stress analysis that were in use in the 1880s and 1890s.

The architects of the nineteenth century were fully aware of the importance of properly working out the foundations for their structures. It is significant that formal testing of soils had been going on since before the mid-nineteenth century. Laboratory analysis had produced data for use by the 1840s. Field sampling for specific projects, however, was limited to only the most complex of buildings, bridges, and dams. Testing of subsurface materials at the site of less than major buildings was not done in a scientific manner. Architects and engineers may simply have augered a hole or holes at foundation locations to depth of foundations to satisfy their questions about water, visual substrata identification, or location of bedrock. Textbooks of the period describe foundations in mostly the same manner. The foundation, as applied to buildings, bridges, and other structures, is considered as that portion of the structure resting on the rock or soil (Hool and Johnson 1929, 347).

While it is true that the level of design sophistication for building foundations was a factor of the importance of the building and the size of the city, the architects and engineers of the time were obviously aware of the fact that large, heavy structures needed special consideration for designing the foundations. It is also true, especially in the smaller towns, that in many instances there was practically no design for foundations for small- to medium-sized buildings. This author has found numerous buildings where the first courses of brick, while laid below frostline, were actually laid directly on the substrata beneath the building. In many cases there was a reverse corbeling or belling out to a thickness of approximately twice the wall thickness in an effort by the designer or builder to lower the unit load of the building on the bearing strata (see Figure 2-1). It appears that the designer/builder was responding to a tradition of following a rule of twice thickness, a system that appears too often and is too widespread geographically to be based on anything but tradition. It appears with equal frequency in areas of good subsurface conditions and areas of poor-quality soils.

Experienced designers of today appreciate the knowledge of the earlier builders. Skill, experience, and intuitive judgment were handed down or taught through the apprentice system. The simplicity of the building system and the compatibility of the materials in use in the nineteenth century was also a contributing factor to the success of buildings. It may be true that only the better constructed buildings have survived for our observation at this point. Past builders, like their modern descendants, must have had failures. We are aware of the nature and complexity of only a few notable early failures. The United States Capitol, the Washington Monument, and the Congressional Library, as well as other structures in the District of Columbia, have inherent foundation problems due to the conditions of the substrata in the area. Chicago is a city built over a thick stratum of damp plastic clay below which is a layer of compacted gravel followed by bedrock. The rule for lightweight buildings was not to penetrate the drier, more compact clay at the upper portion of the strata. The heavier building—the skyscraper—would require piles or caissons that bore directly upon the bedrock. Boston and New York both have areas that have been reclaimed from the sea, and reclaimed land is always a poor place to design building foundations.

In many instances where large stones were available, designers utilized flat stones or cut stones (to the shape desired) of sufficient width to develop a belled out foundation for the building wall or column. They were following the same theory of greater width at the point of loading of the bearing strata producing a lower unit load and thus minimizing settlement, etc. The rule of thumb was to use stones not less than 8 inches in thickness, resulting in a footing of the type shown in Figure 2-2. The rule of a 60° effective spread would keep the cantilevered stone projection within its flexural limits.

The brick building of the seventeenth, eighteenth, and nineteenth centuries was by far the most durable structure of that building era. This was largely a product of

Figure 2-1. Belling out for bearing width. (*From International Correspondence Schools, 1923, sec. 5, 25–28.*)

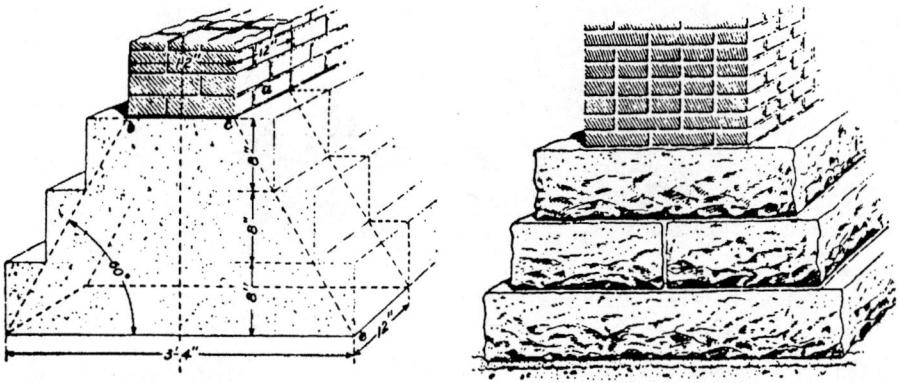

Figure 2-2. Belled out footings of stone. (*From International Correspondence 1923, sec. 5, 2b.*)

the inherent compatibility of the brick and the lime-sand mortars of the period. Even though the brick and the lime-sand mortars were usually manufactured at the site, in the early periods the uniformity of the properties of each of these materials from one region to another is notable. The process for slaking lime produced a near-pure chemical product, quicklime, which was equally available from natural limestone or crushed oyster shells if the region was near the coast and limestone was not quarried nearby. Sand is also a fairly consistent commodity from region to region, and if the builder was careful not to introduce impurities, his mortars were of excellent quality. Mix ratios that were purely the product of the user were also very similar in all parts of the country. One reason for this very compatible mix of product and building is that the size of the building and the magnitude of the loads were also complementary. Products of this era were compatible in that the coefficients of expansion (from temperature extremes) were about the same for hand-made bricks, stones, and the lime-sand mortars. Another reason is that as a wall was built, the thickness of the wall as required for lateral stability during construction gave the wall an inherent strength that was more than adequate for the building. Foundation loads for commercial and residential buildings of this era were also inherently low, partly due to the size and height of the buildings. In addition, when heavy loads were to be supported, the designer was very conservative in his approach.

Reference books on building construction were very scarce in the first half of the nineteenth century in the United States. Most volumes available at this time were printed in England and brought to the United States by craftsman immigrants. Like many other building methods, foundation design was basically an empirical tradition that evolved through the craft apprenticeship system. The architects and master builders of this time used trade rules handed down through the crafts, similar to those laid out by Joseph Moxon of London in 1703 and in later editions (Moxon 1703). Moxon describes how initially the mason should probe the trench of the wall or cellar with a "croe-bar" or "borer" to see whether the foundations are all sound and fit to bear the weight that is to be set upon them. Moxon thus implies a visual and physical assessment. He continues that, if the mason finds any part of the foundations defective, he ought to dig it deeper until he comes to firm ground; or, if it proves to be loose, or made ground (fill) to a great depth, he must take care to make it good and sufficient to carry the loads, which may be done in several ways (ibid., 254):

1. *Stones:* If the foundation is not too loose, it may be made good by ramming in stones (medium to large) with a heavy rammer. Stones should be placed in width about one foot wider than the trench on either side of the wall. Walls should have a "basis" or "footing" at least 4 inches on a side broader than the thickness of the wall. (Here he is describing the reverse corbel or belling out of the base of the brick wall.)
2. *Cribbage:* If the foundation is softer than 1. allows, the trench must be lowered and oak timbers (as long as the trench is wide, including the one foot on either side of the wall) must be laid cross the foundation and rammed into the soft substrata. Upon this foundation, planks are to be laid longitu-

Figure 2-3. Inverted arch foundations. (*From Mahan 1885, 254.*)

dinally upon the cross-timbers and attached with spikes. Upon this cribbing the wall (width including bell) is to be laid.

3. *Piles:* If the foundation stratum is so bad that 2. will not do, piles of heart of oak of such length as will reach ground must be utilized. The diameter of the piles is to be one-twelfth of their length. After the piles are driven, they are to be sawn off level and the planks are to be laid upon them as in 2. The number of piles is to be as high as possible.

Moxon further describes the method of using semicircular, pointed, and inverted arches of brick to span inadequate sections of foundation substrata. Each type of arch shown and described represents a technique that employs the ability of arch geometry to exchange a redirected vertical load and horizontal thrust for the continuous vertical uniform load of a wall. While this inherent characteristic of an arch was and still is known, an intricate geometric design was required to make it function correctly, and the outcome of a successful system was still a set of concentrated loads at points in the wall rather than a continuous uniform load of the wall (see Figure 2-3). Obviously, if there was a narrow section of weak strata passing through the building, this device would provide a method of spanning short sections of wall. The concentrated loads would have had to be dealt with or some assessment made of the degree of loading at those points. A narrow band of weak or soft strata dividing two sections of adequate material almost implies some form of vertical stratification instead of the usual horizontal stratification, the normal way most sedimentatious layering occured. It is not known to how much advantage builders of the first half of the nineteenth century utilized this method of spanning weak sections of substrata with arches. We encounter this type of foundation occasionally when working on historic buildings. Inverted arches occur more frequently on buildings that have rows of columns and/or facades divided into bays, such as storefronts, and result in a series of concentrated loads along the column row or the facade. There are many examples of stone consolidation, as in method 1. above, and the use of timber cribbing and timber piles, as in methods 2. and 3., as a means of strengthening foundations.

It appears that in the United States empirical methods and traditional craftsman responses to foundation designs were used almost exclusively up to the 1860s. Technical advances of the 1870s and 1880s brought about radical changes in building. Larger, taller, heavier structures were constructed beginning at this time, which meant increased loads on walls and columns, and thus on foundations which demanded design solutions. However, empirical methods continued until well into the twentieth century in small to medium-sized cities and towns, due to the problems of communication and technical education. In small towns this phenomenon continues today.

EARLY SOIL TESTING METHODS IN AMERICA

While architects, engineers, and builders relied exclusively on intuitive judgment and experience in estimating the bearing capacity of soil substrata until approximately the mid-nineteenth century, and even later in remote areas, on important buildings in larger cities, architects and engineers would have made investigative borings, drilling (augering) in several locations under the proposed building to ascertain the location of bedrock or the type of material at the various depths under the building. The architect or engineer would physically compress material from the various depths in his hand to determine empirically the approximate moisture content and compactability and estimate the strength or lack of strength for use as foundation-bearing material. If the location of bedrock was too deep for bearing the foundation directly upon the bedrock to be an economical solution, the architect or engineer would determine if he felt that the bearing capacity of the strata just below frostline would carry the load of the building without excessive settlement. If the soil was too moist, he might determine that piles were required. In many cases, other buildings would have been previously constructed in the close proximity, and the experience of other architects and engineers would be utilized. Often, open excavations for other buildings might have revealed the characteristics of the soils to the desired depths, or an "exploration pit" would be dug on the site for visual and physical inspection. Buildings were usually small and the relative weights/loads light, and if the architect or engineer was experienced, conservative, and fortunate, the resulting building would be relatively free of settlement cracks or other foundation-related problems. There were many problems however, as these methods were not always successful. Experience and analysis of earlier failures and the emergence of a more extensive academic approach helped to minimize these nonsuccesses as larger numbers of buildings evolved. The building industry also benefited from the fact that most of the major commissions were obviously occurring in the larger cities, where the more experienced, educated, and scientific architects and engineers lived and practiced. Thus, by circumstance of location and proximity, the more complicated work was usually done by the more competent designers.

With the advent of taller, heavier buildings in the second half of the nineteenth century, the need for a more exact determination of the capacity of the subsurface materials was obvious. A more refined method of evaluating the capacity of slightly compactable to moderately compactable soil strata was needed. Extremely moist, marshy, or extremely stiff soil strata presented fairly obvious solutions, while a soil that compacted and allowed vertical settlement over time was very difficult to identify. These softer clays are very abundant throughout the country, and they will allow subsidence if loaded too heavily. If properly identified, and if the loads are kept within their limits by utilization of wide-based footings, this settlement can be either minimized or eliminated. Fortunately, as the need for spread footings (wide-based footings) was recognized, the technology for soil evaluation was being developed simultaneously. The new heavier, taller buildings also brought about other related technical developments.

Testing in the period 1850–1880 was utilized as a design necessity on only the most important buildings; government buildings and buildings of major size and complexity. In this era, several methods were used in accordance with the desired results or the level of experience of the architect or engineer (Hool and Kinne 1923, 1–20):

Test Pits

Test pits were considered by many to have been the best method to determine the subsurface conditions for smaller lighter-weight to medium-weight buildings that were to bear on subsurface strata 8 to 12 ft below the surface. The pit could have been as large as desired and to a sufficient depth below the level of foundation bearing to give an accurate portrait of the soil condition at bearing point. This method allowed the architect or engineer to see the subsurface conditions in an actual cross-section or profile of the soil layers. The firmness of the material, its water content, its tendency to run horizontally or to collapse or cave in due to a condition of too much or too little moisture, could be analyzed in the pit wall or by physically handling samples from the material removed from the pit.

Rod Test (Sounding Rod)

A 5/8–7/8-in. diameter solid bar of tool steel (tempered), manufactured in sections, was driven into the soil in such a manner as to probe for dense, hard compacted material or bedrock. Each section of the rod was threaded on both ends for couplings, except the lowest section, which was pointed for driving. A special upper-end attachment of a short threaded "driving cap," usually mounted with a hard wood replaceable end, allowed the rod to be driven to depths required. This method required a wood or steel frame above the ground for use in the withdrawing operation. Initially, the sounding rod could be driven into the ground by two or three men using a device that allowed them to apply a twist and downward push on the rod (see Figure 2-4).

As depth was achieved, a 10–12 lb hammer (maul) was used to drive the rod into the earth. Naturally, special devices, exchangeable heads, and withdrawing hooks were used for the driving and withdrawing operations to prevent damage to the threaded ends of the rods. Sections of the rods were normally 4–5 ft long and were added as needed to reach to the depth necessary. The practical maximum depth of this type of sounding was 30–40 ft below the surface. It was impossible to determine with any degree of accuracy the nature of the material penetrated with the sounding rod, or of the hard strata below. The nature of the hard strata was often judged to some extent by the action of the maul on the rod in driving (Hool and Johnson 1929, 3).

Experienced engineers could intrepret the action of the rod during the driving operation, the vertical movement per blow from the maul, etc. If rock was encountered, there would be a sharp rebound and no further penetration. In sand, gravel, or hard clay, the blow would also "ring" or sound "dead," but penetration would continue to occur. In a saturated weak stratum, the blows of the maul would sound distinct and the penetration per blow would be very large. Caution was advised against misintrepretation of results; a large boulder or a thin rock outcrop could be mistaken for bedrock. All authors state that many soundings were required and a comparative analysis of the results would give indications of the validity of the depth of bedrock in certain soundings or groups of soundings, which might have necessitated additional soundings to make further determinations. Bedrock usually runs on a flat or sloped plane and could be plotted on a topographic map of the site and suspicious locations identified. Two to three men could make approximately 10 soundings 20–25 ft deep in an eight-hour day.

Auger Borings

Borings made with an auger that was designed to push its cuttings up the spiral helix by the action of rotation of the auger could provide samples for analysis from the bottom of the test hole. The size of the boring was from 2–6 in. in diameter. Augers were hand-driven to shallow depths or machine-driven where deeper borings were required. Auger-type post hole diggers were sometimes adapted to drill holes as deep as 10–12 ft below grade. One method used to check strata below the wall footing was

Figure 2-4. Rod test with pulling frame. (*From International Correspondence Schools 1923, sec. 1.*)

to hand-auger 4–6 ft below the bottom of footing after excavations for footings had been made and evaluate the material from the bearing strata visually.

This was not in accordance with the principle of preconstruction investigation (since the foundation trench was already excavated); however, it did allow the architect or engineer to determine whether the stratum that supported the proposed foundations/wall footings was thick enough to provide a safe bearing surface. In normal drill operations, three to six men were used to drill test holes 50–60 ft deep (Hool and Kinne 1923). The auger was turned down by the men with levers attached to the top of the rod (see Figure 2-5). If visual inspection of the material was desired at different levels, the auger was rotated enough to embed its full helix into the substrata, then pulled out of the hole and the material carefully cleaned from the auger helix and inspected. This, of course, allowed only for physical inspection and hand squeezing and intrepretation, since the sample was "disturbed" in being taken. In the condition of sand or gravel, or wet material that would not stay in place along the wall of the hole as the operation proceeded to further depths, the boring hole had to be cased. The casing had to be of sufficient size that the auger and drill rods could fit inside the casing. Since the casing is slightly larger than the auger, it had to have a beveled cutting edge (drive shoe) to cut the earth as the casing was driven down. In some cases, when compact sand or gravel or wet material was at the depth of the auger, water jets were utilized to evacuate the loose material.

Wash Borings

The wash boring method of testing became the adopted universal standard by approximately 1910. This method utilized machine-operated well-drilling equipment, usually sinking a casing at the same time as the drill point. Both rotated at about the same level in the hole and moved downward at the same rate. The drill point was kept cool as needed by a steady flow of water that also cleaned out the cuttings. This process could easily drill 150–200 ft in a comparatively short time. This method could be used to obtain relatively undisturbed samples of the material at a specific level as boring progressed by replacement of the drill bit with a thin, sharp-edged section of pipe at the end of the drill rod (see Figure 2-6). The sample pipe was vented at its upper end to prevent compaction as the rotation and downward motion caused the material to enter the pipe section. Upon the rods and pipe section being raised to the surface, the sample was brought to the surface in a relatively undisturbed state, carefully removed from the pipe (later pipe sections were designed to split apart), and placed in a glass bottle to keep its moisture content constant for later inspection. Samples were logged as to which boring and what depth they were obtained from.

Diamond Drill

A diamond drill bit that cut an annular hole in rock was introduced in the twentieth century. It was used on the same equipment as the wash boring and was placed into service when the operator felt that he had encountered a solid layer of stone or bedrock. It was sometimes desirable to know the depth of a rock stratum, or whether a layer of rock was of sufficient depth to support a building. The annular diamond bit would be approximately 10 feet long to allow for a thick section of rock or very hard sample to be obtained before removal of the bit. A special mechanism was provided to break off the core sample at the bottom of the drilled hole before the rods and bit were removed. In the case of wash borings and diamond drill cores, as well as any other samples analyzed by architects and engineers of this period, resulting data were not scientifically tested as they are today. Designers were still using empirical methods, mainly visual, refinement was gradually improving, and laboratory methods were in their infancy throughout the period.

Load Tests

In questionable soils and in some cases of freshly driven or placed piles, load tests were required by local code officials to ensure the safety of the proposed buildings. The New York Building Law allowed the Superintendent of Buildings to order borings

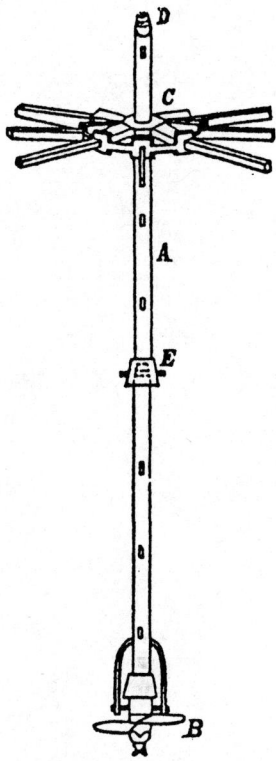

Figure 2-5. Auger boring. (*From Mahan 1885, 199 (upper portion); Trautwire 1888, 627 (lower portion).*)

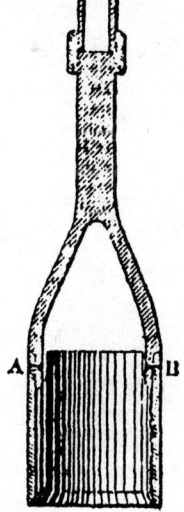

Figure 2-6. Sample pipe. (*From Trautwine 1888, 627.*)

or load tests when doubt existed as to the ability of the soil or the proposed foundation system to carry the loads. When load tests were specified, the New York Building Law had a specific proceedure for testing the bearing resistance of proposed foundation strata: (Huntington, 1929)

> The soil shall be tested in one or more places as shall be determined or conditions warrant, at the level at which the proposed footings are to be placed.
>
> Each test shall be so made as to load the soil over an area of not less than four square feet in any one place.
>
> The accepted safe load shall not exceed two-thirds of the final test load.
>
> The loading of the soil shall proceed as follows:
>
> (a.) The loads per square foot which it is proposed to impose on the soil shall be first applied and allowed to stand for at least 48 hours undisturbed, measurements or readings being taken each 24 hour or oftener in order to determine the settlement if any.
>
> (b.) After the expiration of 48 hours, the additional fifty percent excess load shall be applied and the total allowed to remain undisturbed for a period of at least six days, careful readings and measurements being taken once each 24 hours, or oftener, to determine the settlement.
>
> The test shall not be considered satisfactory or the result acceptable for at least two days, and the total test load shows no settlement for at least four days.

Load tests were performed in basically the same manner in the larger cities throughout the country. The area loaded varied in size from one square foot (12 × 12) (see Figure 2-7) to a full-scale loaded footing (of the actual size footing and loading proposed). In testing in the manner described above, architects and engineers usually performed a second step in the test procedure or a simultaneous evaluation that enabled them to ascertain the effect on the area immediately surrounding the point of loading. The test pit would be surrounded by one or two rows of stakes driven into the ground at points equally spaced radially around the pit and driven to exactly the same elevation for ease in observation. Then, as the test column was loaded (see Figure 2-7), the stakes (shown in this Figure) in the circular rings around the pit and column would be checked at the same intervals at which the column was observed for settlement to see if any uplift or upheaval was occurring in the immediate vicinity due to the load on the column.

Load tests made in preparation for the construction of the Congressional Library Building in Washington, D.C., utilized a frame that rested on four legs (similar to a table), each 12 × 12 in. in size. The frame was moved from place to place along foundation trenches for multiple tests. Many of the authorities of the late nineteenth century stated that small column testing (12 × 12 in.) for purposes of determining the bearing power of the soil stratum was not as reliable as larger column testing.

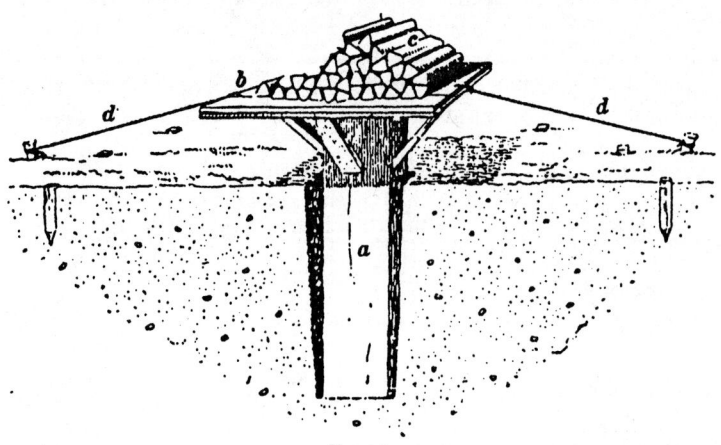

Figure 2-7. Soil test column. (*From International Correspondence Schools 1923, sec. 1, 9.*)

Most said that the area tested should be as large and the test should continue as long as possible (Baker 1910, 335). Evidence indicated that a small area would bear a larger load per unit of area for a short time than would a larger bearing area for a very long period. The science of soil testing and foundation engineering has now evolved into a very intricate system of obtaining undisturbed samples from the site, followed by performing complete laboratory analysis. Methods of, for example, unconfined compression testing, moisture content, and derivation of indexes of plasticity have made the job of designing foundations a very specialized area for all but the most simple structures. Even now, the foundation is still the most intangible part of the overall design procedure. Today, as in the late nineteenth and early twentieth century, many building problems are linked to inadequate foundation designs or ignorance of the actual foundation conditions.

BEARING CAPACITY OF FOUNDATION SUBSTRATA

In the second half of the nineteenth century, Building laws in major cities had maximum allowable bearing values for different soils and bearing strata for use by architects and engineers in designing foundations for their buildings. During the early stages of this period, buildings were relatively small and loads comparatively low. City allowables were naturally conservative, thus minimizing the possibility of overloading. However, identification of the actual soil at the site was difficult; soil contents or types vary within a site, and the actual material encountered would rarely have been the ideal example soil classification as given in the building law. Improper identification and inadequate foundation design were probably the major contributors to the many failures of this early phase of modern building. Failure in this concept means not total building collapse, but a building that develops an unsightly crack or cracks due to differential settlement or improper foundation design. A failure in the context of the designer's criteria would be the lack of total achievement of the intention of the design—not necessarily even a dangerous condition, simply not up to expectations.

Moxon, in 1703, states that after the trenches are dug (just below frostline or below cellar level), an "iron croe" or "rammer," or even a "borer," should be used to see if the foundations are sound and fit to bear the weight of the building. Moxon also states that in soft or "made ground" (fill), wood planks should be laid cross-direction in the foundation trench for the length of the wall. Moxon even mentions piles, inverted arches, and erect arches as a means of making a proper foundation in areas of poor soil (Moxon 1703, 255–256).

Trautwine states that the greatest load that may be safely trusted on an earth foundation shall not exceed 1.5 tons/sq: ft (Trautwine, 1888, 634). Trautwine also gives the following values for safe loads on bearing strata:

- Good Compact Gravel: 2 to 3 Tons per Sq. Ft.
- Good Compact Gravel: 4 to 6 Tons per Sq. Ft. (If a few inches of settlement may be allowed.)
- Pure Clay (Damp): 1 to 2.5 Tons per Sq. Ft. (One Ton equals 2240 Pounds.)
- In the case of soft moist soils which are unable to support building loads, piles are necessary.
- All earth foundations must yield somewhat over time. Vibration (tremor) increases settlements, and causes them to continue for a longer period, especially in weak soils. Foundations in silty soils will probably settle, in years, at the rate of 3to 12 inches per ton (up to 2 tons per Sq. Ft.)
- Equality of pressure is a main point to aim at.

Carnegie Steel Co. (1893) states that the foundations for walls and piers of buildings, if proposed on a "yielding stratum," must be made to produce uniform bearing on the subsurface material. "In case the walls are of different thicknesses and heights, the widths of the foundations must be proportioned according to the different loads resulting therefrom, so that the bearing per unit of ground area will be equal and a uniform settlement of the completed structure is ensured" (ibid., 24). Patton (1893) states that bearing values of 3000 psf should not be exceeded for a soft pliable clay, and 5000 psf for a well-compacted sand that is laterally supported and not subject to any water passage through it. Patton states that if a sand stratum is subject to any water, piles should be used. Interestingly, Patton (ibid., 5) quotes the Bible: "Sand is dangerous to build upon. Sand will hold your structures if you can hold the sand"(Matt 7.26). Beds of gravel and boulders can be relied upon to at least 5000 psf bearing pressure. In each of the above authorities, the bearing capacity is stated as the safe bearing capacity, indicating the values were factored down from the ultimate or crushing capacity. Reference books of this early period are difficult to interpret. Authors of this era were publishing data from their own tests, quoting other authorities, and including values that were only theoretical. Another problem is in the terminology; coordination through professional societies had not yet occurred, and a mixture of such terms as, *safe bearing pressure, actual bearing pressure, maximum strength, and crushing strength* appeared. In most cases, these differences can be sorted out with a certain amount of study and comparison.

Patton provides "actual" bearing power figures from Rankine, one of the known authorities of the period. Rankine's values for "crushing strength" are shown in Table 2-1.

The values shown in Table 2-1 were presented mainly to give figures for buildings bearing upon the stones listed as (1) an intermediate bearing layer between building and soil stratum and (2) a possible bearing stratum in situ. The values of Table 2-1 are the crushing strength and would need to have an appropriate factor of safety applied. Patton states that a factor of safety of from 8 to 10 was considered satisfactory (ibid., 7).

Patton also quotes Baker (obviously from the first edition, 1889), giving the following as the crushing value of stone (see Table 2-2 for these data).

Patton states that we can conclude that good ordinary clay could safely carry 2 tons per sq. ft and good compact sand, free of water, could carry 3–4 tons/sq. ft. In these instances, both Baker and Patton are using 2000-lb tons. Patton's volume is most informative, an excellent narrative accompanying every topic. Many of the solutions are presented in an empirical or practical, rather than quantitative or academic, manner. This is indicative of the transitional state of the science of building; we see later volumes becoming more technical in their analytical solutions.

William H. Birkmire, a pioneer in and authority on high-rise buildings, gives the bearing power of soils and other foundation stratum, as shown in Table 2-3 (Birkmire 1894).

Kidder, one of the most noted authors and designers of the turn of the century, wrote several reference books on construction. *The Architect's and Builder's Pocketbook* was a highly regarded volume from the last quarter of the nineteenth century until well beyond the first quarter of the twentieth century. This book was issued in 18 editions over this period. In the thirteenth edition, 1902, Kidder gives soil and rock bearing values from Rankine, Gaudard, and Baker, three engineers of the period who were specializing in foundation design. The Gaudard and Rankine values are shown in Table 2-4.

Kidder's *Building Construction and Superintendence* (1905) also includes Baker's table on safe bearing power of soils and states that it will enable one to determine the safe bearing power of soils, bedrock, etc. for purposes of constructing buildings and other structures. Baker was the respected authority on foundations in the 1890s.

The American School of Correspondence in Chicago produced a series on construction in the early twentieth century. The *Cyclopedia of Civil Engineering* was a part of the series. The 1908 edition was edited by Frederick E. Turneaure. These volumes are very informative about the period and are nicely illustrated with drawings

Table 2-1. Actual Bearing Power of Strata[a]

Granite	12,861 psi
Sandstone	9,842 psi
Soft sandstone	3,000–3,500 psi
Strong limestone	8,528 psi
Weak limestone	3,050 psi
Clay, sand, and gravel	17–23 psi
Brick	1,100 psi

[a] From Patton 1893, 7.

Table 2-2. Crushing Value of Stone[a]

Granite	12,000–21,000 psi	= 860–1,510 tons/ft²
Marble	8,000–20,000 psi	= 580–1,440 tons/ft²
Limestone	7,000–20,000 psi	= 500–1,440 tons/ft²
Sandstone	5,000–15,000 psi	= 360–1,080 tons/ft²
Brick	674–13,085 psi	= 48–936 tons/ft²
Clay	28–84 psi	= 2–6 tons/ft²
Gravel	112–1,401 psi	= 8–10 tons/ft² [b]

[a] From Patton 1893, 7–8.
[b] The value 1,401 must be in error; it should be 140 to agree with other Baker editions. One ton = 2,000 lb.

and photographs. In the section on foundations, the *Cyclopedia* gives bearing values for ordinary soils, as shown in Table 2-5.

Baker (1910) gives a table of the safe bearing power of soils (see Table 2-6).

Baker cautions designers that some practical considerations must always be applied when a building is designed or constructed:

1. The pressure of the foundation of a tall chimney should be considerably less than that of a low, massive structure. In the case of a tall chimney or spire, a slight inequality of bearing or bearing power and a consequent unequal settling may endanger the stability of the structure.
2. The pressure per unit of area should be less for a light structure subject to the passage of heavy loads, such as a bridge or railroad viaduct; than for a heavy structure subject only to a quiescent load. The shock and vibration of a moving load, especially a heavy moving load is more of a problem on a compressible soil and even more so to a moist compressible soil than a static load with slight increase due to the live loads.[1]
3. A moist soil, moist sands, etc. may require a safeguard against the soils escaping by being pressed out laterally into excavations in the vicinity. In Chicago this condition exists, and several buildings have settled and been damaged due to the flow of plastic clay into open foundations at adjacent sites or even across the street from the building.
4. A large building in New York City (pre-1910) settled due to the pumping of fine sand from an artesian well on the site while obtaining water for the boiler of the building.

In the first quarter of the twentieth century, Milo S. Ketchum, a renowned consulting engineer in the era when the division between the engineer and the architect was totally accepted, published several very detailed and informative reference books on structural engineering. The 1924 edition of *The Structural Engineers Handbook* reproduces Baker's table of safe bearing power of soils (see Table 2-6); however, Ketchum states: "Present practice is more nearly given by the values in Table II" (on allowable bearing on foundations—see Table 2-7). Ketchum also states that foundations should never be placed directly on quicksand. His table is predicated upon the material, the drainage, the amount of lateral support given by adjacent material, the depth of the foundation, and other conditions. Ketchum rightfully considered his values an aid to the judgment of the designer, and warns that sites may vary from location to location and material can vary from one end of a site to the other.

Table 2-3. Bearing Power of Soils and Other Foundation Strata[a]

	Ultimate Crushing Strength	Safe Bearing Power
Stone	180–1800 tons/sq. ft	18–180 tons/ft²
Ordinary clay (saturated by water)		1.5–2 tons/ft²
Ordinary clay (if kept dry)		3–4 tons/ft²
Sand (well cemented with clay, protected from water)		4–6 tons/ft²

[a] From Birkmire 1894, 233–234.

[1] This statement is not entirely correct, and the contemporary designer knows proper design of foundations on compressible clays.

Table 2-4. Soil and Rock Bearing Values (Safe)[a]

Material	Gaudard	Rankine
Rock, hard unyielding	26,000 psf	20,000 psf
Sand, firm compact	16,500 psf	
Firm earth		2,500–3,500 psf
Stiff clay		5,500–11,000 psf

[a] From Kidder 1902, 132–133.

Table 2-5. Bearing Power of Foundation Strata[a]

Ultimate Crushing Strength	Safe Bearing Power
Rock, bedrock (minimum) 186 tons/sq. ft	
Sand, clean-dry	3,000–8,000 psf
Sand, compact, well cemented	8,000–10,000 psf
Gravel, well bedded	6,000–8,000 psf
Gravel, well cemented	12,000–16,000 psf
Clay, soft	2,000–4,000 psf
Clay, medium dry	4,000–8,000 psf
Clay, dry-thick bed	8,000–10,000 psf
Soft semiliquid soils	500–1,500 psf

[a] From Turneaure 1908, 108–109.

Table 2-6. Safe Bearing Power of Soils[a]

| Kind of Material | Safe Bearing Power (tons/ft^2) | |
	Min.	Max.
Rock—the hardest—in thick layers, in native bed	200	
Rock, equal to best ashlar masonry	25	30
Rock, equal to best brick masonry	15	20
Rock, equal to poor brick masonry	5	10
Clay, in thick beds, always dry	6	8
Clay, in thick beds, moderately dry	4	6
Clay, soft	1	2
Gravel and coarse sand, well cemented	8	10
Sand, dry, compact, and well cemented	4	6
Sand, clean, dry	2	4
Quicksand, alluvial soils, etc.	0.5	1

[a] From Baker 1910, 342.

Table 2-7. Allowable Bearing on Foundations[a]

Kind of Material	Tons/ft^2
Soft clay or loam	1
Ordinary clay and dry sand mixed with clay	2
Dry sand and dry clay	3
Hard clay and firm, coarse sand	4
Firm, coarse sand and gravel	6
Shale rock	8
Hard rock	20

[a] From Ketchum 1924, 329.
[b] Ketchum 1921 includes both of the same tables he included in *The Structural Engineer's Handbook*.

Table 2-8. Allowable Soil Pressure (National Board of Fire Underwriters Building Code, 1929)[a]

Kind of Material	Tons/ft²
Soft clay	1
Firm clay, fine sand, or layers of sand and wet clay	2
Clay or fine sand, firm and dry	3
Hard clay, coarse sand, gravel	4
Hardpan	8–15
Rock	15–72

[a] From Huntington 1929, 58.

Huntington (1929) includes a table of allowable soil pressures as taken from the Building Code of the National Board of Fire Underwriters (see Table 2-8).

Urquhart and O'Rourke (1941) also include a more comprehensive table of recommended bearing capacities for construction of footings for buildings, as taken from a later version of the Building Code of the National Board of Fire Underwriters (see Table 2-9).

Tests on soils as performed in the contemporary era—that is, drilling of core samples for laboratory analysis in several locations on the site in preparation for designing the foundations for a building—became commonplace for major buildings in the post-World War II era. Many buildings are still constructed today without the benefit of an analysis of the soil conditions beneath the building site. This is true especially in the smaller to medium-sized cities and towns, predominantly with non-architect-designed buildings. The major buildings of today, 600–1200 ft tall, are so complex in their designs and have such heavy vertical loads with extremely high additive wind loads on their foundations that thorough investigations and special foundation designs are necessary.

FOUNDATIONS OF HISTORIC AMERICAN BUILDINGS 1860–1940

In Europe and the United States, designers of the period knew that not until a very large building was undertaken would loads reach such a magnitude that anything beyond conventional designs would be needed of the foundation system. The rectangular commercial brick building one to two stories tall, which represents the vast majority of the historic building inventory throughout all regions of the United States, would have needed special foundation details only if it was to be built upon problem substrata. Throughout the country, many Georgian and Gothic Revival churches that

Table 2-9. Recommended Bearing Capacities (National Board of Fire Underwriters Building Code, 1940)[a]

Kind of Material	Tons/ft²
Soft clay	1
Firm clay	2
Wet sand	2
Sand and clay, mixed or in layers	2
Fine dry sand	3
Coarse sand	4
Gravel	6
Soft rock	8
Hardpan	10
Medium rock	15
Hard rock	40

[a] Reprinted with the kind permission of McGraw-Hill Publishing Company from L. C. Urquhart and C. E. O'Rourke, *Elementary Structural Engineering*, 1st ed., p. 294. @ 1941 McGraw-Hill Publishing Company.

have tall spires and towers also have very intricate foundation systems to carry these widely varying loads within a structure.

One of the rules that was keenly stressed to designers by their mentors and all of the reference books of the period was the necessity that all sections of the foundation have "equal bearing" pressure on the soil strata (Patton 1893, 3). This refers to a building with continuous wall footings, or one with wall footings and interior columns bearing on footings or inverted arches. Settlement is inevitable in a building constructed on any bearing other than solid rock. All soils or substrata are compressible, some more than others. If footings are proportioned to provide equal bearing, then theoretically and physically the result is even settlement over the entire width and breath of the building. And since the settlement is constant over a period of time (compressible soils of good quality actually take long periods of time to reach maximum compaction), little or no effect is felt by the building. In good-quality strata (called "noncompressible" in the trade), settlement is kept to a minimum, while in softer substrata more settlement will occur and at a faster rate. In any instance, the foundation bearing must be kept well below the frostline or differential movement will occur that will damage the building.

The Building Trades Handbook (International Correspondence Schools 1889) states that "before beginning a structure, the character of the soil should be investigated. "For ordinary work, this can be done by boring, using a 2" auger, at short intervals around the site." The auger will bring up samples sufficient to determine the character of the soil. This is a useful precaution, but it can be dispensed with for the usual run of buildings, the bearing power being judged by experience of loads in adjacent structures, or by examinations of nearby excavations" (Building Trades Handbook, ICS, 1889, 171).

With respect to building foundations, Mahan (1885, 190) classifies soils into three categories:

(1.) First Class, soils which are incompressible, or at least so slightly compressible, as not to affect the stability of the heaviest masses laid upon them, and which, at the same time, do not yield in a lateral direction.[2]
(2.) Second Class, consists of soils which are incompressible, but must be laterally confined to prevent them from spreading out.[3]
(3.) Third Class, consists of all varieties of compressible soils.[4]

Timber Grillage and Platform Foundations and Wood Spread Footings

Mahan discusses methods for placing building loads on each of the soils of the three classes. He suggests that foundations on solid rock be placed on horizontal strata whenever possible. He states that if the stratum slopes to excess, it should be cut off to horizontal sections and stepped to compensate for slope, thus allowing for standard coursing. Designers were cautioned against laying up building walls on sloping substrata that tilted to the degree that horizontal movement of the wall or building might occur. For stony earths, hard clays, and compactable sands, he states that the bed is prepared by digging a trench wide enough to receive the foundation and deep enough to reach compact soil that has not been injured by frost (4–6 ft).

In the case of compressible soils, Mahan states that without extreme care it will be very difficult to guard against unequal settling. In the case of ordinary clay or earth, a trench is dug the proper width, to a level depth below frostline, for the foundation. This foundation may be completed in one of two ways: (1) stones or first courses of brick masonry should be firmly settled in their beds, directly upon the substrata; or (2) a timber grating (grillage) is formed of a course of heavy wood beams laid length-

---

[2] Solid rock, some tufas, compact stony soils, and hard clay belong to this class.
[3] Pure gravel and sand belong to this class.
[4] Ordinary clays, common earths, and marshy soils fit into this class. They are for the most part compressible (some of the clays are slightly incompressible), have high moisture contents, and will yield in any direction.

wise in the trench and connected firmly into cross-pieces into which they are notched, and the whole grillage is firmly pressed into the bed. Mortar or rammed earth is solidly packed between longitudinal and cross-timbers to level at top. A flooring of thick planks to form a platform is spiked to the grillage to receive the lowest course of the wall (see Figure 2-8).

Mahan states that if there is danger from lateral yielding, the area upon which the foundation is to rest must be secured by confining it laterally by means of sheeting piles (see Figure 2-8). There was no quantitative method utilized by early designers to evaluate the allowable bearing pressure of a building wall upon the wood platform and grillage. It was felt that as long as the wood platform and grillage was kept in a moist condition, it would not deteriorate and its capacity to transfer a uniform load to the soil was well within the limits required for buildings of the period 1820–1860.

The use of heavy timber grillages for foundations was prevalent during the Middle Ages, grillages having been found under the walls of Gothic cathedrals. There is also evidence of the use of timber grillages for foundations by the Romans.

In the period after the 1860s, computations were a more common part of the procedure for sizing spread footings of wood timbers for wall and column foundations for buildings. Kidder (1905) provides a method for computing the sizes of transverse timbers in a wood grillage or wood spread footing (see Figure 2-9). Kidder's method is as follows (Kidder 1905, 54):

$$\text{Breadth (width), in.} = \frac{2(W)P^2(S)}{D^2(A)}$$

where:
- W = bearing power of soil strata, psf
- P = projection beyond 3 in. plank, ft
- S = distance between centers of transverse timbers, ft
- D = assumed depth of the transverse beam, in.
- A = constant for strength:
  - Georgia pine = 90
  - Oak = 65
  - Norway pine = 60
  - White pine or spruce = 55

Figure 2-8. Timber grillage. (*From International Correspondence Schools 1923, sec. 5, 23.*)

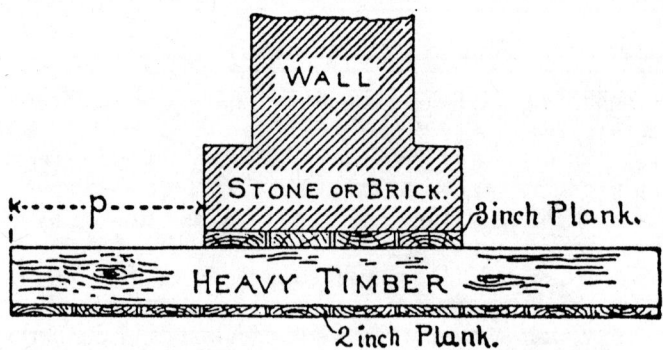

Figure 2-9. Timber footing design. (*From Kidder 1905, 53.*)

Example: Sizing a Timber Spread Footing, Kidder Method
The walls of a building are to impose a load of 20,000 lb/ft upon the proposed foundation. This load has been factored to include the dead loads plus the appropriate live loads. The maximum allowable soil bearing pressure = 2,000 lb/ft to minimize actual settlement. The transverse timbers used were Georgia pine, at 12 in. on centers, with an assumed depth of timber of 10 in. The planking below the timbers is 2 in. thick, and the planking above the timbers is 3 in. thick (the wall bearing point; see Figure 2-18). The masonry width at the contact with the 3-in. planking is 4 ft 0 in. and the area between the transverse timbers is to be filled with a lime-sand aggregate cement (beton or equivalent).

Solution

$$\frac{20,000 \text{ lb/ft}}{2,000 \text{ lb/ft}^2} = 10 \text{ ft 0 in. (width of footing required)}$$

$$\frac{(10 \text{ ft}) - 4 \text{ ft}}{2} = \text{Projection} = 3 \text{ ft 0 in. on each side of the wall base}$$

$$\text{Breadth, in.} = \frac{2(2,000 \text{ lb/ft}^2)(3 \text{ ft})^2(1 \text{ ft})}{(10 \text{ in})^2(90)} = 4 \text{ in. breadth, width of transverse timber}$$

Use a 4 × 10 in. (full-size) Georgia pine timber, oriented with its 10 in. side vertical, in the strong direction

Many utilitarian buildings of the period 1820–1900 were constructed on wood footings. Buildings with moderate to heavy loads may have had a wood footing system composed of transverse timbers laid side by side as a result of the actual computations or by choice of a designer following some tradition (see Figure 2-10). Heavy buildings (skyscrapers and other specific types) of the period required a more specialized construction technique and were very rarely constructed on timber footings. The architects and engineers of that developing period recognized the limitations of the wood footings.

Brick and Stone Footings/Foundations

Smaller buildings of the period 1860 and beyond, especially the lighter structures of the mercantile type in larger cities and the majority of the buildings in smaller towns, were founded directly upon the substrata below the bottom of the load-bearing walls. Tradition required the belling out of the wall thickness, as shown in Figures 2-1 and 2-2. The usual rule was to provide an effective angle of 60° from the horizontal to create a bottom width of twice the thickness of the building wall (see Figure 2-11). An average building, load-bearing masonry, three stories, would transmit a load of approximately 12,000 lb/lineal ft of wall footing (factored for reduced live load). If the lower wall were 24 in. thick, the "twice thickness" rule would indicate a bearing width

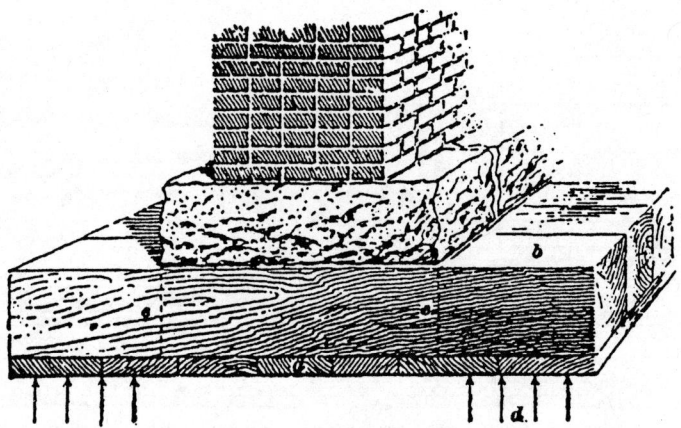

Figure 2-10. Wood/timber spread footing. (*From International Correspondence Schools 1923, sec. 5, 33.*)

of 48 in. ("belled out" thickness) or 4 ft, and an actual load of 3,000 lb/sq. ft. This would be well within the allowable for most slightly compressible soils.

Stones that were used for foundations for heavy loads were called "dimension stones" and were usually granite, bluestone, slate, or some of the hard laminated sandstones and limestones. The bedding was important because the dimension stones were not worked perfectly flat by the masons and were sometimes squared only on the ends and sides and laid on the plane as cut in the quarry. By the turn of the century, bedding was always on lime-cement or a thick bed of mortar.

Stone footings were also sized by rule of thumb methods until approximately the 1890s. The dimension stones to be used for footings were cut to approximately 8 in. thick, and it was customary to use three courses of stone to make up the footing for average buildings. This rule also dictated that the offset between the wall and the first footing course was to be 6 in. on each side, while the successive below were to have an offset of 3 in. on each side at each course (see Figure 2-12).

Building Construction and Superintendence by Kidder (1905) gives a method for computing the proper offset for the various stones, brick, and two types of concrete. The offset is the amount the belling out is stepped between layers of the footing material (see Figure 2-10). Early rules give offsets in brick footings as limited to ¼ brick per course, and stone as above. Kidder gives a table of the safe offsets for various allowable pressures on the bottom of each bearing course. This would allow the architects and engineers of the period to properly size their spread footings based upon assumed allowables for the footing material and the bearing capacity of the founda-

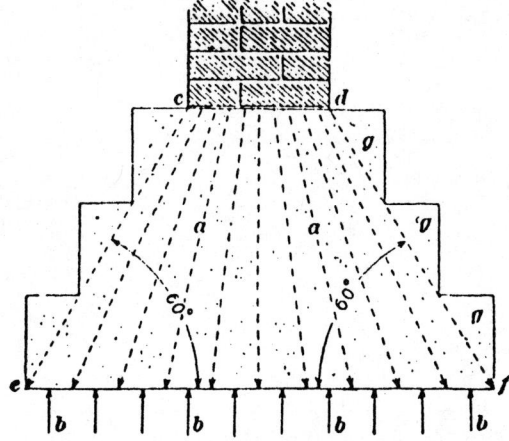

Figure 2-11. Angle of belling out. (*From International Correspondence Schools 1923, sec. 5, 25.*)

Figure 2-12. Stone footings, early designs. (*From International Correspondence Schools 1899, sec. 7, 20–21.*)

tion substrata. In addition, he states that the values given are computed utilizing a factor of safety of 10. As one can see, designers were now able to calculate the proper offset based upon the vertical pressure, the transverse shear allowable of the foundation material, and the thickness of the course of the foundation material.

Kidder gives a table for obtaining the "offsets" for a given allowable soil bearing pressure (see Table 2-10).

Example One: Brick Footing Design
Using Table 2-10, the design of the brick footing was as follows:

    Width of lower-level wall = 24 in.
    Wall load on foundation = 12 tons/ft
    Width of footing = 4 ft, bottom course doubled (from soil allowable bearing capacity)
    Footing to be "best hard brick" in standard 2.5 in. horizontal courses

Table 2-10. Brick Footing Design[a]

Kind of Footing	R (psi)	Offset for a pressure (tors/sq. ft) on the bottom of the course, of					
		0.5	1	2	3	5	10
Bluestone flagging	2,700	Offsets in Terms of the Thickness of the Course					
		3.6	2.6	1.8	1.5	1.2	0.8
Granite	1,800	2.9	2.1	1.5	1.2	1.0	0.7
Limestone	1,500	2.7	1.9	1.3	1.0	0.8	0.5
Slate	5,400	5.0	3.6	2.5	2.2	1.5	1.2
Best hard brick	1,200	2.6	1.8	1.3	1.0	0.8	0.5
Concrete[c] 1 Portland / 2 Sand / 3 Pebbles	150	0.8	0.6	0.4	–	–	–
Concrete[c] 1 Rosendale / 2 Sand / 3 Pebbles	80	0.6	0.4	0.3	–	–	–

[a] From Kidder 1905, 64.
[b] R = modulus of rupture (Baker 1910).
[c] Concrete mixes are proportions by volume.

Load, tons/ft² = $\dfrac{12 \text{ tons/ft}}{4 \text{ ft}}$ = 3 tons/ft²

From Table 2-10, opposite "Best hard brick," constant = 1.0

Allowable offset = 1.0 (2.5 in.) = 2.5 in. on each side, for the next course above

The bottom course (at point of bearing on soil strata) was to be doubled (see Figure 2-13)
Course above the bottom two courses:

Wall load = 12 tons/ft

Width of footing = 48 in.–5 in. = 43 in.

Load, tons/ft² = $\dfrac{12 \text{ tons/ft}}{43 \text{ in.}(12 \text{ in./ft})}$ = 3.35 tons/ft²

By interpolation, the value of the constant for the offset from the table would be 0.97 in. Therefore, 2.5 in. is still the valid offset.

Theoretically, the table requires a varying offset because of the slight increase in bearing pressure (load per square foot), as shown above. This effect is most noticeable in the small thickness of the brick coursing. The table calls for a varying offset at each course level; however, the amount of the variance is so small that it would be physically impossible for the craftsman to lay the bricks to that tolerance. Prudent design would have necessitated that the offset be adjusted at, say, every second or third course to keep the stresses from becoming excessive. Many of the belled-out brick footings were of only a few courses and the difference would have been very small. There was always the factor of safety to compensate for small differences, and theoretically the actual load would decrease slightly due to the fewer courses above the level of the course one was checking.

Example Two: Stone Footing Design
Using the Table 2-10, the design of a stone footing was as follows:

Width of lower-level wall – 24 in.
Wall load on foundation = 12 tons/ft

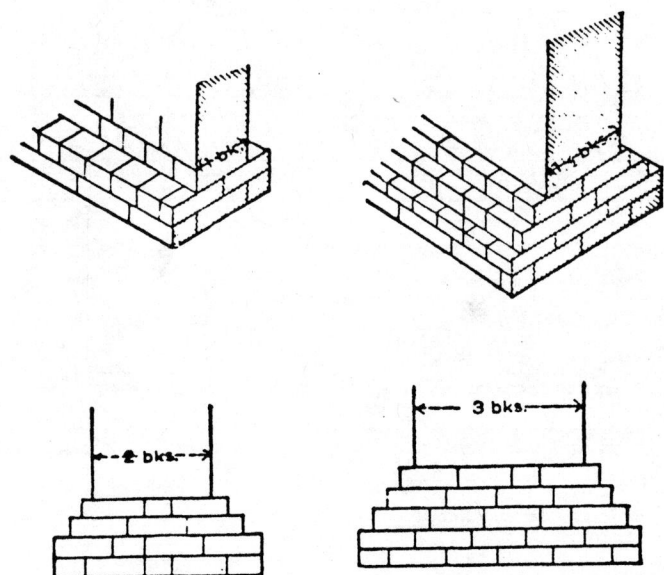

Figure 2-13. Brick footing designs. (*From Kidder 1905, 64.*)

Width of footing = 4 ft (from soil allowable bearing capacity)
Foot to be limestone, course thickness = 10 in.

$$\text{Load, tons/ft} = \frac{12 \text{ tons/ft}^2}{4 \text{ ft}} = 3 \text{ tons/ft (as calculated in the previous example)}$$

From Table 2-10, opposite "Limestone," constant = 1.1
Allowable offset = 1.1(10 in.) = 11 in. on each side, for the next course above
The next course above the bottom course can be 48 in.–22 in. = 26 in. in width.

Kidder states that it is best not to have offsets for stone footings over 6–8 in. and that most building ordinances recommend that the projection of the footings beyond the foundation not exceed 6 in. on each side. He further states that on well-drained sandy soils or a mixture of sand and clay, brick footings are as good as stone for one-, two-, and three-story buildings (ibid., 65).

Inverted arches were used for many years as the solution to distribute the loads of piers or brick or stone wall segments over the length of a continuous footing. Designers of the early periods developed this theory based upon empirical methods and practical theory. They felt that if they could use arches to divide a uniform load from above into concentrated loads passing through columns and piers, they could provide a mirror image of this system at foundation level to redistribute the concentrated loads back to uniform wall loads by means of the inverted arches (see Figure 2-14). By the 1890s, most authorities were convinced that individual spread footings were the preferred method for constructing foundations for a row of piers. Their major complaint against the inverted arch was the condition at the end piers where the horizontal thrust of the end arch was not counteracted by an arch at the opposite side. When there was no restraining arch or tension rod through the end arch (see Figure 2-15), the outside pier would move outward and severe cracks would occur in the outside bay. This, of course, is the same problem that has plagued designers of arches since the beginning of time. It is interesting that in the 1890s authors compared the inverted arch to the individual spread footing, which at that time was the concrete-encased steel beam or rail footing. Their comparison was in the overall depth of each system. The consensus was that the inverted arch system would take up less vertical space than the spread footing. This depth consideration seemed to be a concern of the early designers, who thought in many cases that they did not want to carry their footing excavations to depths that might penetrate one stratum to a weaker one below or get too close to the lower area of a usable stratum.

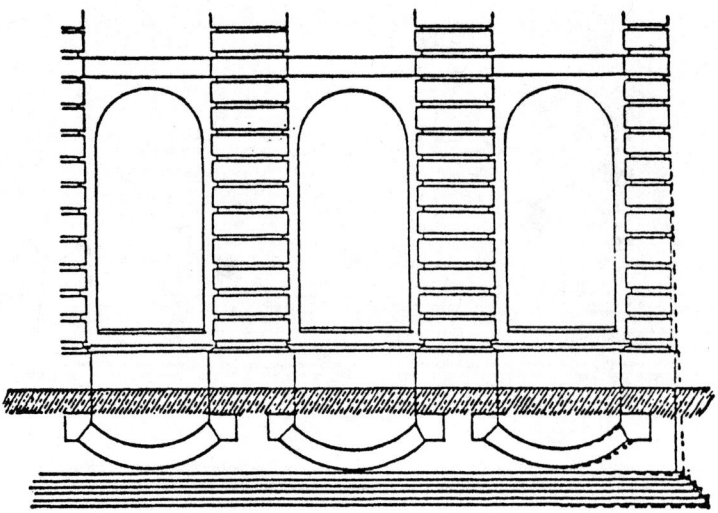

Figure 2-14. Inverted arch foundations. (*From Kidder 1905, 67.*)

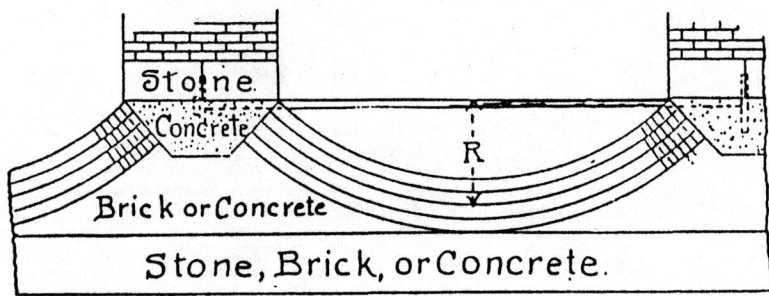

Figure 2-15. Inverted arch foundations. (*From Kidder 1905, 68.*)

Obviously, the state of the art in the design of spread footings was the stacked beam or rail footing, which did consume greater depths (see Figure 2-28) than the reinforced concrete footing of a slightly later period and today. Within approximately 20 years, when the reinforced concrete spread footing was accepted for normal use, the concrete was utilized in compression in the bending section of the footing due to the upward pressure from the soil and resulting in much less depth consumed by the foundation system.

Kidder's method for designing inverted arches for building foundations in the 1890s (see Figures 2-14 and 2-15) was as follows (ibid., 68):

$$\text{Section of arch, in.}^2 = \frac{\text{total load of arch, lb (span, ft)}}{8 \, (R, \text{in.})(10)}$$

Area of tie rods:

$$\text{Area (wrought iron, in.}^2) = \frac{\text{total load of arch, lb (span, ft)}}{8 \, (R, \text{in.})(850)}$$

$$\text{Area (steel, in.}^2) = \frac{\text{total load of arch, lb (span, ft)}}{8 \, (R, \text{in.})(1050)}$$

Example: Kidder's Method—Design of Inverted Arch

Load on wall = 7,000 lb/ft (three-story building)
Piers, stone 4 ft 0 in. long, 14 ft 0 in. center-to-center
Load on pier = 98,000 lb
Span of arch = 10 ft 0 in.
R = Rise of Arch = ⅕ of span = 24 in.

$$\text{Width, footing under arch} = \frac{7{,}000 \text{ lb/ft}}{3{,}000 \text{ lb/ft}^2} = 2 \text{ ft. 4 in. (base width of footing)}$$

$$\text{Section of arch, in.}^2 = \frac{70{,}000 \text{ lb (10 ft)}}{8(24 \text{ in.}) \, 10} = 364 \text{ in.}^2$$

At an arch width of 22 in.: Depth of arch = $\frac{364 \text{ in.}^2}{22 \text{ in.}}$ = 16 in. (four rolocks of brick for arch)

Wrought iron or steel ties (in each outside bay):

$$\text{Area required (wrought iron)} = \frac{70{,}000 \text{ lb (10 ft)}}{8(24 \text{ in.}) \, 850} = 4.289 \text{ in.}^2 \text{ (total area of wrought iron rod)}$$

Diameter = One 2.75 in. diameter rods (root area at threads = 4.62 in.)

= Two 2.00 in. diameter rods (root area at threads = 4.60 in.^{2})

$$\text{Area required (steel)} = \frac{70{,}000 \text{ lb } (10 \text{ ft})}{8(24 \text{ in.}) \ 1050} = 3.472 \text{ in.}^2 \text{ (total area of steel rod)}$$

Diameter = One 2.50 in. diameter rod (root area at threads = 3.72 in.^{2})

= Two 1.76 in. diameter rods (root area at threads = 3.48 in.^{2})

Brick and stone footings were used by early designers in buildings and other structures for literally hundreds of years in an empirical manner. As buildings grew to intermediate size, the need for a more refined design arose with the heavier loads. Enclosures of spaces also increased in size with the use of rows of columns and girders to eliminate interior load-bearing walls, and this brought about the need for the individual column or pier footing. See Figure 2-16 for some examples of brick footings with beton bases.

Beton Bed Foundations

Mahan (1885) states that several failures have occurred in wood grillage and platform systems arising from either compression of the wood from the loads perpendicular to the grain or transverse strain from the soil bearing upward on the outside regions of the transverse timbers of the grillage. He states that engineers at that time now prefer beds formed of an area of beton (a concrete-like mixture of stone, rubble, and mortar) of sufficient thickness to form an adequate foundation. This bed of beton was used over the entire width of the trench and was an obvious beginning of the unreinforced concrete footing that is still in use today (see Figure 2-17). There is no evidence of the early foundations being sized by analytical methods. Rather, it appears that an empirical method was used in which the designers followed the rule that the width of footing was 4.5 in. wider than the belled out wall thickness. Usually the belled-out width was twice the wall width. The depth of the beton would therefore not be critical, since it would not be loaded in transverse flexure. Beton, developed in France in the 1880s, was a monolithic conglomerate of stone, brick, rubble, gravel, sand, mortar (lime or hydraulic lime), and/or Portland cement. "Beton agglomerate" was a name given to a beton of very superior quality, made from only the finest stones, clean sand, top-grade lime, hydraulic lime, and Portland cement. Beton agglomerate, when properly mixed in a mill so as to cause total filling of all voids in the sand, forms a very hard, impervious stone-like material, a predecessor of cast stone or, as it is sometimes called, artificial stone. Mahan also refers to this product being used in "monolithic masonry" buildings in which walls were actually cast in place in forms that were moved up the wall as work progressed. Openings for windows and doors, ornamental cornices, applied ornamentation, and other purposes could be incorporated by manipulation of forms, as could flues, pipe sockets and openings, and other special formable features. Mahan also refers to the common use of wire, rods, iron hooks,

Figure 2-16. Brick and stone footings. (*From International Correspondence Schools 1923, sec. 5, 25.*)

Figure 2-17. Beton foundations. (*From International Correspondence Schools 1923, sec. 5, 27.*)

etc. at junctions of walls, exterior corners, or openings to prevent cracks from occurring due to differential settlement and expansion and contraction due to ordinary changes in temperature. Reinforced concrete or reinforced cast stone structures in 1885 are a very interesting concept.

Pile Foundations on Historic Buildings

Mahan (1885, 195) states that in marshy (moist) soils the principal difficulty consists in forming a bed sufficiently firm to give stability to the structure, owing to the yielding nature of the soil in all directions. Piles were recommended for this condition. This type of foundation was divided into two basic categories:

1. Short wood piles from 6–12 ft long and from 6–9 in. in diameter were driven into the soil as close together as possible, over an area considerably greater than that which the structure was to occupy. The tops of the piles were cut off level to receive a grillage and platform, or a layer of clay from 4–6 ft thick (over the entire area), or a bed of beton over the entire area (beton is mentioned as the best of the choices). (This method utilizes the friction capacity of the large number of piles within the area and the mass of the bed to provide a stable platform as a horizontal structural diaphragm supported by the collection of piles.)
2. Piles to firm bedding: if the unstable stratum overlays a firm stratum or thick layer of bedrock, piles can be driven through the unsuitable stratum to bear on the underlying material. These piles were larger, 9–18 in. in diameter, and were usually placed under walls and columns rather than over the entire building area. The rule or ratio of length to diameter was that the maximum length of pile was no more than 20 times the diameter; thus the maximum length of a 18-in. diameter pile was approximately 30 ft.

These wood piles were fitted with cast iron points and caps (for driving) to prevent damage to the ends of the piles during this phase of the operation (see Figure 2-18). Mahan states that the number of piles should be determined by the weight of the structure. (The weight of the beton bed should be included in the weight of the structure; however, there is no specific statement in the Mahan volume or any others that requires this inclusion. Perhaps it is implied.) Mahan states that when the piles were driven to firm bedding, the allowable load was 1000 psi over the cross-sectional area of pile. When they were driven to only a partial depth of the unsuitable strata, the rule was 200 psi per cross-sectional area of pile. Again, in either type, the piles were to be cut off level and a grillage or a beton bed was to be placed on top of the piles

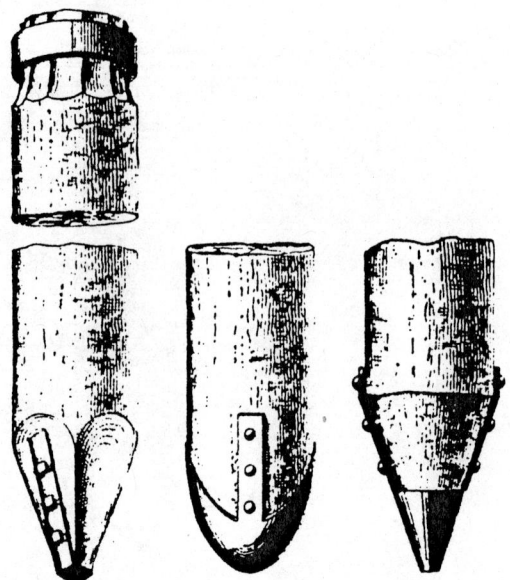

Figure 2-18. Cast iron points and caps used in driving wooden piles. (*From International Correspondence Schools 1908, sec. 47, 13.*)

to act as the foundation to transfer the loads of the building to the piles (see Figure 2-19). Beton was used under both wall and column footings. The width of the beton was 9–12 in. beyond the edge of the brick or stone foundation structure on each side. Normally the brick wall would have been belled out to approximately twice thickness to provide even bearing on the beton and piles.

Piles were driven in several ways in the mid 1880s. Four of the more popular methods were:

1. A "crab engine," a machine used to hoist a weighted driver (ram) to the height required (usually the top of the framework) and allowed to fall on the top of the pile, thusly driving it down a distance.
2. A "Ring Engine" in which men raised the ram repeatedly up a frame and dropped it on the pile. The frame had a ring at the top to allow the rope from the ram to pass through the ring and down to the men who pulled the rope to raise the ram. (This was considered the "softer" method, which was more applicable to cast iron piles.)
3. A "steam pile driver," which would eventually take over the driving duties for steel and concrete piles.
4. A "gun powder pile driver," which used a small charge of gunpowder to drive the pile. See Figures 2-20 and 2-21 for examples of some of these pile-driving machines or simple frames, which were called engines in their time.

Mahan (1885) describes an additional type of pile foundation. This was a technique by which 6 in. diameter piles were driven approximately 6 ft into the soil, then withdrawn, and the resulting open hole then filled with sand and packed tightly. He states that this method should be used only on very moist soils and that the piles should be as close together as possible. The level at the top of the sand piles can then be filled with compacted sand (supported laterally by sheeting piles) or a layer of beton agglomerate approximately 18 in. thick (the preferred option) as foundation for the walls to bear. This method allowed the sand to provide equilibrium and stability as the piles worked together to prevent lateral displacement (see Figure 2-22).

Trautwine (1888, 663–664) states that "The common practice in such cases, of laying planks or wooden platforms in the foundations, for building upon is a very bad one. For if the planks are not constantly kept thoroughly wet, they will decay in a few

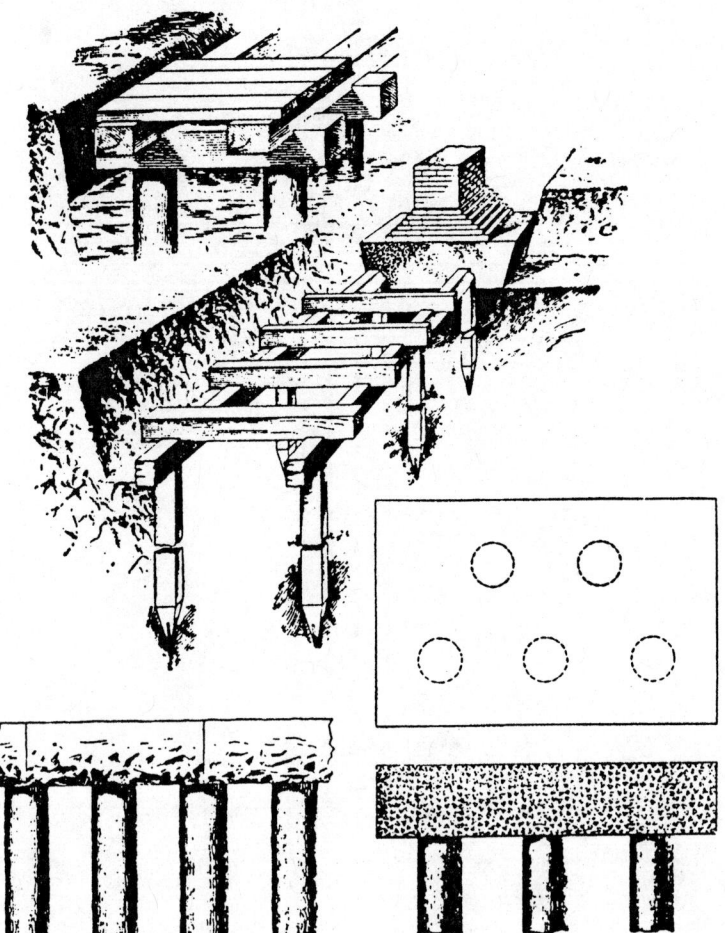

Figure 2-19. Piles, grillage, and platform with stone or beton foundation platform. (*From International Correspondence Schools 1905B, sec. 20, 62.*)

years; causing cracks and settlements in the masonry." Mahan and Patton caution that timber cribbing, piles, etc. must be in constantly moist or wet soils.

W. M. Patton, a professor and noted bridge engineer of the 1880–90s, produced a reference book, *A Practical Treatise on Foundations*, in 1893. It was a very important volume for foundation designs. Patton describes the foundation as that part of the substructure reaching from the foundation bed to the surface of the ground or water and necessarily includes the various means of reaching the foundation bed, such as ordinary excavations on land, driving piles on land or in water, screw pile foundations, Cushing cylinders, coffer dams, pneumatic cylinders, pneumatic caissons, open caissons, Pierre-Perdue foundations on land or in water, sand foundations in swamps, concrete foundations, and rubble stone foundations (Patton 1893, 9). Most of these foundation types are described above. Patton's theories are in most cases consistent with those of Mahan and others that will be discussed. Mahan also includes pneumatic caissons, cylinders, and coffer dams in the sections in his volume that dealt with bridges. Mahan states that screw piles should be used only on lightweight structures, such as houses, while Patton gives examples of some cast and wrought iron screw piles that he has used on piers for small bridges. Cushing cylinders were steel or iron cylinders 4–10 ft in diameter that were sunk around a cluster of 4–12 piles, then filled with concrete, usually to form a bridge pier. Pneumatic caissons and pneumatic cylinders were sometimes used for tall buildings and will be discussed later.

In an earlier paragraph, Mahan presents his method for determining the bearing power of piles. Trautwine (1888, 643) gives a formula for the sustaining power of piles:

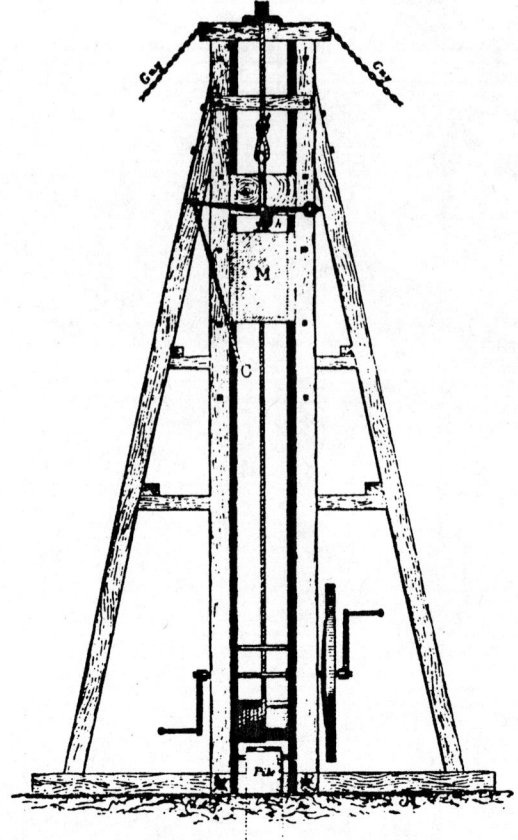

Figure 2-20. Crab engine. (*From Rivington 1900, 233.*)

$$\text{Extreme load, tons} = \frac{\text{cube root of fall, ft (weight of hammer, lbs)(0.023)}}{\text{Last sinking, in.} + 1}$$

$$\text{Safe load, tons} = \frac{\text{extreme load, tons}}{\text{factor of safety}}$$

Factor of safety = 2 for firm soils

6 for marshy soils

Example Solution: (*Mahan Method*)

A 1200 lb hammer is driving piles with a fall of 20 ft. The last blow from the hammer sank the pile 0.75 in. What is the safe load on the pile in marshy soils?

$$\text{Safe load} = \frac{(20 \text{ ft})^{0.333}(1200 \text{ lb})(0.023)}{(0.75 \text{ in.} + 1 \text{ in.})(6)} = 7.128 \text{ tons} = 14{,}250 \text{ lb (each pile)}$$

The total load of the wall on each lineal foot of wall or column on its foundation divided by the safe load per pile will then determine the number of piles that are required in each instance. See Figure 2-19 for a diagram of how a wall footing is placed on a group of wooden piles.

Pneumatic Cylinders and Pneumatic Caissons

Pneumatic cylinders and pneumatic caissons were a phenomenal invention, an engineering application that combined the elements of physics and the principles of engineering to place foundations of this type in very moist, swampy material and in underwater applications. While pneumatic caissons were used mainly on large pro-

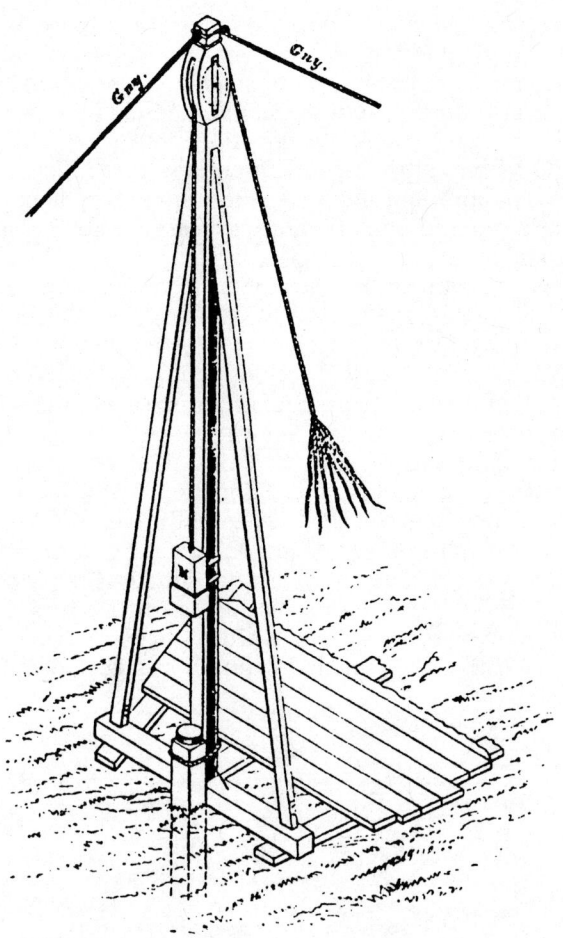

Figure 2-21. Ring engine. (*From Rivington 1900, 231.*)

jects such as bridge piers, pneumatic cylinders were used mainly on building foundations and smaller bridges. Caissons were usually very large objects on which the bridge pier or abutment was constructed; however, there were isolated examples of larger-diameter caissons being used on buildings. In most instances, the caisson used for column foundation for a building was 4–12 ft in diameter. In the pneumatic caisson or cylinder, the principle of differential pressure was utilized to create a water-

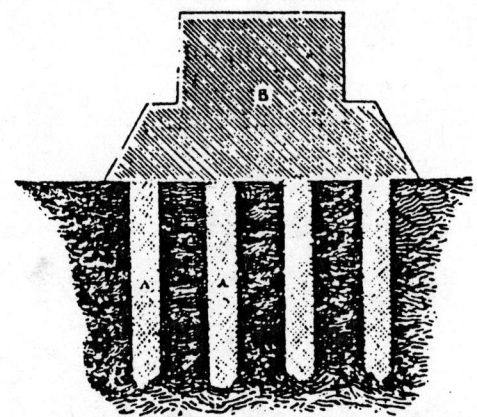

Figure 2-22. Sand piles. (*From Trautwine 1888, 663–664.*)

free workspace below ground and/or below water for men to work. The caisson could be rectilinear, oval, round, or oblong with curved ends; the width and breadth of some designs were as large as 100 × 180 ft. A working chamber could not span these widths, so there were cross-partitions in the form of load-bearing plinths, which complicated the digging out process even more. In flowing rivers, some caissons had a cross-section that came to a point at the upstream end to prevent collection of debris. The principle of the caisson and the cylinder was the same, several common elements in their design being shared and construction procedure following identical steps (see Figures 2-23, 2-24, and 2-25).

The caisson can be compared in principle to a box without a bottom that is placed, open end down, upon the surface of the marshy soil or river bottom. It has a working space surrounded by a frame and a special cutting edge running around the total perimeter of the bottom. The pier or footing is constructed on top of the frame (or on top of the box) and working area. Air is pumped into the working chamber to form a positive pressure zone that keeps all water out of the chamber and allows men to work at the bottom of the caisson. The men must enter through air locks for positive pressure in the working chamber to be maintained. Within the working chamber, during that period, the men dug out the bottom of the caisson with picks and shovels and in the early caissons removed the mud and sand by raising buckets containing the debris through another special chamber that contained air lock doors at top and bottom. Later this method of pulling up the buckets with ropes and pulleys was replaced by a pneumatic conveying system in which air pressure was used to evacuate sand and mud through a pipe by using loading boxes and a set of air valves. Even

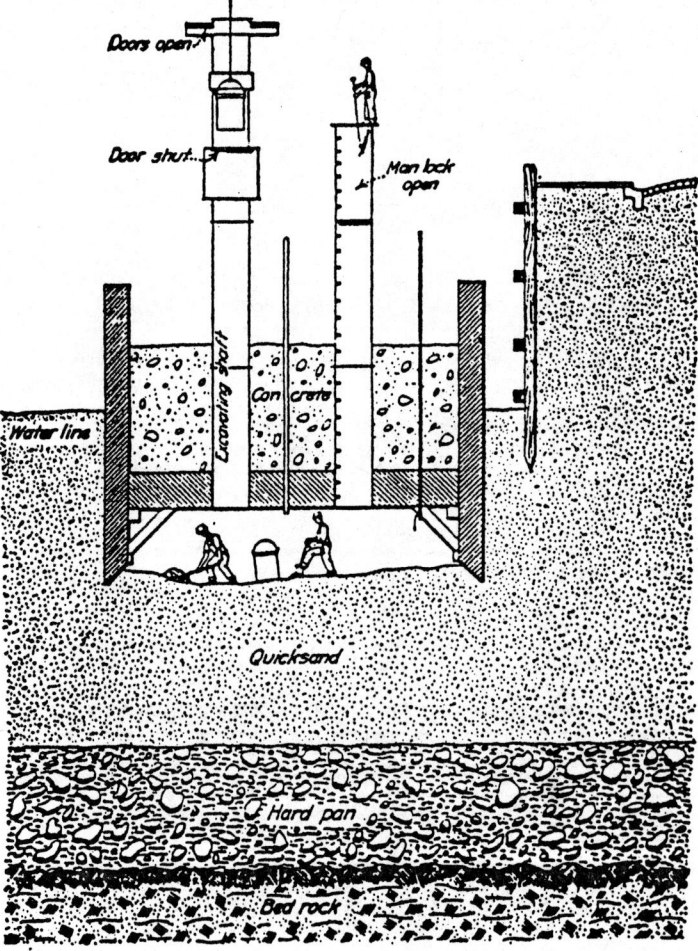

Figure 2-23. Pneumatic caisson: beginning the work. (*From Hool and Kinne 1923, 93; Hool and Johnson 1920, 362–363.*)

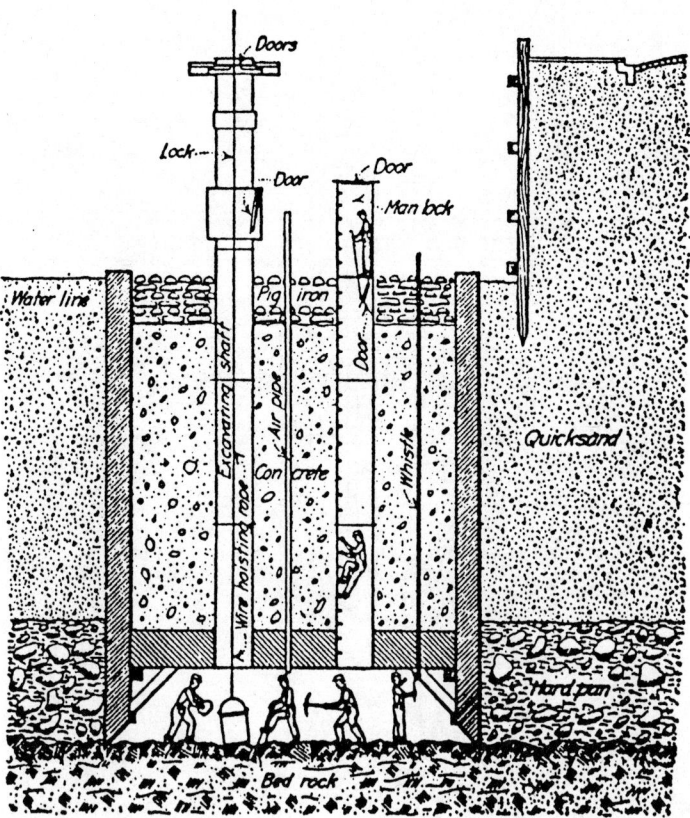

Figure 2-24. Pneumatic caisson: at bearing on bedrock. (*From Hool and Kinne 1923, 93; Hool and Johnson 1920, 362–363.*)

later a vacuum system was employed actually to suck the loose debris from the bottom without having to be physically loaded into wheelbarrows or loading boxes.

The deepest pneumatic caisson was used in St. Louis in 1870, where the east abutment of the Eads Bridge was sunk 109.7 ft below the surface of the Mississippi River. Large bridge caissons were constructed in this manner with the stone or brick masonry pier or abutment above acting as the weight required to sink the caisson as the men dug out the bottom. The weight caused the caisson to slip down, and approximately 1–2 ft per day were excavated on an average. The digging out operation in the working chamber began simultaneously with the construction of the masonry above. The additional weight needed with each slip to overcome the friction on the side of the cutting edge and caisson skin would be acquired by the additional weight on the caisson provided by the amount of stone or brick masonry laid for the pier. When the caisson or cylinder had reached the final bearing stratum and was plumbed correctly, the working chamber was cleaned up, all loose material and tools were removed, and the working chamber and all other voids and shafts were filled with concrete. This ended the operation and a firm, even bearing surface was achieved. In some cases, packed sand was substituted for the concrete filling.

Workers at the bottom of pneumatic caissons or cylinders were subject to special hazards in addition to the danger of working under a structure of this size and weight. The positive pressure could cause a blowout at the caisson wall (cutting edge) and allow the working chamber to lose its positive pressure and fill with water. This blowout could come from a weak section in the stratum, or, if the cutting edge encountered a pothole, either type might fail by allowing the positive pressure inside the chamber to push the material away from the caisson wall. Workers could drown if they could not get up their escape route quickly enough to clear the air lock before the chamber filled with water. Another hazard to the workers was "caisson disease," which was brought about by working at depths in the atmosphere of positive pressure and the entry/return problems associated with men going in and out of the chamber from

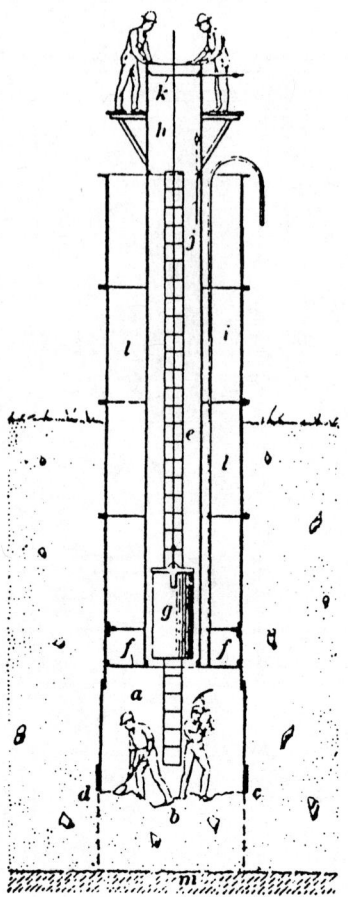

Figure 2-25. Pneumatic cylinder. (*From International Correspondence Schools 1923, sec. 5, 43.*)

one atmosphere to another (Baker 1892, 299–301). This was especially critical as the lower depths were reached. The effect on the men was the same as that on a diver who surfaces too quickly; some texts even refer to it as the "bends," which is the term used by divers. It is also stated that returning to the higher pressure is one way of getting rid of the symptoms at the time, to be followed by reexiting the area of higher pressure at a slower rate to allow the body time to adjust. Most states had laws that required special hazard pay and restricted the time of work in the chambers of higher pressure in a graduated scale in which the time allowed shortened with the increases in depth.

The Manhattan Life Insurance Company Building, New York, 1893, used a system of multiple pneumatic cylinders and rectangular caissons topped with a steel beam platform on which the 17-story building was constructed. A total of 15 caissons and cylinders was used. The smallest cylinder was 9 ft in diameter, while the largest caisson was 25 × 25 ft. Each of the cylinders and caissons was 11 ft high, and they were placed 54 ft below street level. The cylinders and caissons were made of steel, with the upper areas filled with brick and the working chamber filled with concrete (after final positioning). Each of the piers (cylinders or caissons) was topped with a granite cap and iron base for the girders. On some of the rectangular and square caissons the granite caps were followed by a steel bolster (grillage in two directions) as the bearing for the girders. The Manhattan Life Insurance Building was placed on a difficult site in lower Manhattan, with the six-story Consolidated Exchange Building on its south side and a four-story brick building on its north side. The Consolidated Exchange Building had foundations supported by piles that bore on the solid rock bed. The four-story building on the north side had foundations that bore on the strata 28 ft below the street. Each of these buildings was very vulnerable to any disturbance at the site of the Manhattan Life Building. The overwhelming weight of the larger, 17-story Manhattan Life Building would definitely have had a negative effect laterally on the adjacent buildings unless it were founded on the bedrock 54 ft below the surface. New York City law did not allow a building's construction to adversely affect an existing adjacent building. Excavations and foundation construction, especially pile driving in an area near an adjacent building's foundation, would have caused excessive vibration in this stratum, which was in a semiliquid state for its full depth, and would have significantly damaged the foundations of each adjacent building. Only one interior pier on the Manhattan Life Building was constructed of 25 piles (see pier P, Figure 2-26).

Pneumatic caissons and pneumatic cylinders were used on most of the large buildings (skyscrapers) and great bridges of the magnificent building era of the last quarter of the nineteenth century. This was an extraordinary time, following on the development of the new materials (structural steel and curtain wall systems) and the methods of utilizing these new techniques. In addition, as if merely by happenstance, the emergence of a new era of master architects and engineers paralleled the development of the new materials, and their formulation of massive schemes for new build-

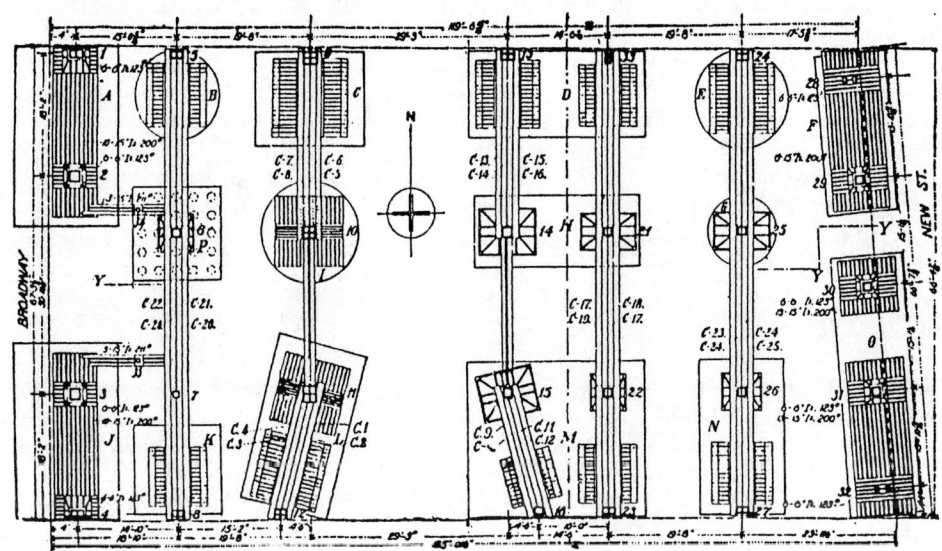

Foundation Plan

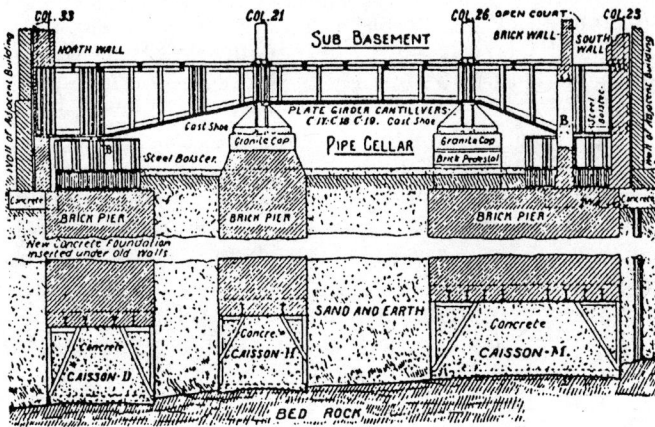

Transverse Section

Figure 2-26. Manhattan Life Building, New York City. (*From Kidder 1905, 59–60.*)

ings was inevitable. Adler and Sullivan, Richardson, Hunt, Gilbert, Burnham and Root, Holabird and Roche, Jenney, and McKim, Meade and White were among the notable professional architects and engineers working on buildings in New York and Chicago, while the great bridge work was being carried out by Roebling and his contemporaries.

The size and weight of these commercial skyscrapers demanded skill and inventiveness on the part of the designers and created special foundation problems that had to be solved. Coincidentally, Chicago and New York were also difficult sites for building with large loads, being located on very unsuitable foundation strata. The use of piles had been widely known for many years, but the pneumatic caisson and pneumatic cylinder were special design solutions that were needed for these new building designs if they were to be built with problem soil conditions. It is not recorded when the idea of working below waterline first occurred and to whom it can be attributed, but the method, from necessity, must have been a collaborative effort on the part of architects and engineers.

Concrete and Rail Footings and Concrete and Beam Footings

Wide footings (spread footings) were used on compressible soils when conditions were such that the anticipated settlement would be minimal. Kidder (1905, 41–42) states that "compressible soils are often met with which will bear from 1–2 tons per square foot with very little settlement, and, as a rule, this settlement is uniform under the same unit pressure (pressure per square foot). In such cases it is often cheaper to spread the foundations so as to reduce the unit pressure to the capacity of the soil rather than to attempt to drive piles." The allowable load on the soil substrata (at foundation bearing level) was estimated or determined by on-site tests. The width of a 1 ft section of footing under a wall was determined by dividing the load on that section of the wall by the allowable soil bearing pressure. The term *allowable* normally meant that the ultimate soil bearing capacity had been reduced by a safety factor; thus, the allowable soil bearing capacity was already factored to a value somewhat lower than the ultimate.

At the turn of the twentieth century, spread footings were constructed of timber grillage, steel rail grillages and steel beam grillages, and concrete and twisted bars. Kidder (1902) describes the use of wood planking for foundation beds. He states that "in erecting buildings on soft ground, where a larger bearing surface is required, planking may be resorted to with great advantage, provided the timber can be kept from decay" (ibid., 144–146). He further states that if the ground is wet, opportunity for decay is minimal, while decay is very possible given the condition of alternating periods of wet and dry. If the latter is the case, he states that the timber is to be prepared by "kyanizing or creosoting." The practicality of timber footings and grillages was limited to lighter loads, and thus there were limits to the size of the building that could utilize this system. In addition, timber was subject to decay unless it was used in moist substrata. The brick or stone footing also had load limitations in that the physical size of the belled out footing for very heavy loads would become excessive. In addition, the actual additional load of the heavy footing system itself could not be ignored.

The architects and engineers of the period understood the concept of dividing the total load of the wall or column by the safe allowable bearing pressure of the soil to determine the width of a wall footing or the size of a column footing. The rule for offsets allowed only small increments of offset for each course or level; the result of this system was that it required a depth that was obviously dependent upon the width of the base of the footing. Buildings that imposed heavy loads upon their foundations would thus require a large vertical dimension for the development of their foundations (see Figure 2-27). The physical dimensions and mass of these large belled out footings would not allow the placement of the bottom of the footing simply 2–4 ft below the ground line or below the floor of the basement of the building, due to the width of the footing system encroaching to within the functional space of the building. This problem was also magnified by the need to keep the bottom of the footing as shallow as possible in many cases so as not to penetrate the stratum upon which the building was to bear.

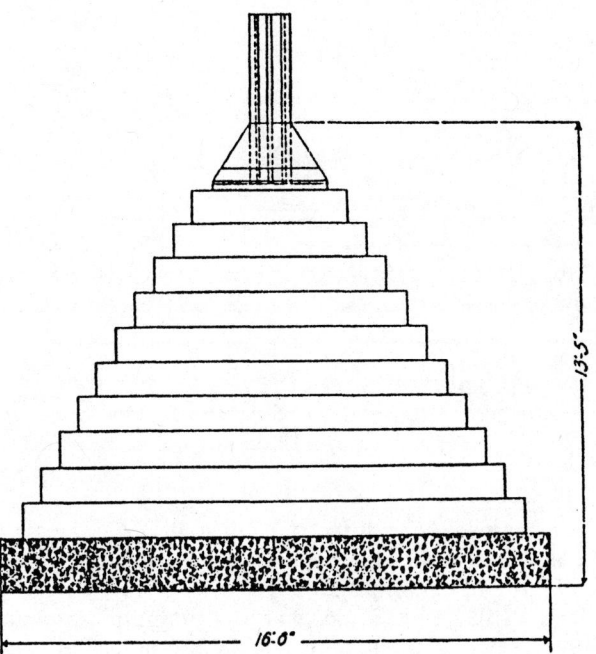

Figure 2-27. Offset column footing: stone, brick or beton. (*From International Correspondence Schools 1905B, sec. 20, 3.*)

As the skyscraper and other special-use buildings evolved in the 1880s as a result of the emergence of structural steel and the new methods of skeletal-frame and curtain wall construction, suddenly buildings were built to the unprecedented heights of 15–20 stories. The resulting foundation loads on these buildings were three to four times as large as the loads on the smaller buildings they superseded. The architects and engineers of the period were aware of the loads that these buildings would place on their foundations and the need for a system that could transmit these loads to the bearing stratum or bedrock.

The concept of using transverse timber sections projecting beyond the edges of the belled out brick walls or column footings was understood by designers in the 1880s, and analytical methods for sizing members were gradually replacing the earlier empirical methods. Steel rails were available in large quantities as a result of their production for railroad and streetcar use and were considered cheaper in cost than the "new sections" I-beams, and wide flanges. Some of these rails were 9 in. deep and were used for joists in fireproof flooring systems and on grillages for building footings. Substituting steel rails for the wood timbers in their spread footings would be a natural transition. It was obvious that steel and iron rails were stronger in flexural bending than most practical sizes of wood timbers. The earliest forms of steel and iron grillages were empirically sized and probably appeared in one form or another in the 1860s in the United States at approximately the same time that the use of rails in fireproof floor construction became fairly common in most larger cities. Architects and engineers were making the transition to rail or beam grillages for their building footings, as the new technology was considered to be a significant saving in space and materials (see Figure 2-28). The rails or beams were placed on top of a beton bed and each layer of rails or beams was then encased in beton for what was then thought to be protection from moisture leading to rust and then loss of structural integrity.

The lack of rational knowledge and of the ability to produce a reinforced concrete footing with the reinforcing bars placed near the bottom of the footing for tension development in the flexural equation as utilized in modern analytical methods is evident. The rails or beams were the only item in the footing being used for resistance in bending. In many instances the grillages were stacked as the architects and engineers were shortening the actual projections at each end of the sections. A second or

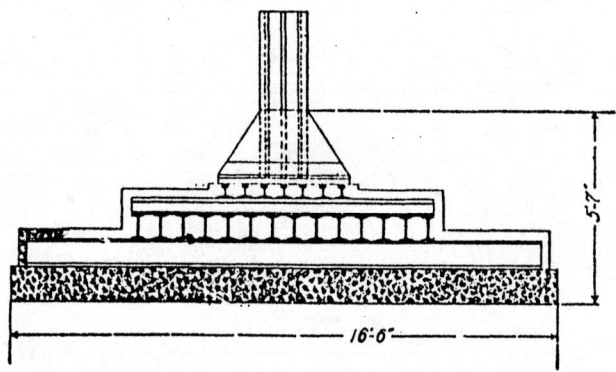

Figure 2-28. Rail and beam grillage footing. (*From International Correspondence Schools 1905, sec. 20, 3.*)

third layer of rails or beams was required to achieve the balance of width and breadth of footing spread, and the additional capacity to resist bending was needed in the computations (see Figures 2-28 and 2-29).

The rail and beam grillage was quickly accepted and its use became widespread. Certain problems arose that would soon cause the architects and engineers to modify their thoughts:

1. It was difficult to get the beton worked into the area between the beams or rails, and spacers had to be used on the rails or beams to keep them erect and spaced uniformly to the desired dimensions (see Figure 2-30).
2. As the rails or beams deflected along their outside edges, the load from the wall or column base above would then become concentrated on the outer edges of the masonry wall footings and thus crush the outer sections of the material.
3. As these problems were encountered, most authorities recommended adding one-third more width to the bottom of the masonry wall, providing more belled out width to minimize this effect (International Correspondence Schools 1905, sec. 20, 1).

The net effect of most of these problems and modifications made the grillage method less attractive to the designers of the period; however, many hundreds of these designs were used in this relatively short period for wall and column footings.

Kidder (1902) has an excellent method for calculating the size and length of steel beams that were to be used in wall and column footings (see Figure 2-31). He provides a table and method of calculation as credited to Carnegie, Phipps & Co.'s *Pocketbook* (Carnegie Steel Co.), to be used in determining the safe length of projections of beams

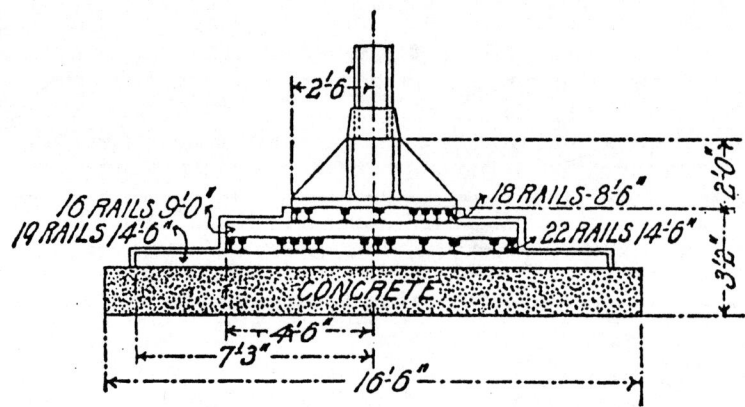

Figure 2-29. Footing of rails and concrete. (*From Freitng 1895, 174.*)

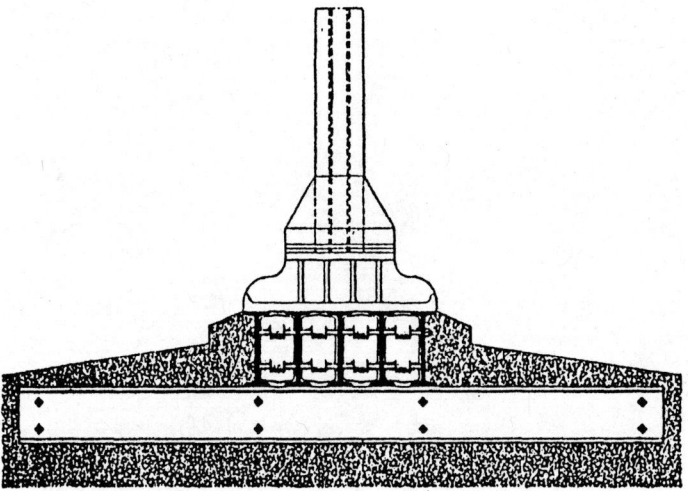

Figure 2-30. Grillage with beam spacers. (*From Kidder 1905, 47–48.*)

from 4–20 in. in depth for soil bearing capacities of from 1–5 tons/sq. ft (see Table 2-11). It is interesting to note that he prefers steel beams to the steel rails first used in early spread footings because of the beams' superior resistance to deflection. This would of course have been especially true in the deeper beam sections. The table shows steel sections and bases its values on a bending stress allowable of 16,000 psi. The deflection problem discussed by Kidder and others was brought about by their using the rail or the steel beams as the only means for the footing to resist deflection through the cantilevered projecting parts. They used the concrete only as an embedment or protecting cover to prevent rust of the steel sections and were not as yet using a footing section with the embedded steel at the lower portion of the footing to take the tension from the flexure while the concrete in the upper portion takes the compression portion of the flexural stress. The solution during the experimental period with steel rail or beam grillages was to stack multiple layers of beams or rails or a combination of the two at 90° to each other in successive layers in an attempt to transfer the concentrated loads (as in the case of a column) from a layer of one width to the next layer of a larger width, and so on. The early designers had their maximum projecting lengths (offsets), and, as could be expected, the only way to spread the load over a larger area was to step the footing outwardly in small increments while utilizing many vertical steps, as had been done in stone and brick. The use of rail and steel beams as footing reinforcement was fortunately confined to the relatively short span of time from the mid-1880s to the mid-1900s. At this point the reinforced concrete footing concept as we know it today was beginning to be developed. The advantages were obvious, and the designers of this period, both architects and engineers, were eager to accept it based upon its shallower depth requirement and the elimination of the deflection problem that came with the use of rails and beams. With the

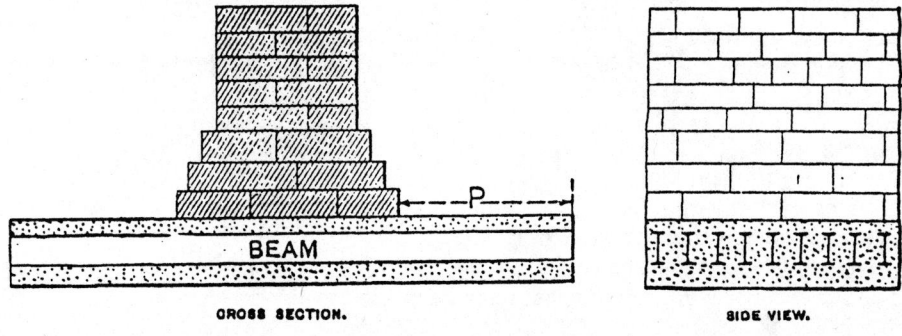

Figure 2-31. Steel beam footing. (*From Kidder 1905, 47–48.*)

Table 2-11. Safe Lengths of Projections, p in feet, for $s = 1$ ft and Values of b Ranging from 1–5 Tons/ft^2 [a]

Depth of Beam (in.)	Weight per Foot (lb)	\multicolumn{11}{c}{b (tons/sq. ft)}
		1	1.25	1.50	2	2.25	2.50	3	3.50	4	4.50	5
20	80	14.0	12.5	11.5	10.0	9.0	9.0	8.0	7.5	7.0	6.5	6.0
20	64	12.5	11.0	10.0	8.5	8.0	8.0	7.0	6.5	6.0	6.0	5.5
15	75	11.5	10.0	9.5	8.0	7.5	7.5	6.5	6.0	6.0	5.5	5.0
15	60	10.5	9.5	8.5	7.5	7.0	6.5	6.0	5.5	5.5	5.0	5.0
15	40	9.5	8.5	8.0	7.0	6.5	6.0	5.5	5.0	5.0	4.5	4.5
15	41	8.5	8.0	7.0	6.0	6.0	5.5	5.0	4.5	4.5	4.0	4.0
12	40	8.0	7.0	6.5	5.5	5.5	5.0	4.5	4.0	4.0	3.5	3.5
12	32	7.0	6.5	5.0	5.0	4.5	4.5	4.0	4.0	3.5	3.5	3.0
10	33	6.5	6.0	5.5	4.5	4.5	4.0	4.0	3.5	3.5	3.0	3.0
10	25.5	5.5	5.0	4.5	4.0	4.0	3.5	3.5	3.0	3.0	2.5	2.5
9	27	5.5	5.0	4.5	4.0	4.0	3.5	3.5	3.0	3.0	2.5	2.5
9	21	5.0	4.5	4.0	3.5	3.5	3.0	3.0	2.5	2.5	2.5	2.5
8	22	5.0	4.5	4.0	3.5	3.5	3.0	3.0	2.5	2.5	2.5	2.0
8	18	4.5	4.0	3.5	3.0	3.0	3.0	2.5	2.5	2.0	2.0	2.0
7	20	4.5	4.0	3.5	3.0	3.0	3.0	2.5	2.5	2.0	2.0	2.0
7	15.5	4.0	3.5	3.0	2.5	2.5	2.5	2.0	2.0	2.0	2.0	1.5
6	16	3.5	3.0	3.0	2.5	2.5	2.0	2.0	2.0	1.5	1.5	1.5
6	13	3.0	3.0	2.5	2.5	2.0	2.0	2.0	1.5	1.5	1.5	1.5
5	13	3.0	2.5	2.5	2.0	2.0	2.0	1.5	1.5	1.5	1.5	1.5
5	10	2.5	2.5	2.0	2.0	1.5	1.5	1.5	1.5	1.5		
4	10	2.5	2.0	2.0	1.5	1.5	1.5	1.5				
4	7.5	2.0	2.0	1.5	1.5	1.5	1.5					

[a] From Kidder 1905, 47. Values given are based on extreme fiber strain of 16,000 psi on steel shape.

advent of reinforced concrete theory and the introduction of Portland cements in the early twentieth century, the reinforcing bars were used in tension and concrete in compression in the concept of flexural bending theory. The capacity to carry load could be met, deflection all but eliminated, and the overall depth cut drastically, and in a simple section. It is therefore understandable that the designers of the period so quickly embraced the new method.

Computations—Wall Footings (Single-Level Grillage). Illustrated below is Kidder's (1905) method for sizing steel beams and the length of projection for spread footings. Table 2-11 was used in the Kidder design of spread footings for walls in the following manner.

Example 1
The building for design has four stories, load-bearing masonry walls.
Total weight of building wall = 30 tons/ft
Bearing capacity of substrata = 2 tons/ft^2
Width of lowest course of masonry = 5 ft
Spacing of transverse beams = 1 ft
Spread footing width = 30 tons/ft divided by 2 tons/ft^2 = 15 ft (the width of bearing required)
Width of projection = (15.0′–5.0′) = 10 ft 0 in. (comb.), projection = 5 ft 0 in. (projection on each side)
From the table using $p = 5′$, $s = 1$ ft, $b = 2$ tons/ft^2
Solution: Beam = 12″ × 32 lb/ft, beam length = 15 ft 0 in., footing width = 16 ft 0 in.

Example 2
Total weight of building wall = 30 tons/ft
Bearing capacity of substrata = 2 tons/ft^2
Width of lowest course of masonry = 5 ft
Spacing of transverse beams = 18 in. = 1.5 ft
Spread footing width = 15 ft (same as Example 1)

Width of projection = 5 ft (each side) (same as Example 1)
From the Table with $p = 5$ ft, $s = 1.5$ ft [$b = 2$ tons/ft² (1.5) = 3 tons/sq.[5] ft]
Beam = 15 in. × 41 lb/ft (use this beam, length[6] = 16 ft 0 in.)

The problems of the beam or rail footings were inherent in the design; the steel sections were merely embedded in the beton or early concrete and were not assisted in any way by the compressive strength of the concrete. Early on, the designers of the period realized that the projecting segments of the steel beams deflected and allowed the wall or column footing to settle as a result. While this settlement was fairly small, it should have been uniform throughout a building, as long as the wall and column footings were designed to the same parameters. The inherent flexibility of the steel section, beam, or rail forming the grillage footing could be a problem if the deflection became excessive or the steel actually buckled at the point of cantilever on each side (see Figure 2-31).

Designers of these early periods were not as well versed in the fields of mechanics and strength of materials as many of today's professionals; and theory was not as complete as it is today. This is quite understandable, as theory and practical use tend to converge as a more thoroughly tested and proven system evolves.

Check of Kidder's Method by Modern Analysis Methodology. Deflection computations for steel beams or rails for footings of this era are not shown in any of the reference manuals. This can be explained partly by the newness of the technology and partly by the fact that deflection computations were not known or used by designers in any segment of the design process. In many instances, such as this design method for the footing grillage, the process itself was the invention of an individual and other designers followed or copied the method without acquiring a thorough understanding of the system. The Kidder method of selecting the proper beams for the grillage requires the selection of members from a table of values derived from computations that were not presented for analysis or interpretation. Apparently the state of the art at that time was to treat the projecting portion of the beam as a simple cantilever. Later volumes give moment maximums that reflect the true maximum bending moment of a steel beam used in a spread footing, as shown in the computations and diagrams below. The true maximum bending moment for the average footing geometry is actually approximately 40 percent higher than the moment obtained by assuming the projection to be a cantilever. This erroneous assumption of a maximum moment equal to that at the edge of the wall may have resulted in members being overstressed. Localized buckling failures may have accounted for many designers' complaints of excessive deflections during this period. In addition, rails, which were used initially as the steel grillages, did not have sufficient depth of section to develop a strong moment of inertia and were therefore not very practical for use in a bending situation.

Modern Analysis: Example No. 1
By modern methods we find the maximum moment in the beams to be 75.0 k-f at the center line of the wall section, and the moment at the edge of the wall is 50.0 k-f (see Figure 2-32).

From the maximum moment and the allowable bending stress on the material we can obtain the section modulus

$$S_x = \frac{M}{F_b} = \frac{(75{,}000 \text{ lb-ft})(12 \text{ in./ft})}{16{,}000 \text{ psi}} = 56.25 \text{ in.}^3$$

The Carnegie Steel Manual gives an I for beam B9, 12 in. × 32 lb = 222.3 in.⁴. From this,

---

[5] Factor to account for beam spacing of 1.5 in. (table is for 1-in. spacing.)
[6] Six in. of concrete outside each end of beam for rust protection.

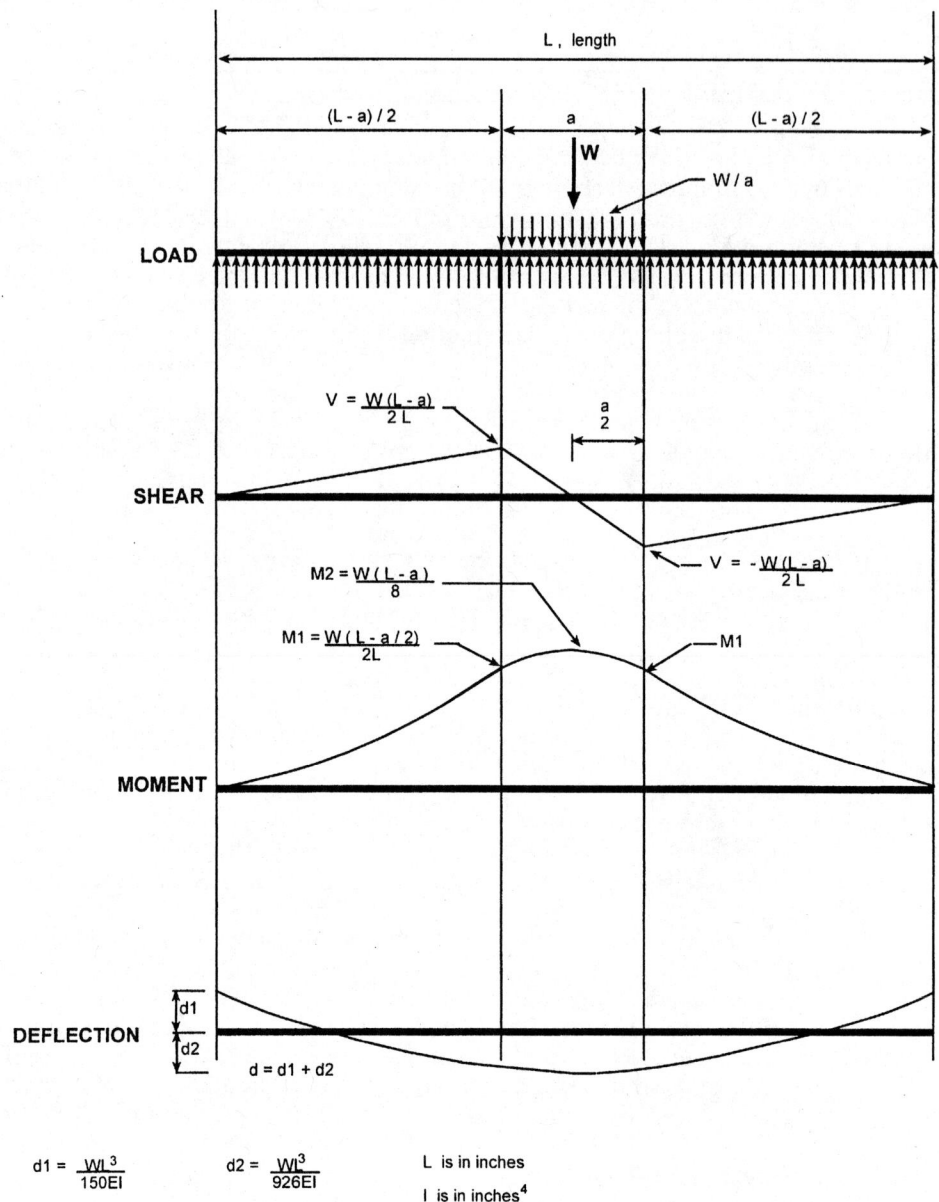

Figure 2-32. Single grillage footing: load, shear, moment, and deflection diagrams.

$$S = \frac{I}{0.5d} = \frac{222.3 \text{ in.}^4}{6 \text{ in.}} = 37.05 \text{ in.}^3$$

As one can see, the section provided by the table is not adequate for the loads that the proposed design was to meet. Reversing the formulae to obtain the actual stress on the member that was provided, we find:

$$F_b = \frac{M}{S} = \frac{75,000 \text{ lb-ft}(12 \text{ in./ft})}{37.05 \text{ in.}} = 24,292 \text{ psi}$$

(a condition of 50 percent overstress according to code but still within the elastic limit of 27,000 psi minimum).

Deflection of the ends of the projected beams will be relatively small = 0.35" (within the standard span/180 allowable. (See deflection computations in Figure 2-32.)

Shear must be checked to confirm that the steel sections can hold the applied loads without failing in vertical shear.

$$\text{Actual shear} = \frac{20{,}000 \text{ lb}}{(2)9.4 \text{ in.}^2} = 1{,}064 \text{ psi}$$

(each beam is in double shear).

The allowable shear on steel sections in 1902 was 10,000 psi.

The moment was calculated at the edge of the wall (the point of cantilever) and was determined to be 50,000 k-f. If the section modulus was calculated from the 50,000 k-f moment (as if it were the design moment), it would be determined to be 37.50 in., which is closer to the 37.05 in. for the Carnegie B9 12 in. × 32 lb section utilized by the Kidder example.

The member as designed for Example 2 will carry 1.5 ft of wall and therefore the total load will be 90,000 lb/ft. The moment at the center of the wall is 112,500 k-f, and at the edge of the wall is 75,000 k-f. In the same manner the design $S_{req'd} = 84.375$ in.^{3} and the section modulus provided by the Carnegie B7 15 in. × 41 lb beam = 56.55 in.^{3}. The actual bending stress on the Carnegie section was 23,873 psi, while the maximum allowable bending stress was 16,000 psi at the time.

The analysis is consistent, and it appears that the Carnegie method as specified by Kidder neglected to consider the actual maximum bending moment in the beam that occurs at the center of the wall. In each example, the actual stated allowable bending stress on the steel section was overstressed by approximately 50 percent. This excess was absorbed by the factor of safety employed in the working stress design method, which set the allowable bending stress at 16,000 psi, while the Carnegie Steel Pocket Companion gives the elastic limit (yield) for "soft steel" as 27,000 psi (Carnegie Steel Co. 1893, 179).

The deflection at the ends of the projected beams that make up the spread wall footings of Examples 1 and 2 calculates to be 0.33 in. Differential movement of the building resulting from this amount of deflection could cause damage to the building in the form of cracks in load-bearing masonry walls. If the design for all footings were consistent with the load to soil-bearing capacity, a minimal amount of differential settlement would have occurred.

Computations—Column Footings (Two or More Layers of Grillage). Concentrated loads from columns required two or more layers of grillage for the point loads to be distributed over an area that was square or basically square. Wood, cast iron, and steel columns were all provided with special bases to transfer their loads to footings. In many instances, loads were of such magnitude that three layers of grillage were required to ensure the proper distribution of the concentrated load over the base of the lower grillage (see Figures 2-33 and 2-34).

The solid bases were normally cast iron shapes that were manufactured to be compatible with the usual column sizes (see Figure 2-35). Columns were normally specified in sizes of 2 in. increments, so the cast iron bases could be manufactured in specific sizes. Standardization may have been creeping into the industry, inasmuch as the steel manufacturers were publishing manuals that suggested column sizes and plate and angle configurations. Many manufacturers were also providing specific shapes that could be combined to make up column sections. Phoenix Z-columns were one of the proprietary series of shapes supplied by the manufacturer, which included the necessary data for its use (see Chapter 3 for details of columns and column loads). In the case of many of the built-up columns that were used for heavy loads, the two- or three-level grillage incorporated a splayed column base section, shown in Figure 2-36. This splayed section actually spread the load from the column to the upper level of grillage by providing a bearing base wide enough to allow a multiple-beam layer as required by the grillage concept.

It is evident that the deflection problem for grillages of two and three layers (used for columns to develop a wider spread of platform under the concentrated load) was even more of a problem than that for a wall footing. Column footings present a special problem due to the fact that they support a concentrated load. The lower grillage is sized to spread the load over a given area of substrata. The area of the lower grillage

92 ◇ Structural Analysis of Historic Buildings

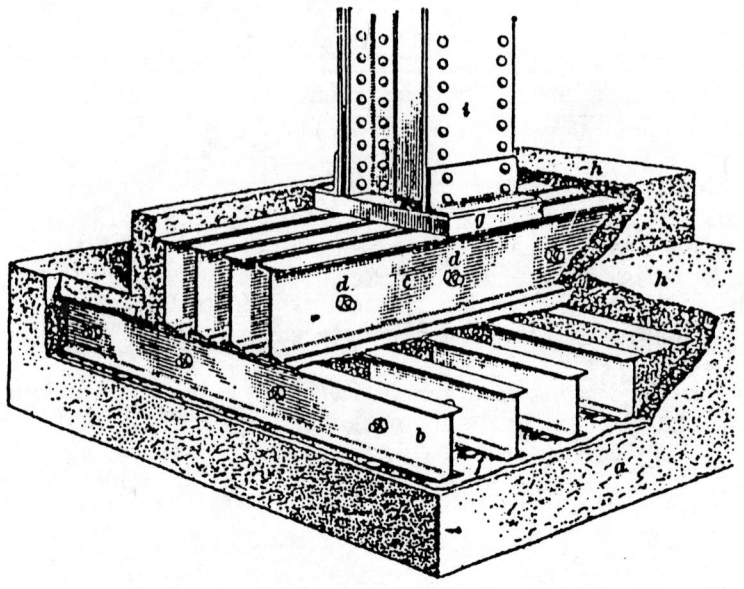

Figure 2-33. Two-layer grillage with column base plate. (*From International Correspondence Schools 1923, sec. 5, 29.*)

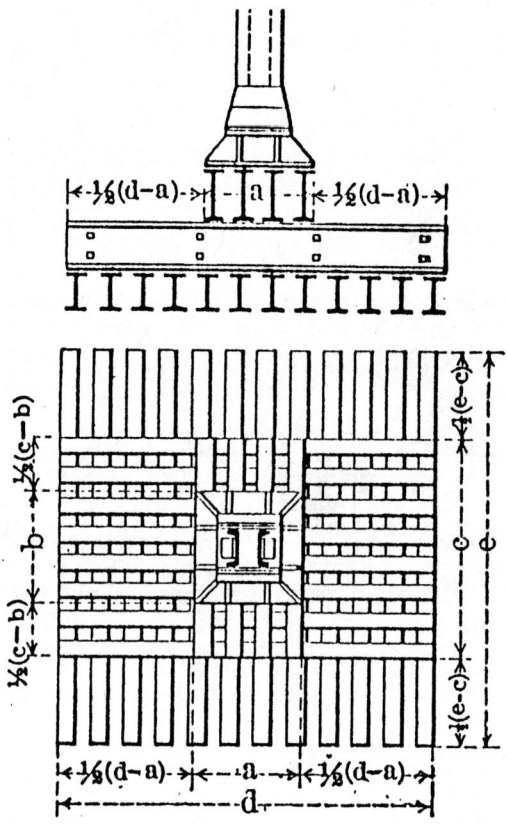

Figure 2-34. Three-layer grillage with column seat. (*From Thackray 1919, 327.*)

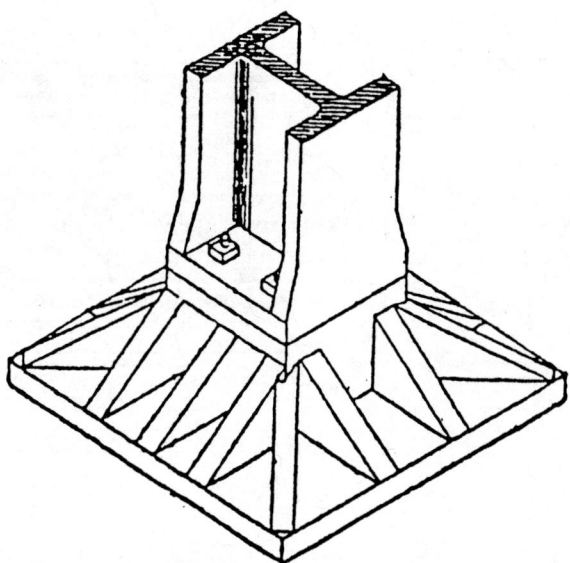

Figure 2-35. Cast iron column bases. (*From Kidder 1902, 255b.*)

corresponds to the allowable bearing pressure. The steel shapes used on multiple layers of grillages are inherently flexible, and the deflections are compounding or additive (see Figure 2-37).

As discussed in the section on wall footings, the designers of this period appear to have interpreted the projecting portion of a beam section used in the grillage as a cantilever. We now know that this produces an error in the design moment. As in the wall footing, the calculated moment at the point of projection is the same as that of a simple cantilever; but the actual maximum moment on the section occurs at the midpoint of the length (see Figure 2-32).

This misinterpretation of a design value resulted in many footings, both wall and column, that contained rail or steel shapes too small for the use intended. The result in most instances would have been an overstressed condition in the steel or rail shape and excessive deflections. The deflections could cause medium to severe cracks in building walls and irregularities in floors, which would be very damaging to the value of a building. Deflection problems from grillage footings would rarely produce actual building failures; however, building owners and users could not tolerate the unsightly cracks in exposed masonry exteriors.

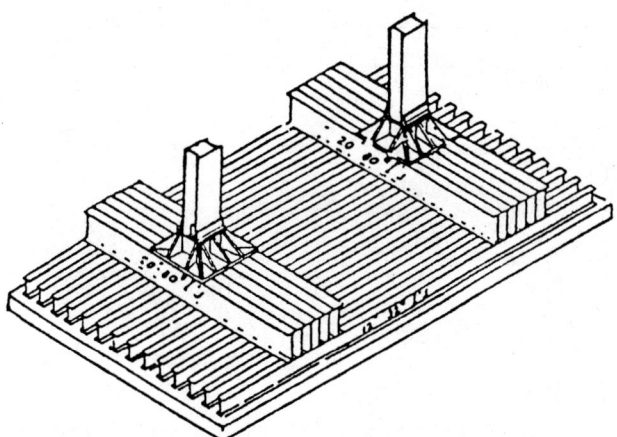

Figure 2-36. Cast iron column bases. (*From Kidder 1905, 51.*)

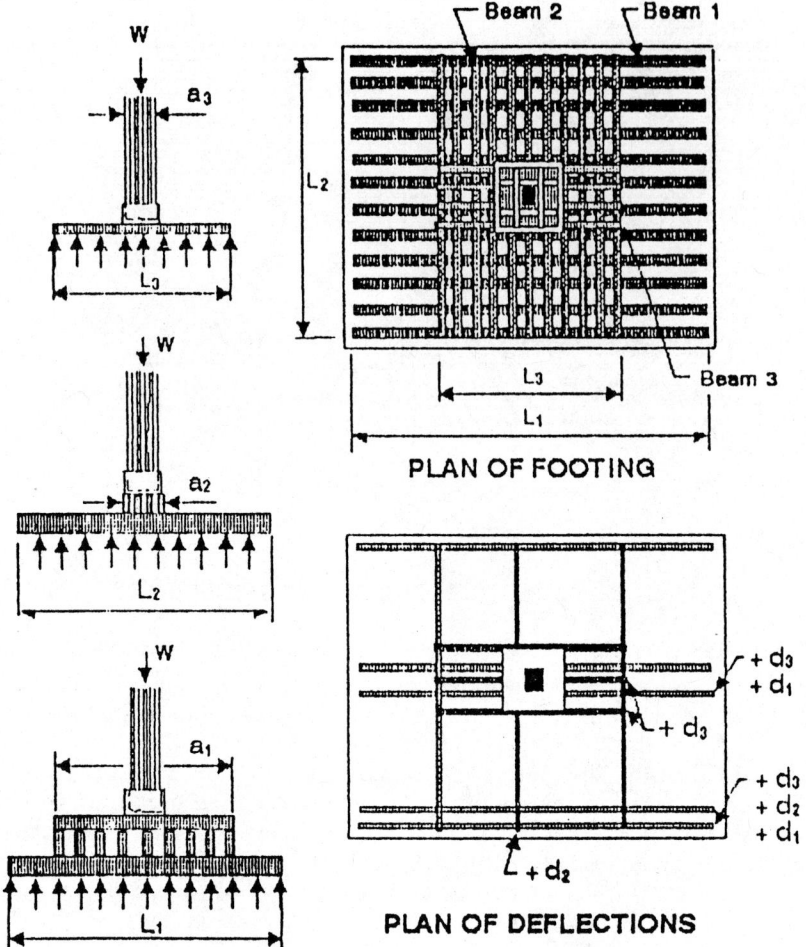

Figure 2-37. Column footing: three-layer grillage.

Example: Column Footing—Using Kidder Method
The grillage footing was designed using the same table shown in Table 2-11 and utilized in the wall footing example.

Column load = 150 tons,
Soil allowable bearing pressure = 1.25 tons/ft²
Footing spread = 150 tons divided by 1.25 tons/ft² = 120 ft² (Use 10 × 12 ft in plan, three-layer grillage)
Footing data:

$L_1 = 12$
$S = 10.5$ in.
$b = 1.25$ tons/ft²
$a_1 = 6$ ft

$L_2 = 10$ ft
$S = 10.5$ in.
$b = 2.5$ tons/ft²
$a_2 = 4$ feet.

$L_3 = 6$ ft $S = 10$ in.
$b = 6.25$ tons/ft²
$a_3 = 2$ ft

Kidder table of beams:

> Beam 1: $S = 0.875$ ft
> Use $b' = (.875)1.25$ tons/ft^2
> $b' = 1.093$ tons/ft^2
> Beam = 6 in. × 16 lb/ft
> Proj. = 3 ft, 11 beams
>
> Beam 2: $S = 0.875'$
> Use $b' = (.875)2.5$ tons/ft^2
> $b' = 2.188$ tons/ft^2
> Beam = 8 in. × 18 lb/ft
> Proj. = 3 ft, 8 beams
>
> Beam 3: $S = 0.833$ ft
> Use $b' = (.833)6.25$ tons/ft^2
> $b' = 5.200$ tons/ft^2
> Beam = 9 in. × 27 lb/ft
> Proj. = 2 ft, 5 beams

Table 2-11 is set up for a spacing of 12 in. on all beams. Kidder states that the table can be used for any spacing by calculating the ratio of the actual spacing to the tabulated spacing of 12 in. and multiplying that ratio by the allowable bearing pressure. The beam to be used is then obtained from the lower bearing pressure or by interpolation between tabulated values. Therefore:

$$\frac{10.5 \text{ in. spacing}}{12 \text{ in. spacing}} = 0.875$$

1.25 tons/ft^2 bearing pressure (0.875) = 1.09375 tons/ft^2

Check of Kidder's Method by Modern Analysis Methodology. As discussed in the preceding section, the early analysis was in error and the full maximum moments were not known to designers. Methods of modern analysis indicate overstress conditions and excessive deflections in the spread footings. This error may have led to the same localized buckling failures shown on wall footings.

Modern Analysis: Example of Three-Layer Grillage
Beam. 3.: Upper Level of Grillage.

> Kidder Method = Steel "W", 9" × 27 lbs./ft. $I_x = 111.29$ in.^{4}
>
> Spacing = 10 inches
>
> Uniform Load from Column = $\dfrac{300 \text{ k}}{(2 \text{ ft.})(4 \text{ ft.})}$ (0.833 ft./beam)
>
> = 31.25 k/ft. on each beam.
>
> Uniform Load from Support = $\dfrac{300 \text{ k}}{(6 \text{ ft.})(4 \text{ ft.})}$ (0.833 ft./beam)
>
> = 10.4167 k/ft. on each beam.

Actual Bending Stresses:

$$f_b = \frac{Mc}{I_x} = \frac{20.833 \text{ k-f}(12''/')(4.5'')}{111.29 \text{ in}^4}$$

$f_b = 10.108$ ksi. at point of projection.

$$f_b = \frac{Mc}{I_x} = \frac{20.833 \text{ k-f}(12''/')(4.5'')}{111.29 \text{ in}^4}$$

f_b = 15.162 ksi. at center under column.

Deflection at Projection End:

$t_{A/C}$ = Moment of Area of M/EI Diagram between A & C about A

= 0.05 inches

Beam 2.: Mid Level of Grillage.

Kidder Method = Steel "W", 8" × 18 lbs./ft. I_x = 57.76 in⁴

Spacing = 10.5 inches

$$\text{Uniform Load from Beam Above} = \frac{300 \text{ K}(.875 \text{ ft./beam})}{6 \text{ ft}(4 \text{ ft})}$$

= 10.9375 k/ft. on each beam.

$$\text{Uniform Load from Support} = \frac{300 \text{ k}(.875 \text{ ft./beam})}{6 \text{ ft.}(10 \text{ ft.})}$$

= 4.375 k/ft. on each beam.

Actual Bending Stresses:

$$f_b = \frac{Mc}{I_x} = \frac{32.8125 \text{ k-f}(12''/')(4'')}{57.76 \text{ inches}^4}$$

f_b = 33.328 ksi. at center of beam length.

$$f_b = \frac{Mc}{I_x} = \frac{19.6875 \text{ k-f}(12''/')(4'')}{57.76 \text{ inches}^4}$$

f_b = 16.360 ksi. at point of projection.

Deflection at Projection End:

$t_{A/C}$ = 0.29 inches

Beam 1.: Lower Level of Grillage.

Kidder Method = Steel "W", 6" × 16 lbs./ft. I_x = 28.64 in⁴

Spacing = 10.5 inches

$$\text{Uniform Load from Beam above} = \frac{300 \text{ k}(.875 \text{ ft./beam})}{6 \text{ ft.}(4 \text{ ft.})}$$

= 4.375 k/ft. on each beam.

$$\text{Uniform Load from Support} = \frac{300 \text{ k}(.875 \text{ ft./beam})}{6 \text{ ft.}(10 \text{ ft.})}$$

= 2.1875 k/ft. on each beam.

Actual Bending Stresses:

$$f_b = \frac{Mc}{I_x} = \frac{9.844 \text{ k-f}(12''/')(3'')}{28.64 \text{ inches}^4}$$

$f_b = 12.382$ ksi. at point of projection.

$$f_b = \frac{Mc}{I_x} = \frac{19.6875 \text{ k-f}(12''/')(3'')}{28.64 \text{ inches}^4}$$

$f_b = 24.764$ ksi. at center of beam length.

Deflection at Projection End:

$t_{A/C}$ = Moment of Area of M/EI Diagram between A & C about A.

= 0.52 inches

The problem inherent in the grillage footing was the result of the designer's mistaken assumption that the projecting portion of the beams or rails was a cantilever. The result was overstress in most beam or rail footings. The resulting deflection, which might be accompanied by compression of the soil, could have caused excessive cracks and foundation settlements. (See Figures 2-38, 2-39, and 2-40.)

The reason for this problem with footings for columns is the compounding or additive characteristics of the deflections at the ends of the projections (see Figure 2-41). As the grillage is compressed by the load from the column acting down, the spread of the grillage footing deflects vertically upward from the force of the soil. The central portion of the grillage footing directly under the column then carries the largest amount of the total load, thus overloading the soil. The soil is compressed beyond its allowable and therefore the pressure either causes a volume change or is compressed and pushed outward beyond the edge of the spread footing, as shown in Figure 2-42.

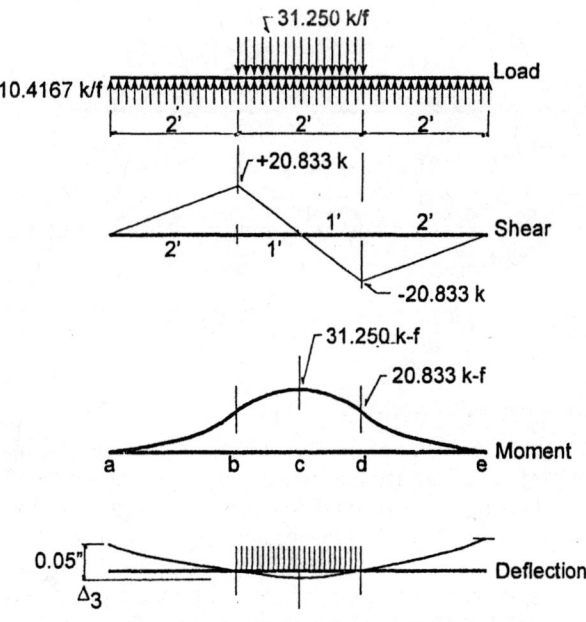

Figure 2-38. Upper level of grillage.

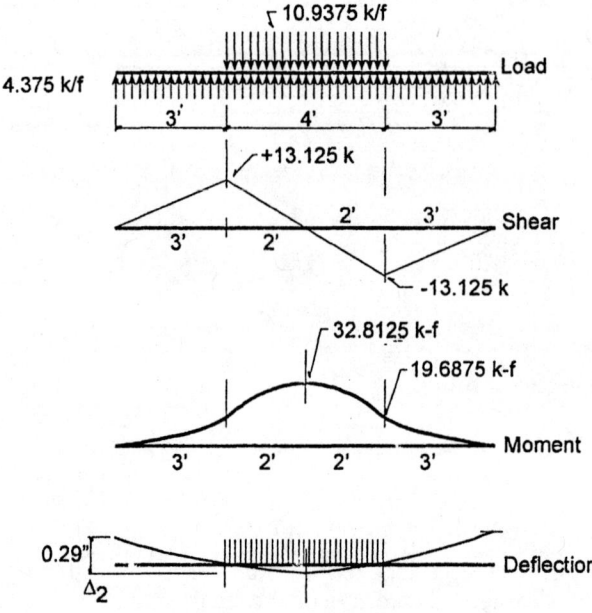

Figure 2-39. Beam 2: middle level of grillage.

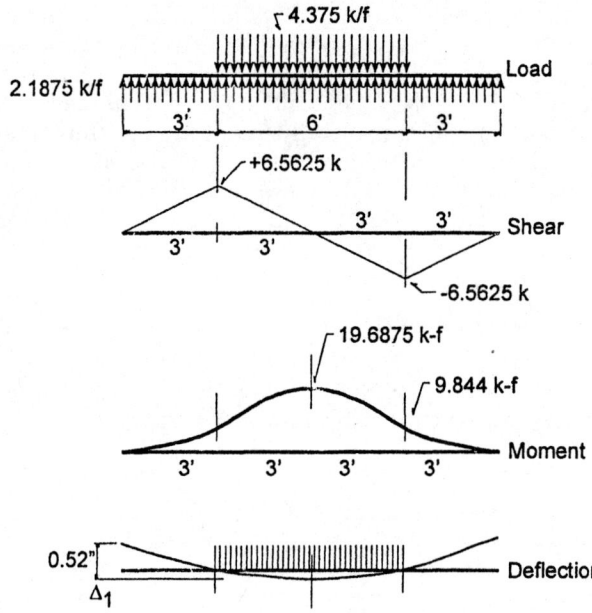

Figure 2-40. Beam 1: lower level of grillage.

Reinforced Concrete Spread Footings

The use of spread footings evolved as the loads from taller and/or skeletal-frame structures became both larger in magnitude and more concentrated in type. The battered footing (the belled out or corbeled masonry footing) and the "timber footing" were adequate for the lower-profile buildings of the mid-nineteenth century. The push for taller buildings beginning in the mid-1870s brought about larger loads per column or wall, and the massive size of the traditional battered footing was unmanageable. Basements were so crowded with battered footings that there was no space for boilers and other mechanical areas; in many instances, they were 20 ft or more wide and taller than they were wide.

Foundation Systems of American Historic Buildings ◇ 99

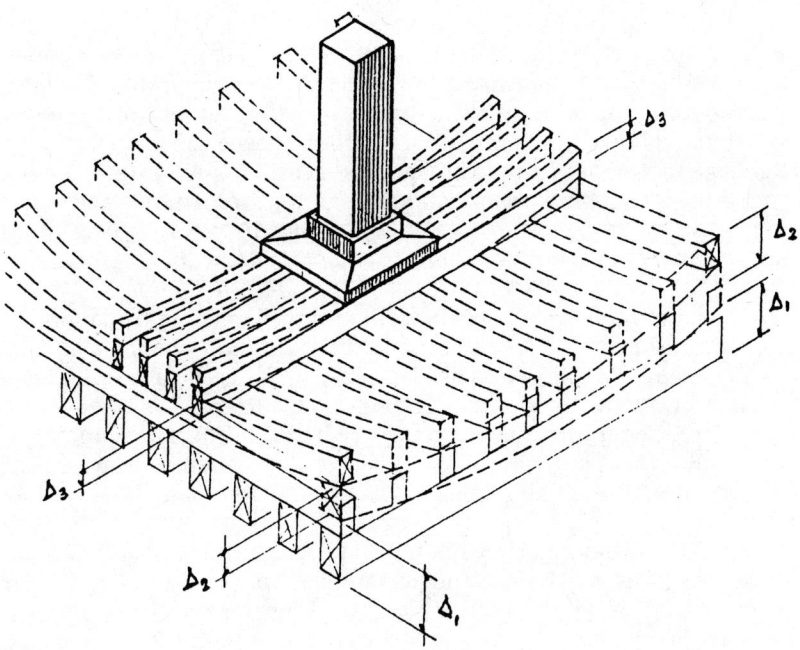

Figure 2-41. Grillage footing: Isometric.

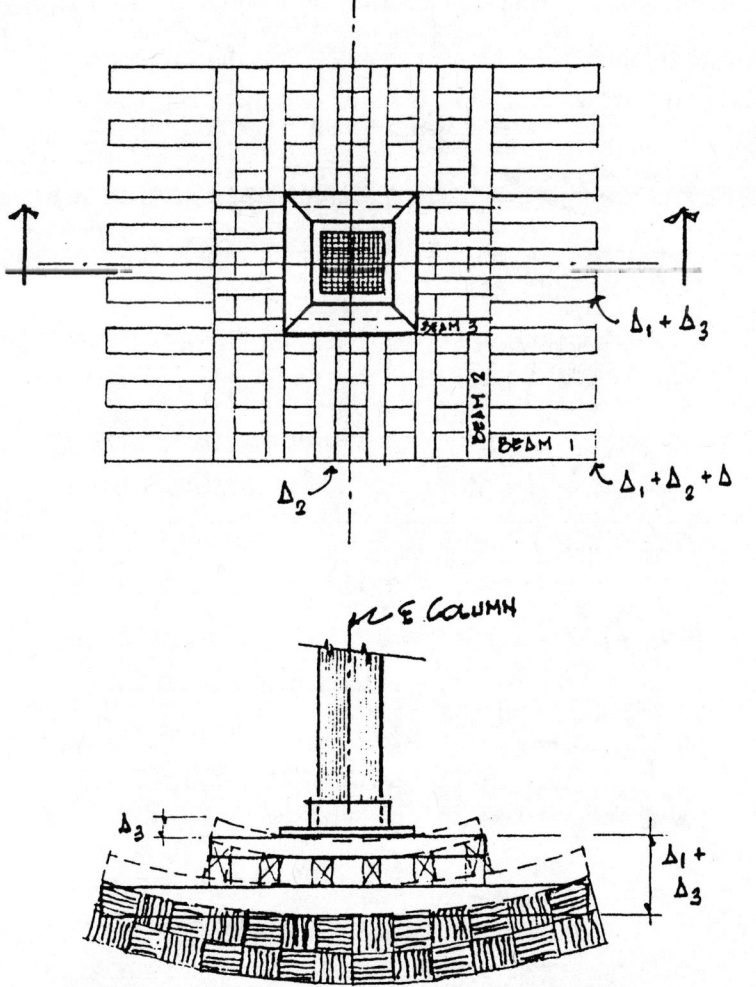

Figure 2-42. Grillage footing: plan and section.

The taller buildings were obviously placing an even heavier load on their foundations than the lower buildings. The battered footing was no longer capable of performing a service, its weight and massive size being a severe problem. The steel rail/beam grillage footing was used on many buildings. However, as has been shown, the designs were often flawed and many of these footings experienced deformations with load. The grillage footing was an advantage when conservatively designed in that the space required for the footing was much less than for the battered masonry footing.

Portland concrete and reinforced spread footings began to appear at about the turn of the century. They were obviously used sparingly at the beginning, as in the application of any new technology. In 1911, in *The Concrete Engineer's Handbook*, the International Correspondence Schools presented very detailed information on weights of materials (live and dead loads), building laws (disposition of loads), and early concrete design. The spread footing was used exclusively for vertical column loads at this time. There is no evidence of any use of the footings to resist moments at the base of the column. The true frame was not yet in use by 1911. The early concrete engineer had a large number of reinforcing bars (or systems) to choose from in his designs. Patented or proprietary bars and special shapes emerged with this new technology (see Figure 2-43).

International Correspondence Schools (1911) designates round, rolled sections as "rods," square sections as "bars," and rectangular sections as "flats." The square twisted bar shown as Bar (a) in Figure 2-43 is the Ransome bar, invented by the early concrete engineer Ransome. This bar was considered superior for two reasons: (1) it gave great resistance to pulling out of the mass of concrete; and (2) when square bars are twisted cold, their elastic limit and ultimate strength are increased from 8 to 25 percent (ibid., 255).

The Ransome bar was generally available throughout the United States and was one of the most popular types with designers. International Correspondence Schools (1911) tabulates properties of the Ransome bars and the Kahn bars (see Tables 2-12 and 2-13).

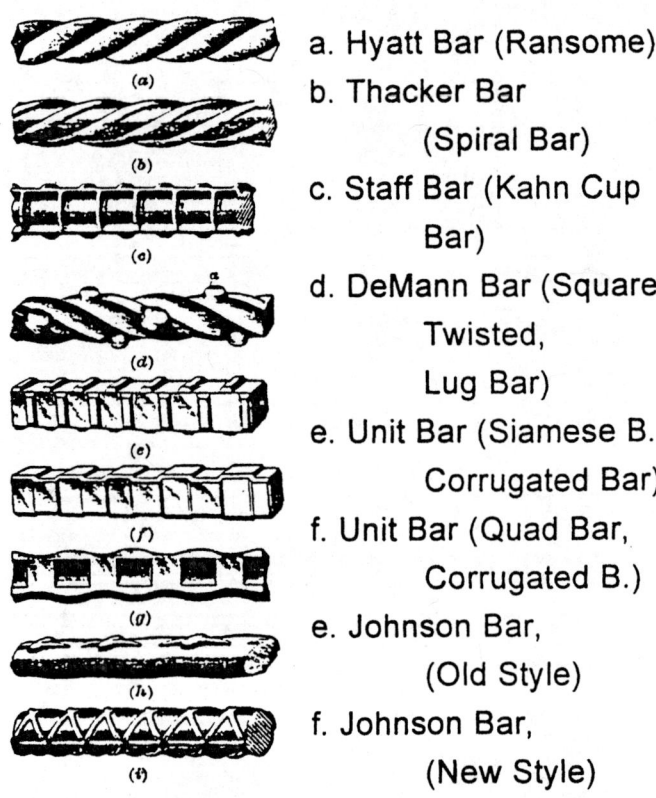

a. Hyatt Bar (Ransome)
b. Thacker Bar (Spiral Bar)
c. Staff Bar (Kahn Cup Bar)
d. DeMann Bar (Square Twisted, Lug Bar)
e. Unit Bar (Siamese B., Corrugated Bar)
f. Unit Bar (Quad Bar, Corrugated B.)
e. Johnson Bar, (Old Style)
f. Johnson Bar, (New Style)

Figure 2-43. Types of steel reinforcement. (*From International Correspondence Schools 1911, 255.*)

Table 2-12. Ransome Bar[a]

Size of Bar (in.)	Number of Twists per Foot of Length	Area of Bar (in.²)	Weight per Foot (lb)	Elastic Limit (psi)	Ultimate Tensile Strength (psi)	Elastic Limit of One Bar (lb)	Ultimate Tensile Strength of One Bar (lb)
¼	5	0.0625	0.213	62,350	86,700	3,897	5,419
⅜	3	0.1406	0.478	61,800	86,600	8,689	12,176
½	2	0.2500	0.850	60,120	86,850	15,030	21,713
⅝	2	0.3906	1.328	57,890	85,820	22,612	33,520
¾	1	0.5625	1.913	56,720	85,240	31,905	47,948
⅞	1	0.7656	2.603	56,150	84,730	42,988	64,869
1	¾	1.0000	3.400	55,760	84,275	55,760	84,275
1¼	½	1.5625	5.312	55,450	83,150	86,641	129,921

[a] From International Correspondence Schools 1911, 255.

Standard square and round bars were available as hot-rolled "metallic reinforcement" for concrete in 1911, according to the Handbook. The square and round bars were available in three grades: mild or soft steel, medium steel, and high-carbon steel. In addition, there was a fourth type, rerolled bars, which were produced by rerolling old steel rails. The areas and weights of the square and round bars are given in Table 2-14.

The reinforced concrete footing design as presented in International Correspondence Schools (1908, sec. 45, 81), is by the "straight line theory" and is as follows:

$$n = \frac{E_s}{E_c} = \frac{30,000,000 \text{ psi}}{2,000,000 \text{ psi}} = 15$$

$M = R(b)(d^2)$ (coefficient from Table 2-15)

$F'_s = 16,000$ psi

$f_c = 650$ psi

where:
  b = width of beam, or unit width of footing
  d = distance from center of steel to top of concrete

This method was the agreed computational analysis as defined by the Joint Committee in their Progress Report published in July 1909. The Committee, which was formed in 1904, consisted of members of the ASCE, ASTM, American Railway Engineering and Maintenance of Way Association, and Association of Portland Cement Manufacturers. The purpose of the Joint Committee was to recommend necessary factors and formulas required in the design of structures where reinforced concrete is to be used (ibid., sec. 45, 91). The value of R, the coefficient, is dependent on the quantity of steel to be used in the design. The Joint Committee set up the criteria for

Table 2-13. Kahn Bar[a]

Nominal Size of Bar (in.)	Sectional Area (in.²)	Weight per Foot (lb)	Elastic Limit of Each Bar (lb)	Ultimate Tensile Strength of Each Bar (lb)
⅜	0.1406	0.502	5,000	9,800
½	0.2500	0.893	9,000	17,500
⅝	0.3906	1.394	14,000	27,300
¾	0.5625	2.008	21,000	39,400
⅞	0.7656	2.733	28,000	53,600
1	1.0000	3.570	37,000	70,000
1⅛	1.2656	4.518	47,000	88,600
1¼	1.5625	5.578	58,000	109,400

[a] From International Correspondence Schools 1911, 257.

Table 2-14. Areas and Weights of Round and Square Bars[a]

	Square		Round	
Size (in.)	Area (in.)	Weight per Foot (lb)	Area (in.)	Weight per Foot (lb)
1/16	0.0039	0.013	0.0031	0.010
1/8	0.0156	0.053	0.0123	0.042
3/16	0.0352	0.120	0.0276	0.094
1/4	0.0625	0.213	0.0491	0.167
5/16	0.0977	0.332	0.0767	0.261
3/8	0.1406	0.478	0.1104	0.376
7/16	0.1914	0.651	0.1503	0.511
1/2	0.2500	0.850	0.1963	0.668
9/16	0.3164	1.076	0.2485	0.845
5/8	0.3906	1.328	0.3068	1.043
11/16	0.4727	1.607	0.3712	1.262
3/4	0.5625	1.913	0.4418	1.502
13/16	0.6602	2.245	0.5185	1.763
7/8	0.7656	2.603	0.6013	2.044
15/16	0.8789	2.989	0.6903	2.347
1	1.0000	3.400	0.7854	2.670
1 1/16	1.1289	3.838	0.8866	3.014
1 1/8	1.2656	4.303	0.9940	3.379
1 3/16	1.4102	4.795	1.1075	3.766
1 1/4	1.5625	5.312	1.2272	4.173
1 5/16	1.7227	5.857	1.3530	4.600
1 3/8	1.8906	6.428	1.4849	5.049
1 7/16	2.0664	7.026	1.6230	5.518
1 1/2	2.2500	7.650	1.7671	6.008
1 9/16	2.4414	8.301	1.9175	6.520
1 5/8	2.6406	8.978	2.0739	7.051
1 11/16	2.8477	9.682	2.2365	7.604
1 3/4	3.0625	10.413	2.4053	8.178
1 13/16	3.2852	11.170	2.5802	8.773
1 7/8	3.5156	11.953	2.7612	9.388
1 15/16	3.7539	12.763	2.9483	10.024
2	4.0000	13.600	3.1416	10.681

[a] From International Correspondence Schools 1911, 253.

the designs and the method for analysis. The ratio p is tabulated in Table 2-15 from 0.02 percent to 2 percent to give the corresponding values of R for three grades of concrete. The ratio of the reinforcement determines which material reaches the working stress first (if the steel ratio is high, the concrete reaches its working stress while the steel is understressed). The "economic steel ratio" is the ratio that brings both the steel and concrete to the level of maximum working stress at a specific bending moment (similar to p balance in today's ultimate strength analysis).

Example: Reinforced Concrete Footing
$\rho = A_s$ divided by $b(d)$ = ratio of reinforcement (see Figure 2-44)

$f'_c = 650$ psi, f

$F'_s = 16,000$ psi, $n = 15$

$A = (0.75 \text{ in.})(3) = 1.69$ in.

$d = 15$ in. $b = 12$ in.

$\rho = 0.0094$, within limits (between .0002 & .02)

Therefore, $R = 114.67$ (from Table 2-15 for 7-15 & $\rho = 0.0094$)

Table 2-15. R Factors for Reinforced Concrete: Table for Special Constants[a]

p	R $n = 15$ $F_s = 16,000$ $F_c = 650$	R $n = 12$ $F_s = 16,000$ $F_c = 500$	R $n = 15$ $F_s = 16,000$ $F_c = 500$	p	R $n = 15$ $F_s = 16,000$ $F_c = 650$	R $n = 12$ $F_s = 16,000$ $F_c = 500$	R $n = 15$ $F_s = 16,000$ $F_c = 500$
0.0002	3.12	3.13	3.12	0.0102	117.01	84.32	90.47
0.0004	6.18	6.20	6.18	0.0104	118.31	84.85	91.01
0.0006	9.20	9.24	9.20	0.0100	119.00	85.37	91.54
0.0008	12.19	12.25	12.19	0.0108	119.68	85.89	92.06
0.0010	15.15	15.24	15.15	0.0110	120.34	86.30	92.57
0.0012	18.10	18.20	18.10	0.0112	120.99	86.89	93.07
0.0014	21.02	21.15	21.02	0.0114	121.64	87.37	93.57
0.0016	23.02	24.08	23.92	0.0116	122.27	87.85	94.05
0.0018	26.81	27.00	26.81	0.0118	122.89	88.32	94.53
0.0020	20.69	29.01	20.69	0.0120	123.50	88.79	95.00
0.0022	32.55	32.80	32.55	0.0122	124.10	89.25	95.46
0.0024	35.40	35.67	35.40	0.0124	124.69	89.70	95.91
0.0026	38.23	38.54	38.23	0.0126	125.27	90.14	96.36
0.0028	41.05	41.40	41.05	0.0128	125.84	90.58	96.80
0.0030	43.87	44.24	43.87	0.0130	126.40	91.01	97.23
0.0032	46.07	47.08	46.67	0.0132	126.96	91.43	97.66
0.0034	49.46	49.91	49.46	0.0134	127.51	91.35	98.08
0.0036	52.24	52.73	52.24	0.0130	128.04	92.26	98.50
0.0038	55.02	55.53	55.02	0.0138	128.57	92.67	98.90
0.0040	57.78	58.34	57.78	0.0140	129.10	93.07	99.31
0.0042	60.54	61.13	60.54	0.0142	129.61	93.47	99.70
0.00426		61.93					
0.0044	63.28	62.74	63.28	0.0144	130.12	93.86	100.09
0.0040	66.02	63.80	66.02	0.0146	130.62	94.24	100.48
0.0048	68.76	64.82	68.76	0.0148	131.12	94.62	100.86
0.00499			71.30				
0.0050	71.48	65.81	71.37	0.0150	131.60	95.00	101.23
0.0052	74.20	66.77	72.37	0.0152	132.08	95.37	101.60
0.0054	76.91	67.70	73.34	0.0154	132.56	95.73	101.97
0.0056	79.61	68.60	74.28	0.0156	133.02	96.09	102.33
0.0058	82.31	69.48	75.19	0.0158	133.49	96.45	102.68
0.0060	85.00	70.33	76.08	0.0160	133.94	96.80	103.03
0.0062	87.69	71.16	76.94	0.0162	134.39	97.15	103.30
0.0064	90.37	71.97	77.78	0.0164	134.83	97.49	103.72
0.0066	93.04	72.76	78.00	0.0166	135.27	97.83	104.06
0.0068	95.71	73.53	79.40	0.0168	135.70	98.16	104.39
0.0070	98.37	74.28	80.17	0.0170	136.13	98.50	104.72
0.0072	101.02	75.01	80.93	0.0172	136.55	98.82	105.04
0.0074	103.68	75.73	81.07	0.0174	136.98	99.15	105.36
0.0076	106.32	76.43	82.39	0.0176	137.38	99.47	105.68
0.00769	107.53						
0.0078	108.02	77.11	83.09	0.0178	137.70	99.78	105.99
0.0080	108.92	77.78	83.78	0.0180	138.19	100.09	106.30
0.0082	109.79	78.44	84.46	0.0182	138.59	100.40	106.60
0.0084	110.65	79.08	85.11	0.0184	138.98	100.71	106.91
0.0086	111.49	79.71	85.76	0.0186	139.37	101.01	107.20
0.0088	112.31	80.33	86.39	0.0188	139.75	101.31	107.50
0.0090	113.11	80.93	87.01	0.0190	140.13	101.60	107.79
0.0092	113.90	81.52	87.61	0.0192	140.50	101.89	108.08
0.0094	114.67	82.10	88.21	0.0194	140.87	102.18	108.36
0.0096	115.43	82.67	88.79	0.0196	141.24	102.47	108.65
0.0098	116.17	83.23	89.36	0.0198	141.60	102.75	108.92
0.0100	116.90	83.78	89.92	0.0200	141.96	103.03	109.20

[a] From International Correspondence Schools 1908, sec. 45, 91.

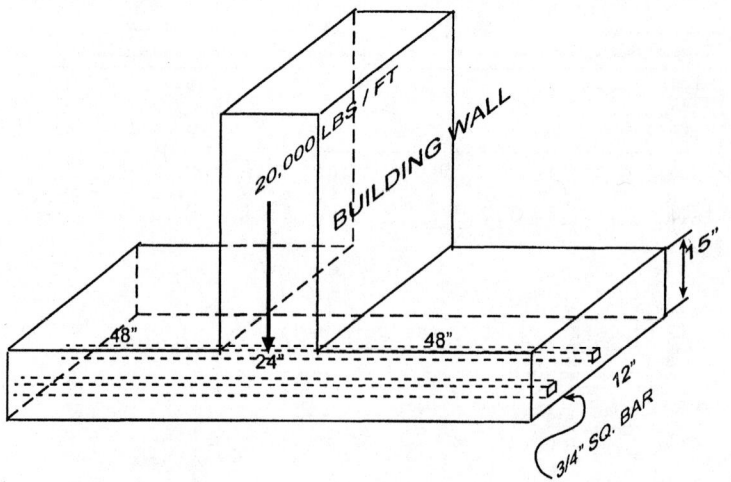

Figure 2-44.

$$M_{all} = R(b)(d^2) = 114.67(12 \text{ in.})(15 \text{ in.})$$
$$= 309{,}609 \text{ lb-in.} = 25{,}800 \text{ lb-ft}$$

$$M_{actual} = \frac{W}{8}(L - a) = \frac{20{,}000 \text{ lbs}}{8}(10 \text{ ft} - 2 \text{ ft}) = 20{,}000 \text{ lb-ft (at center of wall)}$$

$$M_{actual} = 2{,}000 \text{ lb-ft } (4 \text{ ft})\left(\frac{4 \text{ ft}}{2}\right) = 16{,}000 \text{ lb-ft (at edge of wall)}$$

See Figure 2-45.

The maximum applied moment is less than the allowable moment; therefore the design was considered acceptable.

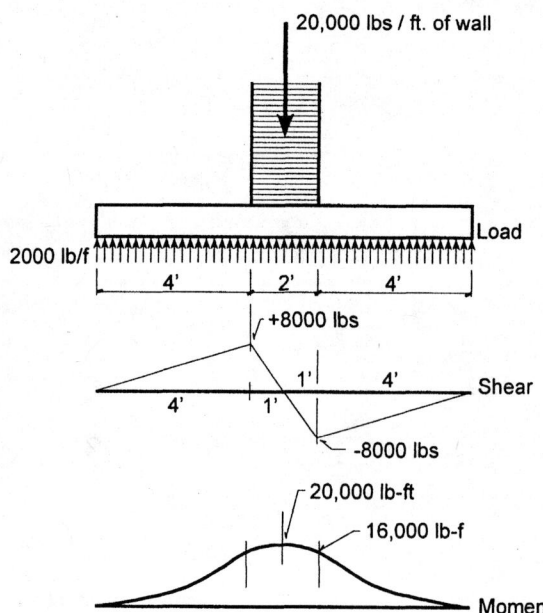

Figure 2-45. The maximum applied moment is less than the computed allowable moment. The design is acceptable.

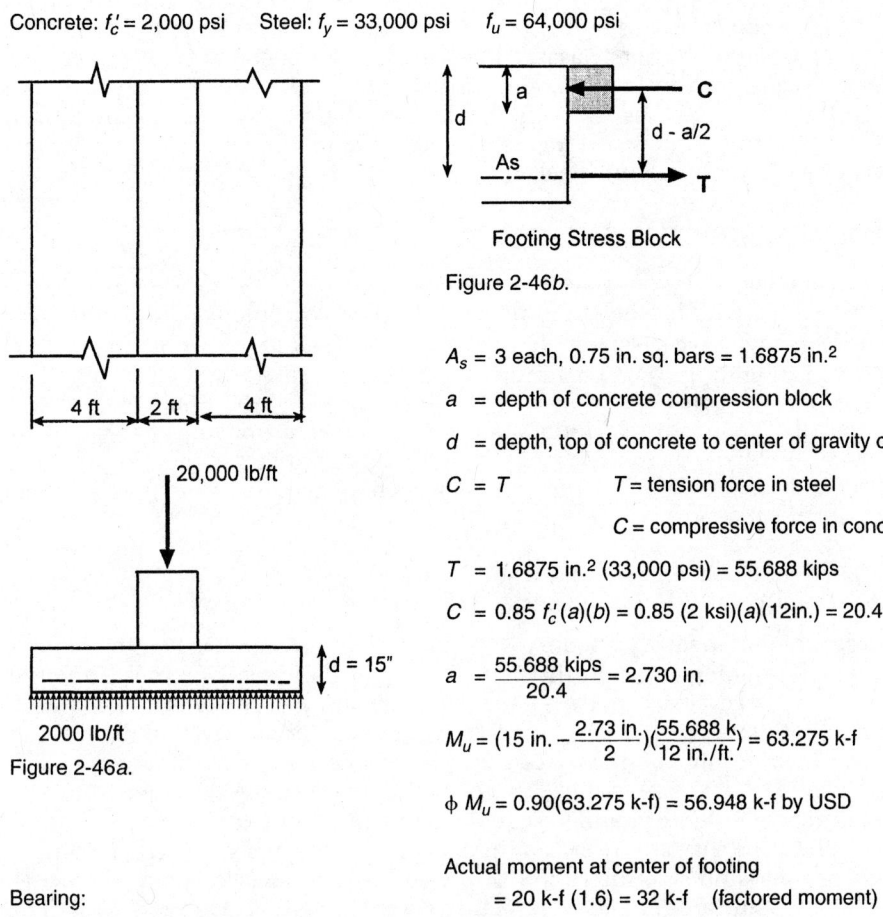

Figure 2-46. Modern analysis by ultimate strength design.

In the design calculation used in Table 2-15, the analysis shown in Figure 2-45 was involved.

The analysis apparently ignored the requirement to check for punching shear (two-way shear) and one-way shear. The example provided is a wall footing; a wall would not produce two-way shear on its footing since the system is designed as if it is continuous. One-way shear, however, would exist and was not considered by the early designers.

Check by Modern Analysis: Reinforced Concrete Spread Footing (Ultimate Strength Design).
By using the same materials as the earlier designers, we can check the early designs by using the analytical design methodology of today. The only assumption that must be made is that the bond of the concrete to the early reinforcing bars was

adequate to develop the stress required for the modern analysis. Bond was considered in early designs and was usually assured by "hooks" at the end of bars.

The analysis shown in Figure 2-46 is based upon the flexure and shear perpendicular to the wall. This was the only direction that the steel was specified in the example. A modern design would include longitudinal reinforcing of at least the minimum required for temperature steel.

FOUNDATION PROBLEMS: ANALYSIS AND REPAIRS

Many problems with buildings are the result of inadequate or improperly designed foundation systems. As has been discussed in this chapter, the early designers went from empirical methods to modest techniques of analysis to fairly detailed methods of analysis as buildings grew from 2 or 4 stories to 12–20 stories at the end of the nineteenth century. Their knowledge of building materials, the availability of technology, and the ability to test foundation materials were evolving throughout the period. Most buildings were designed in a very conservative manner, and in a few instances the designers were working with new technologies that were not thoroughly proven.

Foundation Problems, Problem Indicators, and Methods of Analysis

In the design of load-bearing masonry buildings, the foundation design (the footings) can be the determining factor in the performance of the walls. The most important element is always an accurate assessment of the bearing capacity of the soil at the level of the footings. Proper design requires that the center of gravity of the wall load coincide with the center of gravity of the soil resistance against the bottom of the footing. If the lines of action of these two opposing forces are not collinear, the result is an eccentric condition or moment couple acting on the footing.

The wall will be thrust outwards at midheight if the center of the axial load from the building wall is outside the center of the resisting force (see Figure 2-47a). The wall will be thrust inwards at midheight if the center of the axial load from the building wall is inside the center of the resisting force (see Figure 2-47b).

Designers of the low-rise load-bearing masonry buildings of the late nineteenth century learned to apply this eccentricity to their advantage. They purposely created a condition where the center of the axial load of the wall was inside the line of action of the resisting force from the footing by 1–3 in. to ensure a slight inward thrust to keep the walls in contact with the floor assembly. It appears that the designers were not analyzing the soil pressure under the eccentricity loaded footing by the combined stress equation

$$F_{soil} = \frac{P}{A} \pm \frac{M}{S}$$

where:
  P = axial load on the footing
  A = surface area of the footing
  M = moment on the footing
  S = section modulus of the plan of the footing

Example: Wall Footing: Analysis ca. 1912

When the designer calculated the eccentricity shown in Figure 2-48, his analysis would evaluate the effect of the thrust on the wall. If he considered the inward or outward thrust effect as more desirable, he could adjust the eccentricity by simply shifting the center of the footing in the necessary direction. He could also center the footing under the load rather than under the wall to eliminate the eccentricity and the thrust on the wall.

Modern Analysis

The modern methods of analysis would evaluate the bearing pressure on the soil caused by the footing with the combined axial and bending load. In evaluating the pressure on the soil, the modern designer would calculate the pressure on each side

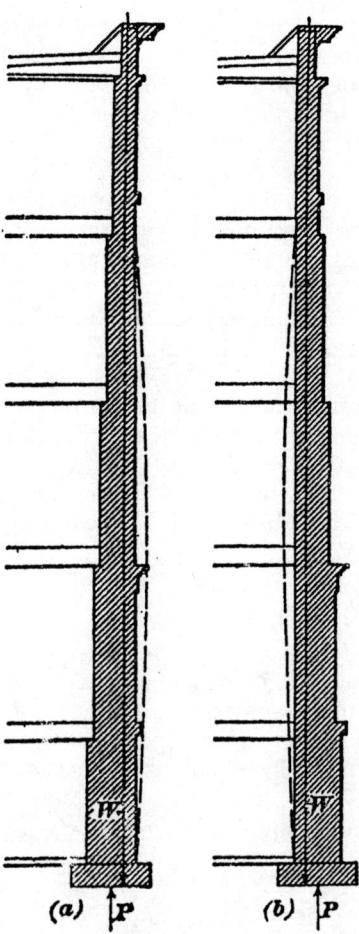

Figure 2-47. Footing eccentricity. (*From International Correspondence Schools 1905B, sec. 18, 25.*)

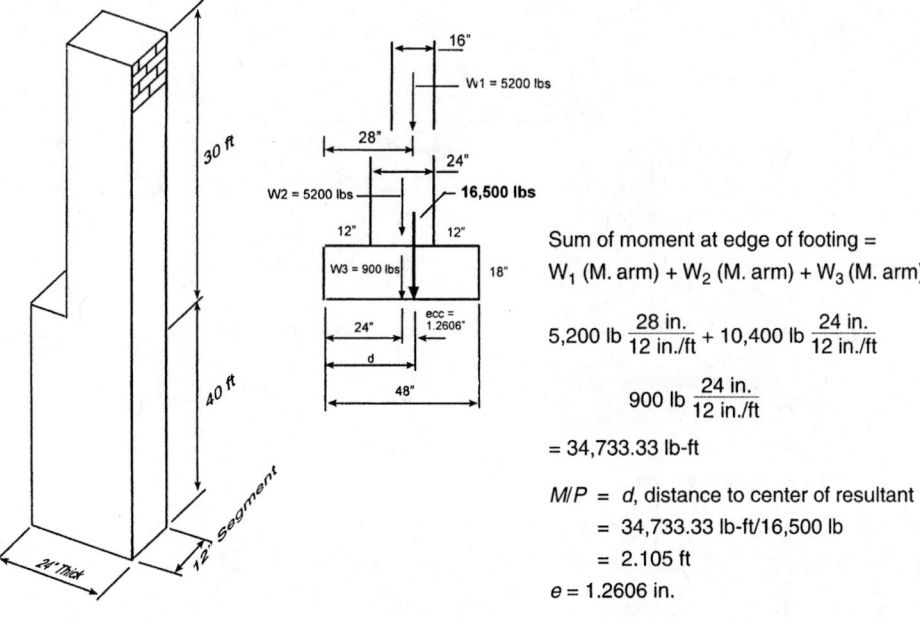

Sum of moment at edge of footing =
W_1 (M. arm) + W_2 (M. arm) + W_3 (M. arm)

$$5{,}200 \text{ lb } \frac{28 \text{ in.}}{12 \text{ in./ft}} + 10{,}400 \text{ lb } \frac{24 \text{ in.}}{12 \text{ in./ft}}$$

$$900 \text{ lb } \frac{24 \text{ in.}}{12 \text{ in./ft}}$$

= 34,733.33 lb-ft

M/P = d, distance to center of resultant
 = 34,733.33 lb-ft/16,500 lb
 = 2.105 ft
e = 1.2606 in.

Figure 2-48.

of the footing. As long as the higher pressure on the positive moment side did not exceed the maximum allowable, and if the pressure was positive under the entire footing, the design would be considered satisfactory. The pressure on the positive moment side of the footing was 4387.5 psf, while the pressure on the opposite side of the footing was 3412.5 psf, as is shown in Figure 2-49, which graphically represents the pressures.

In the era of this type of load-bearing masonry walls, this differential pressure was not understood. The designers did understand the principle of eccentricity and the resulting thrust that the eccentricity caused in the wall. The nonuniform pressures under the footing would cause differential settlements and the support of the wall would be unbalanced. This unbalance in support is the cause of the thrust in the wall that occurs at approximately midheight. The effect of eccentricity caused by the footing that has its center of gravity outside of the center of gravity of the wall in moderately compressible soils would be a thrust that actually made the walls press against the diaphragm of the floor. An eccentricity of the opposite type that caused the walls to spread outwardly and free the diaphragm of the floor and pull the pockets away from joist ends would be highly undesirable and would cause structural problems in the building.

Openings in Continuous Masonry Walls

Continuous footings under walls with large openings, or under structures that produce point loads, such as the tower shown in Figure 2-50, are a problem in that they produce eccentricity under the heavy point loads that results in cracks in the building walls. In Figure 2-50a, the continuous footing under the large opening has caused the eccentricity between the center of the load of the pier and the center of the upward pressure from the footing. This eccentricity has caused the cracks shown on the facade. The eccentricity that occurs in both piers has caused the walls on the two sides of the openings to spread apart.

Figure 2-50b indicates that the individual footings under the piers adjacent to the large openings produce a condition of concentric load and resultant upward footing pressures and no cracks in the building as the walls remain in alignment. The necessity for the designers to provide the exact footing size for consistent soil-bearing pressures throughout the total foundation system of the building is obvious.

It was also necessary for the designers to try to keep the foundation eccentricities at a minimum to control the wall thrusts to keep joists and beams tight in their pockets. Facade problems in masonry buildings, such as cracked mullions and band courses, might also be due to eccentric foundation conditions, as shown in Figure 2-51.

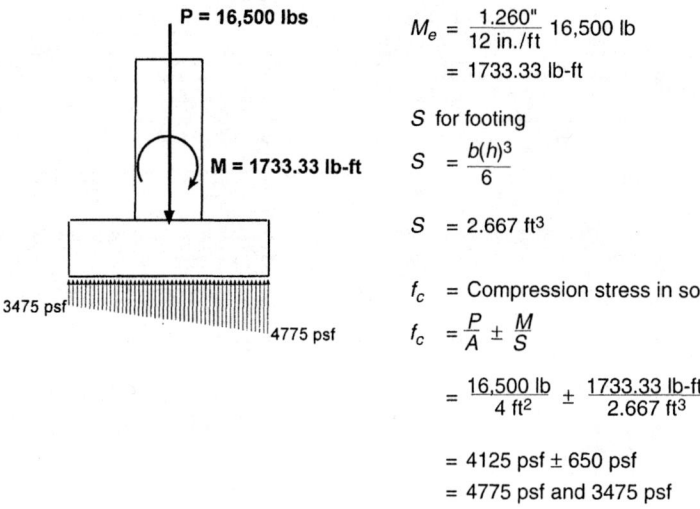

Figure 2-49.

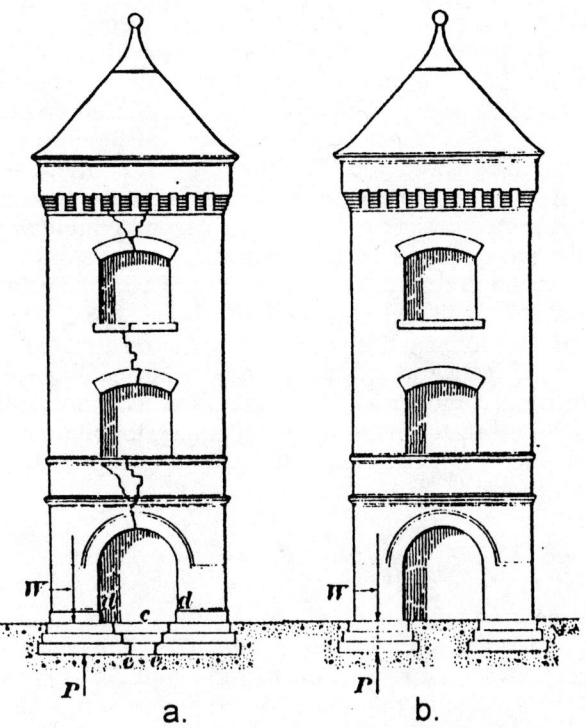

Figure 2-50. Tower footings. (*From International Correspondence Schools 1905, sec. 18, 31.*)

The outside corners of the building are heavily loaded and the continuous footings (continuous at the same width) have caused an eccentricity in the load on the footing that places an outward thrust on each wall, pushing the corner outward and providing vertical cracks in each of the adjacent facades. In this instance the center of gravity of the weight of the building walls at the corner acts along the plane *b-b*

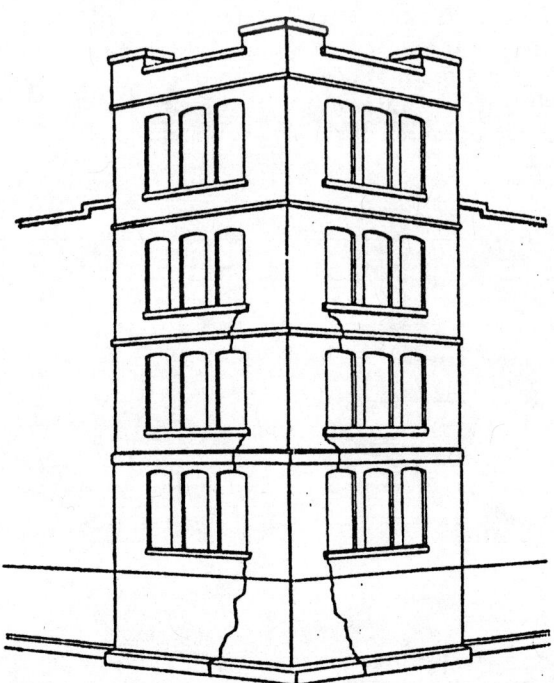

Figure 2-51. Building footings at corner. (*From International Correspondence Schools 1905, sec. 18, 35.*)

and the center of gravity of the foundation pressure acts along the plane *a-a*, as shown in Figure 2-52*a*. This eccentricity at 45° causes an outward thrust at the corner of the building. The solution to the problem of this eccentricity is shown in Figure 2-52*b*, which is a simple crossing of the corner walls and the formation of an eccentricity of the opposite type. This opposite eccentricity would cause an inward thrust in the walls at the corner of the building and keep the building's masonry walls from cracking.

The problem of eccentric footing pressures was often the product of an inner-city building being constructed on its property line adjacent to another site, a preexisting building, or a road right of way. The footing width could then not be centered under a wall. The effect of the eccentricity was minimized when the footing bore on an incompressible stratum such as rock (International Correspondence School 1908, sec. 46, 37). In addition, the lime-sand mortars of the period were compressible, and an eccentricity that would cause any foundation rotation could result in thrusts that would cause excessive compression at the edges of walls and possible wall damage.

Damage to building facades caused by inadequate foundations or any of a series of foundation problems is evidenced in many ways. Figure 2-53 shows three building facades with damage to each facade that is the result of foundation problems.

Analysis of Building Foundation Problems, Degrees of Failure

The masonry load-bearing building was fairly flexible in that minor movements were easily accommodated with only minimal damage. The lime-sand mortars were said to act like cushions between and below the historic brick. The degree of failure is very hard to define from visual analysis. Many determinants are involved in the overall analysis, such as the differential settlement that occurs when foundation loads are not uniform or when an area under a building's footings has zones of soft soils. In the expensive labor period of today, the extent of repairs that can be made on a historic building's foundations reflects the value of the building.

In some cases, the settlement of building foundations that comes from slightly compressible soils occurs in the early or initial period after the building is constructed. Thus, after an initial settlement, the building stabilizes and no further movement normally occurs. At times it is necessary to monitor a building for a period of time, through an annual cycle, to determine whether movements are continuing. In addition, a check of the depth of the footings in relation to the frost line will indicate whether building movements are the result of the freeze-thaw cycle.

Load-bearing masonry buildings with many cracks, mainly vertical and diagonal in direction, have obvious foundation problems. Either settlement from unstable bearing material, inadequate footing design, improper construction, or eccentrically loaded footings has been the cause of the movement and the cracking that resulted. Buildings with excessive numbers of cracks or evidence of differential settlement may

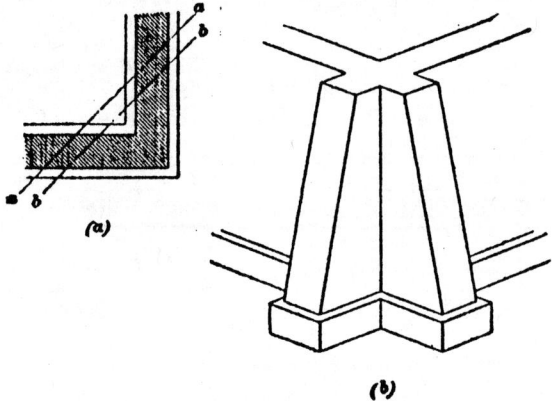

Figure 2-52. Corner eccentricities. (*From International Correspondence Schools 1905, sec. 18, 35.*)

Figure 2-53. Damaged building facades: foundation problems. (*From International Correspondence Schools 1905, sec. 18, 32, 33.*)

need underpinning and the transfer of foundation bearing to a lower, undisturbed bearing stratum. The function of underpinning is to insert a new foundation system below the existing foundations. The typical underpinning of a three- to four-story load-bearing masonry building is shown in Figure 2-54. The beam is inserted in the wall and wedged or jacked as shown to temporarily support the wall while a new foundation is constructed below the original.

"Needle Beam" System. Inserting a temporary beam through a wall to support a section of the wall above the level of the beam is known as "needling." When a series of needle beams is in place to hold up a length of wall, the old foundation system can be removed and the excavation performed to the level of bearing of the new foundation. Needle beams were usually spaced 6–10 feet apart, and the foundation excavation and reconstruction normally could include only three needle beams at one time. Figure 2-54 shows a typical needle beam installation and the construction of the temporary supports for the ends of the needle beams, which in turn support sections of the building wall. In some instances, the need for underpinning was the result of the construction of an adjacent building that required excavations below the foundation of the existing building (see Figure 2-55).

After the needling is completed, the excavation can be made under the wall. At that time, all earthen material and the lower portion of the original wall and foundations may be removed. Between needle beam support points the wall will corbel itself, as shown in Figure 2-54.

The new footing and new lower section of the wall (the underpinning) are then constructed up to the bottom of the old wall. At a point halfway between needles, carefully squared stone blocks of the type shown in Figures 2-56 and 2-54 are installed. Load is removed from the needle beams by driving the steel wedges inward to push up the top stone block and transfer the load from the needle beam to the underpinning or new foundation system. Next, the needle beam is removed, a pair of stones is fitted into the hole, and wedges are driven between the stones to pick up the load as well. After all wedges are driven solidly, the old wall bears on the underpinning system; the spaces between the stones are grouted solid (wedges remain permanently), and the excavation is backfilled.

Figure 2-54. Underpinning: wall foundation. (*From International Correspondence Schools 1923, sec. 3, 24.*)

Brick Piers. Massive brick piers of the type shown in Figure 2-57 are constructed where the soil is firm and where the masonry of the old wall is of reasonably good construction (ibid., sec. 3, 28). The use of brick piers eliminates the need for needle beams and excavations on each side of the wall.

The excavations for the piers are made close enough together to allow the brick wall to corbel itself from pier to to pier to allow the unexcavated sections to be removed. Then the underpinning can be completed between the piers (see Figure 2-58).

The excavations for the piers depended on the masonry wall corbeling above the excavation area (held by the original foundations). The brick piers are massive for large buildings and contain a pair of flat stones and steel wedges, as in the needle

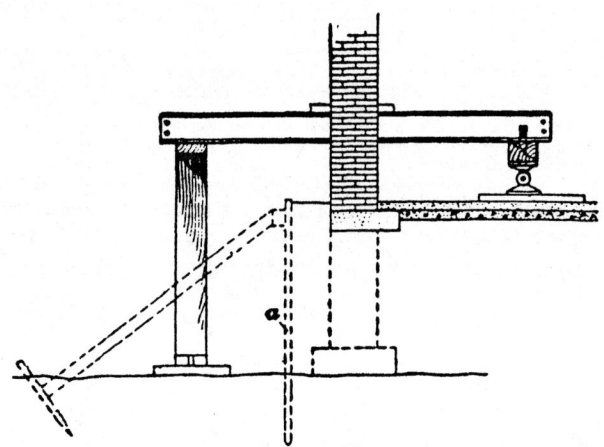

Figure 2-55. Underpinning: wall foundation. (*From International Correspondence Schools 1923, sec. 3, 2b.*)

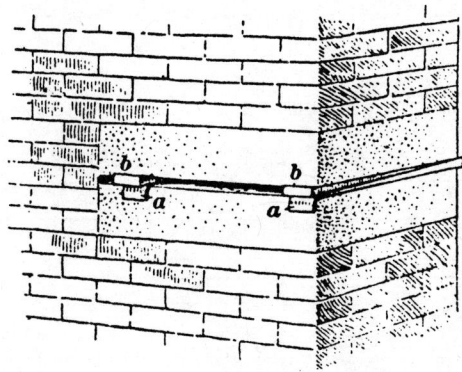

Figure 2-56. Loading blocks. (*From International Correspondence Schools 1923, sec. 3, 27.*)

beam system, to induce the loading (see Figure 2-57). The brick piers were usually constructed at about 10-ft intervals and were prepared for the adjoining walls at each end by means of a lock or shear block to retain wall alignment (ibid., sec. 3, 29). Brick piers are efficient for underpinning buildings to a depth of 10–15 ft. Their mass and size are complimentary to the vertical load that is required to support the section of the building wall. The column action induced in this type of loading is not subject to the normal column buckling because of the mass and size of the pier.

Piles for Underpinning. If a building was constructed with the conventional wall footings for a structure on compressible soil, pilings could be used if a stratum of hard pan or bedrock existed 12–20 ft below the surface. The piles were usually made of steel pipes that were jacked into position in sections and then filled with concrete. The piles were placed into pockets that had been recessed into the inside face of the

Figure 2-57. Underpinning: brick piers. (*From International Correspondence Schools 1923, sec. 3, 29.*)

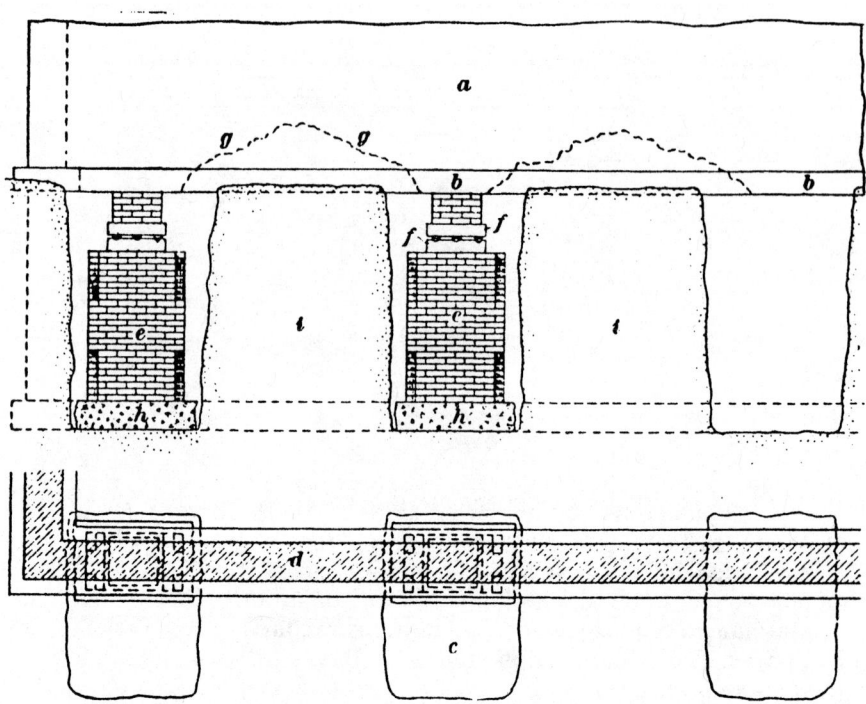

Figure 2-58. Underpinning: brick piers. (*From International Correspondence Schools 1923, sec. 3, 28.*)

walls at footing level (see Figure 2-59). A short section of steel beam was placed at the top of the recessed pocket to be used to jack against to drive the pile into the soft material. The weight of the building was actually used to resist the force of the jack and drive the pile. After the pile was driven to the bearing material, the pile was cleaned out on the inside with compressed air blasts, reinforcing steel was placed in the pile, and the pile was filled with concrete. The pile was finally capped with a steel plate and wedges were utilized to transfer the load of the building to the piles.

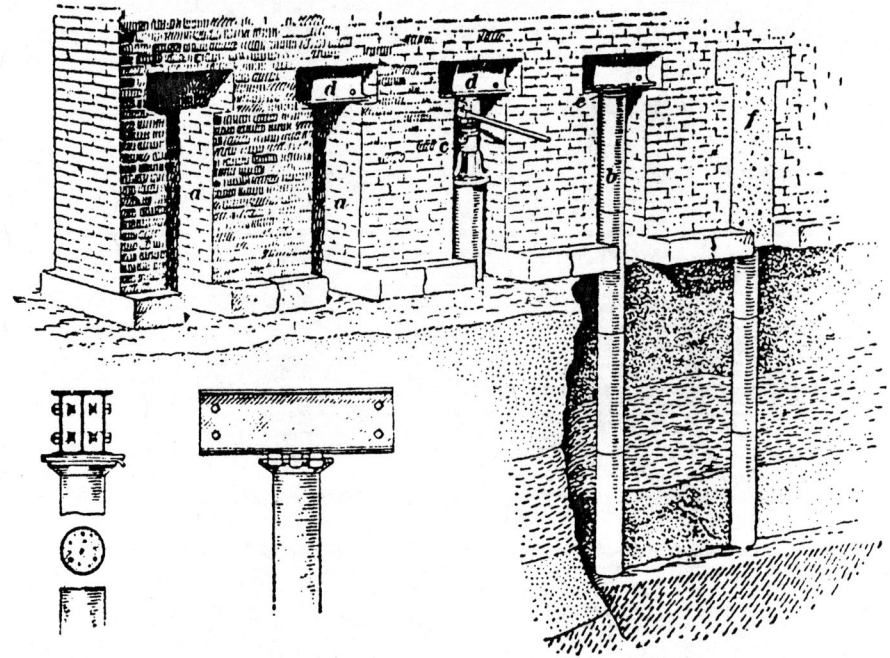

Figure 2-59. Underpinning: piles. (*From International Correspondence Schools 1923, sec. 3, 30.*)

The final phase of the operation was to fill the entire recessed pocket with concrete to solidify the wall. Driving piles by means of jacks against the building weight will work best on moist, noncohesive soils that will not offer much resistance to the driving action. The pile itself is actually converted to a pier by filling with reinforcing and concrete and bearing on hardpan or bedrock. In contrast, the pile would be considered exclusively a pile if it depended upon skin friction to carry the load.

In an historic building, the foundation is probably the most intangible of all areas or systems that a contemporary designer must analyze. If an historic building is in good condition physically, its enclosure systems have continued to be sound over the years, and there is no evidence of water damage, the foundation system will probably be in equally good condition. There should be no differential settlement in the building and no cracks in the masonry as discussed in this chapter. If there has been no negative influence around the building and no new construction near the building that might cause foundation problems for the building, and if there are no indications of problems in the composition of the building (settling, cracks, etc.), the foundation system should remain undisturbed. This chapter has outlined methods of identification, analysis, and repair of existing foundation systems for a building that has experienced any of the array of problems that may be attributed to the building foundation. A thorough diagnostic investigation must be employed before the conclusion can be reached that the foundation is the sole problem and must be revised. In many instances, the building has settled into a state of harmony with its underlaying soil strata and damage could result from disturbing the condition of equilibrium.

The majority of historic buildings were more than adequately designed for the load requirements of contemporary adaptive uses. Most historic buildings (two to three stories) were heavier and more stable than their modern counterparts in weight and in tightness and durability of construction. Early materials had many common material elements, such as coefficients of expansion/contraction with temperature change, that kept the building intact. As heavier buildings (heavier dead load), they inherited a greater dead load to live load ratio, which means that the influence or relative proportion of the live load is less than we experience with contemporary buildings. Therefore, the historic building may be and probably is in better condition than contemporary designers think it is. Professional judgment should influence the decision to revise foundation systems; the question is always in the overall magnitude of the change of the load on the support stratum.

The following chapters explore walls, columns, beams, and lateral support systems as utilized in the design of historic buildings and demonstrate methods and materials of the various periods. Each system in an historic building complements the others, and in the overall sense they are all compatible.

3
Historic American Building Systems: Walls and Columns

BRACED FRAME CONSTRUCTION

The braced frame was the product of medieval Europe and was natural for the early settlers of North America to use due to the abundance of wood on this continent. The virgin timber forests provided large members for braced frame assemblies. Members for the braced frame were empirically sized by craftsmen who had apprenticed in the late-medieval trades tradition. Usually carpenters came to the American colonies as indentured workers and constructed many buildings as partial payment for their passage.

The use of the braced frame, with its mortise and tenon joints and diagonal bracing, continued to be the typical wood construction method until beyond the middle of the nineteenth century. The carpenters of the nineteenth century were mostly products of the American apprenticeship system, and braced frame technology, or building customs, remained the same as in medieval times. Improvements in tools and member sizing allowed joints to be more precise, tighter, and therefore of finer quality. Members were still empirically sized for the most part until into the last quarter of the nineteenth century.

Braced frames were normally used on wooden houses, wooden churches, and wooden barns during the first half of the nineteenth century. The braced or "full" frame, as it was sometimes called, was dependent upon skilled craftsmen and the availability of large timbers (see Figures 3-1 and 3-2). Mortise and tenon joints were extremely durable (if kept dry) and provided a relatively rigid connection. The joint depended upon either the compression transfer through a tight fit or the tenon to hold a tensile load in double shear over its cross-section. Tension members produce a tension pullout perpendicular to the grain of the female member. The ability of the female member to contain the tensile stress is critical to the use of a tension member. In many applications, the mortise and tenon joint was subjected to a moment as well as the axial load. The joint is extremely difficult to analyze for the moment portion of the combined load.

Many buildings during the first three-quarters of the nineteenth century were constructed with braced frame technology. The threat of fire in buildings in urban and semiurban areas caused owners to construct masonry buildings for safety. The westward migration saw first-generation buildings in towns constructed of wood. And as the towns grew, the next generation of buildings would be of masonry (masonry load-bearing walls). The frequency of fire and the fear of the destruction that a fire in an urban city center area could bring made most larger cities regulate against timber buildings in these areas by the 1880s. In residential neighborhoods, wooden houses of balloon frame type were prevalent, especially in the less affluent and working class districts.

Braced frames, wood sills, posts, joists, and girders were often in standard sizes in braced frame buildings in the latter part of the nineteenth century. Kidder (1900) states that a sill can be a 6×6 in. when the sill rests on a brick or stone wall. Further, the sills must have sufficient strength to support the walls and floors when the sill is bearing on posts or when wider than minimum-sized openings exist below the sills (ibid., 51). Kidder also cautions the designer to be aware of the size of the mortise

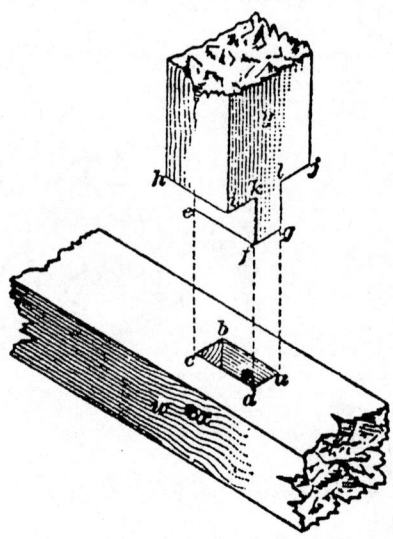

Figure 3-1. Mortise and tenon joint. (*From International Correspondence Schools 1923, sec. 31, 49.*)

and that the area cut away weakens the strength of any member. See Figures 3-3 and 3-4 for details of braced frame buildings. Wood-braced frame structural systems were utilized on houses, commercial and public buildings, and farm buildings during the first three quarters of the nineteenth century. Such systems survive today predominantly in the large wooden barns that are found throughout the middle and eastern sections of the country. The braced frame was the state of the art system for buildings constructed of wood until the fourth quarter of the nineteenth century.

BALLOON FRAME CONSTRUCTION

The need for mass-produced housing in the larger cities and the need for a more efficient system of construction brought the balloon frame into popular use in Chicago after the great fire of 1871. The balloon frame actually had its beginnings in 1832, when George W. Snow used it for a building he was constructing in Chicago (Sprague 1983, 43). Bell (1857) has an entire section on balloon framing in which he describes the new system of construction (Sprague 1983, 43).

The lumber mills could produce studs (vertical members), joists (beams), and decking material in large quantities. Handling and transportation to the job site were easier, but the real savings were in the time and labor required to cut all of the mortises and tenons for the braced frame. As with most of the usual developmental changes in construction, a new process or product such as nail technology was ready for the new construction method. The joints in the balloon frame were simple compared to the mortise and tenon. In the balloon frame, members were cut square and either end-nailed or toe-nailed to the other member. Initially, in the balloon frame of Snow's time, the heavy timber sill with the studs and first-floor joists mortised into place in the sill continued to be used as in the braced frame. Most authors explain this as a reaction to the type of foundation used on these early structures rather than as retention of parts of the old system (ibid., 40). The craftsmen were not sure of the new construction's ability to resist lateral loads, so the one carryover from the braced frame was utilized, the diagonal brace. The diagonal brace was of the same size as the stud in the walls and was "let in" to maintain the surface of the wall. The let-in bracing was necessary to stabilize the wall during construction prior to the placement of the wood sheathing on the surface of all exterior walls. Usually the wood sheathing was 1 × 6 or 1 × 8 in. and was placed in sections that were long enough to extend from the bottom plate to the roof soffit. This sheathing, if nailed properly, would have been sufficient to carry lateral loads, but in the balloon frame method of construction, the structure was erected and fully framed before sheathing was applied.

The balloon frame was the first response to the new industrialized production of wood members. The need for mass-produced housing in and around Chicago pro-

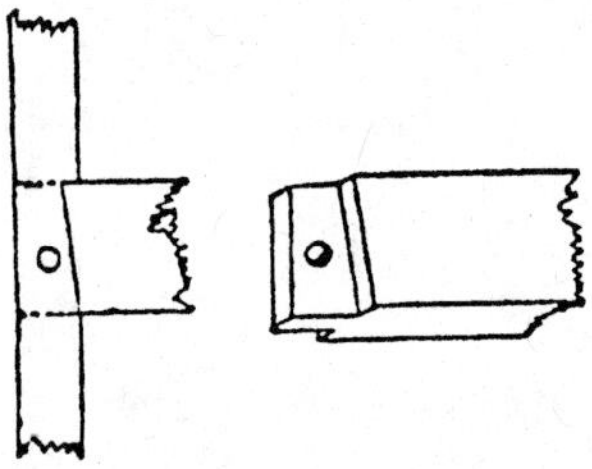

Figure 3-2. Mortise and tenon joint. (*From Kidder 1900, 53.*)

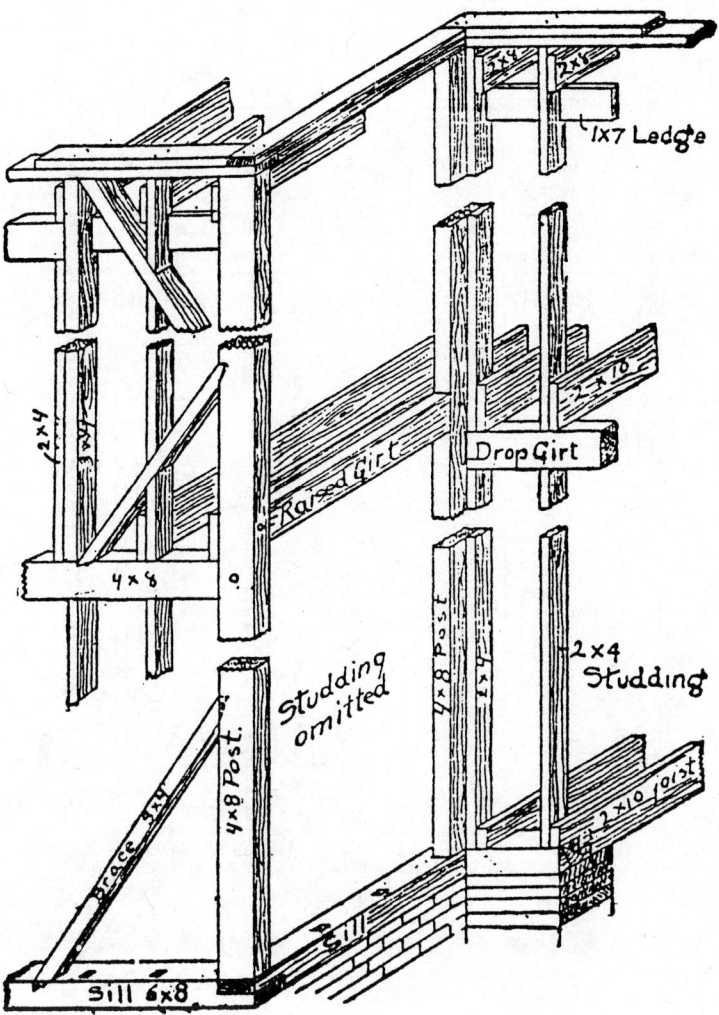

Figure 3-3. Details, braced frame construction. (*From Kidder 1900, 51.*)

vided the stimulus. When combined with the availability of rail transportation to move dimension lumber from mills to areas of need, products and construction methods evolved together. The balloon frame was most efficient for a building or house two stories in height (see Figure 3-5).

PLATFORM OR WESTERN FRAME CONSTRUCTION

The balloon frame was efficient and economical and was the framing method most utilized from the 1850s to approximately the turn of the century, when it was replaced by the platform frame or western frame. Most authors attribute the conversion to convenience and time savings (ibid., 45). Each level of the house consisted of a platform—thus the name. The platform surface enabled the craftsmen to build and raise walls to the next level. This system was not only easier to erect, it allowed for more standardization of member lengths. Studs no longer had to be two stories in length, but could be cut to 8–10 foot standard lengths. It was also very difficult to obtain timber that would allow for studs to be cut to 20-ft and longer lengths.

The western frame had its beginnings in the far West, presumably being developed in the 1850s by craftsmen and architects who had gone to San Francisco at the time of the Gold Rush (ibid.). There is no doubt that it was developed from the balloon frame, the two being quite similar. The western framing system was simple and functional. It made the formation of openings for windows or doors easier: a simple header

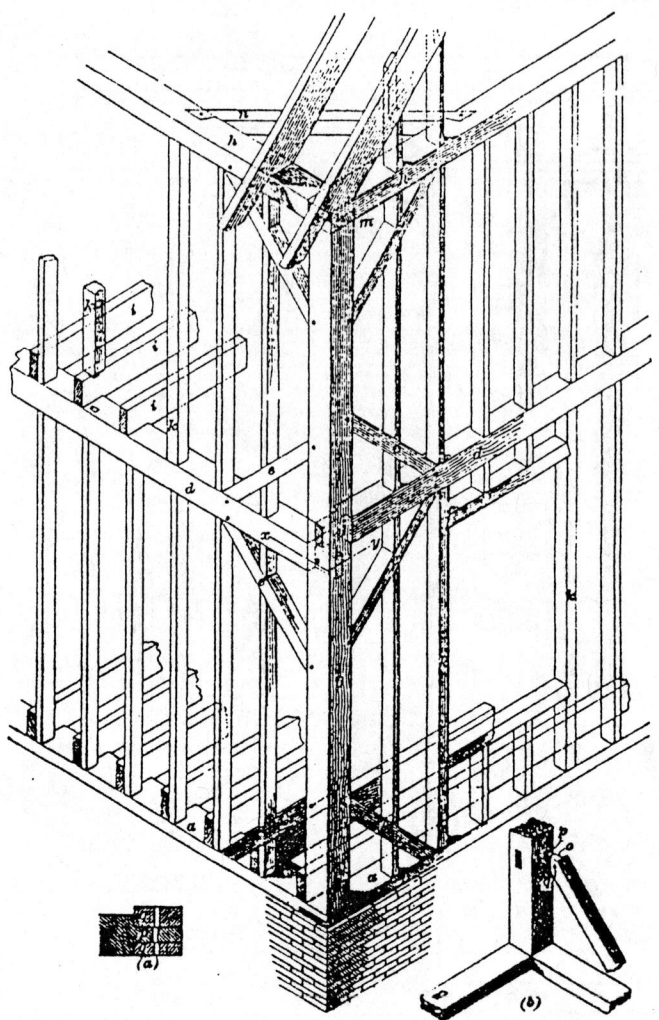

Figure 3-4. Details, braced frame construction. (*From International Correspondence Schools 1923, sec. 31, 51.*)

could be utilized because the platform above spreads loads from an upper floor or roof uniformly to the stud walls below. Thus, window and door openings did not need the conventional trussed header over them because the studs above the openings no longer went above the next upper platform (see Figure 3-6).

The platform is framed by joists and edge band (same size as joists) that bear on a double top plate on the stud wall. The platform is decked with either one by eight planks on diagonal (the early method) or by plywood (the later method). Studs are doubled at each side of the framed opening of the door or window.

Capacity of Studs (Columns on 16 in. Centers)

Historical Method. According to Kidder (1902, 244), "the strength of a column, post or strut (stud) depends, in a large manner, upon the proportion of the length to the diameter or least thickness."

1. When the length of the column (stud) is not more than 12 times the least thickness, the strength of the column must be computed by:

$$\text{Safe load} = \frac{\text{area } (C)}{\text{factor of safety}}$$

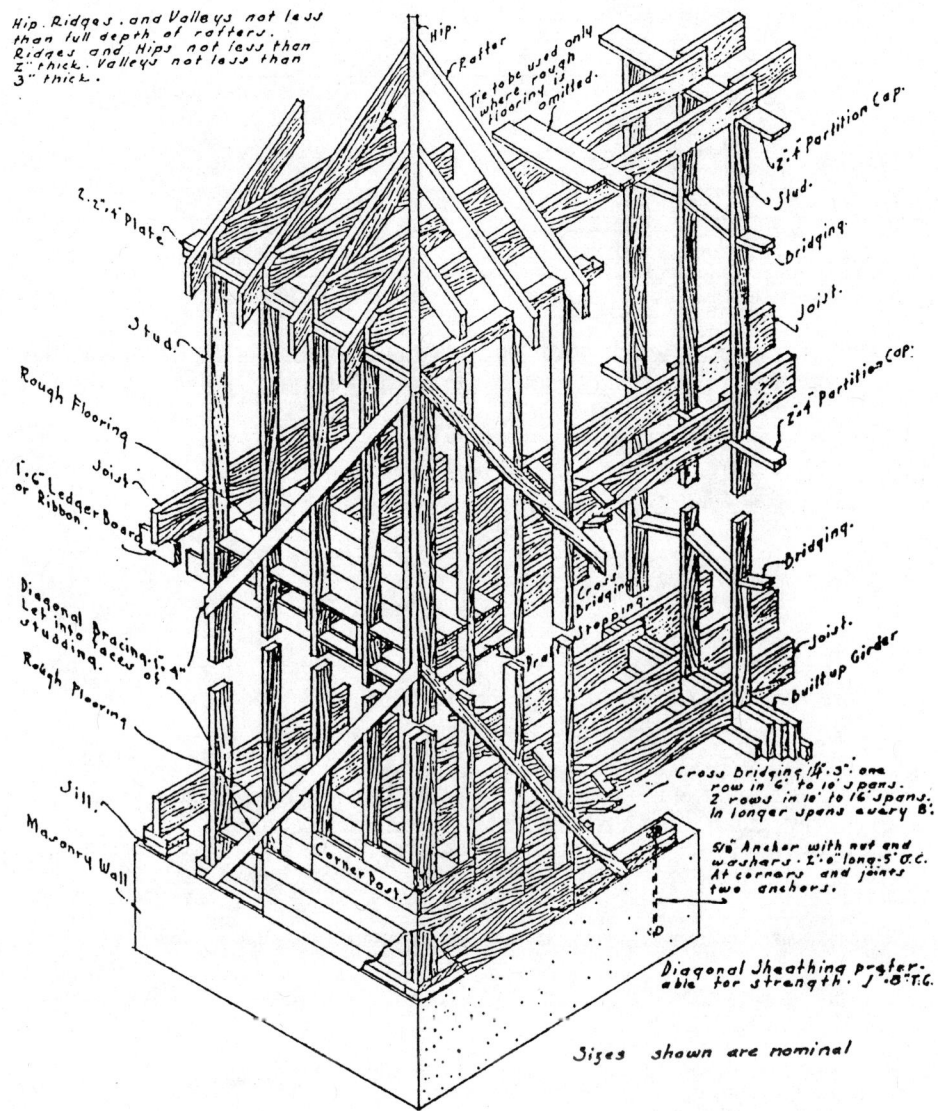

Figure 3-5. Balloon frame construction. (*From Ramsay and Sleeper 1936, 93.*)

where: C = ultimate crushing strength of the column material (see Tables 3-1 and 3-2).
2. When the length of the column (stud) is over 12 diameters (12 least thicknesses).

Professor James H. Stanwood of MIT derived the following formula by plotting the results from a series of extensive tests made at the Watertown, Massachusetts, government testing facility in 1891.

$$\text{Safe load} = \left[1000 - N\left(\frac{\text{length, in.}}{\text{breadth, in.}}\right)\right] (\text{cross sectional area})$$

where:
  Breadth = least side of a rectangular column or diameter of a round column
  N = 10 for yellow pine; 7.5 for oak and Norway pine; 6 for white pine and spruce

The capacity of wooden studs (struts) for building walls would have been calculated in the following manner in 1892 and for a few years thereafter:

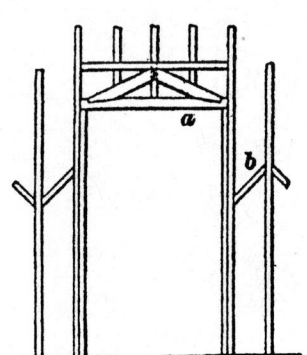

Figure 3-6. Trussed header. (*From International Correspondence Schools 1960, 238.*)

Table 3-1. Average Ultimate Crushing Loads, for Building Materials[a]

Woods (Partial Table)	psi
Oak, white	4000
Pine, Georgia yellow	5000
Pine, Oregon	4500
Pine, white	3500
Spruce	4000

[a] From Kidder 1902, 244.

Stud = 2 × 4 in. (actual size) 1892 through approximately 1920 (specific to location and the usual consideration for time lag to the smaller towns) Southern yellow pine, length = 8 ft, weak axis braced by sheathing

$$\frac{L}{d} = \frac{8 \text{ ft } (12 \text{ in./ft})}{4 \text{ in.}} = 24 \quad \text{(exceeds 12, therefore second formula must be used)}$$

$$\text{Safe load} = \left[1000 - 10\left(\frac{8 \text{ ft } (12 \text{ in./ft})}{4''}\right)\right](8 \text{ in.}^2)$$

$$= 6{,}080 \text{ lb/stud.}$$

Safe load = 4,988 lb/stud (at the latter period, actual size of 1.75 in. × 3.75 in.)

As a comparison, the modern analysis would be as follows (National Forest Products Association 1988). Capacity of stud by NDS modern analysis:

$$P_{all} = F'_c \text{ (area)}$$

F'_c = allowable compressive stress, parallel to the grain

F_c = 1150 psi, southern yellow pine #2, kiln dried, 2 × 4 in., E = 1,600,000 psi

$$\frac{L_e}{d} = \frac{8 \text{ in. } (12 \text{ in./ft})}{4 \text{ in.}} = 24 \qquad K = 0.671\sqrt{\frac{E}{F_c}} = 25.03$$

$$F'_c = F_c\left[1 - 0.333\left(\frac{\frac{L_e}{d}}{k}\right)\right] = 782.45 \text{ psi}$$

$$P_{all} = 782.45 \text{ psi}(8 \text{ in.}^2) = 6259.6 \text{ lb} \qquad (5134.8 \text{ lb, size } 1.75 \times 3.75 \text{ in.})$$

As can be seen, the modern analysis would allow about a 10 percent higher value for the allowable load on the stud. Most experts agree that the strength of a wood member decreases slightly over time. The total decrease varies, usually averaging only 5 percent in 100 years. Checking, moisture infiltration, and insect attack also detract from the strength of any member.

Table 3-2. Factors of Safety (for Formula No. 1)[a]

Material	Factor of Safety
Not very bad knots	5
Badly season-checked, cross-grained, or contains bad knots	6 or 7
Supporting machinery or struts in railway bridges	6–8

[a] From Kidder 1902, 244.

The allowable load on a single stud will be used to check the capacity of stud walls. The usual spacing of studs on historic wood frame structures is 16 in. on centers. The stud will be assumed to be braced about its weak axis by the application of wood sheathing.

Building Example No. 1
Three stories and roof, max. spans = 24 ft

> Roof, $D + L$ = 45 psf
> Floor 3, $D + L$ = 85 psf
> Floor 2, $D + L$ = 85 psf
> Wall floor 1 = actual 2 × 4 in., 8 ft tall
> Actual load on wall of floor No. 1 from floors 2 and 3 and roof is as follows:

215 psf(1.333 ft)(12 ft) = 3439.1 lb/stud

The actual load of 3439.1 lb/stud compares quite favorably with the allowable load capacity of 6259.6 lb − 10% = 5633.6 lb).

Building Example No. 2
Two stories and roof, max spans = 24 ft

> Roof, $D + L$ = 45 psf
> Floor 2, $D + L$ = 85 psf
> Wall Floor 1 = Actual 2 × 4 in., 8 ft tall
> Actual load on wall of floor
> No. 1 from floor 2 and roof is as follows:

130 psf(1.333 ft)(12 ft) = 2080 lb/stud

The actual load of 2080 lb/stud is even lighter than for the three-story frame structure an gives an indication of the conservative nature of these walls.

The stud is structurally oversized for the work that it is required to do in one-, two-, and three-story lightly loaded residential buildings. The balloon frame was constructed with continuous or two floor-height (length) studs. The effect of a continuous member (stud) upon the load-carrying capacity is minimal. The buckling diagram of a two-floor column is an S diagram in which the effective length of each level is equal to the length of each floor (see Figure 3-7). The diaphragm of the floor at intermediate level would serve to stiffen the stud at its midheight, thus allowing the stud to carry more load than the calculated allowable load. The modern analysis could take into account this stiffness at midheight; a k factor for the column design would be approximately 0.85. In later construction, the platform frame with its single floor-height studs was utilized throughout the country. The capacity of a stud wall was shown above.

WOOD COLUMN SIZING

Wood columns were used in buildings of all types. In masonry commercial buildings that were wider than 24–26 ft between bearing walls, columns and girders were introduced to open up spaces and provide larger (wider) spaces for functional reasons. In some instances, the support element between the bearing walls would be a wall (stud wall), actually an intermediate bearing wall of wood.

Wood columns were empirically sized by carpenters for a number of years with moderate success. In the second half of the nineteenth century, loads on building columns increased to the point that analytical methods were required to ensure the design of adequate columns. Wooden railroad bridges and trestles built in the 1840s and 1850s required analytical computations (and large factors of safety) due to the loads of the railroad engines.

Timber column design was investigated by Professor Lanza of MIT through an extensive series of tests of full-size wooden columns. Data generated from these tests

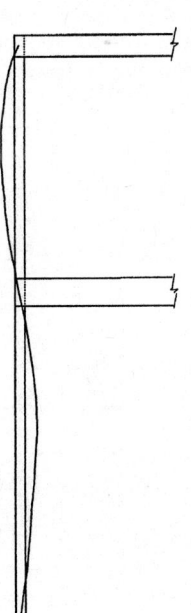

Figure 3-7. Two floor-height studs.

enabled Lanza to determine the following crushing strengths for construction timbers (Thurston 1892, 642):

Average crushing strength of yellow pine	4392 psi*
Average crushing strength of white oak	3323 psi*
Average crushing strength of white wood	3009 psi*

*These values are produced from tests of unselected material and are reported to be very low.

During approximately the same period, the U.S. Government tests at the Watertown Arsenal, Watertown, Massachusetts, produced more usable results (ibid.):

Resistance to Crushing:

Pine, yellow, straight grained, well seasoned	7386 psi
Pine, slow growth	9339 psi
Pine, very green and wet	3015 psi

Thurston (1892) presents timber column design as a group of formulae that had been derived by notable scholars of the time (ibid., 640):

Hodgkinson's formulae: square, flat-ended timber columns (modified Euler's formula) 1 ton = 2240 lb

$$\text{Oak timber } P = 10.95 \left(\frac{d^4}{L^2}\right)$$

where:
 P = Crushing weight, tons
 d = Thickness, in.
 L = Length of column, ft

$$\text{Red pine } P = 7.81 \left(\frac{d^4}{L^2}\right)$$

Where the column is less than 30, and more than 5 diameters in length:

$$W = \frac{P(c)(k)}{P + 0.75(c)(k)}$$

where:
 W = strength of column, tons
 P = crushing weight, tons (from two formulae above)
 C = coefficient of crushing resistance from table below
 K = area, column section, in.^{2}

Coefficients of Resistance in Crushing:

Oak, white	6,500–10,000 psi
Pine, red	6,000–7,500 psi
Pine, white	3,000–6,000 psi
Pine, yellow	6,500–10,000 psi
Maple	5,000–6,000 psi
Cherry	5,000–6,500 psi
Cedar	4,000–6,500 psi

Rankine's Formula:

$$P = \frac{fS}{1 + \frac{L^2}{a(d^2)}}$$

where:
    S = cross-sectional area, in.^{2}
    a, f = constants
    L = length of column, in.
    d = diameter, in.

Rankine's constants:
a = 188 for timber
f = 7200
C. S. Smith's constants:
a = 250 for yellow pine
f = 5000

Morin Equation (from Euler's):

$$P' = A\left(\frac{d^2}{L^2}\right)(\text{area of column})$$

where:
    P' = Load, kg
    d = diameter, cm
    L = length, dm
    A = 160 for pine at a "safe load"

Factors of safety according to Thurston:
Dead loads 5
Moving loads 10
Shock (impact) 10–20

Thurston also gives the following three formulae from *Tredgold's Carpentry* for pillars (columns) above 30 diameters long:

$$W = A\left(\frac{b^4}{L^2}\right)$$

$$W = A\left(\frac{bt^3}{L^2}\right)$$

$$W = A\left(\frac{d^4}{1.7\,L^2}\right)$$

where:
    A = coefficient = 1500 for Beech, ch. elm, and white pine
    b, t, d = breadth, thickness, or diameter, in.
    L = length of column, ft

Evidently some wood columns were bored at their center (center of gravity) to allow "free access of air to all parts of the wood" (ibid., 644). Thurston's "*Strength of Materials*" does not mention the size of the boring, but it does state that the boring should be from one end only to prevent two borings meeting eccentrically at the middle.

Column Tables—Wood

Kidder (1900, 524–525) states that for the safe load on columns for buildings in which the length does not exceed 12 times the least thickness, the following formula applies:

$$P_{safe} = C(A)$$

where:
    P = lb of safe load
    C = 1000 for long-leaf yellow pine
       = 900 for Oregon pine

= 800 for spruce or white oak
= 700 for white pine

For machinery and permanent loads, (brick or stone walls), C must be reduced by 20%.

For columns whose length in inches exceeds 12 times their least dimension, Kidder presents the table (Table 3-3).

The loads in Table 3-3 are the safe loads for yellow pine posts. For other species the tabulated values must be modified by the following factors:

Oregon pine	80%
Oak and Norway pine	75%
Spruce and white pine	62.5%

Trautwine's Civil Engineers Pocketbook, (1888, 458–459) gives the following formula, attributed to Charles Shaler Smith, for square or rectangular columns:

$$P_{breaking\,load} = \frac{5000}{1 + 0.004\left(\frac{L^2}{b^2}\right)}$$

where:
P = breaking load, psi
L = length of column, in.
b = breadth of column, in.

Trautwine also gives a table (see Table 3-4). of breaking loads in tons (2240 lb/ton) for square and rectangular white or yellow pine columns.

The National Lumber Manufacturers Association published *Heavy Timber Mill Construction Buildings* in 1918, in which it lists two formulae for determining the safe loads on timber columns of different lengths (Paul 1918, 66; Voss and Varney 1926, 194).

Table 3-3. Safe Load in Pounds, for Yellow Pine Posts, Round and Square[a]

Size of Post (in.)	Length of Post (ft)								
	8	10	12	14	15	16	18	20	24
4 × 6	18,200	16,800	15,360						
5½ round	19,590	18,760	17,550	16,500					
6 × 6	30,200	28,800	27,400	25,900	25,200	24,500			
6 × 8	40,300	38,400	36,500	34,600	33,600	32,600			
6 × 10	50,400	48,000	45,600	43,200	42,000	40,800			
7½ round	38,540	37,130	35,710	34,300	33,590	32,890			
8 × 8	64,000	54,400	52,500	50,600	49,600	48,600	46,700		
8 × 10	80,000	68,000	65,600	63,200	62,000	60,800	53,400		
8 × 12	96,000	81,600	78,700	76,800	74,400	73,000	70,100		
9½ round	70,900	61,970	60,190	58,350	57,429	56,580	54,800		
10 × 10	100,000	100,000	85,600	83,200	82,000	80,800	78,400	76,000	
10 × 12	120,000	120,000	102,700	99,800	98,400	97,000	94,100	91,200	
10 × 14	140,000	140,000	119,800	116,500	114,800	113,100	109,800	106,400	
11½ round	103,900	103,900	90,912	88,730	87,690	86,550	84,160	82,290	
12 × 12	144,000	144,000	144,000	123,800	122,400	121,000	118,100	115,200	109,440
12 × 14	168,000	168,000	168,000	144,500	142,800	141,100	137,800	134,400	127,680
12 × 16	192,000	192,000	192,000	165,100	163,200	161,300	157,400	153,600	145,920
14 × 14	196,000	196,000	196,000	196,000	170,900	169,100	165,800	162,400	155,800
16 × 16	256,000	256,000	256,000	256,000	229,100	225,300	221,400	217,600	209,900
18 × 18	324,000	324,000	324,000	324,000	324,000	289,400	285,100	280,800	272,160
20 × 20	400,000	400,000	400,000	400,000	400,000	400,000	356,800	852,000	342,400

[a] From Kidder 1900, 525.

Table 3-4. Breaking Loads (in 2240 lb/Tons) for Square and Rectangular White or Yellow Pine Columns[a]

Height (ft)	Side of Square Pine Pillar (in.)												
	1	1¼	1½	1¾	2	2¼	2½	2¾	3	3¼	3½	3¾	4
	BREAKING LOAD												
	Tons	Tons	Tons	Tons	Tons	Tons	Tons	Tons	Tons	Tons	Tons	Tons	Tons
1	1.42	2.54	3.99	5.73	7.80	10.1	12.8	15.7	18.9	22.3	26.1	30.1	34.5
¼	1.17	2.22	3.59	5.26	7.25	9.6	12.2	15.1	18.3	21.7	25.4	29.2	33.7
½	0.97	1.93	3.19	4.80	6.74	9.0	11.6	14.5	17.6	21.0	24.7	28.6	33.0
¾	0.81	1.66	2.81	4.35	6.19	8.4	10.9	13.7	16.8	20.2	23.9	27.8	32.1
2	0.68	1.44	2.48	3.92	5.66	7.8	10.2	12.9	15.9	19.3	23.0	26.9	31.2
¼	0.57	1.24	2.19	3.53	5.17	7.2	9.6	12.3	15.2	18.5	22.0	25.8	30.1
½	0.49	1.07	1.93	3.16	4.70	6.7	8.9	11.5	14.3	17.6	21.1	24.9	29.1
¾	0.42	0.93	1.71	2.85	4.29	6.2	8.3	10.8	13.5	16.7	20.1	23.8	28.0
3	0.36	0.82	1.52	2.55	3.80	5.6	7.6	10.0	12.7	15.8	19.2	22.9	27.0
½	0.28	0.63	1.21	2.07	3.23	4.8	6.6	8.8	11.3	14.2	17.4	20.9	24.8
4	0.22	0.50	0.98	1.70	2.70	4.0	5.7	7.7	9.9	12.7	15.7	19.0	22.7
½	0.18	0.40	0.81	1.42	2.28	3.4	4.9	6.7	8.8	11.4	14.1	17.2	20.7
5	0.15	0.34	0.68	1.19	1.94	3.0	4.2	5.8	7.7	10.0	12.6	15.5	18.8
½	0.12	0.28	0.57	1.02	1.67	2.6	3.7	5.1	6.8	8.9	11.3	14.0	17.1
6	0.10	0.24	0.49	0.86	1.44	2.3	3.3	4.6	6.1	8.0	10.2	12.7	15.6
½	0.09	0.21	0.43	0.74	1.26	2.0	2.9	4.1	5.4	7.2	9.2	11.6	14.2
7	0.08	0.18	0.37	0.66	1.11	1.8	2.6	3.6	4.9	6.5	8.3	10.5	12.9
½	0.07	0.16	0.33	0.59	0.98	1.6	2.3	3.3	4.4	5.9	7.6	9.6	11.8
8	0.06	0.14	0.29	0.52	0.87	1.4	2.0	2.9	3.9	5.2	6.8	8.7	10.8
½	0.05	0.12	0.26	0.47	0.78	1.2	1.8	2.6	3.5	4.8	6.2	7.9	9.9
9	0.03	0.11	0.23	0.42	0.71	1.1	1.6	2.3	3.2	4.3	5.6	7.2	9.1
½		0.10	0.21	0.37	0.64	1.0	1.5	2.1	2.9	3.9	5.1	6.6	8.4
10		0.09	0.19	0.34	0.58	0.93	1.4	2.0	2.7	3.6	4.7	6.1	7.8
½			0.17	0.31	0.53	0.86	1.3	1.8	2.5	3.4	4.4	5.7	7.2
11			0.16	0.28	0.48	0.79	1.2	1.7	2.3	3.1	4.1	5.3	6.7
½			0.14	0.26	0.44	0.72	1.1	1.5	2.1	2.9	3.8	4.9	6.2
12			0.13	0.24	0.41	0.65	1.0	1.4	2.0	2.7	3.4	4.5	5.8
13				0.21	0.35	0.55	0.84	1.2	1.7	2.3	3.1	4.0	5.0
14				0.18	0.31	0.46	0.70	1.0	1.4	2.0	2.7	3.5	4.4
15					0.27	0.41	0.63	0.91	1.2	1.7	2.4	3.1	3.9
16					0.24	0.37	0.57	0.78	1.1	1.5	2.1	2.7	3.5
17							0.50	0.70	1.0	1.4	1.9	2.4	3.1
18							0.45	0.66	0.92	1.3	1.7	2.2	2.8
20									0.76	1.0	1.4	1.8	2.3

Height (ft)	Side of Square Pine Pillar (in.)												
	4¼	4½	4¾	5	5¼	5½	5¾	6	6¼	6½	6¾	7	7¼
	BREAKING LOAD												
	Tons	Tons	Tons	Tons	Tons	Tons	Tons	Tons	Tons	Tons	Tons	Tons	Tons
2	35.8	40.6	45.7	51.1	56.8	62.8	69.0	75.5	82.3	89.4	96.8	104.5	112.4
3	31.4	36.1	41.1	46.2	51.8	57.7	63.8	70.2	76.9	84.0	91.3	98.9	106.7
4	26.8	31.2	35.9	40.8	46.1	51.7	57.6	63.8	70.4	77.3	84.5	92.1	99.9
5	22.6	26.5	30.8	35.4	40.5	45.8	51.5	57.4	63.7	70.3	77.2	84.5	92.1
6	18.9	22.5	26.4	30.5	35.2	40.1	45.4	51.0	57.0	63.3	69.9	76.8	84.1
7	15.8	19.0	22.6	26.2	30.5	35.0	39.9	45.0	50.5	56.5	62.8	69.4	76.3
8	13.3	16.1	19.2	22.5	26.3	30.4	34.9	39.7	45.9	50.4	56.2	62.4	69.0
9	11.3	13.7	16.5	19.5	22.9	26.6	30.7	35.0	39.9	44.8	50.2	56.0	62.1
10	9.7	11.8	14.2	16.9	19.9	23.2	26.9	30.9	35.2	39.9	44.9	50.3	56.0
11	8.3	10.2	12.4	14.8	17.5	20.4	23.8	27.4	31.3	35.6	40.2	45.1	50.4
12	7.2	8.8	10.7	12.9	15.4	18.0	21.1	24.3	27.9	31.8	36.0	40.6	45.5
13	6.2	7.7	9.4	11.4	13.6	16.0	18.8	21.7	24.9	28.5	32.4	36.6	41.1
14	5.5	6.8	8.3	10.1	12.1	14.2	16.7	19.4	22.4	25.7	29.2	33.1	37.3
15	4.8	6.0	7.4	9.0	10.8	12.7	15.0	17.5	20.2	23.2	26.4	30.0	33.9
16	4.4	5.4	6.7	8.1	9.8	11.5	13.6	15.8	18.3	21.0	24.0	27.3	30.8
17	4.0	4.9	6.1	7.3	8.8	10.4	12.3	14.3	16.6	19.1	21.9	24.9	28.3
18	3.6	4.4	5.5	6.6	8.0	9.4	11.2	13.0	15.1	17.4	19.9	22.7	25.8
19	3.3	4.0	5.0	6.0	7.3	8.6	10.2	11.9	13.8	16.0	18.3	20.9	23.7
20	3.0	3.7	4.6	5.5	6.6	7.8	9.3	10.9	12.6	14.6	16.8	19.2	21.8
22	2.5	3.0	3.8	4.6	5.6	6.6	7.9	9.2	10.7	12.4	14.3	16.3	18.6
24	2.1	2.6	3.2	3.9	4.7	5.6	6.7	7.9	9.1	10.6	12.2	14.1	16.0
26	1.8	2.2	2.8	3.4	4.1	4.9	5.8	6.8	7.9	9.2	10.6	12.2	13.9
28	1.5	1.9	2.4	2.9	3.5	4.2	5.1	5.9	6.9	8.0	9.4	10.7	12.2
30	1.3	1.7	2.1	2.6	3.1	3.7	4.4	5.2	6.1	7.1	8.2	9.4	10.8
32	1.2	1.5	1.9	2.3	2.7	3.2	3.9	4.6	5.4	6.3	7.3	8.4	9.6
34	1.1	1.3	1.7	2.0	2.4	2.9	3.5	4.1	4.8	5.6	6.5	7.5	8.6
36	1.0	1.2	1.5	1.8	2.2	2.6	3.1	3.7	4.3	5.0	5.8	6.7	7.7
38	0.9	1.1	1.3	1.6	2.0	2.4	2.8	3.3	3.9	4.5	5.3	6.1	7.0
40	0.8	1.0	1.2	1.5	1.8	2.1	2.6	3.0	3.5	4.1	4.8	5.5	6.3

[a] From Trautwine 1888.

U.S. Department of Agriculture, Division of Forestry Formula:

$$C = \frac{c(700 + 15a)}{700 + 15a + a^2}$$

where:
- C = safe compressive stress, psi
- L = length of column, in.
- d = least side, in.
- c = unit allowable compressive stress
- $a = L/d$

Winslow's Formula (Chicago Building Ordinance):

$$C = c\left[1 - \left(\frac{L}{80(d)}\right)\right]$$

- C = safe compressive stress, psi
- L = length of column, in.
- D = least dim. of column, in.
- c = unit allowable compressive stress

Both formulae are based upon a unit allowable working compressive stress of 1000 psi for a generic column. To modify the formula for a particular species, the safe compressive stress is factored by the safe compressive stress of the wood to be used, divided by 1000 psi. See Table 3-5.

Paul (1918) makes two recommendations on structural columns:

1. If the assumed live load is more than 120 psf and is likely to be permanent, such as for warehouses, no reduction in live load is recommended.
2. For buildings exceeding five stories in height, the designer should use the full live load on the roof and top story and for each succeeding lower floor reduce the live load by 5 percent until 50 percent of the live load is reached for a floor. This ratio of load is to remain constant for all remaining floors. This live load reduction method is discussed in detail in Chapter 1.

Paul gives two tables of safe loads in pounds, one based upon the U.S. Department of Agriculture formula and one upon the Winslow formula (see Tables 3-6 and 3-7).

Milo S. Ketchum was a leading authority on structural theory in the 1920s and wrote several volumes on the subject. His *Structural Engineers Handbook* of 1924 contains timber column design formulae as follows (Ketchum 1924, 379):

Table 3-5. Working Unit Stresses for Wood Columns[a]

Pine, Southern	
Dense grade	1200 psi
Sound	900 psi
Spruce	600 psi
Fir, Douglas	
Dense grade	1200 psi
Sound	900 psi
Oak	900 psi
Hemlock,	
Eastern	700 psi
Western	900 psi

[a] From Paul 1918, 63.

Table 3-6. U.S. Department of Agriculture Formula[a]

Nominal size (in.)	Actual Size (in.)	Area (in.²)	L/d	Length (ft)	Compression Parallel to the Grain (psi) *1000
6 × 6	5½ × 5½	30¼	17.5	8	23,000
"	"	"	21.8	10	20,680
"	"	"	26.2	12	18,580
"	"	"	30.5	14	16,780
8 × 8	7½ × 7½	56¼	12.8	8	47,520
"	"	"	16.0	10	44,220
"	"	"	19.2	12	40,960
"	"	"	22.4	14	37,900
"	"	"	25.6	16	35,040
"	"	"	28.8	18	32,560
"	"	"	32.0	20	30,120
10 × 10	9½ × 9½	90¼	10.1	8	80,580
"	"	"	12.6	10	76,580
"	"	"	15.2	12	72,420
"	"	"	17.7	14	68,120
"	"	"	20.2	16	64,160
"	"	"	22.7	18	60,380
"	"	"	25.3	20	56,640
12 × 12	11½ × 11½	132¼	8.3	8	122,040
"	"	"	10.4	10	117,400
"	"	"	12.5	12	112,420
"	"	"	14.6	14	107,360
"	"	"	16.7	16	102,220
"	"	"	18.8	18	97,200
"	"	"	20.9	20	92,400
14 × 14	13½ × 13½	182¼	7.1	8	171,500
"	"	"	8.9	10	166,380
"	"	"	10.7	12	160,920
"	"	"	12.4	14	155,320
"	"	"	14.2	16	148,260
"	"	"	16.0	18	142,240
"	"	"	17.8	20	137,220
16 × 16	15½ × 15½	240¼	6.2	8	229,140
"	"	"	7.7	10	223,900
"	"	"	9.3	12	217,900
"	"	"	10.8	14	211,600
"	"	"	12.4	16	204,680
"	"	"	14.0	18	197,660
"	"	"	15.5	20	190,980
18 × 18	17½ × 17½	306¼	5.5	8	294,900
"	"	"	6.9	10	289,100
"	"	"	8.2	12	282,960
"	"	"	9.6	14	276,220
"	"	"	11.0	16	268,580
"	"	"	12.3	18	261,520
"	"	"	13.7	20	253,560

[a] From Paul 1918, 67. Safe loads in pounds based on the formula of the U.S. Department of Agriculture, Division of Forestry. Square end bearing and symmetrically loaded.
*1000—Values based upon unit stress of 1000 psi. To use table for a unit stress of other than 1000 multiply values by the ratio of that unit stress to 1000.

For L/d equal to or less than 12:

$$P_{all} = f_c(A)$$

where:
f_c = allowable compressive stress, parallel to the grain
P_{all} = allowable load, lb
A = area of column, in.²
d = least thickness or diameter, in.
L = length of column, in.

Table 3-7. U.S. Department of Agriculture Formula[a]

Nominal size (in.)	Actual Size (in.)	Area (in.²)	L/d	Length (ft)	Compression Parallel to the Grain (psi) *1000
6 × 6	5½ × 5½	30¼	17.5	8	23,640
"	"	"	21.8	10	20,020
"	"	"	26.2	12	20,360
"	"	"	30.5	14	18,700
8 × 8	7½ × 7½	56¼	12.8	8	47,280
"	"	"	16.0	10	45,000
"	"	"	19.2	12	42,740
"	"	"	22.4	14	40,500
"	"	"	25.6	16	38,240
"	"	"	28.8	18	36,000
"	"	"	32.0	20	33,740
10 × 10	9½ × 9½	90¼	10.1	8	78,840
"	"	"	12.6	10	76,060
"	"	"	15.2	12	73,100
"	"	"	17.7	14	70,400
"	"	"	20.2	16	67,680
"	"	"	22.7	18	64,640
"	"	"	25.3	20	61,700
12 × 12	11½ × 11½	132¼	8.3	8	119,020
"	"	"	10.4	10	115,060
"	"	"	12.5	12	111,560
"	"	"	14.6	14	108,200
"	"	"	16.7	16	104,480
"	"	"	18.8	18	101,160
"	"	"	20.9	20	97,860
14 × 14	13½ × 13½	182¼	7.1	8	165,860
"	"	"	8.9	10	162,200
"	"	"	10.7	12	158,540
"	"	"	12.4	14	153,920
"	"	"	14.2	16	149,940
"	"	"	16.0	18	145,800
"	"	"	17.8	20	142,140
16 × 16	15½ × 15½	240¼	6.2	8	221,680
"	"	"	7.7	10	218,000
"	"	"	9.3	12	212,300
"	"	"	10.8	14	207,720
"	"	"	12.4	16	202,880
"	"	"	14.0	18	198,100
"	"	"	15.5	20	193,720
18 × 18	17½ × 17½	306¼	5.5	8	285,100
"	"	"	6.9	10	279,800
"	"	"	8.2	12	275,080
"	"	"	9.6	14	265,500
"	"	"	11.0	16	264,220
"	"	"	12.3	18	259,480
"	"	"	13.7	20	254,180

[a] From Paul 1918, 68. Safe loads in pounds based on the Winslow formula (Chicago Building Ordinance). Square end bearing and symmetrically loaded.
* 1000—Values based upon unit stress of 1000 psi. To use table for a unit stress of other than 1000 multiply values by the ratio of that unit stress to 1000.

For L/d between 12 and 40, L/d shall not go over 40:

$$P_{all} = A(f_c) \left[1 - \frac{L}{60(d)} \right]$$

See Table 3-8.

The 1991 edition of the *National Design Specification for Wood Construction*, by the National Forest Products Association, requires that the design of solid wood columns be handled in the following manner (National Forest Products Assn. 1991, 5). For all solid bolumns, L_e/d shall not exceed 50.

Table 3-8. Allowable Compressive Stress, Parallel to Grain[a]

Species	f_c
White oak	1200 psi
Long leaf yellow pine	1200 psi
White pine and spruce	960 psi
Western hemlock	1200 psi
Douglas fir	1200 psi

[a] From Ketchum 1924.

$$P_{all} = F'_c(A)$$
$$F'_c = F_c C_D C_m C_t C_F C_P$$

where:
- F_c = allowable compressive stress, psi, tabulated value NFPA spec.
- F'_c = adjusted all. comp. stress, psi
- $L_e = k_e L$ = equivalent length, in.
- k_e = effective length factor
- d = least width or breadth, in.

$$\frac{L_e}{d} = \frac{k_e L_1}{d}, \text{ or } \frac{k_e L_2}{d}, \text{ whichever is greater}$$

where:
- C_D = load duration factor
  - = 0.9 permanent dead load
  - = 1.0 live loads, 10 yrs
  - = 1.15 live loads snow, 2 months
  - = 1.25 live loads 7 days
  - = 1.6 live loads wind, earthquake
  - = 2.0 impact load
- C_M = wet service factor
- C_t = temperature factor
  - = 1.0, T = 100°F or less
- c = 0.8 for Sawn Lumber
- F^*_c = allowable comp. stress or F_c times all factors except C_P
- C_F = size factor
- C_P = column stability factor

$C_F = \left(\frac{12}{d}\right)^{0.1111}$ when depth exceeds 12 in.; if not, $C_F = 1$

$C_M = 1$ unless moisture content exceeds 19% for extended period ($C_M = 0.91$)

$$C_P = \frac{1 + \frac{F_{cE}}{F^*_c}}{2c} - \sqrt{\left[\frac{1 + \frac{F_{cE}}{F^*_c}}{2c}\right]^2 - \frac{F_{cE}}{F^*_c(c)}}$$

$$F_{cE} = \frac{k_{cE} E'}{\left(\frac{L_e}{d}\right)^2}$$

- K_{cE} = 0.3 for visually graded lumber
- E = modulus of elasticity
- $E' = E\, C_M C_t C_T$
- C_T = 1 for columns

A column fails in one of three ways: short columns fail by yielding of the material, intermediate-length columns fail by a combination of yielding and buckling, and long columns fail by buckling. The factored equation utilized by the National Forest Products Association today utilizes factors that equate to the three methods of failure. The length of the column is factored by k_e as appropiate in today's building details. See Figure 3-8 for the k_e used in the modern formula.

The column formulae have developed over the years as the product of so-called refinements to the Euler Equation, developed in 1757 by Leonhard Euler, a Swiss mathematician. The Euler Equation determines the critical buckling load for long columns. The capacity of columns of short and intermediate length is determined by modifications to the Euler Equation. Many experts have attempted to improve upon or modify the Euler Equation, but no substantial improvement has ever been made.

The Euler Equation (Fitzgerald 1967, 183):

$$P_{cr} = \frac{\pi^2 EI}{L^2}$$

where:
- P_{cr} = critical buckling load, lb
- E = modulus of elasticity, psi
- I = modulus of inertia, axis of buckling, in.[4]
- L = column length, in.

Modified P_{cr} Equation:

$$F_{cr} = \frac{\pi^2 E}{\left(\dfrac{L}{r}\right)^2}$$

$P_{cr} = f_{cr}(\text{area of column})$

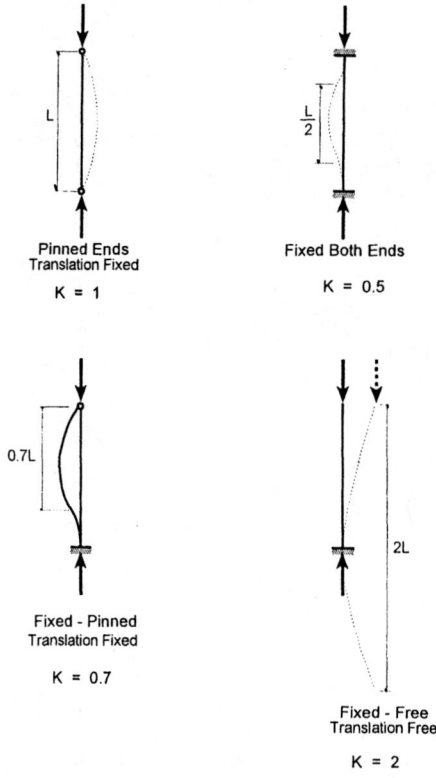

Figure 3-8. Reprinted by permission from National Forest Products Assn., *National Design Specification for Wood Construction*, p. 5. © 1991 National Forest Products Assn.

where:
f_{cr} = compressive stress parallel to grain, at critical buckling, psi
L/r = slenderness ratio about axis of buckling
r = radius of gyration about axis of buckling, in.

The critical buckling load for a column must be adjusted downward by a factor of safety to obtain the allowable load for a column. The factor of safety for the allowable load is 2 to 3.

SUMMARY—TIMBER COLUMN ANALYSIS

Each of the methods outlined in this section will produce an allowable load for a timber column and was utilized at various times by designers of the period. Knowing the date of construction of a particular building will assist the modern designer in applying the formula or tabulated allowable load that the original designer may have used. Verification of the allowable load used by the original designer will assist the modern designer in analyzing the building. The designer can use the comparison between the period allowable load and the contemporary allowable. Table 3-9 was prepared to demonstrate chronologically for a yellow pine, 8 × 8 in. column that is 20 ft long, bearing top and bottom square (hinged connection).

As can be seen in Table 3-9, the early formula produced comparatively similar results for the allowable load on a nominal 8 × 8 in. wood column. The 1991 National Design Specification for Wood Construction revised the method by which solid wood

Table 3-9. Allowable Axial Loads—8 × 8 (7.5″ × 7.5″ actual dimensions) Column—20 ft Long[a]

Formula or Calculation Method	P_{cr}	P_{all}	Remarks
Hodgkinson's formula, 1892 $P = 7.81 \left(\frac{d^4}{L^2}\right)$	120,834 lb	24,166.8 lb	safety factor = 5
Rankine's formula, 1892 $P =$	62,821.8 lb	12,564.3 lb	safety factor = 5
C. S. Smith's formula, 1892 $P =$	55,190.3 lb	11,038.0 lb	safety factor = 5
Morin's formula, 1892 $P = 5670.35$ kg		12,500 lb	
Thredgold's formula, 1892 $W =$		11,865 lb	W is allowable load
Kidder's formula (BC & S), 1900 $P = C(A)$	56,250 lb	18,750	safety factor = 3
Table from Kidder's BC & S	46,700 lb		tabulated L = 18 ft
Trautwine's formula, 1888 $P =$	55,190.3 lb	22,076 lb	safety factor = 2.5
Trautwine's table	48,832 lb	19,533 lb	safety factor = 2.5 (7.25 × 7.25 in.)
Manual, Heavy Timber Mill Construction, Southern Pine Assn., 1918 U.S. Department of Agriculture formula $C = 535.4$ psi $P = C(A)$		30,115.7 lb	safe load
Winslow formula (Chicago Building Ordinance) $C = 600$ psi $P = C(A)$		33,750 lb	safe load
U.S. Dept. of Agriculture (Table 3-6)		30,120 lb	safe load[b]
Winslow formula (Table 3-7)		33,740 lb	safe load[b]
Ketchum's formula, 1924 $P_{all} =$		31,500 lb	safe load
Modern analysis, 1991 NFPA NDS for Wood Construction $P_{all} =$		25,513 lb	safe load

[a] Computations are based upon actual size of 7.5 × 7.5 in. unless noted otherwise.
[b] Computations are based upon an allowable basic stress of 1000 psi.

columns are designed. The method used by the NDS for the preceding 10 years produced an allowable load of 23,000 lb, which is only a few percentage points below the typical allowable load determined by the earlier formulae. The 1991 NDS revised the method of computation to produce an allowable load of 25,500 lb, which is slightly higher than the historic formulae results. The new method here utilizes an allowable stress of 1100 psi for select structural, western hemlock.

The wooden column as utilized in historic buildings was typically a large member, usually 8 × 8 in. (7.5 × 7.5 in., 7.25 × 7.25 in., depending upon surfacing) or even larger. The wood column increased in size by increments of 2 in. from 6 × 6 in. through 18 × 18 in. Wood columns were difficult to assemble with girders, beams, etc., to make up the internal structural frame (nonrigid) of the typical building. The larger columns at lower levels would be holding up girders or beams of approximately the same width as the column thickness. The bottom of the wood column would bear on a cast iron base that spread the load on the foundation. The column would also have a cast iron or steel cap that functioned as the support mechanism for the beam ends. The column cap was also designed to allow an upper column to bear on the cap through a pintle in line with the lower column (see Figures 3-9 and 3-10).

The bases and caps were designed and produced by several manufacturers for use on commercial and industrial buildings. Both were made of cast iron in the earlier periods (1860s–1890s) and steel in the later periods (1880s–1930s) (see Figure 3-11).

Figure 3-12 demonstrates a perspective view of semi-mill construction, showing columns, caps, girders with iron ties, beams with stirrups, and subflooring and fin-

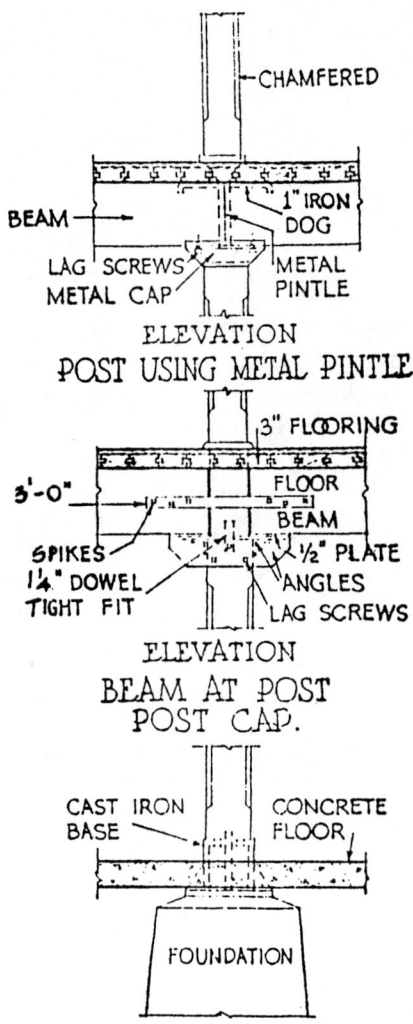

Figure 3-9. Column details. (*From Voss and Varney 1926, 91.*)

HISTORIC AMERICAN BUILDING SYSTEMS: WALLS AND COLUMNS ◇ 135

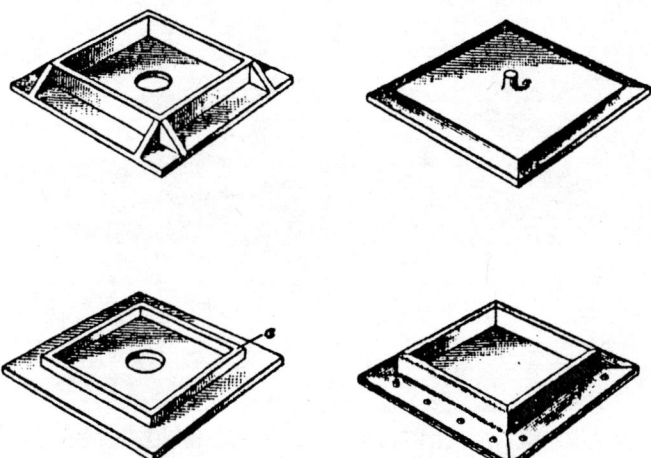

Figure 3-10. Column bases, cast iron. (*From International Correspondence Schools 1904, sec. 15, 20.*)

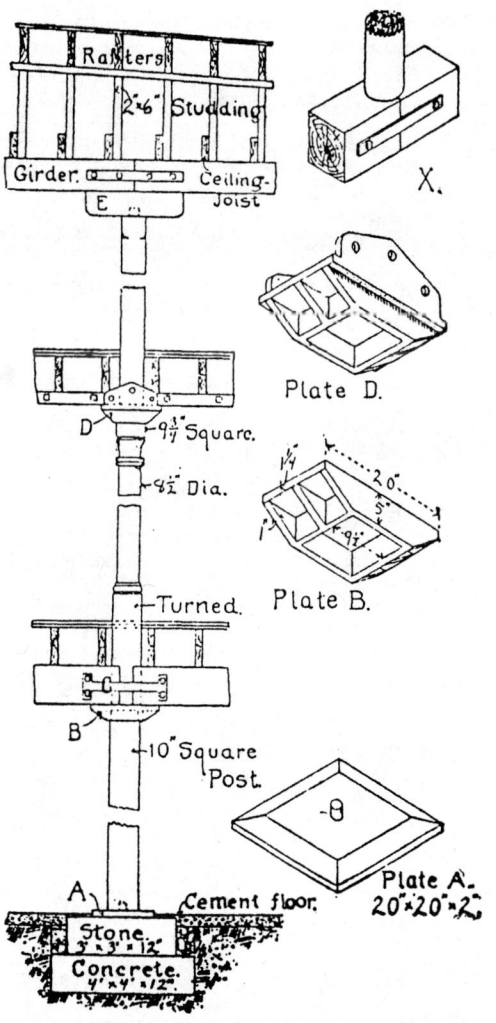

Figure 3-11. Column—base/cap details. (*From Kidder 1900, 429.*)

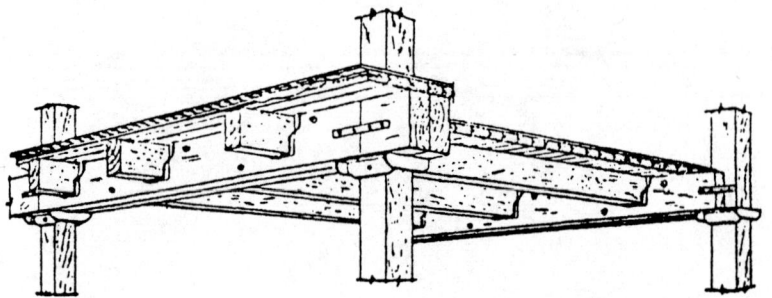

Figure 3-12. Perspective—floor system. (*From Voss and Varney 1926, 112; Hool and Kinne 1924, 108.*)

ished flooring. "Semi-mill" construction was the name given to industrial and commercial buildings that were constructed of relatively large timber members. Beams and columns were capable of carrying relatively heavy loads, and the details of the connections also indicated the construction type.

The problem of too many members intersecting at the same point existed at the top of wooden columns. Recognizing the need for a minimum bearing area requirement on the end of girders, and that a column from above also bears on the column, manufacturers produced a pintle, which transferred load from the upper column to the cap of the lower column while allowing an adequate bearing area for girders (see Figure 3-13).

Some manufacturers recognized the need for a minimum bearing area for girder ends that were framed on a column cap. They produced a cast iron or steel cap including a pintle that transferred the load of an upper column to the cap of the lower column. The pintle was a transfer mechanism to allow the axial load to be passed through the assembly without taking too much space while also allowing two or more girders to frame into the same column cap (see Figure 3-13). The design of a column bracket or column cap at the turn of the century was very much the same as today. The bearing area required at the girder end would depend on the reaction at the end where bearing was being considered and the girder material's allowable compressive stress perpendicular to the grain (bearing stress). The bearing area would be the reaction in pounds divided by the allowable bearing stress in psi (see Figure 3-14).

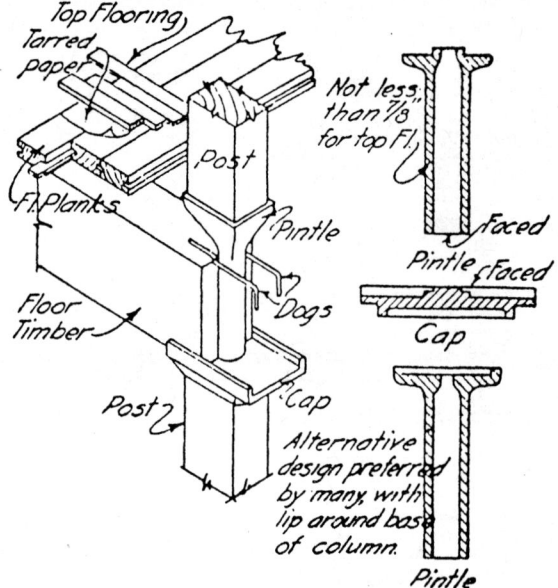

Figure 3-13. Column—pintle assembly. (*From Voss and Varney 1926, 104.*)

Area of Bearing = A_B, in.^{2}

Reaction, R = 10,000 lb

Allowable bearing stress
perpendicular to grain = 325 psi
(for species of girder)

$$A_B = \frac{10,000 \text{ lb}}{325 \text{ psi}} = 30.8 \text{ in.}^2$$

$A_B = a(b) \quad \begin{array}{l} a = 4.25" \\ b = 7.50" \end{array}$

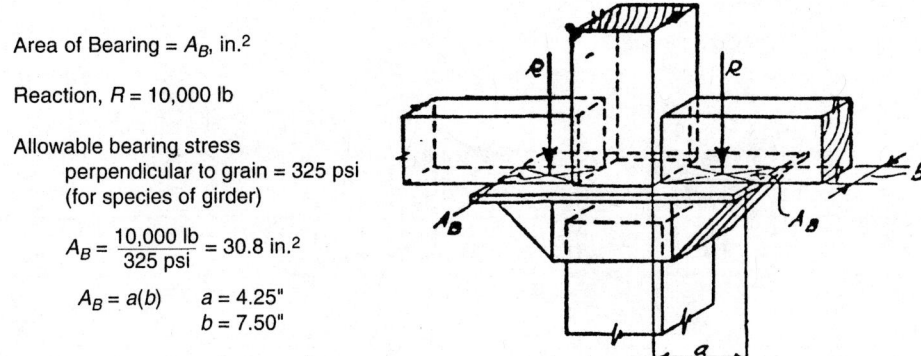

Figure 3-14. Design of column caps.

MILL CONSTRUCTION

The term *mill construction* as used in the late nineteenth century and early twentieth century usually referred to a type of construction that utilized heavy timber girders, beams, and columns as framing inside load-bearing masonry walls. The geometric shape of the building and the early textile mill functions gave it a specific identity that became known as mill construction. Textile mills of the 1870s–90s where three to four stories, as required by the factory process (see Figure 3-15). Later mill buildings were single-story factory structures with trussed roofs and large open areas. Vibrations and heavy machinery made the optimum mill building design a single-story structure with long spans and open spaces for machinery and process. Machinery foundations on grade were easier and cheaper to construct. The best modern practice inclines toward single-floor shops (Ketchum 1924, 7).

The single-story factory building of the later period, into the twentieth century was often a steel-framed structure consisting of a series of bents. These bents, composed of two steel columns and a truss bearing on the columns at its ends, would be subframed with girts on the columns to form metal walls and purlins to support metal roofing. Truss spans would normally be at approximately 60 ft (see Figures 3-16 and 3-17).

Mill buildings, as these industrial buildings became known, were constructed throughout the country and housed almost every type of light manufacturing. Many of these buildings are still in existence; most, however, are now utilized for a different use than the original. The evolution of the mill building took many directions. Traditional mill construction utilizes the least possible number of heavy wooden beams with a double-layered tongued and grooved deck and floor (Hool and Kinne 1924, 113). These buildings also had no concealed spaces, which means that the underside of the floors was exposed, with no false ceiling. The beams and decking layer would be planed very smooth, thereby providing a very high resistance to ignition by fire. The columns and girders would be fire-cut, chamfered corners to eliminate sharp 90° edges that easily ignite. Thus, the name "slow-burning construction" was used interchangeably with "mill construction" and "standard mill construction," which connoted the premium-class construction (see Figure 3-18). In first-class structures, a sprinkler system would have been a requirement. A second type of building, the mill construction with laminated floors, utilized floors of heavy planking laid on edge (see Figure 3-19). This type was not considered as safe as the premium method. A third method was semi-mill construction, which introduced intermediate beams that spanned between the major girders and effectively broke up the span of the flooring (see Figure 3-20). These intermediate beams are what we now know as joists, which are normally spaced between 2–5 ft apart and carry the floor deck and finished floor.

Mill buildings were fairly easy to design. Most or all of the computations were straightforward. Beams, columns, and walls were all of the type called "simple construction," in that there were no cantilevers, continuous members, or rigid-framed connections to complicate the design. Loads were easily determined and quantified,

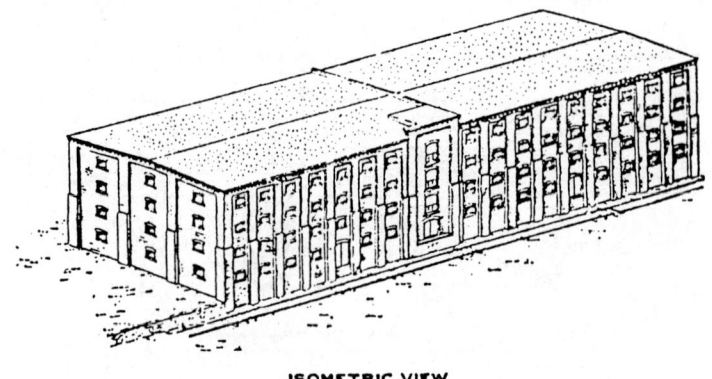

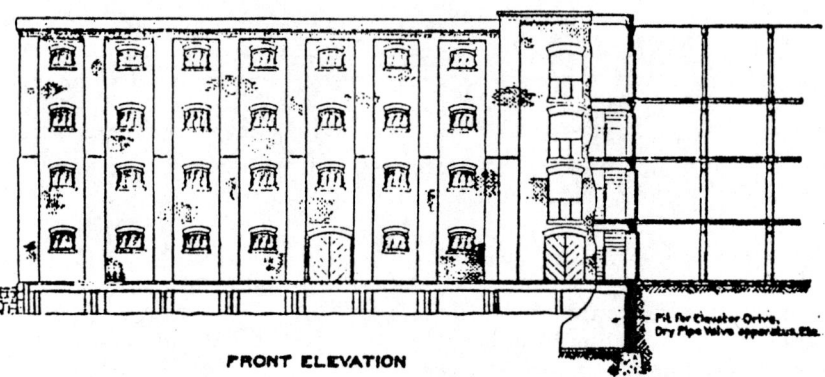

Figure 3-15. Standard mill construction. (*From Paul 1918, 45.*)

so the analysis was routine. Each member, truss, column, or wall would be designed as an individual element.

The design analysis will not be presented here. The components of the masonry mill building will be illustrated on a member-by-member basis. The mill building was the predecessor of many building types and was extremely influential in the design and construction of load-bearing masonry construction. Many existing buildings contain details from mill construction.

MASONRY LOAD-BEARING WALLS

The masonry load-bearing wall is perhaps the oldest of the "permanent" types of construction. Masonry buildings date to prehistoric times. Strangely, thousands of years of masonry construction continued to utilize only empirical methods of design until approximately the turn of the twentieth century. Mesopotamian structures, Greek and Roman buildings, Gothic cathedrals, and Renaissance buildings were apparently constructed by master builders using methods passed on through the apprenticeship system. Master builders understood both the ability of masonry to withstand compressive loads and the stability of arches. These structures depended upon mass and weight to resist the horizontal thrust caused by arches and domes (see Figure 3-21). These early builders knew that the height of a wall depended entirely upon the width or thickness of the wall. The architecture of the modern era depended upon empirical methods throughout almost all of the nineteenth century. As late as the early 1900s, building laws dictated the thickness of masonry load-bearing walls

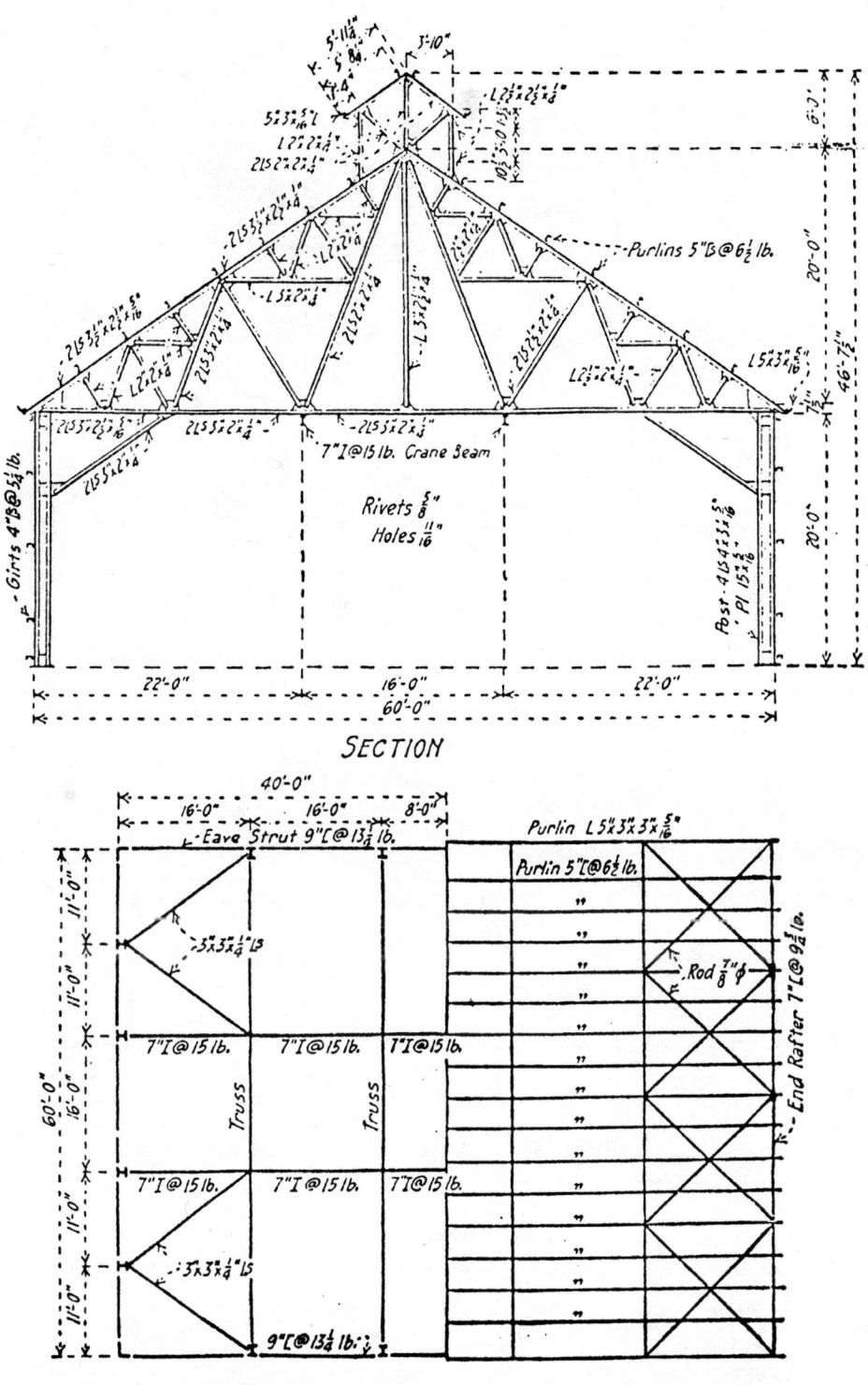

Figure 3-16. Mill Building—early twentieth century. (*From Ketchum 1924, 63.*)

140 ◇ Structural Analysis of Historic Buildings

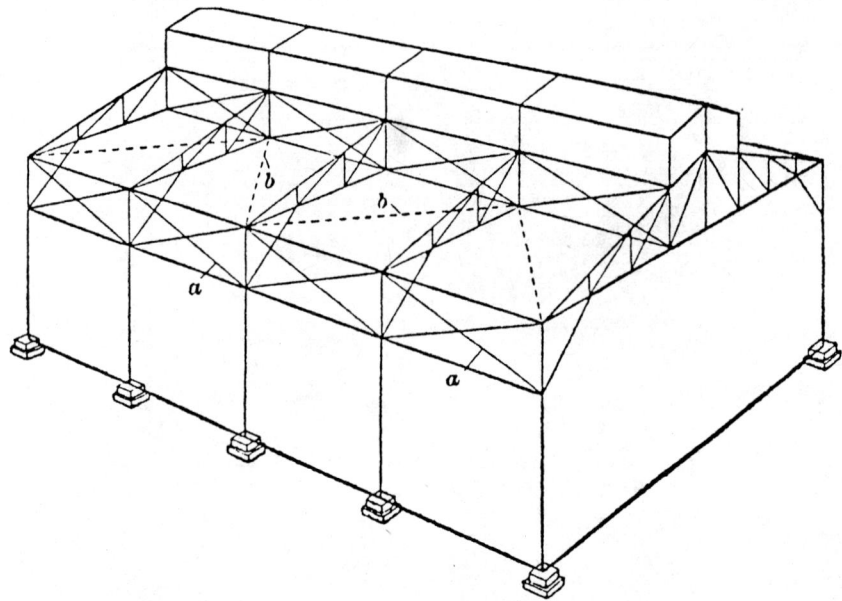

Figure 3-17. Mill building isometric showing wind bracings. (*From International Correspondence Schools 1923, sec. 64, 33.*)

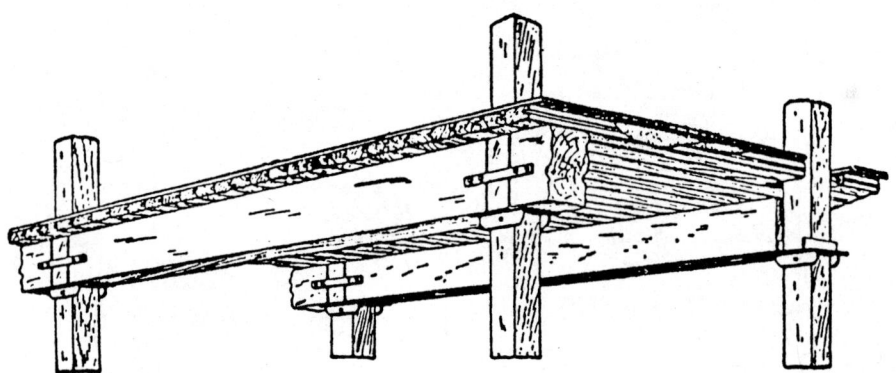

Figure 3-18. Standard mill construction. (*From Hool and Kinne 1924, 113.*)

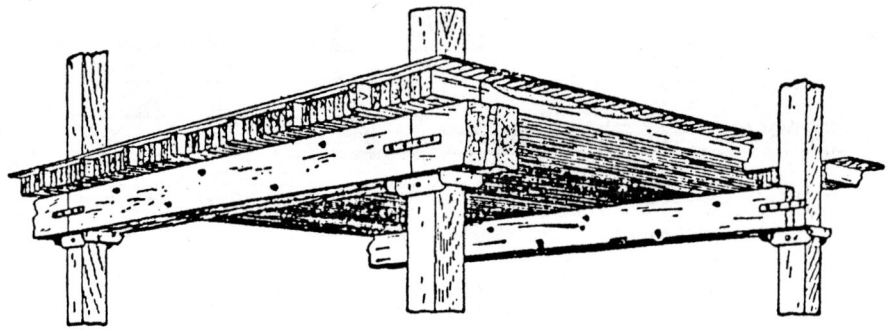

Figure 3-19. Laminated mill construction. (*From Voss and Varney 1926, 101.*)

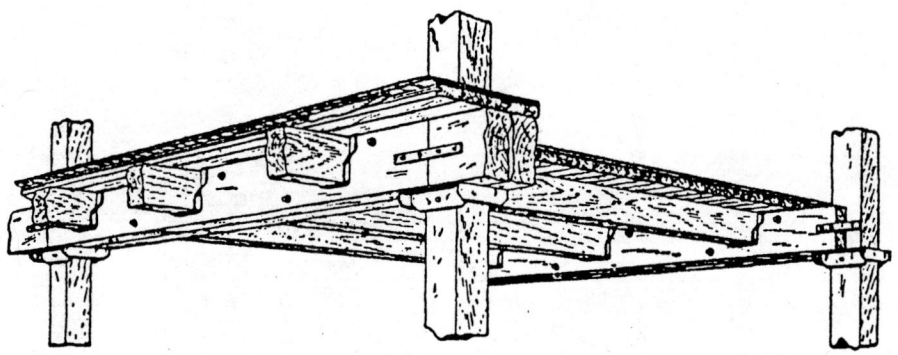

Figure 3-20. Semimill construction. (*From Voss and Varney 1926, 112.*)

for buildings. These building laws used the building height as the determining element in the tabulation of the thickness of the masonry walls (see Chapter 1, and Figure 3-22). Chapter 1 has an in-depth analysis of the various building laws and wall height thickness ratios that governed masonry construction. In this section, compressive stresses, strength of masonry, and modern approaches to analysis of masonry will be considered on a quantitative basis.

Actually, although they did not state it as such, these building laws were dictating the wall thickness to control the maximum compressive stress on the masonry components to within a predetermined allowable. The safe working strengths or allowable strengths are the product of the ultimate strength of a brick and/or a period mortar divided by a factor of safety. See Table 3-10 for a table of the safe strength of brickwork

Figure 3-21. Treasury of Atreus, Mycenae. (*From International Correspondence Schools 1905, sec. 19, 2.*)

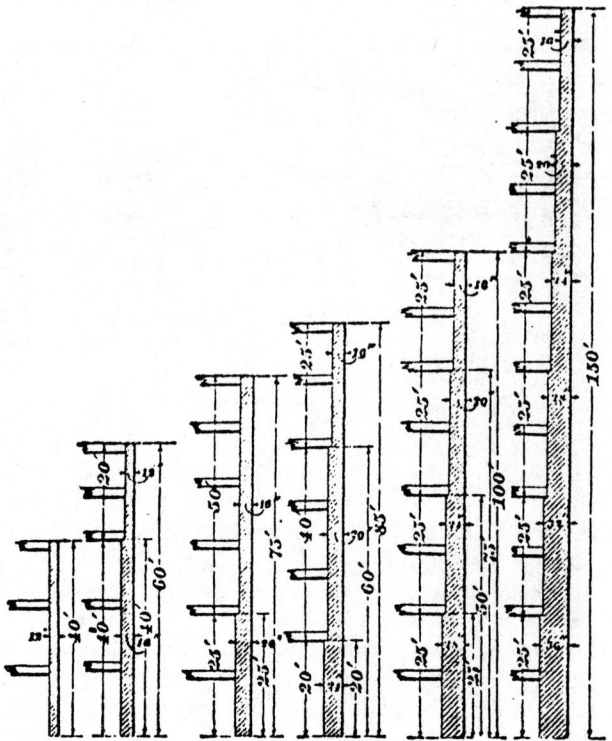

Figure 3-22. Thickness of building walls. (*From International Correspondence Schools 1899, sec. 7, 98.*)

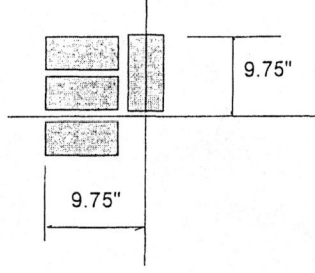

Figure 3-23. Quarter section, 19.5 × 19.5 in. brick pier.

for the average brick used in walls of buildings. The weight of the building components, including the weight of the masonry walls and the superimposed or live loads, make up the total load on the foundation walls or bearing piers.

Kidder (1905, 246) presents a table on safe loads on brickwork (see Table 3-11).

The allowable load on a brick pier would be determined by multiplying the size of the pier in square inches by the allowable stress in psi for the brick and mortar to be used on the project. In normal design, the required load to be carried on the pier is known from the building design. The size of the pier can be determined by dividing the actual maximum design load by the allowable stress of the brickwork (see Figure 3-23).

Example: Brick per Size

$$\text{Size of pier} = \frac{50,000 \text{ lb}}{138 \text{ psi}} = 362.32 \text{ in.}^2 = 19.03 \text{ in.}^2$$

In design, one would use 19.5 × 19.5 in., which is a brick modular dimension.

Walls were designed in much the same way as piers except that many building's walls are subject to a combination of gravity and lateral loads. This combined loading was known to designers and builders of the period, but the method of analysis was more empirical than analytical until well into the twentieth century. Kidder states, "There is no practical rule by which it is possible to calculate the necessary thickness of brick walls, as the Resistance to Crushing, which is the only direct strain, is usually a minor consideration" (ibid., 222).

Kidder further states that wall thickness is the product of experience and that most cities regulate wall thickness according to building height. Baker (1892) states that the optimum use of masonry is when it is in "direct compression," that is, subject to vertical loading only. This would be the vertical live and dead loads from a floor or roof system bearing on a masonry wall plus the actual weight of the wall. Baker

Table 3-10. Safe Stregnth of Brickwork[a]

Materials	Strength	
	psi	tons/ft²
Eastern brick		
Hard-burned brick, laid in good lime mortar	100–140	7–10
Same, laid in 1 to 2 Rosendale cement mortar	195	14
Same, laid in 1 to 3 Rosendale cement and lime mortar	150–165	10.75–11.75
Same, laid in 1 to 2 Portland cement mortar	210	15
Western brick		
Hard-burned brick, laid in 1 to 2 Louisville mortar	145	10.5
Same, laid in 1 to 2 Portland cement mortar	175	12.5

[a]From International Correspondence Schools 1899, sec. 7, 158.

and other authorities of the late 19th century demonstrated methods of obtaining the location of the center of gravity of walls of varying thickness. This relatively simple process was utilized in the design of foundations and in a method of analyzing cracks in masonry walls due to the unbalanced loads (see Chapter 2).

Baker presents a table of the relative strength of brick and brick masonry, the result of tests in Germany in 1883 that were verified by tests performed at the U.S. Arsenal in Watertown, Massachusetts, in 1884 (see Table 3-12). Baker also makes recommendations about the allowable compressive strength of brick masonry in Table 3-13. The allowable stresses in the masonry walls were given as a fraction of the ultimate compressive stresses at this period in the same manner as today (see Table 3-13).

The transverse strength of historic brick masonry walls depends upon the adhesion and cohesion of the mortar. The strength of the mortar and the tensile strength of the brick combine to make the overall transverse strength of the wall. Baker states that "in the case of a high wall whose upper portion is overthrown by a lateral force or pressure of any kind, the failure is due to either (1) to the breaking of the adhesion in the bed-joints and of the cohesion of the side joints, or (2) to the rupture of the mortar in the bed joints alone" (ibid., 167).

Mulligan (1942, 91) states that the use of lime in cement mortar increases the adhesiveness of the mortar to the brick, but at the same time the cohesiveness of the mortar decreases. From this it can easily be concluded that lime and lime cement mortars are more plastic due to the lime and adhere better to the structural units. Plasticity in a mortar makes mason's work much easier.

Brick masonry load-bearing walls were utilized in literally hundreds of thousands of building applications in the United States during the period 1800–1940. During this period many changes occurred in the materials as the brick and the mortars evolved from materials that were handmade at the site to factory-produced, high-quality building components. Early bricks were relatively soft compared to later bricks, which were machine-made and thus denser and stronger. The firing process was also improved with commercial kilns, and this produced a more thoroughly fired brick that was stronger and more uniform. The traditional lime-sand mortars also evolved into a harder mortar made of Portland cement. Brick and mortar improvements seemed to parallel each other: as the brick became stronger or harder, so did the mortars.

As previously mentioned, load-bearing masonry walls are subjected to lateral forces in addition to the normal gravity loads. These lateral forces usually come from wind, horizontal reactions from rafters that bear on walls, and horizontal loads from

Table 3-11. Safe Loads on Brickwork

New England hard-burned brick, in lime mortar	112–138 psi
New England hard-burned brick (1 Rosendale concrete, 2 sand)	166 psi
New England hard-burned brick (1 cement, 3 lime mortar)	194 psi

Table 3-12. Relative Strength of Brick and Brick Masonry[a]

Kind of Brick	Average Crushing Strength of Brick (psi)	Ultimate Strenvth (psi) of Brickwork with Mortar Composed of—			
		1 lime, 2 sand	7 lime, 1 cement, 16 sand	1 cement, 6 sand	1 cement, 3 sand
Ordinary stock	2,930	1,290	1,390	1,610	1,850
Selected stock	3,669	1,620	1,760	2,020	2,320
Clinker stock	5,390	2,370	2,590	2,960	3,410
Porous	2,617	1,150	1,250	1,440	1,650
Porous perforated	1,195	530	570	650	750
Perforated	2,759	1,210	1,320	1,520	1,710
Average strength of the masonry in terms of the strength of the brick		0.44	0.48	0.55	0.63

[a] From Baker 1892, 166.

earthquakes. None of the three lateral forces listed herein is difficult to determine within required accuracies. However, the analytical methods for determination of these forces did not come into common use until sometime in the late 1880s. Also, the methods and assumptions used in early analytical solutions provided designers only with basic approximate answers for the stresses that resulted from combined loadings. The concept of combined stresses, taking into account the fact that bending or flexural stresses are caused in part by lateral pressures, was not fully developed. In fact, this development continues today. With each code revision we are further modifying the method of calculating the forces and/or pressures from lateral influences. Fortunately, we have been operating under a set of fairly conservative numbers for the lateral forces and an equally conservative approach to the determination of working stresses (a fraction, one-half to two-thirds of the ultimate capacity).

International Correspondence Schools (1905) presents a comprehensive analysis by a combination of analytical and graphic statics for the solution of lateral forces on building walls. These forces are the result of the horizontal thrust at rafter ends and the normal force of wind on the windward side of the building. The example they give is shown here in abbreviated form and is followed by a modern analysis. Figure 3-24 shows the building in section and gives its specific dimensions.

Figure 3-25 shows a force diagram with dimensions and member sizes for the second floor and above. Rafter spacing is 2 ft on centers; therefore, the analysis is for a 2 ft segment of the building.

W = 264 lb given, gravity load of roof
P = 648 lb given, wind load perpendicular to the right rafter, position as shown
R_1 = Left vertical reaction from gravity load
H_1 = Left horizontal thrust from gravity load
R_2 = Right vertical reaction from gravity load
H_2 = Right horizontal thrust from gravity load
R_3 = Left reaction from wind on right rafter (line of left rafter, 45° from vertical)
R_4 = Right reaction from wind on right rafter (perpendicular to right rafter).

Figure 3-26 shows further resolution of the forces in the left and right walls. The force polygons represent the forces and reactions for the two foot segments of the walls.

Table 3-13. Allowable Compressive Strength of Brick Masonry[a]

Reasonably good brick in good lime mortar	20 tons/ft^2	275 psi
Best brick in good Portland cement mortar	30 tons/ft^2	415 psi

[a] From Baker 1892, 167. The normal factor of safety against ultimate is between 5 and 6.

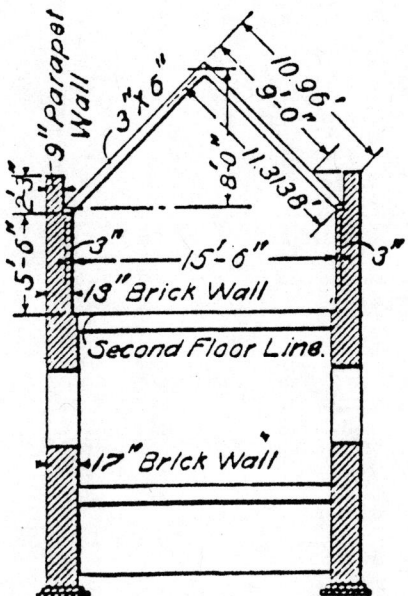

Figure 3-24. Building section. (*From International Correspondence Schools 1905, sec. 18, 59.*)

$R_5 = 700$ lb = resultant of H_1, R_1, and R_3, graphic line of Action, as shown
$R_6 = 440$ lb = resultant of H_2, R_2, and R_4, graphic line of action, as shown
$R_7 = 2430$ lb = resultant of wall weight and R_5, graphic line of action, as shown
$R_8 = 2290$ lb = resultant of wall weight and R_6, graphic line of action, as shown

Reactions R_5 and R_6 are considered to pass through points b on the wall sections. The line of action of these reactions is projected to the point where they intersect the line of the center of gravity of the wall (Figure 3-27), thus, establishing points c_2 and c_3 (see wall sections). Reactions R_7 and R_8 (with wall weights included) lines of action

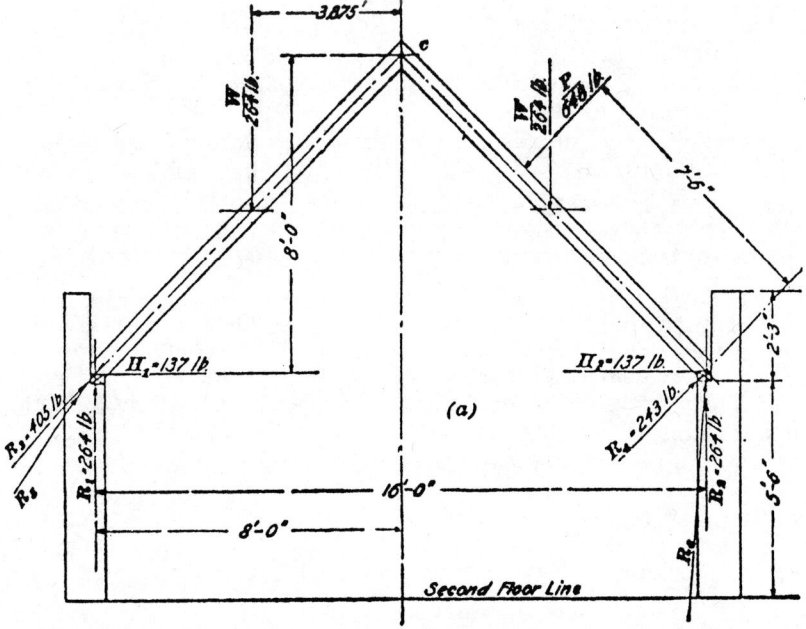

Figure 3-25. Upper section—building. (*From International Correspondence Schools 1905, sec. 18, 61.*)

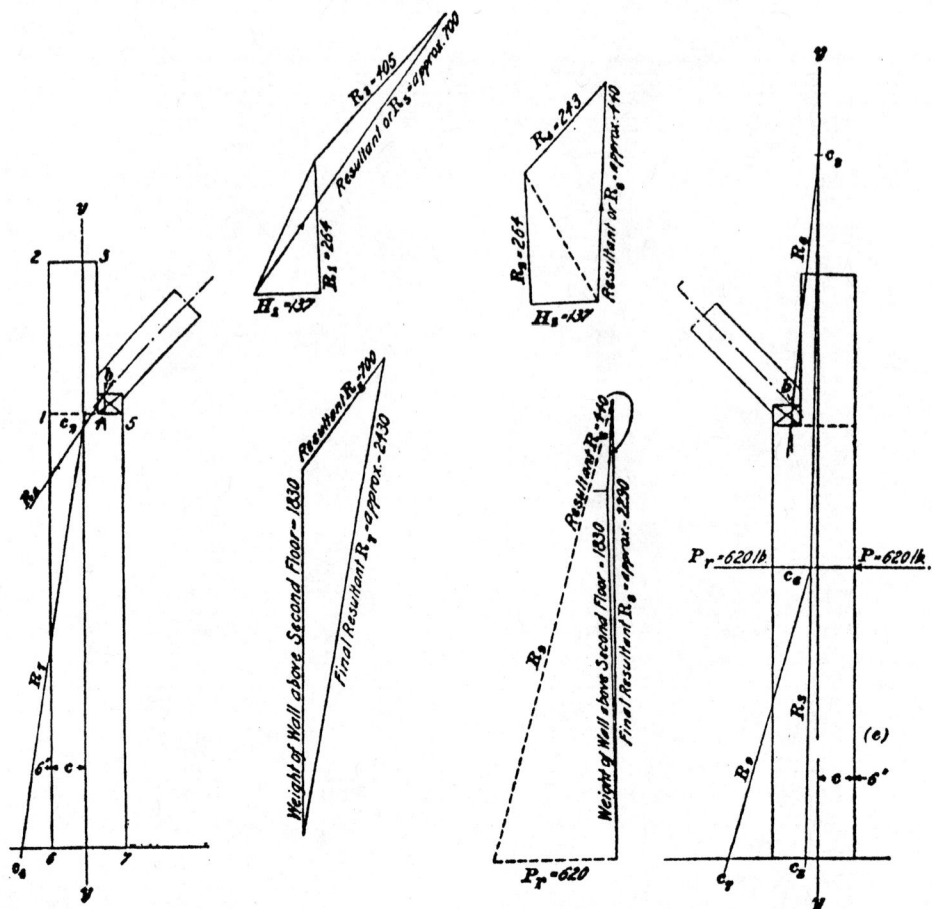

Figure 3-26. Wall sections and force polygons. (*From International Correspondence Schools 1905, sec. 18, 61.*)

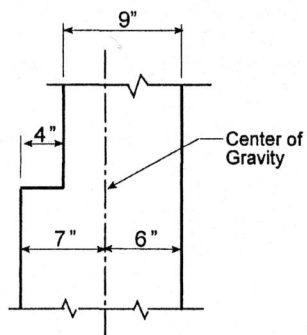

Figure 3-27. Center of gravity—wall.

are projected from points c_2 and c_3 downward toward the second floor line to locate points c_4 and c_5. Point c_6 is the point on the line of action of R_8 at the location of the intersection of the line of action of the equivalent concentrated load of the wind on the wall, 620 lb (at 3875 ft above the second floor line). R_9 line of action is projected from this point to the second floor line at c_7.

Point c_4 (left wall section) falls outside of the wall thickness; therefore, the wall is in tension and compression from bending action from the combination of the gravity loads from the weight of the roof and of the weight of the wall and the wind force on the left rafter. This wall is considered to be unsafe. The fact that point c_4 falls outside of the wall thickness means that the wall is partially in tension. The rule was that the line of action must stay within the wall.

Points c_5 and c_7 (right wall section). Point c_7 falls outside of the wall thickness; therefore, the wall is in both tension and compression from bending action from the combination of the gravity and wind loads and the additive wind load on the right wall (620 lb). Point c_5 falls within the wall section (without the wind load on the wall). This wall is also considered to be unsafe.

The solution presented by the example is to increase the wall thickness to 17 in. by the addition of one wythe for the portion of the wall from the second floor line to the bearing point of the rafters. Figure 3-28 shows the revisions to the wall thickness and the new resolution of forces, which now locate c_4 and c_7 to within the wall thickness. Point c_{10} is the product of the use of a 30 psf wind force for the analysis instead of 40 psf, which is in the original example.

This analysis is very interesting and reflects the state of the analytical methods of the period. The method of analysis has changed over the years and has usually been

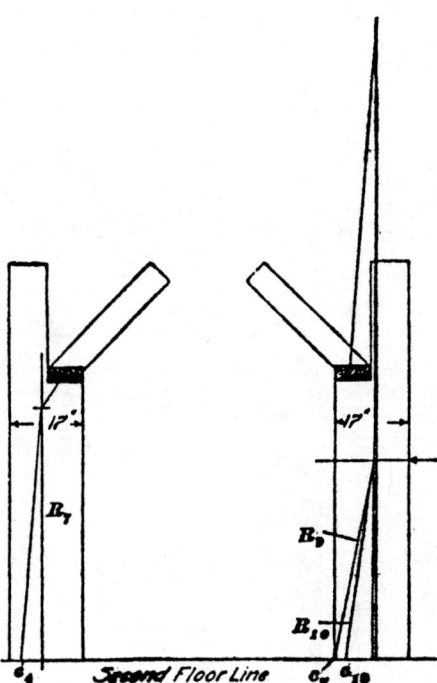

Figure 3-28. Revised wall sections. (*From International Correspondence Schools 1905, sec. 18, 65.*)

driven by the code provisions of the area in which the building is to be constructed. The example problem will now be reworked using Standard Building Code, 1991 methods.

SBC 1991 Wind Criteria

A uniform load of 40 psf (at 0–30 in. above the ground) from wind is generated by a wind velocity of over 130 mph. Modern methods use the geographical location of the building to determine the maximum possible wind velocity on the basis of a 100-year reoccurrence. The wind pressure is classified by heights above the ground, the first category being 0–30 ft above the ground, the second 30–40 ft, and so on. The only geographical locations in the United States that have design wind velocities of 130 mph are the Florida Keys and islands off the coast of North Carolina. It is therefore unreasonable to subject this building to 40 psf wind pressure. As suggested by the ICS in their conclusions, 30 psf is a more reasonable force. The modern analysis will therefore be worked with a wind pressure of 30 psf. The modern analysis also requires leeward suction pressures in addition to reduced windward positive pressures (see Figure 3–29).

Figure 3-30 provides details of the individual walls with the summary of the gravity and wind forces shown on the same diagrams. The forces shown on each section are utilized to compute the moments and stresses at the line of the second floor level and the line of the bottom of the parapet.

The brick or stone wall of a building may have any of several types of mortar joints that were in use at the time. Many of these joints featured a reveal of as much as 0.25–0.5 in. into the wall (see Table 3-14 and Figure 3-32). In an instance where joints of one of these types exist or where the masonry joints have eroded and are not flush with the face of the masonry, a moment of inertia based upon total wall thickness cannot be used. The moment of inertia (I_x) that is utilized in the calculation of the bending stress must be calculated using the net wall thickness rather than the full thickness to the face of the brick. In the analysis of an existing building, the designer must inspect the brick or stone masonry walls to determine the type and

148 ◇ Structural Analysis of Historic Buildings

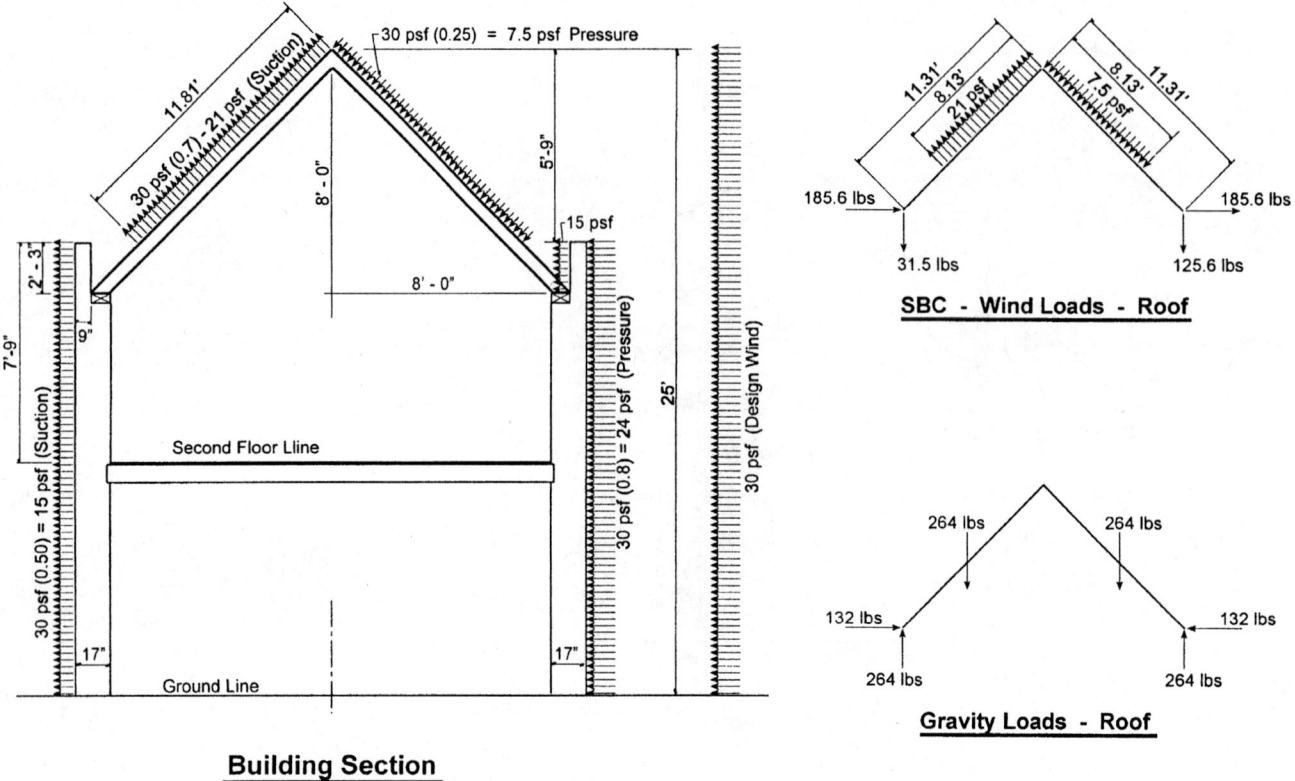

Building Section

Figure 3-29. Wind and gravity loads on building, 1991 Standard Building Code.

condition of the mortar and the type of joint. If the analysis of the wall includes determination of a required flexural strength, as is usually the case, the condition of the wall is an important factor. Cracks in the mortar joints or badly eroded areas must be recorded and considered in the analysis and would effect the flexural capacity of the wall.

PARAPET WALLS, DECORATIVE CORNICES

The stability of the decorative cornice made of corbeled brickwork has always been in question, but within the last 20 years it has become a critical concern. The brick corbeled cornice, which was extensively used on commercial buildings during the period 1875–1925, was an architectural detail required by the style of the buildings. The more important the building, the more decorative and deeper the cornice and the more complicated the detail (see Figure 3-33). Historically, the center of gravity of the weight of the projecting brickwork had to fall within the width of the wall to provide the required stability of the design.

The decorative cornice could be part of a wall that formed the upper area or parapet of a semiflat roof that sloped from front to back of a building. It could also be a freestanding parapet wall with corbeled brick projections. Buildings constructed entirely of load-bearing masonry (without steel frame) would not utilize any steel hooks or bars to hold the cornice in place, and therefore the cornice would depend upon its own weight to remain stable.

Example: Parapet and Cornice
The following example is of a 3 ft (vertical projection) parapet wall, masonry constructed (dimensioned as shown), with a wind force of 15 psf acting in the direction of largest impact (wind acting from behind wall, the most vulnerable side, the over-

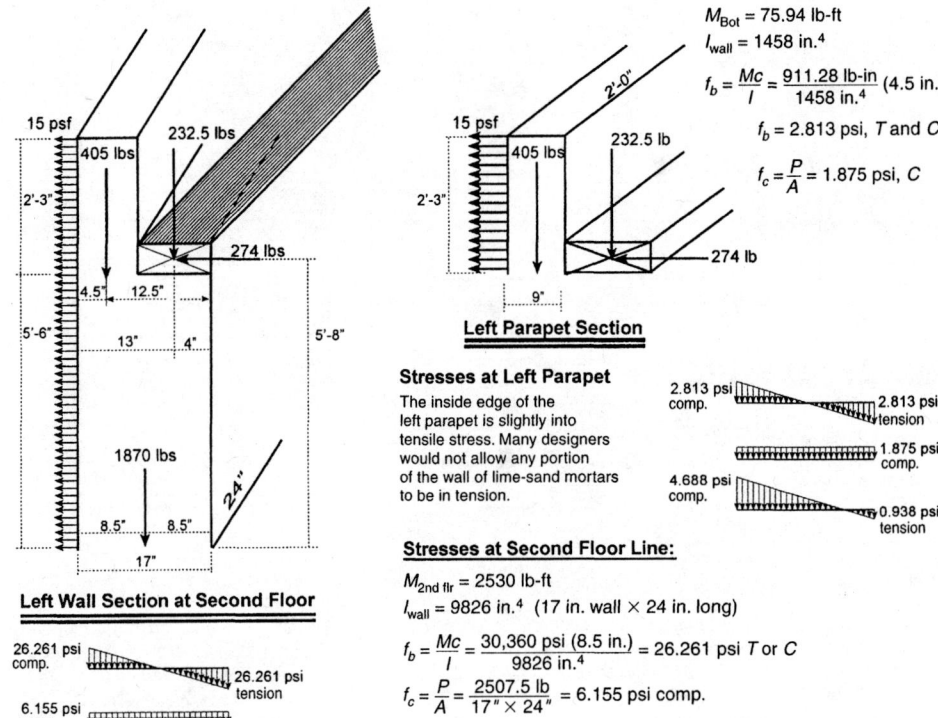

Stresses at Left Parapet

The inside edge of the left parapet is slightly into tensile stress. Many designers would not allow any portion of the wall of lime-sand mortars to be in tension.

Stresses at Second Floor Line:

$M_{2nd\ flr} = 2530$ lb-ft
$I_{wall} = 9826$ in.^{4} (17 in. wall × 24 in. long)

$$f_b = \frac{Mc}{I} = \frac{30{,}360 \text{ psi } (8.5 \text{ in.})}{9826 \text{ in.}^4} = 26.261 \text{ psi } T \text{ or } C$$

$$f_c = \frac{P}{A} = \frac{2507.5 \text{ lb}}{17'' \times 24''} = 6.155 \text{ psi comp.}$$

A segment of the wall at the second floor line is in tension. The usual process was to require the entire section to be in compression to be positively safe on first-class buildings. Some designers of the period would allow 25 psi tensile stress at the face of a wall. In fact, the lime-sand mortar utilized in the construction has cracked and would no longer be in tension. Cracked section analysis, which accounts for the dead zone of the cracked section, does not provide a correct compressive stress in this case as the overall eccentricity places the P outside of the wall. The transverse loading is high in this case relative to the wall weight and its smaller eccentricity.

Critical section:

$$\text{midheight } M_u = \frac{P_g(e_g)}{2} + \frac{w(h)^2}{8} = \frac{1572.5 \text{ lb } (0.03 \text{ in.})}{2} + \frac{2 \text{ ft } (15 \text{ psf}) (7.75 \text{ ft})^2}{8} = 249.07 \text{ ft-lb} = 2988.9 \text{ lb-in.}$$

Overall eccentricity:

$$\text{midheight} = \frac{2988.9 \text{ lb-in.}}{1572.5 \text{ lb}} = 1.9 \text{ in.} \quad \text{kern} = 2.833 \text{ in.} \quad e \text{ is within the kern}$$

Cracked section analysis:

$$\text{2nd floor level } f_c = \frac{4P}{3bt\left(1 - \frac{2e}{t}\right)} = \frac{4(30{,}360 \text{ lb})}{3(17 \text{ in.})(24 \text{ in.})\left(1 - \frac{2(12.108 \text{ in.})}{17 \text{ in.}}\right)} = 233.739 \text{ psi}$$

The adjusted compressive stress of 233.739 psi is not excessive.

Figure 3-30.

turning gravity load working with the wind). (Refer to Figure 3-33 for numerical values; see also Table 3-15.)

The parapet wall is stable. Stage one of the computations (gravity loads, at M_G) indicates that the combined compression and bending stresses are satisfactory. The eccentricity caused by the corbels of the masonry is not sufficient to cause tension in the windward side of the wall. Stage two of the computations (including wind, at M_W) indicates that the combined compression and bending stresses are within reasonable

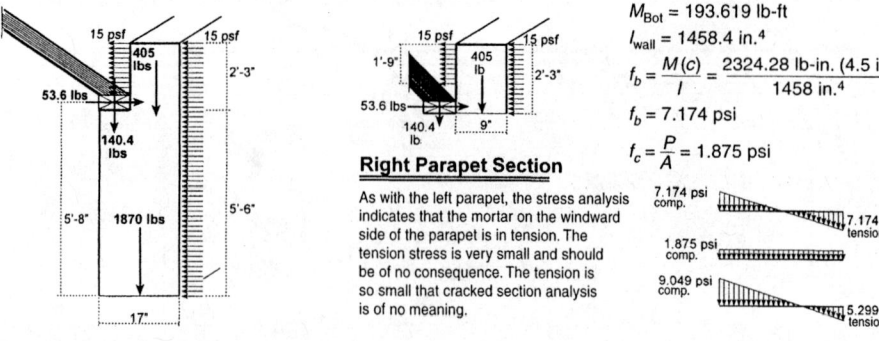

Right Parapet Section

As with the left parapet, the stress analysis indicates that the mortar on the windward side of the parapet is in tension. The tension stress is very small and should be of no consequence. The tension is so small that cracked section analysis is of no meaning.

$M_{Bot} = 193.619$ lb-ft
$I_{wall} = 1458.4$ in.^{4}
$f_b = \dfrac{M(c)}{I} = \dfrac{2324.28 \text{ lb-in. } (4.5 \text{ in.})}{1458 \text{ in.}^4}$
$f_b = 7.174$ psi
$f_c = \dfrac{P}{A} = 1.875$ psi

Right Wall Section at Second Floor

The stress analysis at the second floor line also indicates that the windward edge of the wall is slightly in tension.

$M_{\text{2nd flr}} = 1425$ lb-ft
$I_{wall} = 9826$ in.^{4} (17 in. wall × 24 in. long)
$f_b = \dfrac{M(c)}{I} = \dfrac{17{,}100 \text{ lb-in }(8.5 \text{ in.})}{9826 \text{ in.}^4} = 14.792$ psi T or C
$f_c = \dfrac{P}{A} = 5.920$ psi comp.

The force from the roof member actually reduced the effect of the wind on the wall between the roof and the second floor by acting in the opposite direction from the wind. The lime-sand mortars of the period 1875–1920 varied in quality. However, a small amount of tensile stress was allowed by some of the leading authorities, (see Table 3-14).

The second floor midheight will be checked by cracked section analysis.

$e = 7.080$ in.　　kern $= 2.833$ in.

Cracked section analysis: second floor midheight

$$f_c = \dfrac{4P}{3bt\left(1 - \dfrac{2e}{t}\right)} = \dfrac{4(17{,}100 \text{ lb})}{3(17 \text{ in.})(24 \text{ in.})\left(1 - \dfrac{2(7.080 \text{ in.})}{17 \text{ in.}}\right)}$$

$f_c = 334.51$ psi

The cracked section method indicates that the compression stress at the edge of the wall is high. The 334.51 psi is high but only 10 percent over the allowables in Table 3-14.

Figure 3-31.　Right wall and parapet—general.

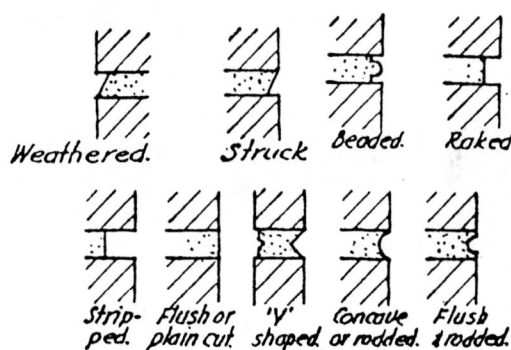

Figure 3-32.　Brick joints. (*From Ramsay and Sleeper 1936, 15.*)

Table 3-14. Strength of Lime-Sand Masonry

The lime-sand mortars utilized in the nineteenth century and into the twentieth century were compatible in material properties with the brick of the period. The allowable compressive stresses and the corresponding allowable tensile stresses given by several texts are as follows:

	Compressive	Tensile
Trautwine 1888[a]	50–75 psi	15–30 psi
Kidder 1902[b]	100 psi	none
Kidder 1905[c]	100 psi	none
International Correspondence Schools 1911	150–300 psi	40–70 psi
International Correspondence Schools 1923	150–300 psi	40–70 psi
Mahan 1885[f]	Bending	40 psi

[a] At 427.
[b] At 180.
[c] At 400.
[d] At 201.
[e] At para. 2, 21.
[f] At 15.

limits. The windward side of the parapet is still slightly in compression. The parapet and cornice are stable. Note that the computations were based upon a 16.625-in. net mortar width in a 17-in. wall. If the mortar is eroded or has deeper joints, a revised moment of interia must be used. (See Figure 3-34.)

MASONRY BUILDING WALLS

The load-bearing masonry commercial building was extensively constructed throughout the United States in the late 19th century. Even into the last quarter of the 19th century, the wall designs were empirically done. The designer or builder would select the bearing wall thickness from wall thickness tables that had been incorporated into the local building laws. The wall thickness depended upon the height of the building. There were no interpretative methods to enable designers to modify a wall thickness, either by increasing or decreasing it, to meet a specific situation. There was no provision for wind or other lateral loads on a building. A designer would depend upon his experience and empirical knowledge as a guide to modifications if they seemed appropriate. Most designers simply chose the wall thickness required by the city's building law and constructed their buildings accordingly. The wall thickness tables were fairly uniform from city to city; where variations occurred, they appeared to be the product of either a loading condition or possibly a foundation condition. See Chapter 1 for many examples of wall thickness tables by building laws in many cities. These tables were quite conservative. There were few to no requirements concerning lateral bracing, and it is not known whether the designers were conscious of the need to provide for this portion of the stability needs of the building. Many buildings were constructed adjacent to each other, and some shared common "party" walls. In most instances, a building was independent of the adjacent structure because each built independent walls. The usual case was to build to the exact line of the property in ownership, and the adjacent building would do the same.

Example
Chicago, Illinois—three-story building, commercial, wall thickness by heights from Table 1-3. First floor = 16 in. thick, height = 18 ft, second floor = 12 in. thick, height = 15 ft, third floor = 12 in. thick, height = 13 ft, 6 in. Parapet = 12 in. thick, height = 3 ft. Loads for the modern analysis areas follow:

$\text{Roof}_{D+L} = 45$ psf
$\text{Third floor}_{D+L} = 75$ psf
$\text{Second floor}_{D+L} = 75$ psf
$\text{First floor}_{D+L} = 75$ psf

Brick and mortar unit weight:

Masonry wall = 120 lb/ft³

Wind force:

Pressure from wind = 15 psf

Constants for gravity and wind loads:

$I_{17} = \frac{bh^3}{12}$ (Use h = 16.625 in. conservative)

$= \frac{12 \text{ in.}(16.625 \text{ in.})^3}{12}$

I_{17} = 4595 in.⁴ per linear ft of wall

Moments at base of corbel ($M(c)$, gravity only):

M = −357.660 lb(8.5 in. − 6.683 in.) = 649.868 lb-in.

$f_b = \frac{M(c)}{I} = \frac{649.868 \text{ lb-in.}(8.3125 \text{ in.})}{4595 \text{ in.}^4}$

f_b = 1.176 psi (tension or compression) per linear ft of wall

$f_c = \frac{P}{A} = \frac{357.660 \text{ lb}}{12 \text{ in.}(16.625 \text{ in.})}$

f_c = 1.793 psi compression

f_c = 2.969 psi compression (outside edge of wall)

f_c = 0.617 psi compression (inside edge of wall)

The parapet is in compression at each edge of wall; there is no problem.

Moment about center of 17 in. wall (at roofline, Mw) (gravity and wind included)

M = −357.660 lb (8.5 in. − 6.683 in.) − 531.140 lb (8.5 in. − 5.240 in.)

= −2381.385 lb-in. (counterclockwise rotation)

$f_b = \frac{M(c)}{I} = \frac{2381.385 \text{ lb-in.}(8.3125 \text{ in.})}{4595 \text{ in.}^4}$

f_b = 4.308 psi (tension or compression) per linear ft of wall

$f_c = \frac{P}{A} = \frac{357.660 \text{ lb} + 531.140 \text{ lb}}{12 \text{ in.}(16.625 \text{ in.})}$

f_c = 4.555 psi compression

f_c = 8.863 psi compression (outside edge of wall)

f_c = 0.247 psi compression (inside edge of wall)

The parapet is in compression at each edge of wall; there is no problem.

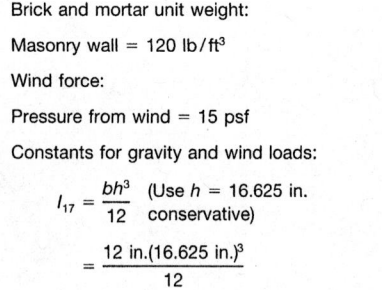

Figure 3-33. Parapet and cornice.

Table 3-15. Summary of Forces, Moments, and Equivalent Forces

Force	Vol. per lin. ft	Force (lb)	M. Arm (in.)	Moment (lb-in.)
F1	0.7839	94.068	6.25	587.925
F2	0.8385	100.620	5.50	583.410
F3	0.3646	43.752	7.00	306.264
F4	0.3465	41.568	7.50	311.760
F5	0.3281	39.372	8.00	314.976
F6	0.3190	38.280	8.25	315.810
	F equival.	357.660	6.683	2390.145
F7	1.549	185.938	8.50	1580.469
Fw		34.200	−17.1	−584.820
FR		−20(0.99563)	+1.0	−19.913
FL		18.000	−14.4	−259.200
FPH		71(1)(0.7)(15)	−32.88	−245.120
FPV		7.455	−10.50	−78.278
	F equival.	531.140	5.240	2783.283

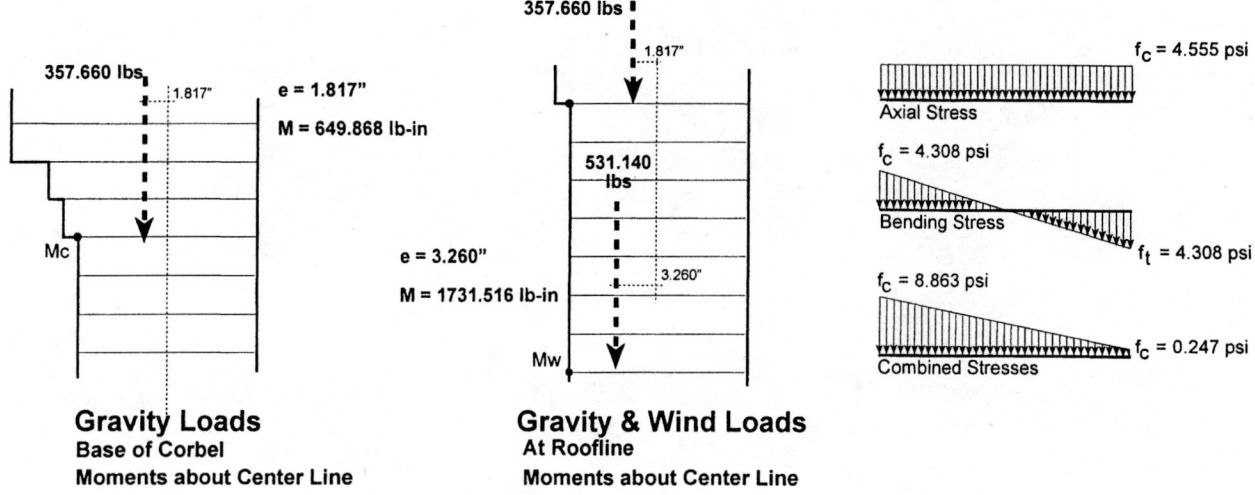

Figure 3-34.

Because the combined stresses for the outside face and inside face of the windward wall are all in compression and within the allowable compressive stress, the upper level of the wall is considered safe for individual wind action. The parapet outside face indicates a small amount of tensile stress on the windward wall. Designers would normally consider this small tensile stress, as not enough to cause concern. The leeward wall would have lower stresses than the windward wall and therefore is not a problem.

The combined lateral and gravity loads must be further analyzed to check the overall lateral effect on the building. If the building is freestanding (vulnerable to wind from any direction), the total lateral wind load on the building must be overcome by the walls from ground level to the second floor joists. The same situation would apply for each level upward. The lower level will produce the highest stresses and is therefore the critical section to check (see Figure 3-37).

The types of failure that were checked in the computations of Figure 3-37 would be typical of an unsupported or 1 ft wide section of wall. Case 1 analyzes the two lower longitudinal walls as if the walls resist moment throughout their entire length. If the walls resist moment at top and bottom, then there is an inflection point at midheight as shown. With the inflection point used as a point of known moment, the shear is determined at the same location. The leeward wall is isolated as a free body diagram to determine the moment at the top of that section of the wall. A combined stress analysis indicates that the maximum compressive stress is 159.73 psi while the maximum tensile stress is 79.57 psi. The stresses are calculated as if there is no diaphragm action from the floor above transferring a part of the lateral load to end walls or trussed partitions. Diaphragm action could occur if the floor is tight and reasonably continuous. Designers worry about counting upon too much shear resistance from inherently weak storefront systems and would rather see this force dissipated in other ways, such as the resistance of the walls. If the building configuration allows, the designer may attribute approximately two-thirds of the tensile stress from the lateral pressure on the wall to be consumed by the shear resistance of the system. Case 2 analyzes the two lower longitudinal walls as if the walls resist moment at the upper end and are pinned at their bases. This method produces a larger moment at the top of the wall than the one developed in Case 1. It is assumed that the actual case for most buildings would be somewhere between these two cases. The shear force shown at the bottom of the section of walls being analyzed would produce a shear stress that would need to be checked.

If the building that is being analyzed has masonry cross-walls (transverse walls) or braced frame partitions perpendicular to the wall being checked, the diaphragm of the floors will transfer the lateral loads to the transverse walls. The transverse walls (masonry walls or braced frame walls) act as shear walls and would consume the lateral loads. The transverse walls or braced frame walls will further transfer the lateral

154 ◇ Structural Analysis of Historic Buildings

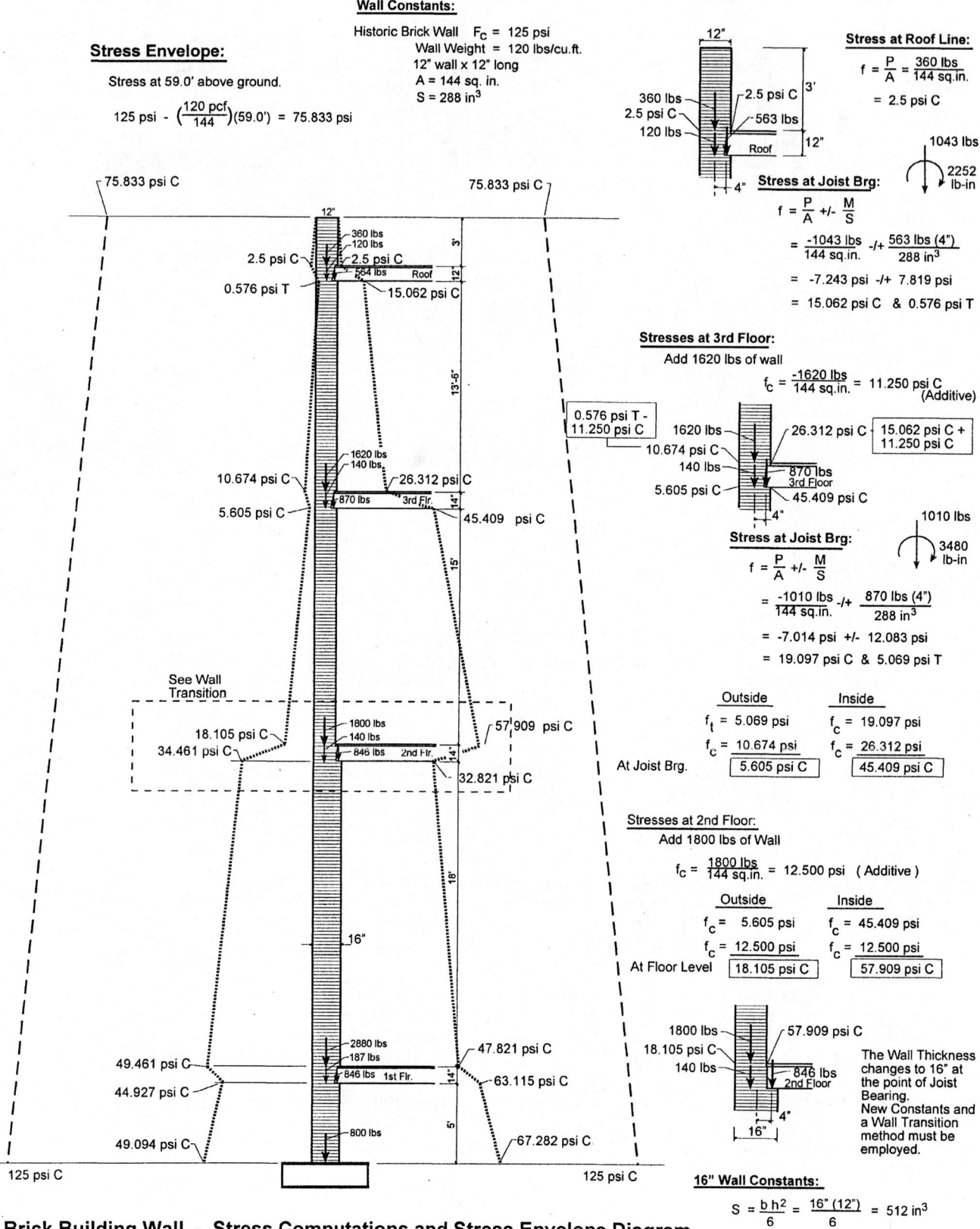

Brick Building Wall - Stress Computations and Stress Envelope Diagram
Loads: Dead Load and Full Live Load - Gravity Only

Figure 3-35. Wall loads.

2nd Floor Thickness Transition
Finding equivalent load at what eccentricity

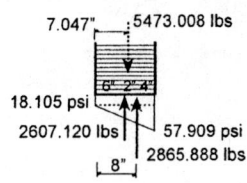

ΣF_v = 2607.120 lbs + 2865.888 lbs
     = 5473.008 lbs

ΣM = − 2607.120 lb (6") − 2865.888 lb (8")
    = − 38,569.824 lb-in

$M_{edge} = \dfrac{38,569.824 \text{ lb-in}}{5473.008 \text{ lbs}} = 7.047"$

e = 8" − 7.047" = 0.953"

$f = \dfrac{-P}{A} -/+ \dfrac{M}{S} = \dfrac{-6459.008 \text{ lbs}}{192 \text{ sq.in.}} -/+ \left(\dfrac{-140 \text{ lbs (2")}}{512 \text{ in}^3} + \dfrac{-5473.008 \text{ lb (0.916 in)}}{512 \text{ in}^3} + \dfrac{846 \text{ lbs (6")}}{512 \text{ in}^3} \right)$

= − 33.641 psi -/+ (+ 0.820 psi)

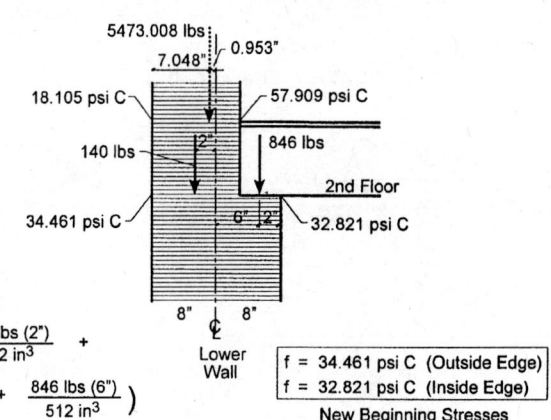

f = 34.461 psi C (Outside Edge)
f = 32.821 psi C (Inside Edge)
New Beginning Stresses

Stresses at 1st Floor

Additional Wall Weight:
$\dfrac{16"}{12"/'}(120 \text{ pcf})(18') = 2880 \text{ lbs/f}$

$f_c = \dfrac{2880 \text{ lbs}}{192 \text{ sq.in.}} = 15.0 \text{ psi}$ (Additive)

Outside	Inside
34.461 psi C	32.821 psi C
15.000 psi C	15.000 psi C
49.461 psi C	**47.821 psi C**

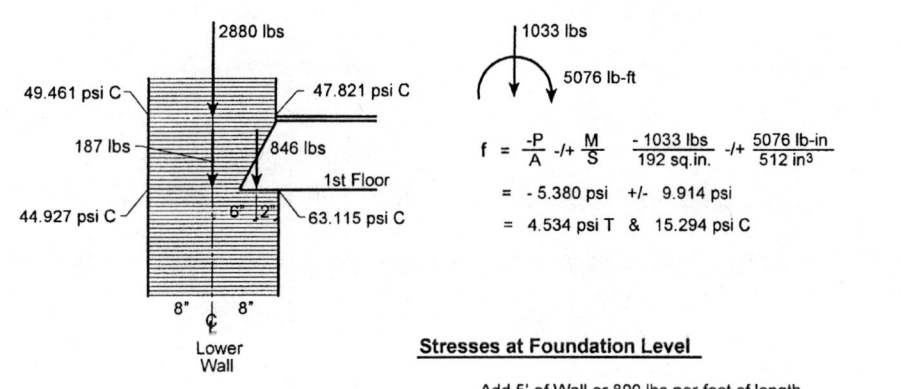

$f = \dfrac{-P}{A} -/+ \dfrac{M}{S}$ $\dfrac{-1033 \text{ lbs}}{192 \text{ sq.in.}}$ -/+ $\dfrac{5076 \text{ lb-in}}{512 \text{ in}^3}$

= − 5.380 psi +/− 9.914 psi
= 4.534 psi T & 15.294 psi C

Outside	Inside
49.461 psi C	47.821 psi C
4.534 psi T	15.294 psi C
44.927 psi C	**63.115 psi C**

Stresses at Foundation Level

Add 5' of Wall or 800 lbs per foot of length.

$f_c = \dfrac{800 \text{ lbs}}{192 \text{ sq.in.}} = 4.167 \text{ psi}$ (Additive)

Outside	Inside
44.927 psi C	63.115 psi C
4.167 psi C	4.167 psi C
49.094 psi C	**67.282 psi C**

Wall Effects on Footing and Soil Pressure

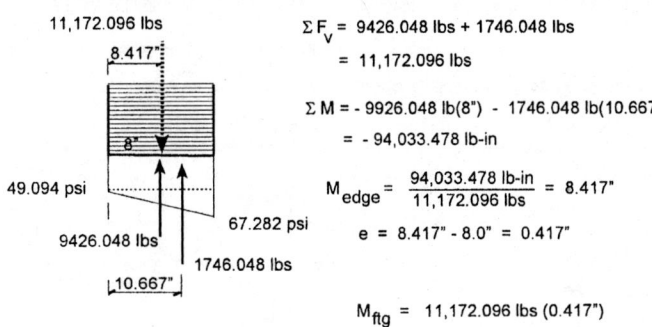

ΣF_v = 9426.048 lbs + 1746.048 lbs
     = 11,172.096 lbs

ΣM = − 9926.048 lb(8") − 1746.048 lb(10.667")
    = − 94,033.478 lb-in

$M_{edge} = \dfrac{94,033.478 \text{ lb-in}}{11,172.096 \text{ lbs}} = 8.417"$

e = 8.417" − 8.0" = 0.417"

M_{ftg} = 11,172.096 lbs (0.417")
     = 4658.764 lb-in

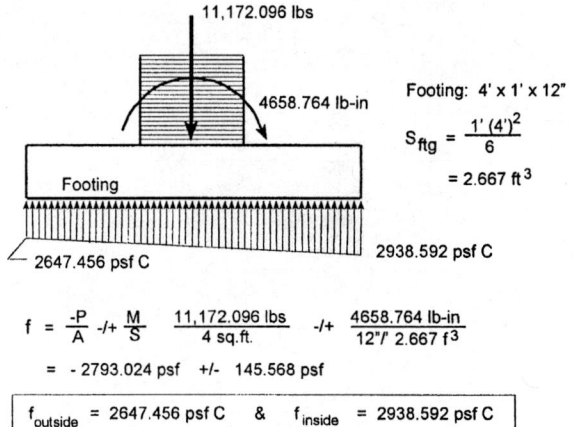

Footing: 4' x 1' x 12"

$S_{ftg} = \dfrac{1' (4')^2}{6} = 2.667 \text{ ft}^3$

$f = \dfrac{-P}{A} -/+ \dfrac{M}{S}$ $\dfrac{11,172.096 \text{ lbs}}{4 \text{ sq.ft.}}$ -/+ $\dfrac{4658.764 \text{ lb-in}}{12"/' \; 2.667 \text{ f}^3}$

= − 2793.024 psf +/− 145.568 psf

$f_{outside}$ = 2647.456 psf C & f_{inside} = 2938.592 psf C

Figure 3-35. (*Continued*).

Brick Building Wall - Stress Computations and Stress Envelope Diagram

Loads: Wind Load, Windward Wall

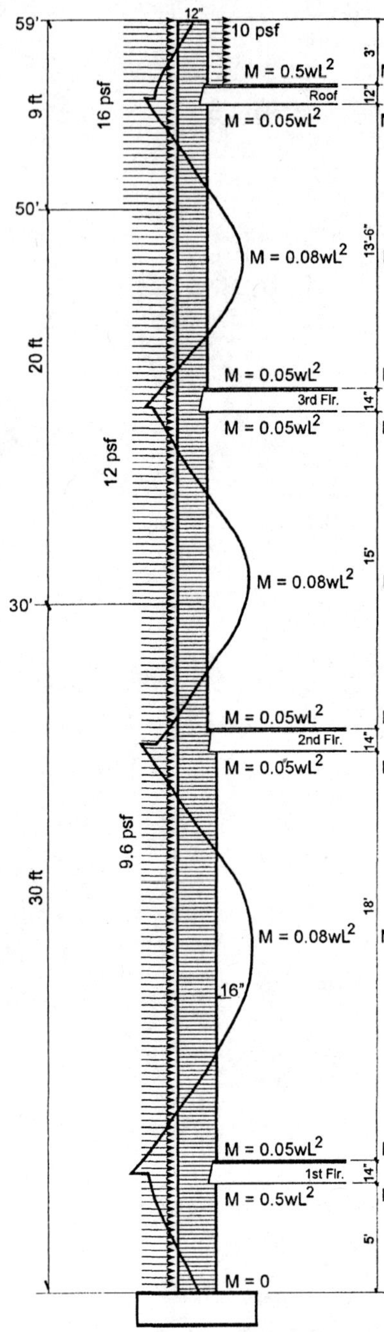

			Wind Loading		Gravity: D + L		Gravity: D Only		Combined W + D + L	
	Moment		Outside Face psi	Inside Face psi	Outside Face psi	Inside Face psi	Outside Face psi	Inside Face psi	Outside Face psi	Inside Face psi
$M = 0.5wL^2$	$M = 117.0$ lb-f	Roof	4.875 T	4.875 C	2.500 C	2.500 C	2.500 C	2.500 C	2.375 T	7.375 C
$M = 0.05wL^2$	$M = 145.8$ lb-f	Joist Brg.	6.075 T	6.075 C	0.576 T	15.062 C	1.417 C	9.083 C	6.651 T	21.137 C
$M = 0.08wL^2$	$M = 233.3$ lb-f	Mid. Ht.	9.721 C	9.721 T	5.625 C	20.687 C	6.907 C	14.708 C	16.628 C	10.966 C
$M = 0.05wL^2$	$M = 145.8$ lb-f	3rd Flr.	6.075 T	6.075 C	10.674 C	26.312 C	12.667 C	20.333 C	6.592 C	32.387 C
$M = 0.05wL^2$	$M = 135.0$ lb-f	Joist Brg.	5.625 T	5.625 C	5.605 C	45.409 C	11.764 C	26.930 C	6.049 C	51.034 C
$M = 0.08wL^2$	$M = 216.0$ lb-f	Mid Ht.	9.000 C	9.000 T	11.855 C	51.659 C	18.014 C	33.180 C	27.014 C	42.659 C
$M = 0.05wL^2$	$M = 135.0$ lb-f	2 nd Flr.	5.625 T	5.625 C	18.105 C	57.909 C	24.264 C	39.430 C	18.639 C	63.534 C
$M = 0.05wL^2$	$M = 155.5$ lb-f	Joist Brg.	3.645 T	3.645 C	34.461 C	32.821 C	37.103 C	14.861 C	33.458 C	36.466 C
$M = 0.08wL^2$	$M = 248.8$ lb-f	Mid. Ht.	5.831 C	5.831 T	41.961 C	39.962 C	44.736 C	22.361 C	50.567 C	34.131 C
$M = 0.05wL^2$	$M = 155.5$ lb-f	1 st Flr.	3.645 T	3.645 C	49.461 C	47.821 C	52.103 C	29.861 C	48.458 C	51.466 C
$M = 0.5wL^2$	$M = 120.0$ lb-ft	Joist Brg.	2.813 T	2.813 C	44.927 C	63.115 C	51.368 C	35.278 C	48.555 C	65.928 C
$M = 0$		Foundation	0	0	49.094 C	67.282 C	55.535 C	39.445 C	55.535 C	67.282 C

Wall Constants:

Historic Brick Wall F_c = 125 psi
Wall Weight = 120 lbs/cu.ft.

12" wall x 12" long
A = 144 sq. in.
S = 288 in^3

16" wall x 12" long
A = 192 sq. in.
S = 512 in^3

Wind Calculations:

Windward Wall:
50 - 100 ft above ground 20 psf (0.8) = 16 psf
30 - 50 ft above ground 15 psf (0.8) = 12 psf
0 - 30 ft above ground 12 psf (0.8) = 9.6 psf

Leeward Wall:
50 - 100 ft above ground 20 psf (0.5) = 10 psf
30 - 50 ft above ground 15 psf (0.5) = 7.5 psf
0 - 30 ft above ground 12 psf (0.5) = 6.0 psf

Wind Calculations:

End and Mid-Height Moments are modified from the suggested constants from ACI. Moments are based upon one-way moment design. Shear walls are far enough apart to make a 2:1 length ratio.

For sections where the Uniform Load from the wind is not constant, use the larger of the two values. If judgement dictates that the higher value is in a very short zone, the lower value may be all that is needed.

Figure 3-36.

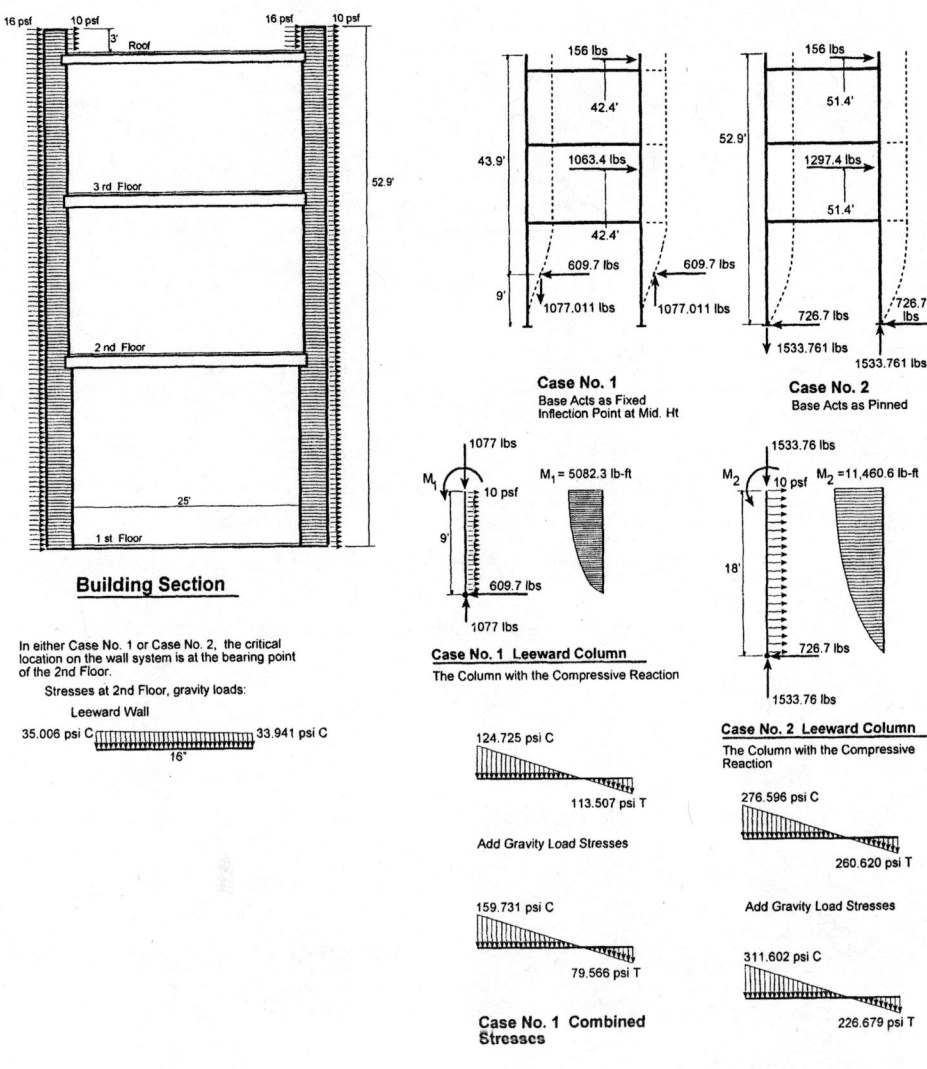

The combined stresses for either case in Figure 3-37 are excessive and demonstrate that a historic masonry building must have shear walls and/or trussed partitions to take the lateral loads. In each case the tensile stress far exceeds any reasonable amount that may be allowed by a designer.

A building of this type will rarely satisfy the stress analysis for gravity plus lateral loads on an overall basis such as is demonstrated here. When buildings of this type are in a contiguous group of several buldings making up a block of buildings, the group can share the lateral loads. Therefore, a contiguous group of historic buildings of approximately the same size and height is interdependent.

Figure 3-37.

loads to the foundations. See Figure 3-38 for an example of a braced frame wall system. If the shear walls are too far apart or if the diaphragm action fails to transfer the loads, the building should not be considered safe. The computations shown in Figure 3-39 provide a check of the shear stress and bending stress on the transverse walls that result as the lateral loads are transferred through diaphragm action from longitudinal to transverse walls. The analysis shown is for an end wall that is solid (no penetrations). The transverse walls would have to be checked with window and/or door penetrations that reduce the actual area and shear capacity of the wall.

The actual shear stress on the transverse wall of 8.208 psi is well below the allowable shear stress of 31.250 psi. The combined axial and bending stresses that result from the weight of the wall (vertical loading) and the lateral forces (wind loading) are

158 ◇ Structural Analysis of Historic Buildings

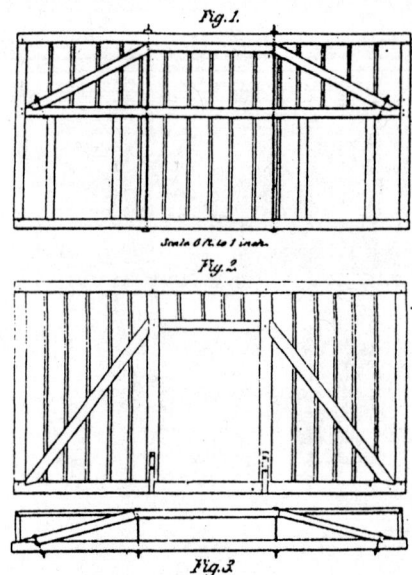

Figure 3-38. Trussed partitions. (*From Bell 1875, 63.*)

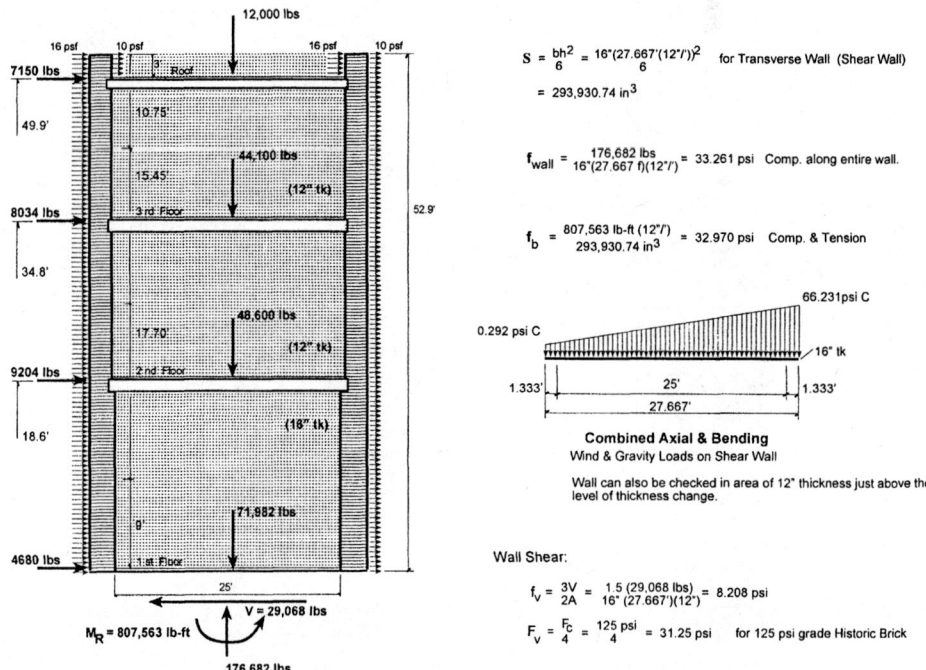

The shear caused by lateral loads on a wall was not analytically determined or checked against allowable stress levels until about the turn of the twentieth century. Merriman and Wiggin (1947, 889) state that "the working shearing unit stress for all stone masonry should be taken as one-quarter of the working compressive unit stresses above given." This is consistent with the thinking of today for all empirically designed masonry.

The average allowable compressive stress for good brick in lime-sand mortar was 125 psi. The allowable shear stress for the same wall would be 31.25 psi.

Figure 3-39. Transverse walls: shear and moment capacity. *Note:* Transverse walls are non-floor load carrying walls. Vertical loads are the weight of the masonry wall.

also well within the allowable. The transverse wall shown was analyzed on the basis of being one of the end walls of a 40-ft-long building with no intermediate transverse walls or trussed walls. The lateral zone of influence from the wind loading would then be 20 ft of wall. the wall was analyzed (determination of stresses) at ground or first floor level, and it was analyzed as a solid wall (with no openings). Moments and stresses at other levels would be lower than those at first floor level (ground floor) and need not to be checked. Walls below the first floor level are considered a part of the foundation system, and that analysis is done within the foundation computations.

In most applications, the transverse walls are not solid, voids for doors and windows exist, and a separate analysis considering the voids would be required. Figure 3-40 illustrates computations for an analysis for the same wall with two separate door and window patterns.

The 6 ft wide opening in the same transverse wall, Case A of Figure 3-40, would only increase the flexural bending stresses of the masonry at first floor level to 33.308 psi T or C. The compressive stress on the masonry due to the gravity load (weight of wall minus the opening size of 6 × 8 ft) would increase to 41.600 psi. In the same manner as in Figure 3-39, the combined stresses would be 7.086 psi C and 73.702 psi C. The shear stress at the first floor level would increase to 10.481 psi. All of these stresses are well within the allowables.

The two 5-ft, 10-in. and one 6-in. wide openings, Case B of Figure 3-40, would have a greater effect on the stresses. The flexural bending stresses would increase to 59.218 psi T or C. The compressive stresses on the masonry due to the gravity loads (weight of wall minus openings) would increase to 157.544 psi C. In the same manner as above, the combined flexural and compressive stresses would be 98.326 psi C and 180.253 psi C. The shear stress would increase to 22.709 psi. The high side flexural stress of 180.253 psi compression does exceed the values of the allowable compressive stress given by some sources. Figure 3-39 gives an average allowable compressive stress for good brick in lime-sand mortar as 125 psi. Figure 3-32 shows an allowable compressive stress of from 100 psi to 150–300 psi. The allowable stresses do seem to

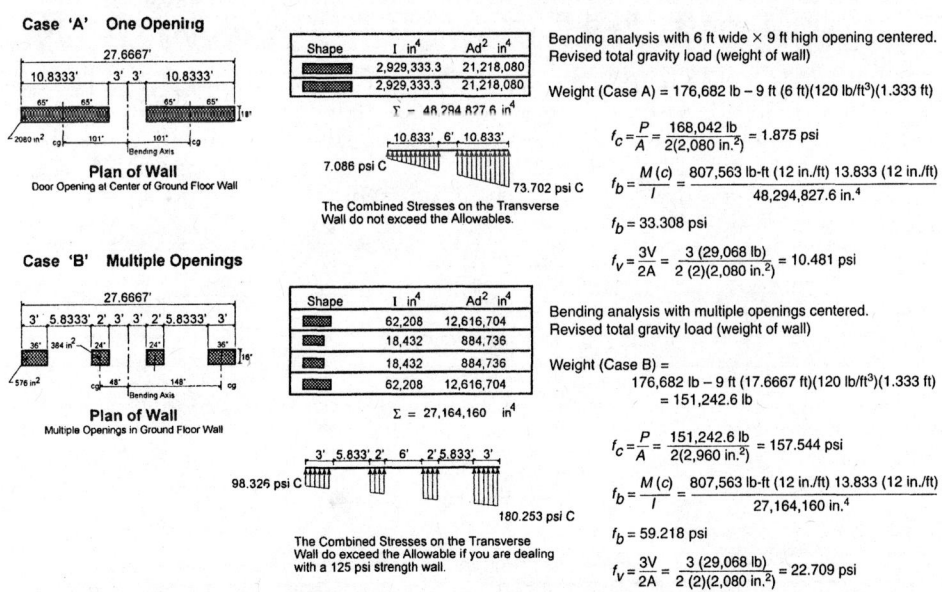

The transverse walls in the example are spaced 40 ft apart. The lateral loads are all transferred to the transverse or shear walls via the masonry from there to the floor diaphragm and thus to the transverse walls.

The combined stresses for Case B with the multiple openings are excessive on the compressive side even though the allowable stresses are increased by one-third when wind and gravity loads are combined. If the allowable were 125 psi, the magnified allowable stress would be 166.7 psi, which is still below the actual compressive stress.

The calculations are based upon the wall being solid masonry above the ground floor, which would almost never be the case. Openings on upper floors would lower the pure compressive stress, which would have an effect on the maximum compressive stress.

Figure 3-40. Analysis: transverse walls with window and door openings.

increase as the text becomes more recent, as from 1902 towards the 1920s, and the level of increase does somewhat reflect the experience component as well as the quality of the brick and the mortar mixing process. Professional judgment must be utilized here. If the building was constructed from the middle of the 19th century to approximately the turn of the century, one should be more conservative. If the building was constructed from about the turn of the twentieth century and later, higher allowables would be justified. If the building is located in a major metropolitan area, then the brick and the mortar would be more uniform at an earlier date. The uniformity and hardness of the brick and the density, uniformity, and condition of the mortar would certainly need to be addressed. In most instances, if the building is sound, a laboratory analysis will not be necessary. A visual inspection will be sufficient.

Masonry Walls, One-Way vs. Two-Way Moments

The masonry wall is usually analyzed as a one-way moment system. A vertical section through the wall, 1 ft wide, is analyzed for lateral loads perpendicular to the wall. The moment resistance of the wall is based upon the ability of the 1-ft section to act as a beam and resist the lateral load through flexural action. Any gravity load on the wall, dead or live, contributes to the level of stress in the wall by providing a constant compressive stress to be combined with the bending stresses. The preceding example of a three-story building in this chapter illustrates a moment combination of that type. This example computed moments and stresses for a wall that was 13.5 ft tall and 40 ft long, with no intermediate supporting system for either direction. The analysis utilized one-way moments as required by the ratio of length of wall to height of wall of 2.963. The principle of two-way moments accepts the fact that there is inherent strength in the wall in its horizontal plane. In concrete, the cutoff for two-way slab design is when the ratio of the length of spans reaches 2 and higher. If the ratio of span to span is less than 2, the slab can resist moment loading in two directions. Many engineers of Eastern and Western Europe utilize the same theory in the analysis of lateral loading on masonry walls. If the length to height ratio is less than 2, the wall can resist moments in two planes, vertical and horizontal. The European method of computing the moments for two-way analysis is the most direct method available (see Figure 3-41).

If the European method is used to check the capacity of masonry walls under lateral load (wind load) and the designer desires to use moment continuity in the wall (not design as simple span), the factors or percentages shown apply to the simple span maximum moment at center of span. The end moments are negative and are the difference between the factored midspan moment and the simple span midspan moment.

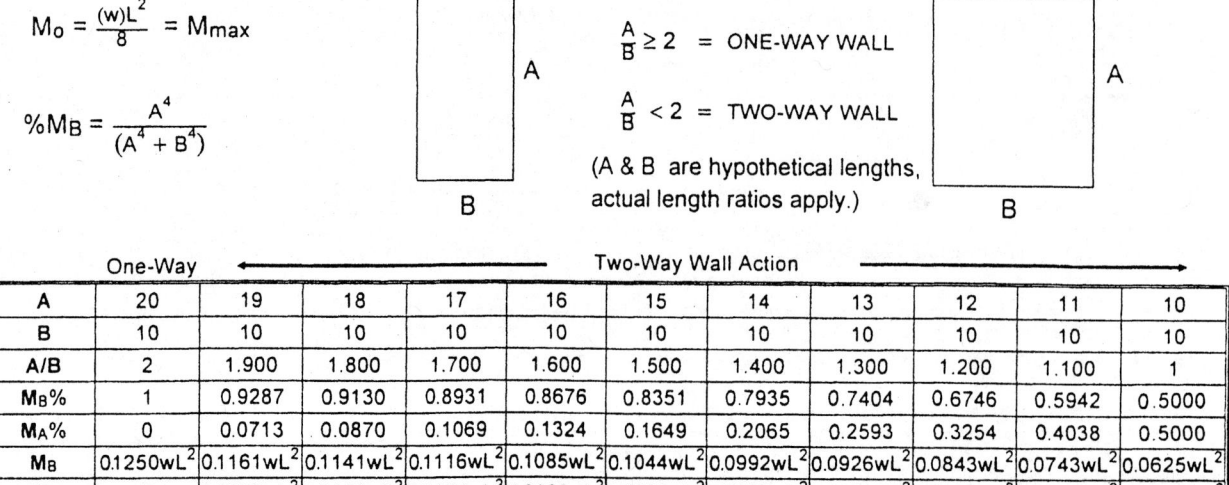

$$M_0 = \frac{(w)L^2}{8} = M_{max}$$

$$\%M_B = \frac{A^4}{(A^4 + B^4)}$$

$\frac{A}{B} \geq 2$ = ONE-WAY WALL

$\frac{A}{B} < 2$ = TWO-WAY WALL

(A & B are hypothetical lengths, actual length ratios apply.)

	One-Way					Two-Way Wall Action					
A	20	19	18	17	16	15	14	13	12	11	10
B	10	10	10	10	10	10	10	10	10	10	10
A/B	2	1.900	1.800	1.700	1.600	1.500	1.400	1.300	1.200	1.100	1
$M_B\%$	1	0.9287	0.9130	0.8931	0.8676	0.8351	0.7935	0.7404	0.6746	0.5942	0.5000
$M_A\%$	0	0.0713	0.0870	0.1069	0.1324	0.1649	0.2065	0.2593	0.3254	0.4038	0.5000
M_B	$0.1250wL^2$	$0.1161wL^2$	$0.1141wL^2$	$0.1116wL^2$	$0.1085wL^2$	$0.1044wL^2$	$0.0992wL^2$	$0.0926wL^2$	$0.0843wL^2$	$0.0743wL^2$	$0.0625wL^2$
M_A	0	$0.0009wL^2$	$0.0109wL^2$	$0.0134wL^2$	$0.0166wL^2$	$0.0206wL^2$	$0.0258wL^2$	$0.0324wL^2$	$0.0407wL^2$	$0.0505wL^2$	$0.0625wL^2$

Figure 3-41. Two-way masonry wall analysis, European method. The computations herein consider the spans to be simply supported.

Example:

$A = 25$ ft
$B = 15$ ft
Building has a floor above and a floor below; therefore, continuity in the wall exists.
Ratio = 1.6667

$$\%M_B = \frac{A^4}{A^4 + B^4} = \frac{25^4}{25^4 + 15^4} = 0.8853$$

$$\%M_A = \frac{B^4}{A^4 + B^4} = \frac{15^4}{25^4 + 15^4} = 0.1147$$

Beam continuity:

$M_{end} = -0.100wL^2$ (Fixed end and center of span moments for the middle span of a continuous member over three spans)
$M_{center} = +0.025wL^2$

Design moment, ends, horizontal direction = $-0.100wL^2(0.1147) = -0.01147wL^2$
Design moment, midspan, horizontal direction = $0.025wL^2(0.1147) = 0.00287wL^2$
Design moment, ends, vertical direction = $-0.100wL^2(0.8853) = -0.0885wL^2$
Design moment, midspan, vertical direction = $+0.025wL^2(0.8853) = +0.0221wL^2$

MASONRY PIERS, AND MASONRY COLUMNS

The capacity of masonry piers and masonry columns is not adequately covered in most early reference books. This type of work, mainly piers and large abutments, was sized empirically or on the basis of allowable compressive stresses only until approximately the turn of the twentieth century. Masonry piers or abutments below grade would not be subject to buckling under the normal axial loading and/or eccentric vertical loading. Above-grade, freestanding masonry piers and columns would have to have been designed with consideration for their height to size ratios. Cast iron columns had been common since the 1870s, but designers relied upon manufacturers to give them the allowable loads that corresponded with the length specified.

Masonry piers and masonry columns had to be designed to sustain the imposed vertical axial loads and/or eccentrically applied vertical loads in the same manner as modern columns. In addition, where applicable, lateral loads such as wind and seismic would produce additional stresses that had to be considered in combination with the vertical loads. Most codes allow the designer to increase the allowable stresses when wind forces are included in the computations. In the case of historic masonry, it is suggested that no increase be allowed when seismic forces are considered in addition to the normal live and dead loading.

By the method presented by the Building Code Committee of the U.S. Department of Commerce, 1925 and utilized by codes, h over $t < 20$ for walls, columns, $t_{min} = 12$ in. for upper 35 ft, plus 4 in. for every 35 ft (Merriman and Wiggin 1947, 889).

Design of Masonry Columns (American Concrete Institute 1988, 33):

$$P_{all} = F'_a A$$

where:
A = area of column, pier, or wall section
$F'_a = 0.25 f'm (1 - h$ over $140r)^2$ for h over $r < 99$
$F'_a = 0.25 f'm (70r$ over $h)^2$ for h over $r > 99$
$0.25f'm = F_a$ for historic masonry
h = height of column, pier, wall = KL
L = length of column, pier, wall
K = effective length factor, usually 1.0 (ends unrestrained)
t = thickness of wall, dimension of pier, column
b = length of wall, width of pier, column
r = radius of gyration of column, pier, wall = $\sqrt{I/A}$

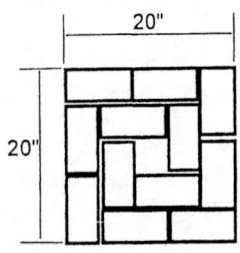

Figure 3-42.

$$I = \frac{bt^3}{12} \quad A = bt$$

Example:
Brick column: good hard brick, lime-sand mortar, F_a = 150 psi (allowable compressive strength) joints raked 0.375"

$b = 20"$

$t = 20"$

$L = 16' = 192"$

$A = (19.25")^2 = 370.563$ in.^{2}

$$I = \frac{(19.25")^4}{12} = 11,443.0 \text{ in.}^4$$

$$r = \sqrt{\frac{11,443.0 \text{ in.}^4}{370.563 \text{ in.}^2}}$$

$r = 5.557$ in.

$\frac{h}{r} = \frac{192 \text{ in.}}{5.557 \text{ in.}} = 34.551$, which is < 99; therefore, the first equation applies for F'_a

$$F'_a = F_a \left(1 - \left(\frac{h}{140r}\right)^2\right)$$

$$= 150 \text{ psi} \left(1 - \left(\frac{192}{778}\right)^2\right) = 140.864 \text{ psi}$$

$P_{all} = F'_a(A) = 140.864$ psi $(370.563$ in.$^2)$

$\quad = 52,211.6$ lb for the 16 ft tall, 20 × 20 in. column

Below-grade piers would not require a stress reduction computation if the soil surrounding the pier was undisturbed and was considered adequate to contain the per, that is, hold the pier in a true vertical or undisturbed vertical axis. And if the full area of the pier 20 × 20 in. were utilized (mortar not raked as in an exposed brickwork) and an adjusted allowable F_a also utilized, P_{all} would be 60,000 lb.

Eccentrically loaded masonry piers, masonry columns, and masonry walls (unreinforced) would require the designer to check the pier, column, or wall for the combined stresses as brought about by the axial load and applied moment on the wall as produced by the eccentricity of the concentrated load. Eccentric axial compression on a pier, column, or wall would produce compressive and bending stresses on the masonry. American Concrete Institute (1988, 530-9) contains the following formulae:

Limiting compressive stress would be $\frac{f_a}{F_a} + \frac{f_b}{F_b} \leq 1$.

(which controls for small eccentricities)

Limiting tensile stress would be $-f_a + f_b$ must be $<F_t$,

the allowable tensile stress (which controls for larger eccentricities)

CAST IRON COLUMNS AND WROUGHT IRON COLUMNS

Cast iron columns were in use in England and France in many applications in the early eighteenth century. Both countries were quick to learn of the disadvantages of cast iron when subjected to tension forces or in bending applications due to the tension zone of the flexural stresses. Cast iron production required little precision and was easy to manufacture. If contained a high quantity of carbon, approximately 4 percent. Wrought iron contains almost no carbon; 0.1 percent is average to high. Wrought iron is therefore very ductile and is capable of carrying tension loads. Cast iron members would be combined with wrought iron members in the construction of many buildings in the second half of the nineteenth century. Designers quickly utilized cast iron for the compression members and wrought iron for tension members in trusses for roof structures, etc. The need for fireproof construction, which became critical in the early part of the last half of the nineteenth century, caused designers to be quick in utilizing these new members and new methods—any noncombustible material would have been embraced with equal enthusiasm. The production of wrought iron rails for railroad trackage was extensive at the midpoint of the nineteenth century, and it was logical that building designers would experiment with rail for floor beams.

Early load-bearing masonry commercial buildings utilized wood joists as floor-carrying members. Building sizes or dimensions between load-bearing walls were limited to the spanning capacity of the wood joists, usually approximately 16–18 ft. Each end of the wood joists would bear on the load-bearing masonry walls. As the need for larger open spaces in these commercial buildings grew, designers began to utilize heavy wood girders and a row of large wood columns at the middle of the space, and the load-bearing walls were now spread to as much as 35 ft or more apart. Joists would then span from a bearing wall on one end to the girder at their other end, the girder being at the center of the space and running parallel to the load-bearing wall. In the larger cities, cast iron columns replaced many of the wood columns as the product became available.

Many foundries were producing cast iron building fronts by the late 1850s. These building fronts were very elaborate, with columns and spandrels that were assembled with bolts and nuts. Cornices at several levels, including the decorative cornice above the roof line, were all of cast iron. The entire fronts of many commercial buildings were of cast iron and glass, which provided merchants with very attractive storefronts. These fronts would carry their own dead loads and could also carry the loads from girder ends at multiple levels. Theoretically, the cast iron front could not be considered a shear wall to provide the stiffness required to brace the front section of the buildings laterally. However, many of the building fronts did just that. In the case of a block of several buildings adjoining each other, the lateral loads would be minimal and would be dissipated over the length of several cast iron fronts.

Cast iron column sections were manufactured at foundries under some measure of control. The columns and other accessories were cast in molds. During the cooling period, cast iron forms a fiber that runs perpendicular to the surface. This is one way of identifying cast iron. Another is the fact that the thickness of adjoining planes is kept relatively constant because thinner planes adjoining each other will cool unevenly and cracking will occur. In addition, corners formed by intersecting planes will always be rounded on cast iron members in the form of a fillet. This is also a matter of crack prevention. The usual cross-section of cast iron columns is shown in Figure 3-43. Decorative bases and capitals would ordinarily be items incorporated into the molds of the round or square sections. The "H," "cross," and "channel" sections would be used in less decorative applications, such as industrial warehouses and other concealed areas. In some instances, these sections would be incorporated into the cast iron storefronts and facades of buildings. Solid columns were rarely used except on small-diameter short sections due to the fact that larger hollow sections would carry more load and use a smaller volume of iron and were thus more economical. Column connections, brackets, and bases were all required as a part of the assembly of the cast iron column. The casting as such cannot be altered after it is cast and cools to ambient temperatures. Therefore, beam seats, brackets, and bolting lugs are drilled at the position required to match the webs of wrought iron beams, etc. See Figures 3-44, 3-45, and 3-46 for examples of these details that were cast separately or cast as a part of the column. Welding cannot be performed on cast iron, and therefore all brackets, lugs, etc. must be designed and cast as a part of the original section.

Reference books of the period and into the twentieth century state that there was practically no limit to the use of cast iron as a building material (International Library

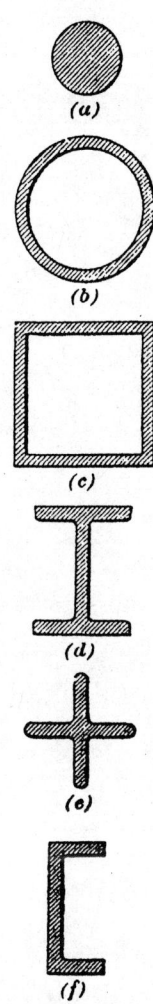

Figure 3-43. Cast iron sections. (*From International Correspondence Schools 1905, para. 15, 33.*)

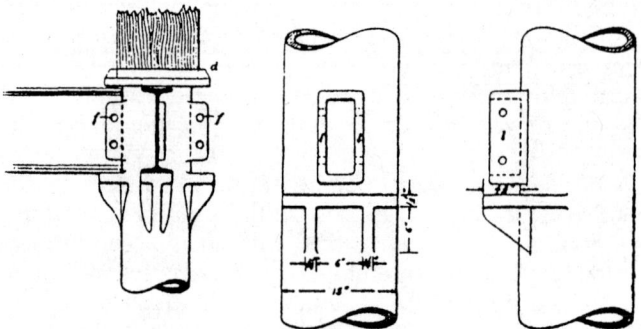

Figure 3-44. Cast iron column connections, brackets. (*From International Correspondence Schools 1905, para. 15, 42.*)

of Technology 1923, para. 47, 29). Except for the problems of tension stress and labor-intensive production that finally eliminated cast iron, the product was extremely durable and could be molded to almost any shape or form. It was easy to provide openings at the back of sections to allow for room to bolt and assemble units for storefronts, window bays, and cornices (see Figure 3-47). Cast iron elements were used on buildings as cornices, finials, window and door hoods, and complete window and door units. The uses of cast iron for purposes other than building details and structural components were vast, from boilers to bathtubs and all of the fittings in between. Cast iron columns were one of the first manufactured building components that became available for purchase for a specific need. A designer would calculate the service load and height of column and, using those parameters, select the type and size required. Many of the scholars of the period tested cast iron columns as derived the formulae used by the manufacturers. Trautwine published *The Civil Engineer's Pocketbook-Book* in 1882, with updates and revised issues through 1888. He presents the following values for the average crushing loads for metals (Trautwine 1888, 438, 439):

Cast iron, average crushing strength = 100,000 psi (compression)
    Average tensile strength = 14,000 psi
    Average elastic limit = 33,600 psi (unit strain = 0.00270 in./in.)
Wrought iron, average elastic limit = 29,120 psi

Trautwine also presents the formula of Professor Lewis Gordon of Glasgow, Scotland, which was derived from a combination of laboratory and full-size loading experiments. F_{cr} as calculated is the compressive stress at failure.

$$F_{cr} = \frac{c}{1 + \frac{L^2}{r^2(a)}}$$

where:
    C = constant, depending upon the material

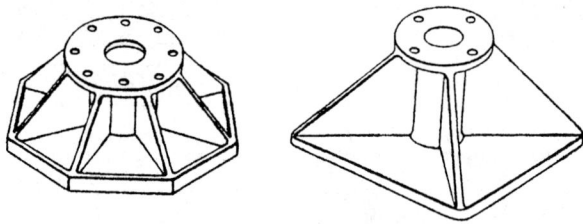

Figure 3-45. Cast iron column bases, connections. (*From International Correspondence Schools 1905, para. 15, 38.*)

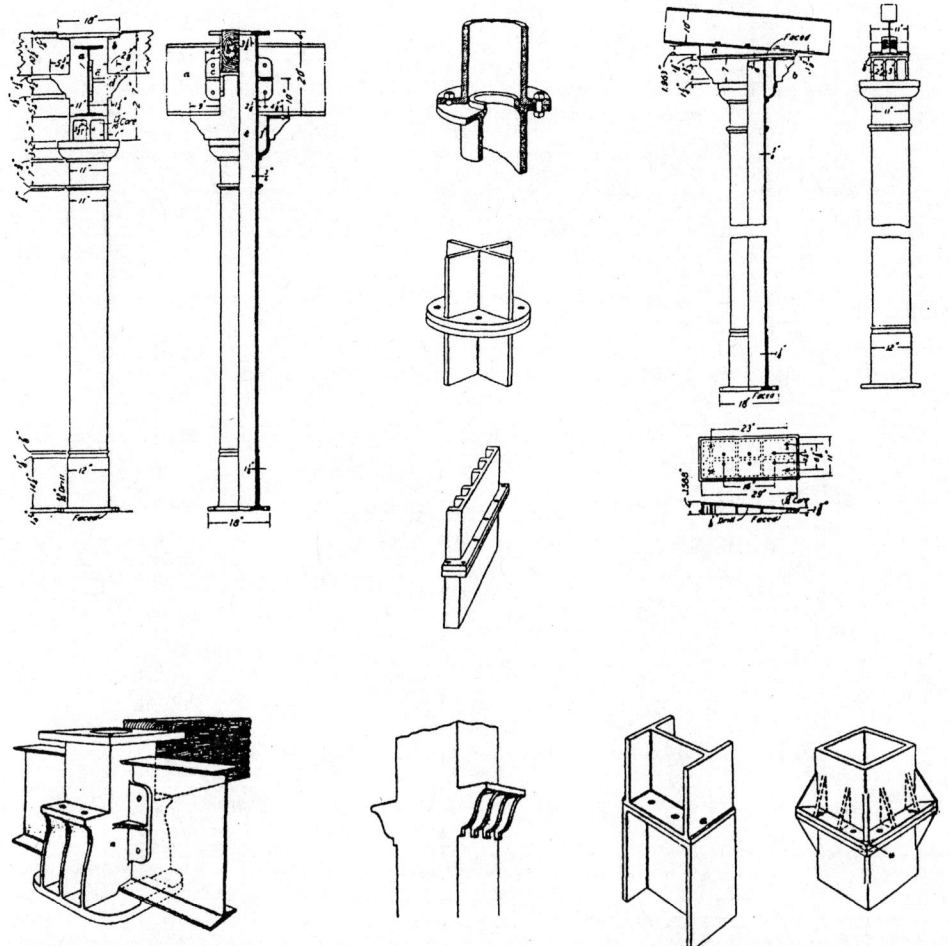

Figure 3-46. Cast iron column details, connections, brackets. (*From International Correspondence Schools 1905, para. 15, 36, 39, 42, 43.*)

L = Unsupported length of column, in.
r = Least radius of gyration, in.
a = end condition constant by material type

The Gordon formula determined the ultimate compressive stress for the material and type of column being designed. For the allowable load on a given column section to be calculated, the ultimate compressive stress must be multiplied by the area of the column and that product divided by the factor of safety (ibid., 439).

For flat ends: factor of safety = 8
For hinged ends: factor of safety = 8

It should be noted that the designers of the day differed widely in their opinions on what the factor of safety should be. Factors in use varied from 15 to 5.

Kidder gives a set of formulae to determine the safe load on cast iron columns. These formulae are known as Gordon's and Rankine's (Kidder 1902, 249).

For solid cylindrical cast iron columns:

$$P_{all} = \frac{13330\,(A)}{1 + \dfrac{L^2}{266\,(d^2)}}$$

For hollow cylindrical cast iron columns:

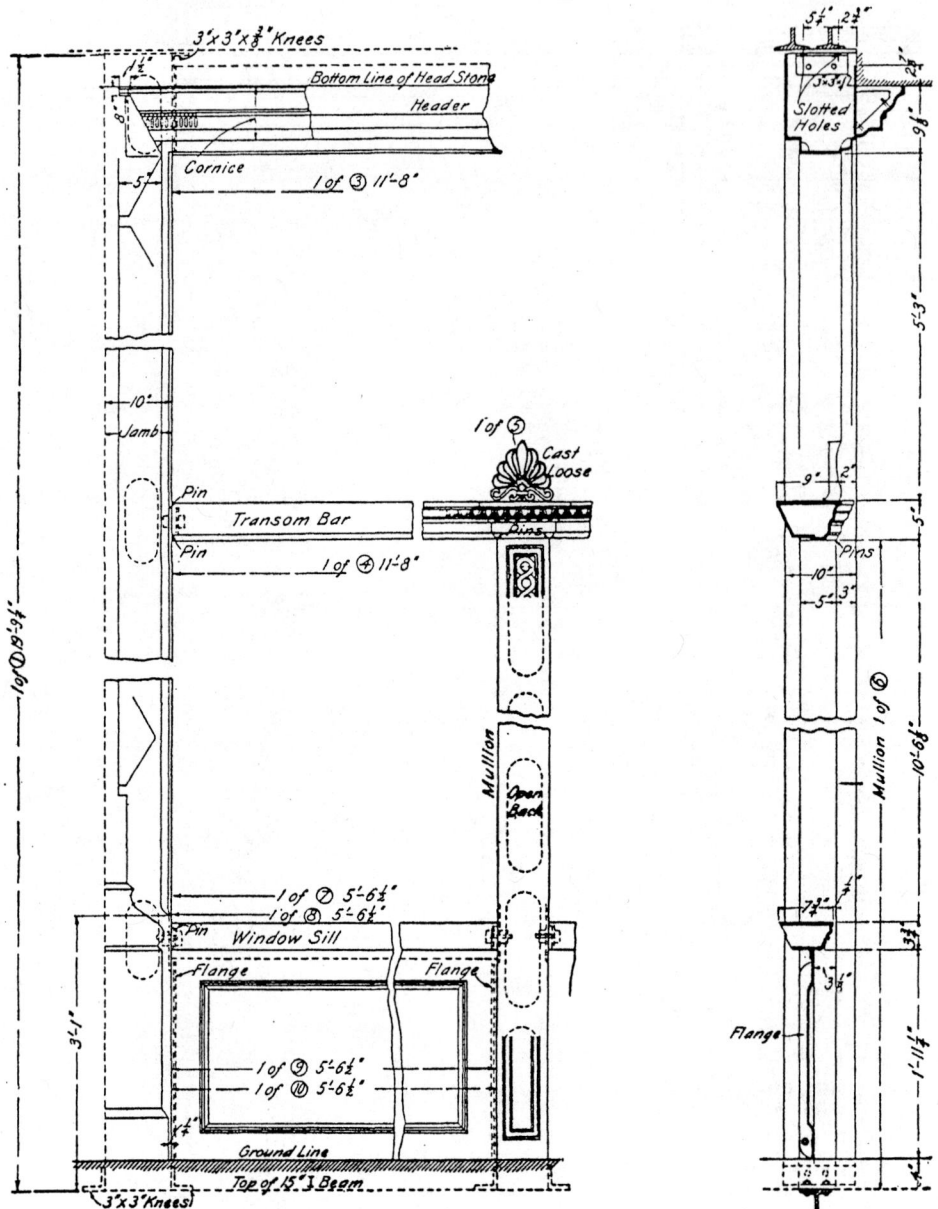

Figure 3-47. Cast iron building fronts. (*From International correspondence Schools 1923, para. 47, 28.*)

Table 3-16.

Column Material	Ends Condition	C	a
Cast iron	Flat ends	80,000	4,750
	Ends hinged	80,000	2,375
	Flat and hinged	80,000	3,375
Wrought iron	Flat ends	40,000	38,000
	Ends hinged	40,000	19,000
	Flat and hinged	40,000	27,000

$$P_{all} = \frac{13330\,(A)}{1 + \frac{L^2}{400\,(d^2)}}$$

For hollow or solid rectangular cast iron columns:

$$P_{all} = \frac{13330\,(A)}{1 + \frac{L^2}{500\,(b^2)}}$$

For cross-shaped (equal leg) cast iron column:

$$P_{all} = \frac{13330\,(A)}{1 + \frac{L^2}{133\,(br^2)}}$$

where, for each type column:
- A = area of column, in.^{2}
- L = length of column, in.
- d = diameter of column, in.
- b = Least side, in.
- br = breadth, in.

Kidder provides a series of tables that provide the tabular load summaries for cast iron columns based upon the formulas from the Pocket-book (ibid., 250).

Table 3-19 is presented in a rather different manner. The left column, "Length Divided by External Breadth or Diameter," requires that the designer perform a com-

Table 3-17. Strengths of Hollow Cylindrical or Rectangular Cast Iron Columns[a]

Length Divided by External Breadth or Diameter	Breaking Weight (psi)		Safe Load (psi)	
	Cylindrical	Rectangular	Cylindrical	Rectangular
5	75,294	76,190	12,549	12,698
6	73,395	74,627	12,232	12,438
7	71,269	72,859	11,878	12,143
8	68,965	70,922	11,494	11,820
9	66,528	68,846	11,088	11,474
10	64,000	66,666	10,666	11,111
11	61,420	64,412	10,236	10,735
12	58,823	62,111	9,804	10,352
13	56,239	59,790	9,373	9,965
14	53,859	57,471	8,976	9,578
15	51,200	55,172	8,533	9,195
16	48,780	52,910	8,130	8,817
17	46,444	50,697	7,741	8,449
18	44,198	48,543	7,366	8,090
19	42,050	46,457	7,008	7,743
20	40,000	44,444	6,666	7,407
21	38,050	42,508	6,341	7,085
22	36,200	40,650	6,033	6,775
23	34,455	38,872	5,742	6,479
24	32,787	37,174	5,464	6,195
25	31,219	35,555	5,203	5,926
26	29,741	34,014	4,957	5,669
27	28,343	32,547	4,724	5,423
28	27,027	31,152	4,504	5,192
29	25,785	29,828	4,297	4,971
30	24,615	25,571	4,102	4,761
31	23,512	27,310	3,918	4,818
32	22,472	26,246	3,745	4,374
33	21,491	25,172	3,581	4,195
34	20,565	24,154	3,427	4,026
35	19,692	23,188	3,282	3,864

[a] From Kidder 1902, 252.

Table 3-18. Safe Loads for Cylindrical Cast Iron Columns[a]

A. Thickness of Shell ¾ in.

Length of Column	Diameter of Column (Outside) (in.)							
	6	7	8	9	10	11	12	13
	Tons	Tons	Tons	Tons	Tons	Tons	Tons	Tons
6	60.6	78.1	94.0	110.8	128.6	144.9	161.7	180.0
7	55.7	72.2	88.9	106.9	124.2	140.1	156.4	176.0
8	50.7	66.3	83.8	101.1	117.7	135.2	151.1	170.3
9	45.8	61.9	78.7	95.2	113.4	130.4	145.8	164.5
10	40.8	56.0	73.5	89.4	106.8	123.2	140.5	158.7
11	37.1	51.5	68.4	83.6	100.1	118.3	135.2	153.0
12	33.4	47.1	63.3	79.7	95.9	113.5	129.9	147.2
13	30.9	44.2	58.1	73.9	89.4	106.3	124.6	141.4
14	27.2	39.8	54.7	70.0	85.0	101.4	119.2	135.6
15	24.7	36.8	49.6	64.1	78.5	96.6	114.0	129.9
16	22.3	33.9	46.2	60.3	71.9	91.8	108.7	124.1
18	—	29.0	41.0	52.5	67.6	84.5	103.4	118.3
20	—	24.4	36.0	44.7	63.3	77.2	98.1	112.5
Metal Area of Cross-Section (in.²)	12.37	14.73	17.10	19.44	21.80	24.15	26.51	28.86

B. Thickness of Shell 1 in.

Length of Column	Diameter of Column (Outside) (in.)							
	6	7	8	9	10	11	12	13
	Tons	Tons	Tons	Tons	Tons	Tons	Tons	Tons
6	77	100	121	143	167	188	211	234
7	71	92	118	138	161	182	204	230
8	64	85	108	131	153	176	197	222
9	58	79	101	123	147	170	190	215
10	52	72	95	116	138	161	183	207
11	47	66	88	108	130	154	175	200
12	42	60	81	102	124	147	169	192
13	39	57	75	95	116	138	162	184
14	35	52	69	90	110	132	155	177
15	31	47	64	83	104	126	148	170
16	28	43	59	78	96	119	142	162
18	25	39	53	68	88	105	128	151
20	22	35	46	58	79	94	114	136
Metal Area of Cross-Section (in.²)	15.71	18.82	22.00	26.14	28.27	31.41	34.58	37.70

C. Thickness of Shell 1¼ in.

Length of Column	Diameter of Column (Outside) (in.)							
	6	7	8	9	10	11	12	13
	Tons	Tons	Tons	Tons	Tons	Tons	Tons	Tons
6	91	119	145	173	203	230	257	286
7	84	111	137	167	196	222	249	281
8	76	102	130	158	186	214	241	272
9	69	94	122	149	179	206	232	263
10	62	86	114	140	168	198	224	253
11	56	79	106	131	158	188	215	245
12	50	73	98	123	151	180	207	235
13	45	68	90	116	141	170	199	226
14	41	62	83	109	134	161	190	217
15	37	56	77	100	127	153	181	208
16	33	52	71	94	117	144	173	198
18	—	45	64	82	106	129	156	184
20	—	—	56	69	96	114	139	166
Metal Area of Cross-Section (in.²)	18.65	22.58	26.52	30.44	34.36	38.29	42.22	46.14

Table 3-18. (Continued)

D. Thickness of Shell 1½ in.

Length of Column	Diameter of Column (Outside) (in.)							
	8	9	10	11	12	13	14	15
	Tons	Tons	Tons	Tons	Tons	Tons	Tons	Tons
6	168	201	236	269	302	336	365	401
7	159	194	228	259	292	330	359	391
8	150	184	219	251	282	320	348	381
9	141	173	211	242	272	309	338	369
10	132	163	196	230	262	298	330	359
11	123	152	184	219	252	287	319	353
12	113	143	176	210	242	276	310	343
13	104	134	164	197	232	265	302	331
14	96	126	156	188	223	255	288	318
15	89	117	148	179	213	244	278	310
16	83	110	136	170	203	233	264	299
18	73	95	124	149	183	217	245	279
20	64	81	112	134	163	195	224	258
Metal Area of Cross-Section (in.²)	30.63	35.34	40.06	44.77	49.48	54.19	58.90	63.62

E. Thickness of Shell 2 in.

Length of Column	Diameter of Column (Outside) (in.)							
	8	9	10	11	12	13	14	15
	Tons	Tons	Tons	Tons	Tons	Tons	Tons	Tons
6	207	251	296	339	383	428	467	514
7	200	242	286	328	371	421	459	502
8	185	229	271	316	358	408	445	490
9	173	215	258	305	345	393	432	474
10	162	202	245	291	333	380	422	462
11	151	189	231	277	320	366	409	450
12	141	178	221	265	307	352	396	438
13	131	167	206	249	295	339	383	424
14	122	158	196	237	283	325	369	412
15	112	145	183	226	270	311	354	398
16	102	136	170	215	257	297	339	384
18	90	119	155	189	232	276	314	359
20	79	101	142	170	207	249	286	332
Metal Area of Cross-Section (in.²)	37.70	43.98	50.266	56.55	62.84	69.11	75.40	81.68

ᵃFrom Kidder 1902, 253–255. Note: If the *breaking load* is desired, multiply the safe load by 6.

putation before entering the table, breadth being the least dimension of the section if rectangular. The table also gives breaking load and save load for the columns. The factor of safety is 6 for the table.

Example

10-in. diameter cast iron column, thickness of column = 1 in.

Length = 15 ft (180 in.)
Determine safe axial load:

$$\frac{L}{d} = \frac{180 \text{ in.}}{10 \text{ in.}} = 18$$

$F_a = 7366$ psi (from Table 3-19)
Safe load $P_a = F_a$ (area) = 7366 psi (28.27 in.²) = 208,236 lb or 104 tons

Check F_a on Table 3-19, check P_a in Table 3-20.

Table 3-19. Safe loads for H-Shaped Cast Iron Columns in Tons of 2000 lb[a]

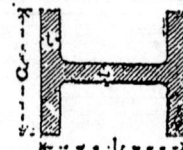

Size of Post (in.)			Area (in.)	Length of Post (ft)							
a.	b.	t.		10	12	13	14	15	16	18	20
6 × 6	×	¾	12½	40	33	31	27				
"	"	1	16	52	43	40	35				
"	"	1¼	19⅜	64	52	48	42				
6 × 8	×	¾	13⅞	45	37	34	30				
"	"	1	18	59	48	45	39				
"	"	1¼	21¼	72	59	54	49				
7 × 7	×	1	19	72	59	57	54	47			
"	"	1¼	23⅛	88	71	69	66	58			
7 × 9	×	1	21	80	65	63	60	52			
"	"	1¼	25⅝	97	79	77	72	64			
8 × 8	×	¾	16½	72	60	55	52	48	45		
"	"	1	22	93	79	72	68	64	59		
"	"	1¼	26⅞	114	96	88	83	78	72		
8 × 10	×	1	24	102	86	79	74	69	65		
"	"	1¼	29⅜	125	105	97	91	85	79		
"	"	1½	34½	146	124	114	107	100	93		
9 × 9	×	1	25	116	100	96	90	82	75	67	
"	"	1¼	30⅝	142	122	118	110	102	92	82	
"	"	1½	36	167	144	138	129	120	108	97	
9 × 10	×	1	26	121	104	100	94	86	78	70	
"	"	1¼	31⅞	148	127	123	115	106	96	86	
"	"	1½	37½	174	150	144	135	125	112	101	
10 × 10	×	1	28	137	123	116	109	102	96	85	
"	"	1¼	34⅜	168	151	142	134	126	118	104	
"	"	1½	40½	198	178	168	158	149	139	123	
"	"	1¾	46⅜	227	204	192	180	170	160	141	
10 × 12	×	1	30	147	132	124	117	110	103	91	
"	"	1¼	36⅞	180	162	153	144	135	127	112	
"	"	1½	43½	213	191	180	169	160	150	132	
"	"	1¾	49⅞	244	219	207	194	183	172	151	
"	"	2	56	274	246	232	218	206	193	170	
12 × 12	×	1	34	180	166	159	152	144	138	125	118
"	"	1¼	41⅞	221	205	196	187	178	170	154	139
"	"	1½	49½	262	242	232	221	210	201	182	164
"	"	1¾	56⅞	301	278	266	254	242	231	209	189
"	"	2	64	339	313	300	286	272	260	235	213
13 × 14	×	1¼	44⅜	235	217	208	198	189	180	163	147
"	"	1½	52½	278	257	246	235	223	213	193	174
"	"	1¾	60⅜	320	295	283	270	257	245	222	201
"	"	2	68	360	332	318	304	288	276	250	225
"	"	2¼	75⅜	398	369	353	337	321	306	277	250

[a] From Kidder 1902, 255c.

1. The area of the 10 in. diameter cast iron column given in the preceding example came from the cross-sectional areas for a 1 in. thick column, Table 3-20.

    Area, 1 in. thick = 28.27 in.^{2}

2. Table 3-20 gives the safe loads for cylindrical cast iron columns by material thickness. A check of the preceding example is given by looking in the 1 in. thick portion of the table for a 10 in. diameter column, 15 ft long.

Table 3-20. Safe Loads for Hollow Rectangular Cast Iron Columns, in Tons of 2000m lb[a]

Outside Dimensions (in.)	Thickness of Shell (in.)	Sectional Area (in.)	10	12	13	14	15	16	18	20
6 × 6	¾	15¾	58	48	44	40				
" "	1	20	74	61	56	51				
" "	1¼	23¾	87	73	65	61				
6 × 8	¾	18¾	69	58	52	48				
" "	1	24	88	74	67	62				
" "	1¼	28¾	106	88	80	74				
6 × 10	¾	21¾	80	67	61	56	51			
" "	1	28	103	80	78	72	66			
" "	1¼	33¾	124	104	94	87	80			
7 × 7	¾	18¾	78	67	62	58	53			
" "	1	24	100	86	80	74	68			
7 × 9	¾	21¾	91	78	72	67	62			
" "	1¼	28	117	100	93	86	79			
8 × 8	¾	21¾	100	87	81	76	71	65		
" "	1	28	128	112	105	98	92	84		
" "	1¼	33¾	155	135	126	118	110	101		
8 × 10	¾	24¾	113	99	92	86	80	74		
" "	1	32	147	128	120	112	105	96		
" "	1¼	38¾	178	155	145	135	125	116		
8 × 12	¾	27¾	127	111	104	97	90	83		
" "	1	36	165	144	135	126	117	108		
" "	1¼	43¾	201	175	164	153	144	125		
10 × 10	¾	27¾	143	130	123	117	111	105	94	
" "	1	36	186	169	160	151	144	136	122	
" "	1¼	43¾	226	205	194	184	175	166	148	
" "	1½	51	263	239	227	215	204	193	173	
10 × 12	¾	30¾	159	144	137	129	122	116	104	
" "	1	40	206	188	178	168	160	152	136	
" "	1¼	48¾	252	229	217	205	195	185	165	
" "	1½	57	293	267	253	240	228	216	193	
10 × 14	¾	33¾	174	158	150	142	135	128	114	
" "	1	44	227	206	196	185	176	167	149	
10 × 16	1	48	248	225	214	202	192	182	163	
10 × 18	1	52	268	244	231	219	208	197	176	
10 × 24	1	64	330	300	285	270	256	243	217	
12 × 12	¾	33¾	187	174	168	161	154	148	136	124
" "	1	44	244	227	219	210	201	193	177	162
" "	1¼	53¾	298	278	267	256	246	236	217	198
" "	1½	63	349	325	312	300	289	277	254	233
12 × 14	¾	36¾	203	189	182	175	168	161	148	136
" "	1	48	266	248	239	229	220	211	193	177
12 × 16	1	52	288	268	258	248	238	228	210	192
12 × 24	1	68	374	351	338	325	312	299	274	251
14 × 16	1	56	330	308	297	288	278	268	250	235
16 × 16	1	60	354	336	330	324	318	310	291	275
16 × 16	1	64	377	358	352	345	239	330	314	293
18 × 18	1	68	414	401	391	380	374	367	346	333
18 × 24	1	80	488	472	460	448	440	432	408	392

[a] From Kidder 1902, 255d.

P_{all} = 104 tons (which agrees with the manual calculations)

Kidder also provides data on H-shaped cast iron columns and includes details of connections and bases and tables of the safe loads on H-shaped columns. He states, "For skeleton construction, when the height of the building exceeds twice its width, it seems unquestionable that the riveted steel column, 'breaking point' in alternate stories, and with riveted steel connections with the beams and girders, is much the best; but for the larger proportion of the buildings in which iron posts are used, cast-iron possesses advantages which the author believes are not exceeded by the riveted post" (ibid., 253, 254). The H-shaped cast iron columns provided the designers with some advantages over the cylindrical and rectangular hollow sections. Casting flaws were much easier to locate, and all surfaces could be painted. Connections were easier, and the beam loads could in some cases be placed on the columns much closer to the axis of symmetry. Lugs and brackets for connections could be placed on the H section much more easily than on the circular sections. And finally, the H column easily met new requirements for fireproofing of all exposed structural elements. Bolted connections would provide some stability; however, failures in the cast iron sections would occur at low levels of tensile stress. H section cast iron columns were introduced in 1894 and were quickly introduced into the building industry. See Tables 3-21 and 3-22 for H section and hollow rectangular column load tables. Figures 3-48 and 3-49 show additional cast iron column connections and column base details. Connections were the weakness of the cast iron sections. Axial compressive load capacity was not a problem. The cast iron sections had to be cast (fabricated) specially for each beam connection. It was impossible to weld or otherwise alter the connections after casting. The rigidity of the connections was in question from their beginning. Wrought steel sections and structural steel sections were easily fabricated, and connection details were very simple. Semifixity of the connections allowed designers to design for some rigidity.

The Carnegie Steel Company of Pittsburgh, Pennsylvania, was a producer of cast iron, wrought iron, and steel shapes. Carnegie also distributed books containing design data and tables to assist designers in the selection of columns and beams. These Pocket Companions were edited by the chief engineer for the company and were used by designers throughout the distribution area. The Carnegie *Pocket Companion* of 1893 contained the following table on cast iron columns (see Table 3-23).

International Correspondence Schools (1912, 98) gives the following formula for determining the breaking loads of cast iron columns:

$$S = \frac{U}{1 + \frac{L^2}{3600\, R^2}}$$

where:
S = breaking strength of column, psi
U = Ultimate comp. strength of cast iron, psi
L = length of column, in.
R = least radius of gyration

The safe or allowable load on a column is found by multiplying the breaking strength S by the area of the column and dividing by 6 (the factor of safety).

International Correspondence Schools (1912) also has two tables giving the breaking loads for cast iron columns (see Table 3-24).

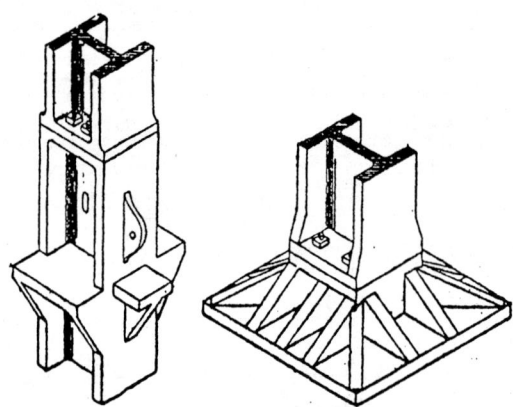

Figure 3-48. Cast iron column details. (*From Kidder 1902, 255b.*)

Figure 3-49. Cast iron column details. (*From International Correspondence Schools 1905, para. 18, 14.*)

Table 3-21. Ultimate Strength of Hollow Cylindrical and Hollow Rectangular Cast iron Columns[a]

Ultimate Strength (psi)

	Cylindrical Columns			Rectangular Columns		
Square Bearing:	Pin & Square:	Pin Bearing:	Square Bearing:	Pin & Square:	Pin Bearing:
$\dfrac{80000}{1 + \dfrac{(12L)^2}{800 d^2}}$	$\dfrac{80000}{1 + \dfrac{3(12L)^2}{1600 d^2}}$	$\dfrac{80000}{1 + \dfrac{(12L)^2}{400 d^2}}$	$\dfrac{80000}{1 + \dfrac{3(12L)^2}{3200 d^2}}$	$\dfrac{80000}{1 + \dfrac{9(12L)^2}{6400 d^2}}$	$\dfrac{80000}{1 + \dfrac{3(12L)^2}{1600 d^2}}$

L = Length of column (ft)
d = External diameter or least side of rectangle (in.)

$\dfrac{L}{d}$	Cylindrical Columns Ultimate Strength (psi)			Rectangular Columns Ultimate Strength (psi)		
	Square Bearing	Pin and Square	Pin Bearing	Square Bearing	Pin and Square	Pin Bearing
1.0	67,800	62,990	58,820	70,480	66,520	62,990
1.1	65,690	60,300	55,730	68,790	64,260	60,300
1.2	63,530	57,600	52,690	67,000	61,940	57,600
1.3	61,340	54,930	49,740	65,140	59,600	54,960
1.4	59,140	52,310	46,900	63,260	57,270	52,320
1.5	56,940	49,770	44,200	61,350	54,960	49,760
1.6	54,760	47,300	41,630	59,450	52,680	47,300
1.7	52,620	44,940	39,210	57,550	50,460	44,960
1.8	50,530	42,670	36,930	55,670	48,300	42,670
1.9	48,490	40,510	34,790	53,800	46,230	40,510
2.0	46,510	38,460	32,790	51,940	44,200	38,460
2.1	44,600	36,520	30,920	50,160	42,260	36,520
2.2	42,750	34,680	29,180	48,400	40,400	34,680
2.3	40,980	32,940	27,540	46,670	38,630	32,950
2.4	39,280	31,310	26,030	44,990	36,930	31,310
2.5	37,650	29,770	24,620	43,390	35,310	29,760
2.6	36,090	28,320	23,300	41,820	33,770	28,320
2.7	34,600	26,950	22,070	40,320	32,310	26,950
2.8	33,180	25,670	20,930	38,870	30,920	25,670
2.9	31,820	24,460	19,860	37,470	29,600	24,460
3.0	30,530	23,320	18,870	36,120	28,340	23,320
3.1	29,310	22,250	17,940	34,830	27,150	22,250
3.2	28,140	21,250	17,070	33,580	26,030	21,250
3.3	27,030	20,300	16,260	32,390	24,960	20,300
3.4	25,970	19,410	15,500	31,240	23,940	19,410

Table 3-21. (Continued)

Outside Diameter (in.)	Thickness of Metal	Safe Loads, in Tons of 2,000 lb, for Hollow Cylindrical Cast Iron Columns[b]								Sectional Area (in.)	Weight (lb) of Columns per ft of Length
		Length of Columns (ft)									
		8 Tons	10 Tons	12 Tons	14 Tons	16 Tons	18 Tons	20 Tons	22 Tons	24 Tons		
6	½	26.2	23.0	20.1	17.5	15.2	13.2	11.5			8.6	26.95
6	¾	37.5	33.0	28.8	25.0	21.7	18.9	16.5			12.4	38.59
6	⅞	42.7	37.6	32.8	28.5	24.7	21.5	18.8			14.1	43.95
6	1	47.6	41.9	36.5	31.8	27.6	24.0	21.0			15.7	49.01
6	1⅛	52.2	46.0	40.1	34.8	30.2	26.3	23.0			17.2	53.76
7	¾	47.7	43.1	38.5	34.3	30.4	26.9	23.9	21.2	18.9	14.7	45.96
7	1	61.1	55.2	49.3	43.8	38.9	34.4	30.6	27.1	24.2	18.9	58.90
7	1⅛	67.2	60.8	54.3	48.3	42.8	37.9	33.7	29.9	26.7	20.8	64.77
8	¾	57.9	53.3	48.6	44.1	39.7	35.8	32.2	28.9	26.1	17.1	53.29
8	1	74.6	68.7	62.5	56.7	51.1	46.0	41.4	37.3	33.6	22.0	68.64
8	1¼	89.9	82.8	75.5	68.4	61.7	55.5	49.9	44.9	40.5	26.5	82.71
9	¾	68.1	63.6	58.9	54.2	49.6	45.2	41.2	37.5	34.1	19.4	60.65
9	1	88.0	82.3	76.2	70.0	64.1	58.4	53.2	48.4	44.1	25.1	78.40
9	1¼	106.6	99.6	92.2	84.8	77.6	70.8	64.4	58.7	53.4	30.4	94.94
9	1½	123.8	115.7	107.1	98.5	90.1	82.2	74.8	68.1	62.0	35.3	110.26
9	1¾	139.6	130.5	120.8	111.1	101.6	92.7	84.4	76.8	69.9	39.9	124.36
10	1	101.4	95.9	89.8	83.6	77.4	71.5	65.8	60.5	55.5	28.3	88.23
10	1¼	123.3	116.5	109.1	101.6	94.1	86.8	79.9	73.4	67.5	34.4	107.28
10	1½	143.7	135.8	127.3	118.5	109.7	101.2	93.2	85.6	78.7	40.1	124.99
10	1¾	162.7	153.8	144.1	134.1	124.2	114.6	105.5	97.0	89.1	45.4	141.65
11	1	114.8	109.4	103.5	97.3	91.0	84.8	80.2	73.1	67.7	31.4	98.03
11	1¼	139.9	133.3	126.1	118.6	110.9	103.3	97.8	89.4	82.5	38.3	119.46
11	1½	163.5	155.9	147.5	138.6	128.7	120.8	114.3	104.1	96.4	44.8	139.68
11	1¾	185.7	177.1	167.5	157.5	147.3	137.2	129.8	118.3	109.5	50.9	158.68
11	2	206.6	196.9	186.3	175.1	163.8	152.6	144.4	131.5	121.8	56.6	178.44
12	1	128.0	122.9	117.2	111.0	104.7	98.4	92.2	86.1	80.4	34.6	107.51
12	1¼	156.4	150.1	143.1	135.7	127.9	120.2	112.6	105.2	98.2	42.2	131.41
12	1½	183.3	175.9	167.7	159.0	149.9	140.9	132.0	123.3	115.1	49.5	154.10
12	1¾	208.7	200.4	191.0	181.1	170.7	160.4	150.3	140.5	131.1	56.4	175.53
12	2	232.7	223.4	213.0	201.9	190.4	178.9	167.6	156.6	146.1	62.8	195.75
13	1	141.2	136.3	130.7	124.7	118.5	112.1	105.8	99.5	93.5	37.7	117.53
13	1¼	172.8	166.8	160.0	152.7	145.0	137.2	129.4	121.8	114.4	46.1	143.86
13	1½	203.0	195.9	187.9	179.3	170.3	161.1	152.0	143.1	134.3	54.2	168.98
13	¾	231.6	223.6	214.5	204.7	194.4	183.9	173.5	163.3	153.3	61.9	192.88
13	2	258.9	249.9	239.7	228.7	217.3	205.5	193.9	182.5	171.3	69.1	215.56
14	1	154.3	149.6	144.3	138.5	132.3	125.9	119.5	113.1	106.8	40.8	127.60
14	1¼	189.2	183.4	176.9	169.7	162.2	154.4	146.5	138.6	131.0	50.1	156.31
14	1½	222.6	215.8	208.1	199.7	190.8	181.7	172.3	163.1	154.1	58.9	183.67
14	1¾	254.4	246.7	237.9	228.3	218.1	207.6	197.0	186.5	176.2	67.4	210.00
14	2	284.8	276.2	266.4	255.6	244.2	232.4	220.6	208.8	197.2	75.4	235.12
15	1	167.4	162.9	157.8	152.1	146.0	139.7	133.3	126.8	120.4	44.0	137.28
15	1¼	205.5	200.0	193.7	186.7	179.3	171.5	163.6	155.7	147.9	54.0	168.48
15	1½	242.1	235.7	228.2	220.0	211.2	202.1	192.8	183.5	174.2	63.6	198.74
15	1¾	277.2	269.8	261.3	251.9	241.9	231.4	220.7	210.1	199.5	72.9	227.45
15	2	310.8	302.5	293.0	282.5	271.2	259.5	247.5	235.5	223.6	81.7	254.90

[a] From Carnegie Steel Co. 1893.
[b] Based on factor of safety of 10 in Gordon's formula.

Carnegie Steel Co. (1913, 280) gives two tables of allowable loads for cast iron columns, based upon the New York City Building Law formula (date of adoption unknown).

Allowable unit stress = $9000 - 40(L/r)$ (in psi)

where:
L = length of column, in.
r = least radius of gyration

The tables in the New York City Building Law are reproduced in Table 3-25.

Table 3-22. Breaking Loads for Cast Iron Columns[a]

Dimension of Side (in.)	Thickness (in.)	Breaking Loads for Square Cast Iron Columns In thousand lb U = 81,000 Length of Column (ft)						
		8	10	12	14	16	18	20
6	½	594						
6	⅝	700	584					
6	⅖	820	685	567				
6	⅞	929	770	638	535			
7	½	775	664	573	494	430		
7	⅝	840	725	625	532	460	395	
7	¾	1,090	935	810	688	589	505	440
7	⅞	1,220	1,060	915	770	660	570	491
7	1	1,368	1,180	1,020	860	780	625	540
8	⅝	1,162	1,040	915	805	708	621	540
8	¾	1,380	1,218	1,060	930	815	710	582
8	⅞	1,550	1,370	1,220	1,065	930	814	700
8	1	1,730	1,530	1,340	1,169	1,018	882	770
8	1⅛	1,900	1,660	1,450	1,248	1,098	950	835
9	⅝	1,391	1,262	1,138	1,013	908	800	715
9	¾	1,643	1,473	1,338	1,192	1,060	940	831
9	⅞	1,861	1,678	1,510	1,340	1,180	1,060	938
9	1	2,100	1,880	1,690	1,500	1,320	1,176	1,040
9	1⅛	2,300	2,070	1,840	1,630	1,450	1,276	1,135
10	¾	1,934	1,784	1,584	1,467	1,328	1,190	1,064
10	⅞	2,184	2,000	1,820	1,643	1,485	1,334	1,200
10	1	2,440	2,250	2,040	1,850	1,660	1,490	1,345
10	1⅛	2,890	2,650	2,400	2,140	1,930	1,730	1,550
10	1¼	3,040	2,710	2,470	2,225	1,992	1,780	1,600
11	⅞	2,510	2,340	2,150	1,950	1,783	1,623	1,484
11	1	2,810	2,650	2,430	2,200	2,020	1,821	1,680
11	1⅛	3,100	2,920	2,670	2,440	2,230	2,030	1,840
11	1¼	3,430	3,180	2,930	2,660	2,440	2,200	1,990
12	1	3,150	2,970	2,770	2,570	2,370	2,180	1,980
12	1⅛	3,520	3,300	3,007	2,850	2,620	2,420	2,230
12	1¼	3,870	3,620	3,370	3,120	2,870	2,640	2,430
12	1⅝	4,170	3,900	3,600	3,350	3,080	2,820	2,570
13	1⅛	3,890	3,690	3,460	3,250	3,140	2,780	2,560
13	1¼	4,290	4,070	3,820	3,580	3,340	3,060	2,830
13	1⅜	4,650	4,400	4,130	3,940	3,570	3,000	3,040
13	1½	5,000	4,740	4,420	4,130	3,830	3,520	3,240
14	1¼	4,740	4,530	4,280	4,000	3,760	3,510	3,260
14	1⅜	5,060	4,870	4,620	4,340	4,050	3,770	3,520
14	1½	5,520	5,270	4,960	4,680	4,870	4,060	3,800
14	1⅝	5,900	5,610	5,300	5,980	4,650	4,330	4,030

[a] From International Correspondence Schools 1912, 101, 102. Both tables are for hollow sections.

The Carnegie Steel Company continued to carry the same two column tables in subsequent editions. The same tables appear in their 1923 edition. The Cambria Steel Company of Philadelphia and Johnstown, Pennsylvania, issued a publication that provides information and data on their shapes and methods of utilizing their shapes in the construction of buildings. The 12th edition, published in 1919, was comprehensive in its treatment of data, loads, and load capacities for the shapes provided by Cambria. George E. Thackray, C. E., the "special engineer" for Cambria, was responsible for the data. These Handbooks were issued to assist designers in the use of the manufacturers products, but each volume contained much more than just the data on shapes and design formulae. The Handbooks averaged about 400 pages and contained a vast array of mathematical and trigonometric tables as well as the live and dead load requirements and allowable unit stresses for cities. The two tables on allowable loads on cast iron columns are reproduced in Table 3-26.

Carnegie Steel Co. (1923) includes the two tables reproduced in Table 3-27 on allowable loads on cast iron columns. The tables are virtually unchanged from the 1913 edition.

Table 3-22. (Continued)

		Breaking Loads for Round Cast Iron Columns In thousand lb U = 81,000						
Dimension of Col. (in.)	Thickness (in.)	Length of Column (ft)						
		8	10	12	14	16	18	20
6	½	418						
6	⅝	505	412					
6	¾	570	465	374				
6	⅞	651	525	423	347			
7	½	560	474	393	332	270		
7	⅝	655	566	476	402	332	287	
7	¾	820	658	544	456	384	324	280
7	⅞	880	732	610	510	425	360	313
7	1	971	810	670	559	470	396	339
8	⅝	850	735	631	542	462	400	352
8	¾	1,000	910	743	640	545	468	405
8	⅞	1,120	960	810	710	620	529	455
8	1	1,260	1,082	930	785	675	578	502
8	1⅛	1,360	1,185	1,005	865	732	625	545
9	⅝	1,040	923	810	710	620	535	480
9	¾	1,220	1,073	950	820	720	628	560
9	⅞	1,360	1,200	1,050	918	805	595	614
9	1	1,550	1,362	1,192	1,039	900	758	682
9	1⅛	1,710	1,500	1,300	1,130	958	851	751
10	¾	1,420	1,280	1,141	1,020	900	798	705
10	⅞	1,640	1,495	1,318	1,170	1,036	910	820
10	1	1,830	1,580	1,460	1,292	1,143	1,010	895
10	1⅛	2,030	1,810	1,470	1,310	1,160	1,110	980
11	¾	1,640	1,510	1,363	1,260	1,120	1,000	895
11	⅞	1,840	1,680	1,540	1,385	1,250	1,113	990
11	1	2,100	1,923	1,740	1,570	1,420	1,258	1,119
11	1⅛	2,300	2,110	1,920	1,720	1,550	1,370	1,217
12	¾	1,869	1,740	1,600	1,440	1,310	1,200	1,080
12	⅞	2,130	1,960	1,800	1,630	1,490	1,360	1,223
12	1	2,380	2,180	2,020	1,840	1,670	1,500	1,360
12	1⅛	2,580	2,350	2,180	1,990	1,800	1,620	1,470
13	¾	2,020	1,900	1,740	1,620	1,480	1,363	1,265
13	⅞	2,400	2,250	2,100	1,920	1,760	1,590	1,465
13	1	2,660	2,480	2,320	2,120	1,940	1,745	1,610
13	1⅛	2,920	2,720	2,540	2,310	2,120	1,900	1,750
14	⅞	2,600	2,450	2,310	2,160	1,970	1,820	1,670
14	1	2,930	2,700	2,580	2,390	2,140	2,090	1,880
14	1⅛	3,250	3,090	2,850	2,670	2,440	2,300	2,070
14	1¼	3,560	3,360	3,140	2,910	2,660	2,440	2,250

The manufacturers of cast iron columns were the developers of the technology. Each manufacturer had its own staff of engineers and metallurgists working on the materials and their use. In addition, each manufacturer hired a leading authority or engineer to develop design procedures, formulae, and tables to assist the designers in the building discipline. Initially each manufacturer worked independently in the development of its products in a competitive environment with little or no interaction with other manufacturers. Cast iron columns and their accessories were among the first building materials to be specified by the designer according to the manufacturer's stated capacity and then purchased by the contractor for the actual point of use. An association of these manufacturers, with standardized elements, would evolve in a relatively short period of time. The sizes of columns, thickness of casting, and other factors were being brought into some conformity at the turn of the century, but each manufacturer had items at the top end or lower end of the product line that were specific to that one manufacturer. A summary of the allowable loads for cast iron columns by manufacturer is presented in a tabular format to allow comparisons between manufacturers and between analytical methods, see Table 3-18.

Table 3-23. Carnegie Cast Iron Column Tables[a]

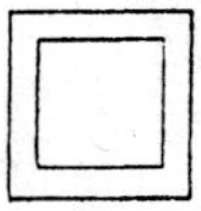

A. Square Cast Iron Columns
ALLOWABLE LOADS IN THOUSANDS OF lb.
By Proposed New York Building Law Formula
(in.²) Weights do not include details

Outer Width (in.)	Thickness (in.)	Area (in.²)	Weight per Foot (lb)	Least Radius (in.)	Effective Length of Column in (ft)										
					8	10	12	14	16	18	20	22	24	26	28
6	½	11.00	34.4	2.26	80	76	71								
	⅝	13.44	42.0	2.21	98	92	86								
	¼	15.75	49.2	2.17	114	107	100								
	⅞	17.94	56.1	2.12	129	121	113								
7	⅝	15.94	49.8	2.62	120	114	108	103							
	¾	18.75	58.6	2.57	141	134	127	120							
	⅞	20.44	63.9	2.53	153	145	137	130							
	1	24.00	75.0	2.48	179	170	160	151							
8	¾	21.75	68.0	2.98	168	161	154	147	140						
	⅞	24.94	77.9	2.93	192	184	175	167	159						
	1	28.00	87.5	2.89	215	205	196	187	178						
	1⅛	30.94	96.7	2.84	237	226	216	205	195						
9	⅞	27.44	85.8	3.34	215	208	200	192	184	176					
	1	32.00	100.0	3.29	251	241	232	223	213	204					
	1⅛	35.44	110.8	3.25	277	267	256	246	235	225					
	1¼	38.75	121.1	3.21	302	291	279	268	256	244					
10	1	36.00	112.5	3.70	287	277	268	259	249	240	231				
	1⅛	39.94	124.8	3.65	317	307	296	286	275	265	254				
	1¼	43.75	136.7	3.61	347	336	324	312	301	289	277				
	1⅜	47.44	148.3	3.57	376	363	350	338	325	312	299				
11	1⅛	44.44	138.9	4.06	358	347	337	326	316	305	295	284			
	1¼	48.75	152.3	4.01	392	380	369	357	345	334	322	310			
	1⅜	52.94	165.4	3.97	425	412	400	387	374	361	348	336			
	1½	57.00	178.1	3.93	457	443	429	416	402	388	374	360			
12	1¼	53.78	168.1	4.42	437	426	414	402	391	379	367	356	344		
	1⅜	58.44	182.6	4.37	475	462	449	436	423	410	398	385	372		
	1½	63.00	196.9	4.33	511	497	483	469	455	441	427	413	399		
	1⅝	67.44	210.8	4.29	547	532	516	501	486	471	456	441	426		
13	1⅜	63.94	199.8	4.78	524	511	498	486	473	460	447	434	421	409	
	1½	69.00	215.6	4.74	565	551	537	523	509	495	481	467	453	439	
	1⅝	73.94	231.1	4.69	605	590	575	560	544	529	514	490	484	469	
	1¾	78.75	246.1	4.65	644	627	611	595	579	562	546	530	514	497	
14	1½	75.00	234.4	5.14	619	605	591	577	563	549	535	521	507	493	479
	1⅝	80.44	251.4	5.10	663	648	633	618	603	588	572	557	542	527	512
	1¾	85.75	267.9	5.05	707	690	674	658	641	625	609	593	576	560	544
	1⅞	90.94	284.2	5.01	749	731	714	696	679	662	644	627	609	592	574
15	1⅝	86.94	271.7	5.50	722	707	691	676	661	646	631	616	600	585	570
	1¾	92.75	289.8	5.46	769	753	737	721	704	688	672	655	639	623	606
	1⅞	98.44	307.6	5.41	816	799	782	764	746	729	711	694	676	659	642
	2	104.00	325.0	5.37	862	843	824	806	787	769	750	731	713	694	676
16	1¾	99.75	311.7	5.86	832	816	800	783	767	751	734	718	702	685	669
	1⅞	105.94	331.1	5.82	884	866	849	831	814	796	779	761	744	726	709
	2	112.00	350.0	5.77	934	915	896	878	859	840	822	803	785	766	747
	2⅛	117.94	368.6	5.73	982	963	943	923	903	883	864	844	824	804	785

[a]From Carnegie Steel Co. 1913, 281, 282.

Table 3-23. (Continued)

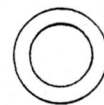

B. Round Cast Iron Columns
ALLOWABLE LOADS IN THOUSANDS OF lb.
By Proposed New York Building Law Formula
(in.²) Weights do not include details

Outer Diam. (in.)	Thickness (in.)	Area (in.²)	Weight per Foot (lb)	Least Radius (in.)	\multicolumn{11}{c}{Effective Length of Column in (ft)}											
					8	10	12	14	16	18	20	22	24	26	28
6	½	8.64	27.0	1.95	61	56									
	⅝	10.55	33.0	1.91	74	68									
	¾	12.37	38.7	1.88	86	80									
	⅞	14.09	44.0	1.84	97	90									
7	⅝	12.52	39.1	2.27	92	86	81								
	¾	14.73	46.0	2.23	107	101	95								
	⅞	16.84	52.6	2.19	122	115	107								
	1	18.85	58.0	2.15	136	128	119								
8	¾	17.08	53.4	2.58	128	122	116	109							
	⅞	19.59	61.2	2.54	147	139	132	124							
	1	21.99	68.7	2.50	164	156	147	139							
	1⅛	24.30	75.9	2.46	181	171	162	152							
9	⅞	22.34	69.8	2.89	171	164	157	149	142						
	1	25.13	78.5	2.85	192	184	175	167	158						
	1⅛	27.83	87.0	2.81	212	203	193	184	174						
	1¼	30.43	95.1	2.78	232	221	211	200	190						
10	1	28.28	88.4	3.20	221	212	204	195	187	178					
	1⅛	31.37	98.0	3.16	244	235	225	216	206	197					
	1¼	34.36	107.4	3.13	267	257	246	235	225	214					
	1⅜	37.26	116.4	3.09	289	277	266	254	243	231					
11	1⅛	34.90	109.1	3.51	276	266	257	247	238	228	219				
	1¼	38.29	119.7	3.48	302	292	281	271	260	250	239				
	1⅜	41.58	129.9	3.44	328	316	305	293	281	270	258				
	1½	44.77	139.9	3.40	352	340	327	314	302	289	277				
12	1¼	42.22	131.9	3.83	338	327	316	306	295	285	274	264			
	1⅜	45.90	143.4	3.79	367	355	343	332	320	308	297	285			
	1½	49.48	154.6	3.75	395	382	369	357	344	331	319	306			
	1⅝	52.97	165.5	3.71	422	408	394	381	367	353	340	326			
13	1⅜	50.22	156.9	4.14	405	394	382	370	359	347	336	324	312		
	1½	54.19	169.4	4.10	437	424	412	399	386	374	361	348	335		
	1⅝	58.07	181.5	4.06	468	454	440	427	413	399	385	372	358		
	1¾	61.85	193.3	4.03	498	483	468	454	439	424	409	395	380		
14	1½	58.91	184.1	4.45	479	467	454	441	429	416	403	390	373		
	1⅝	63.18	197.4	4.41	514	500	486	472	459	445	431	417	404		
	1¾	67.35	210.5	4.38	547	532	518	503	488	473	459	444	429		
	1⅞	71.42	223.2	4.34	580	564	548	532	516	501	485	469	453		
15	1⅝	68.29	213.4	4.76	560	546	532	518	504	491	477	463	449	436	
	1¾	72.85	227.6	4.73	597	582	567	552	537	523	508	493	478	463	
	1⅞	77.31	241.6	4.69	632	617	601	585	569	553	538	522	506	490	
	2	81.68	255.3	4.65	668	651	634	617	600	583	565	550	533	516	
16	1¾	78.34	244.8	5.08	646	631	616	601	587	572	557	542	527	513	498
	1⅞	83.20	260.0	5.04	685	670	654	638	622	606	590	574	559	543	527
	2	87.97	274.9	5.00	724	707	690	673	657	640	623	606	589	572	555
	2⅛	92.63	289.5	4.96	762	744	726	708	690	672	654	636	619	601	583

Table 3-24. Cambria Cast Iron Column Tables[a]

SAFE LOADS IN THOUSANDS OF POUNDS FOR HOLLOW ROUND CAST IRON COLUMNS. SQUARE ENDS.

Based on Gordon's Formula $P = \dfrac{10,000}{1 + \dfrac{L^2}{800\,d^2}}$.

P = safe load in pounds per square inch.
L = length of column in inches.
d = outside diameter of column in inches.

Ultimate compressive strength = 50,000 psi. Safety factor 8. Safe loads for other safety factors than that of the tables may be obtained as follows:

New safe load = Safe load from table $\times \dfrac{8}{\text{New factor}}$.

Outside Diameter (in.)	Thickness (in.)	Length of Column (ft)										Area of Metal (in.²)	Weight per ft (lb)
		6	8	10	12	14	16	18	20	22	24		
6	¾	105	94	82	72	62	54	47	41	36	32	12.4	38.7
	⅞	119	107	94	82	71	62	54	47	41	36	14.1	44.0
7	¾	130	119	108	96	86	76	67	60	53	47	14.7	46.0
	⅞	149	136	123	110	98	87	77	68	61	54	16.8	52.6
8	¾	155	145	133	122	110	99	89	80	72	65	17.1	53.4
	⅞	178	166	153	139	126	114	104	92	83	75	19.6	61.2
	1	200	186	172	158	142	128	115	103	93	84	22.0	68.7
9	⅞	207	196	183	169	156	142	130	118	108	98	22.3	69.8
	1	233	220	206	190	175	160	146	133	121	110	25.1	78.5
	1⅛	258	244	228	211	194	177	162	147	134	122	27.8	87.0
10	⅞	235	225	212	199	185	172	158	146	134	123	25.1	78.4
	1	265	254	240	224	209	194	178	164	151	139	28.3	88.4
	1⅛	294	281	266	249	232	215	198	182	168	154	31.4	98.0
	1¼	323	308	291	273	254	235	217	200	184	169	34.4	107.4
11	1	298	287	273	259	243	227	212	197	183	169	31.4	98.2
	1⅛	330	319	304	287	270	253	235	219	203	188	34.9	109.1
	1¼	363	350	333	315	296	277	258	240	223	206	38.3	119.7
	1⅜	395	380	361	342	322	301	280	261	242	224	41.6	129.9
12	1⅛	368	356	342	326	309	291	274	256	239	223	38.4	120.1
	1¼	404	391	375	358	339	320	300	281	263	245	42.2	131.9
	1⅜	439	425	408	389	369	348	327	306	287	267	45.9	143.4
	1½	473	458	440	419	397	375	352	330	308	288	49.5	154.6
13	1⅛	404	393	379	364	347	330	312	294	277	260	42.0	131.2
	1¼	444	432	417	400	382	363	343	323	304	286	46.1	144.2
	1⅜	484	470	454	435	415	395	373	352	331	311	50.2	156.9
	1½	522	507	490	470	448	426	403	380	358	336	54.2	169.4
14	1¼	485	473	459	442	424	405	386	366	347	327	50.1	156.5
	1⅜	528	515	499	482	462	441	420	399	378	357	54.5	170.4
	1½	570	556	540	520	499	477	454	431	408	385	58.9	184.1
	1⅝	612	597	579	558	535	511	487	462	437	413	63.2	197.4
15	1⅜	573	560	545	528	509	489	467	446	424	406	58.9	183.9
	1½	618	605	589	570	550	528	505	482	459	439	63.6	198.8
	1⅝	664	650	632	612	590	567	542	517	492	471	68.3	213.4
	1¾	708	694	675	653	630	605	579	552	525	502	72.8	227.6
16	1½	666	654	638	620	600	679	557	533	510	486	68.3	213.5
	1⅝	716	702	686	666	645	622	598	573	548	522	73.4	229.3
	1¾	764	750	732	711	689	664	638	611	584	558	78.3	244.8
	1⅞	811	796	777	756	731	705	678	649	621	592	83.2	260.0

[a] From Thackray 1919, 302, 303.

Table 3-24. (Continued)

SAFE LOADS IN THOUSANDS OF POUNDS FOR HOLLOW ROUND CAST IRON COLUMNS. SQUARE ENDS.

Based on Gordon's Formula $P = \dfrac{10{,}000}{1 + \dfrac{L^2}{800\, d^2}}$.

P = safe load in pounds per square inch.
L = length of column in inches.
d = outside diameter of column in inches.

Ultimate compressive strength = 80,000 psi. Safety factor 8. Safe loads for other safety factors than that of the tables may be obtained as follows:

New safe load = Safe load from table × $\dfrac{8}{\text{New factor}}$.

Outside Diameter (in.)	Thickness (in.)	Length of Column (ft)										Area of Metal (in.²)	Weight per ft (lb)
		14	16	18	20	22	24	26	28	30	32		
18	1 5/8	754	732	708	684	659	633	608	596	557	533	83.6	261.2
	1 3/4	806	782	757	732	704	677	650	637	596	569	89.3	279.2
	1 7/8	857	832	805	777	749	720	691	677	633	605	95.0	296.8
	2	907	880	852	823	792	762	731	717	670	641	100.5	314.2
20	1 3/4	922	900	876	850	824	797	769	742	714	687	100.3	313.6
	1 7/8	981	957	932	905	877	848	819	789	760	731	106.8	333.6
	2	1,039	1,014	987	958	929	898	867	836	805	774	113.1	353.4
	2 1/8	1,097	1,070	1,041	1,011	980	948	915	882	849	817	119.3	372.9
22	1 7/8	1,105	1,082	1,058	1,032	1,005	976	947	918	888	859	110.5	370.5
	2	1,171	1,147	1,122	1,094	1,065	1,035	1,004	974	941	910	125.7	392.7
	2 1/8	1,239	1,213	1,186	1,157	1,126	1,094	1,062	1,029	996	962	132.9	415.3
	2 1/4	1,301	1,275	1,246	1,215	1,183	1,150	1,116	1,081	1,046	1,011	139.6	436.3
24	2	1,303	1,280	1,241	1,229	1,201	1,171	1,141	1,110	1,079	1,047	138.2	432.0
	2 1/8	1,376	1,352	1,311	1,298	1,268	1,238	1,206	1,173	1,140	1,106	146.0	456.4
	2 1/4	1,449	1,423	1,380	1,367	1,335	1,303	1,269	1,235	1,200	1,165	153.7	480.4
	2 3/8	1,520	1,494	1,448	1,434	1,402	1,367	1,332	1,296	1,259	1,222	161.4	504.2
26	2 1/8	1,515	1,492	1,467	1,440	1,412	1,382	1,351	1,319	1,286	1,252	159.4	498.1
	2 1/4	1,596	1,572	1,546	1,517	1,487	1,456	1,423	1,389	1,354	1,319	167.9	524.6
	2 3/8	1,675	1,650	1,623	1,593	1,562	1,528	1,494	1,458	1,422	1,385	176.3	550.9
	2 1/2	1,754	1,728	1,699	1,668	1,635	1,600	1,564	1,527	1,489	1,450	184.6	576.8
28	2 1/4	1,742	1,719	1,694	1,667	1,638	1,608	1,576	1,542	1,508	1,474	182.0	568.8
	2 3/8	1,829	1,806	1,780	1,751	1,721	1,689	1,655	1,620	1,584	1,548	191.2	597.5
	2 1/2	1,917	1,892	1,864	1,834	1,802	1,769	1,734	1,697	1,660	1,622	200.3	625.9
	2 5/8	2,002	1,967	1,948	1,917	1,883	1,848	1,811	1,773	1,734	1,694	209.3	653.9
30	2 3/8	1,982	1,961	1,936	1,909	1,879	1,848	1,816	1,782	1,747	1,711	206.1	644.1
	2 1/2	2,078	2,055	2,028	2,000	1,969	1,937	1,903	1,867	1,830	1,793	216.0	675.0
	2 5/8	2,172	2,148	2,119	2,090	2,058	2,024	1,989	1,952	1,913	1,874	215.8	705.5
	2 3/4	2,265	2,240	2,210	2,180	2,147	2,111	2,074	2,035	1,995	1,954	235.4	735.7
32	2 1/2	2,239	2,217	2,192	2,165	2,135	2,104	2,071	2,036	2,000	1,963	231.7	724.0
	2 5/8	2,341	2,318	2,292	2,264	2,233	2,200	2,165	2,129	2,092	2,053	242.2	757.0
	2 3/4	2,442	2,418	2,391	2,361	2,329	2,295	2,259	2,221	2,182	2,141	252.7	789.7
	2 7/8	2,542	2,517	2,489	2,458	2,424	2,389	2,351	2,312	2,271	2,229	263.1	822.1
34	2 5/8	2,511	2,488	2,463	2,436	2,406	2,374	2,341	2,306	2,272	2,232	258.7	808.6
	2 3/4	2,620	2,596	2,570	2,542	2,511	2,478	2,441	2,406	2,370	2,329	270.0	843.7
	2 7/8	2,728	2,703	2,676	2,646	2,614	2,580	2,544	2,505	2,468	2,425	281.1	878.5
	3	2,835	2,810	2,781	2,750	2,717	2,681	2,643	2,604	2,565	2,520	292.2	913.0
36	2 3/4	2,796	2,774	2,749	2,721	2,692	2,660	2,626	2,591	2,553	2,515	287.3	897.7
	2 7/8	2,913	2,889	2,863	2,834	2,803	2,770	2,735	2,698	2,659	2,619	299.2	935.0
	3	3,028	3,003	2,976	2,946	2,904	2,880	2,849	2,805	2,765	2,723	311.0	971.9

Table 3-25. Allowable Loads for Cast Iron Columns[a]

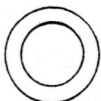

Round Cast Iron Columns
ALLOWABLE LOADS IN THOUSANDS OF POUNDS
By New York Building Law, 1917
Weights do not include details

Outer diameter (in.)	Thickness (in.)	Area (in.²)	Weight per Foot (lb)	Least Radius (in.)	Effective Length of Column in (ft)											
					8	10	12	14	16	18	20	22	24	26	28	
6	½	8.64	27.0	1.95	61	56										
	⅝	10.55	33.0	1.91	74	68										
	¾	12.37	38.7	1.88	86	80										
	⅞	14.09	44.0	1.84	97	90										
7	⅝	12.52	39.1	2.27	92	86	81									
	¾	14.73	46.0	2.23	107	101	95									
	⅞	16.84	52.6	2.19	122	115	107									
	1	18.85	58.9	2.15	136	128	119									
8	¾	17.08	53.4	2.58	128	122	116	109								
	⅞	19.59	61.2	2.54	147	139	132	124								
	1	21.99	68.7	2.50	164	156	147	139								
	1⅛	24.30	75.9	2.46	181	171	162	152								
9	⅞	22.34	69.8	2.89	171	164	157	149	142							
	1	25.13	78.5	2.85	192	184	175	167	158							
	1⅛	27.83	87.0	2.81	212	203	193	184	174							
	1¼	30.43	95.1	2.78	232	221	211	200	190							
10	1	28.28	88.4	3.20	221	212	204	195	187	178						
	1⅛	31.37	98.0	3.16	244	235	225	216	206	197						
	1¼	34.36	107.4	3.13	267	257	246	235	225	214						
	1⅜	37.26	116.4	3.09	289	277	266	254	243	231						
11	1⅛	34.90	109.1	3.51	276	266	257	247	238	228	219					
	1¼	38.29	119.7	3.48	302	292	281	271	260	250	239					
	1⅜	41.58	129.9	3.44	328	316	305	293	281	270	258					
	1½	44.77	139.9	3.40	352	340	327	314	302	289	277					
12	1¼	42.22	131.9	3.83	338	327	316	306	295	285	274	264				
	1⅜	45.90	143.4	3.79	367	355	343	332	320	308	297	285				
	1½	49.48	154.6	3.75	395	382	369	357	344	331	319	306				
	1⅝	52.97	165.5	3.71	422	408	394	381	367	353	340	326				
13	1⅜	50.22	156.9	4.14	405	394	382	370	359	347	336	324	312			
	1½	54.19	169.4	4.10	437	424	412	399	386	374	361	348	335			
	1⅝	58.07	181.5	4.06	468	454	440	427	413	399	385	372	358			
	1¾	61.85	193.3	4.03	498	483	468	454	439	424	409	395	380			
14	1½	58.91	184.1	4.45	479	467	454	441	429	416	403	390	378			
	1⅝	63.18	197.4	4.41	514	500	486	472	459	445	431	417	404			
	1¾	67.35	210.5	4.38	547	532	518	503	488	473	459	444	429			
	1⅞	71.42	223.2	4.34	580	564	548	532	516	501	485	469	453			
15	1⅝	68.29	213.4	4.76	560	546	532	518	504	491	477	463	449	436		
	1¾	72.85	227.6	4.73	597	582	567	552	537	523	508	493	478	463		
	1⅞	77.31	241.6	4.69	632	617	601	585	569	553	538	522	506	490		
	2	81.68	255.3	4.65	668	651	634	617	600	583	566	550	533	516		
16	1¾	78.34	244.8	5.08	646	631	616	601	587	572	557	542	527	513	498	
	1⅞	83.20	260.0	5.04	685	670	654	638	622	606	590	574	559	543	527	
	2	87.97	274.9	5.00	724	707	690	673	657	640	623	606	589	572	555	
	2⅛	92.63	289.5	4.96	762	744	726	708	690	672	654	636	619	601	583	

[a]From Carnegie Steel Co. 1923, 248, 249.

Table 3-25. (Continued)

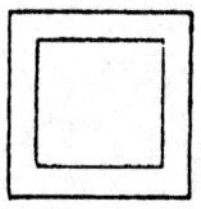

Square Cast Iron Columns
ALLOWABLE LOADS IN THOUSANDS OF POUNDS
By New York Building Law, 1917
Weights do not include details

Outer Width (in.)	Thickness (in.)	Area (in.²)	Weight per Foot (lb)	Least Radius (in.)	8	10	12	14	16	18	20	22	24	26	28
6	½	11.00	34.4	2.26	80	76	71								
	⅝	13.44	42.0	2.21	98	92	86								
	¾	15.75	49.2	2.17	114	107	100								
	⅞	17.94	56.1	2.12	129	121	113								
7	⅝	15.94	49.8	2.62	120	114	108	103							
	¾	18.75	58.6	2.57	141	134	127	120							
	⅞	21.44	63.9	2.53	153	145	137	130							
	1	24.00	75.0	2.48	179	170	160	151							
8	¾	21.75	68.0	2.98	168	161	154	147	140						
	⅞	24.94	77.9	2.93	192	184	175	167	159						
	1	28.00	87.5	2.89	215	205	196	187	178						
	1⅛	30.94	96.7	2.84	237	226	216	205	195						
9	⅞	27.44	85.8	3.34	215	208	200	192	184	176					
	1	32.00	100.0	3.29	251	241	232	223	213	204					
	1⅛	35.44	110.8	3.25	277	267	256	246	235	225					
	1¼	38.75	121.1	3.21	302	291	279	268	256	244					
10	1	36.00	112.5	3.70	287	277	268	259	249	240	231				
	1⅛	39.94	124.8	3.65	317	307	296	286	275	265	254				
	1¼	43.75	136.7	3.61	347	336	324	312	301	289	277				
	1⅜	47.44	148.3	3.57	376	363	350	338	325	312	299				
11	1⅛	44.44	138.9	4.06	358	347	337	326	316	305	295	284			
	1¼	48.75	152.3	4.01	392	380	369	357	345	334	322	310			
	1⅜	52.94	165.4	3.97	425	412	400	387	374	361	348	336			
	1½	57.00	178.1	3.93	457	443	429	416	402	388	374	360			
12	1¼	53.78	168.1	4.42	437	426	414	402	391	379	367	356	344		
	1⅜	58.44	182.6	4.37	475	462	449	436	423	410	398	385	372		
	1½	63.00	196.9	4.33	511	497	483	469	455	441	427	413	399		
	1⅝	67.44	210.8	4.29	547	532	516	501	486	471	456	441	426		
13	1⅜	63.94	199.8	4.78	524	511	498	486	473	460	447	434	421	409	
	1½	69.00	215.6	4.74	565	551	537	523	509	495	481	467	453	439	
	1⅝	73.94	231.1	4.69	605	590	575	560	544	529	514	499	484	469	
	1¾	78.75	246.1	4.65	644	627	611	595	579	562	546	530	514	497	
14	1½	75.00	234.4	5.14	619	605	591	577	563	549	535	521	507	493	479
	1⅝	80.44	251.4	5.10	663	648	633	618	603	588	572	557	542	527	512
	1¾	85.75	267.9	5.05	707	690	674	658	641	625	609	593	576	560	544
	1⅞	90.94	284.2	5.01	749	731	714	696	679	662	644	627	609	592	574
15	1⅝	86.94	271.7	5.50	722	707	691	676	661	646	631	616	600	585	570
	1¾	92.75	289.8	5.46	769	753	737	721	704	688	672	655	639	623	696
	1⅞	98.44	307.6	5.41	816	799	782	764	746	729	711	694	676	659	642
	2	104.00	325.0	5.37	862	843	824	806	787	769	750	731	713	694	676
16	1¾	99.75	311.7	5.86	832	816	800	783	767	751	734	718	702	685	669
	1⅞	105.94	331.1	5.82	884	866	849	831	814	796	779	761	744	726	709
	2	112.00	350.0	5.77	934	915	896	878	859	840	822	803	785	766	747
	2⅛	117.94	368.6	5.73	982	963	943	923	903	883	864	844	824	804	785

Table 3-26. Allowable Loads for Cast Iron Columns, by Manufacturer

Summary of Analytical Methods and Their Results:	10-in. Diameter, Hollow, 1 in. Thick Length = 15 ft, Flat End Conditions $r = 3.202$ in., $A = 28.274$ in.^{2}
Trautwine, 1882 $F_{cr} = 48{,}039.8$ psi, factor of safety = 8 (Gordon formula)	$P_{all} = 169{,}785$ lb
Kidder 1902 (from formula)	$P_{all} = 208{,}228$ lb
Kidder 1902 (from table)	$P_{all} = 208{,}000$ lb
Carnegie 1893 (from formula)	$P_{all} = 160{,}992$ lb
Carnegie 1893 (from table)	$P_{all} = 161{,}000$ lb
ICS 1912, factor of safety = 6 (from formula)	$P_{all} = 203{,}268$ lb
ICS 1912, factor of safety = 6 (from table)	$P_{all} = 202{,}917$ lb
Carnegie 1913 (from formula)	$P_{all} = 190{,}889$ lb
Carnegie 1913 (from table)	$P_{all} = 191{,}000$ lb
Cambria 1919, factor of safety = 8 (from formula)	$P_{all} = 201{,}238$ lb
Cambria 1919, factor of safety = 8 (from table)	$P_{all} = 201{,}500$ lb
Carnegie 1923 (from formula)	$P_{all} = 190{,}889$ lb
Carnegie 1923 (from table)	$P_{all} = 191{,}000$ lb

The more reliable values for the allowable loads are the later ones. It appears that in earlier periods the factor of safety was variable, thus producing erratic allowable loads. If the P_{cr} value of 1,358,276 lb for the Trautwine column were divided by a factor of safety of 7, the allowable load for the column would be 194,000 lb, which is practically the same as the Carnegie 1923 allowable load. The differences in the remainder of the handbooks are minimal. The handbooks also covered wood and steel sections. Steel producers began a very close cooperative effort toward uniform shapes and sizes as early as 1923, with the 1st edition of the A.I.S.C. *Steel Construction Manual*.

WROUGHT IRON COLUMNS

Wrought iron sections and early structural steel sections were coming into use at approximately the same time. A couple of foundries or mills produced wrought iron sections in the 1870s, while most of the others entered production in the 1880s. Steel sections began to be produced in the mid- to late 1880s. In addition, the wrought and steel sections became available approximately 20 years after cast iron members were introduced. At first, the uses and height of buildings appeared to dictate which type of column would be used in a building. Later, approaching the turn of the twentieth century, steel sections for both columns and beams became increasingly popular with designers. Steel provided the advantage of having the same properties in tension and compression and was the best material for flexural bending. Steel was also reliable for connections having a relatively high shear stress. The American Institute of Steel Construction (AISC) produced a volume that gives the dates, dimensions, and properties of steel and wrought iron beams and columns from 1873 to 1952. The volume was first printed in 1953 subsequently reprinted. The sections are catalogued by size and date and manufacturer, and therefore the work can be used as a resource to determine the date at which a shape entered the market. The work also includes a summary of the "unit stresses" (allowable stresses) that were recommended by the manufacturers. The "unit stresses" tabulated for wrought iron sections are as follows (Ferris 1978, 5):

1873	Carnegie Kloman & Co. (factor of safety = 3)	14,000 psi
1874	New Jersey Steel & Iron Co.	12,000 psi
1881–84	Carnegie *Brothers & Co., Ltd.*,	12,000 psi
	″ ″	10,000 psi
1884	Passaic Rolling Mill Co.	12,000 psi
	″ ″	10,000 psi
1885	Phoenix Iron Co.	12,000 psi
1885–87	Pottsville Iron & Steel Co.	12,000 psi
1889	Carnegie, Phipps & Co.	12,000 psi
	″ ″	10,000 psi

Trautwine (1888) contains a comprehensive section on wrought iron columns and the various formulae and tables that were in use in the period. J. D. Whitmore produced a "simple formula" "which was found to agree very closely with the results of experiments on Phoenix Columns." It is shown below (ibid., 442).

$$F_{cr} = [(1{,}200 - H)(30)] + \frac{525{,}000}{H^2}$$

where:
F_{cr} = ultimate stress on column, psi
$H = \dfrac{L, \text{in}}{D, \text{in}}$ = length of pillar/D, outside diameter (both same units)

The factor of safety (F.S.) for wrought iron and steel columns as developed by a Mr. Christie for the Pencoyd Iron Works in 1884 is as follows (ibid.):

For flat and fixed ends:

$$\text{F.S.} = 3 + \left[0.01 \left(\frac{\text{Length, in.}}{r, \text{in.}} \right) \right]$$

For hinged and round ends:

$$\text{F.S.} = 3 + \left[0.015 \left(\frac{\text{Length, in.}}{r, \text{in.}} \right) \right]$$

where:
r = least Radius and Gyration

Trautwine also provides a set of formulae for the four main types of columns and tables of ultimate stresses for wrought iron and early steel columns. These formulae were derived by Charles Shaler Smith, and the table in Table 3-29 is by C. L. Gates. The Phoenix Iron Company produced some very interesting segmental columns, which were rolled in the standard mill method and then riveted together to form their final column shape. A segmental column of four segments with section properties is shown in Table 3-30.

Trautwine also provides tables to indicate the breaking loads or P_{cr} for hollow cylindrical wrought iron columns (see Table 3-31.)

The factor of safety for wrought iron columns by Gates is as follows:

$$\text{F.S.} = 4 + 0.05\,H$$

where:
$H = \dfrac{\text{Length, in}}{d, \text{in}}$ = length between end bearings/least diameter, d

Birkmire (1894) describes the beginnings of the new skyscrapers being constructed in Chicago and New York City. They were the product of the new method of construction, the skeletal steel frame. Birkmire volume lists the prominent buildings that utilized cast iron columns and those that utilized wrought iron and steel columns (see Table 3-32).

It is difficult to differentiate in the literature of the period which column shapes would have been made of wrought iron and which of steel, as well as whether certain shapes were made only of wrought iron and until what date. The handbooks and reference books are not clear as to which sections are of which material, apparently because the two metals closely resemble each other and the only true way of determining if a member is wrought iron or steel is to have the material tested by a commercial laboratory. Wrought iron is more malleable and contains less carbon than steel. The Brinnell and Rockwell hardness tests can be performed on in situ columns or beams without causing harm to the material (Merriman and Wiggin 1947, 613). The Brinnell test produces a scale of relative hardness that can be used to distinguish

Table 3-27. Wrought Iron Column Details[a]

	A	B	C	D
Flat ends	$\dfrac{38{,}500}{1 + \dfrac{H^2}{5{,}820}}$	$\dfrac{42{,}500}{1 + \dfrac{H^2}{4{,}500}}$	$\dfrac{36{,}500}{1 + \dfrac{H^2}{3{,}750}}$	$\dfrac{36{,}500}{1 + \dfrac{H^3}{2{,}700}}$
One pin end	$\dfrac{38{,}500}{1 + \dfrac{H^2}{3{,}000}}$	$\dfrac{40{,}000}{1 + \dfrac{H^2}{2{,}250}}$	$\dfrac{36{,}500}{1 + \dfrac{H^2}{2{,}250}}$	$\dfrac{36{,}500}{1 + \dfrac{H^2}{1{,}500}}$
Two pin ends	$\dfrac{37{,}800}{1 + \dfrac{H^2}{1{,}900}}$	$\dfrac{36{,}600}{1 + \dfrac{H^2}{1{,}500}}$	$\dfrac{36{,}500}{1 + \dfrac{H^2}{1{,}750}}$	$\dfrac{36{,}500}{1 + \dfrac{H^2}{1{,}200}}$

Ultimate and safe loads in psi, of the above four pillars, with flat ends, and equally loaded. Coefficient of safety = 4 + 0.05 H. By C. L. Gates, C. E.

H:	A. Square Column		B. Phoenix Column		C. American Column		D. Common Column	
	Ultimate	Safe	Ultimate	Safe	Ultimate	Safe	Ultimate	Safe
15	37,067	7,822	40,476	8,521	34,434	7,249	33,693	7,093
16	36,876	7,683	40,212	8,377	34,167	7,118	33,339	6,946
18	36,470	7,443	39,645	8,091	33,597	6,856	32,589	6,651
20	36,024	7,205	39,030	7,806	32,982	6,596	31,790	6,358
22	35,544	6,970	38,373	7,524	32,327	6,338	30,952	6,069
25	34,767	6,622	37,317	7,110	31,285	5,959	29,639	5,646
30	33,344	6,063	35,424	6,440	29,435	5,352	27,375	4,977
35	31,806	5,531	33,406	5,810	27,512	4,789	25,108	4,367
40	30,198	5,033	31,352	5,226	25,584	4,264	22,919	3,820
45	28,562	4,570	29,310	4,690	23,701	3,792	20,857	3,337
50	26,932	4,143	27,321	4,203	21,900	3,369	18,952	2,916
55	25,833	3,728	25,415	3,765	20,203	3,004	17,214	2,559
60	23,787	3,398	23,611	3,373	18,621	2,660	15,643	2,235

[a]From Trautwine 1888, 443. Trautwine's note reads: "Ultimate crippling strengths in lbs per sq inch of metal section of the four wrought iron pillars below. These formulas are deduced by Chs. Shaler Smith, from many tests by G. Bouscaren, C. E., of large pillars of good American iron. The lower Table is an abridgment of the full ones by C. L. Gates; C. E., in the Trans. Am. Soc. C. E., Oct., 1880."

$H = \dfrac{\text{length between end bearings}}{\text{least diameter } d}$ both in the same measure; and is to be squared.

For safety take from 1/3 to 1/8, according to circumstances.

between wrought iron and steel. In the Brinnell test, a hardened steel ball (10 mm in diameter) at the end of the instrument is held against the material and struck with a standard force. The penetration of the ball is read through the instrument, which is calibrated to directly indicate the hardness. Cast iron is harder than structural steel, which is harder than wrought iron.

In investigating a building to determine the structural capacity of beams, joists, and columns for an adaptive reuse project; the first step is to identify the structural materials that frame the building. In some instances, the date of the building will provide some indication of the probable material; however, it is best to check the composition of the material to make the proper analysis. In a typical building, only a small amount of semidestructive investigation will be required to determine the material. Actually, the investigation required to analyze the connections and the condition of the rivets, plates, etc., would allow the material to be checked by the Brinnell test.

Many hundreds of buildings were constructed with wrought iron columns and beams. Many still remain in use today, serving their original function. Wrought iron is more malleable than steel, and therefore it may be more vulnerable to tearing shear

Table 3-28. Phoenix Columns[a]

Mark	Thickness D (in.)	Diameters (in.)			Area of Cross-section (in.²)	One Column Weight per ft Run (lb)	Least Radius of Gyration (in.)	Size of Rivet
		d	D	D'				
A	3/16	3 5/8	4	6 1/16	3.8	12.6	1.45	1 1/8 × 3/8
A	1/4	3 5/8	4 1/8	6 3/16	4.8	16.	1.50	1 1/4 × 3/8
A	5/16	3 5/8	4 1/4	6 5/16	5.8	19.3	1.55	1 3/8 × 3/8
A	3/8	3 5/8	4 3/8	6 7/16	6.8	22.6	1.59	1 1/2 × 3/8
B¹	1/4	4 13/16	5 5/16	8 1/16	6.4	21.3	1.92	1 5/8 × 1/2
B	3/8	4 13/16	5 9/16	8 1/4	9.2	30.6	2.02	1 3/4 × 1/2
B	1/2	4 13/16	5 13/16	8 7/16	12.	40.	2.11	1 7/8 × 1/2
B	5/8	4 13/16	6 1/16	8 5/8	14.8	49.3	2.20	2 1/8 × 1/2
B²	1/4	5 15/16	6 7/16	9 1/8	7.4	24.6	2.34	1 5/8 × 1/2
B	3/8	5 15/16	6 11/16	9 5/16	10.6	35.3	2.43	1 3/4 × 1/2
B	1/2	5 15/16	6 15/16	9 1/2	13.8	46.	2.52	1 7/8 × 1/2
B	5/8	5 15/16	7 3/16	9 11/16	17.	56.6	2.61	2 1/8 × 1/2
C	1/4	7 3/16	7 11/16	11 9/16	10.	33.3	2.80	1 7/8 × 5/8
C	1/2	7 3/16	8 3/18	11 15/16	18.	60.	2.98	2 1/4 × 5/8
C	3/4	7 3/16	8 11/18	12 3/16	25.2	84.	3.16	2 5/8 × 3/4
C	1	7 3/16	9 3/16	12 9/16	33.2	110.6	3.34	3 × 3/4
C	1 1/4	7 3/16	9 11/16	12 15/16	41.2	137.3	3.52	3 1/4 × 3/4
E	1/4	11	11 1/2	15 7/16	16.8	56.	4.18	2 × 3/4
E	1/2	11	12	15 7/8	26.4	88.	4.36	2 3/8 × 3/4
E	3/4	11	12 1/2	16 5/16	37.8	126.	4.55	2 3/4 × 3/4
E	1	11	13	16 3/4	49.8	166.	4.73	3 × 3/4
E	1 1/4	11	13 1/2	17 3/16	61.8	206.	4.91	3 1/4 × 3/4
G	5/16	14 3/8	15	19 1/8	24.	80.	5.45	1 7/8 × 5/8
G	1/2	14 3/8	15 3/8	19 7/16	36.	120.	5.59	2 1/4 × 5/8
G	3/4	14 3/8	15 7/8	19 7/8	52.	173.3	5.77	2 5/8 × 3/4
G	1	14 3/8	16 3/8	20 3/8	68.	226.6	5.95	3 × 3/4
G	1 3/8	14 3/8	17 1/8	21	92.	306.6	6.23	3 3/8 × 3/4

[a]From Trautwine 1888, 449. G columns have 8 segments, E, 6 segments, all others, 4 segments.

at connections. Wrought iron has a lower shear stress than steel. In wrought iron the direction of fibers gives it properties different from steel. Allowable stresses in directions perpendicular to the fiber are reduced by a factor of 12 percent. The oxidation process for wrought iron and steel is more rapid than for cast iron. Buildings must be checked to ascertain that the roof or curtain walls are not allowing water or water vapor to get to the structural elements and the area of the connections.

Wrought iron latticed columns utilizing channels and bars were very popular built-up columns during the period when wrought iron was used. In addition, built-up columns were constructed of channels and plates (square columns), angles and plates (common column), Phoenix columns and Z-bar columns. All but the last can be seen in Figures 3-67 and 3-68. Z-bar columns of wrought iron did exist; the relative numbers of these columns that were used in buildings is not known. Birkmire (1894, 34) states that C. L. Strobel had tested iron Z-bar columns in which the central web plates were replaced by lattice bars and reported the tests in the Transactions of the American Society of Civil Engineers in 1888.

Birkmire presents a number of types of compound columns, which were made up of rolled shapes, plates, and rivets (see Figure 3-50). The section shown as "Fig. 17," which is the heaviest of the compound columns shown, represents a column of the 12-story Venetian Building in Chicago. The column 13.25 × 21 in. × 27.2 ft long. Birkmire also produces results of a number of tests on full-size Phoenix columns that were performed at the Watertown Arsenal in 1881 by Clark of the consulting firm of Clark, Reeves & Company (ibid., 33).

Table 3-29. Breaking Loads for Hollow Cylindrical Wrought Iron Columns[a]

Length (ft)	Wrought Iron, Thickness ⅛ in. Outer Diameter (in.)									
	¾	1	1¼	1½	1¾	2	2¼	2½	2¾	3
	Breaking Load (2000 lb Tons)									
	Tons	Tons	Tons	Tons	Tons	Tons	Tons	Tons	Tons	Tons
1	3.64	5.27	6.88	8.50	10.1	11.7	13.2	14.8	16.4	18.0
2	2.94	4.64	6.32	8.00	9.6	11.2	12.8	14.5	16.1	17.8
3	2.30	3.86	5.57	7.28	8.9	10.6	12.2	13.9	15.6	17.3
4	1.77	3.13	4.74	6.36	8.1	9.9	11.6	13.3	15.0	16.7
5	1.36	2.51	4.07	5.66	7.3	9.1	10.8	12.5	14.2	16.0
6	1.04	2.03	3.46	4.91	6.6	8.3	9.9	11.6	13.4	15.2
7	0.81	1.65	2.91	4.24	5.7	7.4	9.1	10.8	12.6	14.4
8	0.61	1.36	2.46	3.67	5.1	6.7	8.3	9.9	11.7	13.5
9	0.50	1.05	2.03	3.18	4.5	6.0	7.5	9.1	10.8	12.6
10	0.41	0.95	1.75	2.77	4.0	5.4	6.9	8.4	10.1	11.8
11	0.34	0.81	1.52	2.41	3.6	4.8	6.2	7.7	9.3	11.0
12	0.29	0.70	1.34	2.14	3.2	4.3	5.6	7.0	8.6	10.2
13	0.24	0.60	1.16	1.88	2.8	3.9	5.2	6.5	8.0	9.5
14	0.21	0.53	1.03	1.69	2.5	3.5	4.7	6.0	7.4	8.9
15	0.19	0.47	0.91	1.50	2.3	3.2	4.3	5.5	6.9	8.3
16	0.18	0.42	0.84	1.38	2.1	2.9	4.0	5.1	6.4	7.7
18	0.14	0.33	0.67	1.11	1.7	2.4	3.4	4.4	5.6	6.8
20		0.27	0.55	0.91	1.4	2.0	2.8	3.7	4.7	5.8
25					0.9	1.4	2.0	2.6	3.4	4.2
Weight of 1 ft of length of pillar (lb)										
	0.820	1.15	1.47	1.80	2.13	2.45	2.78	3.11	3.43	3.77
Area of ring of solid metal (in.²)										
	0.246	0.344	0.442	0.540	0.638	0.736	0.835	0.933	1.03	1.13

Length (ft)	Wrought Iron, Thickness ¼ in. Outer Diameter (in.)									
	2	2¼	2½	2¾	3	3½	4	4½	5	5½	6
	Breaking Load (2000 lb Tons)									
	Tons	Tons	Tons	Tons	Tons	Tons	Tons	Tons	Tons	Tons	Tons
1	21.9	25.4	28.3	31.4	34.5	40	47	53	60	66	72
2	21.1	24.3	27.6	30.7	33.9	40	47	53	60	66	72
3	19.9	23.1	26.4	29.7	33.0	39	46	52	59	65	71
4	18.6	21.8	25.3	28.5	31.9	38	45	51	58	64	71
5	17.0	29.4	23.5	27.3	30.7	37	44	50	57	63	70
6	15.4	18.8	22.1	25.7	29.2	36	43	49	56	62	69
7	13.9	17.3	20.5	23.8	27.8	34	41	47	54	61	68
8	12.5	15.6	19.1	22.3	25.9	32	40	46	53	60	67
9	11.2	14.2	17.5	20.6	24.3	30	38	44	51	58	65
10	10.0	13.0	16.1	19.1	22.7	29	37	43	50	57	64
11	9.0	10.7	15.7	17.6	21.1	27	35	41	48	55	62
12	8.1	10.6	13.5	16.4	19.6	26	33	40	46	54	61
13	7.3	9.6	12.4	15.1	18.2	24	31	38	44	52	59
14	6.6	8.8	11.3	14.0	17.0	23	30	36	43	51	57
15	6.0	8.0	10.4	12.9	15.8	21	28	34	41	49	55
16	5.5	7.3	9.5	12.0	14.6	20	27	33	40	47	54
18	4.5	6.0	8.0	10.3	12.7	18	24	30	37	43	50
20	3.8	5.1	6.8	8.7	11.0	16	21	27	34	40	47
25					7.9	12	16	21	27	33	39
30							13	17	22	27	32
35							10	14	18	22	27
40									14	18	23
45									11	15	19
50									8	12	16
Weight of 1 ft of length of pillar (lb)											
	4.30	5.23	5.90	6.53	7.20	8.50	9.83	11.1	12.4	13.7	15.0
Area of ring of solilid metal (in.²)											
	1.38	1.57	1.77	1.96	2.16	2.55	2.95	3.34	3.73	4.12	4.51

Table 3-29. (Continued)

Length (ft)	Wrought Iron, Thickness ½ in. Outer diameter in.											
	5	5½	6	6½	7	7½	8	8½	9	10	11	12
	Breaking Load (2000 lb Tons)											
	Tons	Tons	Tons	Tons	Tons	Tons	Tons	Tons	Tons	Tons	Tons	Tons
2	112	125	139	152	166	177	189	201	214	238	263	290
4	110	123	136	149	163	174	186	199	212	237	262	289
6	106	119	132	145	158	171	184	197	210	235	261	288
8	101	114	127	140	154	167	181	194	207	232	258	284
10	95	108	123	136	149	162	176	189	203	228	254	280
12	89	102	116	129	143	157	171	185	199	224	250	276
14	82	95	109	122	137	151	165	179	194	219	245	272
16	76	89	103	117	131	145	160	173	187	213	240	268
18	70	83	97	110	124	138	153	166	180	207	235	263
20	64	77	91	104	117	131	145	159	173	201	227	257
22	58	70	83	96	109	123	138	151	165	192	220	250
25	52	64	76	89	102	115	129	143	157	183	212	241
30	42	52	63	74	87	100	113	127	141	167	195	224
35	34	43	53	64	75	87	99	112	125	151	178	207
40	27	35	44	53	64	75	86	98	110	135	163	190
45	23	30	38	46	55	65	76	87	98	123	148	174
50	19	24	32	38	47	56	66	76	87	109	133	158
60	15	19	24	29	36	43	51	60	69	88	109	132
70	11	14	18	23	28	34	40	48	56	73	91	111
80	9	11	14	18	22	27	32	37	44	57	74	93
90	7	9	11	14	18	22	26	31	36	49	63	78
100	6	7	9	12	15	18	22	26	30	41	53	66
Weight of 1 ft of length of pillar (lb)												
	23.6	26.2	28.8	31.4	34.0	36.6	39.3	42.0	44.7	49.7	55.0	60.3
Area of ring of solid metal (in.²)												
	7.07	7.85	8.64	9.43	10.2	11.0	11.8	12.6	13.4	14.9	16.5	18.1

Length (ft)	Wrought Iron, Thickness 2 in. Outer diameter (in.)										
	13	14	15	16	17	18	20	22	24	26	28	30
	Breaking Load (2000 lb Tons)											
	Tons	Tons	Tons	Tons	Tons	Tons	Tons	Tons	Tons	Tons	Tons	Tons
1	603	653	704	753	805	854	955	1056	1157	1257	1357	1458
10	588	638	691	742	795	846	949	1049	1149	1248	1354	1457
20	543	595	651	702	759	810	913	1016	1120	1223	1327	1430
30	479	538	594	645	699	758	866	973	1077	1186	1289	1364
40	415	470	528	584	636	691	806	912	1027	1130	1237	1348
50	335	405	462	516	570	627	740	848	961	1067	1179	1294
60	300	348	400	452	505	559	669	781	891	1005	1115	1228
70	256	300	348	398	448	499	606	715	824	936	1046	1160
80	215	255	298	344	392	440	543	649	757	868	978	1092
90	185	222	261	303	347	392	489	590	694	800	910	1023
100	157	190	225	262	303	345	436	532	631	735	843	955
110	134	162	193	227	264	302	386	474	568	666	770	877
125	111	135	162	192	225	259	336	416	505	598	697	799
130	82	101	122	145	171	198	262	328	405	485	574	666
175	62	78	95	112	133	155	208	266	331	400	478	560
200	49	60	74	89	106	124	168	216	269	328	395	467
Weight of 1 ft of length of pillar (lb)												
	126	136	147	157	168	178	199	220	241	262	263	344
Area of ring of solid metal (in.²)												
	37.7	40.8	44.0	47.1	50.3	53.4	59.7	66.0	72.3	78.5	84.8	91.1

[a] From Trautwine 1888, 447, 448. Trautwine's note reads: "Table 4, of breaking loads in tons of hollow cylindrical wrought iron pillars, with flat ends, perfectly true, and firmly fixed, and the loads pressing equally on every part of the top. Calculated by Gordon's formula. No pains have been taken to have the last figure of the laods perfectly correct in every case."

Table 3-30. Buildings of Cast Iron, Wrought Iron, and Steel Colums to 1894[a]

Cast Iron Columns (New York City)	Cast Iron Columns (Chicago)	Wrought Iron and Steel Columns (NYC)	Wrought Iron and Steel Structures (Chicago)
Postal Telegraph Building	The Rookery	The New Netherlands	Rand McNaly Building
Decker Brothers Building	Home Insurance Building	Havemeyer Building	The Ashland Block
The Waldorf	The Monon Block	Lancashire Building	Venetian Building
Jackson Building	Western Bank Note Building	World Building	The Kearsarge
Scott & Browne Building	Tacoma Building Cold Storage Building	Home Life Insurance Co. Building	The Fair
D., L. & W.R.R. Building	The Auditorium	Hotel Majestic	Masonic Temple
The Western Union Annex	The Chamber of Commerce	Mail and Express	German Theatre
Lincoln Building	Manhattan Building	Mutual Reserve Fund Building and others	The Pontiac
McIntyre Building	Unity Building		Northern Hotel
Mutual Life Annex (wall columns)	Owens Building		Woman's Temple and others

[a] From Birkmire 1894, 20.

The tests at the Watertown Arsenal gave interesting data concerning the net change in length due to compression at intervals, elastic limit (load and stress), and ultimate strength (failure, in load and stress) and a comparison between the ultimate strength from the tests and the calculated ultimate strength from Gordon's formula for wrought iron columns.

Gordon's formula (ibid., 33):

$$\frac{P}{A} = \frac{36{,}000}{1 + \dfrac{L^2}{36{,}000\, r^2}}$$

where:
P = breaking weight, lbs
A = area of section

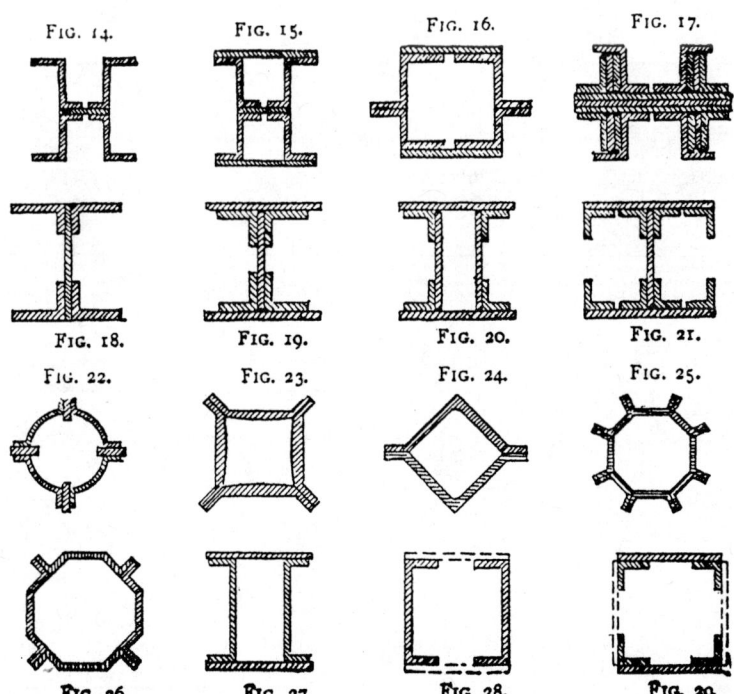

Figure 3-50. Wrought iron and steel columns: compound sections. (*From Birkmire 1894, 22, 23.*)

L = length of column, in.
r = least radius of gyration

Birkmire also includes tables of the results of tests on wrought iron columns made up of channels spaced 8 in. apart and latticed at different intervals, four Z-bars latticed in the web, and box columns of channels and plates (see Table 3-34).

The latticed columns of channels and lattice bars are fairly clear in the presentation of test results. It is unclear whether the channels are turned inward or outward, but they must have been turned inward because a portion of the bars is shown to be 6 in. long. Testing the columns with different-length lattice bars would produce different results because the unbraced length of column sections would vary with spacing. In this case, however, it is not as easily analyzed because the cross-sectional area of the channels changed at the same time the spacing of lattice bars changed. The test results appear to be well defined.

The Z-bar columns in the second table indicate the ultimate strength of the test columns (breaking loads) and then compare the results of the tests with results of the theoretical ultimate strengths of the same columns as computed by formula:

$$F_a = 46{,}000 - 125 \left(\frac{L}{r}\right)$$

where:

F_a = ultimate comp. stress, psi
L = length of column, in.
r = least radius of gyration, in.

The preceding tables indicate by test results and by formula theory that the length of a column and the radius of gyration are both factors in combination that have great effect on the column's ability to carry load. The publications of the time were gradually pointing to the fact that the column fails in a combination of crushing and buckling for shorter lengths and in buckling only for longer lengths. The cross-sectional area of the shape, end conditions, length, and radius of gyration (which is associated directly with the area) determine the manner in which a column reacts to loads. Johnson et al. (1894) brings the effect of the end conditions into the design

Table 3-31. Phoenix Columns: Clark's Tests, 1881, at Watertown Arsenal[a]

No. of Experiment	Length of Column (ft)	Ratio of Diameter to Length	Weight (lb)	Sectional Area (in.²)	Total Compression under Loads Lb 200,000	Total Compression under Loads Lb 300,000	Elastic Limit Total lb	Elastic Limit psi	Ultimate Strength Total lb	Ultimate Strength psi	Total Ultimate Strength (lb) by Gordon's Formula
1	28	42	1,142	12.062	0.190				424,000	35,150	330,146
2	28	42	1,153	12.181	0.186				416,000	34,150	333,459
3	25	37½	1,034	12.233		0.255	342,000	27,960	431,500	35,270	352,013
4	25	37½	1,023	12.100	0.168	0.264			424,000	35,040	348,119
5	22	33	920	12.371	0.160	0.243			440,000	35,570	372,837
6	22	33		12.311	0.152	0.236			423,000	34,360	371,043
7	19	28½	773	12.023		0.198			425,200	35,365	377,955
8	19	28½	777	12.087	0.139	0.213	354,000	29,290	446,000	36,900	380,197
9	16	24	650	12.000	0.120				439,000	36,580	391,701
10	16	24	650	12.000	0.116				439,000	36,580	391,701
11	13	19½	536	12.185	0.092	0.142	342,000	28,890	449,000	36,857	410,660
12	13	19½	531	12.009	0.091				449,000	37,200	406,886
13	10	15	415	12.248		0.110	330,000	26,940	446,800	36,480	423,886
14	10	15	418	12.339		0.109	350,000	28,360	449,100	36,397	427,047
15	7	10½	291	12.265	0.054		360,000	29,350	468,000	38,157	433,021
16	7	10½	284	11.962			354,000	29,590	517,000	43,300	469,324
17	4	6	164	12.081	0.031				598,000	49,500	432,132
18	4	6	164½	12.119	0.025	0.042	340,000	28,050	621,000	51,240	433,507

[a] From Birkmire 1894, 33.

Table 3-32. Wrought Iron Column Tables: Latticed Channels, Z-Bars, and Box Columns[a]

	Size of Bars	Length		Sectional Area	Lattice Spacing	Ultimate Strength		Manner of Failure
						Actual	psi	
	in.	ft	in.	in.²	in.	lb	psi	
Flat ends	6	10	0	4.760	18	174,800	36,720	Channels buckled
Flat ends	6	10	0	4.670	18	165,000	35,330	Channels buckled
Pin ends	6	12	0	4.600	18	159,800	34,740	Horizontal deflection
Pin ends	6	15	0	4.480	18	151,500	33,820	Horizontal deflection
Pin ends	6	17	6	4.660	18	152,600	32,750	Horizontal deflection
Pin ends	6	20	6	4.660	18	136,000	29,180	Horizontal deflection
Pin ends	6	22	6	4.570	18	139,800	30,590	Horizontal deflection
Pin ends	6	25	0	4.710	18	110,000	23,350	Horizontal deflection
Pin ends	6	27	6	4.690	18	102,500	21,850	Horizontal deflection
Pin ends	6	30	0	4.700	18	69,300	14,740	Horizontal deflection
Pin ends	8	13	4	7.520	18	261,800	34,810	Defl. upward; ch. bars buckled
Pin ends	8	16	8	7.480	18	254,100	33,970	Defl. horizon; ch. bars buckled
Pin ends	8	20	0	7.550	18	246,200	32,610	Defl. horizon; ch. bars buckled
Pin ends	8	23	4	7.990	18	257,500	32,230	Defl. horizon; ch. bars buckled
Pin ends	8	26	8	7.780	18	243,900	31,350	Defl. horizon; ch. bars buckled
Pin ends	8	30	0	7.810	18	194,100	24,850	Defl. horizon; ch. bars buckled
Pin ends	10	12	6	9.680	22	344,120	35,550	Channel bars buckled
Pin ends	10	16	8	9.550	22	323,200	33,840	Channel bars buckled
Pin ends	10	20	10	9.740	22	330,000	33,880	Channel bars buckled
Pin ends	10	25	0	10.040	22	342,700	34,130	Channel bars buckled
Pin ends	10	29	2	9.300	22	299,300	32,180	Deflection horizontally
Pin ends	12	20	0	11.980	22	411,600	34,360	Channel bars buckled
Pin ends	12	25	0	12.144	22	400,000	32,940	Channel bars buckled
Pin ends	12	25	0	11.910	22	407,800	34,240	Channel bars buckled
Pin ends	12	30	0	12.180	22	385,000	31,610	Channel bars buckled
Pin ends	12	30	0	12.540	22	393,000	31,340	Deflection horizontally

Section of columns: 4 Z-bars, 2¼ × 3 × 2¼ in. (latticed)
Radius of gyration (latticed bars not considered) = 2.05 in.

Length of Column	Sectional Area (in.²)	Ultimate Strength by Actual Tests (psi)	Ratio of Length to Least Radius of Gyration	Ultimate Strength by Formula $46{,}000 - 125\frac{L}{r}$
15'-0"	9.480	34,600	88	35,000
15'-0"	9.280	36,600	88	35,000
19'-0¾"	9.241	33,800	112	32,200
19'-0¾"	10.104	33,700	112	32,200
22'-0"	9.286	30,700	129	29,900
22'-0"	9.286	29,500	129	29,900
22'-0"	9.286	30,700	129	29,000
25'-0"	9.156	28,100	146	27,750
25'-0"	9.456	28,000	146	27,750
25'-0"	9.516	28,400	146	27,750
28'-0"	9.375	27,700	164	25,500
28'-0"	9.643	28,000	164	25,500
28'-0"	9.375	27,600	164	25,500

solution. The length of column divided by the least diameter or least radius of gyration has been a consideration in most formulas. Johnson et al. indicate that for "a perfectly ideal column" (lower limit of failing by bending only) $L/r = 100$ for wrought iron and $L/r = 85$ for mild steel. Any larger ratios would certainly produce column failure in pure bending. Johnson et al. also compare the theories of the day to the basic research of Euler in 1759 (ibid., 146). Euler's Equation and the application of the principle of end conditions, which is still utilized today in intact form, are shown in Figure 3-51.

The Johnson et al. point toward a generic set of formulae to be used for the design of built-up columns. The designer could elect to utilize a built-up or compound column made up of angles and plates, channels and plates, etc. The variable in the design

Table 3-32. (Continued)

			Wrought Iron Box Columns with Flat Ends		
Style of Column	Total Length	Sectional Area (in.²)	Ultimate Strength		Manner of Failure
			Total lb	psi	
Two 6 in. channels 5.5 in. apart, flanges turned out with two ¼ in. cover-plates	10' 7.9" 10' 7.9"	12.08 11.11	383,200 372,900	31,722 33,564	Plates buckeled between the rivets
Two 8 in. channels 7.6 in. apart, flanges turned out with two ⁵⁄₁₆ in. cover-plates	13' 11.8" 13' 11.8"	17.01 17.80	594,500 633,600	34,950 35,595	Same as above Triple flexure
Four plates connected with four angles forming a box 7 in. × 7½ in. inside	13' 11.9"	15.74	517,000	32,846	Buckling plates
Plates and angles all ⁵⁄₁₆ in. thick	13' 11.6" 20' 7.63" 20' 7.80"	15.84 15.68 15.56	555,200 517,500 536,900	35,050 33,003 34,505	Buckling plates Deflecting upward Buckling plates
Single web columns with 3⅛ in. pin-ends. One ⁵⁄₁₆ in. web 8 in. wide with four angles, and 8 in. channels used in place of cover-plates, flanges onward	13' 4"	15.34	47,500	30,945	Deflecting upward in plane of pin

ᵃFrom Birkmire 1894, 34, 35.

would be the cross-sectional area and the radius of gyration. Johnson et al. include a very interesting graph of the result of the Euler Equation when modified for end conditions and material elastic modulus (see Figure 3-52).

Each of the authors quoted in the section on wrought-iron columns recommends that a factor of safety of 4 be applied to the ultimate load or stress to determine the allowable load or stress.

The Carnegie Steel Co. (1893) contains a table of the values of the ultimate stress of wrought iron columns for three conditions (see Table 3-35).

STRUCTURAL STEEL COLUMNS

Structural steel sections were initially produced by the same mills and foundries that produced wrought iron and cast iron sections for the building industry. From the late 1880s until into the early twentieth century, the individual mills produced catalogue of information and data and gave the unit stresses (allowable stresses) for their products. Ferris (1953–78, 5) provides a summary of unit stresses recommended by each manufacturer:

1887	Pottsville Iron & Steel Co.	15,600 psi
1889–1893	Carnegie, Phipps & Co.	16,000 psi
1893–1908	Jones & Laughlins Steel Co.	16,000 psi
1896	Carnegie Steel Co., Ltd.	16,000 psi
1898–1919	Cambria Steel Co.	16,000 psi
1900–1903	Carnegie Steel Co.	16,000 psi
1907–1911	Bethlehem Steel Co.	16,000 psi
	Moving Loads	12,500 psi
1915	Lackawanna Steel Co.	16,000 psi

The steel industry began to standardize shapes and sizes to conform to American Standard Beams under the Association of American Steel Manufacturers in 1893. The

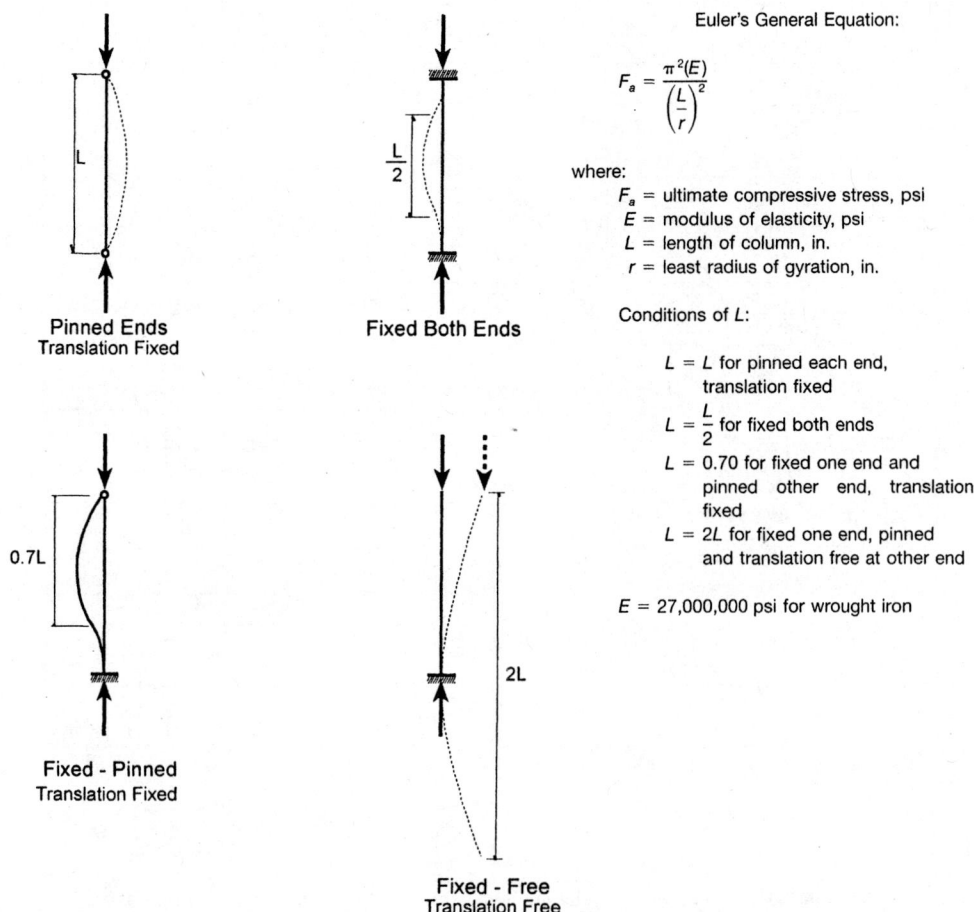

Figure 3-51. Euler's Equation and end conditions. (*From Fitzgerald 1967, 184, 185.*)

Association also brought about standardized testing procedures and finally conformance to the American Society for Testing of Materials standardized specifications (herein referred to as ASTM Specifications). Ferris also summarizes the history of the ASTM and AISC structural steel types by date, ASTM specification, ultimate tensile strength and minimum yield point tensile stress for designer uses, as shown in Table 3-36.

The strength of structural steels of a certain period or date is very difficult to be exact about. Prior to 1900, the specifications for the manufacture of steel were not adopted by all mills; each mill would issue technical data on its own product. The technology was new, and each mill would develop its own metallurgy and perhaps slightly different compositions for its steel. There would also logically be some variance in the strength of the material. Fortunately, designers of the period were fairly conservative in their approaches to the material, as were the engineers who worked as specialists for the mills. The transition from load-bearing masonry to skeletal construction and the use of curtain walls was actually fairly rapid, considering the period. Many advances were occurring almost simultaneously, either by a masterful job of coordination or by necessity. Skeletal frames, elevators, grillage footings, pneumatic Cassions, and fireproof Construction etc. all seemed to be interdependent. It must have been a very exciting period to be involved in construction.

Early experiments in structural steel for columns indicated that the strength of steel sections was about 20 percent higher than that of the same wrought iron section for lengths up to 90 radii of gyration. Beyond this length, the excess of strength diminished to the point where steel and wrought iron were approximately the same strength at lengths of about 200 radii of gyration (Birkmire 1894, 35). This comparison

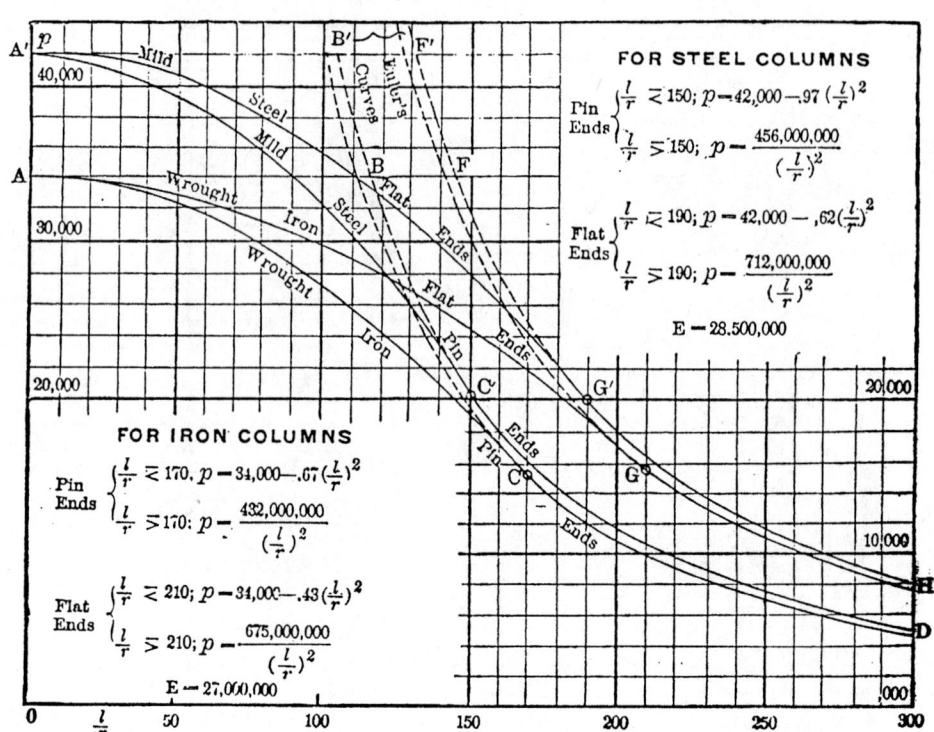

Figure 3-52. Graphs of Euler's Equation: ultimate axial stresses on columns. (*From Johnson et al. 1894, 149.*)

was made in 1894, at the point when steel was beginning to take precedence over wrought and cast irons.

Kidder (1902, 270) gives the following formula for determining the allowable stress for steel columns and struts in buildings.

For steel columns:

$$F_a = 17{,}100 - 57\frac{L}{r}$$

For steel struts in trusses:

$$F_a = 13{,}750 - 60\frac{L}{r}$$

where:

L = length of column, in.

r = least radius of gyration

For steel columns length less than $90(r)$:

$F_a = 12{,}000\text{–}14{,}000$ psi

For steel struts where L/r is less than 50:

$F_a = 10{,}750$ psi

Table 3-33. Carnegie Wrought Iron Columns[a]

Ultimate Strength of Wrought Iron Columns

For different proportions of length in feet ($= L$)
To least radius off gyration in inches ($= r$)

Ultimate strength in psi =

Column Square Bearing:	Column Pin and Square Bearing:	Column Pin Bearing:
$\dfrac{40{,}000}{1 + \dfrac{(12L)^2}{36{,}000 r^2}}$	$\dfrac{40{,}000}{1 + \dfrac{(12L)^2}{24{,}000 r^2}}$	$\dfrac{40{,}000}{1 + \dfrac{(12L)^2}{18{,}000 r^2}}$

To obtain safe resistance:

For quiescent loads, as in buildings, divide by 4.
For moving loads, as in bridges, divide by 5.

$\dfrac{L}{r}$	Ultimate Strength (psi)			$\dfrac{L}{r}$	Ultimate Strength (psi)		
	Square	Pin and Square	Pin		Square	Pin and Square	Pin
3.0	38,610	37,950	37,310	11.0	26,950	23,170	20,330
3.2	38,430	37,680	36,970	11.2	26,640	22,820	19,960
3.4	38,230	37,400	36,610	11.4	26,320	22,470	19,610
3.6	38,030	37,110	36,240	11.6	26,000	22,130	19,270
3.8	37,820	36,810	35,860	11.8	25,690	21,800	18,930
4.0	37,590	36,500	35,460	12.0	25,380	21,460	18,590
4.2	37,360	36,170	35,050	12.2	25,070	21,130	18,260
4.4	37,120	35,840	34,640	12.4	24,770	20,810	17,940
4.6	36,870	35,500	34,210	12.6	24,470	20,490	17,620
4.8	36,620	35,140	33,770	12.8	24,170	20,180	17,310
5.0	36,360	34,780	33,330	13.0	23,870	19,860	17,000
5.2	36,090	34,420	32,890	13.2	23,570	19,560	16,710
5.4	35,820	34,050	32,440	13.5	23,140	19,110	16,280
5.6	35,540	33,670	31,980	13.8	22,700	18,670	15,850
5.8	35,260	33,280	31,520	14.0	22,420	18,380	15,580
6.0	34,970	32,890	31,060	14.2	22,150	18,100	15,310
6.2	34,670	32,500	30,590	14.5	21,740	17,690	14,920
6.4	34,370	32,110	30,130	14.8	21,320	17,290	14,530
6.6	34,060	31,710	29,670	15.0	21,050	17,020	14,290
6.8	33,750	31,310	29,200	15.2	20,790	16,760	14,040
7.0	33,440	30,910	28,740	15.5	20,290	16,390	13,690
7.2	33,130	30,510	28,270	15.8	20,020	16,010	13,350
7.4	32,810	30,110	27,820	16.0	19,760	15,770	13,120
7.6	32,490	29,710	27,360	16.2	19,510	15,540	12,910
7.8	32,170	29,310	26,910	16.5	19,150	15,190	12,590
8.0	31,850	28,900	26,460	16.8	18,790	14,850	12,280
8.2	31,520	28,500	26,010	17.0	18,550	14,630	12,080
8.4	31,190	28,100	25,570	17.2	18,320	14,410	11,880
8.6	30,870	27,700	25,130	17.5	17,980	14,100	11,590
8.8	30,540	27,310	24,700	17.8	17,640	13,790	11,320
9.0	30,210	26,920	24,270	18.0	17,420	13,590	11,140
9.2	29,880	26,530	23,850	18.2	17,200	13,390	10,960
9.4	29,550	26,140	23,430	18.5	16,880	13,100	10,700
9.6	29,230	25,760	23,030	18.8	16,570	12,820	10,450
9.8	28,900	25,370	22,620	19.0	16,370	12,630	10,290
10.0	28,570	25,000	22,220	19.2	16,170	12,450	10,130
10.2	28,250	24,630	21,830	19.5	15,870	12,190	9,890
10.4	27,920	24,260	21,440	19.8	15,570	11,930	9,670
10.6	27,600	23,890	21,060	20.0	15,380	11,760	9,520
10.8	27,270	23,530	20,690	20.2	15,200	11,600	9,380
				20.5	14,920	11,360	9,170
				20.8	14,650	11,120	8,970

[a] From Carnegie Steel Co. 1893, 149, 150.

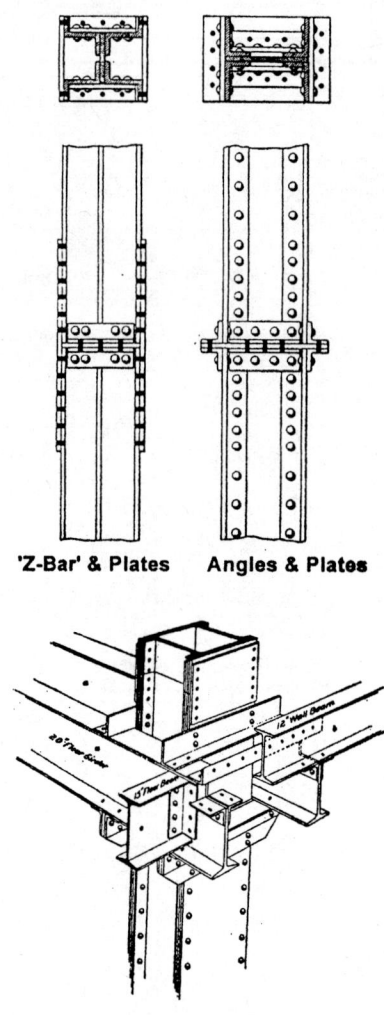

Figure 3-53. Wrought iron column details. (*From International Correspondence Schools 1905, para. 31, 47; 1905, para. 16, 46.*)

Carnegie's Steel Co. (1893, 132) gives the following formulae for steel Z-bar columns:

For columns of lengths of 90 radii of gyration (least):

$F_a = 12,000$ psi

For columns of lengths above 90 radii of gyration:

$$F_a = 17,100 - 57 \frac{L}{r}$$

where:
    L = length of column, in.
    r = least radius of gyration

Carnegie Steel Co. (1893) also includes a set of tables of the safe loads on steel Z-bar columns, a portion of which is shown in Table 3-37.

Table 3-34. Structural Steel Stress Chronology by Dates and Specifications

Date	Specification	Ultimate Tensile Strength, psi	Minimum Yield Point psi
1900	ASTM, A9, buildings, rivet steel	50,000–60,000	30,000
	buildings, medium steel	60,000–70,000	35,000
1901–1908	ASTM, A9, buildings, rivet steel	50,000–60,000	0.5 T.S.
	buildings, medium steel	60,000–70,000	0.5 T.S.
1909–1913	ASTM, A9, buildings, rivet steel	48,000–58,000	0.5 T.S.
	buildings, structural steel	55,000–65,000	0.5 T.S.
1914–1923	ASTM, A9, buildings, rivet steel	46,000–56,000	0.5 T.S.
	buildings, structural steel	55,000–65,000	0.5 T.S.
	AISC, allowable basic working stress	18,000 psi	
1924–1931	ASTM, A9, buildings, rivet steel	46,000–56,000	0.5 T.S. or ≥ 25,000 psi
	buildings, structural steel	55,000–65,000	0.5 T.S. or ≥ 30,000 psi
	AISC, allowable basic working stress	18,000 psi	
1932	ASTM, A140-32T, buildings, shapes bars and plates	60,0000–72,000	0.5 T.S. or ≥ 33,000 psi
	ASTM, A140-32T, buildings, eyebar flats, unannealed	67,000–82,000	0.5 T.S. or ≥ 36,000 psi
	ASTM, A141-32T, rivet steel	52,000–62,000	0.5 T.S. or ≥ 28,000 psi
	AISC, allowable basic working stress	18,000 psi	
1933	ASTM, A140-32T discontinued		
	ASTM, A9, buildings, revised structural steel	55,000–65,000	0.5 T.S. or ≥ 30,000 psi
	ASTM, A9-33T, buildings structural steel	60,000–72,000	0.5 T.S. or ≥ 33,000 psi
	ASTM, A141-32T, rivet steel	52,000–62,000	0.5 T.S. or ≥ 28,000 psi
	AISC, allowablle basic working stress	18,000 psi	
1934–1938	ASTM, A9-34, buildings, adopted as a standard structural steel	60,000–72,000	0.5 T.S. or ≥ 33,000 psi
	ASTM, A141-33, rivet steel	52,000–62,000	0.5 T.S. or ≥ 28,000 psi
1936	AISC revised allowable basic working stress	to 20,000 psi	
1939–1948	ASTM, A7-39, (revises ASTM A7-34 and ASTM A9-34 to one specification for bridges and buildings)	60,000–72,000	0.5 T.S. or ≥ 33,000 psi
	ASTM, A141-39, rivet steel	52,000–62,000	0.5 T.S. or ≥ 28,000 psi
	AISC, allowable basic working stress	20,000 psi	

AISC Historical Shapes, 1953.

Carnegie Steel Co. (1893) states that their tables are based upon a maximum fiber strain of 12,000 psi (column tables) and a factor of safety of 4. They also state that from this point forward (1893) their products will be exclusively of steel.

Birkmire (1894) also provides several tables of safe loads on steel Z-bar columns for designers (see Table 3-38).

Birkmire does not state the actual limit stresses utilized for the calculations of the tables, but an analysis of the tables and the formulae shown with them indicates that 12,000 psi is the allowable stress. This time (1894) is still within the period when manufacturers designated the ultimate and allowable stresses for their products. Therefore, the stresses utilized in the Carnegie and Birkmire tables do not correlate with the AISC stresses.

Table 3-35. Carnegie's Safe Loads for Steel 'Z-Bar' Columns in Tons (2,000 lb)[a]

Z-Bar Columns
Square Ends

Allowed strains psi safety factor 4 $\begin{cases} 12{,}000 \text{ lb, for lengths of 90 radii or under} \\ 17{,}100 - 57\frac{l}{r}, \text{ for lengths over 90 radii} \end{cases}$

14 in. Z-bar Columns
Section: 4 Z-Bars 6¹⁄₁₆ × ¹³⁄₁₆ in. 1 Web Plate 8 × ¹³⁄₁₆ in. 2 Side Plates 14 in. wide

Length of Column (ft)	14 × ³⁄₈ Plates = 185.6 lb/ft area = 54.6 in.² r (min.) = 3.73	14 × ⁷⁄₁₆ Plates = 191.5 lb/ft area = 56.3 in.² r (min.) = 3.74	14 × ½ Plates = 197.5 lb/ft area = 58.1 in.² r (min.) = 3.75	14 × ⁹⁄₁₆ Plates = 203.4 lb/ft area = 59.8 in.² r (min.) = 3.76	14 × ⅝ Plates = 209.4 lb/ft area = 61.6 in.² r (min.) = 3.77	14 × ¹¹⁄₁₆ Plates = 215.3 lb/ft area = 63.3 in.² r (min.) = 3.78	14 × ¾ Plates = 221.3 lb/ft area = 65.1 in.² r (min.) = 3.78	14 × ¹³⁄₁₆ Plates = 227.2 lb/ft area = 66.8 in.² r (min.) = 3.79	14 × ⅞ Plates = 233.2 lb/ft area = 68.6 in.² r (min.) = 3.80
26 and under	327.5	338.0	348.5	359.0	369.5	380.0	390.5	401.0	411.5
28	326.7	337.5	348.5	359.0	369.5	380.0	390.5	401.0	411.5
30	316.7	327.2	337.7	348.3	358.9	369.5	380.0	390.6	401.1
32	306.6	318.0	327.2	337.4	347.7	358.0	368.2	378.5	388.8
34	296.6	306.6	316.6	326.5	336.5	346.5	356.4	366.4	376.4
36	286.7	296.4	306.0	315.7	325.3	335.0	344.7	354.3	364.0
38	276.7	286.0	295.4	304.8	314.2	323.6	332.9	342.3	351.7
40	266.6	275.7	284.8	293.9	303.0	312.1	321.2	330.3	339.3
42	256.6	265.5	274.3	283.0	291.8	300.6	309.4	318.2	327.0
44	246.6	255.2	263.6	272.2	280.6	289.2	297.6	306.1	314.6
46	236.6	244.9	253.0	261.3	269.5	277.7	285.8	294.0	302.3
48	226.7	234.6	242.5	250.4	258.3	266.2	274.1	282.0	290.0
50	216.6	224.3	231.9	239.5	247.1	254.8	262.3	269.9	277.6

14 in. Z-bar Columns
Section: 4 Z-Bars 6⅛ × ⅞ in. 1 Web Plate 8 × ⅞ in. 2 Side Plates 14 in. wide

Length of Column (ft)	14 × ³⁄₈ Plates = 197.8 lb/ft area = 58.2 in.² r (min.) = 3.71	14 × ⁷⁄₁₆ Plates = 203.8 lb/ft area = 59.9 in.² r (min.) = 3.72	14 × ½ Plates = 209.7 lb/ft area = 61.7 in.² r (min.) = 3.73	14 × ⁹⁄₁₆ Plates = 215.7 lb/ft area = 63.4 in.² r (min.) = 3.74	14 × ⅝ Plates = 221.6 lb/ft area = 65.8 in.² r (min.) = 3.75	14 × ¹¹⁄₁₆ Plates = 227.6 lb/ft area = 66.9 in.² r (min.) = 3.76	14 × ¾ Plates = 233.5 lb/ft area = 68.7 in.² r (min.) = 3.77	14 × ¹³⁄₁₆ Plates = 239.5 lb/ft area = 70.4 in.² r (min.) = 3.77	14 × ⅞ Plates = 245.4 lb/ft area = 72.2 in.² r (min.) = 3.78
26 and under	349.1	359.6	370.1	380.6	391.1	401.6	412.1	422.6	433.1
28	347.4	358.3	369.1	380.0	390.9	401.6	412.1	422.6	433.1
30	336.7	347.2	357.9	368.4	378.9	389.5	400.1	410.7	421.2
32	326.0	336.3	346.6	356.8	367.1	377.3	387.6	397.9	408.2
34	315.3	325.2	335.2	345.2	355.1	365.2	375.2	385.1	395.1
36	304.5	314.2	324.0	333.6	343.3	353.0	362.7	372.4	382.0
38	293.8	303.2	312.6	322.0	331.4	340.8	350.2	359.6	369.0
40	283.1	292.2	301.3	310.4	319.5	328.6	337.7	346.8	355.9
42	272.3	281.2	290.0	298.8	307.6	316.4	325.2	334.0	342.8
44	261.6	270.2	278.7	287.2	295.7	304.2	312.7	321.2	329.8
46	250.9	259.1	267.4	275.6	283.8	292.1	300.3	308.5	316.7
48	240.2	248.1	256.1	264.0	272.0	279.8	287.8	295.7	303.6
50	229.5	237.1	244.8	252.4	260.0	267.6	275.3	283.0	290.6

Table 3-35. (Continued)

Z-Bar Column Dimensions

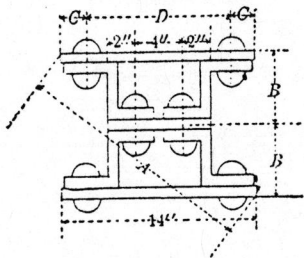

14 in. Columns.
4 Z-bars 6¹/₁₆ × ¹³/₁₆ in.
1 web plate 8 × ¹³/₁₆ in.
2 side plates 14 in. wide

Thickness of Side Plates	Diameter of Bolt or Rivet, ⅞ in.			
	A	B	C	D
⅜	19⁹/₁₆	6²⁷/₃₂	1¹³/₁₆	10⅜
⁷/₁₆	19⅝	6²⁹/₃₂	1¹³/₁₆	10⅜
½	19¾	6³¹/₃₂	1¹³/₁₆	10⅜
⁹/₁₆	19⅞	7¹/₃₂	1¹³/₁₆	10⅜
⅝	19¹⁵/₁₆	7³/₃₂	1¹³/₁₆	10⅜
¹¹/₁₆	20¹/₁₆	7⁵/₃₂	1¹³/₁₆	10⅜
¾	20⅛	7⁷/₃₂	1¹³/₁₆	10⅜
¹³/₁₆	20³/₁₆	7⁹/₃₂	1¹³/₁₆	10⅜
⅞	20¼	7¹¹/₃₂	1¹³/₁₆	10⅜

14 in. Columns.
4 Z-bars 6⅛ × ⅞ in.
1 web plate 8 × ⅞ in.
2 side plates 14 in. wide

Thickness of Side Plates	Diameter of Bolt or Rivet, ⅞ in.			
	A	B	C	D
⅜	19¾	6¹⁵/₁₆	1⅞	10½
⁷/₁₆	19¹³/₁₆	7	1⅞	10¼
½	19⅞	7¹/₁₆	1⅞	10¼
⁹/₁₆	20	7⅛	1⅞	10¼
⅝	20¹/₁₆	7³/₁₆	1⅞	10¼
¹¹/₁₆	20⅛	7¼	1⅞	10¼
¾	20¼	7⁵/₁₆	1⅞	10¼
¹³/₁₆	20⁵/₁₆	7⅜	1⅞	10¼
⅞	20⁷/₁₆	7⁷/₁₆	1⅞	10¼

ᵃFrom Carnegie Steel Co. 1893, 139, 140.

During this period, cities also designated column design formulae through their building laws, and the leading engineers of the period would derive formulae through tests and analytical derivations and advocate their implementation by both cities and manufacturers. Gordon's formula, adopted by the New York City Building Law, is as follows (Birkmire 1894, 26):

$$F_{a(ult)} = \frac{S}{1 + K\left(\frac{L^2}{d^2}\right)}$$

where:
S = comp. unit stress, 40,000 psi for steel
K = 0.000333 for Steel

Table 3-36. Birkmire's Safe Loads for Steel Z-Bar Columns in Tons (2000 lb)[a]

Steel Z-Bar Columns, Square Ends
Allowed strains per square inch for steel, safety factor 4:
12,000 lb for lengths of 90 radii or under

$$17,100 - 57\frac{l}{r} \text{ for lengths over 90 radii}$$

10 in. Steel Z-Bar Columns
Section: 4 Z-bars 5 in. deep and 1 web plate 7 in. × thickness of Z-bars

Length of Column (ft)	5/16 Metal = 53.7 lb = 15.8 in.² r (min.) = 3.08	3/8 Metal = 64.7 lb = 19.0 in.² r (min.) = 3.13	7/16 Metal = 75.8 lb = 22.3 in.² r (min.) = 3.18	1/2 Metal = 83.3 lb = 24.5 in.² r (min.) = 3.10	9/16 Metal = 94.2 lb = 27.7 in.² r (min.) = 3.15	5/8 Metal = 105.2 lb = 30.9 in.² r (min.) = 3.21	11/16 Metal = 111.0 lb = 32.7 in.² r (min.) = 3.13	3/4 Metal = 122.8 lb = 35.8 in.² r (min.) = 3.18	12/16 Metal = 132.6 lb = 39.0 in.² r (min.) = 3.25
22 and under	94.7	114.2	133.9	147.0	166.2	185.6	196.0	214.9	234.0
24	92.8	112.6	133.1	144.6	164.8	185.3	193.6	213.9	234.0
26	89.3	108.6	128.3	139.2	158.7	178.7	186.5	206.2	226.6
28	85.8	104.4	123.5	133.8	152.7	172.1	179.3	198.5	218.4
30	82.3	100.2	118.7	128.4	146.7	165.5	172.2	190.8	210.2
32	78.8	96.1	113.8	123.0	140.7	158.9	165.0	183.1	202.0
34	75.3	91.9	109.1	117.6	134.7	152.3	157.9	175.4	193.8
36	71.8	87.8	104.3	112.2	128.7	145.7	150.7	167.8	185.6
38	68.3	83.6	99.5	106.8	122.7	139.1	143.6	160.0	177.4
40	64.8	79.4	94.7	101.4	116.7	132.5	136.5	152.3	169.1
42	61.3	75.3	89.9	96.0	110.6	125.9	129.4	144.6	160.9
44	57.7	71.1	85.1	90.6	104.6	119.3	122.2	136.9	152.7
46	54.2	67.0	80.3	85.2	98.6	112.7	115.1	129.2	144.5
48	50.7	62.8	75.5	79.8	92.6	106.1	107.9	121.5	136.3
50	47.2	58.6	70.7	74.4	86.6	99.5	100.8	113.8	128.1

12 in. Steel Z-Bar Columns
Section: 4 Z-bars 6 in. deep and 1 web plate 8 in. × thickness of Z-bars

Length of Column (ft)	3/8 Metal = 72.7 lb = 21.4 in.² r (min.) = 3.67	7/16 Metal = 85.2 lb = 25.0 in.² r (min.) = 3.72	1/2 Metal = 97.8 lb = 28.8 in.² r (min.) = 3.77	9/16 Metal = 106.2 lb = 31.2 in.² r (min.) = 3.70	5/8 Metal = 118.5 lb = 34.8 in.² r (min.) = 3.75	11/16 Metal = 130.9 lb = 38.5 in.² r (min.) = 3.73	3/4 Metal = 137.8 lb = 40.5 in.² r (min.) = 3.68	13/16 Metal = 149.9 lb = 44.1 in.² r (min.) = 3.66	7/8 Metal = 162.1 lb = 47.7 in.² r (min.) = 3.64
26 and under	128.3	150.3	172.6	187.3	209.1	231.0	243.0	264.5	286.1
28	127.0	149.7	172.5	186.0	208.9	230.3	240.8	261.4	282.1
30	123.0	145.1	167.6	180.2	202.5	223.3	233.2	253.2	273.2
32	119.0	140.5	162.4	174.5	196.1	216.3	225.7	245.0	264.2
34	115.8	135.9	157.2	168.7	189.8	209.2	218.2	236.7	255.2
36	111.1	131.3	152.0	162.9	183.4	202.1	210.6	228.4	246.3
38	107.1	126.7	146.8	157.1	177.0	195.1	203.1	220.2	237.3
40	103.1	128.1	141.5	151.4	170.7	188.0	195.6	211.9	228.3
42	99.1	117.5	136.3	145.5	164.4	180.9	188.0	203.7	219.4
44	95.1	112.9	131.1	139.8	158.0	173.9	180.5	195.5	210.4
46	91.2	108.3	126.2	134.0	151.6	166.8	172.9	187.2	201.4
48	87.2	103.6	120.7	128.2	145.3	159.8	165.4	179.0	192.4
50	83.2	99.1	115.5	122.4	138.9	152.7	157.9	170.7	183.5

L = length of column, ft
d = least side, in.

Factor of safety = 4, as designated by New York City Building Law

Birkmire gives the following formula for determining the ultimate loads on Phoenix steel columns (ibid., 46):

Table 3-36. (Continued)

Safe Loads for Stress Z-Bar Columns (Tons, 2000 lb)

Plates = 14″ × 3/8″   14″ × 7/16″
Weight = 166.6 lbs/ft
Area = 49.0142
min = 3.804

Steel Z-Bar Columns, Square Ends
Allowed strains per square inch for steel, safety factor 4:
12,000 lb for lengths of 90 radii or under

$$17,100 - 57\frac{l}{r} \text{ for lengths over 90 radii}$$

14 in. Steel Z-Bar Columns
Section: 4 Z-bars 6 1/8 × 11/16 in.   1 web plate 8 × 11/16 in.   2 side plates 14 in. wide

Length of Column (ft)	14 × 3/8 Plates = 166.6 lb/ft area = 49.0 in.² r (min.) = 3.80	14 × 7/16 Plates bE 172.6 lb/ft area = 50.8 in.² r (min.) = 3.81	14 × 1/2 Plates = 178.5 lb/ft area = 52.5 in.² r (min.) = 3.82	14 × 9/16 Plates = 184.5 lb/ft area = 54.3 in.² r (min.) = 3.82	14 × 5/8 Plates = 190.4 lb/ft area = 56.0 in.² r (min.) = 3.83	14 × 11/16 Plates = 196.4 lb/ft area = 57.8 in.² r (min.) = 3.84	14 × 3/4 Plates = 202.3 lb/ft area = 59.5 in.² r (min.) = 3.85	14 × 13/16 Plates = 208.4 lb/ft area = 61.3 in.² r (min.) = 3.85	14 × 7/8 Plates = 214.2 lb/ft area = 63.0 in.² r (min.) = 3.85
28 and under	294.0	304.5	315.0	325.5	336.0	346.5	357.0	367.5	378.0
30	286.6	297.2	307.7	318.3	328.9	339.5	350.0	360.4	370.9
32	277.8	288.1	298.3	308.6	318.9	329.2	339.4	349.5	359.7
34	269.0	278.9	288.9	298.9	308.9	318.9	328.8	338.6	348.6
36	260.1	269.8	279.5	289.2	298.9	308.6	318.2	327.7	337.4
38	251.3	260.7	270.1	279.5	289.0	298.3	307.6	316.8	326.2
40	242.5	251.6	260.7	269.7	278.9	288.0	297.0	306.0	315.0
42	233.7	242.5	251.3	260.1	269.0	277.8	286.4	295.1	303.8
44	224.9	233.3	241.9	250.4	258.9	267.4	275.8	284.2	292.6
46	216.0	224.3	232.4	240.7	249.0	257.2	265.2	273.3	281.5
48	207.2	215.1	223.0	230.9	238.9	246.9	254.6	262.4	270.3
50	198.4	206.0	213.6	221.3	229.0	236.5	244.0	251.5	259.1

14 in. Steel Z-Bar Columns
Section: 4 Z-bars 6 × 3/4 in.   1 web plate 8 × 3/4 in.   2 side plates 14 in. wide

Length of Column (ft)	14 × 3/8 Plates = 173.4 lb/ft area = 51.0 in.² r (min.) = 3.75	14 × 7/16 Plates = 179.4 lb/ft area = 52.8 in.² r (min.) = 3.76	14 × 1/2 Plates = 185.3 lb/ft area = 54.5 in.² r (min.) = 3.77	14 × 9/16 Plates = 191.3 lb/ft area = 56.3 in.² r (min.) = 3.78	14 × 5/8 Plates = 197.2 lb/ft area = 58.0 in.² r (min.) = 3.79	14 × 11/16 Plates = 203.2 lb/ft area = 59.8 in.² r (min.) = 3.80	14 × 3/4 Plates = 209.1 lb/ft area = 61.5 in.² r (min.) = 3.80	14 × 13/16 Plates = 215.2 lb/ft area = 63.3 in.² r (min.) = 3.81	14 × 7/8 Plates = 221.0 lb/ft area = 65.0 in.² r (min.) = 3.82
28 and under	306.0	316.5	327.0	337.5	348.0	358.5	369.0	379.5	390.0
30	296.7	307.2	317.8	328.3	338.9	349.4	359.9	370.5	381.1
32	287.4	297.6	307.9	318.2	328.4	338.7	348.9	359.1	369.4
34	278.1	288.0	298.0	308.0	318.0	327.9	337.8	347.8	257.8
36	268.8	278.4	288.2	297.9	307.4	317.2	326.8	336.4	346.1
38	259.5	268.8	278.3	287.7	297.0	306.4	315.7	325.1	334.5
40	250.2	259.3	268.4	277.5	286.5	295.6	304.7	313.7	322.8
42	240.9	249.7	258.5	267.3	276.1	284.8	293.6	302.4	311.2
44	231.6	240.1	248.6	257.1	265.6	274.1	282.5	291.0	299.6
46	222.4	230.5	238.7	246.9	255.1	263.4	271.5	279.7	287.9
48	213.0	220.9	228.8	236.8	244.7	252.6	260.4	268.3	276.2
50	203.7	211.3	219.0	226.6	234.2	241.8	249.4	257.0	264.6

Table 3-37. Phoenix Steel Columns: Dimensions and Allowable Stresses[a]

Length (ft)	Safe Loads for Phoenix Columns, in lb/sq. in. of Sectional Area Square-End Bearings					
	Col. A	Col. B^1	Col. B^2	Col. C	Col. E	Col. G
10	9,323	9,833	10,024	10,195	10,351	10,411
12	8,885	9.564	9.830	10,067	10,288	10,371
14	8,420	9,267	9,607	9,924	10,215	10,326
16	7,943	8,944	9,364	9,783	10,131	10,275
18	7,463	8,610	9,105	9,575	10,037	10,216
20	6,997	8,260	8,830	9,386	9,935	10,152
22	6,526	7,906	8,541	9,185	9,824	10,082
24	6,090	7,550	8,250	8,973	9,705	10,005
26		7,201	7,955	8,755	9,580	9,926
28		6,860	7,660	8,527	9,450	9,841
30		6,527	7,366	8,297	9,314	9,750
32			7,075	8,070	9,170	9,654
34				7,837	9,021	9,555
36				7,604	8,870	9,441
38				7,375	8,717	9,341
40				7,147	8,561	9,235

Least Radius of Gyration Equals $D \times 0.3636$

One Segment		Diameters (in.)			One Column			
Thickness (in.)	Weight (lb/yard)	d Inside	D Outside	D_1 Over Flanges	Area of Cross-section (in.²)	Weight per ft (lb)	Least Radius of Gyration (in.)	Size of Rivets (in.)
3/16	9½	A—3⅝	4	6 1/16	3.8	12.6	1.45	3/8 × 1 1/8
1/4	12		4 1/8	6 3/16	4.8	16.0	1.50	1 1/4
3/26	14½		4 1/4	6 5/16	5.8	19.3	1.55	1 3/8
2/3	17		4 3/8	6 7/16	6.8	22.6	1.59	1 1/8
1/4	16		5 5/16	8 1/16	6.4	21.3	1.92	1/2 × 1 5/8
‰	19½		5 7/16	8 1/8	7.8	26.0	1.96	1 3/4
2/3	23		5 9/16	8 1/4	9.2	30.6	2.02	1 3/4
7/16	26½	B^1—4 13/16	5 11/16	8 3/8	10.6	35.3	2.07	1 7/8
1/2	30		5 13/16	8 7/16	12.0	40.0	2.11	1 7/8
9/16	33½		5 15/16	8 1/2	13.4	44.6	2.16	2
5/8	37		6 1/16	8 5/8	14.8	49.3	2.20	2 1/8
1/4	18½		6 7/16	9 1/8	7.4	24.6	2.34	1/4 × 1 3/4
5/16	22½		6 9/16	9 1/4	9.0	30.0	2.39	1 3/4
3/8	26½		6 11/16	9 5/16	10.6	35.3	2.43	1 3/4
7/16	30½	B^2—5 15/16	6 13/16	9 3/8	12.2	40.6	2.48	1 7/8
1/2	34½		6 15/16	9 1/2	13.8	46.0	2.52	1 7/8
9/16	38½		7 1/16	9 5/8	15.4	51.3	2.57	2
5/8	42½		7 3/16	9 11/16	17.0	56.6	2.61	2 1/8

$$F_{a(ult)} = \frac{42,000}{1 + \frac{L^2}{50,000\, r^2}}$$

where:
Factor of safety = 4 (to be used)
L = length of column, ft
r = least radii of gyration

Birkmire also provides a set of tables of dimensions and safe loads (as allowable compressive stresses) for phoenix steel columns (derived from the above formula) (see Table 3-39).

Phoenix Iron Co. (1906) (using their original name) provides a table of the dimensions of Phoenix steel columns (including safe loads, tons, for 16-ft-long columns)

Table 3-37. (Continued)

One Segment		Diameters (in.)			One Column			Size of Rivets (in.)
Thickness (in.)	Weight (lb/yard)	d Inside	D Outside	D^1 Over Flanges	Area of Cross-section (in.^{2})	Weight per ft (lb)	Least Radius of Gyration (in.)	
1/4	25		7 11/16	11 9/16	10.0	33.3	2.80	5/8 × 1 7/8
5/16	30		7 13/16	11 5/8	12.0	40.0	2.85	2
3/8	35		7 15/16	11 11/16	14.0	46.6	2.90	2 1/8
7/16	40		8 1/16	11 3/4	16.0	53.3	2.94	2 1/4
1/2	45		8 3/16	11 13/16	18.0	60.0	2.98	2 5/8
9/16	48		8 5/16	11 7/8	19.2	64.0	3.03	2 1/2
5/8	53	C—7 3/16	8 7/16	12	21.2	70.6	3.08	3/4 × 2 5/8
11/16	58		8 9/16	12 1/16	23.2	77.3	3.12	2 3/4
3/4	63		8 11/16	12 3/16	25.2	84.0	3.16	2 7/8
13/16	68		8 13/16	12 5/16	27.2	90.6	3.21	3
7/8	73		8 15/16	12 7/16	29.2	97.3	3.26	2 3/4
1	83		9 3/16	12 9/16	33.2	110.6	3.34	2 7/8
1 1/8	93		9 7/16	13 3/4	37.2	124.0	3.43	3
1 1/4	103		9 11/16	13 15/16	41.2	137.3	3.52	
1/4	28		11 1/2	15 7/16	16.8	56	4.18	5/8 × 2
5/16	32		11 5/8	15 9/16	19.2	64	4.23	2 5/8
3/8	36		11 3/4	15 11/16	21.6	72	4.28	2 1/8
7/16	40		11 7/8	15 13/16	24.0	80	4.32	2 1/4
1/2	44		12	15 7/8	26.4	88	4.36	2 3/8
9/16	48		12 1/8	16	28.8	96	4.40	2 3/8
5/8	53		12 1/4	16 1/16	31.8	106	4.45	2 1/2
11/16	58	E—11	12 3/8	16 3/16	34.8	116	4.50	3/4 × 2 5/8
3/4	63		12 1/2	16 5/16	37.8	126	4.55	2 5/8
13/16	68		12 5/8	16 7/16	40.8	136	4.60	2 3/4
7/8	73		12 3/4	16 5/8	43.8	146	4.64	2 3/4
1	83		13	16 3/4	49.8	166	4.73	2 7/8
1 1/8	93		13 1/4	17	55.8	186	4.82	3
1 1/4	103		13 1/2	17 3/16	61.8	206	4.91	3 1/8
5/16	30		15	19 1/8	24	80.0	5.45	5/8 × 2
3/8	35		15 1/8	19 1/4	28	93.3	5.50	2
7/16	40		15 1/4	19 3/8	32	106.6	5.55	2 1/8
1/2	45		15 3/8	19 7/16	36	120.0	5.59	2 1/4
9/16	50		15 1/2	19 1/2	40	133.3	5.63	2 3/8
5/8	55		15 5/8	19 5/8	44	146.6	5.68	2 1/2
11/16	60		15 3/4	19 3/4	48	160.0	5.72	3/4 × 2 5/8
3/4	65	G—14 3/8	15 7/8	19 7/8	52	173.3	5.77	2 5/8
13/16	70		16	20	56	186.6	5.82	2 3/4
7/8	75		16 1/8	20 1/8	60	200.0	5.87	2 7/8
1	85		16 3/8	20 3/8	68	226.6	5.95	3
1 1/8	95		16 5/8	20 5/8	76	253.3	6.04	3 1/8
1 1/4	105		16 7/8	20 3/4	84	280.0	6.14	3 1/4
1 3/8	115		17 1/8	21	90	306.6	6.23	3 5/8

[a] From Birkmire 1894, 46–48.

(see Table 3-40), as well as a table of the ultimate strength of columns, medium steel (ultimate stress), which can be applied to any built-up or compound column of different shapes and plates (see Table 3-41).

The Gray steel column was patented by J. H. Gray and was immediately very popular with designers due to its versatility. The configuration was extremely strong while economically utilizing only angle sections and steel tie-plates either 8 or 9 in. long. The weight and thus the strength could be significantly increased by utilization of thicker and/or unequal leg angles. Larger angles did not increase the physical size by very much, and connections for beams, girders, and bracing were easily made. Cover plates on each face would also add strength without a large increase in size. Long or continuous columns were easily fabricated, and butt joints were made in line with tie plates. The butt joints of the angles did not have to occur at the same point vertically on the column; it was logical to stagger the angle joints unless a change in

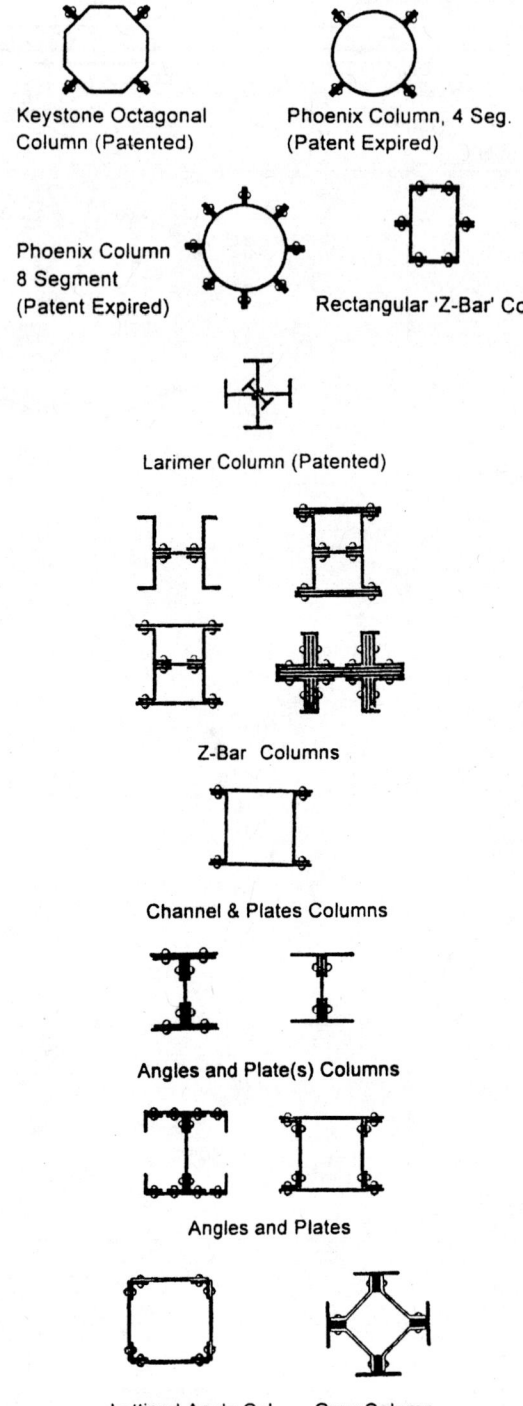

Figure 3-54. Steel column sections used in early skeletal frame construction. (*From International Correspondence Schools 1905, para. 16, 7–9.*)

size at or slightly above a floor was to take place. See Figure 3-55 for sections of Gray columns, and Table 3-42 for a table of safe loads for Gray columns given by Kidder (1902).

The Gray steel column was very popular with designers in spite of its being patented. It was considered extremely economical and very well suited for load carrying, there being a theory at the time that most of the steel area should be as far away from the center of the section as possible. Interior columns would also be symmet-

Table 3-38. Phoenix Segmental Columns[a] [A, B1, B2 and C = 4 segments, t = 6 segments, G = 8 segments]

One Segment		Diameters (in.)			Area of Cross-Section (in.²)	Weight per ft (lb)	Least Radius of Gyration (in.)	Safe Load, in Net Tons for 16 Lengths
Thickness (in.)	Weight per ft (lb)	d Inside	D Outside	D^1 Over Flanges				
3/16	3.2		4	6 1/16	3.8	12.9	1.45	18.1
1/4	4.1	A	4 1/8	6 3/16	4.8	16.3	1.50	23.5
5/16	4.9	3 5/8	4 1/4	6 5/16	5.8	19.7	1.55	29.1
3/8	5.8		4 3/8	6 7/16	6.8	23.1	1.59	34.7
1/4	5.4		5 3/8	8 1/8	6.4	21.8	1.95	36.8
5/16	6.6		5 1/2	8 3/16	7.8	26.5	2.00	45.0
3/8	7.8	B1	5 5/8	8 5/16	9.2	31.3	2.04	54.0
7/16	9.0	4 7/8	5 3/4	8 7/16	10.6	36.0	2.09	62.9
1/2	10.2		5 7/8	8 1/2	12.0	40.8	2.13	72.0
9/16	11.4		6	8 9/16	13.4	45.6	2.18	80.4
5/8	12.6		6 1/8	8 11/16	14.8	50.3	2.23	88.8
1/4	6.3		6 9/16	9 1/4	7.4	25.2	2.39	44.4
5/16	7.6		6 11/16	9 3/8	9.0	30.6	2.43	54.0
3/8	9.0	B2	6 13/16	9 7/16	10.6	36.0	2.48	63.6
7/16	10.4	6 1/16	6 15/16	9 1/2	12.2	41.5	2.52	73.2
1/2	11.7		7 1/16	9 5/8	13.8	46.9	2.57	82.8
9/16	13.1		7 3/16	9 3/4	15.4	52.4	2.61	92.4
5/8	14.4		7 5/16	9 13/16	17.0	57.8	2.66	102.0
1/4	8.5		7 13/16	11 11/16	10.0	34.0	2.84	60.0
5/16	10.3		7 15/16	11 3/4	12.1	41.3	2.88	72.6
3/8	12.0		8 1/16	11 13/16	14.1	48.0	2.93	84.6
7/16	13.7		8 3/16	11 7/8	16.0	54.6	2.97	96.0
1/2	15.3		8 5/16	11 15/16	18.0	61.3	3.01	108.0
9/16	17.0		8 7/16	12	20.0	68.0	3.06	119.4
5/8	18.3	C	8 9/16	12 1/16	21.9	74.6	3.11	131.4
11/16	20.7	7 5/16	8 11/16	12 3/16	24.3	82.6	3.16	145.8
3/4	22.7		8 13/16	12 5/16	26.6	90.6	3.20	159.6
13/16	24.3		8 15/16	12 7/16	28.6	97.3	3.24	171.6
7/8	26.0		9 1/16	12 1/2	30.6	104.0	3.29	183.6
1	29.7		9 5/16	12 5/8	34.9	118.6	3.34	208.8
1 1/8	33.0		9 9/16	12 13/16	38.8	132.0	3.48	232.8
1 1/4	36.3		9 13/16	13	42.7	145.3	3.57	256.2
1/4	9.3		11 9/16	15 1/2	16.5	56.0	4.20	99.0
5/16	10.8		11 11/16	15 5/8	19.1	65.0	4.25	114.6
3/8	12.3		11 13/16	15 3/4	21.7	74.0	4.29	130.2
7/16	14.0		11 15/16	15 7/8	24.7	84.0	4.34	148.2
1/2	15.7		12 1/16	15 15/16	27.6	94.0	4.38	165.6
9/16	17.3		12 3/16	16 1/16	30.6	104.0	4.43	183.6
5/8	19.0	E	12 5/16	16 3/16	33.5	114.0	4.48	201.0
11/16	20.7	11 1/16	12 7/16	16 5/16	36.4	124.0	4.52	218.4
3/4	22.7		12 9/16	16 7/16	40.0	136.0	4.56	240.0
13/16	24.3		12 11/16	16 9/16	43.0	146.0	4.61	258.0
7/8	26.0		12 13/16	16 11/16	45.9	156.0	4.66	275.4
1	29.3		13 1/16	16 13/16	51.7	176.0	4.73	310.2
1 1/8	32.7		13 5/16	17 1/16	57.6	196.0	4.84	345.6
1 1/4	36.0		13 9/16	17 5/16	63.5	216.0	4.93	381.0

rical, and that would be of advantage. Actually, the loads from the beams would be transmitted into the double angles in a manner that would place the loads near the center of gravity of the angles, thus forming a sort of subcolumn. One building that utilized Gray columns was the Reliance Building, Chicago, 1890–94, by Burnham & Root Architects. Kidder states that "In the Reliance Building, Chicago, there is a Gray column 290 feet long, built in one piece at the shop" (ibid., 289c).

Kidder also gives a list of the notable skyscrapers of the period just before the turn of the century that gives the building, the architect, the number of stories and the kind of column used in the construction (see Table 3-43).

American School of Correspondence (1908) provides a table of the allowable stresses for materials and column formulas as specified by the building laws of various cities (see Table 3-44).

Table 3-38. (Continued)

One Segment		Diameters (in.)			Area of Cross-Section (in.²)	Weight per ft (lb)	Least Radius of Gyration (in.)	Safe Load, in Net Tons for 16 Lengths
Thickness (in.)	Weight per ft (lb)	d Inside	D Outside	D¹ Over Flanges				
5/16	10.3		15¼	19⅜	24.3	82.6	5.54	145.2
3/8	12.0		15⅜	19½	28.2	96.0	5.59	168.6
7/16	13.7		15½	19⅝	32.1	109.3	5.64	192.0
½	15.3		15⅝	19¹¹⁄₁₆	36.0	122.6	5.68	216.0
9/16	17.0		15¾	19¾	40.0	136.0	5.73	239.4
5/8	18.7		15⅞	19⅞	43.9	149.3	5.77	262.8
11/16	20.3	G	16	20	47.8	162.6	5.82	286.2
3/4	22.0	14⅝	16⅛	20⅛	51.7	176.0	5.88	310.2
13/16	23.7		16¼	20¼	55.7	189.3	5.91	333.6
7/8	25.3		16⅜	20⅜	59.6	202.6	5.95	357.6
1	28.7		16⅝	20⅝	67.4	229.3	6.04	404.4
1⅛	32.0		16⅞	20⅞	75.3	256.0	6.13	451.8
1¼	35.3		17⅛	21	83.1	282.6	6.27	498.6
1⅜	38.7		17⅜	21¼	90.9	309.3	6.32	545.4

ᵃFrom Phoenix Iron Co. 1906, 43, 44. A, B1, B2, and C = 4 segments, E = 6 segments, G = 8 segments; least radius of gyration equals D × 0.3636.

Carnegie Steel Co. (1913) contains the Standard Specifications for Structural Steel for Buildings, adopted August 16, 1909, by the American Society for Testing Materials (ASTM), which were at this time the standard for the industry, at least for all manufacturers that subscribed to the ASTM. There is no mention of the ASTM A9 notation called for in the AISC Historic Shapes Volume (Ferris 1953–78), but the tensile strength specification is the same, 55,000 − 65,000 psi ultimate. The yield Point is stated to be 0.5 of the ultimate (Carnegie Steel Co. 1913, 10–14).

Carnegie Steel Co. (1913) also includes a section on the Specifications for Steel Structures by the American Bridge Company. An important predecessor to the Standard AISC Specifications. These specifications, adopted in 1912, contain design, fabrication, and construction information. There are important provisions in these specifications that appear for the first time, as far as this researcher has been able to find (Carnegie Steel Co. 1913, 126–127):

1. Cranes and moving machinery are stated to require an additional 25 percent for impact moving load (no mention of lateral force from moving loads).
2. Wind pressure, 20 psf sides and ends of buildings and on vertical projection of roofs (no mention of wind pressure increases for height above the ground). Wind pressure, 30 psf on all vertical surfaces during construction.
3. Columns:

$$F_a = 19,000 - 100 \frac{L}{r} \text{ (in psi)} \quad \text{(but not to exceed 13,000 psi)}$$

where:

L = length of column, in.
r = least radii of gyration Main members, L/r shall not exceed 120. Secondary members, L/r shall not exceed 200. Provision must be made for eccentric loading of columns.

Carnegie Steel Co. (1913) also provides a table of comparisons of the various compression formulae adopted by the American Bridge Company, The American Railroad Engineers Association, Gordon, and the cities of New York, Philadelphia, and Boston (see Table 3-45).

Freitag (1909, 201) gives an adapted form of the Gordon or Rankine formula for eccentrically loaded columns:

Table 3-39. Phoenix Built-up or Compound Columns[a]

Ultimate Strength of Columns
Medium Steel

For different proportions of length in ft (= L).
To least radius of gyration in in. (= r).

Ultimate strength in psi:

Column Square Bearing: $\dfrac{50,000}{1 + \dfrac{(12L)^2}{36,000 r^2}}$;

Column Pin and Square Bearing: $\dfrac{50,000}{1 + \dfrac{(12L)^2}{24,000 r^2}}$;

Column Pin Bearing: $\dfrac{50,000}{1 + \dfrac{(12L)^2}{18,000 r^2}}$;

To obtain safe resistance:
  For quiescent loads, as in buildings, divide by 4.
  For moving loads, as in bridges, divide by 5.

$\dfrac{l}{r}$	Ultimate Strength (psi)		
	Square	Pin and Square	Pin
3.0	48,260	47,440	46,640
3.2	48,030	47,110	46,210
3.4	47,790	46,760	45,770
3.6	47,540	46,390	45,300
3.8	47,270	46,010	44,820
4.0	46,990	45,620	44,330
4.2	46,710	45,210	43,820
4.4	46,410	44,800	43,300
4.6	46,100	44,370	42,760
4.8	45,780	43,930	42,220
5.0	45,460	43,480	41,670
5.2	45,120	43,020	41,110
5.4	44,780	42,560	40,540
5.6	44,430	42,080	39,970
5.8	44,070	41,600	39,400
6.0	43,710	41,120	38,820
6.2	43,340	40,630	38,240
6.4	42,960	40,140	37,660
6.6	42,580	39,640	37,080
6.8	42,200	39,140	36,500
7.0	41,810	38,640	35,920
7.2	41,410	38,140	35,340
7.4	41,020	37,640	34,770
7.6	40,620	37,130	34,200
7.8	40,210	36,630	33,630
8.0	39,810	36,130	33,070
8.2	39,400	35,630	32,510
8.4	38,990	35,130	31,960
8.6	38,590	34,630	31,410
8.8	38,180	34,140	30,870
9.0	37,760	33,650	30,340
9.2	37,350	33,160	29,810
9.4	36,940	32,680	29,290
9.6	36,530	32,200	28,780
9.8	36,120	31,720	28,280
10.0	35,710	31,250	27,780
10.2	35,310	30,780	27,290
10.4	34,900	30,320	26,810
10.6	34,500	29,870	26,330
10.8	34,090	29,420	25,870

Table 3-39. (Continued)

$\dfrac{L}{r}$	Ultimate Strength (psi)		
	Square	Pin and Square	Pin
11.0	33,690	28,970	25,410
11.2	33,290	28,530	24,960
11.4	32,900	28,090	24,510
11.6	32,510	27,670	24,080
11.8	32,110	27,240	23,650
12.0	31,730	26,820	23,230
12.2	31,340	26,410	22,820
12.4	30,960	26,010	22,420
12.6	30,580	25,610	22,030
12.8	30,210	25,210	21,640
13.0	29,830	24,830	21,260
13.2	29,460	24,450	20,890
13.5	28,930	23,890	20,350
13.8	28,380	23,340	19,810
14.0	28,030	22,980	19,470
14.2	27,680	22,630	19,130
14.5	27,180	22,110	18,650
14.8	26,650	21,610	18,170
15.0	26,320	21,280	17,860
15.2	25,990	20,950	17,550
15.5	25,500	20,490	17,110
15.8	25,020	20,020	16,680
16.0	24,700	19,720	16,400
16.2	24,390	19,420	16,130
16.5	23,940	18,990	15,740
16.8	23,490	18,560	15,350
17.0	23,190	18,290	15,100
17.2	22,900	18,020	14,850
17.5	22,480	17,630	14,490
17.8	22,050	17,240	14,150
18.0	21,780	16,980	13,920
18.2	21,510	16,740	13,700
18.5	21,100	16,380	13,380
18.8	20,720	16,020	13,060
19.0	20,460	15,790	12,860
19.2	20,210	15,570	12,660
19.5	19,840	15,240	12,360
19.8	19,470	14,920	12,090
20.0	19,230	14,710	11,910
20.2	19,000	14,500	11,730
20.5	18,650	14,200	11,460
20.8	18,310	13,910	11,210

[a] From Phoenix Iron Co. 1906, 88, 89.

$$F_a(\text{ult}) = \dfrac{E}{1 + a\left(\dfrac{L^2}{r^2}\right) + X_o\left(\dfrac{y_1}{r}\right)}$$

where:

F_a = ultimate stress, in compression, psi
E = elastic limit of material
a = constant, end conditions
L = length of column, in.
r = radius of gyration

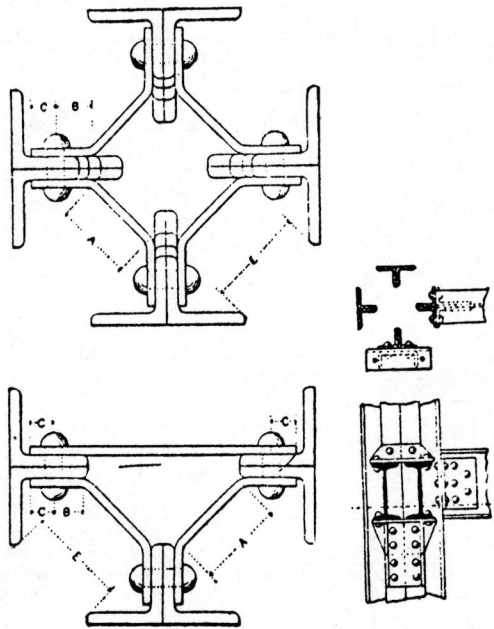

Figure 3-55. Gray steel columns. (*From Kidder 1902, 289d.*)

x_o = eccentricity, in.
y_1 = distance, C.G. to extreme fiber of section, in.

Milo S. Ketchum was one of the leading authorities of structural design around 1920. He was Professor and Dean of the College of Engineering of the University of Illinois and the author of several volumes on structural engineering. Ketchum wrote a set of General Specifications for Steel Frame Buildings in which he describes in detail the requirements for designing and specifying structural steels for building construction. The third edition of the specifications, 1914, contains the following formulae for the design of steel columns (Ketchum 1910, 57; 1921, 604, 605):

$$F_a \text{ (allowable)} = 16000 \text{ psi} - 70 \frac{L}{r}$$

where:
F_a = allowable unit stress, psi
L = length of column, in.
r = least radius of gyration

Columns subjected to combined axial and eccentric loading would be designed by the following formula:

$$F_a \text{ (allowable)} = \frac{P}{A} + \frac{My_1}{I \pm \frac{P(L^2)}{10E}}$$

where:
P = axial load, lb
A = area of section, in.^{2}
M = (ecc. load) (eccentricity), lb-in.
y_1 = distance, N.A. to extreme fiber (bending direction)
I = moment of inertia (bending direction)
E = modulus of elasticity, psi

Table 3-40. Load Tables, Gray Steel Columns[a]

Safe Loads for Gray Columns

Computed by the Formula $W = 17{,}100 - 57\dfrac{l}{r}$

12 in. Square Columns

No. of Pieces	Dimensions of Angles — Length of Legs	Thick	Area (in.²)	I	r	12 ft	16 ft	20 ft	30 ft
b { 8	2½ × 2½	5/16	11.76	172	3.8	175	165	160	140
8	2½ × 2½	3/8	13.84	202	3.8	205	195	185	160
8	3 × 2½	5/16	12.96	198	3.9	195	185	175	150
8	3 × 2½	3/8	15.36	234	3.9	230	220	205	180
8	3 × 3	5/16	14.24	206	3.8	210	200	190	165
8	3 × 3	3/8	16.88	241	3.8	250	240	225	195
8	3 × 3	7/16	19.52	276	3.8	290	275	260	225
8	3 × 3½	5/16	15.44	209	3.7	230	215	205	180
8	3 × 3½	3/8	18.40	245	3.7	270	260	245	210
8	3 × 3½	7/16	21.20	282	3.7	315	300	280	245
8	3 × 3½	½	24.00	318	3.7	355	340	320	275
8	3 × 4	5/16	16.72	213	3.5	245	230	220	185
8	3 × 4	3/8	19.84	249	3.5	290	275	260	220
8	3 × 4	7/16	22.96	285	3.5	335	320	300	255
8	3 × 4	½	26.00	321	3.5	380	360	340	290
8	3 × 4	9/16	28.96	357	3.5	425	405	380	325
c { 8	3 × 5	3/8	22.88	255	3.3	335	315	295	250
8	3 × 5	7/16	26.48	291	3.3	385	365	340	290
8	3 × 5	½	30.00	327	3.3	435	410	385	325
8	3 × 5	9/16	33.44	363	3.3	485	460	430	365

14 in. Square Columns

No. of Pieces	Length of Legs	Thick	Area (in.²)	I	r	12 ft	16 ft	20 ft	30 ft
b { 8	2½ × 2½	5/16	11.76	238	4.5	180	170	165	145
8	2½ × 2½	3/8	13.84	280	4.5	210	200	195	170
8	3 × 2½	5/16	12.96	274	4.6	195	190	180	160
8	3 × 2½	3/8	15.36	325	4.6	235	225	215	190
8	3 × 3	5/16	14.24	286	4.5	215	205	200	180
8	3 × 3	3/8	16.88	336	4.5	255	245	235	210
8	3 × 3	7/16	19.52	386	4.5	295	285	275	245
8	3 × 3½	5/16	15.44	293	4.3	235	225	215	190
8	3 × 3½	3/8	18.40	340	4.3	280	265	255	225
8	3 × 3½	7/16	21.20	386	4.3	320	305	295	260
8	3 × 3½	½	24.00	433	4.3	365	350	330	295
c { 8	3 × 3½	9/16	26.72	479	4.3	405	385	370	330
8	3 × 3½	5/8	29.36	526	4.3	445	425	405	360
8	3 × 3½	11/16	32.00	572	4.3	485	465	445	395
8	3 × 3½	3/4	34.48	619	4.3	520	500	480	425
8	3 × 3½	13/16	36.96	666	4.3	560	535	515	455
b { 8	3 × 4	5/16	16.72	300	4.2	250	240	230	205
8	3 × 4	3/8	19.84	348	4.2	300	285	275	240
8	3 × 4	7/16	22.96	396	4.2	345	330	315	280
8	3 × 4	½	26.00	444	4.2	390	375	360	315
c { 8	3 × 4	9/16	28.96	491	4.2	435	420	400	350
8	3 × 4	5/8	31.84	539	4.2	480	460	440	385
8	3 × 4	11/16	34.72	587	4.2	525	500	480	420
8	3 × 4	3/4	37.52	635	4.2	565	540	520	455
8	3 × 4	13/16	40.24	688	4.2	605	580	555	490

[a] From Kidder 1902, 289e, f.
[b] tie-plates 8 in., 2 ft 6 in. C. to C.
[c] tie-plates 9 in., 2 ft 6 in. C. to C.

Table 3-41. Columns used in Principal Office Buildings[a]

Building	City	Architect	Number of Stories	Kind of Columns
Manhattan	Chicago	W. L. B. Jenney	16	Cast
The Fair	Chicago	W. L. B. Jenney	9	Z-bar
Y.M.C.A.	Chicago	W. L. B. Jenney	13	Z-bar
Isabella	Chicago	W. L. B. Jenney	10	Z-bar
New York Life	Chicago	Jenney & Mundie	12	Steel plates and angles
Fort Dearborn	Chicago	Jenney & Mundie	12	Channels and plates
Tacoma	Chicago	Holabird & Roche	13	Cast
Pontiac	Chicago	Holabird & Roche	14	Z-bar
Venetian	Chicago	Holabird & Roche	13	Z-bar
Monadnock (New)	Chicago	Holabird & Roche	17	Z-bar
Old Colony	Chicago	Holabird & Roche	17	Z-bar and Phoenix
Champlain	Chicago	Holabird & Roche	15	Z-bar
Marquette	Chicago	Holabird & Roche	16	Z-bar
Auditorium	Chicago	Adler & Sullivan	10 and 17	Cast iron
Schiller Theatre	Chicago	Adler & Sullivan	13 and 17	Z-bar and Phoenix
Stock Exchange	Chicago	Adler & Sullivan	13	Z-bar
Rookery	Chicago	Burnham & Root	12	Cast
Woman's Temple	Chicago	Burnham & Root	13	Z-bar
Masonic Temple	Chicago	Burnham & Root	20	Plates and angles
Ashland	Chicago	Burnham & Root	16	Z-bar
Reliance Bldg.	Chicago	D. H. Burnham & Co.	15	Gray
Fisher	Chicago	D. H. Burnham	18	Gray
Great No. Theatre	Chicago	D. H. Buurnham	16	Gray
Title & Trust	Chicago	Henry Ives Cobb	16	Phoenix
Owings	Chicago	Henry Ives Cobb	14	Cast
American Surety	New York	Bruce Price	21	Angles & plates Z-bar
Manhattan Life Insurance	New York	Kimball & Thompson	18	Cast, 5 stories plates and angles
Meyer, Jonasson	New York	Geo. B. Post	14	Plates and angles
Havemeyer (Ct. St.)	New York	Geo. B. Post	15	Plates and angles
New York World	New York	Geo. B. Post	22	Phoenix
Union Trust	New York	Geo. B. Post	10	Phoenix
Postal Telegraph	New York	Harding & Gooch	14	Cast
Hotel Waldorf	New York	H. J. Hardenberg	12	Cast
American Tract Society	New York	R. H. Robertson	20	Riveted

[a] From Kidder 1902, 289j.

L = length of member, in.

Shedd (1934, 213–225) gives a comparison of the best known formulae in use:

$$F_{a(all)} = F_{(max)} - K\frac{L}{r}$$

where:

K = a constant, set by Shedd
L = length of column, in.
r = least radius of gyration
$F_{(max)}$ = maximum allowable fiber stress

$$F_a = 16{,}000 - 70\frac{L}{r} \quad \text{(max = 14,000 psi) A.R.E.A., 1910 and others.}$$

$$F_a = 15{,}000 - 50\frac{L}{r} \quad \text{(max = 12,500 psi) A.R.E.A., 1920, 23, 25, 31}$$

$$F_a = 19{,}000 - 100\frac{L}{r} \quad \text{(max = 13,000 psi) American Bridge Co., 1912}$$

$$F_a = 16{,}000 - 60\frac{L}{r} \quad \text{(fixed ends) J. L. Waddell, 1916}$$

Table 3-42. Allowable Stresses, Building Laws of Cities[a]

	New York 1900	Chicago 1900	Philadelphia 1903	Boston 1900
Compression, direct				
Rolled steel	16,000 psi		16,250 psi	
Cast steel	16,000 psi		16,250 psi	
Wrought iron	12,000 psi		12,500 psi	
Steel pins and rivets (bearing)	20,000 psi	20,000 psi		18,000 psi
Wrought iron pins and rivets, bearing	15,000 psi	15,000 psi		15,000 psi
Columns				
Mild steel	$15{,}200 - 58\frac{L}{r}$	15,000	$\dfrac{14{,}500}{1 + \dfrac{L^2}{13{,}500\, r^2}}$	12,000
Medium steel	$15{,}200 - 58\frac{L}{r}$	15,000	$\dfrac{16{,}250}{1 + \dfrac{L^2}{11{,}000\, r^2}}$	12,000
Wrought iron	$14{,}000 - 80\frac{L}{r}$	12,000	$\dfrac{12{,}500}{1 + \dfrac{L^2}{15{,}000\, r^2}}$	10,000
Cast iron	$11{,}300 - 30\frac{L}{r}$	10,000	$\dfrac{17{,}500}{1 + \dfrac{L^2}{400\, r^2}}$	

[a]From Turneaure 1908, 53.

$$F_a = 16{,}000 - 80\frac{L}{r} \quad \text{(hinged ends) J. L. Waddell, 1916}$$

Rankine-Gordon formula:

$$F_{a\,(all)} = \frac{16{,}000}{1 + \dfrac{1}{13{,}500}\left(\dfrac{L}{r}\right)^2} \quad \text{maximum } L/r = 40$$

$$F_{a\,(all)} = \frac{18{,}000}{1 + \dfrac{1}{18{,}000}\left(\dfrac{L}{r}\right)^2} \quad \text{maximum } L/r = 60 \quad \text{(A.I.S.C. column formula)}$$

Shedd also presents formula for designers to use in the determination of the stresses in a column subjected to a combination of axial plus eccentric loading:

$$F_a = \frac{P}{A} + \frac{M_c}{I}\left(1 - \frac{1}{\dfrac{PL^2}{10\,EI}}\right)$$

where:
- P = Axial load, lb
- A = area of section, sq. in.^{2}
- M = moment from eccentricity ($P_1 \times e_1$)
- c = distance, N.A. to extreme fiber, in.
- P_1 = eccentric load, lb
- e_1 = Eccentricity, in.
- I = Moment of inertia, in.^{4}
- E = modulus of elasticity of material, psi
- L = length of column, in.

The American Institute of Steel Construction was formed as an association of steel manufacturers and quickly undertook the work of promoting uniform practice in the

Table 3-43. Steel Column Allowable Stresses[a]

Comparison of Compression Formulas
Allowable Unit Stresses in psi

$\dfrac{L}{r}$	A. B. Co. See Construction Specifications	A. R. E. Association Chicago $16{,}000 - 70\dfrac{L}{r}$ 14,000 max.	Gordon $\dfrac{12{,}500}{1 + \dfrac{L^2}{36{,}000\,r^2}}$	New York $15{,}200 - 58\dfrac{L}{r}$	Philadelphia $\dfrac{16{,}250}{1 + \dfrac{L^2}{11{,}000\,r^2}}$	Boston $\dfrac{16{,}000}{1 + \dfrac{L^2}{20{,}000\,r^2}}$
0	13,000	14,000	12,500	15,200	16,250	16,000
5	13,000	14,000	12,490	14,910	16,215	15,980
10	13,000	14,000	12,460	14,620	16,100	15,920
15	13,000	14,000	12,420	14,330	15,925	15,820
20	13,000	14,000	12,365	14,040	15,680	15,690
25	13,000	14,000	12,285	13,750	15,375	15,515
30	13,000	13,900	12,195	13,460	15,020	15,310
35	13,000	13,550	12,090	13,170	14,620	15,075
40	13,000	13,200	11,970	12,880	14,185	14,815
45	13,000	12,850	11,835	12,590	13,725	14,530
50	13,000	12,500	11,690	12,300	13,240	14,220
55	13,000	12,150	11,530	12,010	12,745	13,900
60	13,000	11,800	11,365	11,720	12,240	13,560
65	12,500	11,450	11,185	11,430	11,740	13,210
70	12,000	11,100	11,000	11,140	11,240	12,850
75	11,500	10,750	10,810	10,850	10,750	12,490
80	11,000	10,400	10,615	10,560	10,275	12,120
85	10,500	10,050	10,410	10,270	9,810	11,755
90	10,000	9,700	10,205	9,980	9,360	11,390
95	9,500	9,350	9,995	9,690	8,930	11,025
100	9,000	9,000	9,785	9,400	8,510	10,670
105	8,500	8,650	9,570	9,110	8,115	10,315
110	8,000	8,300	9,355	8,820	7,740	9,970
115	7,500	7,950	9,140	8,530	7,380	9,630
120	7,000	7,600	8,930	8,240	7,035	9,300
125	6,750	7,250	8,715		6,715	
130	6,500	6,900	8,510		6,405	
135	6,250	6,550	8,300		6,115	
140	6,000	6,200	8,095		5,840	
145	5,750	5,850	7,890			
150	5,500	5,500	7,690			
155	5,250		7,495			
160	5,000		7,305			
165	4,750		7,120			
170	4,500		6,935			
175	4,250		6,755			
180	4,000		6,580			
185	3,750		6,410			
190	3,500		6,240			
195	3,250		6,080			
200	3,000		5,920			

Name of Formula	Abbreviation	Maximum Ratio of l/r Main Members	Maximum Ratio of l/r Bracing Struts
American Bridge Company	A. B.	120	200
American Railway Engineering Association	A. R. E.	100	120
Chicago Building Law	C.	120	150
Gordon	G.	..	..
New York Building Law	N. Y.	120	..
Philadelphia Building Law	P.	140	..
Boston Building Law	B.	120	..

[a] From Carnegie Steel Co. 1913, 254.

industry. Uniform practice would not only include the manufacture of the steel, it would also set standards for the design, fabrication, and erection of structural steel for buildings. The work to develop such standard specifications began in the early 1920s and concluded with the completion of the Standard Specification for Structural Steel for Buildings adopted by the American Institute of Steel Construction in 1923. This specification has been carefully updated over the years and is still the code in use today. The American Institute of Steel Construction developed and produced the First edition of the manual *Steel Construction* in 1923. It contains the specifications, tables, and data required for the use of structural steels, including formulae for the design of steel columns (American Institute of Steel Construction 1930, 9):

$$F_a = \frac{18,000}{1 + \dfrac{\left(\dfrac{L}{r}\right)^2}{18,000}}$$

where:
  L = unsupported length of column, in.
  r = least radius of gyration, in.
    Main compression members: L/r shall not exceed 120.
    Bracing and secondary members: L/r shall not exceed 200.

Bethlehem Steel Co. (1926) gives the same basic column formula as the AISC (above), however, Bethlehem gives a maximum allowable stress on an unsupported column of 15,000 psi while also allowing main member L/r not to exceed 120, secondary member L/r not to exceed 150 (ibid., 97). Steel manufacturers continued to produce their own handbooks for a number of years after the AISC volumes were produced, and still do, but they contain mostly data and descriptions of specialty or exotic items not included in the AISC Manual.

The American Institute of Steel Construction (1939) specification provides the following formulae for steel columns (ibid., 264, 265, 263):

For L/r up to and not exceeding 120:

$$F_a = 17,000 - .485 \left(\frac{L}{r}\right)^2$$

where:
  L = length of column, in.
  r = least radii of Gyration, in.

For L/r values greater than 120:

$$F_a = \frac{18,000}{1 + \dfrac{\left(\dfrac{L}{r}\right)^2}{18,000}}$$

For members subjected to both axial and bending stresses (eccentrically loaded columns):

$$\frac{f_a}{F_a} + \frac{f_b}{F_b} \cdots \text{shall not exceed unity}$$

where:
  f_a = actual axial stress, psi
  F_a = allowable axial compressive stress, psi
  f_b = actual bending stress, psi
  F_b = allowable bending stress, psi

If an eccentricity was applied to both axes, then the second term would actually appear a second time.

The AISC Specifications/Code remained basically the same through the 1950s. All column formulae were loose derivations based upon Euler's original work of 1759, many attempts to correlate test results with mathematical formulae have always proven that the elder work of 1759 was and is still the best and most accurate. The present Code (American Institute of Steel Construction 1989) contains the same formulae for the design of steel columns that have been in the Code since the sixth edition in 1964. The present method for determining the allowable axial stress in a given steel section is through the use of a formula that closely resembles Euler's work (ibid., pt 5, sec. E, 5-42).

On the gross section of axially loaded columns:

when $k\frac{L}{r} < C_c$ and where $C_c = \sqrt{\frac{2\pi^2 E}{F_y}}$

$$F_a = \frac{\left[1 - \frac{(Kl/r)}{2C_c^2}\right] F_y}{\frac{5}{3} + \frac{3(Kl/r)}{8C_c} - \frac{(Kl/r)^3}{8C_c^3}}$$

when $k\frac{L}{r} > C_c$

$$F_a = \frac{12 \pi^2 E}{23 \left(\frac{KL}{r}\right)^2}$$

KL/r shall not exceed 200 for compression members. Most designers still limit main column slenderness ratios to $kL/r = 120$.

The AISC Specification, 1989 also includes the modern formula for determining the stresses on a member that is subjected to both axial compression and bending (ibid., pt. 5, sec. H, 5-54). Members must satisfy the following two equations:

$$\frac{f_a}{F_a} + \frac{C_{mx} f_{bx}}{\left(1 - \frac{f_a}{F'_{ex}}\right) F_{bx}} + \frac{C_{mx} f_{bx}}{\left(1 - \frac{f_a}{F'_{ey}}\right) F_{by}} \leq 1 \tag{1}$$

$$\frac{f_a}{0.60 F_y} + \frac{f_{bx}}{F_{bx}} + \frac{f_{by}}{F_{by}} \leq 1 \tag{2}$$

If f_a/F_a is equal to or less than 0.15, Equation 3 is permitted in lieu of Equation 1.

$$\frac{f_a}{F_a} + \frac{f_{bx}}{F_{bx}} + \frac{f_{by}}{F_{by}} \leq 1 \tag{3}$$

f_a = Actual axial stress.
F_a = Allowable axial stress.
f_b = Actual bending stress x or y direction.
$F'_e = \dfrac{12 \pi^2 E}{23 \left(\dfrac{KL_b}{r_b}\right)}$
E = Modulus of Elasticity
K = Effective length factor

L_b = Unbraced Length, in.
r_b = Radius of Gyration, in.
C_m = A coefficient.
1. Sidesway C_m = .85
2. Restrained Mbrs. $C_m = 0.6 - 0.4\left(\dfrac{M_1}{M_2}\right)$ but not less than 0.4
3. Braced Frames w/transverse loading.
   Restrained $C_m = 0.85$
   Unrestrained $C_m = 1.0$

The steel column has had an complicated but interesting evolution. In the early periods and through the 1910s, every manufacturer produced its own design methodology. Many engineering authorities also provided the designer with formulae for their use in the sizing of steel columns. Not until the Standard Specification for Structural Steel for Buildings was adopted and accepted by designers in the 1920s did any uniformity in analytical design become a reality. It is difficult to check the capacity of a column through the methodology of the period due to the lack of knowledge concerning which system the designer was utilizing on the building. Therefore, it is necessary to take a building column section, date the construction to determine the allowable stresses of the period, check the column capacity by utilizing several formulae of the period, and then check the capacity utilizing a combination of the modern formulae and the allowable stresses of the period of manufacture of the steel section. Through this procedure, the modern method of analysis should verify the capacity of the column. This method will determine whether the results of the earlier computations were conservative. We further assume that the formulae and computational analysis of today provide accurate while modestly conservative solutions. This is further enhanced by the knowledge that the work of Euler and his original experiments of the 1760s and beyond has never really been improved upon. The condition of the column would be an additional variable; however, semidestructive investigations at several points should determine whether any deterioration has occurred over the period from construction to date. If the building has been properly maintained, with water intrusion and other factors controlled, there should not be a problem of deterioration.

The first column to be checked by modern analysis is the steel Z-bar column. The dates of the publications that provide a table of safe loads are 1894 (Birkmire) and 1893 (Carnegie Steel Co.). The Column is a 14-in. Z-bar, a composite section composed of four 6.125 × 0.6875 in. Z-bars, one web plate 8 × 0.6875, and two side plates 14 × 0.375 in. The total area is 49.00 sq in. and the least radii of gyration is 3.80 in.

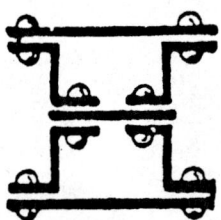

Figure 3-56.

P_{all} = 286.6 tons (2,000) for a 30 ft column (from table, Birkmire and Carnegie).
See Figure 3-5.

Proof of allowable load. Formula $F_a = 17,100 - 57\left(\dfrac{L}{r}\right)$

F_a = 11,700 psi (from the formula)

P_{all} = 11,700 psi(49.00 in.^{2}) = 573,300 lb = 286.65 tons (table checks)

Check of column utilizing modern formulae (this chapter, AISC 1989, ASD).

F_a = 12,837.7 psi (from modern equation) (Steel 1894, F_a = 16,000 psi, assume F_y = 32,000 psi)

Modulus of elasticity = 29,000,000 psi

P_{all} = (49.00 in.^{2})12,837.7 psi = 629,047 lb.= 314.52 tons

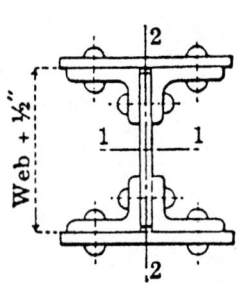

Figure 3-57.

The above computations indicate that by the modern method, the allowable load on the column is 314.5 tons. This difference is understandable because the modern formulae are slightly less conservative than the earlier formulae.

The second column to be checked is a compound column made up of steel plates and angles. The date is 1913 and the column is 30 ft long with the equivalent of pinned ends. Carnegie Steel Co. (1913) has a table of the safe loads for these plate and angle columns. The column is made up of four 6 × 4 × 0.5 in. angles, a web plate of 12 × 0.5 in., and two flange plates of 14 × 0.5 in. The area of the column is 39.00 sq. in. The least radii of gyration is 3.18 in. (similar to Figure 3-57).

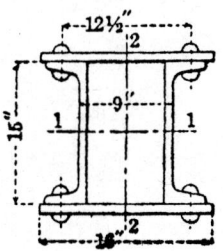

Figure 3-58.

P_{all} = 299 kips for a 30' column (from Carnegie Steel Co. 1913). See Figure 3-37.

Proof of allowable load. Formula $F_a = 19,000 - 100 \left(\dfrac{L}{r}\right)$ (with max. of 13,000 psi)

F_a = 9551.2 psi (from the formula)

P_{all} = 7679.25 psi × 39.00 in.² = 299.49 kips

Check of column utilizing modern formulae (this Chapter, AISC 1989, ASD).

F_a = 10,458.24 psi (from modern equation)

(steel 1913, F_{ult} = 55–60 ksi, F_y = 30,000 psi)

P_{all} = 39.00 in.² × 10,548.24 psi = 407.87 kips

This computation provides a much more liberal allowable load than does the formula utilized in 1913. The loads shown and calculated herein do not make any allowance or reduction of load for the rivet holes that appear in the section. It was not customary to consider rivet holes in compression members. However, rivet holes were an important factor in members that were in tension or had tension zones due to flexural bending or other factors.

Carnegie Steel Co. (1923) contains exactly the same load table for plate and angle columns. The same section on column formulae also appeared in the construction specifications.

The third column to be checked is a compound column made up of two channels and two plates. The date is 1923 and the column is 30 ft long with the equivalent of pinned ends. Carnegie Steel Co. (1923) has a table of safe loads for these channel and plate columns. The column is made up of two 15 in. × 35 lb channels and two 18 in. × 0.75 in. plates. The area of the column is 47.46 sq. in. The least radii of gyration is 5.53 in. (similar to Figure 3-58).

P_{all} = 593 kips for a 30-in. column (from Carnegie Steel Co. 1923). See Figure 3-58.

Proof of allowable load. Formula $F_a = 19,000 - 100 \left(\dfrac{L}{r}\right)$ (with max. of 13,000 psi)

F_a = 12,490 psi (from the formula)

P_{all} = 12,490 psi × 47.46 in.² = 592.78 kips (table checks)

Check of column utilizing modern formulae (this Chapter, AISC 1989, ASD).

F_a = 14,569.5 psi (from modern formula)

F_{ult} = 55,000–65,000 psi, F_y = 30,000 psi) 1923 steel

P_{all} = 47.46 in.² × 14,569.5 psi = 691.47 kips

The allowable load by modern methods is much higher than the allowable load of the period.

The fourth column to be checked is a Phoenix steel column, C type with four segments, 0.75 wall thickness, 7.3125 inside diameter, 8.8125 outside diameter (see Figure 3-59). The total weight is 90.6 lb/ft and the area of the section is 26.6 sq. in. The least radii of gyration is 3.20 in. The tabulated load in the Phoenix Handbook of 1906 is 159.6 tons (2,000 lb) or 319,200 lb for a 16 ft-long column.

P_{all} = 319.2 kips for a 16-in. column (From table, Phoenix Iron Co. 1906)

Proof of allowable load. Phoenix Iron Co. 1906,

F_a = 12,000 psi for $\frac{L}{r}$ = ≤ 90 $\left(\frac{L}{r} = 192/3.2 = 60 \text{ for this use}\right)$

P_{all} = 12,000 psi × 26.6 in.² = 319.2 kips (table checks)

Check of column utilizing modern formulae (this chapter, AISC 1989, ASD).

F_a = 15,782 psi (from modern formula) (steel 1906, F_a = 16,000 psi, assume F_y = 32,000 psi)

Modulus of elasticity = 29,000,000 psi

P_{all} = 15,782 psi × 26.6 in.² = 419.8 kips

Figure 3-59.

The modern method allows a higher allowable load than the formulae utilized in 1906.

The last column to be checked is the 12-in. rolled column section from American Institute of Steel Construction (1930). The column is 30 ft long with the equivalent of pinned ends. The AISC Manual has a table of allowable concentric loads for rolled columns of Bethlehem sections. The column section is a 12 in. × 99.5 lb Bethlehem rolled column, which was one of the standard shapes of the period. The area of the section is 29.21 sq. in. and the least radii of gyration is 3.04 in. See Figure 3-60.

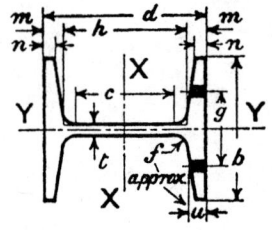

Figure 3-60.

P_{all} = 296 kips for a 30 ft column) (from table, AISC 1930)

Proof of allowable load. Formula $F_a = \dfrac{18,000}{1 + \dfrac{\left(\dfrac{L}{r}\right)^2}{18,000}}$ (with max. of 15,000)

F_a = 10,117.6 psi. (from the formula)

P_{all} = 10,117.6 psi × 29.21 in.² = 295.53 kips (table checks)

Check of column utilizing modern formulae (this chapter, AISC 1989, ASD)

F_a = 9937.94 psi (from modern formula) (steel 1930, ASTM A7, min. F_y = 30,000 psi)

P_{all} = 9937.94 psi × 29.21 in.² = 290.3 kips

The modern method is slightly more conservative than the formula of 1930.

Figure 3-61 gives typical details of steel columns of the period 1890–1930. When the building that the section or detail is taken from is known it is noted.

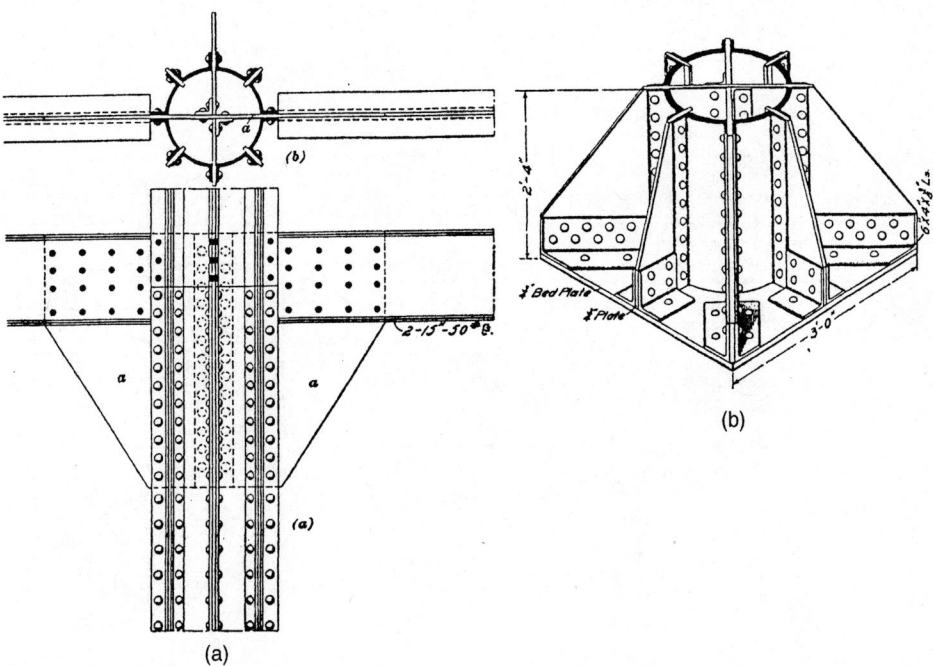

Figure 3-61. Column details. (a) Phoenix column: elevation and plan section. (*From International Correspondence Schools 1905, para. 16, 44.*) (b) Phoenix column: base detail. (*From id., 41.*)

FIREPROOFING—FERROUS STRUCTURAL SYSTEMS

The use of cast iron, wrought iron, and structural steels in buildings was seen to be of immediate importance because the materials were noncombustible. They were safer than wood, stronger, more durable, and had no inherent problems such as rot and insects. Other materials within and in buildings were still combustible, and buildings would still burn. The early buildings that utilized the new iron and steel structural sections were not fireproofed; at least there is no mention of fireproofing until the 1880s when "fireproof buildings" were being discussed. And "fireproof building" was initially a label meant in that there were no wood structural elements in the building. True fireproofing in the sense that a building could sustain a fire with no major structural damage or failures was not wholly in place until it was learned that the iron and steel elements had to be protected from the heat of the fire. In the days when wood structural elements were utilized extensively with masonry load-bearing walls, the wood floor systems would burn and fail and often take down walls or parts of walls in their collapse. Massive wood girders and columns would not burn through in normal fires that received some abatement measures while burning or simply burned themselves out in a relatively short period of time. In many instances, these massive wood girders and columns would only char, would not suffer any structural damage, and could be reused in situ as the reconstruction took place.

What must have been learned very early on, or was already known by technicians and metallurgists, was that all ferrous structural elements lose their strength significantly in the presence of fires of most magnitudes. American Institute of Steel Construction (1930, 61) provides a temperature chart for use of riveters that has additional data defining the subject of heat and steels. Structural steels are still capable of carrying their full loading until approximately 1000°F (538°C). Above that point they lose capacity at a very rapid rate as the temperature increases. Steel melts at approximately 2800°F (1524°C); however, it becomes very plastic and has no structural capacity at approximately 2200°F (1200°C). Cast iron melts at approximately 2300°F (1260°C) and becomes more brittle than plastic at temperatures above 1000°F (538°C). Iron and steel columns are extremely vulnerable to "cold water shock" when they are subjected

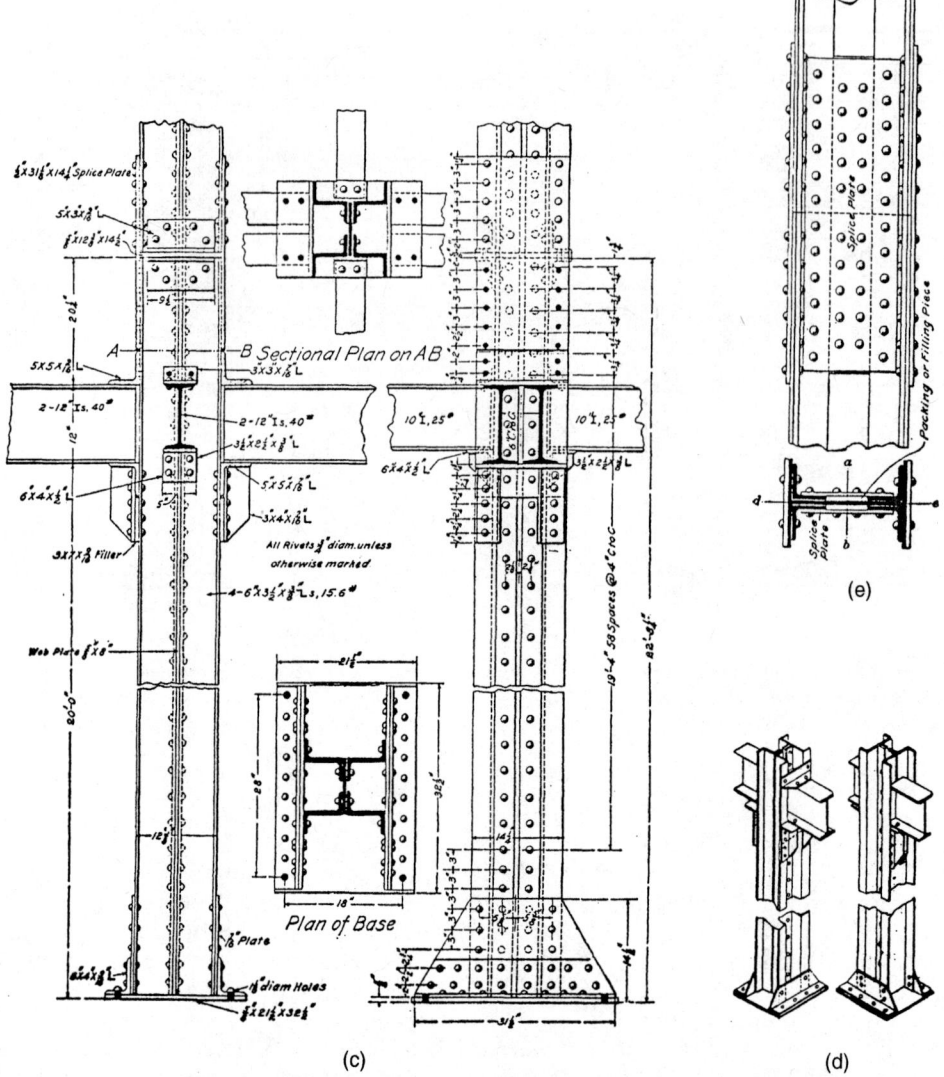

Figure 3-61. *(Continued).* (c) Z-bar. *(From id., 43.)* (d) Z-bar. *(From Birkmire 1894, 35.)* (e) Angle and plate column: typical splice. *(From International Correspondence Schools 1905, para. 16, 6.)*

to water by firefighters during a fire when they are at these elevated temperatures (Kidder 1902, 471). The sudden shock of rapid temperature lowering from the cold water causes contraction of the steel on the cold side, and the column bends and loses it's structural capacity and fails under load.

When a building burns extensively, temperatures can reach and exceed 2000°F (1093°C), and thus failures in steel sections are commonplace if the steel is not protected. The standard protection methods derived in the 1880s and 1890s included covering the columns with brick and/or terra cotta shapes designed expressly to encase columns. In most instances, the brick or terra cotta covering was then plastered over to provide the finished surface. The covering protected the iron or steel for a period of time that would keep the temperature from rising to critical points during the life cycle of the fire.

Many designers chose concrete encasement of the structural steel members in their buildings as their means of fireproofing. Concrete was almost as effective as brick and/or terra cotta shapes. It was not quite as good an insulator nor quite as good in the presence of fire as the fired clay products; however, it was easier to apply

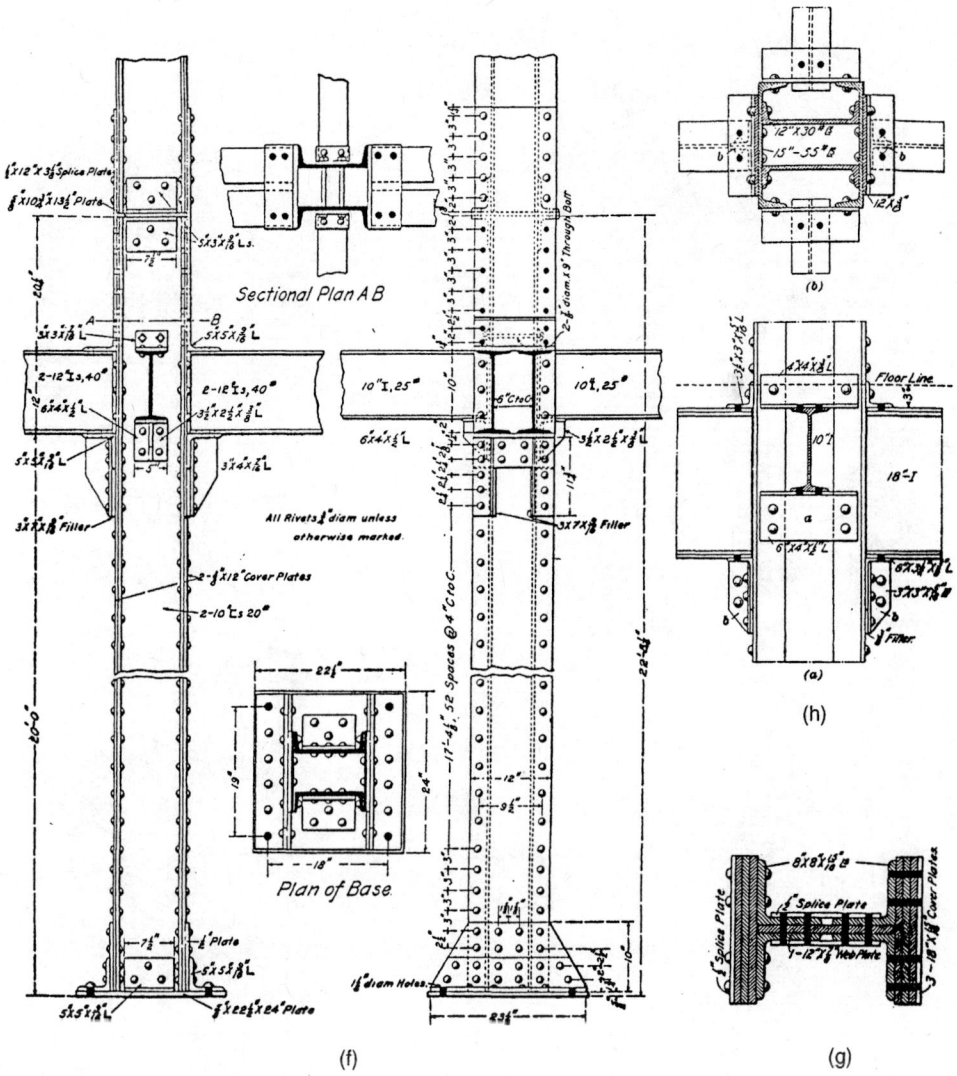

Figure 3-61. (*Continued*). (*f*) Channel and plate. (*From* id., *42.*) (*g*) Angle and plate: column section. (*From* id., *40.*) (*h*) channels and plates: elevation and plan section. (*From* id.)

in that it was cast as a viscous material held in place during curing by formwork. Figure 3-62 shows two types of wrought iron or steel columns that were encased in concrete. The columns are shown with formwork and concrete and with formwork removed or partially removed.

Concrete encasement for fireproofing of columns was utilized most often on warehouse buildings where the concrete could be left exposed and usually would not have any further finishing material. Forms were manufactured to fit standard column sizes and be reusable (see Figure 3-62).

Molded terra cotta shapes for round, square, and rectangular columns were manufactured in a range of sizes to make exact enclosures for any size column. The outer surface of the terra cotta would then be plastered and finished as needed for proper architectural treatment. The terra cotta shapes, which were kiln-fired clay castings, were manufactured in the same manner as bricks. The shapes were therefore cured, and they would resist the heat of fire and were of a higher efficiency than the concrete encasements. In addition, the shapes were hollow, which provides a partial thermal break and thereby enhances their insulating ability. Most shapes for column fireproofing were manufactured in 2 and 3 in. thicknesses.

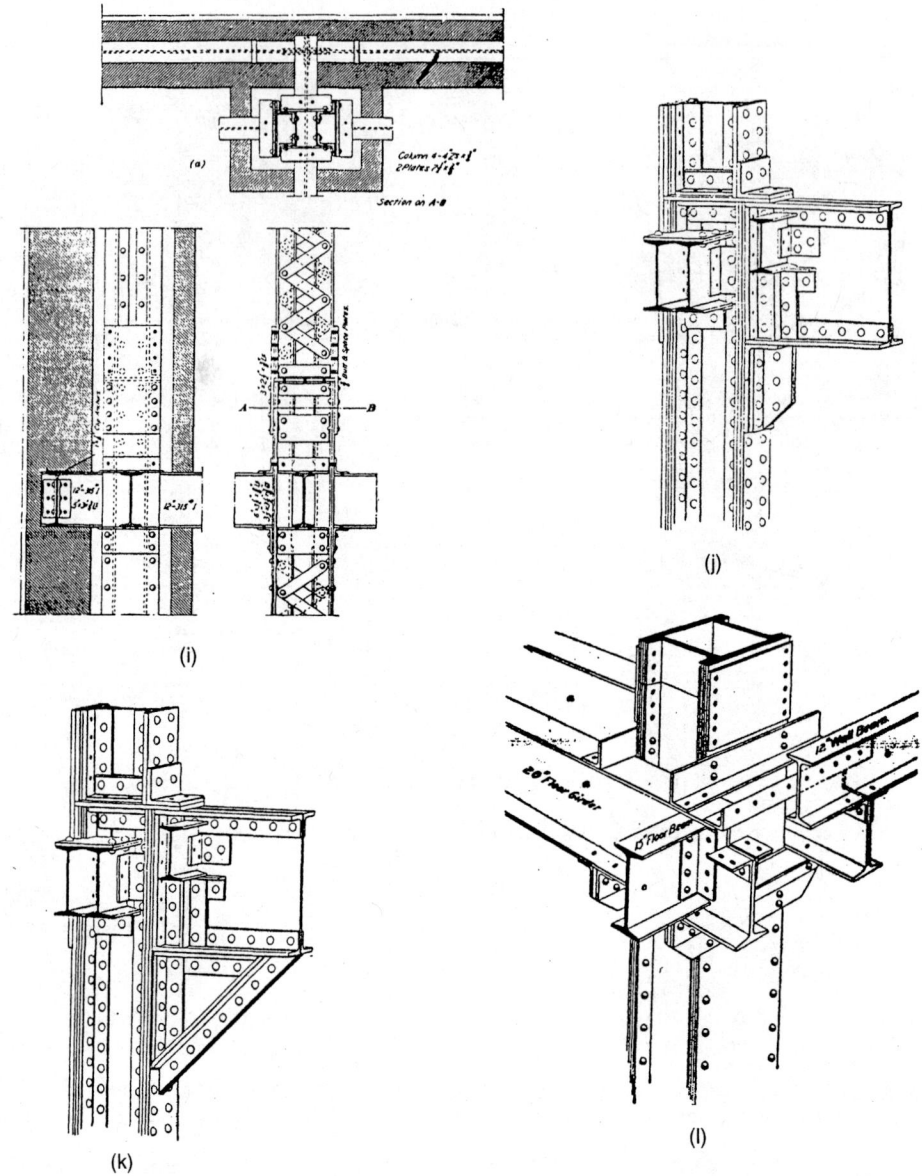

Figure 3-61. (Continued). (i) Z-bar column with lattice, plan section, elevations. (From id., 45.) (j) Typical column details. (From International Correspondence Schools 1905, para. 31, 2.) (k) Angle and plate column with beams and braces. (From id., 4.) (l) Channel and plate column with beam connections. (From International Correspondence Schools 1905, para. 16, 46.)

Terra cotta shapes were also cast as rectangular blocks in several thickness and were used as partition blocks in the construction of nonloadbearing partitions in buildings. These partitions would be plastered to provide the required surface finishes. They would also be capable of providing "rated walls" to give fire protection to adjacent spaces. Figure 3-63 illustrates the typical terra cotta column encasement shapes and their uses.

Terra cotta encasements for columns to meet fireproofing requirements were extensively used in buildings from the 1890s to well into the 1940s. Structural reinforced concrete construction began to gain momentum in the 1920s and gradually took part of the market for the terra cotta product. World War II halted construction for a period, and the technology seemed to fade during that time. As commercial and housing construction restarted after the war, terra cotta encasements for fireproofing were

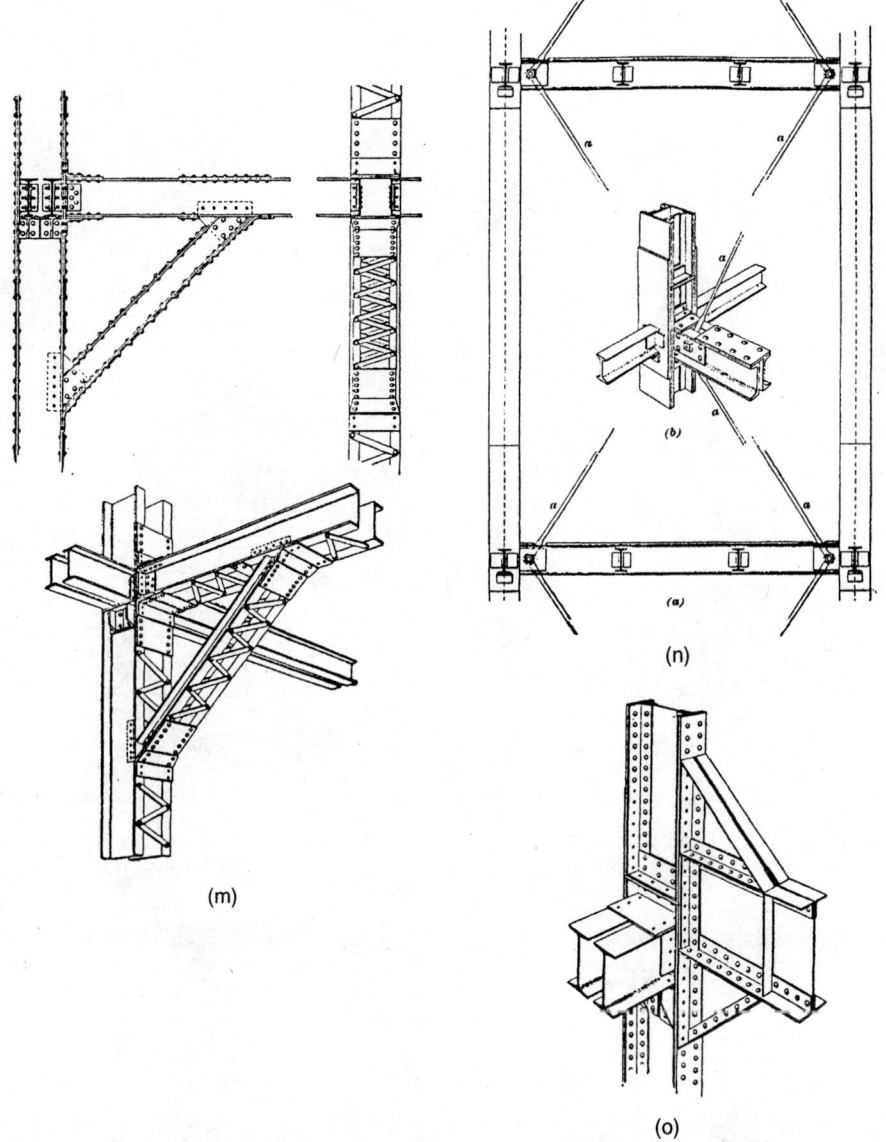

Figure 3-61. (*Continued*). (*m*) Column with knee brace. (*From International Correspondence Schools 1905, para. 31, 5.*) (*n*) Columns with sway rods or diagonal bracing. (*From id., 7.*) (*o*) Column with knee braces, top and bottom. (*From id., 6.*)

used less and less. New products and new technology replaced the type of fireproofing that the terra cotta provided.

Fireproofing using ordinary brick laid in the usual stretcher courses to form rectangular enclosures around columns was utilized by some designers. This method was as effective as the terra cotta shapes; it is a closely related product. Metal lath and plaster became an acceptable means of "protecting" the steel column; however, the term *protecting* is a poor choice because the method really involved covering the column with a metal lath with metal corner beads and applying the standard three-coat plaster system to the lath to form round or rectilinear finished shapes. The plaster was thin, usually no more than 3/4 in. of total built-up thickness, and would not provide much resistance to fires. A sprayed-on insulating covering of the structural steel column was quickly developed to provide the fire-rating requirements demanded by the building codes. Figure 3-64 shows details of columns that utilized metal lath and plaster coverings.

224 ◇ Structural Analysis of Historic Buildings

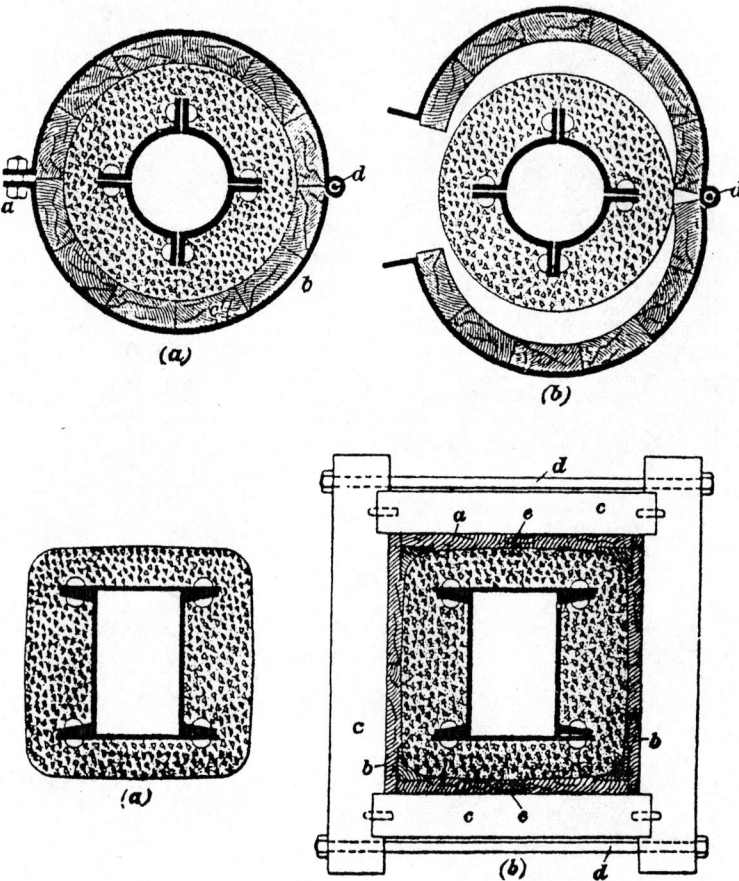

Figure 3-62. Fireproofing on columns. (*From International Correspondence Schools 1905, para. 23, 42, 43.*)

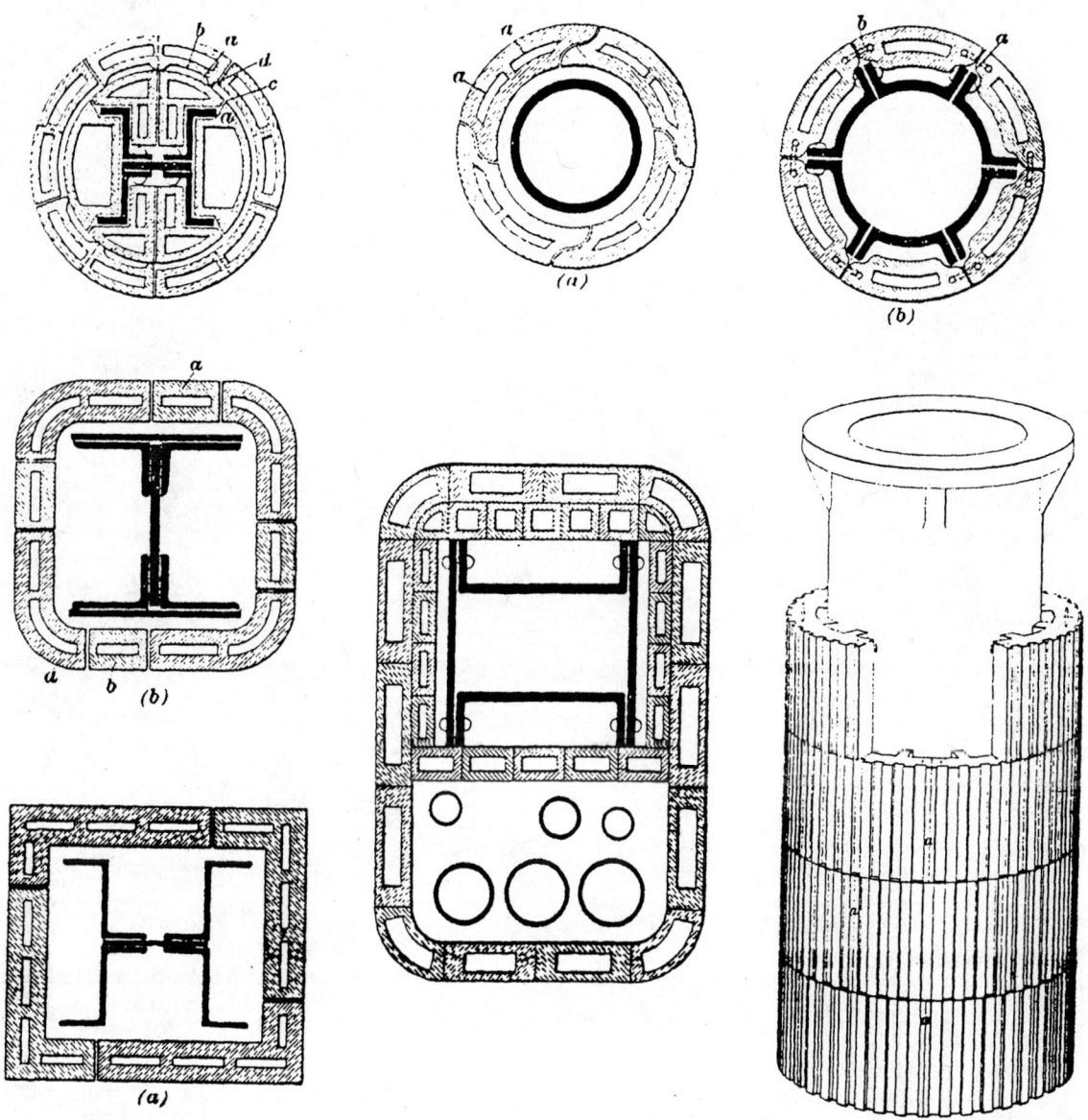

Figure 3-63. Terra cotta encasements for columns. (*From International Correspondence Schools 1905, para. 23, 45, 40, 39, 38; Lowndes 1936, 44, 45; Freitag 1909, 157.*) (The figure from Lowndes is reprinted by permission from W. S. Lowndes, *Hollow Tile and Fireproofing.* © 1936 International Textbook Company.)

226 ◇ Structural Analysis of Historic Buildings

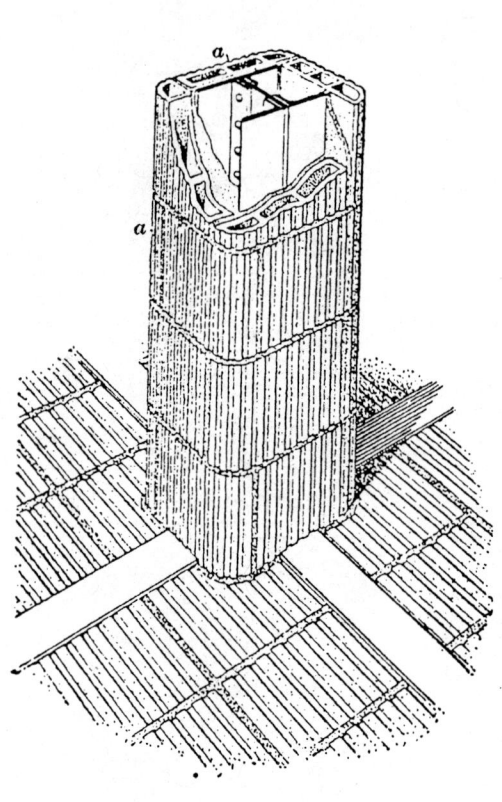

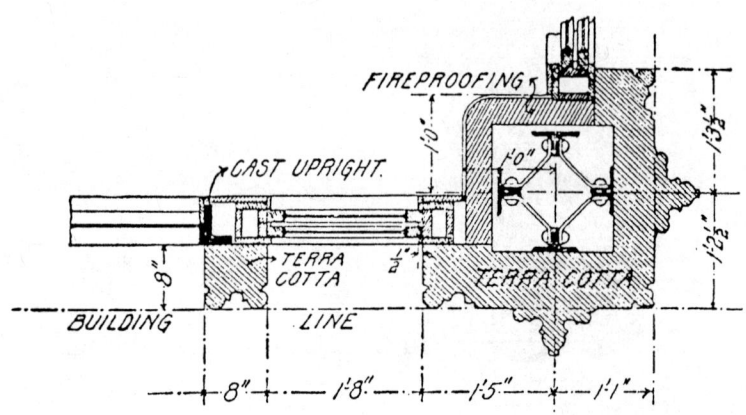

Detail of Corner Pier and Column.   Reliance Building.

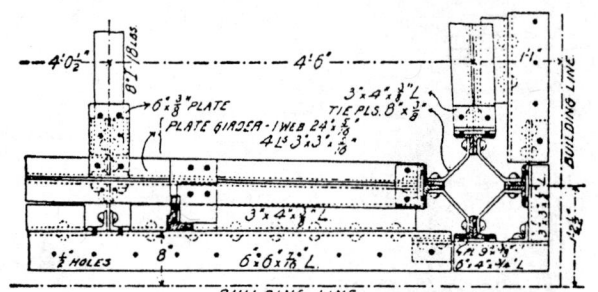

Detail of Wall Girders and Corner Column.   Reliance Building.

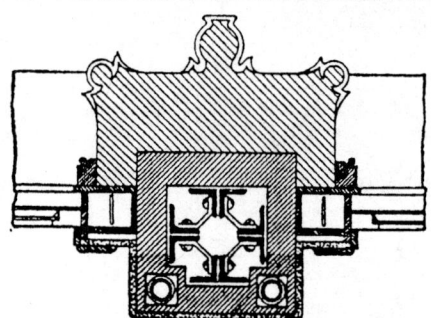

Detail of Columns in Exterior Walls.   Fisher Building.

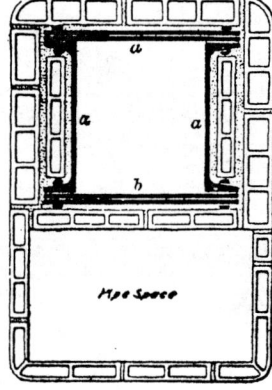

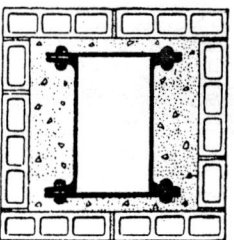

Figure 3-63.   (Continued).

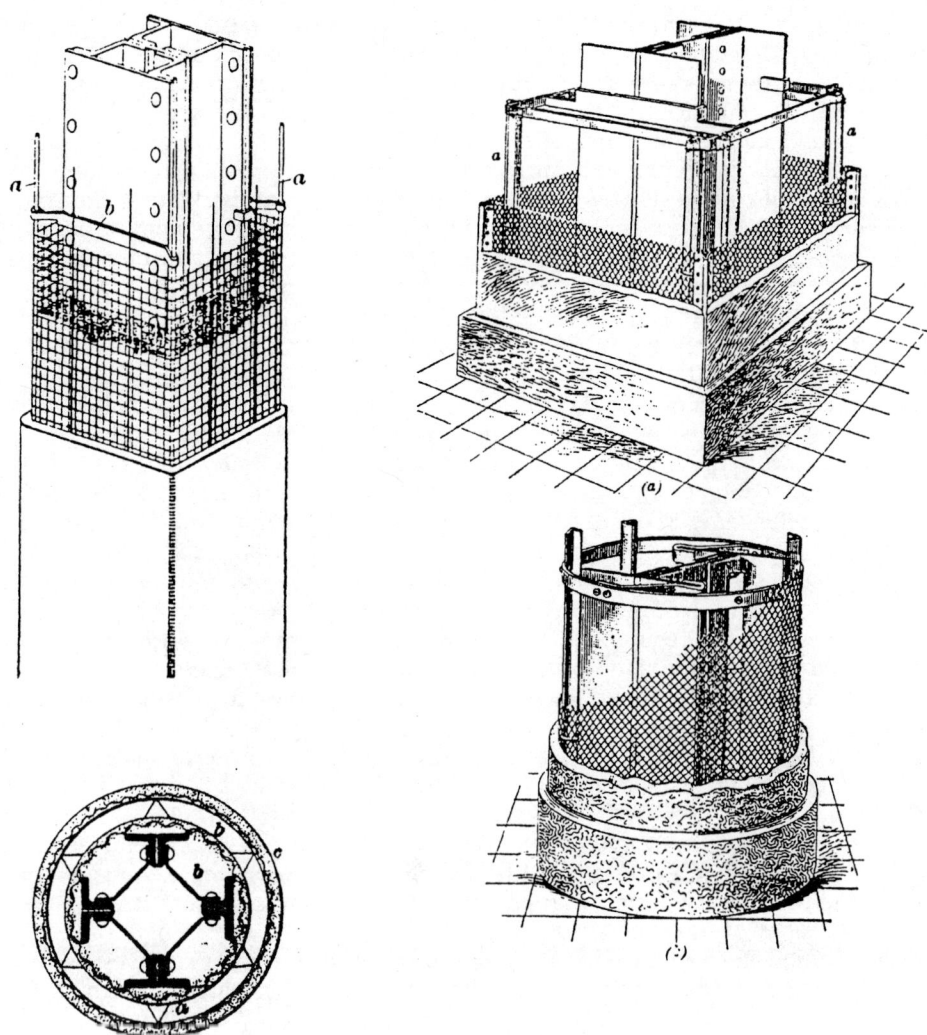

Figure 3-64. Metal lath and plaster encasements for columns. (*From International Correspondence Schools 1905, para. 23, 42, 41.*)

REINFORCED CONCRETE COLUMNS

Reinforced concrete structural elements—beams and columns—were developed in France and England in the mid- to late 19th century. French and German engineers continued to improve the early techniques to about 1900, when the United States took the lead in the production and use of concrete. The first cement plant in the United States was founded in Coplay, Pennsylvania, in 1875. The first use of reinforced concrete in America was in the development of spread footings with steel rails as the reinforcement in "grillage footings" (see Chapter 2). Reinforced concrete was sometimes called "concrete-steel," "ferro-concrete," or "armored concrete" of the turn of the twentieth century (Reid 1907, 3). The first building of reinforced concrete in America was constructed by W. E. Ward in Port Chester, New York in 1875, and the first bridge of reinforced concrete was an arch bridge at Golden Gate Park in San Francisco built in 1889 by Ransome and Smith (ibid., 5).

In the early twentieth century, the use of concrete and reinforced concrete increased tremendously, with such names as Hyatt, Jackson, Kahn, Hennebique, Ransome, and Thacher providing the early process developments. By 1905, engineers were ready to use reinforced concrete in foundations, floors, girders, walls, columns, roofs, and stairways. Reference books of the period quoted a 3–5 percent savings in the cost of construction when reinforced concrete was utilized as the structural system. The Ingalls Office Building, 1904, Cincinnati, Ohio, 16 stories, was the first office building of notable height that was constructed of reinforced structural concrete. The development of reinforced concrete construction paralleled the development of structural steels in the same period. However, the concrete researchers and engineers of the period were not as attached to the manufacturers as were the early steel designers. The concrete engineers developed and patented many types of reinforcing and reinforcing cages during the period prior to standardization in the industry.

A number of early tests were made on concrete mixes at the Watertown Arsenal in Watertown, Massachusetts, from 1899–1904. The amounts of cement, sand, stone (course aggregate), and water were varied under careful controls to determine the effect on the strength of concrete. The concrete was cast in 6 and 12 in. cubes, cured under different conditions, and loaded by calibrated machine to compression failure. Table 3-46 shows the results of the tests for the crushing strength of 12 in. cubes.

Table 3-47 shows the results of the tests for the crushing strength of 6 × 6 × 36 in. prisms. Comparison of the two tables indicates a slight difference in the two methods of testing. Each method utilized "unconfined" compression tests; therefore, the larger section would produce slightly higher results. Table 3-48 shows the effect of the size of stone (course aggregate) on the strength of the concrete for a constant mix of one part alpha cement, one part sand, and three parts stone. The size and type of the stone varied, as did the curing time. This test, which utilized several sizes of stone, also tabulated results of determinations of the value of the modulus of elasticity for those early concretes. Note that the modulus of elasticity was determined by monitoring tests during the segment of 100 psi and 600 psi, which would be well within the so-called straight-line zone of the load-deformation plot. These tests appear to have been quite accurately conducted and provided valuable early design information on the strength of concretes.

Turneaure and Mauer (1914) provide a table of results of Watertown Arsenal tests conducted in 1899 for five of the cement brands commercially available of 1899 (see Table 3-49).

The tensile strength of concrete was determined in the early twentieth century as approximately one-tenth to one-twelfth of the compressive strength (ibid., 15). The test method of the early period was said to be by pure tension rather than by transverse tests. Today we use one-tenth of the ultimate compressive strength of the concrete as the tensile strength. The ACI Code provides a formula that gives a slightly different number for the tensile strength, but most designers simply use the rule of thumb. Turneaure and Mauer also provide results from five different engineers who published the results of concrete tests conducted in their own laboratories (see Table 3-50).

The shearing strength of concrete was not known to engineers in that there was no standard test method at the turn of the century. Several methods were utilized at this time to determine a shear value. The most accepted form of the test to determine the ultimate shear strength of a concrete cylinder 5 in. in diameter was to clamp the

Table 3-44. Crushing Strength of 12 in. Cubes (from Watertown Arsenal Tests, 1901)[a]

Composition			Age		Ultimate Resistance (psi)
Cement	Sand	Broken Stone	Years	Months	
1	2	4 ⅝ in trap.	1		3,187
1	3	6 ⅝ in trap.	1		2,070
1	4	8 ⅝ in trap.	1		1,499
1	5	10 ⅝ in trap.	1		949
1	6	12 ⅝ in trap.	1		791
1	2	4 1½ and 2½ in. trap	2		2,789
1	2	4 1 and 2½ in. trap	2		2,549
1	2	4 2½ in. trap	2		2,466
1	2	7 2½ in. trap	1	2	2,406
1	2	4 1½–3 in. pebbles	1	2	3,589

[a] From Reid 1907, 3.

Table 3-45. Crushing Strength of 6 × 6 × 36 in. Prisms (from Watertown Arsenal Tests, 1901)[a]

Volume Ratios and Aggregate Sizes			Number of Specimens Tested	Ultimate Crushing Strength (psi)
Cement	Sand	Stone		
1 part	2½ part	4 part ½ in.–2 in. diameter pebbles.	8	2,326
1 part	2½ part	4 part ¼ in.–2½ in. diameter gravel.	6	3,363
1 part	2½ part	4 part 1 in.–2½ in. diameter hard	6	3,886

[a] From Reid 1907, 189.

ends in cylindrical bearings and move the clamps horizontally to shear the cylinder. Professor C. M. Spofford produced test results in 1906 for the ultimate shear strength of concrete (5-in. cylinders) as 63 percent of the compressive strength for 1:2:4 mix, 89 percent of the compressive strength for 1:3:5 mix, and 104 percent of the compressive strength for 1:3:6 mix (ibid., 17).

Table 3-46. Effect of Size of Stone on Strength of Concrete[a]

A. From Watertown Arsenal Tests, 1898 12-inch cubes, 1:1:3 alpha cement						
Size and Kind and Aggregate	Compressive Strength in psi at age in days				Coefficient of Elasticity in psi. between Loads of 100 and 600 psi	
	7–8	19–23	29–34	61–76	About 1 month	About 2 months
½ in. trap rock	1,391	2,220	2,800	5,021	3,571,000	4,167,000
¾ in. trap rock	1,900	2,769	3,200		8,333,000	
1 in. trap rock	3,390	4,254	4,917	5,272[b]	6,250,000	8,333,000
1½ in. trap rock	3,189	4,006	4,562	2,583		
2½ in. trap rock	2,400	4,143	4,140	4,523	5,000,000	12,500,000
½ in. trap rock 1 part / 2½ in. trap rock 2 parts	2,800	3,786	4,340	4,544[b]	8,333,000	6,250,000
½ in. trap rock 1 part / 1 in. trap rock 1 part / 2½ in. trap rock 1 part	2,800	4,156	4,800	5,542[b]	8,333,000	8,333,000
Mean strength trap rock	2,553	3,619	4,110	4,581		
Gravel, ⅜ in.	1,298	2,600	2,992	3,870	4,167,000	3,125,000
Gravel, 1½ in.	2,276	3,186	3,817	4,018	4,167,000	2,778,000
Gravel ⅜ in., 1 part / Gravel, 1½ in. 2 parts	1,994	3,023	3,800	3,490	4,167,000	5,000,000
Gravel, ⅛ in., 1 part / Gravel, ⅜ in., 1 part / Gravel, 1½ in., 1 part	1,486	2,676	3,000	3,800	3,125,000	3,125,000
Mean strength gravel	1,764	2,871	3,402	3,794		

B. From Watertown Arsenal Tests, 1904						
Mixture			Compressive Strength			
Cement	Sand	Gravel	½ in trap	2½ in trap	¼ Gravel	⅜ in Gravel
1		1	6,400			
1		1			4,800	
1		1				4,360
1		1		4,200		
1	1	2	4,180	2,200		2,600
1	2	3	3,700	1,680		2,060
1	2	4	1,480	1,210		1,700
1	3	6	1,410	790		580

[a] From Reid 1907, 189, 190.
[b] Not fractured.

Table 3-47. Compressive Strength of Concrete (Watertown Arsenal, 1899)[a]

Mixture	Brand of Cement	Strength (psi)			
		7 Days	1 Month	3 Months	6 Months
1:2:4	Saylor	1,724	2,238	2,702	3,510
	Atlas	1,387	2,428	2,966	3,953
	Alpha	904	2,420	3,123	4,411
	Germania	2,219	2,642	3,082	3,643
	Alsen	1,592	2,269	2,608	3,612
	Average	1,565	2,399	2,896	3,826
1:3:6	Saylor	1,625	2,568	2,882	3,567
	Atlas	1,050	1,816	2,538	3,170
	Alpha	892	2,150	2,355	2,750
	Germania	1,550	2,174	2,486	2,930
	Alsen	1,438	2,114	2,349	3,026
	Average	1,311	2,164	2,522	3,088

[a] From Turneaure and Maurer 1914, 12.

The Hennebique column utilized a system of round longitudinal bars and flat plates with holes to keep the longitudinal bars in place (see Figure 3-65). In the Hennebique system, concrete columns of any type—round, square, or rectangular—were initially divided into two types for purpose of design: (1) columns with light reinforcement and (2) columns with a larger quantity of reinforcement (reinforcement area greater than 1 percent of the gross area of the column).

The Hennebique column was utilized on a number of buildings, the most notable being the Ingalls Office Building, 1904, Cincinnati, Ohio. The column detail shown in Figure 3-66 is similar to the columns used on the Ingalls Office Building. Reid (1907) states that the column was "modified"; however, therearе no details about the type of modification. The Hennebique column was one of the first of the early reinforced concrete column types to utilize straight round vertical rods (longitudinal steel), which were restrained in a manner so as to maintain true alignment. The flat bars that were used as ties were included in the early designs, while later revisions or modifications included turning the flat bars vertical and forming a hoop, and then later utilizing

Table 3-48. Compressive and Tensile Tests on Concrete[a]

	Compressive Strength (psi)	Tensile Strength (psi)
Whitney		
1:2:4 concrete, 28 days	1940	189
Henby		
1:2:4 concrete	3000	180
1:3:6 concrete	1800	115
Hatt		
1:2:4 (broken stone), 30 days	–	311
1:2:5 (broken stone), 90 days	2413	359
1:2:5 (broken stone), 28 days	2290	237
1:5 (gravel), 90 days	2804	290
1:5 (gravel), 28 days	2400	253
Woolsen		
1:2:4 5–7 weeks old	1753 (average)	161
Talbot		
1:3:6 concrete	–	178
(50–84 days old)		160
		170

[a] From Turneaure and Maurer 1914, 15, 16.

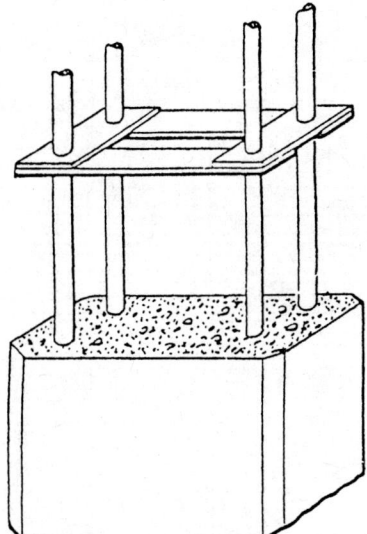

Figure 3-65. Hennebique column reinforcing. (*From Reid 1907, 266.*)

0.375 in. thick, square bands to form hoops, later to be known as ties. The vertical rods were as large as 3 in. in diameter and were usually four to a square column.

The design formula for axially loaded reinforced concrete columns as presented by Reid is as follows for lightly reinforced columns (ibid., 189):

$$P = f_c(A_g + nA_s)$$

where:
- f_c = allowable working stress of concrete, psi (compression)
- A_g = sectional area of concrete, in.^{2}
- n = ratio of modulus of elasticity of steel to that of concrete
- A_s = area of longitudinal reinforcing, in.^{2}

and, for heavily reinforced columns (ibid., 189, 190):

$$P = f_c[A + (n - 1)A_s] \quad L \text{ is not to exceed } 25\,d \text{ or diameter)}$$

where:
- A = gross area of column, in.^{2}
- A_c = area of concrete, in.^{2}
- A_s = area of longitudinal steel, in.^{2}
- E_c = modulus of elasticity, concrete, psi
- y = distance, center of gravity of column to center of longitudinal bar, in.
- b = width of column, in.
- d = depth or diameter of column, in.

and for heavily reinforced long columns:

$$P = \frac{39.48\left(\frac{bd^3}{12} + 9A_sY^2\right)E_c}{L^2} \quad \text{for rectangular columns, ultimate strength}$$

$$P = \frac{39.48(0.0491d^4 + 9A_sY^2)E_c}{L^2} \quad \text{for round columns, ultimate strength}$$

Factor of safety (ibid., 195) (recommended) = 5 to 6, buildings with no vibration or impact = 4

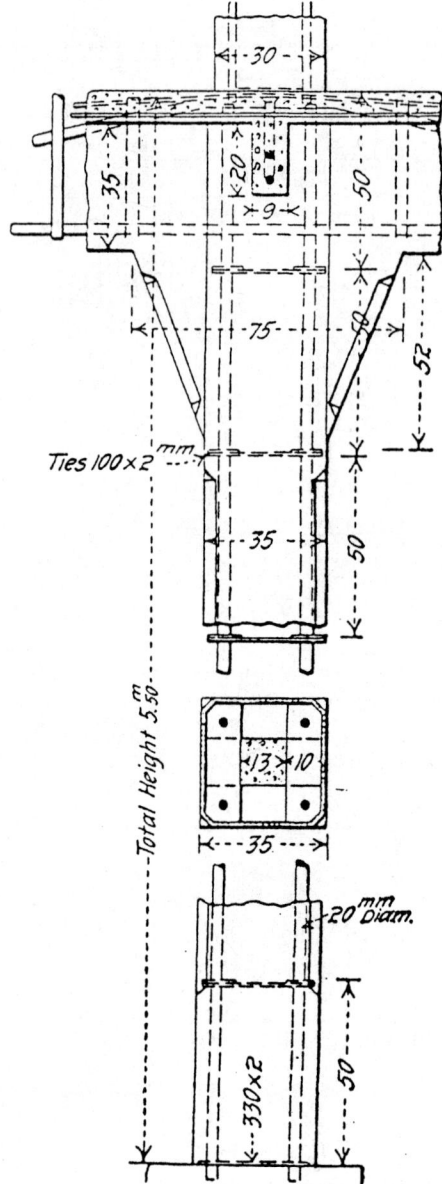

Figure 3-66. Hennebique column detail. (*From Reid 1907, 467.*)

Cities enacted laws to govern the use of concrete and reinforced concrete within their jurisdiction. The laws usually determined the allowable working stresses for concrete that would be utilized in building construction. Table 3-51 shows the allowable unit stresses for concrete for the various cities as given in International Correspondence Schools (1911). It is interesting to note that the cities do not appear to have dictated the type of concrete upon which they based their allowable stresses. It appears that the concrete mix they might have used as a basis was a 1:2:4 mix that produced a 2000 psi ultimate strength at about 30 days' common curing. The "safe grip or bond" of concrete on steel bars with plain surfaces is stated to be 50 lb/in. of surface of the metal in contact with the concrete (ibid., 239). The same volume also provides a table of the average ultimate strength of concrete made from portland cement, sand, and crushed stone, as attributed to W. Purves Taylor, who based his work on more than 600 tests made in the Philadelphia Municipal Testing Laboratory (see Table 3-52). Reinforcing rods (bars) in the early periods of structural reinforced concrete were mostly patented shapes (see Figure 3-67).

Reinforcing bars were produced in three basic grades of steel (ibid., 252):

Table 3-49. Allowable Stresses for Concrete[a]

City	Direct Compression (psi)	Shear (psi)	Unit Compressive Stress under Bending Loads (psi)
New York	350	50	500
Philadelphia	250	50	600
Cleveland	400	50	500
San Francisco	450	75	500
Buffalo	350	50	500
Toronto	340	50	500

[a] From International Correspondence Schools 1911, 240.

1. Mild or soft steel, 52,000–62,000 psi ultimate tensile strength, (yield point being 50 percent of ultimate), elongation 25 percent in 8 in.
2. Medium steel, 60,000–70,000 psi ultimate tensile strength (yield point being 50 percent of ultimate), elongation 22 percent in 8 in.
3. High-carbon steel, 80,000–100,000 psi ultimate tensile strength yield point being 50 percent of ultimate), elongation 5–15 percent in 8 in. (brittle to bending beyond 90°).

The early reinforcing bars were sometimes twisted while they were hot and coming off the roll mill, and sometimes after they had cooled. In either case, the twisted bars were considered to be 8–25 percent stronger than in the nontwisted state. (ibid., 254, 256). The twisted bars were generally known as Ransome bars (see Figure 3-67). Twisted bars were considered to be superior to plain bars in their resistance to pulling out of the concrete, bond, or grip. Some of these bars were made from rerolled rail stock (railroad rails).

Early reinforcing bars were also made from flat bars. A look at the nomenclature concerning reinforcing steel for concrete is necessary. In the early manuals, round reinforcing sections were known as "rods," square sections as "bars," and rectangular sections as "flats." In addition, several other bars were special and did not fit into the above categories (see Figure 3-68).

This array of reinforcement was the product of the patent system, and the effect was a different area and weight for each type. See Table 3-53 and Figure 3-69 for the properties, areas, and weights of these types of reinforcement.

In the earliest stages of reinforced concrete columns there were two general types, straight reinforcement and hooped reinforcement. Straight reinforcement implies straight vertical bars, as in the modern reinforced concrete column. Hooped reinforcement could be one of two types: hoops (individual hoops) and spiral hoops, or con-

Table 3-50. Ultimate Strength of Concrete[a]

Average Ultimate Strength of Concrete Made from Portland Cement, Sand, and Crushed Stone													
Proportion of Ingredients			Tension (psi)				Compression (psi)				Shear (psi)			
Cement	Sand	Stone	7 days	1 month	3 months	6 months	7 days	1 month	3 months	6 months	7 days	1 month	3 months	6 months
1	2.0	4	160	210	240	250	1,600	2,150	2,400	2,500	200	269	300	313
1	2.5	5	143	195	225	235	1,430	1,950	2,250	2,350	179	244	281	294
1	3.0	6	125	180	210	220	1,250	1,800	2,100	2,200	156	225	263	275
1	3.5	7	110	166	196	208	1,100	1,660	1,960	2,080	138	208	245	260
1	4.0	8	98	152	182	195	980	1,520	1,820	1,950	123	190	228	244
1	4.5	9	85	140	169	184	850	1,400	1,690	1,840	106	175	211	230
1	5.0	10	75	126	155	172	750	1,260	1,550	1,720	94	158	194	215
1	5.5	11	65	112	142	160	650	1,120	1,420	1,600	81	140	178	200
1	6.0	12	60	100	130	150	600	1,000	1,300	1,500	75	125	163	188

[a] From International Corrrespondence Schools 1911, 241.

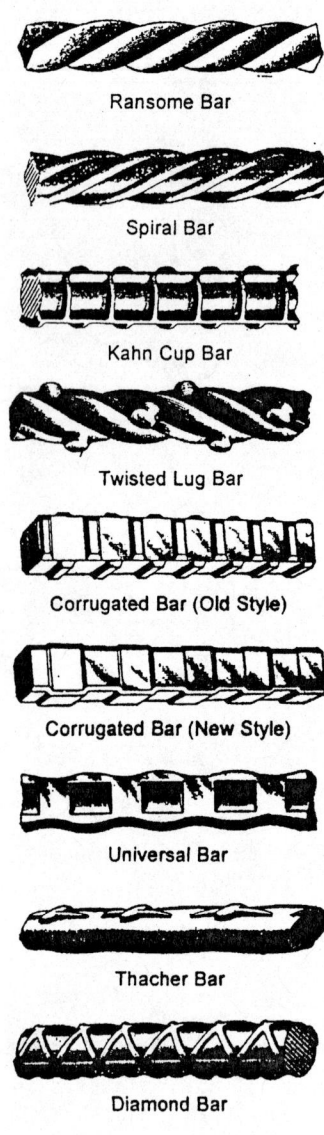

Figure 3-67. Reinforcing bars. (*From International Correspondence Schools 1911, 255.*)

tinuous spirals. Both types are shown in Figure 3-70. As the names imply, the hooped or spiral columns had no vertical bars and depended upon the circular hoop to contain the core of the concrete. Hooped or spiral columns were most often round, hexagonal or octagonal. It was possible to use the hoop or spiral system for square columns; however, no examples have been found. Another variation of the hoop or spiral system was expanded metal hooping. This system normally utilized a double layer of encasement reinforcing around a vertical rebar or rod and metal lathing on the outside of the expanded metal (see Figure 3-71). The expanded metal hooping system could also be used on square and rectangular columns. The Monolith Steel Company of Washington, D.C., utilized a "monolith bar" and a combination of a single hoop made from a flat bar and a spiral made from a narrow flat bar (see Figure 3-72).

In either case, the hooped column was good only for pure vertical axial load. It was a fairly efficient method of constructing a short, fairly heavily loaded column. However, the hooped column was not capable of carrying any moment, as flexural bending results in tension and compression stresses in the column section. If the pure compression stresses from the axial loads are not sufficient to prevent the combined stresses from having tension in one face of the column, the result will be failure. The tensile capacity of concrete is only one-tenth of the ultimate compressive strength, and therefore the column without longitudinal reinforcing will be capable of carrying only a very small moment. Thus, the hooped column, with no longitudinal reinforcing, was quickly discarded in favor of the longitudinally reinforced concrete column.

The quality of concrete produced and mixed in the early twentieth century was comparable to that of today. Many producers were very scientific in their formulations and produced very consistent concrete. There is little reason to suspect that early concrete or reinforced concrete would deteriorate or lose strength with age. A thorough inspection would be necessary to determine the condition of the concrete, and tests to determine the strength of the concrete would be necessary.

In the initial periods of use of columns of structural concrete, designers utilized plain concrete columns or reinforced concrete columns. Plain concrete columns, as the name implies, were columns of concrete with no reinforcing of any kind. Reinforced concrete columns would include several types of reinforcing, including round steel hoops spaced at intervals and steel continuous spirals, both of which provided only containment of the inner core of the column. That containment allowed the core to resist the axial stress without failing in shear by providing a volume of horizontally restrained concrete to resist the vertical force. It is worthy of note that this is the same principle utilized by the modern reinforced concrete column in that the cage formed by the longitudinal bars and the horizontal ties reacts to load in the same manner. The early designers were also very aware of the need to consider eccentricity of loads and the effect that the eccentricity had on the analysis or design of the column.

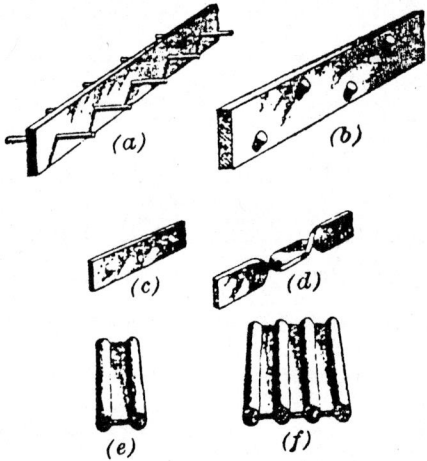

Figure 3-68. Plain bar iron: (a) Hyatt bar; (b) Thacher bar; (c) staff bar; (d) DeMann bar; (e) Siamese bar; (f) quad bar. (*From International Correspondence Schools 1911, 254.*)

Table 3-51. Physical Properties, Areas, and Weights and Reinforcing[a]

A. Physical Properties of the Ransome Bar

Size of Bar (in.)	Number of Twists per ft of Length	Area of Bar (in.^{2})	Weight per ft (lb)	Elastic Limit (psi)	Ultimate Tensile Strength (psi)	Elastic Limit of One Bar (lb)	Ultimate Tensile Strength of One Bar (lb)
1/4	5	0.0625	0.213	62,350	86,700	3,897	5,419
3/8	3	0.1406	0.478	61,800	86,600	8,689	12,176
1/2	2	0.2500	0.850	60,120	86,850	15,030	21,713
5/8	2	0.3906	1.328	57,890	85,820	22,612	33,520
3/4	1	0.5625	1.913	56,720	85,240	31,905	47,948
7/8	1	0.7656	2.603	56,150	84,730	42,988	64,869
1	3/4	1.0000	3.400	55,760	84,275	55,760	84,275
1 1/4	1/2	1.5625	5.312	55,450	83,150	86,641	129,921

B. Properties of the Kahn Cup Bar

Nominal Size of Bar (in.)	Sectional Area (in.^{2})	Weight per Foot (lb)	Elastic Limit of Each Bar (lb)	Ultimate Tensile Strength of Each Bar (lb)
3/8	0.1406	0.502	5,000	9,800
1/2	0.2500	0.893	9,000	17,500
5/8	0.3906	1.394	14,000	27,300
3/4	0.5625	2.008	21,000	39,400
7/8	0.7656	2.733	28,000	53,600
1	1.0000	3.570	37,000	70,000
1 1/8	1.2656	4.518	47,000	88,600
1 1/4	1.5625	5.578	58,000	109,400

C. Properties of Twisted Lug Bars

Size of Bar (in.)	Net Sectional Area (in.^{2})	Weight per ft (lb)	Safe Working Stress for Each Bar (lb)
1/4	0.0625	0.222	1,250
3/8	0.1406	0.492	2,810
1/2	0.2500	0.870	5,000
5/8	0.3906	1.350	7,810
3/4	0.5625	1.940	11,250
7/8	0.7656	2.640	15,310
1	1.0000	3.450	20,000
1 1/8	1.2656	4.350	25,310
1 1/4	1.5625	5.370	31,250

D. Size, Net Section and Weight of Corrugated, or Johnson, Bars

Size of Bars (in.)	Weight per ft (lb)	Net Section (in.^{2})
1 1/4	0.78	.19
3/4	1.56	.38
7/8	2.25	.55
1	2.90	.70
1 1/4	4.56	1.10

E. Net Sections and Weights of Universal Bars

No.	Size of Bar (in.)	Weight per ft (lb)	Net Section (in.^{2})
1	1/4 × 1	0.73	0.19
2	3/8 × 1 3/8	1.35	0.41
3	3/8 × 1 3/4	1.97	0.54
4	3/8 × 2	2.27	0.65
5	3/8 × 2 1/2	2.85	0.80

Table 3-51. (Continued)

F. Size, Weight, and Ultimate Tensile Strength of Thacher Bars (Medium Steel)

Diameter of Bar (in.)	Weight per ft (lb)	Area of Net Section (in.)	Average Ultimate Tensile Strength for Each Bar (lb)
1/4	0.16	0.047	3,000
3/8	0.34	0.10	6,400
1/2	0.61	0.18	11,500
5/8	0.95	0.28	17,900
3/4	1.39	0.41	26,200
7/8	1.87	0.55	35,200
1	2.41	0.71	45,400
1 1/8	3.06	0.90	57,600
1 1/4	3.74	1.10	70,400
1 3/8	4.49	1.32	84,500
1 1/2	5.30	1.56	99,800
1 5/8	6.15	1.81	115,800
1 3/4	7.07	2.08	133,100
1 7/8	7.99	2.35	150,400
2	9.01	2.65	169,600

G. Size, Area, and Weight of Diamond Bars

Nominal Size of Bar (in.)	Area of Section (in.^{2})	Weight per ft (lb)
1/4	0.062	0.22
3/8	0.14	0.48
7/16	0.19	0.65
1/2	0.25	0.85
5/8	0.39	1.33
3/4	0.56	1.91
7/8	0.76	2.60
1	1.00	3.40
1 1/4	1.56	5.31

H. Areas and Weights of Square and Round Bars

Size (in.)	Square Area (in.)	Square Weight per ft (lbs)	Round Area (in.)	Round Weight per ft (lb)
1/16	0.0039	0.013	0.0031	0.010
1/8	0.0156	0.053	0.0123	0.042
3/16	0.0352	0.120	0.0276	0.094
1/4	0.0625	0.213	0.0491	0.167
5/16	0.0977	0.332	0.0767	0.261
3/8	0.1406	0.478	0.1104	0.376
7/16	0.1914	0.651	0.1503	0.511
1/2	0.2500	0.850	0.1963	0.668
9/16	0.3164	1.076	0.2485	0.845
5/8	0.3906	1.328	0.3068	1.043
11/16	0.4727	1.607	0.3712	1.262
3/4	0.5625	1.913	0.4418	1.502
13/16	0.6602	2.245	0.5185	1.763
7/8	0.7656	2.603	0.6013	2.044
15/16	0.8789	2.989	0.6903	2.347
1	1.0000	3.400	0.7854	2.670
1 1/16	1.1289	3.838	0.8866	3.014
1 1/8	1.2656	4.303	0.9940	3.379
1 3/16	1.4102	4.795	1.1075	3.766
1 1/4	1.5625	5.312	1.2272	4.173
1 5/16	1.7227	5.857	1.3530	4.600
1 3/8	1.8906	6.428	1.4849	5.049
1 7/16	2.0664	7.026	1.6230	5.518
1 1/2	2.2500	7.650	1.7671	6.008
1 9/16	2.4414	8.301	1.9175	6.520
1 5/8	2.6406	8.978	2.0739	7.051
1 11/16	2.8477	9.682	2.2365	7.604
1 3/4	3.0625	10.413	2.4053	8.178
1 13/16	3.2852	11.170	2.5802	8.773
1 7/8	3.5156	11.953	2.7612	9.388
1 15/16	3.7539	12.763	2.9483	10.024
2	4.0000	13.600	3.1416	10.681

[a] From International Correspondence Schools 1911, 253, 256–260; *id.*, 1907, para. 45, 20.

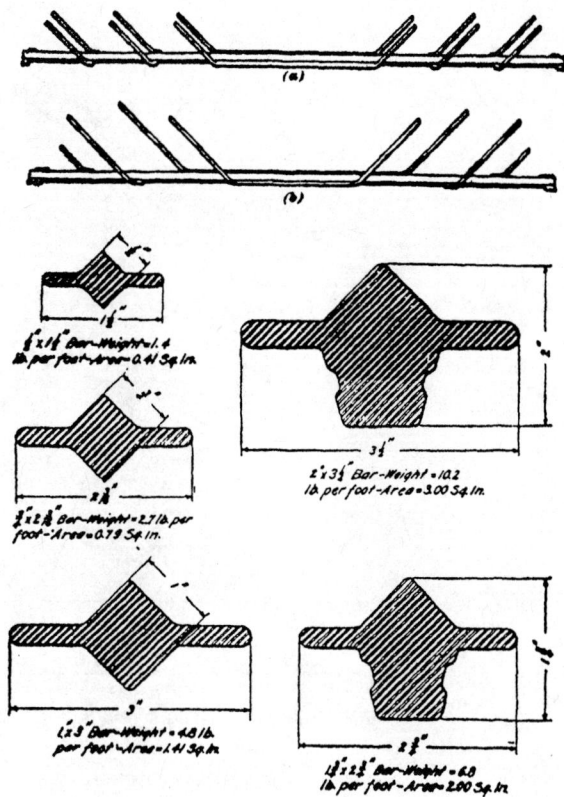

Figure 3-69. Kahn trussed bars: details and sections. Elastic limit 33,000–35,000 psi. From International Correspondence Schools 1907, para. 45, 20.

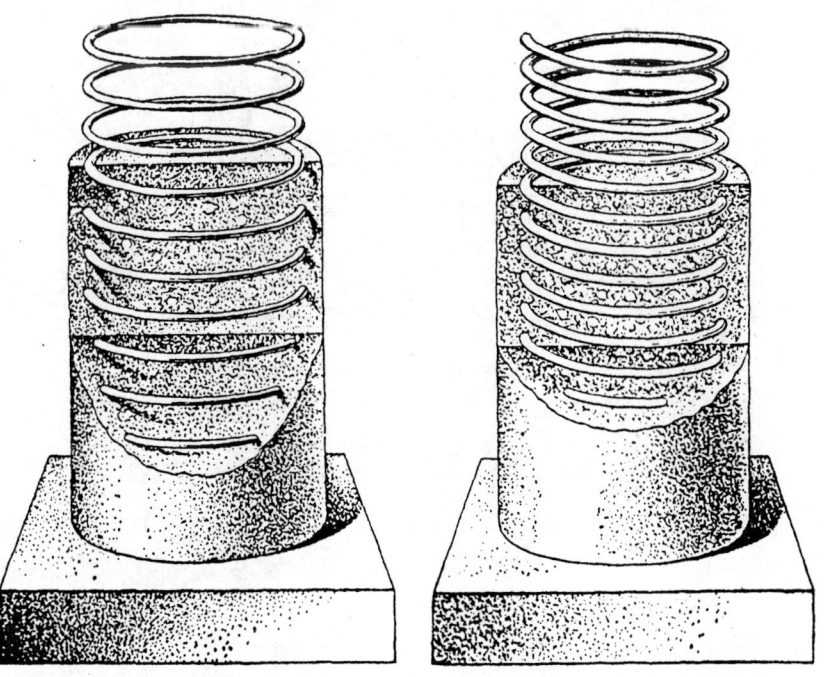

Figure 3-70. Concrete columns, hooped and spiral. (*From International Correspondence Schools 1907, para. 45, 20.*)

238 ◇ Structural Analysis of Historic Buildings

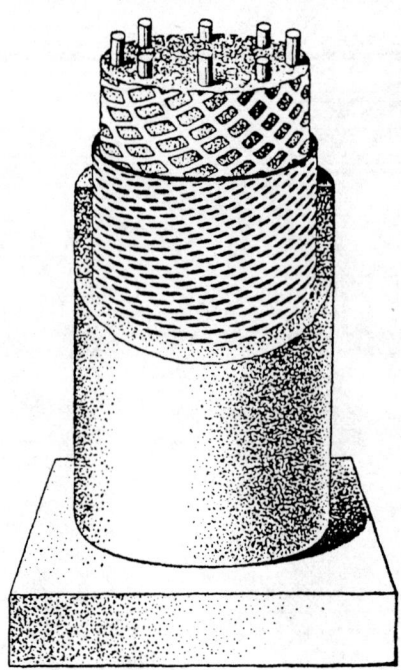

Figure 3-71. Expanded metal hooping. (*From Reid 1907, 497.*)

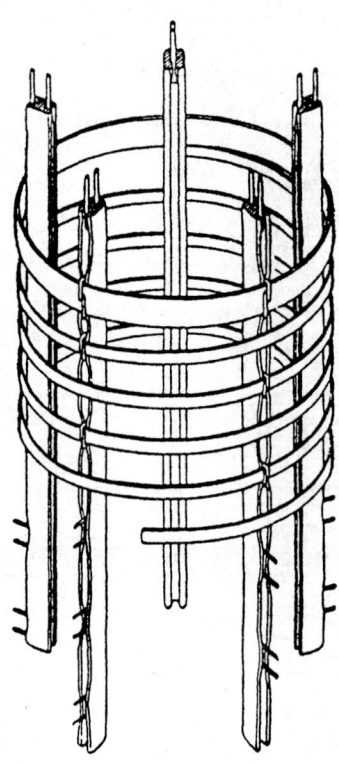

Figure 3-72. Monolith hooped column. (*From International Correspondence Schools 1911, 285.*)

The Concrete Engineer's Handbook (The International Correspondence Schools contains several design formulae that enable a recreation of the design methodology of the period.

Plain concrete columns, axially loaded (ibid., 285) (height of column limited to 12(d)):

$$P_{all} = (f_c)_{all}(\text{area of column})$$

where:
d = diameter or least dimension of column

Plain concrete columns, eccentrically loaded (ibid., 303, 304, 305) (height of column limited to 12(d)):
Circular columns:

$$(f_c)_{all} = \frac{P}{A} + \frac{8(e)P_e}{Ad}$$

Rectangular columns:

$$(f_c)_{all} = \frac{P}{A} + \frac{6(e)P_e}{Ad}$$

where:
f_c = allowable compressive stress in concrete, psi or ksi
e = eccentricity, in.
P = total load on column, $P_c + P_e$, lb or kips
P_c = pure axial load, lb or kips
P_e = eccentric load, lb or kips
A = area of column, gross, in.^{2}
d = diameter of column, gross, in.

Reinforced concrete columns, concentrically loaded (ibid, 304, 305) (height of column limited to 15(d)):

Straight longitudinal reinforcement:

$$P_{all} = f_c(A_c + nA_s)$$

where:
f_c = allowable compressive stress in concrete, psi or ksi
f_s = allowable compressive stress in steel, psi or ksi
A_c = net area of column, gross area of concrete minus the area of longitudinal steel, in.^{2}
E_c = modulus of elasticity, concrete, psi or ksi
E_s = modulus of elasticity, steel, Psi or ksi
n = ratio E_s/E_c
f_c = 450 psi
n = 15 for this equation, as required by the Joint Committee of ASCE, ASTM, AEMWA, and AAPCM

Hooped reinforcement (hoops or spirals only, no Longitudinal reinforcement): Hooped reinforcement shall not be less than 1% of the volume enclosed (within the circumference of the hoop) (1 in. of hoop reinforcement equals in volume 1% of the volume of the concrete area within the hoop × one hoop spacing). The Joint Committee recommended a concrete allowable Compressive Stress of 450 psi on the concrete area within the hoop diameter.

$$P_{all} = 450 \text{ psi}(A) \qquad A = \text{area of concrete within the hoop diameter}$$

Hooped reinforcement with longitudinal steel: For columns containing the standard 1% by volume of hooped reinforcement that also contain 1–4% of longitudinal reinforcement, the Joint Committee recommends a concrete allowable compressive stress of 650 psi on the concrete within the hoop diameter.

$P_{all} = 650 \text{ psi}(A)$ where:
A = area of concrete within hoop diameter
No specific allowance for the longitudinal steel.

Empirical rules for straight reinforcement (ibid.) (bars, rods, or structural shapes)
Straight reinforcement, concentric loading: (vertical reinforcing bars, patented type):

$P_{all} = f_c(A)$ where:
f_c = 500 psi, allowable compressive stress of concrete
A = area of concrete column minus cover; A_s included in A

Rolled structural sections encased in concrete (steel core sections):

$P_{all} = f_s(A_s)$ where:
f_s = allowable compressive stress of rolled steel section (usually taken as 16,000 psi)
A_s = cross-sectional area of steel section, in.^{2}

The concrete cover was assumed to prevent the structural steel section from buckling under the column loads. It was also assumed that the use of a set of longitudinal bars around the perimeter of the column would also help the concrete prevent buckling in the structural steel section. (No allowance was made for the strength of the concrete or longitudinal steel in the computations.)

It is not known how many columns were sized through empirical formulae. Probably they were sized this way only for a short period, until the early technology advancements allowed the designers of the period to accept and understand the analytical methods.

Reinforced Concrete Columns, eccentrically loaded (ibid., 306):
Reinforced concrete columns with concrete beams and girders constructed integrally to the columns would by method of construction almost always be eccentrically loaded. The ICS *Concrete Engineer's Handbook* states that the eccentrically loaded or moment column must be designed the same as a beam in that the tensile reinforcement in beam action carries the tensile force of the couple. The compression load, as it exists from above, would cause a pure compressive stress distributed equally over the whole column section. The combined bending and axial stresses would sum algebraically for the maximum stresses in the column section to be analyzed. The Handbook states that because this method is approximate, low allowable stresses should be utilized.

Hool et al. (1918) gives designers the latest analytical methods for the period and further specifies that concrete columns are of four principal types:

1. Plain concrete columns or piers
2. Columns reinforced with longitudinal rods only
3. Columns reinforced with hoops and longitudinal rods
4. Columns reinforced with structural steel shapes

The Handbook utilizes design specifications as recommended by the Joint Committee, which consisted of members from a U.S. Government committee, the ASCE

Board of Directors, and Committee C-1 of the ASTM. These Standard Specifications became effective January 1, 1917.

<u>Plain Rectangular Concrete Columns or Piers</u> (No Steel Reinforcement) (ibid., 371).
  Compression members in which the ratio of the unsupported length to least width is 4 or less are referred to as *piers*.
Rectangular and square columns:

$$f_c = \frac{P}{A}\left(1 + 6\frac{x_o}{t}\right)$$ where:

P = total load, lb or kips
A = area of column, in.^{2}
X_o = eccentricity, in.
t = breadth of column, in. (in direction of eccentricity)
f_c = max. concrete compressive stress, psi or ksi

If eccentricity = 0, $f_c = \frac{P}{A}$

Reinforced concrete columns with longitudinal reinforcement only (ibid., 371, 372):

$P_{all} = f_c(A)[1 + (n - 1)\rho]$ formula allows for the concrete and the reinforcing to carry load)

where:
A = total net area, in.^{2}
A_c = area of concrete, in.^{2}
A_s = area of longitudinal steel, in.^{2}
ρ = steel ratio, A_s/A
n = Ratio of modules of elasticity, E_s/C_c
f_c = comp. allowable stress, concrete, psi or ksi

<u>Reinforced concrete columns with hooped (spiral) and longitudinal reinforcement</u> (ibid., 372).
  Hool et al. state that the addition of bands or spirals to columns having longitudinal reinforcement does not have much positive effect up to the point of failure without hooping. They further state that the effect of the hooping raises the ultimate strength of the concrete (by providing a confined mass) and thus raises the load-carrying capacity of the column (above that without hoops).

p = percentage of hoop steel (spiral steel) = 1% recommended amount

$A_s = \frac{spd_1}{4}$ S = pitch of spiral or distance between hoops, in.

d_1 = diameter of spiral or hoop, in.

$S = \frac{A_s}{0.0025d_1}$ A_s = sectional area of spiral or hoop, in.^{2}

<u>Concrete columns reinforced with structural shapes</u> (ibid., 372, 373).
  Hool et al. state that structural steel shapes that are fireproofed by concrete encasement cannot rightfully be considered reinforced concrete columns. They are to be considered steel columns and are subject to only the allowable loads on the steel section. The concrete does not properly bond to the steel and therefore the concrete does not provide adequate lateral support for the steel section.

The effective end of the plain concrete column with encasement of structural steel section came with the 1917 Joint Committee recommendations. The use of the same effective area of longitudinal reinforcing rods with ties or spirals (hoops) was said to be more efficient by the committee than a structural steel section encased in concrete.

Reduction formula for long reinforced concrete columns (ibid., 381) (any column type, shape). The Joint Committee determined that columns with L/d equal to or less than 15 shall be considered short or standard columns to be designed by the formulae as presented herein. Columns with an L/d between 15 and 30 shall be designed as long columns. In the case of a long column, the following is the allowable stress reduction factor.

$$R = 1.6 - 0.04 \left(\frac{L}{d}\right) \quad \text{for } L/d \text{ between 15 and 30}$$

where:

R = allowable stress reduction factor
L = unsupported length of column, in.
D = least width or diameter of column, in.

Plain concrete columns with eccentric loadings (ibid., 385) (with Axial and eccentric loading)

$$f_c = \frac{P}{A} + \frac{P_1 X_0 X_1}{I_x}$$

where:
P = axial load or axial + eccentric load, lb or kips
P_1 = eccentric load, lb or kips
X_o = eccentricity, in.
X_1 = distance C.G. to extreme fiber, in.
I_x = moment of inertia about C.G., in direction of eccentricity, in.^{4}
A = area of column, in.^{2}
F_{cr} = 50 psi or less (conc tension)

Reinforced concrete columns with eccentric loadings (with axial and eccentric loadings) (ibid., 386):

$$f_c = \frac{P}{A_c + nA_o} + \frac{P_1 X_0 X_2}{I_c + n I_s}$$

where:
P = axial + eccentric load, lb or kips
P_1 = eccentric load, lb or kips
A_c = area of concrete (bt), in.^{2}
I_c = moment of inertia, concrete in direction of eccentricity
I_s = moment of inertia, steel in direction of eccentricity
A_o = area of steel, $A_s + A'$, in.^{2}
A_s = area of tension steel, in.^{2}
A' = area of compression steel, in.^{2}
A = total area of column = $bt + n(A_s + A')$
 in.^{2} concrete + transformed area of longitudinal reinforcement.

If the tensile steel and the compression steel are equal in area, then the C.G. of the concrete and the C.G. of the steel are the same point. If the steel is unbalanced, the composite I_x must be calculated.

A_c is the area of the concrete (bt) with no allowance for displacement of any concrete by the longitudinal reinforcing steel.

As with any new analytical method that is based upon theory, gradual realizations and evolutions in theory sharpened the calculation methods through a series of mod-

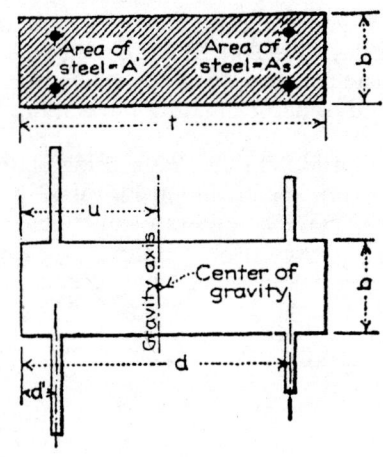

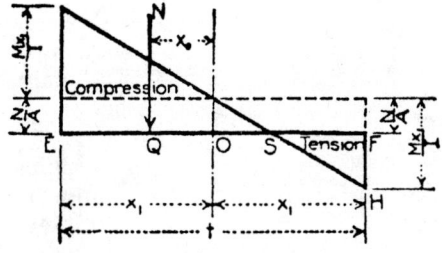

Figure 3-73.

ifications and developments. From 1907 through the 1940s concrete theory was in this process.

Reinforced concrete columns, compression over entire section with eccentric load (ibid., 387)

$$f_c = \frac{P_1}{bt}\left[\frac{1}{1+np_o} \pm \frac{6x_o t}{t^2 + 12np_o r^2}\right]$$

where:
P_1 = eccentric load

$$p_o = p + p' \qquad p = \frac{A_s}{A_g} \qquad p' = \frac{A'}{A_g} \qquad r = \frac{4}{2} - d'$$

In 1927–1928, Committees from the American Concrete Institute and the Concrete Reinforcing Steel Institute met as a combined Committee known as Joint Committee E-1. Committee E-1 produced a set of Reinforced Concrete Building Regulations intended for adoption as a part of a general building code. The Regulations were

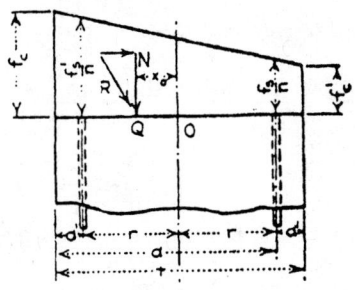

Figure 3-74.

adopted by almost all cities and codes and soon became the standard that governed all concrete building construction. This Building Regulation divided reinforced concrete columns into two basic categories: standard-length columns and long columns.

Design provisions of the Code of Building Regulation, 1928:

Standard-length columns (Lord 1928, 238–240): Principal columns in buildings shall have a minimum diameter or dimension of 12 in. Noncontinuous columns (floor to floor) shall have a minimum diameter or dimension of 6 in. The unsupported length of columns shall be the distance between the floor (bottom) and the underside of beams (shallowest), capitals (for slab construction).

Spiral columns:

$$P = A_c[1 + (n - 1)\rho]f_c$$

$$f_c = [300 + (0.10 + 4\rho)]f'_c$$

where:

f_c = allowable comp. strength, concrete, psi, ksi
f'_c = ultimate comp. strength, concrete, psi, ksi
$n = E_s/E_c$
E = modulus if elasticity
ρ = ratio, area of longitudinal steel to area of concrete core, between 1% and 6% of area of concrete.
Longitudinal steel = min. 6 Bars, 0.5 in. square column min. 3 per side
Spacing, 2.5 × diameter of bar
$A_{s\,spirals}$ = 25% in. minimum of longitudinal reinforcement

Columns with lateral ties:

$$P = 0.225f'_c A_G[1 + (n - 1)\rho]$$

where:

A_G = gross area of tied columns, outside dimensions, in.^{2}
ρ = ratio of longitudinal steel, $0.2A_G$,
$A_{s\,min}$ = 4-5/8 in. bars, cover 2 in.

Splices: deformed bars 24 in. min.
Smooth bars 30 in. min.
Ties: 0.25 in. diameter min. spacing 12 in. max.

Bending in standard-Length columns:

wall columns (flat slab const.) $M_{max} = \dfrac{WL^2}{35}$

(counter moments from D. L. eccentricities may be deducted)

<u>Axial and bending loads on standard-length reinforced concrete columns</u> (ibid., 240).

Until the 1928 Code, columns were designed as axially loaded with one set of formulae and columns with eccentric loadings by another set of formulae. There were formulae for each type and shape of column. In the 1928 Code, columns with axial and bending moment from eccentric loading were to be designed by analysis of combined sections with compressive stresses not to exceed allowable values and tension (if found) carried by the reinforcing and tension stresses in the reinforcing not to exceed the allowable for steel.

Limits of axial plus bending (combined) allowable stresses by 1928 Code.
Columns with spiral reinforcement:

$$f_c = [300 + (0.10 + 4p)f'_c] + 0.15f'_c$$

Columns with lateral ties:

$$f_c = 0.3f'_c \quad A_s < 0.04(A_{gross})$$

Allowable tensile stress in longitudinal rebars = 16,000 psi:
Check combined stresses through analysis of force couple on section of column at point of maximum moment.

Long columns (ibid., 242): (reinforced concrete long columns)

Spiral columns: allowable load reduction factor

$$P' = \left(1.50 - \frac{L}{100r}\right)P$$

Tied columns: allowable load reduction factor

$$P' = \left(1.33 - \frac{L}{120r}\right)P$$

where:
L = unsupported length of Column, in.
   L is equal to or greater than $50r$
r = radius of gyration, in. (column core)
P = allowable load on column, lb or kips
P' = Factored allowable load on long column, lbs or kips

r to be calculated on the basis of the concrete area used in design, and the transformed area of the longitudinal steel

Composite reinforced concrete column (steel or cast iron section encased inside longitudinally reinforced spiral or tied Column) (ibid., 214, 215). The concrete column, spiral or tied, meets normal requirements for longitudinal steel and concrete Loads (minus the area of steel or cast iron section, plus:

Steel section:

$$f_a = \frac{18,000}{1 + \frac{L^2}{18,000\, r^2}} \leq 15,000 \text{ psi}$$

Cast iron section:

$$f_a = 12,000 - 60\frac{l}{r} \leq 9000 \text{ psi}$$

Combination columns (straight steel column encased with wire fabric and concrete (ibid., 240):

$$P\left(1 + \frac{A_c}{100\, A_s}\right)$$

where:
A_c = net area of concrete, within wire fabric core, in.^{2}
A_s = area of structural section, in.^{2}
P' = factored load, lb or kips

P = calculated allowable load, steel column in standard steel method, lb or kips

The 1928 Reinforced Concrete Building Regulations written by Joint Committee E-1, ACI, and the CRSI Committee on Engineering Practice published a table of the allowable unit stresses in concrete and the allowable unit stresses in reinforcement, (see Table 3-54).

The 1937 Code also incorporated the use of both allowable stress design and ultimate strength design as alternative methods of analysis for the use of the designer. The ultimate strength method of 1937 was not the same as that used today in that the 1937 method calculated the ultimate strength of columns and then utilized a factor of safety of 5.5 to determine the safe allowable load on the column. Today's code

Table 3-52. Allowable Unit Stresses (1928)[a]

Description	Allowable Unit Stresses			
	For any Strength of Concrete as Fixed by Test in Accordance with Sec. 303 $n = \dfrac{30{,}000}{f'_c}$	When Strength of Concrete Is Fixed by the Water-Cement Ratio in Accordance with Sec. 302		
		$f'_c =$ 2000 lb $n = 15$	$f'_c =$ 2500 lb $n = 12$	$f'_c =$ 3000 lb $n = 10$
Flexure: fc				
Extreme fiber stress in compression (f_c)	$0.40f'_c$	800	1,000	1,200
Extreme fiber stress in compression adjacent to supports of continuous or fixed beams or of rigid frames (f_c)	$0.45f'_c$	900	1,125	1,350
Shear: v				
Beams with no web reinforcement and without special anchorage of longitudinal steel (v_c)	$0.02f'_c$	40	50	60
Beams with no web reinforcement, but with special anchorage of longitudinal steel (v_c)	$0.03f'_c$	60	75	90
Beams with properly designed web reinforcement, but without special anchorage of longitudinal steel (v)	$0.06f'_c$	120	150	180
Beams with properly designed web reinforcement and with special anchorage of longitudinal steel (v) For conditions determining the use of greater shear values see Sec 903(e).	$0.09f'_c$	180	225	270
Flat slabs at distance d from edge of column cap or drop panel (v_c)	$0.03f'_c$	60	75	90
Footings where longitudinal bars have no special anchorage (v_c)	$0.02f'_c$	40	50	60
Footings where longitudinal bars have special anchorage (v_c)	$0.03f'_c$	60	75	90
Bond: u				
In beams and slabs and one-way footings:				
Plain bars (u)	$0.04f'_c$	80	100	120
Deformed bars (u)	$0.05f'_c$	100	125	150
In two-way footings:				
Plain bars (u)	$0.03f'_c$	60	75	90
Deformed bars (u) (Where special anchorage is provided (see Sec. 903), double these values in bond may be used.)	$0.0375f'_c$	75	94	112
Bearing: fc				
Where a concrete member has an area at least twice the area in bearing (f_c)	$0.25f'_c$	500	625	750
Axial Compression: fc				
In columns with lateral ties (f_c)	$0.225f'_c$	450	563	675

Table 3-52. (Continued)

Description	Allowable Unit Stresses			
	For any Strength of Concrete as Fixed by Test in Accordance with Sec. 303 $n = \dfrac{30{,}000}{f'_c}$	When Strength of Concrete Is Fixed by the Water-Cement Ratio in Accordance with Sec. 302		
		$f'_c =$ 2000 lb $n = 15$	$f'_c =$ 2500 lb $n = 12$	$f'_c =$ 3000 lb $n = 10$
In columns with continuous spirals enclosing a circular core:[b]				
Ratio of longitudinal reinforcement $\begin{cases} p = 0.01 \\ 0.02 \\ 0.03 \\ 0.04 \\ 0.05 \\ 0.06 \end{cases}$	$300 + 0.14f'_c$ $300 + 0.18f'_c$ $300 + 0.22f'_c$ $300 + 0.26f'_c$ $300 + 0.30f'_c$ $300 + 0.34f'_c$	580 660 740 820 900 980	650 750 850 950 1,050 1,150	720 840 960 1,080 1,200 1,320
(Spiral reinforcement not to be less than ¼ the longitudinal.)				

Allowable Unit Stresses in Reinforcement:
The following unit stresses in reinforcing steel shall not be exceeded:

Tension:
- Intermediate grade billet steel — $(f_s) = 20{,}000$ psi
- Rail steel bars — $(f_s) = 20{,}000$ psi
- Web reinforcement — $(f_v) = 16{,}000$ psi
- Structural steel shapes — $(f_s) = 18{,}000$ psi
- Other steel reinforcement 50 percent of the yield point stress, but not to exceed — $(f_s) = 20{,}000$ psi

Compression:
- Bars — nf_c
- Structural steel section in composite columns — 15,000 psi
- Cast iron section in composite columns — 9,000 psi

[a] Reprinted by permission from A. R. Lord, *A Handbook of Reinforced Concrete Building Design, in Accordance with 1928 Joint Standard Building Code*, 1st ed., pp. 238–240. @ 1928 American Concrete Institute.
[b] Unit stress in spirally reinforced columns = $[300 + (0.10 + 4p)f'_c]$.

method utilizes a system of load factoring to artificially raise the load for computations against the ultimate strength of concrete, thus introducing a factor of safety. That factor of safety is about 1.6, as occurs in the modern ultimate strength analysis. This appears to be a more radical shift than it actually is. It should be noted that over 50 years of concrete technology have passed to today.

The 1937 Joint Committee of the ACI and others produced a set of Standard Specifications for Concrete Reinforced Concrete. This Code is the basis of the ACI 318 Building Code that we now use today and is revised about every five years to remain up to date.

Design of tied columns, 1937 Code method (Dunham 1939, 145–147) (short columns, length ≤ 10 × least lateral dimension or diameter):

$P = A_G f_c [1 + (n - 1)\rho_G]$ (formula based upon allowable stresses)

$P' = A_G[1 - \rho_G]f_c + \rho_G f'_s$ (formula based upon ultimate strength design, safety factor = 5.5)

$P = 0.18 f'_c (A_G - A_s) + 0.8 A_s f_s$ (formula for allowable load based upon ultimate stresses of concrete and steel)

where:
A_G = gross effective area of concrete core, in.²
$\rho_G = A_s/A_G$ $f_s = n(f_c)$

f_s = allowable stress, steel, psi or ksi
f_c = allowable stress, concrete, psi or ksi
P = allowable load on Column, lb or kips
P' = ultimate load on column, lb or kips
f_s = yield stress, steel, psi or ksi
f_c = ultimate strength, concrete, psi or ksi
f_s = 16,000 psi

Design of spiral columns, 1937 Code Method (ibid., 150, 151):

$$P = A_G[0.225 f'_c(1 - \rho_G) + \rho_G f_s]$$

Spirals $\rho' = 0.45(R - 1)\dfrac{f'_c}{f'_s}$

where:
$f_s = 0.45 f'_s$ or 0.45 yield
ρ' = volume of Spiral/volume of concrete within spiral
$R = A_G/A_c$

Long columns, 1937 Code method: L greater than 10 times least lateral dimension or diameter (ibid., 159):

$$P' = P\left(1.3 - 0.03\left(\dfrac{L}{d}\right)\right)$$

where:
L/d must not exceed 20
L = unsupported length of column, in.
d = least lateral dimension, or diameter, in.

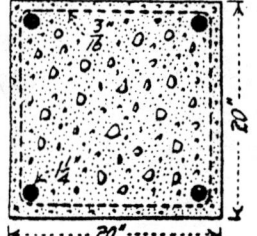

Figure 3-75. Ingals column. (*From Reid 1907, 470.*)

Reinforced concrete columns were designed with proprietary reinforcing systems from 1905 through the 1930s and perhaps beyond. Round and square bars were used in large quantity by some designers, avoiding the proprietary materials. Gradually, "diamond" bars became more uniformly used in the late 1930s and early 1940s. In addition, the embedment of structural steel shapes became less frequently used by the 1940s. As with most technological advances, the more remote areas embraced newer technologies more slowly than the centers of population. The Ingals Building, 1907, Cincinnati, Ohio, the first office building of reinforced concrete, used round bars at the corners of square columns with bar ties very much like the conventional arrangement in today's design methodology (see Figure 3-75).

The ACI Building Code Requirements for Reinforced Concrete (ACI 318-89), published by the American Concrete Institute, are the building code for concrete construction that has been adopted by all code-writing and enforcement agencies in America. In 1956, The Code introduced ultimate strength design as an alternative to the working stress design methods that had existed during the period since reinforced concrete construction began in America. This was not a signal to start a mass movement toward ultimate strength design, but all of the designers of the period realized that this new technology was coming. In the 1963 Code, ACI Standard 318-63, dated June 1963, ultimate strength design received equal weight in the section on analysis and design of structural concrete members. By the mid 1970s, the Code was all but totally converted to ultimate strength design. Working stress design was by then only an alternative method of analysis.

ACI 318-89 Reinforced Concrete Columns, USD:

Tied column, pure axial load: (American Concrete Institute 1989, secs. 9.3.2, 10.3.5.1):

$$P_u = \phi 0.80[0.85(f'_c)(A_G - A_s) + f_y(A_s)]$$

Spiral column, pure axial load (ibid.):

$$P_u = \phi 0.85[0.85(f'_c)(A_G - A_s) + f_y(A_s)]$$

where:
ϕ = 0.75 for spiral column; 0.70 for tied column
P_n = axial load capacity, lb or kips
ρ_G = Between 0.01 and 0.08 (A_G)
f'_c = ultimate strength concrete, psi or ksi
f_y = yield stress of re-bars, psi or ksi
A_G = gross area of concrete, in.^{2}
A_s = area of longitudinal steel, in.^{2}

Columns with eccentric loading (with and without additional axial loads) are analyzed by use of the standard interaction diagram which demonstrates the interactive strength of a column by plotting axial load vs. moment and showing graphically the ability of the column to carry axial load plus bending (see Figure 3-77).

Reinforcing for concrete is now almost exclusively Grade 60 deformed (diamond) bars, which have a yield strength of 60,000 psi. Some designers have utilized the earlier (1970s) Grade 40 bars, which have a yield strength of 40,000 psi.

Modern designers prefer square or round columns and either spiral or tied horizontal bars to restrain the longitudinal reinforcement. The conventional method of arranging ties is to attempt to ensure that most longitudinal bars are restrained by a 90° corner in the horizontal reinforcement at each required spacing point. It is felt that in this manner the maximum potential for restraint against any buckling of the longitudinal reinforcement can be achieved (see Figure 3-77).

In addition to the interaction diagram plotted for each column at each steel strength and configuration, size, etc., and concrete strength, a tabular form of the data for the interaction diagram is quite popular with designers (see Table 3-55). By the use of these data, it is possible to correlate the design axial load and moment on the member, from floor beams above, by looking at the eccentricity at the point of load. The column interaction diagram for an 18 × 18-in. concrete column utilizing 5000 psi concrete and eight #9 reinforcing bars as shown in Figure 3-76, can be examined and cross-checked by the tabular data shown in Table 3-55.

Any load at a given eccentricity produces an equivalent axial load plus an applied moment.

Example:
500 kips load at 6 in. eccentricity equates to 500 kips axial load plus 250 k-f moment. This falls within the shaded area of the interaction diagram, and therefore load and moment are within the allowables.

A check of the historic plain or reinforced concrete column by modern analysis is made by first identifying the time of construction, column type, size, method of reinforcing, and original method of analysis. The early analytical method or formulae in common use at a given time will be utilized to determine the column capacity, and the column will be checked by the method utilized by today's code. The original column capacity (by original calculations) will be checked against the modern calculations as much as is applicable.

Plain concrete column, 1911 (method of ICS, *Concrete Engineer's Handbook*) 18 × 18 in. square column (plain, no reinforcing)

$$f_c = 300 \text{ psi} \quad \text{(this chapter)}$$

$$P_{all} = f_c(A) = 97{,}200 \text{ lb. (limit of height = 10 ft.)}$$

Reinforced concrete column, 1907 (lightly reinforced, Reid) 18 × 18 in. square column with 8 each 0.75 in., square, twisted Ransome bars

250 ◇ Structural Analysis of Historic Buildings

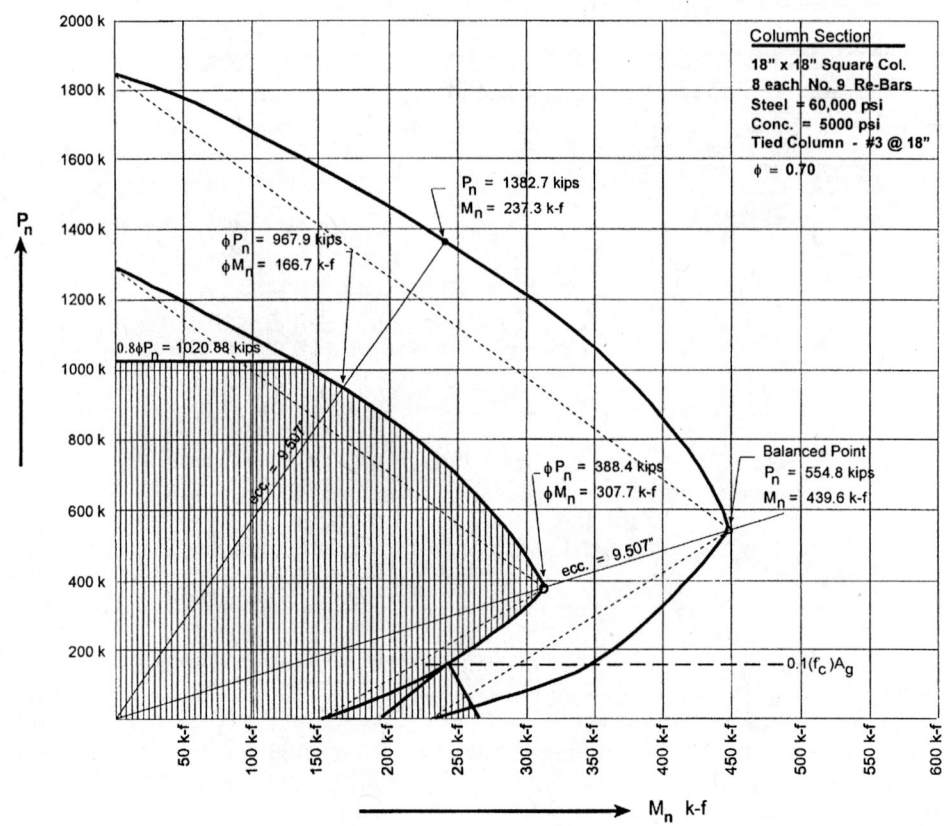

Figure 3-76. Column interaction diagram.

Table 3-53. C.R.S.I. Column Design Tables[a]

Bars Grade 60		Square Tied Columns 16" × 16" Short columns; no sidesway [1] Bars symmetrical in 4 faces											0.10f'_cA_g = 128 kips [2]				
Concrete f'_c = 5,000 psi		P_u (kips)—Ultimate Usable Capacity													For P_u of [2] 128		
		M_u/P_u = e (in.) (φ = 0.70)											OT[0]	Balance		k [0]	O[0]
Bars	p %	0	0.1t	2"	3"	4"	6"	8"	12"	16"	20"	24"	28"	e (in.)	e (in.)	P_u (k)	e (in.)	M_u (k-ft.)
4-#8	1.23	885	707	666	569	483	355	262	134	0	0	0	0	2.87	7.23	302	12.34	93
4-#9	1.56	918	732	690	591	504	376	294	158	0	0	0	0	2.96	7.89	300	13.77	114
4#10	1.98	960	764	721	619	531	401	319	188	0	0	0	0	3.07	8.74	296	15.58	141
4-#11	2.43	1,005	797	752	646	555	422	336	212	141	0	0	0	3.18	9.71	284	17.26	167
4-#14	3.51	1,113	877	828	714	617	476	383	270	187	140	0	0	3.36	12.12	268	21.51	229
4-#18	6.25	1,386	1,075	1,016	878	763	598	482	343	266	218	176	147	3.63	20.03	217	31.59	374
		Square Tied Columns 18" × 18"													0.10f'_cA_g = 162 kips			
4-#9	1.23	1,120	896	873	763	662	501	394	213	0	0	0	0	3.17	8.13	388	14.10	134
4-#10	1.56	1,162	929	906	793	691	529	421	250	0	0	0	0	3.28	8.90	385	15.80	165
4-#11	1.92	1,207	963	938	822	717	554	444	282	183	0	0	0	3.41	9.71	378	17.40	197
4-#14	2.77	1,315	1,044	1,019	895	785	615	500	356	241	177	0	0	3.62	11.68	369	21.51	272
4-#18	4.93	1,588	1,247	1,217	1,073	947	754	622	455	355	277	222	184	3.97	17.33	331	31.25	450
8-#6	1.08	1,101	874	850	736	630	464	341	189	0	0	0	0	2.90	7.11	397	13.25	121
8-#7	1.48	1,151	911	886	768	661	494	382	222	0	0	0	0	2.98	7.79	394	15.21	159
8-#8	1.95	1,210	954	929	806	697	528	417	258	178	0	0	0	3.07	8.60	392	17.28	204
8-#9	2.46	1,276	1,002	976	848	734	561	448	293	205	0	0	0	3.15	9.50	388	19.48	252
8-#10	3.13	1,360	1,064	1,035	901	782	603	486	335	237	182	0	0	3.24	10.66	384	22.17	312
8-#11	3.85	1,451	1,127	1,096	953	828	640	519	372	265	204	166	0	3.35	11.92	375	24.57	370
8-#14	5.55	1,666	1,282	1,246	1,082	942	735	602	437	336	261	213	179	3.49	15.06	360	30.81	488
12-#10	4.70	1,558	1,209	1,176	1,021	888	690	559	403	303	237	194	164	3.31	13.06	376	28.41	428
12-#11	5.77	1,694	1,303	1,267	1,098	954	743	603	437	338	265	218	184	3.40	15.00	362	31.69	498
16-#10	6.27	1,757	1,355	1,320	1,145	997	778	635	461	360	286	236	201	3.39	15.38	375	34.40	543

[a] From Concrete Reinforcing Steel Institute 1978, sec. 3-29.

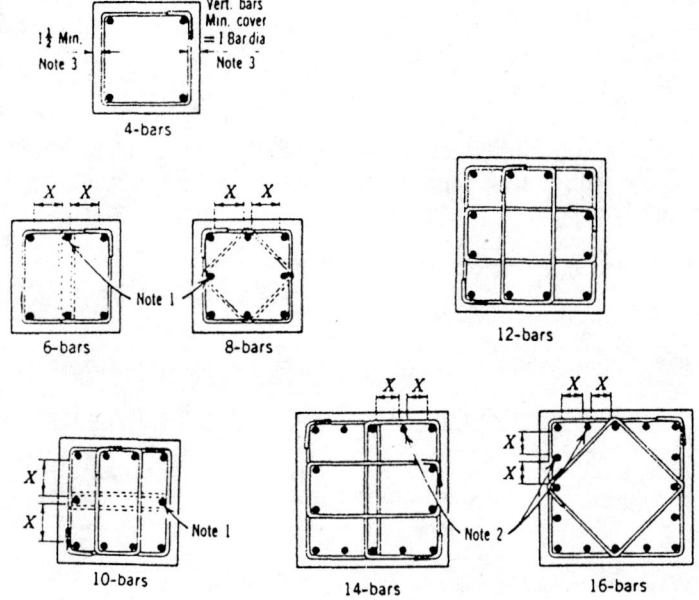

Figure 3-77. Column reinforcing and ties. (*From Ferguson 1973, 561.*)

$P_{all} = f_c[A_G + nA_s]$ this chapter

$= 350 \text{ psi } [324 \text{ in.}^2 + 10(4.5 \text{ in.}^2)]$

$= 129{,}150 \text{ lb}$

Reinforced concrete column, 1907 (heavily reinforced, reid) 18 × 18 in. square column with 8 each 0.25 in., square, twisted Ransome bars

Table 3-54. Concrete Column Capacities

	f_c or f'_c (psi)	Column (in.)	Re-bar	A_s (in.²)	P_{all} (lb)
Plain concrete column, 1911	300	18 × 18	None	None	97,200
Reinforced concrete column, 1907	350	18 × 18	8—0.75 in., sq.	4.500	129,150
Reinforced concrete column, 1907	350	18 × 18	8—1.25 in., sq.	12.500	152,775
Reinforced concrete column, 1911	450	18 × 18	8—1.25 in., sq.	12.500	196,425
Plain column, Hooped or Spiral, 1911	450	18 diameter	None	None	69,270
Reinforced concrete column, H or S, 1911	650	18 diameter	8—1.25 in., sq.	12.500	100,000
Empirical rule, reinforced column, 1911	500	18 × 18	8—1.25 in., sq.	12.500	98,000
Plain concrete column, 1918	450	18 × 18	None	None	145,800
Reinforced Concrete column, 1918	450	18 × 18	8—1.25 in., sq.	12.500	189,375
Reinforced concrete column + eccentric, 1918	435	18 × 18	8—1.25 in., sq.	12.500	100,000 50,000
Reinforced concrete column, 1928	1,068	18 × 18	8—1.25 in.,sq.	12.500	222,100
Reinforced concrete column, 1928	675	18 × 18	8—1.25 in., sq.	12.500	295,460
Reinforced concrete column, 1928 900 psi concrete and 16,000 psi steel T	900				
Reinforced concrete column, 1937	650	18 × 18	8—1.25 in., sq.	12.500	284,520
Reinforced concrete column, 1937, USD Factor of safety = 1.6	3,000	18 × 18	8—1.25 in., sq.	12.500	243,350
Reinforced concrete column, 1989, USD Factor of safety = 1.6	3,000	18 × 18	8—1.25 in., sq.	12.500	418,000
Plain concrete column, 1989, USD Factor of safety = 1.6	3,000	18 × 18	None	None	194,700

$P_{all} = f_c[A_G + (n - 1)A_s]$ (this chapter)

$= 350 \text{ psi}[324 \text{ in.}^2 + (9)12.5 \text{ in.}^2]$

$= 152{,}775 \text{ lb}$

Reinforced concrete column, concentrically loaded, 1911 (straight longitudinal reinforcement) 18 × 18 in. square column with 8 each 1.25 in., square, twisted, Ransome bars

$P_{all} = f_c[A_c + nA_s]$ $f_c = 450$ psi (this chapter)

$= 450 \text{ psi}[311.5 \text{ in.}^2 + 10(12.5 \text{ sq in.}^2)]$

$= 196{,}425 \text{ lb}$

Hooped or spiral concrete column, 1911 (hooped or spiral reinforcement only, ICS method)
Hooped reinforcement shall not be less than 1% of volume enclosed.
18 in. diameter column, hooped reinforcement, hoop diameter 14 in. (2 in. cover.)

$P_{all} = 450 \text{ psi}(A_G)$ (this chapter)

$= 450 \text{ psi}[0.7854(14 \text{ in})^2 \text{ in.}^2$

$= 69{,}270 \text{ lb}$

Reinforced concrete Column, 1911 (hooped or spiral reinforcement only, ICS method)
Hooped spiral reinforcement shall not be less than 1% of volume enclosed.
18 × 18 in. square column, hoop diameter 14 in.

$P_{all} = 650 \text{ psi}(A_G)$ (this chapter)

$= 650 \text{ psi}[0.7854(14 \text{ in.})^2]$

$= 100{,}060 \text{ lb}$

Empirical rules for straight reinforcement, 1911 (ICS method)
Straight reinforcement, concentric loading, (patented type longitudinal bars)
No specific allowance made for reinforcing, reinforcing size, etc., except in the larger value for f_c $f_c = 500$ psi
18 × 18 in. square column, with Patented Rebar, (longitudinal)

$P_{all} = f_c(A)$ (this chapter)

$= 500 \text{ psi}(14 \text{ in.})^2$

$= 98{,}000 \text{ lb}$

where:
A = area of concrete minus cover

Plain rectangular concrete columns, 1918 (Hool et al., Joint Committee)
18 × 18 in. square column, no reinforcing, eccentricity = 0.
L is limited to $\leq 4d$ or 4 diameter

$P_{all} = f_c(A_G)$ (this chapter)

$= 450 \text{ psi}(18 \text{ in.})^2$

$= 145{,}800 \text{ lb}$

where:
$f_c = 450$ psi, average in 1918.

A column this short is limited to use as a Pier.

Reinforced concrete column with longitudinal reinforcement only, 1918, Joint Committee
18 × 18 in. square column with 8 each 1.25 in. square, twisted ransome bars.

$P_{all} = f_c A[1 + (n - 1)p]$ (this chapter)

$= 450 \text{ psi}(311.5 \text{ in.}^2)[1 + (10 - 1)(0.039)]$

$= 189{,}375 \text{ lb}$

where:

$$p = \frac{A_s}{A_G} = \frac{12.5}{324} = 0.039$$

$$A = A_c - A_s = 311.5 \text{ in.}^2$$

Reinforced concrete column, with eccentric loading, 1918 (Joint Committee) 18 × 18 in. square column with 8 each 1.25 in. square, twisted Ransome bars

$$f_c = \frac{P}{A_c + nA_s} + \frac{P_1 \, ; x_0 \, X_1}{I_c + nI_s} \quad \text{(this chapter)}$$

$$= \frac{100{,}000 \text{ lb}}{324 \text{ in.}^2 + 10(12.5 \text{ in.}^2)} + \frac{50{,}000 \text{ lb}(6")(9")}{8{,}748 \text{ in.}^4 + 10 \, (396 \text{ in.}^4)} = 222.7 \text{ psi} + 212.5 \text{ psi}$$

$= 435.2 \text{ psi (within the 450 psi allowable)}$

where:
$f_c = 450 \text{ psi, max}$
$P = 100{,}000 \text{ lb axial load}$
$P_1 = 50{,}000 \text{ lb eccentric load}$
$x_o = \text{eccentricity} = 6.0 \text{ in.}$

This formula utilizes the combination of axial load and eccentric loads on the column to determine the actual stress on the column to compare to the allowable maximum stress.

Reinforced concrete column, standard length, 1928 (Joint Committee) spiral column 18 in. diameter, 8 each 1.25 in. square, twisted Ransome bars

$P_{all} = A_c[1 + (n - 1)p]f_c$ (this chapter)

$f_c = [300 \text{ psi} + (0.10 + 4p)f'_c]$

$= [300 \text{ psi} + (0.10 + 4(0.039))3000 \text{ psi}]$

$= 1068 \text{ psi}$

where:
$p = 0.039$

$P_{all} = 0.7854(14 \text{ in.})^2[1 + 9(0.039)]1068 \text{ psi}$
$= 222{,}100 \text{ lb}$

Reinforced concrete column, standard length, 1928 Joint Committee Tied Column, 18 × 18 in. square column, 8 each 1.25 in. square, twisted Ransome bars

$P_{all} = 0.225 f'_c(A_G)[1 + (n - 1)p]$

$= 0.225(3000 \text{ psi})(324 \text{ in}^2)[1 + 9(0.39)]$

$= 295{,}460 \text{ lb}$

Reinforced concrete Column, standard length, axial and bending loads,

1928 (Joint Committee) Tied column, 18 × 18 in. square column, 8 each 1.25 square, twisted, Ransome bars

$f_{c\,max} = 0.3f'_c = 0.3(3000 \text{ psi}) = 900 \text{ psi}$

Tension in longitudinal reinforcing, A_s = 16,000 psi max.

The 1928 Code states that designers are to utilize the conventional method of designing for axial and bending loads and make the above allowable stresses the maximum bending stresses in the member. The member would need to be checked for actual bending stresses and add the compressive stress from the axial load to determine the compressive and tension maximums. In modern codes we analyze by the interaction diagram for the column. To check a column with these allowable stresses, an interaction diagram based upon allowable stresses should be developed.

Reinforced concrete Columns, long Columns by the 1928 Joint Committee Code. The standard length column was limited to a length equal to or less than 11 times the least dimension or the diameter. Any length of column above that control point must have a factor applied to the allowable load to compensate for the additional length. The factor is a variable that reduces itself as the length of column increases. See formulae, this chapter.

At the same time that reinforced concrete columns were being developed and increasingly being used, cast Iron, wrought iron, and steel columns were still being used in conjunction with concrete encasement systems. The concrete encasement was not calculated into the capacity of the column unless there was either longitudinal reinforcement with ties or spirals or a wire fabric or mesh containing a core of concrete. See this chapter for formulae for this type of column.

Reinforced concrete column, standard length, (L equal to or less than 10 (least dimension) or 10 diameter), 1937 (Joint Committee)
<u>Tied column: 18 × 18 in. square column, 8 each 1.25 in. square, twisted Ransome bars</u>

$P_{all} = A_G f_c[1 + (n - 1)p_G]$

$= 324 \text{ in.}^2 (650 \text{ psi})[1 + (10 - 1)0.039]$

$= 284{,}520 \text{ lb}$

$P_{ult} = A_G[(1 - p_G)f'_c + p_G f'_s]$

$= 324 \text{ in.}^2[(1 - 0.039)3000 \text{ psi} + 0.039(32{,}000 \text{ psi})]$

$= 1{,}338{,}444 \text{ lb}$

Formula for allowable load based upon ultimate stress of concrete

$P_{all} = 0.18 f'_c(A_G - A_s) + 0.8(A_s) \text{ in.}^2\ .8A_s(f_s)$

$= 0.18(3000 \text{ psi})(324 \text{ in.}^2 - 12.5 \text{ sq. in.}^2) + 0.8(12.5 \text{ psi})(16{,}000 \text{ psi})$

$= 328{,}210 \text{ lb}$

<u>Reinforced concrete columns standard length, 1989 (ACI 318-89) Check of Historic 18 × 18 in. square tied column, 8 each 1.25 in. square, twisted Ransome bars</u>

$$P_u = 0.70(0.80)[0.85f'_c(A_G - A_s) + f_y A_s]$$
$$= 0.70(0.80)[0.85(3000 \text{ psi})(324 \text{ in.}^2 - 12.5 \text{ in.}^2) + 32,000 \text{ psi}(12.5 \text{ in.}^2)]$$
$$= 668,822 \text{ lb}$$

$$P_{all} = \frac{P_{ult}}{1.6} = \frac{668,822 \text{ lb}}{1.6}$$
$$= 418,000 \text{ lb}$$

Plain concrete column, 1989 (ACI 318-89, as applies) (shear across inclined plane in column would govern) (American Concrete Institute 1989, see 11.7.5). ACI does not allow plain concrete columns which have a length of over three times the least dimension d or diameter d.

$$P_u = 0.70(0.80)[0.85f'_c(A_G)]$$
$$= 0.70(0.80)[0.85(3000 \text{ psi.})(324 \text{ in.}^2)]$$
$$= 462,672 \text{ lbs} \quad (\text{with } A_s = 0)$$
$$P_{all} = 462,672 \text{lb}/1.6 = 289,170 \text{ lb}$$

Table 3-55 is a summary of the allowable loads on concrete columns, arranged chronologically for purposes of comparison.

As can be seen by inspection of the table, the early concrete formulae were very conservative, as would be expected due to the lack of development of the technology. The condition of the concrete column would determine the percentage one would allow of the allowable load that governed at the time of construction. It would be difficult to core (extract a sample) a column that had to remain in service, or to determine a location that would not include reinforcing. If the top of the column could be exposed, the core of the column could be cored (sample) and the void filled with grout to return the section to its original density. The ACI 318-89 Commentary states that core tests indicate about 85 percent of the strength of laboratory-cured cylinders (ibid., sec. R20-2). An in-place concrete testing instrument such as the Kelly Ball test equipment would give a close approximation of the strength of the concrete; but Code requirements dictate that the building be load tested to determine the performance of the building, frame, or column, as an indication not only of strength but of serviceability. A professional engineer must be in charge of the load testing operations, and the Code official responsible for the area in which the building is located must be informed and offered the choice of being in attendance at the testing location.

4
Historic American Floor Systems—Beams

WOOD JOISTS AND BEAMS

Wood has been used as a spanning device, probably for a crude form of roofing, since long before recorded history. It has been used for everything from tools to furniture to building products. Probably the original huts with wood spanning members for roof framing were built of stone directly upon the ground and utilized the ground as the floor. Many historians state that the first dwellings were near lakes and other sources of water, such as swamps or rivers, and were elevated above the waterline for the obvious reason, to keep dry, and for a not so obvious reason, to keep out of the reach of most animals. If that is so, then a floor system had to be devised, which had to span between posts (columns or piers) and support the weight of the occupants. During that era and beyond, the sizing of these members was empirical. The availability of timber, its size (size and length of logs), and the ability to shape these members would have had a great bearing on the look and sophistication of the floor system. Medieval man had been shaping members and had invented the system of members and submembers at some point probably within a range of years before or after the tenth century. Stone, stone arches, and floors bearing on arches (where level was achieved by filling the lower portions of the voids with sand and rubble) were utilized in many areas of the world for permanence and protection. Wood would have been used for doors and shutters and for furniture prior to being used as a structural element. Joseph Moxon of London was one of the first to publish a reference or guidebook on the subject of building with his *Mechanick Exercises or the Doctrine of Handy-Works*, "applied to the arts of Smithing, Joinery, Carpentry, Turning, and Bricklayery" in 1703. Moxon quotes building laws of the City of London that were set by Parliment after the Great Fire of 1666. Moxon gives sizes for timbers for houses of the "first sort" and houses of the "other two sorts" as follows (Moron 1703, 138, 139):

"First sort" (first-class) construction:
   Floor: summer beams, 15, 12 in, and 8 in.
   Wall plates, 7 in. and 5 in.

Roof: principal rafters,

$$15 \text{ ft} \left(\frac{\text{st foot—8}}{\text{st top—5}} \right) 6 \text{ in.}$$

   Single rafters, 4 in. and 3 in.

   Joists: to 10 ft, 3 in. thickness, 7 in. depth
   Garret floors (attics), 3 ft, ? thickness, 6 in. depth

This is the only mention of first-class construction and the members required in this section of Moxon's volume.

"Other two sorts" of construction:
Floor members: summers and girders which bear in length from
- 10–15 ft, 11 in. breadth and 8 in. depth; joists, 6 in. depth
- 15–18 foot, 13 in. breadth and 9 in. depth; joists, 7 in. depth
- 18–21 foot, 14 in. breadth and 10 in. depth; josits, 7 in. depth
- 21–24 foot, 16 in. breadth and 10 in. depth; joists, 8 in. depth
- 24–26 foot, 17 in. breadth and 14 in. depth; joists, 8 in. depth

Joists, all 3 in. thick, which bear 10 ft (see above)
Principal discharges upon piers, 13 in. and 12 in. In the first story in the fronts, 15 in. and 13 in.
Binding joists with their trimming joists, 5 in. thickness, depth = floor
Wall plates, or railing pieces and beams, 10 in. and 6 in.
                                             8 in. and 6 in.
                                             7 in. and 5 in.
Lintels of oak in the first and second Story, 8 in. and 6 in. Third story, 5 in. and 4 in.

For the roof:

Principal rafters, from 15–18 ft, 9 in. at foot, 7 in. at top, thickness 7 in.
  from 18–21 ft, 10 in. at foot, 8 in. at top, thickness 8 in.
  from 21–24 ft, 12 in. at foot, 9 in. at top, thickness 8.5 in.
  from 24–26 ft, 13 in. at foot, 9 in. at top, thickness 9 in.
Single rafters, not exceeding in length, 9 ft, 5 in. thick, 4 in. depth not exceeding in length 6 ft, 6 in. thick, 3.5 in. depth.
Purlins from 15–18 ft, 9 in. thickness, 8 in. depth from 18–21 ft, 12 in. thickness, 9 in. depth

Moxon's terms and sizes are difficult to understand in the context of today's terms and design styles. They reflect the thinking of early times, which was that the thickness was required to be as much as the depth or even more. Rafters are shown as tapered, in an attempt to lighten the weight of roofs. There would be no trouble with lateral bracing to resist torsional buckling with those thicknesses. This thinking may have been the result of carpenters experiencing buckling problems with spanning devices. Also, nails were not available as they are today. Members would be joined by mortise and tenons. The forming of mortise and tenon joints was very time-consuming and required certain skills. Carpenters were responsible for planning and sizing of all members. Moxon mentions that all members are of "good oak," and at this period most, if not all, of the member sizing would be by adze, axe and froe. Small members might have been formed by splitting logs, and very little sawing of large sizes would occur. See Figure 4-1 for a plan layout of a floor system from Moxon's volume.

Early experiments in determining the scientific aspects of wood flexural members were done by army engineers and college faculty. Mahan (1885) gives some of the early analytical and experimental testing methods for wood members "transversely loaded" (flexure). Mahan includes a table of the results of some testing of timber sections that averaged about 2 × 2 or 5 in., simply supported and loaded to failure at midspan. The materials were a range of forest products from England and North America (see Figure 4-2).

The state of the art in the United States in 1885 was still very empirical for everyone except the educated designers and builders of larger commercial buildings.

Correlation of the results of these tests with the numbers and constants of today is relatively simple. Using today's conventional formulae for the moment of inertia, the moment of this type of loading, the average modulus of elasticity, and the standard equation for the deflection of a member loaded in this manner gives us a verification of Mahan's tabular quantities that result from the early testing.

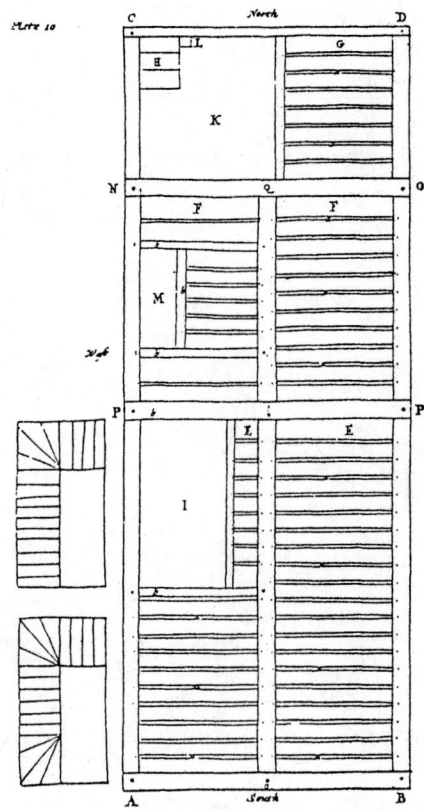

Figure 4-1. Plan: timber floor framing. (*From Moxon 1703, 138, 139.*)

Verification of the Tabular Results in the Mahan Volume:

$$I = \frac{bh^3}{12} = \frac{(2 \text{ in.})^4}{12} = 1.333 \text{ in.}^4 \text{ (the moment of inertia)}$$

$$M = \frac{Pb}{4} = \frac{637 \text{ lb } (84 \text{ in.})}{4} = 13{,}377 \text{ lb-in. (the maximum moment at the center span)}$$

$$F_b = \frac{M_c}{I} = \frac{13{,}377 \text{ lb-in. } (1 \text{ in.})}{1.333 \text{ in.}^4} = 10{,}032.75 \text{ psi (the maximum flexural stress)}$$

$$E = 1{,}500{,}000 \text{ psi (the modulus of elasticity for oak) (Trautwine 1888, 434e)}$$

$$\delta_E = \frac{PL^3}{48EI} = \frac{200 \text{ lb } (8 \text{ in.}^3)}{48(1{,}500{,}000)(1.333 \text{ in}^4)} = 1.23 \text{ in.}$$

(the deflection at the center of span (the maximum point) for a concentrated load of 200 lb located at the center of span)

As can be seen from a comparison of Table 4-2, the deflection given for the 200 lb load at midspan is given as 1.280 in. from the tests. This verification is actually closer than that justified by the data. The choice of a modulus of elasticity of 1,500,000 psi was from an average according to Trautwine; any other choice would have seriously affected the value of the deflection.

Trautwine gives formulae for the determination of the load required to take the stress to the elastic limit of the material. For wood, he states that for knots and irregular grain of material, the designer should utilize only one-third of the amount given in the equation and then utilize only one-third to one-eighth of that amount for actual

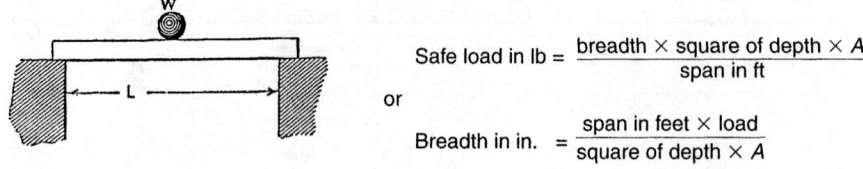

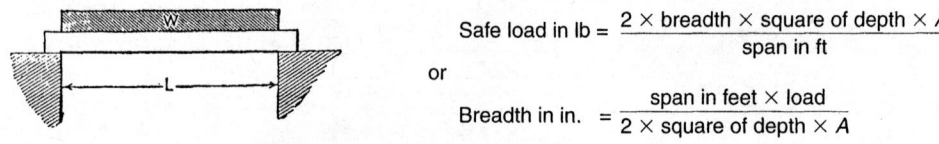

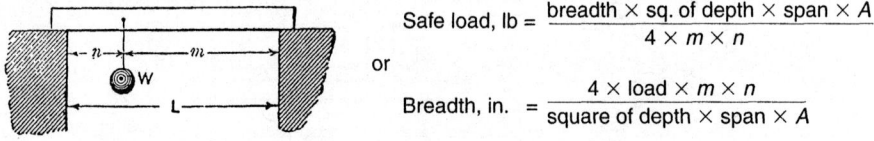

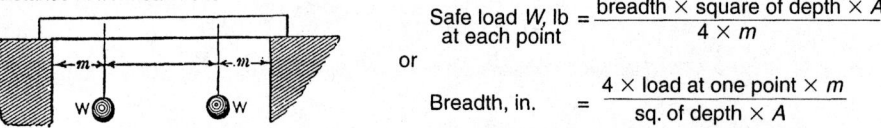

Note: In the last two cases the lengths denoted by m and n should be taken in ft, the same as the spans.

Figure 4-2. Rectangular beam equations for safe loads. *(From Kidder 1902, 372.)*

practice. The formula is based upon a constant for the species or type of material used (see Table 4-2).

The Trautwine Formula (Ibid., 492):

$$W = \frac{\text{breadth (in.)} \times (\text{depth})^2 \text{ in.}^2}{\text{clear span, ft}} (\text{constant})$$

Table 4-1. Early Experiments with Flexural Strains in Wood[a]

Description of Wood	Specific Gravity	Values of L	Values of b	Value of d	Value of f	Value of w	Value of W	Authors of Experiments
		Inches	Inches	Inches	Inches	lb	lb	
Oak (English)	0.934	84	2	2	1.280	200	637	Prof. Barlow
Oak (Canadian)	0.872	84	2	2	1.080	225	673	Prof. Barlow
Pine (American)	—	84	2	2	0.931	150	—	Prof. Barlow
Oak (English)	—	30	1	1	0.5	137	—	Tredgold
White spruce (Canadian)	0.465	24	1	1	0.5	180	285	Tredgold
White pine (American)	0.455	85.2	2.75	5.55	0.177	777	5,189	Lieut. Brown
Black spruce (American)	0.490	85.2	2.75	5.55	0.177	892	5,646	Lieut. Brown
Southern pine, (American)	0.872	85.2	2.75	5.54	0.177	1,175	9,237	Lieut. Brown

[a] From Mahon 1885, 125.

Table 4-2. Trautwine's Flexural Constants[a]

Woods[b]	
Ash, English	650
Ash, white (Author)	650
Ash, Swamp	400
Ash, Black	600
Arbor Vilæ, Amer.	250
Baleam, Canada	350
Beech,	850
Beech, Amer.	550
Birch, Amer. Black	550
Birch, Amer. Yellow	850
Cedar, Bermuda	400
Cedar, Guadaloupe	600
Cedar, Amer. White or arbor vitæ	250
Chestnut	450
Elm, Amer. white	650
Elm, Rock, Canada	800
Hemlock	500
Hickory, Amer.	800
Hickory, bitter nut	800
Iron, Wood, Canada	600
Locust	700
Lignum, Vitæ	650
Larch	400
Mahogany	750
Mangrove, White	650
Mangrove, Black	550
Maple, Black	750
Maple, Soft	750
Oak, English	550
Oak, Amer. white (by author)	600
Oak, Amer. red, black, basket	850
Oak, Live	600
Pine, Amer. white (by author)	450
Pine, Amer. yellow (by author)	500
Pine, Amer. pitch (by author)	550
Pine Georgia	850
Poplar	550
Poon	700
Spruce (by author)	450
Spruce, Black	550
Sycamore	500
Tumarack	400
Teak	750
Walnut	550
Willow	350

Metals	
Brass	850
Iron, cast, 1500–2700 average	2,100
Iron, cast, common pig	2,000
Iron, cast, castings from pig	2,300
Iron, cast, employed in our tables	2,025
Iron, cast, for castings 2½ or 3 in. thick	1,800
Iron, wrought, 1900–2600	2,250
Wrought iron does not *break*; but at about the average of 2,250 lb its elas. limit is reached	
Steel, hammered, or rolled; elas. destroyed by 3,000–7,000. Under heavy loads hard steel snaps like cast iron, and soft steel bends like wrought iron.	5,000

Stones, etc.	
Blue stone flagging, Hudson River	125
Brick, common, 10–30 average	20
Cedar, good Amer. pressed, 30 to 50 average	40
Clæn Stone	25
Cement, Hydraulic, English Portland. artificial,	
7 days in water	30
1 year in water	50
Cement, Hydraulic, Portland Kingston, N.Y.,	
7 days in water	30
Cement, Hydraulic, Saylor's Port.,	
7 days in water	26
Cement, Hydraulic, Common U.S. cements, 7 days in water	5
The following hydraulic cements were made into prisms, in vertical moulds, under a pressure of 32 psi, and were kept in sea water for 1 year.	
Portland, Cement, English, pure, 1 year old	64
Roman Cement, Scotch, pure	23
American Cement, pure, average about	25
Granite, 50–150, average	100
Granite, Quincy	100
Glass, Millville, N. Jersey, thick flooring (by author)	170
Mortar, of lime alone, 60 days old	10
Mortar, 1 measure of slacked lime in powder, 1 sand	8
Mortar, 1 measure of slacked lime in powder, 2 sand	7
Marble, Italian, White (Author)	116
Marble, Manchester, Vt, White (Author)	95
Marble, East Dorset, Vt, White (Author)	111
Marble, Lee, Mass, White (Author)	86
Marble, Montg'y Co. Pa, Gray (Author)	103
Marble, Montg'y Co. Pa, Clouded (Author)	142
Marble, Rutland Vt, Gray (Author)	70
Marble, Glenn's Falls, N.Y., Black (Author)	155
Marble, Baltimore, Md, White, coarse	102
Oolites, 20 to 50	35
Sandstones, 20 to 70 average	45
Sandstones, Red of Connecticut and New Jersey	45
Slate, laid on its bed, 200–150 average	825

[a] From Trautwine (1888), 493.
[b] In wooden beams in practice deduct one-third part, to allow for knots, crooked grain, etc.

where:
W = breaking load at center of span

Using the Previous Example:

$$W = \frac{2 \text{ in.} (2 \text{ in.})^2}{7 \text{ ft}} (550) = 628.6 \text{ lb}$$

The calculated value of 628.8 lb compares quite nicely with the value in the Mahan table of 637 lb.

Trautwine also includes a table of the safe loads of 1 in. wide beams of white pine or spruce supported at both ends and loaded at the center and includes the corresponding deflection for the corresponding loads. The table is set up for beams of various depths and spans from 4–40 ft. When the user was designing a member more than 1 in. wide, the value in the table was multiplied by the actual width in inches over 1 in. (see Table 4-3). The table uses constants to convert its values to other species of wood. For wite oak and southern pine, use a factor of 1.25 times the values shown. The table is said also to be safe against "shearing and crushing at the ends of the beams." This is an early mention of the recognition that there were other parameters beyond flexure in the design of wooden beams and girders.

The Carnegie Steel Company included design data for wood structural members in their *Pocketbook* that have been referenced often in this work. In the 1893 edition of the *Pocketbook,* a section on wooden beams refers to extensive testing by Professor Lanza on wooden beams for both flexural capacity and horizontal shear. Lanza's work centered around his determination of a modulus of rupture for each species of wood that was used for structural members. The modulus of rupture was equal to the moment of the forces causing rupture divided by the moment of resistance of the cross-section (see Table 4-4). The modulus of rupture is the same as the maximum bending stress in the extreme fiber of a rectangular beam at the time of failure (Kidder 1902, 330). The Carnegie *Pocketbook* reduces the values of Professor Lanza by about 35 percent and says that the allowable unit must contain a safety factor of 4 (see Table 4-5). The *Pocketbook* also mentions horizontal shear as a mode of failure for some of the tests conducted by Lanza. It states that the "mean intensity of the shearing strains, (horizontal shear), for beams that failed in this manner, was 191 lbs., and for yellow pine 248 lbs." Carnegie Steel Co. 1893, 185, 186) ("lbs." must mean psi).

Johnson (1905) also gives values for the modulus of rupture and quotes these values as being from the U.S. Forestry Circular, No. 15. Johnson gives the average modulus of rupture for white pine at 12 percent moisture content as 7900 psi, his results being on the basis of 120 tests. Johnson does not give results from yellow pine; however, he does give the results of tests on "red pine," which is what some authors of the time called yellow pine. These results were based upon 95 tests on 12 percent moisture material, and the modulus of rupture was determined to be 9,100 psi.

Johnson states that the first comprehensive quantitative tests on wood in the United States were the United States Timber Tests of 1891, conducted by Dr. B. E. Fernow, Chief of the Forestry Division of the U.S. Agricultural Department. Fernow was assisted by Dr. Charles Mohr, Filibert Roth, and J. B. Johnson, among many others, in laboratories in Washington and St. Louis, and at the U.S. Forestry Department. In all, over 40,000 tests were made and the results were published in 1897 in Bulletin No. 15 of the U. S. Forestry Department.

Verification of the two previous tables from Trautwine and Carnegie Steel Co.:
The Trautwine table (concentrated load at center of span): 10 ft span, 12 in. depth of beam

$$W = \frac{\text{breadth (in.)} \times (\text{depth, in.})^2}{\text{span, ft}} (\text{constant}) \quad (\text{constant of Table 4-2})$$

$$W = \frac{1 \text{ in.} (12 \text{ in.})^2}{10 \text{ ft}} (450) = 6480 \text{ lb} \quad (\text{breaking load at center of span for white pine})$$

$$W = \frac{\text{breaking load}}{\text{factor of safety}} = \frac{6480 \text{ lb}}{6} = 1080 \text{ lb} \quad (\text{as tabulated in Table 4-3})$$

Table 4-3. Safe Loads: 1 in. Wide White Pine or Spruce—Concentrated Load Placed at the Center of Span[a]

Depth of Beam (in.)	Span 4 ft load lb	Span 4 ft def. in.	Span 6 ft load lb	Span 6 ft def. in.	Span 8 ft load lb	Span 8 ft def. in.	Span 10 ft load lb	Span 10 ft def. in.	Span 12 ft load lb	Span 12 ft def. in.	Span 14 ft load lb	Span 14 ft def. in.	Span 16 ft load lb	Span 16 ft def. in.	Weight of 10 ft of beam lb
1	19	0.39	13	0.92	10	1.8	8	3.0	6	4.4					2
2	75	0.22	50	0.45	38	0.82	30	1.3	25	1.9	21	2.7	19	3.7	4
3	170	0.13	114	0.30	85	0.53	67	0.84	57	1.3	48	1.7	42	2.3	6
4	300	0.10	200	0.22	150	0.39	120	0.63	100	0.92	86	1.3	75	1.7	8
5	469	0.08	312	0.18	234	0.31	187	0.50	156	0.72	134	1.0	117	1.3	10
6	675	0.06	450	0.15	337	0.26	270	0.41	225	0.60	193	0.83	168	1.1	12
7	919	0.06	612	0.12	460	0.22	367	0.35	306	0.51	262	0.70	230	0.93	14
8	1,200	0.05	800	0.11	600	0.19	480	0.31	400	0.45	343	0.61	300	0.81	16
9	1,520	0.04	1,014	0.10	760	0.17	607	0.27	507	0.40	434	0.54	380	0.72	18
10	1,875	0.04	1,250	0.09	937	0.16	750	0.24	625	0.35	536	0.49	468	0.64	20
11	2,270	0.04	1,514	0.08	1,135	0.14	907	0.22	757	0.32	648	0.44	567	0.58	22
12	2,700	0.03	1,800	0.07	1,350	0.13	1,080	0.20	900	0.29	772	0.40	675	0.53	24
14	3,675	0.03	2,450	0.06	1,837	0.11	1,470	0.17	1,225	0.25	1,050	0.34	918	0.45	28
16	4,800	0.02	3,200	0.05	2,400	0.10	1,920	0.15	1,600	0.22	1,372	0.30	1,200	0.40	32
18	6,075	0.02	4,050	0.05	3,037	0.09	2,430	0.14	2,025	0.20	1,736	0.27	1,518	0.35	36
20	7,500	0.02	5,000	0.04	3,750	0.08	3,000	0.12	2,500	0.18	2,145	0.24	1,875	0.31	40
22	9,075	0.02	6,050	0.04	4,537	0.07	3,630	0.11	3,025	0.16	2,593	0.22	2,268	0.29	44
24	10,800	0.02	7,200	0.04	5,400	0.06	4,320	0.10	3,600	0.15	3,088	0.20	2,700	0.26	48

Depth of Beam (in.)	Span 18 ft load lb	Span 18 ft def. in.	Span 20 ft load lb	Span 20 ft def. in.	Span 26 ft load lb	Span 26 ft def. in.	Span 30 ft load lb	Span 30 ft def. in.	Span 35 ft load lb	Span 35 ft def. in.	Span 40 ft load lb	Span 40 ft def. in.	Weight of 10 ft of beam lb
6	150	1.4	135	1.8	108	2.9	90	4.5	77	6.5	67	9.2	12
7	204	1.2	184	1.5	147	2.5	122	3.9	105	5.8	92	7.6	14
8	267	1.0	240	1.3	192	2.1	160	8.2	137	4.6	120	6.4	16
9	338	0.92	304	1.2	243	1.9	202	2.8	174	4.0	152	5.5	18
10	417	0.82	375	1.0	300	1.7	250	2.5	214	3.5	188	4.9	20
11	505	0.74	454	0.93	363	1.5	302	2.2	259	3.2	227	4.3	22
12	600	0.68	540	0.85	432	1.4	360	2.0	308	2.9	270	3.9	24
14	817	0.58	735	0.72	588	1.2	490	1.7	420	2.4	367	3.2	28
16	1,067	0.50	960	0.63	768	1.0	640	1.5	548	2.1	480	2.8	32
18	1,350	0.45	1,215	0.56	972	0.90	810	1.3	694	1.8	607	2.5	36
20	1,666	0.40	1,500	0.50	1,200	0.79	1,000	1.2	857	1.6	750	2.2	40
22	2,017	0.37	1,815	0.45	1,452	0.72	1,210	1.1	1,037	1.5	907	2.0	44
24	2,400	0.33	2,160	0.41	1,728	0.65	1,440	0.96	1,234	1.3	1,080	1.8	48
26	2,817	0.31	2,526	0.38	2,018	0.60	1,684	0.88	1,449	1.2	1,263	1.6	52
28	3,267	0.28	2,940	0.35	2,352	0.55	1,960	0.81	1,680	1.1	1,470	1.5	56
30	3,750	0.26	3,375	0.33	2,700	0.50	2,250	0.76	1,928	1.1	1,687	1.4	60
32	4,267	0.25	3,840	0.30	3,072	0.45	2,560	0.71	2,194	1.0	1,920	1.3	64
34	4,817	0.23	4,335	0.29	3,468	0.44	2,890	0.67	2,477	0.92	2,167	1.2	68
36	5,400	0.22	4,860	0.27	3,888	0.43	3,240	0.63	2,777	0.86	2,430	1.1	72

[a] From Trautwine 1888, 499, 500. Trautwine's note reads:
  White oak, and best Southern pitch pine will bear loads ¼ greater.
  For cast iron, mult the loads in the table by 4.5; and for wrought by 5.3. For these new loads, mult the defs by 0.4 for cast; and by 0.3 for wrought.
  If the load is equally distributed over the span, it may be twice as great as the center one, and the defs will be 1¼ times those in the table. If the loads in the table be equally distributed along the whole beam, the defs will be but five-eighths as great as those in the table. See Art. 26, p 505b. When more accuracy is reqd, half the wt of the beam itself must be deducted from the center load; and the whole of it from an equally distributed load. The wt of the beam, in the last column, supposes the wood to be but moderately seasoned, and therefore to weigh 28.8 lbs per cub ft.

Modern Analytical Method:

$$P = \frac{4\,(F_b)(s)}{\text{span}} = \frac{4\,(750 \text{ psi})(24 \text{ in.}^3)}{120 \text{ in.}} = 600 \text{ lbs}$$

(conservative, F_b probably higher for early white pine)

where:
  F_b = Allowable bending stress, southern pine, select structural, psi
  S = section modulus, in.^{3}
  L = simple span of beam, in.

Table 4-4. Modulus of Rupture, Wood Beams[a]

Kind of Timber	Modulus of Rupture = $\frac{M}{R}$ = $\frac{\text{(Moment of forces causing rupture)}}{\text{(Moment of resistance of cross-section)}}$		
	Maximum	Minimum	Mean
Spruce	5,878	2,995	4,884
White pine	6,415	3,438	4,808
Oak	7,659	4,984	6,075
Yellow pine	11,360	5,092	7,292

[a] From Carnegie Steel Co. 1893, 185. The original note reads:
   The following is a general summary of the results obtained by Prof. Lanza from numerous experiments upon wooden beams. They were of an average section of about 12 × 4 inches and were tested for mean span lengths of about 18 feet.

Table 4-5. Carnegie Steel Co.: Safe Loads, Uniformly Distributed, for Rectangular Spruce or White Pine Beams (1 in. thick)[a]

Span (ft)	Depth of Beam (in.)										
	6	7	8	9	10	11	12	13	14	15	16
5	600	820	1,070	1,350	1,670	2,020	2,400	2,820	3,270	3,750	4,270
6	500	680	890	1,120	1,390	1,680	2,000	2,350	2,730	3,120	3,560
7	430	580	760	960	1,190	1,440	1,710	2,010	2,330	2,680	3,050
8	380	510	670	840	1,040	1,260	1,500	1,760	2,040	2,340	2,670
9	330	460	590	750	930	1,120	1,330	1,560	1,810	2,080	2,370
10	300	410	530	670	830	1,010	1,200	1,410	1,630	1,880	2,130
11	270	370	490	610	760	920	1,090	1,280	1,490	1,710	1,940
12	250	340	440	560	690	840	1,000	1,180	1,360	1,560	1,780
13	230	310	410	520	640	780	930	1,080	1,260	1,440	1,640
14	210	290	380	480	590	720	860	1,010	1,170	1,340	1,530
15	200	270	360	450	560	670	800	940	1,090	1,250	1,420
16	190	260	330	420	520	630	750	880	1,020	1,180	1,330
17	180	240	310	400	490	590	710	830	960	1,100	1,260
18	170	230	290	370	460	560	670	780	910	1,040	1,190
19	160	210	280	360	440	530	630	740	860	990	1,130
20	150	200	270	340	420	510	600	710	820	940	1,070
21	140	190	260	320	390	480	570	670	780	890	1,020
22	140	190	240	310	380	460	540	640	740	850	970
23	130	180	230	290	360	440	520	610	710	810	920
24	130	170	220	280	350	420	500	590	680	780	890
25	120	160	210	270	330	410	480	560	660	750	860
26	110	160	210	260	320	390	460	540	630	720	820
27	110	150	200	250	310	370	440	520	610	690	790
28	110	140	190	240	300	360	430	500	580	670	760
29	110	140	180	230	290	350	410	490	560	640	740

[a] From Carnegie Steel Co. 1893, 185, 186. For oak, increase values in table by ⅓; for yellow pine, increase values by ⅔. To obtain the safe load for any thickness: multiply values for 1 in. by thickness of beam. To obtain the required thickness for any load: divide by safe load for 1 in. The original note reads:
   Owing to the wide ranges of the results obtained and the generally erratic behavior of timber subjected to strains, Prof. Lanza recommends the following values for moduli of rupture to be adopted in practice:

Spruce and White pine,	3,000 lbs.
Oak,	4,000 lbs.
Yellow pine,	5,000 lbs.

These values are lower than heretofore in use and a safety factor of 4, on the basis of these values, may be assumed as ample for all cases.
   The table has been calculated for extreme fibre strains of 750 lbs. per square inch.:

It is difficult to ascertain the design conditions to utilize the NFPA 1991 Design Specifications and Member Data. Continuous lateral support was assumed and a moderately low material specification was chosen in the Mixed Southern Pine, Select Structural category. Even in this selection, modern codes give their design data based upon nominal 2 in. thick material.

The Carnegie Steel Co. 1893 (uniform load over span)
10 ft span, 12 in. depth of beam

$$M = F_b S \quad \text{and} \quad M = \frac{wL^2}{8}$$

$$w = \frac{8 F_b S}{L^2} = \frac{8(750 \text{ lb/in.}^2)(24 \text{ in.}^3)}{(10 \text{ ft})^2(12 \text{ in./ft})} = 120 \text{ lb/ft}$$

$W = 1200$ lb (the total load on the beam)

Modern analytical method: under the same conditions, the formulae are the same and the allowable bending stress on the Mixed Southern Pine, is 750 psi. The allowable load is the same.

As can be seen from the modern analysis, the member can carry almost twice as much load as the design method of the period 1893 would allow. This points out the conservative nature of the design of the period. However, the modern analysis did consider the member as having continuous lateral bracing, but no other adjustment factors were applied to the formula for the allowable bending stress.

Kidder (1902) defines a method for sizing wooden beams that utilizes a constant A as a part of a formula for the safe load in beams. Each type of loading has its own formula, which makes the system cumbersome. This method incorporates the statics of the type of loading into the design formulae through the use of the constant A. In modern analysis, we determine the moment on the member through statics and the design through a general design formula that applies for any moment. The Kidder beam types and formula with constant A are shown in Figure 4-2.

The coefficient A for beams as developed in Kidder is the modulus of rupture divided by 18 (see Table 4-6). Kidder also states that the coefficients are one-third of the breaking weight of timbers of the same size and quality as that used in first-class buildings and that the beams produced by this method are sufficient for dwelling houses, small halls, etc., but where large loads or vibration is possible, in buildings such as gymnasiums and public halls, a factor of four-fifths must be applied to the coefficient.

Table 4-7 and the formula for the uniformly loaded member were reduced to a table of results that allowed designers to select beams from a chart that gave the allowable loads by span and depth of a 1 in. wide beam.

Verification of the tables from Kidder (1902):
Table 4-7: uniformly distributed load
Span = 10 ft, 1 in. wide × 12 in. deep member

$$\text{Safe load (lb)} = \frac{2(\text{breadth, in.})(\text{depth, in.})^2 (A)}{\text{span, ft}} = \frac{2 \text{ in. } (12 \text{ in.})^2(100)}{10 \text{ ft}} = 2880 \text{ lb}$$

The number read from the table or from the formulae gives the total load in pounds for the member. The uniform load w is the total load divided by the span, which is 288 lb/foot.

Modern Analytical Method:

$$w = \frac{8(F_b)(S)}{(L)^2} = \frac{8(1900 \text{ lb/in.}^2)(24 \text{ in.}^3)}{(10 \text{ ft})^2 (12 \text{ in./ft})} = 304 \text{ lb/ft}$$

$W = 3040$ lb (the total load on the member)

Table 4-6. Kidder's Beam Coefficient A[a]

Material	A lb
Cast iron	308
Wrought iron	666
Steel	888
American woods:	
Chestnut	60
Hemlock	55
Oak, white	75
Pine, Georgia yellow	100
Pine, Oregon	90
Pine, red or Norway	70
Pine, white, Eastern	60
Pine, white, Western	65
Pine, Texas yellow	90
Spruce	70
Whitewood (popular)	65
Redwood (California)	60
Bluestone flagging (Hudson River)	21
Granite, average	17
Limestone	15
Marble	17
Sandstone	8
Slate	50

[a] From Kidder 1902, 374.

The modern method utilizes the allowable stress for a southern pine member, select structural, from the 1991 NFPA Specifications for Wood Construction. It was felt that the material used in 1902 would have been equal to that quality. Again, by the 1991 Code method, no adjustment factors for service conditions were applied to the allowable stress.

In each of the preceding examples, the conservative nature of the design of the early periods is adequately shown. The age of the wooden member would not affect its capacity to work as required unless outside interference such as water infiltration or other man-inflicted damage has endangered the integrity of the member. The modern solutions verifying the formulae from the previous periods did not include any of the in-service factors (both positively and negatively affecting the allowable stress),

Table 4-7. Kidder's Table for Uniformly Loaded Wood Beams[a]

Depth of Beam	Span (ft)												
	6	8	10	12	14	15	16	17	18	20	22	24	25
in.	lb	lb	lb	lb	lb	lb	lb	lb	lb	lb	lb	lb	lb
6	1,200	900	720	600	514	480							
7	1,633	1,225	980	816	700	653	612						
8	2,133	1,600	1,280	1,066	914	853	800	753					
9	2,700	2,025	1,620	1,350	1,157	1,080	1,012	963	900				
10	3,333	2,500	2,000	1,666	1,428	1,333	1,250	1,176	1,111	1,000			
12	4,800	3,600	2,880	2,400	2,056	1,920	1,800	1,694	1,600	1,440			
14	6,533	4,900	3,920	3,266	2,800	2,613	2,450	2,306	2,177	1,960	1,782		
15	7,500	5,633	4,500	3,750	3,214	3,000	2,816	2,653	2,500	2,250	2,045	1,875	1,800
16	8,583	6,400	5,120	4,266	3,656	3,412	3,200	3,012	2,844	2,560	2,327	2,133	2,048

[a] From Kidder 1902, 377. Table of safe quiescent loads for horizontal rectangular beams of Georgia yellow pine, 1 in. broad, supported at both ends, load *uniformly distributed*. For *concentrated* load at center, *divide by two*. For *permanent* loads (such as masonry), reduce by 10 percent. Loads above and to the right of heavy line will crack plastered ceilings.

nor have any of the other design criteria, such as deflection, horizontal shear, bearing stress, or stress at point of loading, been checked. Since the exact conditions of all of the parameters were not described in the author's examples, they were also not verified. Part of this is owing to the state of the art at the time an example or design computation was created. Later examples will go into more detail as the design process evolves.

Kidder (1902) also includes an extensive section on the construction of mill floors. These floors were made of wood and heavy timber supporting beams and girders. The design of spanning floor members is similar to the design of beams, except that unlike most beams, floor planking spans over more than one clear opening; usually they are continuous over two, three, or four clear openings. Kidder provides a formula for the thickness of flooring members (Kidder 1902, 385):

$$\text{thickness of plank} = \sqrt[3]{\frac{\text{weight/ft}^2 \, (L^3)}{19.2(e)}}$$

where:
 e = constant for deflection
 L = span and member, ft

Weight per square foot is the total of live and dead loads.

The constant e used in the deflection computations in the design of beams is as shown in Table 4-8. Two sets of computations existed, one allowing one-fortieth of an inch of deflection per foot of span and one allowing one-thirtieth of an inch per foot of span. These values correspond to the limits used today, but today's designers use different terms. In today's terms, in the first computation, the deflection cannot exceed the span over 480, and in the second, the span over 360.

Floor planking is of equal importance to beam design, especially in mill construction systems, in that the floor joists or beams are usually spaced 6–10 ft apart. In this sense, the joists are really beams as they carry heavy loads and the flooring would span a long enough distance to require specific design consideration. Joists usually imply that there are many of them carrying a relatively thin system of floor deck and finished flooring. Mill buildings had thick flooring, many as thick as 6 in. or more.

Carnegie Steel Co., (1913) gives a comprehensive table of the allowable stresses of wood structural members, using the system of notation that is still in use today. Carnegie states that most of their information comes from the American Railway En-

Table 4-8. Values of the Deflection Constant e^a

Material	E	$F = \dfrac{E}{432}$	$e = \dfrac{E}{17,280}$	$e_1 = \dfrac{E}{12,960}$
Cast iron	15,700,000	36,300	907	1,210
Wrought iron	26,000,000	60,000	1,500	2,000
Steel	31,000,000	71,760	1,794	2,358
Yellow pine	1,780,000	4,120	103	137
Spruce	1,294,000	3,000	75	100
White oak	1,240,000	2,870	72	95
White pine	1,073,000	2,480	62	82
Hemlock	1,045,000	2,420	60	80
Whitewood	1,278,000	2,960	74	98
Chestnut	944,000	2,180	54	72
Ash	1,482,000	3,430	86	114
Maple	1,902,000	4,400	110	146
Oregon pine	1,425,000	3,300	82	110

[a] From Kidder 1902, 385.
E = modulus of elasticity, psi.
F = constant for deflection of beam, loaded at midspan.
e = constant, allowing a deflection of $\frac{1}{40}$ in. per ft. of span.
e_1 = constant, allowing a deflection of $\frac{1}{30}$ in. per ft of span.

Table 4-9. Strength of Solid Timber and Plank Floors[a]

Superficial Load	Weight per Square Foot of Floor			Dimensions of Beams			Thickness of floorplank (in.)
	Weight of beam (lb)	Weight of floorplank	Total	Depth (in.)	Breadth (in.)	Span (ft)	
50	3.00	6.07	59.07	12	6	20.95	2.43
	4.08		60.15	14	7	26.16	
	5.33		61.40	16	8	31.63	
75	3.00	7.40	85.40	12	6	17.42	2.96
	4.08		86.48	14	7	21.82	
	5.33		87.73	16	8	26.46	
100	3.00	8.55	111.55	12	6	15.25	3.42
	4.08		112.63	14	7	19.12	
	5.33		113.88	16	8	23.23	
125	3.00	9.55	137.55	12	6	13.73	3.82
	4.08		138.63	14	7	17.23	
	5.33		139.88	16	8	20.96	
150	3.00	10.45	163.45	12	6	12.59	4.18
	4.08		164.53	14	7	15.82	
	5.33		165.78	16	8	19.25	
175	3.00	11.26	189.26	12	6	11.71	4.51
	4.08		190.34	14	7	14.70	
	5.33		191.59	16	8	17.91	
200	3.00	12.05	215.05	12	6	10.98	4.82
	4.08		216.13	14	7	13.80	
	5.33		217.38	16	8	16.81	
225	3.00	12.75	240.75	12	6	10.38	5.11
	4.08		241.83	14	7	13.06	
	5.33		243.08	16	8	15.90	
250	3.00	13.45	266.45	12	6	9.86	5.38
	4.08		267.53	14	7	12.40	
	5.33		268.78	16	8	15.08	
275	3.00	13.55	291.55	12	6	9.43	5.62
	4.08		292.63	14	7	11.86	
	5.33		293.88	16	8	14.46	
300	3.00	14.72	317.72	12	6	9.03	5.89
	4.08		318.80	14	7	11.36	
	5.33		320.05	16	8	13.85	

[a] From Kidder 1902, 435.

gineering Association Manual of 1911 (see Table 4-9). The railroad engineers were very advanced in wood design, utilizing timbers in the construction of railroad bridges and buildings. The weights of the locomotives and impact loadings of the moving loads required accurate design methods, and the railroad engineers had specifications that contained allowable stresses for species of wood, the time of the year that the tree was felled, the age of the tree, and even the method of sawing. The moisture content and method of curing were also important to the strength of the members. Table 4-10 gives the Working Unit Stresses for Structural Timber adopted by the American Railway Engineering Association, 1911.

Table 4-11 gives an example of the Uniform Load Tables for longleaf pine based upon a maximum bending stress of 1300 psi. These values and tables from Carnegie were extensively used by designers, the Pocketbooks being distributed free to every designer that company representatives could locate. Carnegie was in the business of selling steel products, but they considered it good for steel and their business to include all types of information in their manuals. With the wide distribution of these, it is logical that we see the almost universal use of the design criteria.

Table 4-10. Carnegie Steel Co.: Working Unit Stresses for Structural Timber (Adopted by the American Railway Engineering Association)[a]

	Unit Stresses (psi)[b]											
	Bending			Shearing				Compression				Working Stresses for Columns
	Extreme Fiber Stress		Modulus of Elasticity	Parallel to the Grain		Longitudinal Shear in Beam		Perpendicular to the Grain		Parallel to the Grain		
Kind of Timber	Average Ultimate	Working Stress	Average	Average Ultimate	Working Stress	Average Ultimate	Working Stress	Elastic Limit	Working Stress	Average Ultimate	Working Stress	Length under 15 × d	Length over 15 × d
Douglas fir	6,100	1,200	1,510,000	690	170	270	110	630	310	3,600	1,200	900	1,200(1−l/60d)
Longleaf pine	6,500	1,300	1,610,000	720	180	300	120	520	260	3,800	1,300	975	1,300(1−l/60d)
Shortleaf pine	5,600	1,100	1,480,00	710	170	330	130	340	170	3,400	1,100	825	1,100(1−l/60d)
White pine	4,400	900	1,130,000	400	100	180	70	290	150	3,000	1,000	750	1,000(1−l/60d)
Spruce	4,800	1,000	1,310,000	600	150	170	70	370	180	3,200	1,100	825	1,100(1−l/60d)
Norway pine	4,200	800	1,190,000	590*	130	250	100		150	2,600*	800	600	800(1−l/60d)
Tamarack	4,600	900	1,220,000	670	170	260	100		220	3,200*	1,000	750	1,000(1−l/60d)
Western hemlock	5,800	1,100	1,480,000	630	160	270*	100	440	220	3,500	1,200	900	1,200(1−l/60d)
Redwood	5,000	900	800,000	300	80			400	150	3,300	900	675	900(1−l/60d)
Bald cypress	4,800	900	1,150,000	500	120			340	170	3,900	1,100	825	1,100(1−l/60d)
Red cedar	4,200	800	800,000					470	230	2,800	900	675	900(1−l/60d)
White oak	5,700	1,100	1,150,000	840	210	270	110	920	450	3,500	1,300	975	1,300(1−l/60d)

[a] From Carnegie Steel Co. 1913, 310. The original note reads:
    The working unit stresses given in the table are intended for railroad bridges and trestles. For highway bridges and trestles, the unit stresses may be increased 25 per cent. For buildings and similar structures, in which the timber is protected from the weather and practically free from impact, the unit stresses may be increased 50 per cent. To compute the deflection of a beam under long continued loading instead of that when the load is first applied, only 50 per cent of the corresponding modulus of elasticity given in the table is to be employed.
[b] Unit stresses are for green timber and are to be used without increasing the live load stresses for impact. Values noted * are for partially air-dry timbers.
    In the formula given for columns. l = length of column, in. and d = least side or diameter, in.

Example: Beam Design, Longleaf Pine, Span = 20 ft, Load = 1000 lb/ft:

allowable uniform load, 20 ft span, 18 in. depth = 2340 lb per 1 in. wide beam

$$\text{breadth} = \frac{20{,}000 \text{ lb}}{2340 \text{ lb}} = 8.5 \text{ in. (use } 9 \times 18 \text{ in.)}$$

$$\text{deflection} = \frac{19.38}{18} = 1.077 \text{ in.}$$

The designs of 1913 do not mention the lateral bracing of the member. The criteria for lateral bracing of the compression zone of the beam to prevent buckling and loss of strength by alteration of the section modulus are not mentioned until later periods. The decking or flooring on members would reduce the tendency to buckle; however, many designers require a positive type of solid or diagonal bridging. Modern design methods require that the allowable stresses on the beam be reduced in accordance with the maximum length of a lateral unbraced section.

The deflections for beams were still computed from coefficients in 1913, rather than individually calculated as is done today. The deflection table given in Table 4-12 is for uniformly loaded simple span beams and is based upon a deflection limit of span (inches) over 360 to prevent cracking of plastered ceilings. For the deflection of a simple span with a concentrated load located at midspan, the deflection constant must be multiplied by a factor of 0.8 to adjust the factor.

The deflections as defined by the coefficients are very liberal. The deflection as shown in the example problem from the coefficient and method of the period was 1.077 in. The designers of the period would experience actual deflections of approx-

Table 4-11. Rectangular Wooden Beams (1 in. thick), Longleaf Pine

Span (ft)	Allowable Uniform Load (lb) Maximum Bending Stress, 1,300 psi											
	Depth of Beam (in.)											
	2	4	6	8	10	12	14	16	18	20	22	24
2	<u>320</u>											
	289											
3	193	<u>640</u>										
4	144	578										
5	116	462										
			<u>960</u>									
6	96	385	867									
7	83	330	743	<u>1,280</u>								
8	72	289	650	1,156								
9		257	578	1,027	<u>1,600</u>							
10		231	520	924	1444							
						<u>1,920</u>						
11		210	473	840	1,313	1,891						
12		193	433	770	1,204	1,733	<u>2,240</u>					
13			400	711	1,111	1,600	2,178					
14			371	660	1,032	1,486	2,022	<u>2,560</u>				
15			347	616	963	1,387	1,887	2465				
16			325	578	903	1,300	1,769	2,311	<u>2,880</u>			
17				544	850	1,224	1,665	2,175	2,753			
18				514	802	1,156	1,573	2,054	2,600	<u>3,200</u>		
19				487	760	1,095	1,490	1,946	2,463	3,041	<u>3,520</u>	
20				462	722	1,040	1,416	1,849	2,340	2,889	3,496	
21					688	991	1,348	1,761	2,229	2,751	3,329	<u>3,840</u>
22					657	945	1,287	1,681	2,127	2,626	3,178	3,782
23					628	904	1,231	1,608	2,035	2,512	3,040	3,617
24					602	867	1,180	1,541	1,950	2,407	2,913	3,467
25						832	1,132	1,479	1,872	2,311	2,796	3,328
26						800	1,089	1,422	1,800	2,222	2,689	3,200
27						770	1,049	1,370	1,733	2,140	2,589	3,082
28						743	1,011	1,321	1,671	2,064	2,497	2,971
29							976	1,275	1,614	1,992	2,411	2,869
30							944	1,233	1,560	1,926	2,330	2,773
31							913	1,193	1,510	1,864	2,255	2,684
32							885	1,156	1,463	1,806	2,185	2,600
33								1,121	1,418	1,751	2,119	2,521
34								1,088	1,377	1,699	2,056	2,447
35								1,057	1,337	1,651	1,998	2,377
36								1,027	1,300	1,605	1,942	2,311
37									1,265	1,562	1,890	2,249
38									1,232	1,521	1,840	2,189
39									1,200	1,482	1,793	2,133
40									1,170	1,444	1,748	2,080

[a] From Carnegie Steel Co. 1913, 323. Horizontal lines indicate the limit for resistance to shear in the horizontal direction of the grain.

imately half this amount. The modern analysis of the member (the parameters checked) indicates that the member section is adequate for flexure, provided the allowable bending stress is 1300 psi. The stress reduction formulae for a beam of this size may determine an allowable even higher than the one used in the 1913 Carnegie Pocketbook.

Check of Previous Example: Modern Methods
N.F.P.A. 1991 Specification Method (National Forest Products Assn. 1991, 8–10)

$$L_e = 1.63 L_u + 3d = 391.2 + 54 = 445.2 \text{ in.}$$

$$E_{0.16} = E - E(1.0)(COV_E) = 1.6 \times 10^6 - 1.6 \times 10^6(1.0)(0.25) = 1,200,000 \text{ psi}$$

Table 4-12. Coefficients of Deflections for Uniform Loads[a]

Span (ft)	White Oak	Longleaf Pine	Shortleaf Pine, Western Hemlock	White Pine, Douglas Fir	Spruce
1	0.06	0.05	0.05	0.05	0.05
2	0.23	0.19	0.18	0.19	0.18
3	0.52	0.44	0.40	0.43	0.41
4	0.92	0.78	0.71	0.76	0.73
5	1.44	1.21	1.12	1.19	1.15
6	2.07	1.74	1.61	1.72	1.65
7	2.81	2.37	2.19	2.34	2.24
8	3.67	3.10	2.85	3.06	2.93
9	4.65	3.92	3.61	3.87	3.71
10	5.74	4.85	4.46	4.77	4.58
11	6.95	5.86	5.40	5.78	5.54
12	8.27	6.08	6.42	6.87	6.60
13	9.70	8.19	7.54	8.07	7.74
14	11.25	9.50	8.74	9.36	8.98
15	12.92	10.90	10.04	10.74	10.31
16	14.69	12.40	11.42	12.22	11.73
17	16.59	14.00	12.89	13.79	13.24
18	18.60	15.70	14.45	15.47	14.84
19	20.72	17.49	16.10	17.23	16.53
20	22.96	19.38	17.84	19.09	18.32
21	25.31	21.37	19.67	21.05	20.20
22	27.78	23.44	21.59	23.10	22.17
23	30.37	25.63	23.59	25.25	24.23
24	33.06	27.91	25.69	27.49	26.38
25	35.88	30.28	27.88	29.83	28.63
26	38.80	32.75	30.15	32.27	30.96
27	41.85	35.32	32.51	34.80	33.39
28	45.00	37.99	34.97	37.42	35.91
29	48.27	40.75	37.51	40.14	38.52
30	51.66	43.61	40.14	42.96	41.22
31	55.16	46.56	42.86	45.87	44.01
32	58.78	49.61	45.67	48.88	46.90
33	62.51	52.76	48.57	51.98	49.88
34	66.35	56.01	51.56	55.18	52.95
35	70.32	59.35	54.64	58.47	56.11
36	74.39	62.79	57.80	61.86	59.36
37	78.58	66.33	61.06	65.34	62.70
38	82.89	69.96	64.40	68.92	66.14
39	87.31	73.69	67.84	72.60	69.66
40	91.84	77.52	71.36	76.37	73.28

[a] From Carnegie Steel Co. 1913, 321.

$$E_{0.04} = E - E(1.65)(COV_E) = 1.6 \times 10^6 - 1.6 \times 10^6(1.65)(0.25) = 940,000 \text{ psi}$$

Average $E' = 1,070,000$ psi

$$R_B = \sqrt{\frac{L_e(d)}{b^2}} = \sqrt{\frac{445.2 \text{ in. } (18 \text{ in.})}{(9 \text{ in.})^2}} = 9.947$$

$$F_{bE} = \frac{K_{bE} E'}{(R_B)^2} = \frac{0.438(1,070,000 \text{ lb/in.}^2)}{(9.947)^2} = 4737 \text{ psi}$$

$$C_L = \frac{1 + \frac{F_{bE}}{F_b^*}}{1.9} - \sqrt{\left[\frac{1 + \frac{F_{bE}}{F_b^*}}{1.9}\right]^2 - \frac{F_{bE}}{0.95 F_b^*}}$$

$$= 1.9510 - \sqrt{(1.9510)^2 - 2.8943} = 0.973$$

$F_b^* = F_b$(all adjustment factors except C_{fu}, C_v, and C_L)

C_D, C_M, C_T, C_F, C_V, C_{fu}, C_r, C_C, C_f are all equal to 1.0

$F_b' = F_b(C_D)(C_M)(C_T)(C_F)(C_V)(C_{fu})(C_r)(C_C)(C_f)(C_L) = 1300 \text{ psi }(0.973) = 1264.9 \text{ psi}$

$$\text{Design moment} = \frac{wL^2}{8} = \frac{1000 \text{ lb/ft } (20 \text{ ft})^2}{8} = 50{,}000 \text{ lb-ft}$$

$$\text{Section modulus required} = \frac{M_{design}}{F'_b} = \frac{50{,}000 \text{ lb-ft } (12 \text{ in./ft})}{1264.9 \text{ lb/in.}^2} = 474.3 \text{ in.}^3$$

The section modulus of a 9×18 in. beam is $bh^2/6 = 486$ in.^{3}

The section modulus is calculated with the full dimension of 9×18 in. because lumber was not yet being dressed in 1913.

$$\text{Deflection, } \delta = \frac{5 wL^4}{384 EI} = \frac{5(1000 \text{ lb/ft})(20 \text{ ft})^4(1728 \text{ in.}^3/\text{ft}^3)}{384 (1{,}070{,}000 \text{ lb/in.}^2)(4374 \text{ in.}^4)} = 0.769 \text{ in.}$$

As can be seen from an inspection of the results of the example, the early designs were actually very conservative. The same service today would require a smaller member—8.5×16 in. would probably be sufficient. As all parameters in the N.F.P.A. Specification would change, it is not possible to positively say the 8.5×16 in. would work without going through the whole procedure. In a real design situation, you would "run it through" to see if it would satisfy. The timber available now is also not as good as the timber cut at and just beyond the turn of the century, and in the earlier periods the allowables on this superior material were lower than those of today. The National Lumber Manufacturers' Association, formed in the last half of the nineteenth century, published a design manual, in 1916 that defined mill construction as the type of construction that originated in the cotton and woollen mills of New England (Paul 1916, 5). The use of masonry walls and heavy timber columns, beams, and floors reduced the losses from fire in these manufacturing buildings. Slow burning construction was different from mill construction in that it could utilize smaller, thinner, more combustible members and protect them with metal lath and plaster. The mill construction technique involved more than just utilization of larger, thicker members that were difficult to ignite or took so much time to ignite that firefighters could extinguish almost any fire before the structure would be damaged. Mill construction was also under certain dictates of the Associated Factory Mutual Fire Insurance Companies, which included stairways and elevators enclosed in fireproof shafts, floors with no openings or with all openings protected by fireproof covers, automatic fire doors, fireproof window shutters, wire glass, and automatic sprinklers. These measures were very popular and easily understood and slowly became utilized in practically all kinds of large buildings.

There are three general types of framing that make up mill buildings. They may be classified as follows (ibid., 7):

1. *Standard mill construction:* floors of heavy plank laid flat upon large girders spaced 8–11 ft apart on centers. These girders are supported by wood posts or columns spaced 16–25 ft apart.
2. *Mill construction with laminated floors:* floors of heavy plank laid on edge and supported by girders spaced from 12–18 ft apart on centers. These girders are supported by wood posts or columns spaced 16 ft or over apart, depending upon the design of the structure.
3. *Semimill construction:* floors of heavy plank laid flat upon large beams spaced 4–10 ft part on centers and supported by girders spaced as far apart as the loading will allow. These girders are supported by wood posts or columns located as far apart as consistent with the general design of the building. A spacing of 20–25 feet is not uncommon for columns in this class of framing where the loading is not excessive.

See Figure 4-3 for a visual comparison of the three types of mill floors that were normally constructed and Figure 4-4 for a section through a building of standard mill construction.

Paul (1916) presents formulae for the design of girders and beams in mill construction. These formulae followed the typical types of mill construction (ibid., 59, 60):

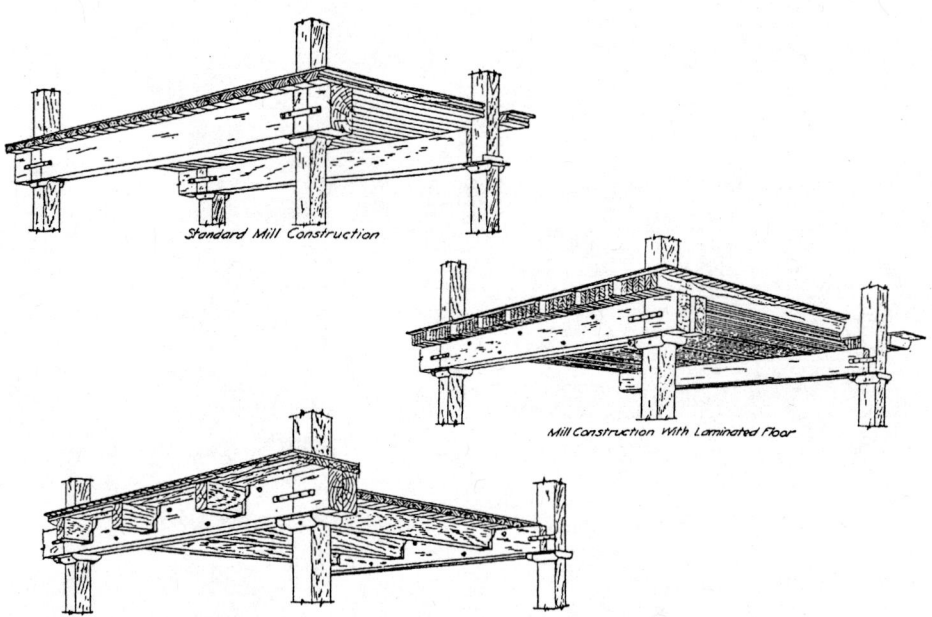

Figure 4-3. Types of mill construction. (*From Voss and Varney 1926, vol. 2, bkl, 95, 101, 112.*)

Uniform Load on Simple Beam:

Bending: $d^2 = \dfrac{9\,WL}{f(b)} \quad W = \dfrac{f(b)(d^2)}{9L}$

Shear: $W = \dfrac{4bds}{B}$

Deflection: $\dfrac{270\,WL^3}{Ebd^3}$

Concentrated Load at Center of Girder:

Bending: $d^2 = \dfrac{18\,PL}{f(b)} \quad P = \dfrac{f(b)(d^2)}{18b}$

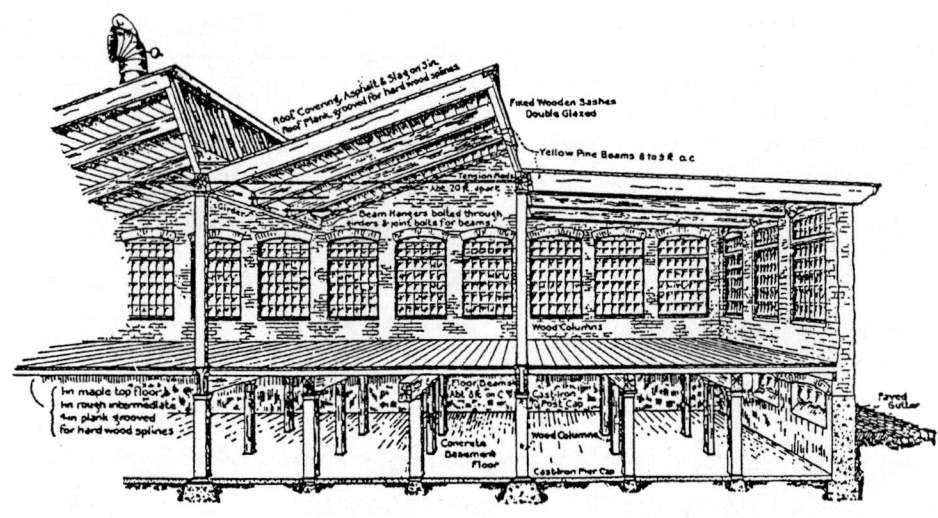

Figure 4-4. Section: Building of standard mill construction. (*From Paul 1916, 34.*)

Shear: $P = \dfrac{4\,bds}{3}$

Deflection: $\dfrac{432\,P(L^3)}{Ebd^3}$

Two equal concentrated loads at the third points of girder:

Bending: $d^2 = \dfrac{24\,PL}{f(b)}$ $P = \dfrac{f(b)d^2}{24\,L}$

Shear: $p = \dfrac{2\,bds}{3}$

Deflection: $D = \dfrac{184.4\,P(L^3)}{Ebd^3}$

where:
- W = total load on the member, lb
- D = depth of member, in.
- b = breadth of member, in.
- s = allowable shear stress, psi
- P = concentrated load, lb
- w = uniform load, lb/ft
- L = length of span, in.
- f = allowable bending stress, psi
- E = modulus of elasticity, psi
- D = deflection, in.

The formulae given above are manipulations of the standard formulae for moments, section modulus, etc., as can be seen below:

Uniform Load:

$M = \dfrac{wL^2}{8}$ and $S = \dfrac{M}{F_b}$ $S = \dfrac{bd^2}{6}$

$S = S \rightarrow \dfrac{M}{F_b} = \dfrac{bd^2}{6} \rightarrow M = \dfrac{F_b bd^2}{6}$

$M = M \rightarrow \dfrac{WL}{8}(12\text{ in./ft}) = \dfrac{F_b\,bd^2}{6} \rightarrow d^2 = \dfrac{9\,W(L\text{ in.})}{b\,(F_b)}$

Paul (1916) provides tables of safe working stresses and average values of the modulus of elasticity of structural-size timbers of the common species of wood used in construction in the United States (see Tables 4-13 and 4-14).

Paul presents a table of the safe loads (uniformly distributed) on heavy timber beams, with the limiting loads based upon the maximum horizontal shear allowed by the cross-section. This is a particularly handy table horizontal shear governs many of the spans of heavy timber beams (see Table 4-17).

The shear values are calculated in the same manner as today. The formulae are rearranged, as are the other parameters (see Fig. 4-5).

Modern Formula: Horizontal Shear:

$F_v = \dfrac{3V}{2\,bd}$

$V = \dfrac{wL}{2} = \dfrac{W}{2}$

Table 4-13. Average Values of the Modulus of Elasticity, Structural Timber[a]

Species of Timber	Modulus of Elasticity in Bending (psi)
*Pine, southern	
Dense grade	1,600,000
Sound grade	1,200,000
Spruce	1,000,000
Tamarack	1,200,000
*Fir, Douglas	
Dense grade	1,600,000
Sound	1,200,000
Hemlock, western	1,400,000
Oak	1,300,000
Pine, eastern white	1,300,000
Pine, Norway	1,400,000

[a] From Paul 1916, 63

$$F_v = \frac{3W}{4bd}$$

$$W = \frac{F_v \, 4bd}{3} \quad \text{(N.L.M.A. formula)}$$

Voss and Varney (1926) contains plates of details and sections of the three types of mill construction. The plate on standard mill construction shows details of the method of beam-to-column connections and beam-to-wall connections (see Figure 4-5).

Large wood members are difficult to connect to each other. The historic method of mortise and tenon connections requires close time-consuming details and weakens the girder by the mortise being cut into the member. When semimill construction is used, the joists are either "stacked" (bear on top of beams) or framed into the beams (tops at same elevation). In the former case, the building is increased in height and expense, and in the latter, the problem of the butt connection is difficult. Designers

Table 4-14. Allowable Unit Stressess for Structural Timbers[a]

	Bending		Compression	
Species of Timber	Stress in Extreme Fiber (psi)	Horizontal Shear Stress (psi)	Parallel to Grain "Short Columns" (psi)	Perpendicular to Grain (psi)
*Pine, Southern				
Dense, grade	1,600	125	1,200	350
Sound	1,300	85	900	300
Spruce	900	70	600	200
Tamarack	1,200	95	900	350
*Fir, Douglas				
Dense, grade	1,600	100	1,200	350
Sound	1,300	85	900	300
Hemlock, eastern	1,000	70	700	300
Hemlock, western	1,300	75	900	300
Oak	1,400	125	900	400
Pine, eastern white	900	80	700	250
Pine, Norway	1,100	85	800	300

[a] From Paul 1916, 64.

Table 4-15. Maximum Spans for Timber Mill Floors[a]

Nominal Thickness	Actual Thickness	Fiber Stress (psi)	(Actual Thickness) Made of Matched and Dressed Plank Span (ft) Live Load (psi)											
			50	100	125	150	175	200	225	250	275	300	350	400
3"	2⅝"	1,200	13' 8"	10' 1"	9" 1"	8' 4"	7' 9"	7' 3"	6'10"	6' 6"	6' 3"	6' 0"	5' 7"	5' 2"
"	"	1,300	14' 3"	10' 6"	9' 6"	8' 8"	8' 1"	7' 7"	7' 2"	6'10"	6' 6"	6' 3"	5' 9"	5' 5"
"	"	1,500	15' 4"	11' 3"	10' 2"	9' 4"	8' 8"	8' 2"	7' 8"	7' 4"	7' 0"	6' 8"	6' 2"	5'10"
"	"	1,600	15'10"	11' 8"	10' 6"	9' 7"	8'11"	8' 4"	7'11"	7' 7"	7' 2"	6'11"	6' 5"	6' 0"
"	"	1,800	16' 9"	12' 4"	11' 2"	10' 3"	9' 6"	8'11"	8' 5"	8' 0"	7' 8"	7' 4"	6' 9"	6' 4"
"	"	Deflection	9' 0"	7' 4"	6'11"	6' 6"	6' 2"	5'11"	5' 8"	5' 6"	5' 4"	5' 2"	4'11"	4' 9"
4"	3⅝"	1,200	18' 5"	13' 8"	12' 4"	11' 5"	10' 7"	10' 0"	9' 5"	9' 0"	8' 7"	8' 3"	7' 7"	7' 2"
"	"	1,300	19' 2"	14' 3"	12'11"	11'10"	11' 0"	10' 4"	9'10"	9' 4"	8'11"	8' 7"	7'11"	7' 5"
"	"	1,500	20' 7"	15' 4"	13'10"	12' 9"	11'10"	11' 2"	10' 6"	10' 0"	9' 7"	9' 2"	8' 6"	8' 0"
"	"	1,600	21' 3"	15'10"	14' 4"	13' 2"	12' 3"	11' 6"	10'11"	10' 4"	9'11"	9' 6"	8'10"	8' 3"
"	"	1,800	22' 7"	16' 9"	15' 2"	13'11"	13' 0"	12' 2"	11' 7"	11' 0"	10' 6"	10' 1"	9' 4"	8' 9"
"	"	Deflection	12' 3"	10' 1"	9' 5"	8'11"	8' 6"	8' 2"	7'10"	7' 7"	7' 4"	7' 2"	6'10"	6' 6"
5"	4⅝"	1,200	22'10"	17' 8"	15' 7"	14' 5"	13' 5"	12' 7"	11'11"	11' 4"	10'10"	10' 5"	9' 8"	9' 1"
"	"	1,300	23'10"	17'11"	16' 3"	14'11"	13'11"	13' 1"	12' 5"	11'10"	11' 4"	10'10"	10' 1"	9' 5"
"	"	1,500	25' 7"	19' 3"	17' 5"	16' 1"	15' 0"	14' 1"	13' 4"	12' 8"	12' 2"	11' 8"	10'10"	10' 2"
"	"	1,600	26' 5"	19'11"	18' 0"	16' 7"	15' 6"	14' 7"	13' 9"	13' 1"	12' 6"	12' 0"	11' 2"	10' 6"
"	"	1,800	28' 0"	21' 1"	19' 1"	17' 7"	16' 5"	15' 5"	14' 7"	13'11"	13' 4"	12' 9"	11'10"	11' 1"
"	"	Deflection	15' 4"	12' 9"	11'11"	11' 3"	10' 9"	10' 4"	10' 0"	9' 8"	9' 4"	9' 1"	8' 8"	8' 4"
6"[b]	5⅝"[b]	1,200		20' 8"	18' 9"	17' 4"	16' 2"	15' 3"	14' 5"	13' 9"	13' 2"	12' 8"	11' 9"	11' 0"
"	"	1,300		21' 6"	19' 6"	18' 0"	16'10"	15'10"	15' 0"	14' 3"	13' 8"	13' 1"	12' 2"	11' 5"
"	"	1,500		23' 1"	21' 0"	19' 4"	18' 1"	17' 0"	16' 1"	15' 4"	14' 8"	14' 1"	13' 1"	12' 3"
"	"	1,600		23'10"	21' 8"	20' 0"	18' 8"	17' 7"	16' 7"	15'10"	15' 2"	14' 7"	13' 6"	12' 8"
"	"	1,800		25' 3"	23' 0"	21' 2"	19'10"	18' 8"	17' 8"	16'10"	16' 1"	15' 5"	14' 4"	13' 6"
"	"	Deflection		15' 4"	14' 5"	13' 8"	13' 0"	12' 6"	12' 1"	11' 8"	11' 4"	11' 0"	10' 6"	10' 1"

[a] From Paul 1916, 61. Fiber stress 1,200, 1,300, 1,500, 1,600, and 1,800 psi; modulus of elasticity 1,620,000 psi. The sum of the live load and the weight of the floor was used in calculating the spans. In the line marked "Deflection" is given the span that has a deflection of 1/30 in. per ft of span.
[b] Use for laminated floors when made of 2 × 6 and 4 × 6 pieces.

The tables give maximum spans for timber mill floors, based upon a modulus of elasticity of 1,620,000 psi. In each of the tables, if the material to be used has a lower modulus of elasticity, the values of the lengths in the tables are to be multiplied by a factor as shown below.

E = 1,600,000 psi: Use tables as they stand.
E = 1,400,000 psi: Use a factor of 0.952 times the lengths in the tables.
E = 1,300,000 psi: Use a factor of 0.929 times the lengths in the tables.
E = 1,200,000 psi: Use a factor of 0.904 times the lengths in the tables.

have devised a series of "joist hangers" or "stirrup irons" for the framing of joists into girders. (see Figure 4-7).

Beams or girders began to be designed utilizing the same methods and criteria as with modern analysis by approximately the mid-1920s. The nomenclature and methods of application of the design procedure are almost the same. Modern analysis has taken on a new procedure since 1991, when a series of design factors was introduced to either reduce or magnify the allowable stress on a member. Voss and Varney (1976) establishes the method used in the pre-modern era. The determination of member size by computation of a design moment and use of that with the allowable stress to determine the section modulus of the member and thus the depth and breadth of the section is most similar to the modern method. In addition, Voss and Varney contains a nomograph for the selection of beams or girders (see Figure 4-8).

Table 4-16. Maximum Spans for Timber Laminated Floors[a]

(Actual Thickness) Made of Planks on Edge, Laid Close
Span (ft)
Live Load (psi)

Nominal Thickness	Actual Thickness	Fiber Stress (psi)	100	125	150	175	200	225	250	275	300	350	400
6"	5½"[b]	1,200	20' 3"	18' 4"	16'11"	15'10"	15' 1"	14' 1"	13' 5"	12'10"	12' 4"	11' 6"	10' 9"
"	"	1,300	21' 1"	19' 1"	17' 8"	16' 5"	15' 8"	14' 8"	14' 0"	13' 4"	12'10"	11'11"	11' 2"
"	"	1,500	22' 7"	20' 9"	18'11"	17' 8"	16'10"	15' 9"	15' 0"	14' 4"	13' 9"	12'10"	12' 0"
"	"	1,600	23' 4"	21' 3"	19' 7"	18' 3"	17' 5"	16' 4"	15' 6"	14'10"	14' 3"	13' 3"	12' 5"
"	"	1,800	24' 9"	22' 6"	20' 9"	19' 4"	18' 5"	17' 3"	16' 5"	15' 9"	15' 1"	14' 0"	13' 2"
"	"	Deflection	15' 0"	14' 1"	13' 4"	12' 9"	12' 3"	11' 9"	11' 5"	11' 1"	10' 9"	10' 3"	9'10"
8"	7½"	1,200	26'10"	24' 6"	22' 8"	21' 2"	20' 0"	19' 0"	18' 1"	17' 4"	16' 7"	15' 6"	14' 7"
"	"	1,300	27'11"	25' 6"	23' 7"	22' 1"	20'10"	19' 9"	18'10"	18' 0"	17' 4"	16' 1"	15' 2"
"	"	1,500	30' 0"	27' 5"	25' 4"	23' 9"	22' 4"	21' 2"	20' 3"	19' 4"	18' 7"	17' 4"	16' 3"
"	"	1,600	31' 0"	28' 3"	26' 2"	24' 6"	23' 1"	21'11"	20'10"	20' 0"	19' 2"	17'10"	16' 9"
"	"	1,800	32'10"	30' 0"	27' 9"	26' 0"	24' 6"	23' 3"	22' 2"	21' 2"	20' 4"	19' 0"	17'10"
"	"	Deflection	20' 1"	19' 4"	17'11"	17' 2"	16' 6"	15'11"	15' 5"	15' 0"	14' 7"	13'11"	13' 4"
10"	9½"	1,200									20'10"	19' 5"	18' 3"
"	"	1,300									21' 9"	20' 3"	19' 1"
"	"	1,500									23' 4"	21' 9"	20' 5"
"	"	1,600									24' 1"	22' 5"	21' 2"
"	"	1,800									25' 7"	23'10"	22' 5"
"	"	Deflection									18' 4"	17' 6"	16'10"
12"	11½"	1,200											22' 0"
"	"	1,300											22'11"
"	"	1,500											24' 7"
"	"	1,600											25' 4"
"	"	1,800											26'11"
"	"	Deflection											20' 3"

[a] From Paul 1916, 62. Fiber stress 1,200, 1,300, 1,500, 1,600 and 1,800 psi; modulus of elasticity 1,620,000 psi. The sum of the live load and the weight of the floor was used in calculating the spans. In the line marked "Deflection" is given the span that has a deflection of 1/30 in. per ft of span.
[b] Use for 2½ × 6, 3 × 6, and 6 × 6 pieces; for 2 × 6 and 4 × 6, use Table 4-16 for mill floors.

Example: Beam Design

Southern Pine, $F_b = 1500$ psi (Method by Voss and Varney)

Load/ft = 4 ft. × 110 lb/ft² + 20 lb (beam weight) = 460 lb/ft

$$M_{Des} = \frac{wL^2}{8} = \frac{460 \text{ lb/ft} (16 \text{ ft})^2 (12 \text{ in./ft})}{8} = 176{,}640 \text{ in.-lb} = 14{,}700 \text{ ft-lb}$$

$M_{Des} = F_b S_x$, which is 176,600 lb/in. = 1,500 lb/in.² $\left(\frac{bd^2}{6}\right)$ in.³

and $bd^2 = 706.5$ in.³ if $b = 5.5$ in. then $d = 11.333$ in. use nom. 6 × 14 in. Actual dim. beam would require a 5.5 × 13.25 in (nom. 6 × 14 member)

$S = 160.9$ in.³

$$\text{Deflection}_{6 \times 14} = d = \frac{5 wL^3}{384 \, EI} = \frac{5(460 \text{ lb/ft})(16 \text{ ft})^4 (1728 \text{ in.}^3/\text{ft}^3)}{384(1{,}620{,}000 \text{ lb/ft})(1066.2 \text{ in.}^4)} = 0.391 \text{ in.}$$

$$\text{Deflection allowable} = \frac{\text{span, in.}}{360} = \frac{16 \, (12 \text{ in./ft})}{360} = 0.533 \text{ in.}$$

Table 4-17. Safe Loads in Pounds, Uniformly Distributed for Timber Beams: Limited by Resistance to Horizontal Shear Along the Neutral Axis[a]

Nominal Size	Actual Size	(Actual Size) Horizontal Shearing Stress (psi)				
		100	125	150	175	200
6 × 10	5½ × 9½	6,966	8,707	10,449	12,190	13,932
8 × 10	7½ × 9½	9,500	11,875	14,250	16,625	19,000
10 × 10	9½ × 9½	12,032	15,040	18,048	21,056	24,064
6 × 12	5½ × 11½	8,432	10,540	12,648	14,756	16,864
8 × 12	7½ × 11½	11,500	14,375	17,250	20,125	23,000
10 × 12	9½ × 11½	14,566	18,207	21,849	25,490	29,132
12 × 12	11½ × 11½	17,632	22,040	26,448	30,856	35,264
6 × 14	5½ × 13½	9,900	12,375	14,875	17,325	19,800
8 × 14	7½ × 13½	13,500	16,875	20,250	23,625	27,000
10 × 14	9½ × 13½	17,100	21,375	25,650	29,925	34,200
12 × 14	11½ × 13½	20,700	25,875	31,050	36,225	41,400
14 × 14	13½ × 13½	24,300	30,375	36,450	42,525	48,600
6 × 16	5½ × 15½	11,366	14,207	17,049	19,890	22,732
8 × 16	7½ × 15½	15,500	19,375	23,250	27,125	31,000
10 × 16	9½ × 15½	19,634	24,542	29,451	34,359	39,268
12 × 16	11½ × 15½	23,766	29,707	35,649	41,590	47,532
14 × 16	13½ × 15½	27,900	34,875	41,850	48,825	55,800
16 × 16	15½ × 15½	32,032	40,040	48,048	56,056	64,064
6 × 18	5½ × 17½	12,834	16,042	19,251	22,459	25,668
8 × 18	7½ × 17½	17,500	21,875	26,250	30,625	35,000
10 × 18	9½ × 17½	22,166	27,707	33,249	38,790	44,332
12 × 18	11½ × 17½	26,834	33,542	40,251	46,959	53,668
14 × 18	13½ × 17½	31,500	39,375	47,250	55,125	63,000
16 × 18	15½ × 17½	36,166	45,207	54,249	63,290	72,332
18 × 18	17½ × 17½	40,832	51,040	61,248	71,456	81,664
20 × 20	19½ × 19½	50,700	63,375	76,050	88,725	101,400

[a] From Paul 1916, 65. Note: to use table for values of the horizontal shearing stress less than 100 psi, multiply the safe load in the 100 column by the ratio of the unit stress used to 100. For example: for 85 psi, use 0.85 of the load in the 100 column.

Wood shapes for joists, planking, and heavy timber members were all sawn to nominal sizes. The actual size would be slightly smaller due to shrinkage during the kiln drying and final surfacing of the material. Figure 4-9 gives the sizes of larger wood members commercially produced for bridge and building construction. Voss and Varney state that the usual rule for the allowance from nominal to actual is 0.25 in. for sizes of 4 in. width and under and 0.5 in. for sizes of 6 in. width and over, and that all depths are to have an allowance of 0.5 in. All sizes in Figure 4-9 are based upon southern pine members except for the sizes shown in brackets [ ], which are Douglas fir and other Pacific Coast species. Bold-faced sizes were the most common for beam sections. Sizes in parentheses ( ) were not commonly available.

Voss and Varney (1926) also discusses the effect of lateral bracing on the design of beams. This is the first mention that this author has found concerning the laterally unsupported sections on the actual beam design. Voss gives the following formula for

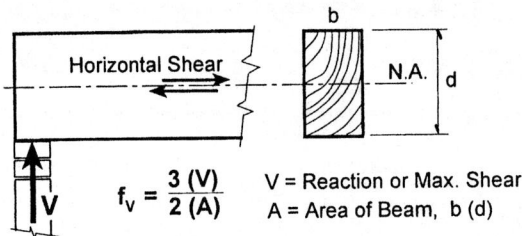

Figure 4-5. Horizontal shear on wood beam

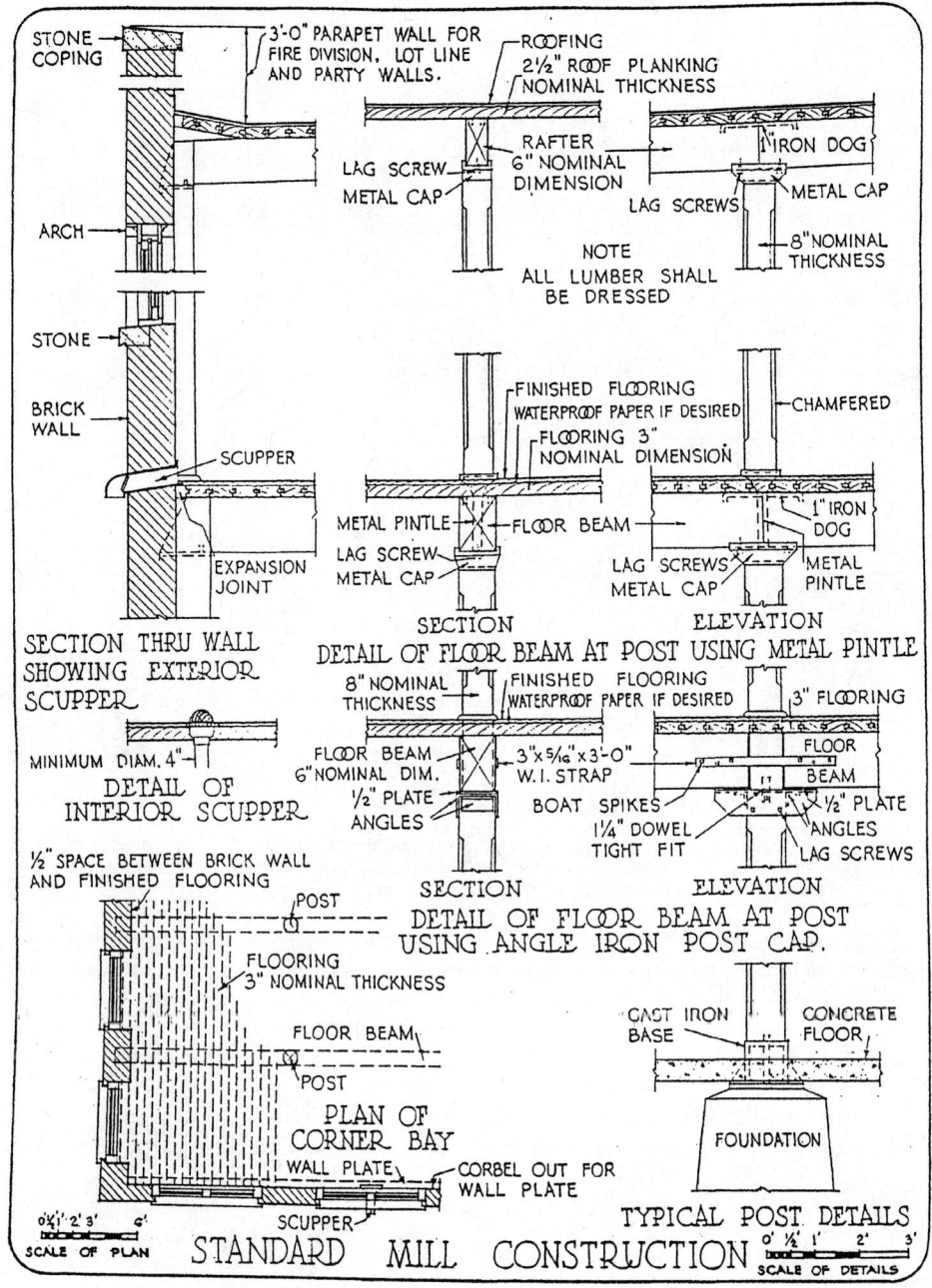

Figure 4-6. Standard mill construction. (*From Voss and Varney 1926, 91.*)

the reduction of the allowable bending stress for members that have laterally unbraced sections that exceed 20 times the width of the member (Voss and Varney 1926, 13–15).

$$F_{bu} = F_b \left(1 - \frac{L_u}{90b}\right)$$

where:
  F_{bu} = maximum allowable bending stress considering maximum unbraced length of member
  F_b = allowable bending stress for species of wood

280 ◇ Structural Analysis of Historic Buildings

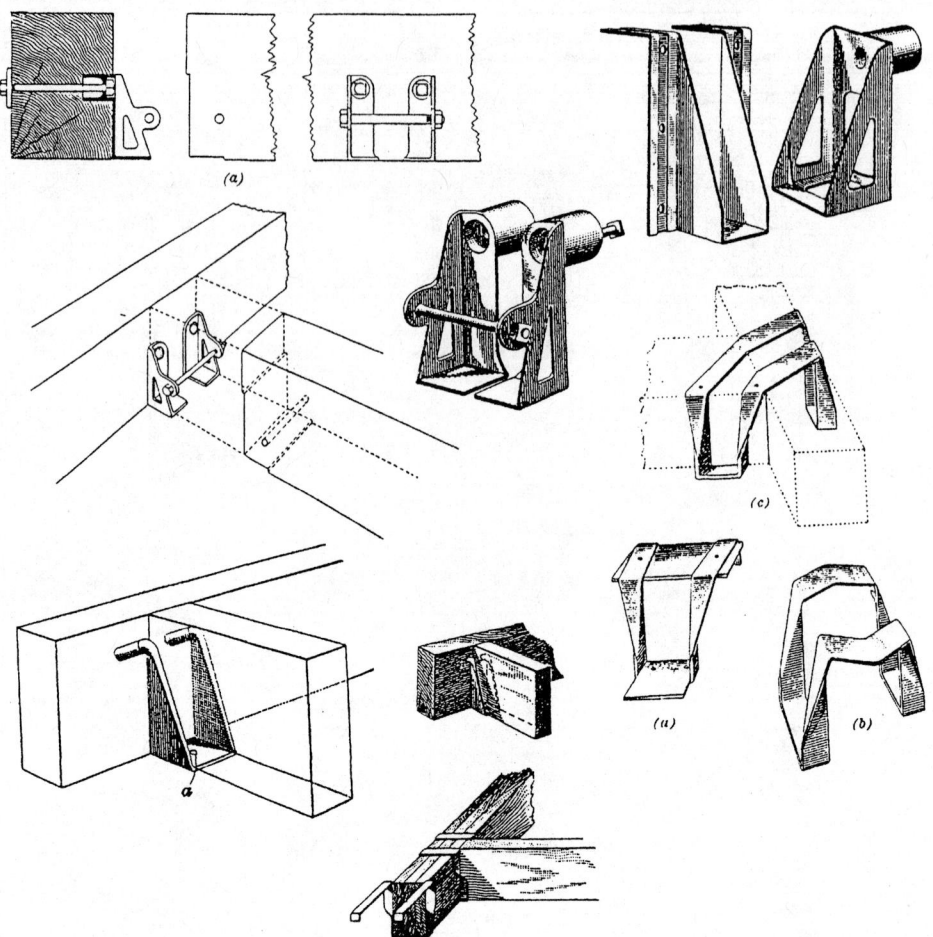

Figure 4-7. Joist hangers: stirrup irons. (*From International Correspondence Schools 1905A, para. 13, 13, 16; id. 1923, para. 31, 53.*)

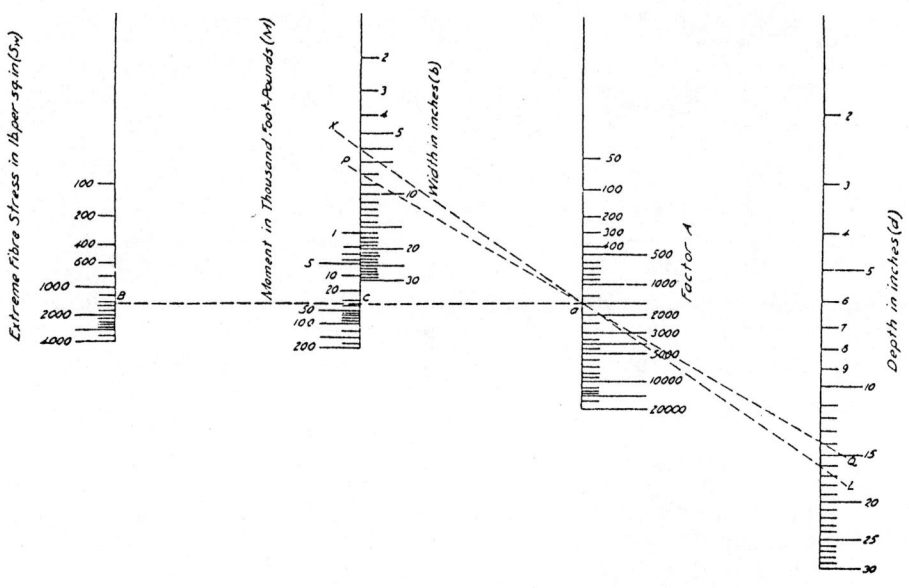

Figure 4-8. Monograph for simple wood beams, 1926. (*From Voss and Varney 1926, 11.*)

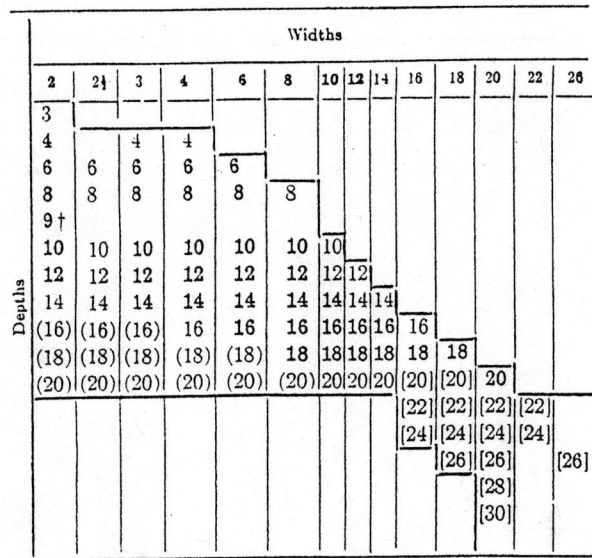

Figure 4-9. Commercial timber sizes. (*From Voss and Varney 1926, 12.*)

L_u = maximum unbraced length of member, in.
b = breadth of beam, in.

Voss and Varney also describe and quantitatively discuss horizontal shear.

$$F_v = \frac{3V}{2bd}$$

where:
V = maximum vertical shear on member, lb
b = breadth of member, in.
d = depth of member, in.

The design of uniformly loaded beams is controlled by horizontal shear if the span in feet is less than the depth of the beam in inches.

Voss and Varney also state that the ends of beams must be checked to see that there is sufficient bearing area to ensure that the allowable compressive stress (perpendicular to the grain) of the member is not exceeded.

Actual bearing stress, $f_{br} = \dfrac{R}{ba}$

where:
R = reaction at the support, lb
b = breadth of member at bearing, in.
a = length of bearing, in.

Voss and Varney include a table of the allowable stresses for structural timbers that were suggested for use during the period following the publication of their work, (see Table 4-18).
The table contains special notations and explanations of some of the values:

Grades of the species are in conformance with Southern Pine Association Standards and the Standards of the U.S. Forest Products Laboratory.
  Grade 1 = Dense material
  Grade 2 = Sound material
When the analysis of a structure includes the lateral loads from wind and live and dead gravity loads, the allowable stresses may be increased by 50 percent.

Table 4-18. Allowable Stresses for Structural Timbers[a]

Species	Grade	Bending — In Extreme Fiber lb/in.²	Bending — Horizontal Shear lb/in.²	Compression — Parallel to Grain "Short Columns" lb/in.²	Compression — Perpendicular to Grain lb/in.²	Modulus of Elasticity 1,000 lb/in.²
Ceder, western red	1	900	80	700	200	1,000
	2	600	53	467	200	
Cedar, northern white	1	750	70	550	175	800
	2	500	47	384	175	
Chestnut	1	950	90	800	300	1,000
	2	,633	60	533	300	
Cypress	1	1,300	100	1,100	350	1,400
	2	867	67	733	350	
Douglas fir§	1	1,500	90	1,100	325	1,600
	2	1,000	60	750	300	
Douglas fir (Rocky Mountain region)	1	1,100	85	800	275	1,200
	2	767	57	533	275	
Fir, balsam	1	900	70	700	150	1,000
	2	600	47	467	150	
Gum, red	1	1,100	100	800	300	1,200
	2	767	67	533	300	
Hemlock, western	1	1,300	75	900	300	1,400
	2	867	50	600	300	
Hemlock, eastern	1	1,000	70	700	300	1,100
	2	667	47	467	300	
Larch, western	1	1,200	100	1,100	325	1,300
	2	800	67	733	325	
Maple, sugar or hard	1	1,500	150	1,200	500	1,600
	2	1,000	100	800	500	
Maple, silver or soft	1	1,000	100	800	350	1,100
	2	667	67	533	350	
Oak, white or red	1	1,400	125	1,000	500	1,500
	2	933	83	667	500	
Pine, southern yellow	1	1,500	110	1,100	325	1,600
	2	1,000	70	750	300	
Pine, eastern white, western white, and western yellow	1	900	85	750	250	1,000
	2	600	57	500	250	
Pine, Norway	1	1,100	85	800	300	1,200
	2	733	57	533	300	
Spruce, red, white, and Sitka	1	1,100	85	800	250	1,200
	2	733	57	533	250	
Spruce, Engelmann	1	750	70	600	175	800
	2	500	47	400	175	
Tamarack, eastern	1	1,200	95	1,000	300	1,300
	2	800	63	667	300	

[a] From Voss and Varney 1926, 8.

The construction of mill buildings and other types of buildings that utilized heavy timbers declined significantly in the 1930s with all of the problems of the Great Depression, which was followed by the World War II and the special concentration by all industry on the production of war-related materials. The postwar construction boom was mostly concentrated on housing and the use of lighter timber building components. Post war industrial and commercial construction utilized mainly steel and concrete, and the relative ease of use of these types of construction served to practically eliminate the use of heavy timbers in the building industry.

COMPOUND WOOD BEAMS: COMBINATION WOOD AND STEEL BEAMS

Compound beams were utilized during the last half of the nineteenth century and the first third of the twentieth century. Designers looked for a method to make larger or stronger beams than were available or could be acquired for a required service.

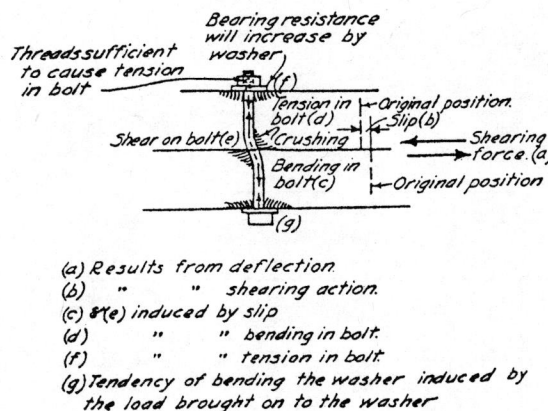

Figure 4-10. Diagram of bolted joint. (*From Voss and Varney 1926, 22.*)

Generally the compound beam consisted of two basic types: one was composed of two beams, one over the other, and one was composed of two beams, one beside the other. The strength of the combination was the element that designers were looking for. A single beam would have one unit of strength. Combining two beams side by side would only double the strength. (In the equation for section modulus, $b(h^2)/6$, the two beams side by side would only double the value of b, and therefore the strength would be doubled.) Stacking the two beams one over the other would increase the strength by the factor of the square of the doubled depth over the square of the single depth (true only if the connection between the two members is capable of preventing any differential movement or horizontal slip at the juncture of the two members). The difficulty in obtaining that positive connection was the concern of the designers of the period.

Designers of the period considered the ratio of the strength of a compound beam to that of a solid beam of the same dimensions as the "efficiency of the section." The capacity of the stacked beams would depend totally upon the means of attachment and the ability to resist the lateral movement at the juncture. Mechanical fasteners such as bolts and pins are not totally positive connections—slipping occurs, even with relatively small loads. The free surface of the holes slips as load is applied and causes a bearing stress buildup as the bolt or pin is pressed against the surface of the wood (see Figure 4-10).

As load is applied, the round section applies a bearing pressure against the projected area of the member: b (the breadth of the member) times d (the diameter of the pin or bolt).

The designers of the period felt that the bearing against the wood member could be replaced by a parallel and a normal component (see Figure 4-11). They felt that the parallel component tended to split the member and the perpendicular component caused tension across the fibers of the member. The assumption that the pressure produces the components made the designers choose a working stress of the average of the allowable end and cross-bearing stresses. Southern yellow pine would therefore have the following average bearing stress:

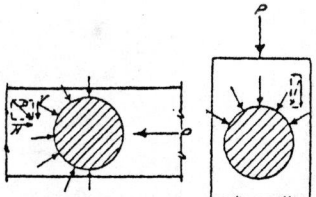

Figure 4-11. Pressure on wood member from round pin or bolt. (*From Voss and Varney 1926, 23.*)

Southern yellow pine (from Table 4-18):
    Allowable cross-bearing = 325 psi
    Allowable end bearing = 1100 psi
    Average bearing = 710 psi for a pin or bolt pressure on the projected area of the member

Another form of the compound, all wood beam is the "keyed" beam, as shown in Figure 4-12. The design is sound in principle. It allows the wedged keys to prevent the slip (from the horizontal shear) while the bolts hold the two members together to make a single beam. Kidder (1900) provides design information for keyed beams. The object of the keyed beams is to obtain longer spans and heavier loads, which requires much deeper beams. The keys were usually of oak or cast iron and were made as

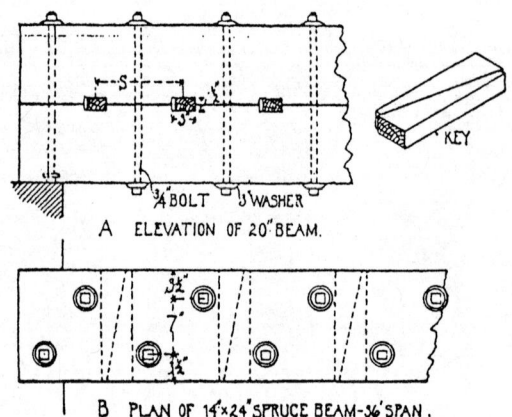

Figure 4-12. Keyed beam details. (*From Kidder 1900, 448.*)

wedges so they could be driven tightly in the key and thus hold the position firmly. Kidder cites Professor Kidwell's tests on these members as rating the efficiency of the beams as 75 percent when oak keys are used and 80 percent when cast iron keys are used. The keys work in shear against the material of the beam and are spaced closer in the region of the beam where the greater shear value exists, (see Figure 4-13).

Kidder provides two design tables for the direct selection of compound keyed beams. Table 4-19 shows the safe distributed loads (uniform loads) in pounds (the total load) for compound keyed beams, and Table 4-20 shows the number of oak keys required on each side of the center of the beam. The depths given in the tables are the total depth of the combined section. Kidder states that the breadth of the compound beams should not be less than two-fifths of the depth. There is no mention of lateral bracing or the reduction of bending stress being based upon the laterally unbraced length. The time period was too early for the design to include this parameter. The number of keys required is not affected by the length or breadth of the beam if the beam is figured at the full safe load (ibid., 448). The deflection of compound keyed beams (oak keys) is about 25 percent more than a solid beam, and that of cast iron keys is about the same as that of the solid beam.

Also, for beams loaded at the center, the spacing of the keys should be uniform from X to Y. Y being one-eighth of the span from the center. If the distance between the keys, center to center, works out to be less than the minimum spacing, the safe load should be correspondingly reduced or the breadth (thickness) of the beam increased.

Kidder states that if the beam is not over 10 in. wide, the designer may use a single row of bolts as shown on the spruce beam in Figure 4-13 if 12 in. wide or wider, the bolts shall be staggered as shown on the hard pine beam of Figure 4-13. Kidder suggests the spacing of keys (beginning at the reaction end) for a uniformly loaded simple span (see Table 4-21).

The authors of the period all appear to agree that the most widely used form of compound beam is the beam that is formed by nailing diagonal boards on each side

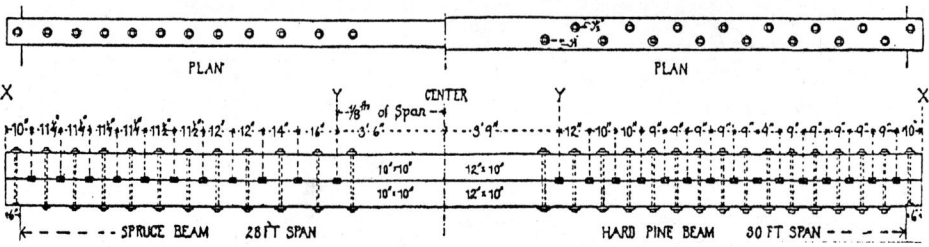

Figure 4-13. Compound keyed beams. (*From Kidder 1900, 46.*)

Table 4-19. Safe Distributed Loads (lb) Compound Keyed Beams[a]

Size of Beam		Span of Beam (ft)					
		20	24	28	30	32	36
1 × 16	White pine	1,152	960	823	768	720	
	Spruce	1,344	1,120	960	896	840	
	Oregon pine		1,440	1,234	1,152	1,080	
	Georgie pine		1,600	1,371	1,280	1,200	
1 × 20	White pine	1,800	1,500	1,285	1,200	1,125	
	Spruce		1,750	1,500	1,400	1,312	
	Oregon pine		2,250	1,928	1,800	1,687	1,500
	Georgia pine			2,142	2,000	1,875	1,666
1 × 24	White pine		2,160	1,851	1,728	1,620	1,440
	Spruce		2,520	2,160	2,016	1,890	1,680
	Oregon pine			2,777	2,592	2,430	2,160
	Georgia pine			3,085	2,880	2,700	2,400
1 × 28	White pine			2,520	2,352	2,205	1,960
	Spruce			2,744	2,572	2,286	
	Oregon pine			3,528	3,307	2,940	
	Georgia pine			3,920	3,675	3,266	

To find safe loads for any given thickness of beam, multiply the load in the table by breadth of beam in inches.
For center loads, take one-half those in table.
Beams should not be used for shorter or longer spans than those for which safe loads are given, except that 28 in. beams may be used up to 40 ft.
   16 and 20 in. beams to have 1½ × 3 in. oak keys, ¾ in. bolts, 3 in. washers.
   24 in. beam to have 2 × 4 in. oak keys, ⅞ in. bolts, 3½ in. washers.
   28 in. beam to have 2¼ × 4½ in. oak keys, ⅞ in. bolts, 3½ in. washers.

[a] From Kidder 1900, 447.

of the two members that make up the beam, as shown in Figure 4-14. The diagonals, usually 1.25 × 6–8 in. wide, were nailed to the members as shown. None of them discuss the direction of the diagonals—whether one direction is better than the other or whether they should be reversed at midspan. This author would intuitively want to keep the diagonals the same directions on either side of the members (the opposite of what is shown in Figure 4-14). It seems that perhaps a torsional consideration would dictate this. Kidder states that these beams should be designed utilizing an allowable stress of the beam only 65 percent of what it would be for a solid member. He makes no mention of including the 1.25 in. members in the section modulus or moment of inertia, and it would seem inappropiate to do so.

    Voss and Varney (1926) have a section on this type of beam; they refer to it as the Clark method. They state that the efficiency of this type of beam is about 75 percent of that of the solid beam and that the deflection is about twice that of the solid beam. Voss and Varney state that the boarding is kept at 45° and that the strips

Table 4-20. Number of Oak Keys Required Each Side of Center of Beam[a]

	For Beams of			
	White Pine	Spruce	Oregon Pine	Georgia Pine
16 in. beams 1½ × 3 in. keys	7	8	11	12
20 in. beams 1½ × 3 in. keys	9	11	13	15
24 in. beams 2 × 4 in. keys	8	9	12	13
28 in. beams 2¼ × 4½ in. keys	9	10	12	14
Minimum spacing of keys				
1½ × 3 in. keys	11¼ in.	11¼ in.	9 in.	9 in.
2 × 4 in. keys	15 in.	15 in.	11½ in.	11½ in.
2¼ × 4½ in. keys	17 in.	17 in.	13 in.	13 in.

[a] From Kidder 1900, 448.

Table 4-21. Spacing of Keys on Compound Beams[a]

16 in. spruce beam, 32 ft span	10, 12, 12, 16, 19, 24, 32
20 in. spruce beam, 32 ft span	10, 11½, 11½, 11½, 12, 12, 13, 15, 18, 24
24 in. spruce beam, 36 ft span	13, 15, 15, 15, 15, 16, 18, 20, 30
28 in. spruce beam, 36 ft span	15, 17, 17, 17, 17, 17, 17, 17, 17, 17

[a] From Kidder 1900, 449.

should be placed at opposite inclinations on the two sides of the beam to form a "shear resisting couple." This is not clearly understood and may not be valid. The authors provide no description or example of the effect of this shear resisting couple. They further state that the designer may use either 1.25 in. or 2 in. stock for the diagonal cladding and specify the type of nails for each. The nails are to be concentrated at the center of the diagonals near the plane of slippage and at the outside ends of the diagonal. If 1.25 in. cladding, use 10d or 12d nails; if 2 in. cladding, use 20d or 30d nails. Bolts are also highly recommended for the beams with diagonal cladding, and Voss and Varney say that they are to be placed on about 2 ft centers (see Figure 4-15 and Table 22).

If the breadth of the members is large, staggered bolts may be used. Bolts should never be closer than 2 in. from the edge of the members. Voss and Varney state that the bolts are not counted on for part of the resistance of the member slippage; they are only for holding the two members together. The bolt sizes are as follows:

For overall depths less than 16 in., use 0.5 in. diameter bolts.
For overall depths 16–20 in., use 0.625 in. diameter bolts.
For overall depths over 20 in., use 0.75 in diameter bolts.

Diagonal cladding would be an effective measure to reinforce or retain the original strength of wooden members that have severely checked and cannot any longer carry their design loads. It would be comparatively easy to install such a system on an existing beam and in situ. This author has used a good grade of plywood with two layers on each side of the member (to stagger plywood joints) very successfully in several instances. Purist preservationists argue against such a measure in any situation where the beam is exposed because it practically encases the beam and is unsightly.

The International Correspondence Schools (1905A), in its chapter on beams and girders, details the most popular compound beams and girders used by designers at the turn of the twentieth century. It does not give any explanation of the design of these beams; however, it does provide a table of the relative efficiencies of these sections. The beams described by the I.C.S. are shown in Figure 4-16.

Kidder (1902) calls the compound beam a "solid-built beam." Kidder recognizes the keyed system and describes the keys as hardwood, solid, or wedges for easy readjustment as minor shrinking of the timber occurs. Kidder describes the Threadgold rule, which says the breadth of the key should be twice its depth and the sum of the depths should be equal to one and one-third the total depth of the beam. The second type of solid built beam is the "indented" or "scarfed" beam, which is held together

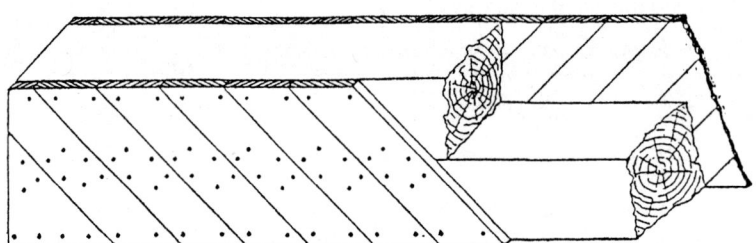

Figure 4-14. Compound beam: diagonal cladding. (*From Kidder 1900, 445.*)

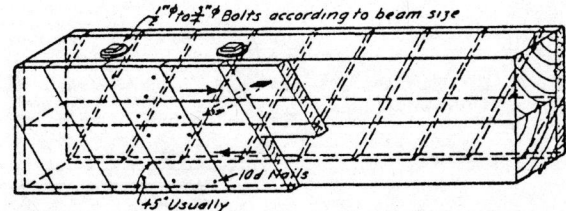

Figure 4-15. Clark's compound beam. (*From Voss and Varney 1926, 31.*)

by bolts or straps. The indented or scarfed beam would be slightly tapered from the middle to the ends so that the wrought iron bands could be slipped over the ends and driven into place. See Figure 4-17 for details of the two types of solid-built beams presented by Kidder. The upper section is the keyed girder and the lower section is the tapered, scarfed beam. The tapering could have had a very significant effect in that it provided a natural camber to the beam that would level out, or tend to, as load was applied to the beam.

Compound beams were also made in other ways. One popular way was to produce a "flitch-plate girder," which was a steel plate sandwiched between two wood members to form a stiffened beam. The plate was usually 0.25–1.75 in. thick, times the full depth and of the same length as the wood members. The section was bolted together to form a very efficient girder (see Figure 4-18).

Kidder (1902) gives two formulae for the design of flitch-plate girders. These formulae are for a uniformly loaded beam and a beam with a concentrated load at the center of the beam. The concept behind the flitch-beam is to have the steel section take the first, major portion of the load and the wood sections take a small portion of the load or the remainder. In this operation, the designer does not want either material to be overstressed. Kidder's formulae for the safe loads on these flitch-girders are as follows (ibid., 402).

Simple Span Flitch Beams (Kidder Method)

safe load at center, P, lb $= \dfrac{d^2}{b} (f + 750\, t)$

safe uniform load, W, lb $= \dfrac{2\, d^2}{L} (f + 750\, t)$

d for the concentrated load at the center $= \sqrt{\dfrac{PL}{f(b) + 1500(t)}}$

d for the uniformly distributed load $= \sqrt{\dfrac{WL}{2f(b) + 1500(t)}}$

Table 4-22. Efficiency of Built-up Wooden Beams[a]

Kind of Beam	Deflection	Efficiency (percent)
Indented	2.00	69.5
Clark's	2.00	76.0
Piped	1.70	84.6
Oak keys	1.25	90.6
Flat iron keys	1.50	78.6
Square iron keys	1.50	89.5

[a] From International Correspondence Schools 1905A, para. 13, 12.

288 ◇ Structural Analysis of Historic Buildings

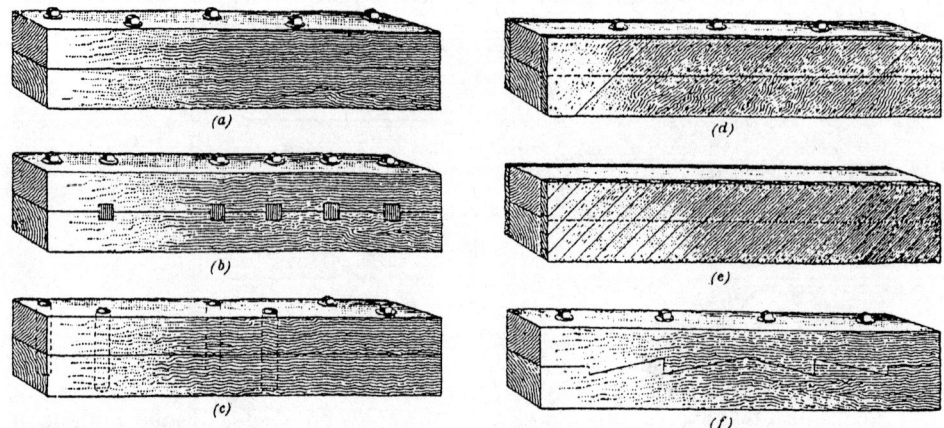

Figure 4-16. I.C.S. compound beams and girders: (a) horizontal joint—bolted; (b) keyed compound beam; (c) horizontal joint, bolted and pipe inserts; (d) diagonal cladding with bolts; (e) diagonal cladding without bolts; (f) "scarfed" or "indented" beam.

where:
- P = concentrated load, lb
- W = uniform load, lb
- L = simple span, ft
- t = thickness of steel plate, in.
- f = 100 lb for hard pine
-     = 73 lb for spruce
- d = depth of beam, in.

Example: Kidder Method (Southern yellow pine = hard pine)

span = 25 ft, uniform load = 1,250 lb/ft
Total load = 31,500 lb
breadth of wood members = 6 in.
thickness of steel = 0.75 in.

$$d = \sqrt{\frac{31{,}500(25)}{2(100)(12) + 1{,}500(0.75)}} = \sqrt{223.4} = 14.94 \text{ in.}$$

Use: $d = 16$ in. (within this range, beams come in even depths only)

Kidder does not mention the size of bolts nor the required number of bolts either by maximum spacing of bolts, etc. This detail would have been left to the designer.

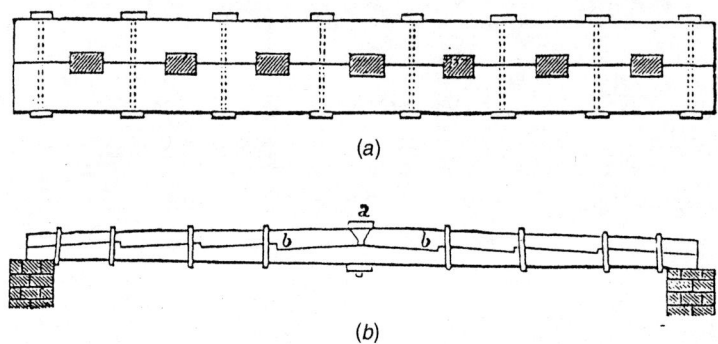

Figure 4-17. Kidder's Solid-Built Beams: (a) Keyed solid-built beam; (b) "scarfed" or "indented" beam. The top part of (b) is two pieces, b and b, which abut against a broad, flat iron bolt, a, termed a "king bolt." Some builders are said to have preferred compound girders to conventional members due to the difficulty in obtaining quality members of the full size. (From Kidder 1902, 382.)

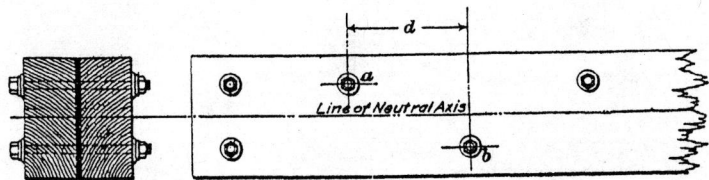

Figure 4-18. Flitch-plate beams and girders. (*From International Correspondence Schools 1924, para. 7, 87.*)

Modern Analysis: (southern yellow pine, F_b = 1500 psi, E = 1,600,000 psi) (Steel, 1900, F_b = 16,000 psi, E = 29,000,000 psi)

$n = \dfrac{29}{1.6} = 18.125$ Steel, t_{eff} = 18.125(0.75 in.) = 13.594 in.

t_{total} = 13.594 in. + 12 in. = 25.594 in.

$M = \dfrac{1250 \text{ lb/ft } (25 \text{ ft})^2}{8}$

M = 97,656.25 lb-ft

$S_{\text{Eq. Sec.}} = S_{x\text{EQ.SEC.}} = \dfrac{bh^2}{6} = \dfrac{25.594 \text{ in. } (16 \text{ ft})^2}{6}$

$= 1092 \text{ in.}^3$

$f_{b \text{ wood}} = \dfrac{M}{S_{\text{Eq. Sec.}}} = \dfrac{97,656.25 \text{ lb-ft}(12 \text{ in./ft})}{1092 \text{ in.}^3} = 1073.15 \text{ lb/in.}^2$

M capacity (wood members) $= \dfrac{F_b S_{\text{wood}}}{12 \text{ in./ft}} = \dfrac{1500 \text{ lb/in.}^2 (512 \text{ in.}^3)}{12 \text{ in./ft}} = 64,000 \text{ lb-ft}$

Moment carried by the steel = 97,656.25 lb-ft. − 64,000 lb-ft = 33,656.25 lb-ft

$f_{b \text{ steel}} = \dfrac{M}{S_{\text{steel}}} = \dfrac{33,656.25 \text{ lb-ft}(12 \text{ in./ft})}{32 \text{ in.}^3} = 12,621 \text{ lb/in.}^2$

The steel is stressed below the allowable bending stress of 16,000 psi. Therefore the flitch-girder design is o.k.

There were other shapes of flitch beams at the time of the turn of the century. A common one was two steel plates sandwiching a wood section; another was a pair of channels sandwiching a wood section (back of channels against wood section); and a difficult beam to make (and not much utilized) was a wood section with steel plates at top and bottom or at either top or bottom.

COMPOUND BEAMS—TRUSSED GIRDERS

Trussed girders were by far the most popular method of reinforcing (strengthening) wood beams and girders at the turn of the twentieth century (see Figure 4-19). Many buildings were designed for larger than average loads, while the need existed to keep the column spacing at a reasonable length for interior functions. At the time, wood was still the framing tool for many designers. Steel sections were new, and details of construction, experience, and costs were factors that kept the multitude of designers from making the change immediately. The skyscrapers in the large cities designed by the major architects of the period were made with the steel frame and were on the cutting edge. It is understandable that for average and minor buildings designers would continue to use the traditional methods and materials.

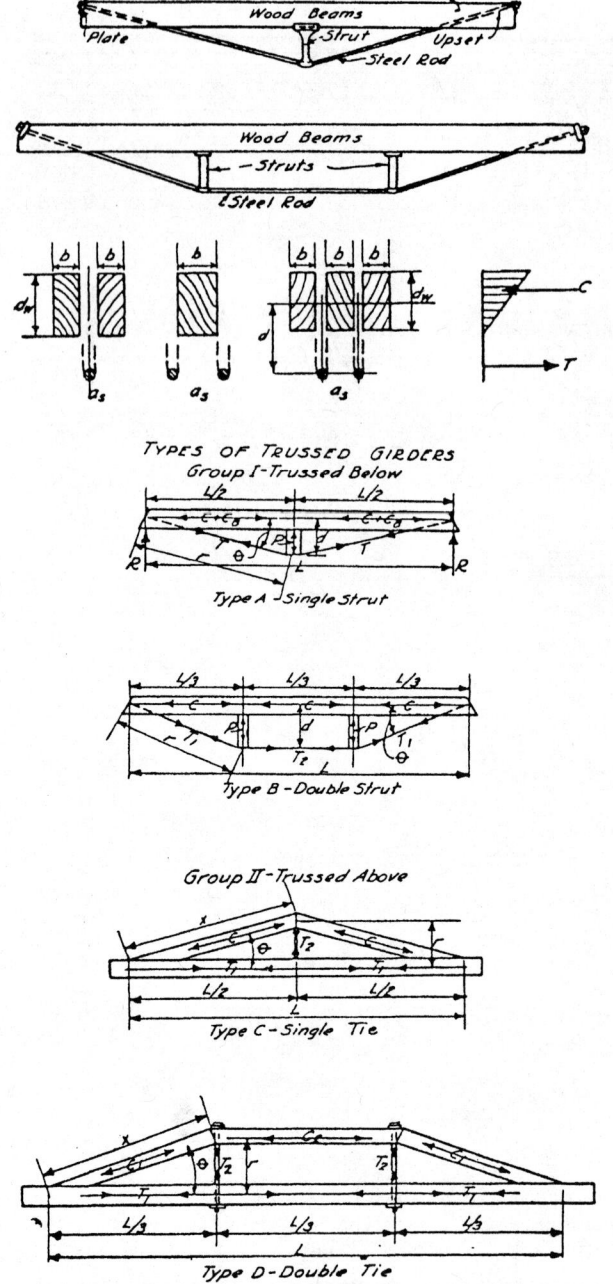

Figure 4-19. Trussed girders. (*From Voss and Varney 1926, 36, 40.*)

Trussed girders began to be used in the 1880s as a solution for longer-span beams. Kidder (1902) contains a brief section on the design of trussed girders. Kidder states that "whenever we wish to support a floor upon girders having a span of more than thirty feet, we must use either a trussed girder, a riveted iron plate girder, or two or more iron beams. The cheapest and most convenient way is, probably, to use a large wooden girder, and truss it" (ibid., 404). Kidder calls trussed girders "belly-rod trusses." Trussed girders utilize a tension rod to support a vertical strut at midspan or two vertical struts at one-third points, utilizing the principle that the tension in the rods would provide a vertical uplift on the strut and thus support the beam at these points. Voss and Varney (1926) have a detailed section on the design and layout of

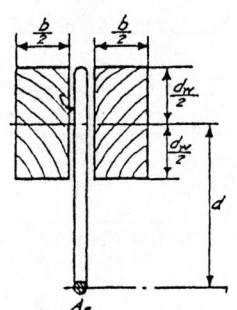

The usual minimum values of this distance are as follows:

(1) $d \geq 1\text{-}1/2 \cdot d_w$
(2) $b = 1/2$ to $2/3$ of d_w.
(3) b and d_w must be even widths and depths to match commercial timber sizes.
(4) d_w varies from 10" to 16" (averaging about 1/2" per foot of span).

Figure 4-20. Truss girder parameters. (*From Kidder 1902, 404.*)

trussed girders. They provide a set of parameters concerning minimum dimensions and member sizes for trussed girders (see Figure 4-20). The designers of the period were quick to realize that the deeper the strut dimension, the greater the angle of the rods with the horizontal and the larger the value of the upward force.

Kidder (1902) was one of the most influential volumes; every designer had a copy of this book due to its versatility in dealing with almost every aspect of building. Kidder's formulae for the design of trussed girders would therefore have been the basis of the design for members in many buildings (see Figures 4-21 and 4-22).

Kidder also provides criteria and formulae for the selection of the members and rods that make up these truss girders for use by designers after they had determined the forces in the components.

$$\text{Area of cross-section of strut } C \text{ or } R = \frac{\text{tension or compression in strut}}{\text{constant } C}$$

$$\text{Diameter of single tie rod} = \sqrt{\frac{\text{tension in rod}}{9{,}425}} \quad \text{(for allowable } f_t \text{ in rod of 12,000 psi)}$$

$$\text{Area of cross-section of beam } B = \frac{\text{Tension}}{t} \text{ or } \frac{\text{Comp.}}{C}$$

In single-strut trusses:

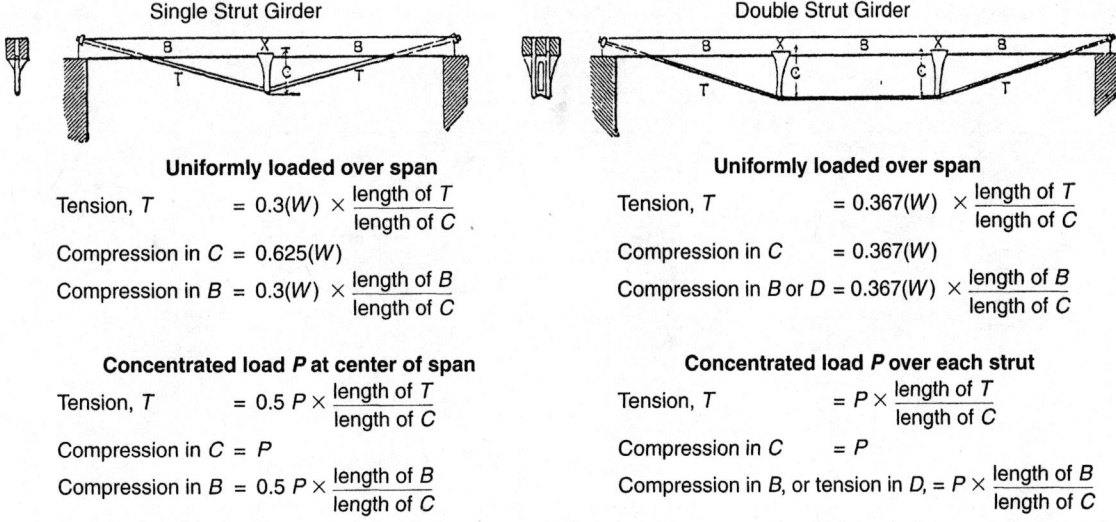

Figure 4-21. Kidder's formulae: trussed girders. (*From Kidder 1902, 405.*)

Single Strut Girder

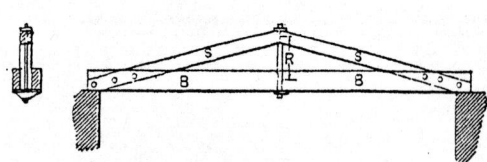

Uniformly loaded over span

Compression in $S = 0.3(W) \times \dfrac{\text{length of } S}{\text{length of } R}$

Tension in $R \quad = 0.625(W)$

Tension in $B \quad = 0.3(W) \times \dfrac{\text{length of } B}{\text{length of } R}$

Concentrated load P at center of span

Compression in $S = 0.5(P) \times \dfrac{\text{length of } S}{\text{length of } R}$

Tension in $R \quad = P$

Tension in $B \quad = 0.5(P) \times \dfrac{\text{length of } B}{\text{length of } R}$

Double Strut Girder

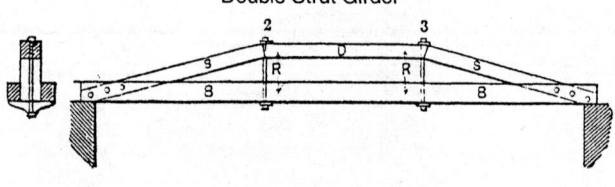

Uniformly loaded over span

Compression in $S = 0.367(W) \times \dfrac{\text{length of } S}{\text{length of } R}$

Tension in $R \quad = 0.367(W)$

Tension in B or compression in $D = 0.367(W) \times \dfrac{\text{length of } B}{\text{length of } R}$

Concentrated load P over each strut

Compression in $S = P \times \dfrac{\text{length of } S}{\text{length of } R}$

Tension in $R \quad = P$

Tension in B or compression in $D = P \times \dfrac{\text{length of } B}{\text{length of } R}$

Figure 4-22. Kidder's formulae: trussed girders. (*From Kidder 1902, 406.*)

Breadth of B (as a beam) $= \dfrac{\text{total load } (L)}{2\,(D)^2 A}$

In double strut trusses:

Breadth of B (as a beam) $= \dfrac{\text{total load } (L)(2)}{5\,(D)^2 A}$

Constants for the above formulae:
C = 1,000 psi for hard pine (southern yellow pine)
 = 800 psi for spruce and white oak
 = 700 psi for white pine
 = 13,000 psi for cast iron

T = 2,000 psi for hard pine and oak
 = 1,600 psi for spruce
 = 1,400 psi for white pine
 = 10,000 psi for wrought iron
 = 12,500 psi for steel

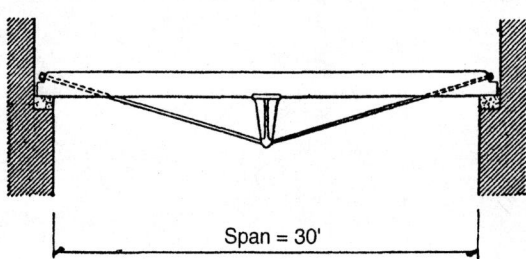

Figure 4-23. The design includes:

Three members for the girder, $5\tfrac{1}{2} \times 12$ in. (actual dimensions southern yellow pine)
Two "belly rods $1\tfrac{7}{8}$ in. diameter, wrought iron or steel
Cast iron strut 1×2 in. at small end

HISTORIC AMERICAN FLOOR SYSTEMS—BEAMS ◇ 293

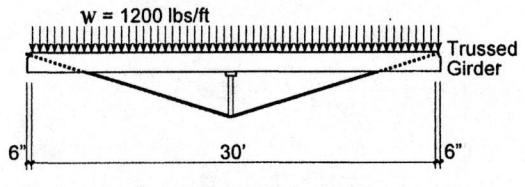
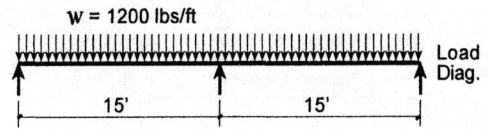
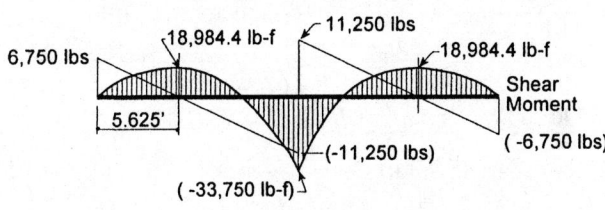
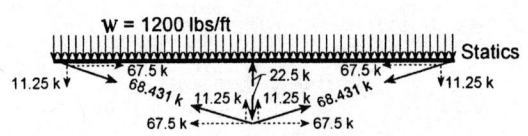
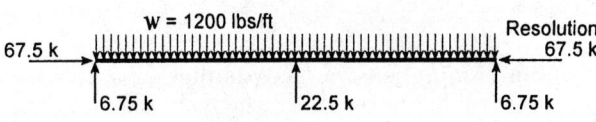

The Modern Analysis would also check for bending and compressive combined stresses.
The Girder is actually a two-span member if the rods and strut combine to act as a column or support at the center of the member.

Checking the Combined Stresses:

$$f_a = \frac{\text{Compressive Force in Mbr.}}{\text{Area of Member}} = \frac{67{,}500 \text{ lbs}}{3\,(5.5")(12")}$$

$$= 340.9 \text{ psi (Compression)}$$

$$f_b = \frac{M_{max}}{S} = \frac{33{,}750 \text{ lb-ft }(12"/')}{396 \text{ in}^3} \quad S = \frac{b(h)^2}{6} = \frac{3(5.5")(12")^2}{6}$$

$$= 1022.7 \text{ psi (T \& C)} \quad = 396 \text{ in}^3$$

The Combined Stress Formulas:

$$\frac{f_a}{F_a} + \frac{f_b}{F_b} \le 1$$

$$\frac{340.9 \text{ psi}}{1100 \text{ psi}} + \frac{1022.7 \text{ psi}}{1500 \text{ psi}} \le 1$$

$$0.310 + 0.682 = 0.992$$

The Allowable Stresses in Compression and Bending for Southern Yellow Pine are from the Table of Allowable Stresses, Table 4-18.

The Allowable Stresses do not need to be reduced for reasons of Lateral Bracing. Column Buckling, Etc. since the Joists or decking are providing that type of support.

The Design Checks and is O.K.

Figure 4-24.

A = 100 psi for hard pine (yellow pine)
  = 90 psi for Oregon pine
  = 70 psi for spruce
  = 60 psi for white pine

Kidder provides an example of the design of a trussed girder utilizing his method:

Single strut girder: southern yellow pine, wrought iron rods, cast iron strut
  Simple span = 30 ft, girders 12 ft on centers
Load = 100 psf = 1200 lb/in. ft.
  Use a three-strut beam with two tension rods.
  Beam depth = 12 in. Struts go 2 ft below beams.
  Dimension C = 30 in. (center of beam to apex of rods)
  Dimension B = 15 ft Dimension T = 15 ft, 2 in. or 182.5 in.

$$\text{Tension in rod} = 0.3(W)\left(\frac{\text{length of }T}{\text{length of }C}\right) = 0.3(36{,}000 \text{ lb})\left(\frac{182.5 \text{ in.}}{30 \text{ in.}}\right)$$

$$= 65{,}700 \text{ lb}$$

$$\text{Diameter of each rod} = \sqrt{\frac{\text{tension in single rod}}{9{,}425}} = \sqrt{\frac{32{,}850 \text{ lb}}{9{,}425}} = 1.867 \text{ in.} = 1\tfrac{7}{8} \text{ in.}$$

Compression in Beam:

$$B_C = 0.30(W) \frac{\text{length of } B}{\text{length of } C} = 0.3(36{,}000 \text{ lb}) \left(\frac{180 \text{ in.}}{30 \text{ in.}}\right) = 64{,}000 \text{ lb}$$

$$\text{Compression in each member} = \frac{64{,}800 \text{ lb}}{3} = 21{,}600 \text{ lb/member}$$

$$\text{Member size for compression} = \frac{21{,}600 \text{ lb}}{1{,}000 \text{ lb/in.}^2}$$
$$= 21.6 \text{ in.}^2 = 2 \times 12 \text{ in. each minimum}$$

$$\text{Breadth of beam from load} = \frac{18{,}000 \text{ lb}}{3} = 6{,}000 \text{ lb (load of each horizontal strut)}$$

$$\text{Breadth} = \frac{W(L)}{2(d^2)A} = \frac{6{,}000 \text{ in. (15 ft)}}{2(12 \text{ in.})^2(100 \text{ lb/in.}^2)} = 3.125 \text{ in.}^2 = 3\tfrac{1}{2} \text{ in.}$$

The total breadth of the three members is 5.5 in. (each member) (the addition of 2 in. and 3½ in.)

Design of strut at center: C, strut = 0.625(36,000) = 22,500 lb

$$\text{Cast iron, area} = \frac{22{,}500 \text{ lb}}{13{,}000 \text{ lb/in.}^2} = 1.8 \text{ in.}^2$$

$$\text{Wood, oak, area} = \frac{22{,}500 \text{ lb}}{1{,}000 \text{ lb/in.}^2} = 22.5 \text{ in.}^2 \text{ (if strut is constructed of oak)}$$

The Kidder method appears to be a comprehensive analysis that separates the compressive and bending forces on the three beam sections, which should allow the combined stresses to stay within the allowable.

The example problem that was worked by the Kidder method will now be checked by statics and modern analytical methods. The modern analysis would also check for bending and compressive stresses. The girder is actually a two-span member if the rods and strut combine to act as a column or support at the center of the member.

> Statics: We know from beam statics that the reaction at the center of a two-span continuous member uniformly loaded over the entire spans is 0.625(W) and that the reaction at the ends is 0.1875(W) at each end. Therefore, the compression in the vertical strut at the center is 0.625(W). In this case, 0.625(36,000 lb) or 22,500 lb. From that:
> The vertical strut is in compression and actually acts as a column holding up the center of the beam. The strut is held in place by (or supported by) the tension rods that go from the bottom of the strut to the outside ends of the beam.
> The diagonal tension of the rods produces a vertical component of force that provides the support for the strut. The diagonal also produces a horizontal component of force which results in the compression of the beams.

The Design checks and is o.k. Kidder (1912) has a revised set of formulae for the solution of trussed girders. In each of the formulae for the tension or compression in the truss chords of uniformly loaded single strut girders (both the 'underbelly' and the 'overhead' type), the factor preceeding W is changed to 0.50. There is also a new statemeent that if the horizontal chord is "jointed at midspan" the force in the strut is 0.5W (removing the continuity). In double strut trussed girders, the factor of 0.367 preceeding W (in each formula) in the uniformly loaded trusses is changed to 0.333(W). W is the total load over the 30 ft span.

The forces in the members generated by the Kidder (1902) formulae are a small amount less than the actual member forces as generated by the statics method. The revised-formulae from 1912 give member forces that are quite a lot higher than the statics method provides.

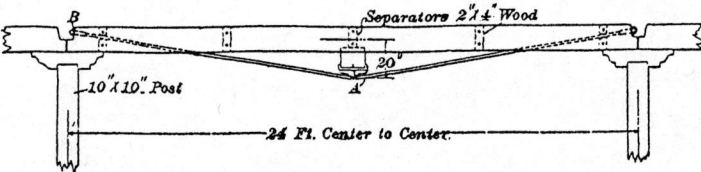

Figure 4-25. Trussed girders with member separators. (*From International Correspondence Schools 1924, para. 7, 92.*)

Table 4-23. Rods and Turnbuckles for Trussed Girders[a]

Upset Screw Ends for Round Bars
American Bridge Company Standard

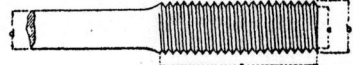

Pitch and Shape of Thread A. B. Co. Standard

Bar		Weight per ft (lb)	Upset				Area	
Diameter d (in.)	Area, (in.²)		Diameter b (in.)	Length a (in.)	Additional Length for Upset +10% (in.)	Diameter at Root of Thread c (in.)	At Root of Thread (in.²)	Excess Over Area of Bar (%)
¾	0.442	1.50	1	4	4	0.838	0.551	24.7
⅞	0.601	2.04	1¼	4	5	1.064	0.890	48.0
1	0.785	2.67	1⅜	4	4	1.158	1.054	34.2
1⅛	0.994	3.38	1½	4	4	1.283	1.294	30.2
1¼	1.227	4.17	1⅝	4	4	1.389	1.515	23.5
1⅜	1.485	5.05	1¾	4	4	1.490	1.744	17.5
1½	1.767	6.01	2	4½	4½	1.711	2.300	30.2
1⅝	2.074	7.05	2⅛	4½	4	1.836	2.649	27.7
1¾	2.405	8.18	2¼	5	4	1.961	3.021	25.6
1⅞	2.761	9.39	2⅜	5	4	2.086	3.419	23.8
2	3.142	10.68	2½	5½	4	2.175	3.716	18.3
2⅛	3.547	12.06	2⅝	5½	3½	2.300	4.156	17.2
2¼	3.976	13.52	2⅞	6	4½	2.550	5.108	28.4
2⅜	4.430	15.06	3	6	4½	2.629	5.428	22.5
2½	4.909	16.69	3¼	6½	5½	2.870	6.509	32.6
2⅝	5.412	18.40	3¼	6½	4½	2.879	6.509	20.3
2¾	5.940	20.19	3½	7	5½	3.100	7.549	27.1
2⅞	6.492	22.07	3¾	7	6	3.317	8.641	33.1
3	7.069	24.03	3¾	7	5	3.317	8.641	22.2
3⅛	7.670	26.08	4	7½	6	3.567	9.993	30.3
3¼	8.296	28.21	4	7½	5	3.567	9.993	20.5
3⅜	8.946	30.42	4¼	8	5½	3.798	11.330	26.6
3½	9.621	32.71	4¼	8	5	3.798	11.330	17.8
3⅝	10.321	35.09	4½	8½	5½	4.028	12.741	23.4
3¾	11.045	37.55	4¾	8½	6	4.255	14.221	28.8
3⅞	11.793	40.10	4¾	8½	5½	4.255	14.221	20.6

Table 4-23. *(Continued)*

Turnbuckles and Sleeve Nuts
American Bridge Company Standard
All Dimensions in Inches

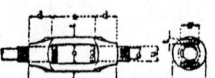

Turnbuckles | Sleeve Nuts

2 = 6", 2 = 9" for turnbuckles marked[b].
Pitch and shape of thread, A. B. Co. Standard.

Pitch and shape of thread, A. B. Co. Standard

Diam. of Screw u	Standard Dimensions						Weight (lb)	Diam. of Screw u	Standard Dimensions						Weight (lb)
	d	l	c	t	g	b			d	l	a	b	c	t	
3/8	9/16	7 1/8	9/16	3/16	1/2	1 1/16	1								
7/16	21/32	7 5/16	5/8	1/4	5/8	1 3/8	1								
1/2	3/4	7 1/2	5/8	1/4	5/8	1 3/8	1								
9/16	27/32	7 11/16	13/16	5/16	3/4	1 9/16	1 1/2								
5/8	15/16	7 7/8	13/16	5/16	3/4	1 9/16	1 1/2								
3/4	1 1/8	8 1/4	1 1/16	11/32	7/8	2	2								
7/8	1 3/16	8 5/8	1 1/4	3/8	1	2 1/4	3	7/8	1 1/2	7	1 5/8	1 7/8	1 1/8	1/4	3
1	1 1/2	9	1 5/16	7/16	1 1/4	2 7/16	4	1	1 1/2	7	1 5/8	1 7/8	1 1/8	1/4	3
1 1/8	1 11/16	9 3/8	1 7/16	1/2	1 1/4	2 9/16	5	1 1/8	1 3/4	7 1/2	2	2 5/16	1 3/8	5/16	4
1 1/4	1 7/8	9 3/4	1 9/16	1/2	1 1/2	2 3/4	6	1 1/4	1 3/4	7 1/2	2	2 5/16	1 3/8	5/16	4
1 3/8	2 1/16	10 1/8	1 11/16	1/2	1 5/8	3 1/16	7	1 3/8	2	8	2 3/8	2 3/4	1 5/8	3/8	5
1 1/2	2 1/4	10 1/2	1 3/4	5/8	1 3/4	3 3/16	8	1 1/2	2	8	2 3/8	2 3/4	1 5/8	3/8	6
1 5/8	2 7/16	10 7/8	2	5/8	1 7/8	3 1/2	10	1 5/8	2 1/4	8 1/2	2 3/4	3 3/16	1 7/8	7/16	8
1 3/4	2 5/8	11 1/4	2 1/8	5/8	2	3 3/4	11	1 3/4	2 1/4	8 1/2	2 3/4	3 3/16	1 7/8	7/16	9
1 7/8	2 13/16	11 5/8	2 3/16	11/16	2 1/8	3 7/8	12	1 7/8	2 1/2	9	3 1/8	3 5/8	2 1/8	1/2	10
2	3	12	2 3/8	11/16	2 1/4	4 1/4	14	2	2 1/2	9	3 1/8	3 5/8	2 1/8	1/2	11
2 1/8	3 3/16	12 3/16	2 1/2	22/32	2 1/2	4 1/2	17	2 1/8	2 3/4	9 1/2	3 1/2	4 1/16	2 3/8	9/16	14
2 1/4	3 3/8	12 3/4	2 11/16	13/16	2 1/2	4 3/4	20	2 1/4	2 3/4	9 1/2	3 1/2	4 1/16	2 3/8	9/16	15
2 3/8	3 9/16	13 1/8	2 3/4	13/16	2 3/4	4 7/8	22	2 3/8	3	10	3 7/8	4 1/2	2 5/8	5/8	18
2 1/2	3 3/4	13 1/2	3 1/16	27/32	3	5 3/8	25	2 1/2	3	10	3 7/8	4 1/2	2 5/8	5/8	19
2 3/4	4 1/2	14 1/4	3 1/4	15/16	3 1/4	5 3/4	33	2 3/4	3 1/4	10 1/2	4 1/4	4 13/16	2 7/8	11/16	23
2 7/8	4 9/16	14 5/8	3 7/16	1 1/32	3 1/4	6 1/16	36	2 7/8	3 1/2	11	4 5/8	5 3/8	3 1/8	3/4	27
3	4 1/2	15	3 5/8	1 1/32	3 1/2	6 3/8	40	3	3 1/2	11	4 5/8	5 3/8	3 1/8	3/4	28
3 1/4	4 7/8	15 3/4	3 7/8	1 1/16	4	6 3/4	50	3 1/4	3 3/4	11 1/2	5	5 13/16	3 3/8	13/16	35
3 1/2	5 1/4	16 1/2	4 1/4	1 7/32	4	7 1/4	65	3 1/2	4	12	5 3/8	6 1/4	3 5/8	7/8	40
3 3/4	5 5/8	17 1/4	4 7/16	1 5/16	5	8 1/4	95	3 3/4	4 1/4	12 1/2	5 3/4	6 11/16	3 7/8	15/16	47
4	6	18	4 5/8	1 7/16	5	8 3/4	108	4	4 1/2	13	6 1/8	7 1/16	4 1/8	1	55
4 1/4[b]	6 1/4	21 1/2	4 5/8	1 5/8	5 5/32	9 1/4	140	4 1/4	4 3/4	13 1/2	6 1/2	7 1/2	4 3/8	1 1/16	65
4 1/2[b]	6 3/4	22 1/2	5 1/2	1 3/4	6 1/2	10 3/4	195	4 1/2	5	14	6 7/8	7 13/16	4 3/4	1 1/16	75
4 3/4[b]	7 1/4	23 1/2	5 5/8	2	6 1/2	11 1/4	205								
5	7 1/2	24	6	2 1/4	6 1/2	11 7/8	250								

[a] From Carnegie Steel Co. 1920, 149, 153.
[b] Special.

The forces computed by the 1912 formula are 60 percent higher than those found using the statics method, and thus the members chosen to respond to the magnified forces are larger than required. Thus, if the formulae of 1912 were widely adopted, the resulting truss girders would be very conservative and our calculations checking those trusses would indicate that.

The Kidder method checks when compared to a modern method of analysis; however, both methods simplify the solution, and some other considerations are necessary. Voss and Varney (1926) utilize the Kidder method, but they state that the design method is only approximate because the configuration of the system changes

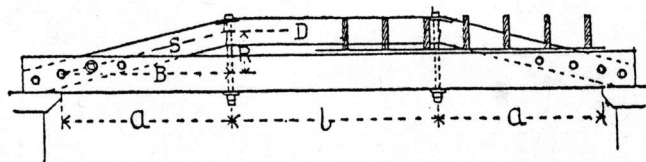

Figure 4-26. Trussed girders. (*From Kidder 1900, 451.*)

under load and the analysis may be inaccurate after the deflection of the girder occurs. The girders tend to act like simple spans rather than a two-span continuous member, assuming that the strut does not move (meaning the overall beam does not deflect). The ends of the members must be checked for bearing compressive stresses perpendicular to the grain of the wood at the points of support. The same applies at midpoint, where the strut provides the support at midspan. As long as the designs are conservative, the trussed girders are very good design solutions.

The three members that make up the girder are separated to allow the wrought iron rods to pass between them (see Figure 4-19), and they should have spacers along their length to keep the alignment constant to allow the compression and bending loads to be carried by a contiguous unit (see Figure 4-25).

The belly rods or tension rods would have been wrought iron until well after the turn of the century. The rods, whether wrought iron or steel, would be sized by calculation of their actual diameter and cross-sectional area, and the threaded portions would be done on upset sections so as not to have the root diameter of the threads actually reduce the effective diameter of the rod. The rods specified in the Kidder design are wrought iron, and the allowable tensile stress is 12,000 psi. The rods used for tension rods in trussed girders were the same ones used for diagonal bracing of buildings and roof trusses (see Table 4-23).

Hundreds of buildings were constructed with trussed girders of every type. These very innovative systems were excellent for that midrange of spans that were accompanied by relatively medium-sized loads and where costs were considered and the need for steel skeletal construction was not warranted. Another type of trussed girder that was often used had a compression chord above the girder and thus the girder

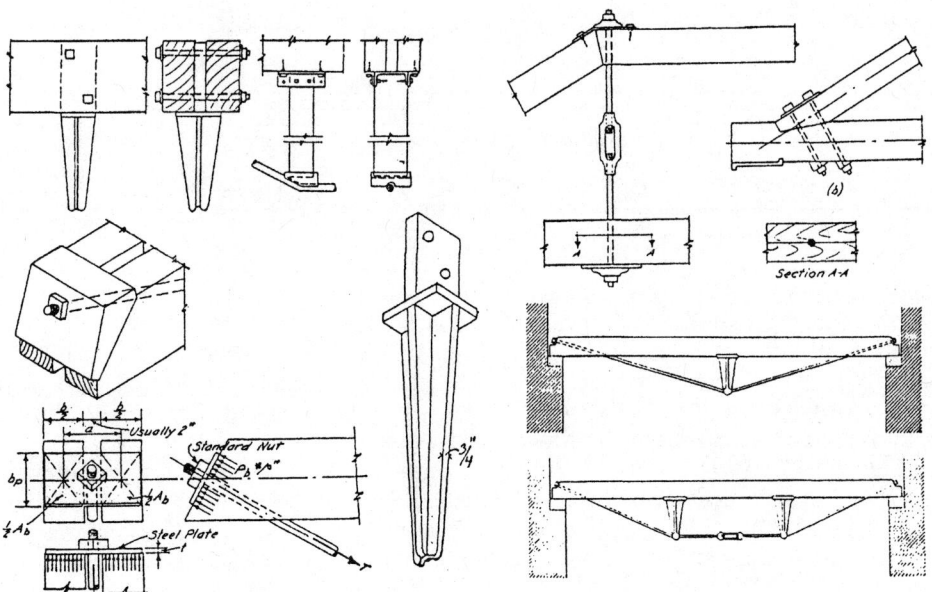

Figure 4-27. Trussed girder details. (*From Kidder 1900, 450; International Correspondence Schools 1899B, para. 5, 101, 102; Voss and Varney 1926, 49, 50.*)

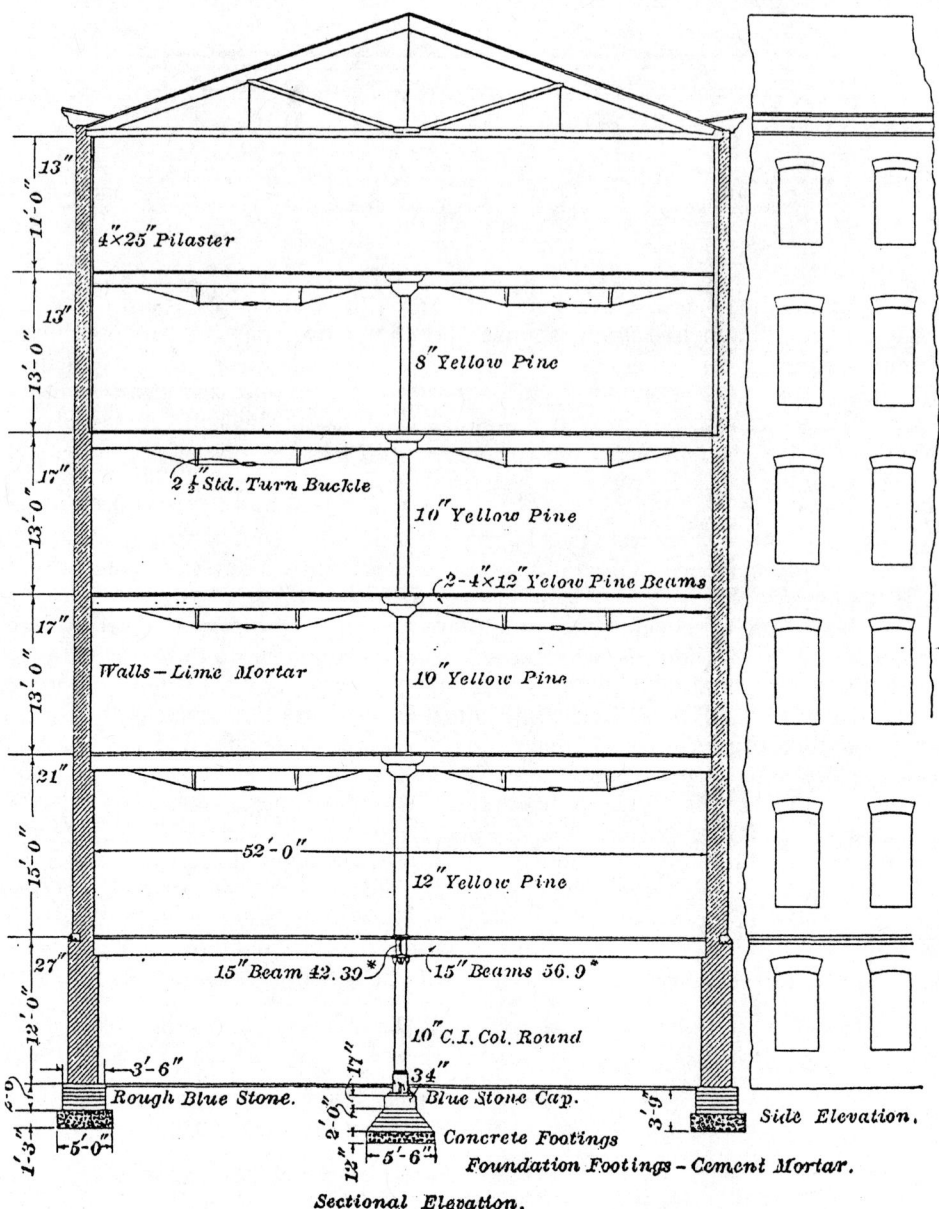

Figure 4-28. Section through a building, showing walls, framing, trussed girders, etc. The 15 in. beams at the first level are fireproof construction (see Figure 4.30). (*From Huntington 1929, 6.*)

members (minimum two) would be in tension when under the influence of load. Figure 4-26 shows a very compact type of this trussed girder. Note that the floor joists and the vertical height of the upper chord (tops of chord and joists) are coordinated to conserve space both above and below the trussed girder.

Figure 4-27 shows details of many of the truss girder connections, separators, rod end plates, and vertical struts that were used by designers from the 1880s through the 1930s, when for all practical purposes the use of these systems ceased. Figures 4-28 and 4-29 are large sections through buildings that used trussed girders.

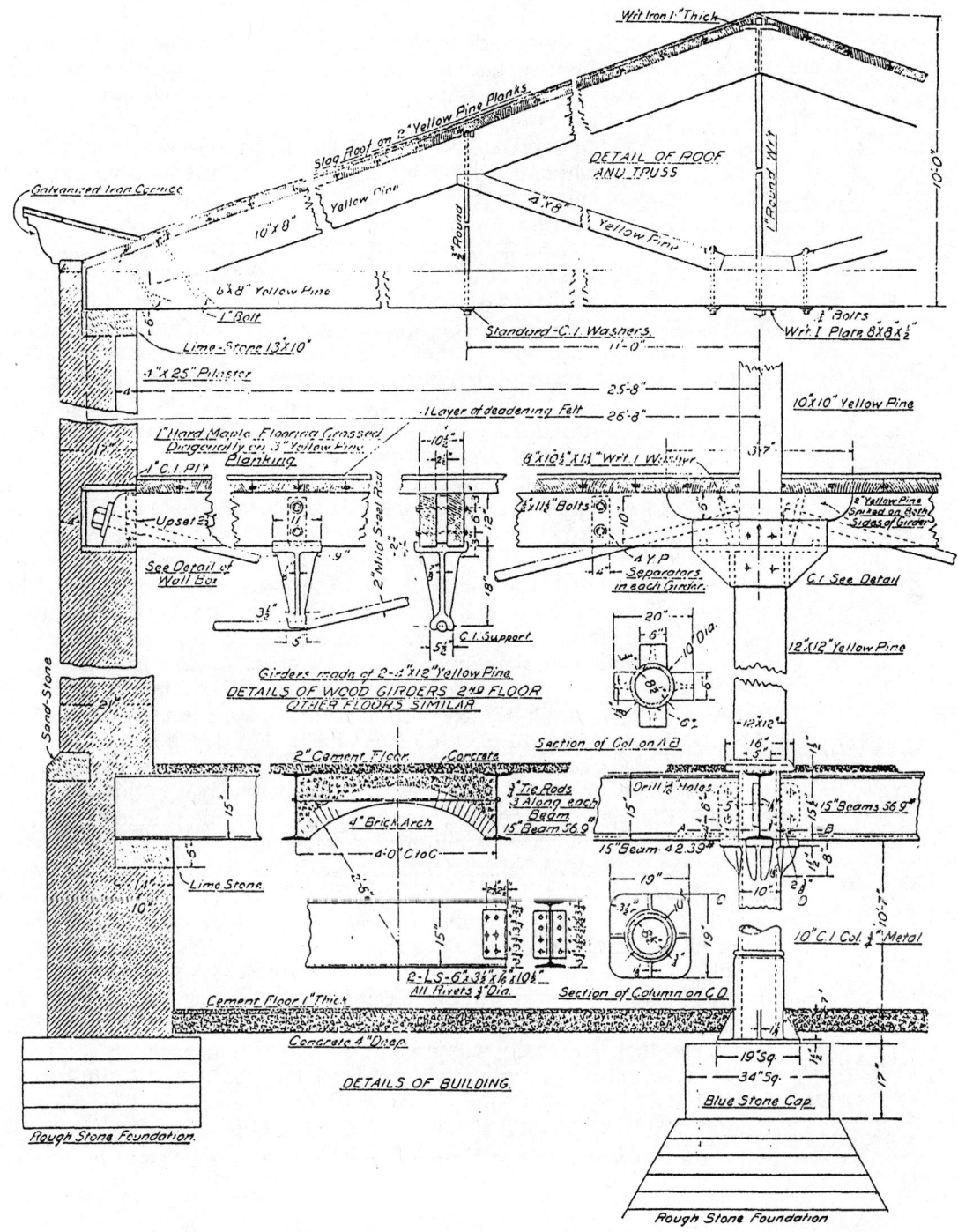

Figure 4-29. Details, building section of Figure 4-29. (*From Trautwine 1888, 552.*)

FIREPROOF CONSTRUCTION: FIREPROOF FLOORS

Fireproof construction encompasses many material types and construction systems. The need for "fireproof" buildings was created by the great losses in money and lives to fires in buildings. Fires have been a problem in buildings since almost the beginning of construction, with buildings being made of the most flammable construction material, wood. Fire was so feared that in the original Colonial Capitol at Williamsburg, 1707, fireplaces and candlelight were both forbidden by law. The loss of land records and other documents was feared. In the mid-1720s, the records were found to be deteriorating due to dampness. Fireplaces were added to the Capitol, and it burned in 1747. It was rebuilt and later burned again. Every city in America has had one or more major fires. The Chicago Fire of 1871 caused probably the largest loss of property and materials of any fire in the world. Most of the buildings that burned were constructed of wood or wood and brick. The losses were so great that concerned building owners and city officials wanted the buildings reconstructed in a manner that would eliminate the hazard of total destruction in the future. The spread of fire was as much a concern to those involved in the design, construction, ownership, and administration of buildings as fire itself. Many things were happening in parallel at this time, and for the next 10–20 years building developments were so interconnected that one would not have developed without the other. The growing use of wrought iron and steel and the tall buildings, 12–20 stories of metallic frame, required protection of the frame elements, columns, and beams. A frame that would be the entire means of support for a tall building would cause a complete building failure if an intense fire at a lower level attained a temperature of 1300°F or higher, as wrought iron and steel lose half their strength at that temperature. Most fires of any magnitude can reach 2000°F within a relatively short span of time.

Fireproofing is not totally attainable. It would dictate that no combustible materials could be within the building in the construction materials, no matter how minor. Wood windows, window casings, door casings, trim, and wood and textile furniture could not be allowed. The elimination of all combustibles would not be possible as the principal material of the office function is paper. Huntington (1929) states that fireproof construction became known as that type of construction that includes buildings of masonry, steel, or reinforced concrete construction. No wood or other combustible materials could be used as structural elements in fireproof buildings, except wooden finish flooring and interior doors and windows with their frames, trim and casings. In the best class of fireproof construction, no wood whatsoever is used, the floors being of fireproof material and all doors, door and window frames, casings, etc. being of metal (ibid., 6). Floors were the most vulnerable to fire, especially in lighter buildings of 3–12 stories, which is exactly the type of building that predominated in Chicago and most other cities and towns in the United States in the 1870s and 1880s. These buildings would have multiple floor joists and two to three layers of 1 in. (usual) thick decking and flooring, creating a high degree of fire potential. The heavy timbers of the mill construction types would not be as inclined to fire because of their mass. The first fireproofing attempts werein floor construction, where designers used wrought iron beams and brick arches (see Figures 4-30 and 4-31).

Brick arches and corrugated iron arches were the systems employed in the first fireproof floors. It is not known which came first, but probably it was the brick because designers would look first to a traditional material where it would fit the requirement because of their familiarity with and knowledge of that material. The first fireproof arches consisted of I-beams placed on about 5 ft centers with 4 in. brick arches sprung between the I-beams. The upper surfaces of these arches were then filled with beton (a form of early concrete made of a rich mixture of lime-sand-aggregate), leveled to a surface just below finished floor. If the finished floor was to be wood, sleepers

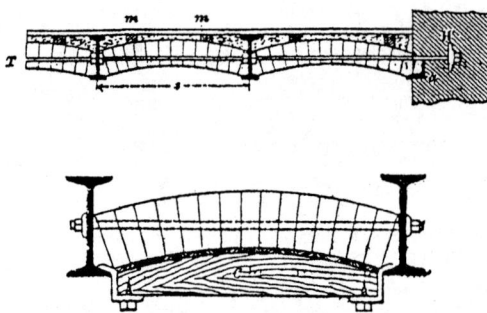

Figure 4-30. Fireproof floor systems. (*From Trautwine 1888, 522.*)

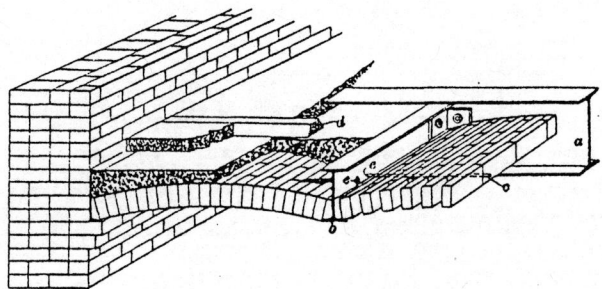

Figure 4-31. Details of brick arch fireproof floor. (*From International Correspondence Schools 1905, para. 22, 15.*)

(nailing strips of wood) were used as a combination leveling strip and nailing strip. Corrugated sheet metal forms, which could be sprung into the space between I-beams, would serve two useful purposes: they would eliminate the false formwork or centering that had to be used on brick arches. Corrugated iron, which is sheet steel, would eliminate the need for the brick. The beton could be poured directly on the corrugated iron to form the floor in the same way as with brick arches (see Figure 4-32).

These early arches had one common problem, their self-weight. Arched floors were much heavier than wooden floors, usually four to five times as heavy. The safety and fire resistance of the system made it an instant success, and designers quickly embraced it. The cost of this type of construction was considered to be quite high, as would be expected. A larger labor factor was involved, and the actual cost of materials was also much more than with wood. Wood had the positive cost factor in that the product itself was structural as well as very easy to finish. Brick and beton were fire-resistant, and that value could not be discounted. The early arches were fireproof and would prevent the flow or spread of fire, but the solution needed by designers and owners of buildings was still not at hand. Each of the systems presented to this point, and a couple yet to be presented, had one feature that made it imperfect: the exposed (and vulnerable) bottom flanges of the I-beams. A fire in a space could be intense enough to bring down the arch system if the heat buildup was enough to reduce the strength of the I-beams. The weight of the floor systems could cause a severe sag or actual failure of the floor when temperatures melted the tension flanges of the I-beams.

Reduction of the weight of the arches was paramount, and designers and manufacturers worked to devise schemes to make the systems more efficient. Also, the overall depth of some of the early systems would be of some concern. Terra cotta shapes were emerging in partition construction as well as the exterior cladding of buildings. The ability to have strong sections of terra cotta tiles that had hollow cavities would significantly reduce the weight of the overall floor system. The first terra cotta flat arches were used in the Equitable Building, Chicago, 1872, a little over a year after the great fire (see Figure 4-33). As with most new products of this type, they were crude and underdeveloped at first, and they were patented by the developers.

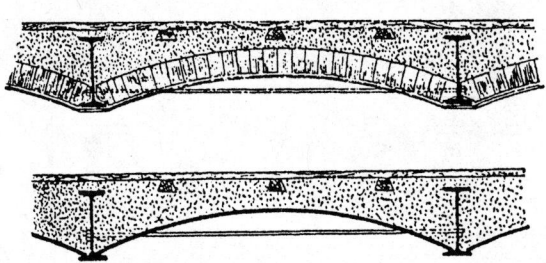

Figure 4-32. Brick arch and corrugated iron arch. (*From Freitag 1909, 90.*)

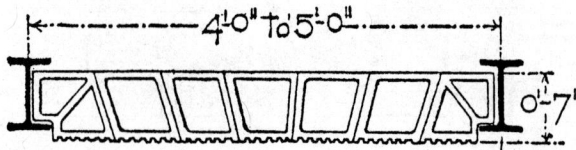

Figure 4-33. Terra cotta flat arch: Equitable Building, Chicago, 1872. (*From Freitag 1909, 9.*)

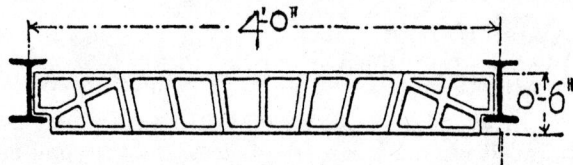

Figure 4-34. Flat arches: Montauk Building, Chicago, 1881. (*From Freitag 1909, 9.*)

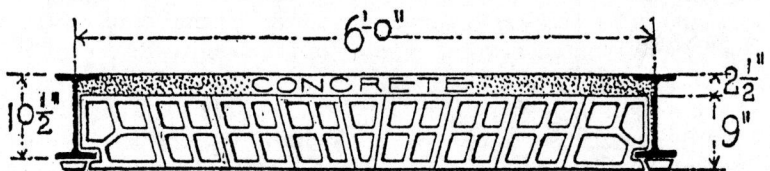

Figure 4-35. Terra cotta arch: Home insurance Building, Chicago, 1884. (*From Freitag 1909, 9.*)

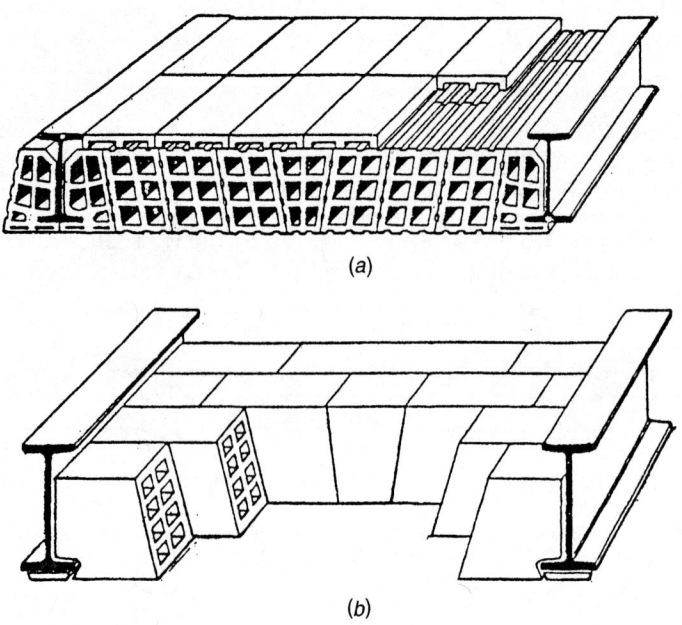

Figure 4-36. Types of terra cotta elements: (*a*) side construction terra cotta arch; (*b*) end construction terra cotta arch. (*From Freitag 1909, 97, 99, 101.*)

Table 4-24. Side Construction Terra Cotta Arches

Depth (in.)	Width of Span (ft)	Weight/ft² (lb) Hd. Burn	Weight/ft² (lb) Porous
6	3–4	27	25
7	4–4.5	29	26
8	4.5–5	32	28
9	5–6	37	32
10	6–6.5	40	36
12	6.5–7	44	40

Table 4-25. End Construction Terra Cotta Arches

Depth (in.)	Maximum Span (ft)	Weight/ft² (lb)
6	4.5	29
8	5.5	31
9	6	32
10	6.5	33
12	7	39
15	8	46

This system did not quite expose the tension flanges of the I-beams, as there was a plaster coating placed on the bottom surface of the tiles and through and across the area of the beam flanges. The plaster would provide only limited protection of the members, as it would crack and fall away from the ceiling in the locations of the I-beam flanges. Another building, the Post Office Building in New York City, utilized a different type of terra cotta arch at approximately the same time as the Equitable in Chicago. These early forms of brick, corrugated iron arches, terra cotta systems were obviously being developed prior to the great Chicago Fire. There is almost no evidence of any masonry or beton systems in use in Chicago prior to the fire. The Montauk Building, Chicago, 1881, was said to be the "first building of modern design in Chicago" (Freitag 1909, 90). The Montauk Building's arches were slightly improved and more perfected, but they still left only a thick plaster coating on the bottom flanges of the I-beams (see Figure 4-34). The weight of the arches was critical. The earliest brick arches, shown in Figures 4-30, 4-31, and 4-32, weighed approximately 75 psf, while the terra cotta tile arches averaged approximately half that amount. This was a very significant reduction in the dead load, and it made buildings much more reasonable in their dead loads and significantly reduced the costs. The difference in the dead loads was significant in that these weights reflected in the column sizes and cost and the footing sizes and cost. The first terra cotta arch to afford protection to the bottom flanges of the I-beams was in the Home Insurance Building, Chicago, in 1884. This system included a beveled bottom section of the arch segment against the I-

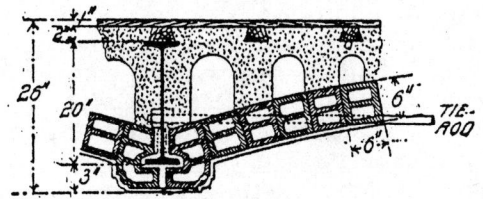

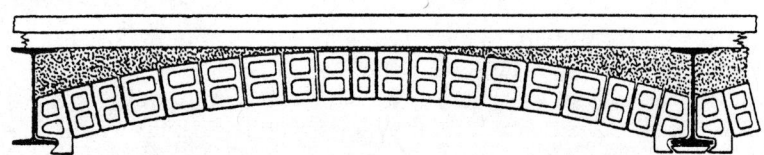

Figure 4-37. Segmental terra cotta arch. (*From Freitag 1909, 102, 103.*)

Table 4-26. Spacing of Tie Rods for Floor Arches[a]

Span of Arch (ft)	Diameter of Tie Rods (in.)	Spacing, in Feet, for a Uniform Load of 100 psf, in Addition to the Weight of the Arch							
		Nominal Depth of Arch (in.)							
		6	8	9	10	12	14	15	16
		Effective Depth or Rise of Arch (in.)							
		3.6	5.6	6.6	7.6	9.6	11.6	12.6	13.6
3	5/8	7.70	11.43	13.07	14.62	17.47	20.29	21.75	23.18
3	3/4	11.50	17.08	19.54	21.86	26.12	30.33	32.52	34.66
3	7/8	16.00	23.76	27.18	30.40	36.32	42.18	45.23	48.20
4	5/8	4.33	6.43	7.35	8.22	9.83	11.41	12.24	13.04
4	3/4	6.47	9.61	10.99	12.30	14.69	17.06	18.29	19.50
4	7/8	9.00	13.36	15.29	17.10	20.43	23.73	25.44	27.11
5	5/8	2.77	4.11	4.71	5.26	6.29	7.30	7.83	8.35
5	3/4	4.14	6.15	7.03	7.87	9.40	10.92	11.71	12.48
5	7/8	5.76	8.55	9.78	10.94	13.08	15.19	16.28	17.35
6	5/8		2.86	3.27	3.66	4.37	5.07	5.44	5.80
6	3/4		4.27	4.89	5.46	6.53	7.58	8.13	8.66
6	7/8		5.94	6.79	7.60	9.08	10.55	11.31	12.05
7	5/8			2.40	2.69	3.21	3.73	4.00	4.26
7	3/4			3.59	4.01	4.80	5.57	5.97	6.37
7	7/8			4.99	5.58	6.67	7.75	8.31	8.85
8	5/8				2.06	2.46	2.85	3.06	3.26
8	3/4				3.07	3.67	4.27	4.57	4.87
8	7/8				4.28	5.11	5.93	6.36	6.78

[a] Reprinted with permission from Bethlehem Steel Co., *Bethlehem Structural Shapes,* Catologue S-18, pp 189–191

beam that was set to retain an insert in the form of a section with beveled edges, which was simple but effective in the protection of the flange (see Figure 4-35). Most of the arches and flat arches, sometimes called "jack arches," were fitted around the flange and into and against the web of the beams that framed the floor. The tops of the terra cotta were usually set 2–3 in. below the top flange of the beam for a concrete topping slab. In addition, all arches have the common characteristic of exerting outward thrust on the points that they bear. The arched flooring systems require tension rods between beams to prevent the horizontal thrust from separating the beams and opening up the span and causing the arch to fail. Figures 4-30, 4-31, and 4-32 show the tension rods and how the last one is secured in a wall. Between beams, the rods are staggered for easier adjustment. The rods are permanent and remain built into the flooring system.

The terra cotta arch factories were constantly working to improve their product. The desire was to maintain the required strength, which is practically empirical, and lighten the weight of the sections. One notable improvement was to reduce the density of the actual material itself by mixing sawdust and finely cut straw into the clay that was used to form the terra cotta. In the process of firing the clay shapes, the sawdust and straw would burn away in the extreme temperatures and leave a porous lighter element. Manufacturers were always working to improve their systems and attract more of this market. Many plants manufactured brick and terra cotta, some chose to produce end construction-type arch elements, and others produced side construction arch elements. Each type required several precise shapes and types (see Figure 4-36. Manufacturers pushed their products by advertising their weight, safe span, and ease of construction. Many had form systems that were quickly assembled, attached to the steel framing, and easily relocated to another section for the next placement of arches. Manufacturers also had systems that allowed leveling of surfaces with top pieces as shown in Figure 4-36. Hardwood floors were popular at that time, and therefore the most utilized system was to embed floor nailers in the topping slabs to allow for the placement of the hardwoods. In the modern era, designers would take the opportunity

Table 4-27. Bethlehem Tables of Safe Loads for Terra Cotta Arches[a]

Bethlehem Steel Company	187

Example: What will a 10 in. arch carry with a factor of safety of 7 for a span of 6 ft 6 in., the keys having a thickness of ⅝ in. in their shells and webs, and being three cells in depth?

⅝ × 12 × 4 = 30 in.² net = actual area of key
Area of 10 in. tile in table = 36 in.²
30 ÷ 36 = 0.833 ratio of actual area to tabular area
Safe load in table = 198 lb × 0.833 = 165 lb
= safe live and dead load for the actual area

Assuming 38 lb as the weight of arch, then 165 − 38 = 127 lb, which is the net safe live load.

Safe Loads for Segmental Floor Arches of Hollow Tile
As Given by Manufacturers of this Material
Safety factor = 7
Weight of arch blocks not included

Arch Spans		Depth of Arch = 6 in.					
		Area of Arch = 36 in.²					
		Rise of Arch (in.)					
		¾	1	1¼	1½	1¾	2
ft	in.	Safe Loads (psf)					
4	0	902	1,148	1,485	1,740	1,986	2,233
4	6	792	1,044	1,313	1,539	1,775	1,975
5	0	709	951	1,172	1,379	1,592	1,773
5	6	641	864	1,062	1,266	1,439	1,619
6	0	585	788	969	1,154	1,315	1,476
6	6	551	724	902	1,058	1,218	1,358
7	0	508	669	834	981	1,127	1,264
7	6	471	621	774	920	1,049	1,176
8	0	439	588	724	859	987	1,099
8	6	411	551	678	806	926	1,037
9	0	386	518	645	758	871	977
9	6	364	489	608	721	823	923
10	0	344	462	576	683	784	879

to provide Spanish tile or bluestone flagstone, which would be perfect for the arched masonry.

The arches were tabulated as to their depth, span width, and weight, as shown in Tables 4-24 and 4-25.

Segmental terra cotta arches were also quite popular. They were designed to span larger distances than the conventional flat arches or jack arches. Segmental arch floor systems were considered the systems to use on warehouses, factories, or breweries, where heavy floor loads were required and ceiling appearances were not a concern. Segmental arches were most often side-construction elements, 4, 5, 6, or 8 in. square × 12 in. long (see Figure 4-37).

Designers had to figure out how to size and space the tie rods for the arch systems. For a period at the beginning of the use of these types of flooring systems, the tie rods were empirically sized and spaced. Later there were methods and formulae for the design of the tie rods. Arches and the thrust from arches are still a difficult statical problem, and the empirical methods have not been dramatically improved upon. See Table 4-26.

Table 4-27. (Continued)

| | 188 | Bethlehem Steel Company | | | | |

Safe Loads for Flat Floor Arches of Hollow Tile
As Given by Manufacturers of this Material
Safety factor = 7
Total dead and live load

Arch		Depth (in.)					
		6	8	9	10	12	15
Spans		Net Sectional Area (in.²)					
		27	27	27	36	36	36
ft	in.	Safe Loads (psf)					
3	0	420	560	630	933	1,120	1,400
3	3	357	477	537	795	954	1,193
3	6	308	411	462	685	823	1,028
3	9	268	358	403	597	716	895
4	0	236	315	354	525	630	786
4	3		279	314	465	558	697
4	6		249	279	415	497	622
4	9		223	251	372	447	558
5	0		201	227	336	402	504
5	3		182	205	305	365	457
5	6			187	277	333	417
5	9			171	254	305	381
6	0			157	233	280	350
6	3				214	258	322
6	6				198	238	298
6	9					221	276
7	0					206	257
7	6					178	223
8	0					157	197
8	6						174
9	0						155
9	6						140
10	0						126

[a] Reprinted with permission from Bethlehem Steel Co., *Bethlehem Structural Shapes*, Catalogue S-18, pp. 187, 188.

Spacing of tie rods for terra cotta arches:

$$S = \frac{12,000 \, (a)(R)}{W(L)^2}$$

$$18,000a = S(p) = (S)\frac{3(W)(L^2)}{2R}$$

where:
　　Tension in one rod = $18,000(a)$
　　　　Arch thrust = p (per ft of arch) (see thrust formula below)
　　Rod spacing = S ft
　　　　A = area of one rod, in.²
　　　　L = span of arch, ft
　　　　R = rise of segmental arch, in.
　　　　w = total load, psf

Table 4-28. Safe Loads for Terra Cotta Arches[a]

Segmental Terra Cotta Arches
Manufacturers Standard
Safe Loads in psf
Factor of safety = 7

Span of Arch (ft-in.)	Rise of Arch (in.)	Depth of Arch Blocks (in.) 6 — Area 36	8 — 43	10 — 47	Span of Arch (ft-in.)	Rise of Arch (in.)	6 — 36	8 — 43	10 — 47	Span of Arch (ft-in.)	Rise of Arch (in.)	6 — 36	8 — 43	10 — 47	Span of Arch (ft-in.)	Rise of Arch (in.)	6 — 36	8 — 43	10 — 47
4-0	¾	902	1,078	1,178	7-0	¾	508	606	662	10-0	¾	344	411	449	15-0	¾	225	268	293
	1	1,184	1,414	1,545		1	669	799	873		1	462	552	603		1	302	361	394
	1¼	1,485	1,774	1,939		1¼	834	996	1,089		1¼	576	688	751		1¼	377	450	491
	1½	1,740	2,079	2,272		1½	981	1,171	1,280		1½	683	816	892		1½	447	534	583
	1¾	1,986	2,373	2,593		1¾	1,127	1,346	1,471		1¾	784	937	1,024		1¾	515	616	673
	2	2,233	2,667	2,915		2	1,264	1,510	1,650		2	879	1,050	1,147		2	577	690	754
4-6	¾	792	946	1,034	7-6	¾	471	563	615	10-6	¾	331	396	432	16-0	¾	209	249	272
	1	1,044	1,247	1,363		1	621	741	810		1	438	523	572		1	281	336	367
	1¼	1,313	1,568	1,713		1¼	774	925	1,011		1¼	546	652	713		1¼	353	421	460
	1½	1,539	1,838	2,009		1½	920	1,099	1,201		1½	648	774	846		1½	419	500	546
	1¾	1,775	2,121	2,318		1¾	1,049	1,253	1,369		1¾	744	889	972		1¾	481	575	628
	2	1,975	2,359	2,578		2	1,176	1,405	1,536		2	832	994	1,086		2	540	645	705
5-0	¾	709	847	926	8-0	¾	439	525	573	11-0	¾	315	376	411	17-0	¾	194	232	254
	1	957	1,143	1,249		1	588	703	768		1	421	503	550		1	265	316	345
	1¼	1,172	1,400	1,530		1¼	724	864	944		1¼	519	621	678		1¼	330	394	430
	1½	1,379	1,647	1,800		1½	859	1,026	1,122		1½	617	737	805		1½	392	468	512
	1¾	1,592	1,902	2,078		1¾	987	1,179	1,288		1¾	709	847	925		1¾	452	540	590
	2	1,773	2,118	2,315		2	1,099	1,312	1,434		2	794	948	1,036		2	506	605	661
5-6	¾	641	766	837	8-6	¾	411	491	536	12-0	¾	285	341	372	18-0	¾	182	218	238
	1	864	1,032	1,128		1	551	658	719		1	383	458	500		1	248	296	324
	1¼	1,062	1,269	1,387		1¼	678	810	885		1¼	477	569	622		1¼	310	370	404
	1½	1,266	1,512	1,652		1½	806	963	1,052		1½	566	676	738		1½	370	442	482
	1¾	1,439	1,719	1,879		1¾	926	1,106	1,208		1¾	649	776	848		1¾	425	507	554
	2	1,619	1,933	2,113		2	1,037	1,239	1,354		2	727	869	949		2	477	570	623
6-0	¾	585	699	764	9-0	¾	386	461	504	13-0	¾	261	312	341	19-0	¾	173	206	225
	1	788	941	1,028		1	518	619	677		1	351	419	458		1	233	279	304
	1¼	969	1,157	1,265		1¼	645	770	842		1¼	437	522	570		1¼	293	350	382
	1½	1,154	1,379	1,507		1½	758	906	990		1½	519	620	677		1½	348	416	455
	1¾	1,315	1,570	1,716		1¾	871	1,041	1,137		1¾	596	712	778		1¾	402	480	524
	2	1,476	1,763	1,927		2	977	1,167	1,275		2	670	801	875		2	451	539	589
6-6	¾	551	658	719	9-6	¾	364	435	475	14-0	¾	240	287	313	20-0	¾	163	194	212
	1	724	864	944		1	489	584	638		1	326	390	426		1	221	265	289
	1¼	902	1,077	1,177		1¼	608	726	793		1¼	406	485	530		1¼	277	331	361
	1½	1,058	1,264	1,382		1½	721	862	942		1½	482	575	620		1½	330	395	431
	1¾	1,218	1,455	1,590		1¾	823	983	1,074		1¾	553	661	722		1¾	381	455	497
	2	1,358	1,622	1,772		2	923	1,102	1,204		2	619	740	808		2	427	510	558

Table 4-28. (Continued)

Terra Cotta Arches for Floor Load of 150 psf

Flat Arch — Typical Construction: Bottom of arch below bottom of beam

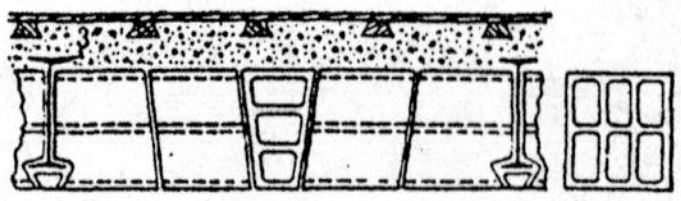

Segmental Arch — Typical Construction: Top of arch level with top of beam

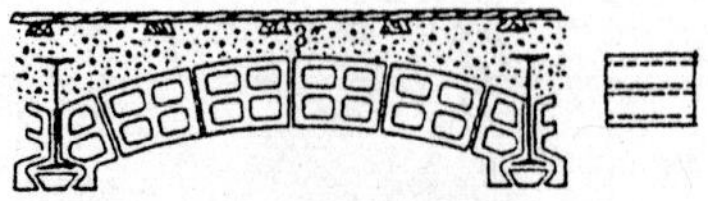

Depth of Beam (in.)	Depth of Arch Blocks (in.)	Depth of Floor (in.)	Span of Arch (ft)	Approx. Weight (psf)						Depth of Beam (in.)	Depth of Arch Blocks (in.)	Rise of Arch (in.)	Span of Arch (ft)	Approx. Weight (psf)					
				Steel	Terra Cotta	Concrete	Flooring	Ceiling	Total					Steel	Terra Cotta	Concrete	Flooring	Ceiling	Total
6	6	11	5¼	6	22	30	4	5	67	6	4	¾	4½	7	20	27	4	5	63
7	6	12	5¼	7	22	38	4	5	76	7	4	1	5	7	20	28	4	5	64
8	6	13	5¼	8	22	45	4	5	84	8	4	1¼	5½	7	20	29	4	5	65
7	7	12	6	8	24	30	4	5	71	9	4	1½	6	8	20	30	4	5	67
8	7	13	6	8	24	38	4	5	79	8	6	¾	5	8	26	27	4	5	70
9	7	14	6	8	24	45	4	5	86	9	6	1	5½	8	26	28	4	5	71
8	8	13	6½	8	27	30	4	5	74	10	6	1¼	6	9	26	29	4	5	73
9	8	14	6½	8	27	38	4	5	82	12	6	1½	6½	9	26	30	4	5	74
10	8	15	6½	8	27	45	4	5	89	10	8	¾	5½	9	31	27	4	5	76
9	9	14	7½	8	29	30	4	5	76	12	8	1	6	9	31	28	4	5	77
10	9	15	7½	9	29	38	4	5	85	12	8	1¼	6½	10	31	29	4	5	79
12	9	17	7½	9	29	53	4	5	100	15	8	1½	7	10	31	30	4	5	80
10	10	15	8	9	31	30	4	5	79	12	10	¾	5¾	10	34	27	4	5	80
12	10	17	8	9	31	45	4	5	94	12	10	1	6½	11	34	28	4	5	82
12	12	17	9½	10	35	30	4	5	84	15	10	1¼	7	11	34	29	4	5	83
15	12	20	9½	10	35	53	4	5	107	15	10	1½	7½	12	34	30	4	5	85
15	15	20	11	12	42	30	4	5	93										

[a] From Carnegie Steel Co. 1923, 267.

Thrusts from Arches:

$$P = \frac{3wL^2}{2R} \quad \text{Uniform load } w, \text{psf}$$

$$P = \frac{3PL}{R} \quad \text{Load } P \text{ at center, lb}$$

Bethlehem Steel Co. (1926) provides tables of safe loads for segmental and flat floor arches of terra cotta tile (see Table 4-27).

Carnegie Steel Co. (1923) contains three tables of safe loads for terra cotta arches, prefaced by the words "Manufacturers' Standard," indicating that there was an association of manufacturers setting standards and providing engineering data and load tables for the sizes and types of arches being produced (see Tables 4-28 and 4-29).

The one problem with all terra cotta arches is the vulnerability of the bottom flange of the I-beam or girder to the heat of the flames. Efforts to protect the flanges were successful in eliminating the problem (see Figures 4-38–4-40, Table 4-30, and Figure 4-41).

Table 4-29. Flat Terra Cotta Arches, Safe Loads[a]

Span of Arch (ft-in.)	Depth of Arch Blocks (in.) Manufacturer's Standard Safe Loads in psf Factor of safety = 7						
	6	7	8	9	10	12	15
	Area of Arch Blocks (in.²)						
	31	34	37	40	43	49	58
3-0	458	588	735	901	1,084	1,487	2,210
3-3	386	496	622	763	916	1,262	1,877
3-6	330	424	531	653	785	1,083	1,612
3-9	284	365	459	565	679	938	1,398
4-0	247	318	399	493	593	820	1,223
4-3	216	278	350	433	521	722	1,079
4-6	190	245	309	382	461	640	951
4-9	168	217	274	340	410	571	855
5-0	149	193	244	304	367	511	767
5-3		172	218	272	330	460	691
5-6		154	196	245	297	416	626
5-9		139	176	222	269	378	569
6-0			159	201	244	344	518
6-3			144	183	222	314	474
6-6			131	166	203	287	435
6-9				152	186	264	400
7-0				139	170	243	369
7-6					144	206	315
8-0						177	272
8-6						153	236
9-0						132	205
9-6							180
10-0							158

[a] From Carnegie Steel Co. 1923, 266. The original note reads:
The weight of the terra cotta arch has been deducted from the safe load given in the tables, so that only the dead load of the concrete fill, plastering, etc., must be deducted to obtain the net safe live load for any arch and span; blocks of different areas and for other factors of safety are calculated as follows:

EXAMPLE.—Required the load per square foot for a 5'-6" span and 8 inch arch blocks with three horizontal and four vertical webs, ¾ inch thick, set in end construction, cross-section through webs of blocks parallel to webs of beams.

Sectional area of the blocks is 8" × ¾" × 4 + (12" − 4 × ¾") × ¾" × 3 = 44.25 sq. in. at 0.06 pounds per cu. in., the weight is 44.25 × 12 × 0.06 = 32 pounds.

The net safe load of the 8 inch block given in the table is 196 pounds. Adding the weight of the block, 37 × 12 × 0.06 = 26 pounds, the total safe load is 222 pounds. The net safe load for blocks with an area of 44.25 sq. in. and a safety factor of 5 is (44.25 ÷ 37 × 222 × 7/5) − 32 = 340 pounds per sq. ft.

310 ◇ Structural Analysis of Historic Buildings

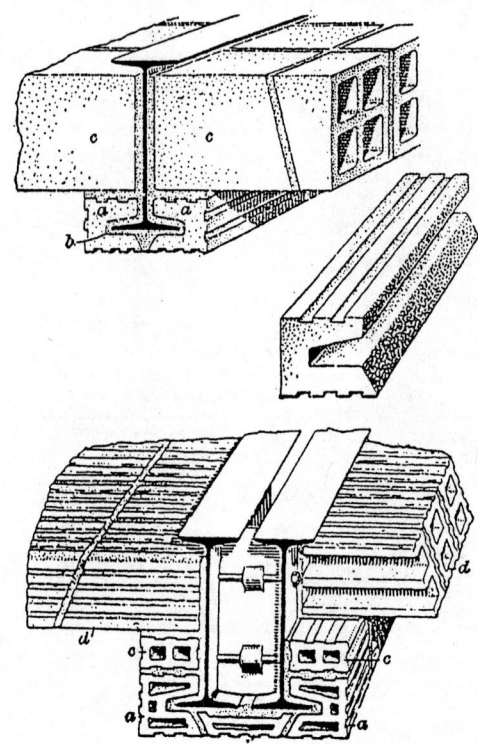

Figure 4-38. Fireproofing beams, details. Reprinted with permission from C. E. White, *Hollow Tile*, Part 1 of *Hollow Tile and Fireproofing*, ed. W. S. Lowndes, pp. 14, 15.

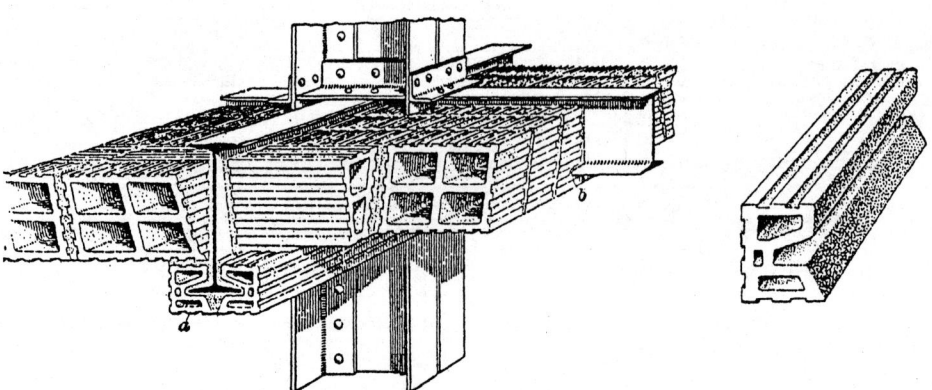

Figure 4-39. Fireproof floor systems: terra cotta arch details.

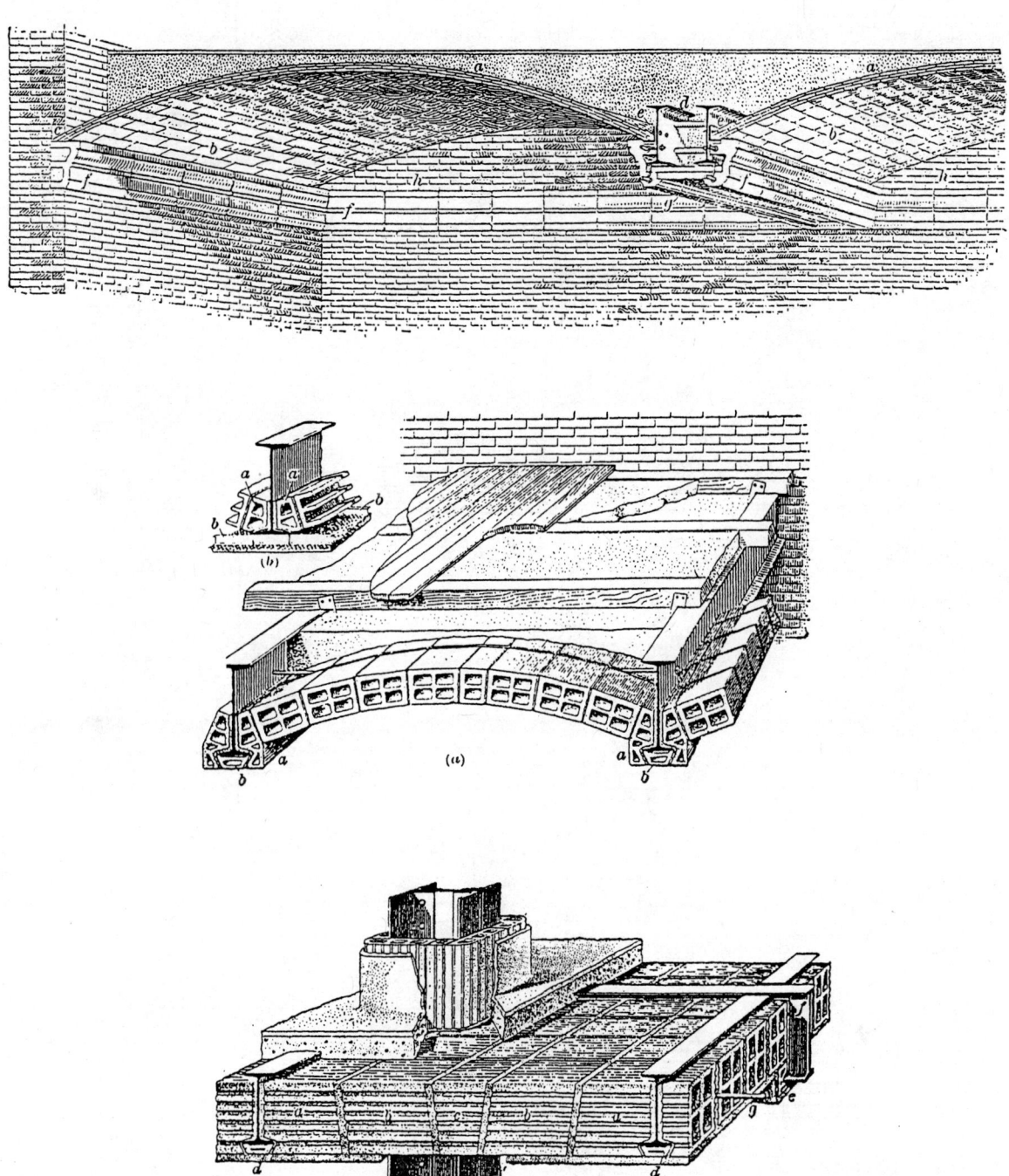

Figure 4-40. Miscellaneous details: fireproof floor systems; terra cotta arches. International Correspondence Schools 1905B, para. 22, 21.

312 ◇ Structural Analysis of Historic Buildings

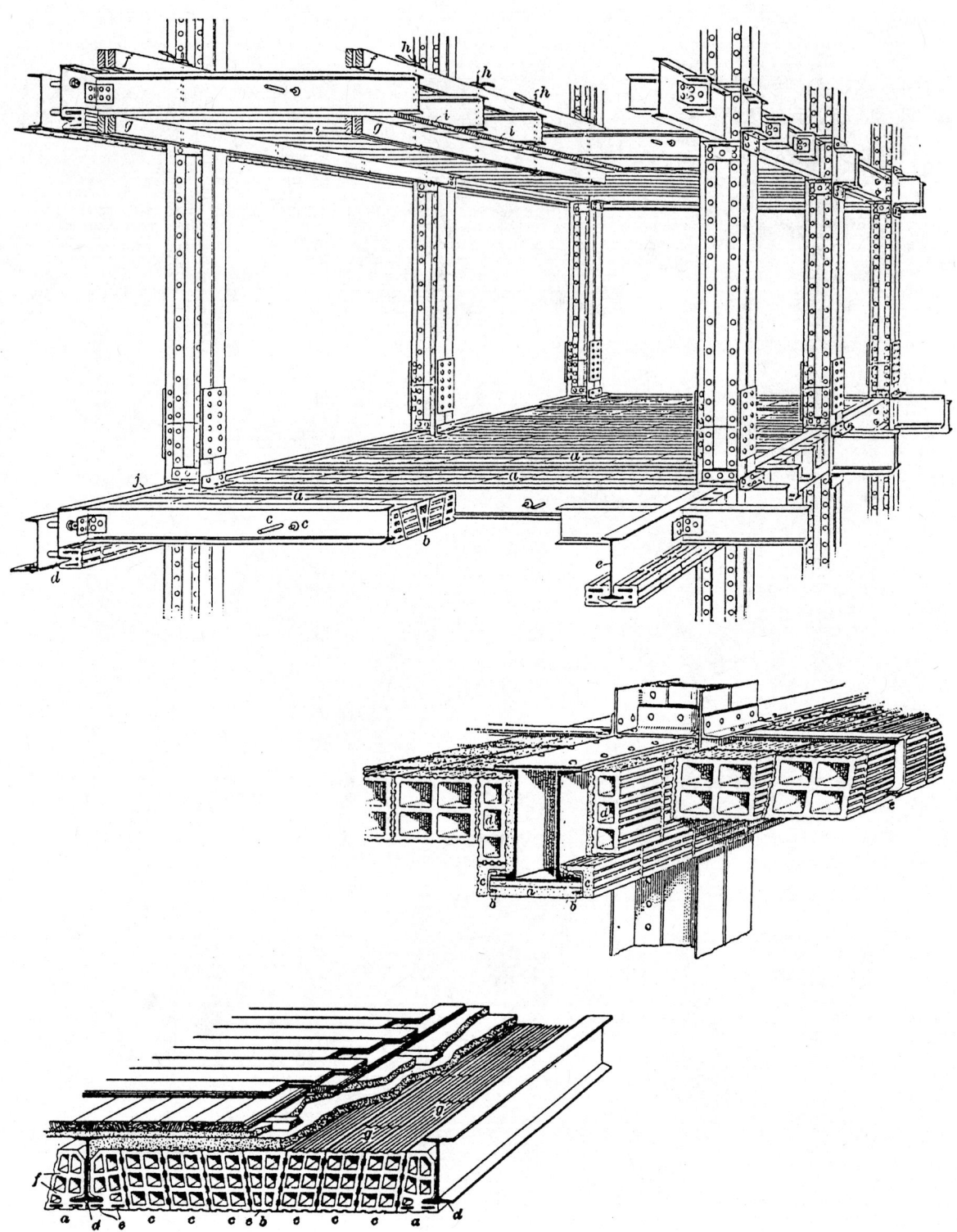

Figure 4-40. (Continued).

Table 4-30. Miscellaneous Tables: Fireproof Floor Systems; Terra Cotta Arches[a]

A. Safe Loads for End-Construction and Combination-Construction Flat Arches[b]

Depth of Arches	6 in.	7 in.	8 in.	9 in.	10 in.	12 in.	15 in.
Weight of Arch (psf)	26	29	32	35	38	42	50
Span (ft, in.)	psf	psf	psf	psf	psf	psf	psf
3-0	482						
3-3	410	525					
3-6	354	453	563				
3-9	308	394	491	597			
4-0	271	347	431	525	627		
4-3	240	307	382	465	555		
4-6	214	274	341	414	495		
4-9	192	246	306	372	444	608	
5-0	173	222	276	336	401	548	
5-3	157	201	250	304	364	497	
5-6	143	183	228	277	331	453	
5-9	131	168	208	254	303	415	614
6-0		154	191	233	278	381	563
6-3		142	176	215	256	351	519
6-6		131	163	198	237	324	480
6-9			151	184	220	301	445
7-0			140	171	204	280	414
7-6			122	149	178	243	360
8-0				131	156	214	317
8-6				116	138	190	281
9-0					123	169	250
9-6						152	225
10-0						137	203
10-6						124	184
11-0							167
11-6							153

B. Safe Loads for Side-Construction Flat Arches[b]

Depth of Arches	6 in.	7 in.	8 in.	9 in.	10 in.	12 in.	15 in.
Weight of Arch (psf)	26	29	32	36	38	44	54
Span (ft, in.)	psf	psf	psf	psf	psf	psf	psf
3-0	439	515					
3-3	370	434	500				
3-6	314	371	427	509			
3-9	270	320	368	439	514		
4-0	234	276	319	382	448		
4-3	204	243	280	335	393	543	
4-6	178	213	246	296	347	480	
4-9	158	188	218	262	308	427	549
5-0	140	167	193	233	275	382	491
5-3	124	149	173	209	246	343	442
5-6		133	155	188	221	309	399
5-9		118	139	169	200	279	362
6-0		107	125	153	181	253	329
6-3			113	138	164	231	300
6-6			102	125	149	210	274
6-9			92	114	136	192	251
7-0				104	124	176	231
7-6				86	104	148	196
8-0					87	126	167
8-6						107	144
9-0						91	124
9-6						78	107

Table 4-30. (Continued)

C. Safe Loads foor Segmental Arches of Hollow Tile (psf)[c]															
Span (ft)	Rise (in.)	4 in. (lb)	6 in. (lb)	8 in. (lb)	10 in. (lb)	Span (ft)	Rise (in.)	4 in. (lb)	6 in. (lb)	8 in. (lb)	10 in. (lb)	Span (ft)	Rise (in.)	4 in. (lb)	6 in. (lb)	8 in. (lb)	10 in. (lb)
4	¾	702	902	1,078	1,178	7	¾	394	508	606	662	11	1	327	421	503	550
	1	920	1,184	1,414	1,545		1	520	609	799	873		1¼	404	519	621	676
	1¼	1,155	1,485	1,774	1,939		1¼	648	834	996	1,089		1½	479	617	737	805
	1½	1,353	1,740	2,079	2,272		1½	762	981	1,171	1,280		1¾	551	709	847	925
	1¾	1,545	1,986	2,373	2,593		1¾	876	1,127	1,346	1,471		2	617	794	948	1,030
	2	1,736	2,233	2,667	2,915		2	983	1,264	1,510	1,650						
4½	¾	615	792	946	1,034	8	¾	341	439	525	573	12	1	297	383	458	600
	1	812	1,044	1,247	1,363		1	457	588	703	768		1¼	370	477	569	622
	1¼	1,020	1,313	1,568	1,713		1¼	562	724	864	944		1½	439	566	676	738
	1½	1,196	1,539	1,838	2,009		1½	668	859	1,026	1,122		1¾	505	649	776	848
	1¾	1,381	1,775	2,121	2,318		1¾	767	987	1,179	1,288		2	565	727	869	949
	2	1,536	1,975	2,359	2,578		2	854	1,099	1,312	1,434						
5	¾	551	709	847	926	9	¾	300	386	461	504	13	1	272	351	419	456
	1	744	957	1,143	1,249		1	403	518	619	677		1¼	339	437	522	570
	1¼	911	1,172	1,400	1,530		1¼	501	645	770	842		1½	403	519	620	677
	1½	1,072	1,379	1,647	1,800		1½	590	758	906	990		1¾	463	596	712	778
	1¾	1,238	1,592	1,902	2,078		1¾	677	871	1,041	1,137		2	521	670	801	873
	2	1,379	1,773	2,118	2,315		2	759	977	1,167	1,275						
6	¾	455	585	699	764	10	¾	267	344	411	449	14	1	253	326	390	426
	1	612	788	941	1,028		1	359	462	552	603		1¼	315	406	485	530
	1¼	753	969	1,157	1,265		1¼	447	576	688	751		1½	374	482	575	629
	1½	898	1,154	1,379	1,507		1½	531	683	816	892		1¾	430	533	661	722
	1¾	1,022	1,315	1,570	1,716		1¾	610	784	937	1,024		2	481	619	740	808
	2	1,148	1,476	1,763	1,927		2	683	879	1,050	1,147						
												15	1	234	302	361	394
													1¼	292	377	450	491
													1½	347	447	534	583
													1¾	401	515	616	673
													2	449	577	690	754

[a] From Peters 1987, 14–23.
[b] Materials semiporous, factor of safety 7; National Fire Proofing Co.
[c] The weight of the arch tile has been deducted in table so that only the dead load of concrete fill, plastering, etc. must be deducted to obtain net live load.

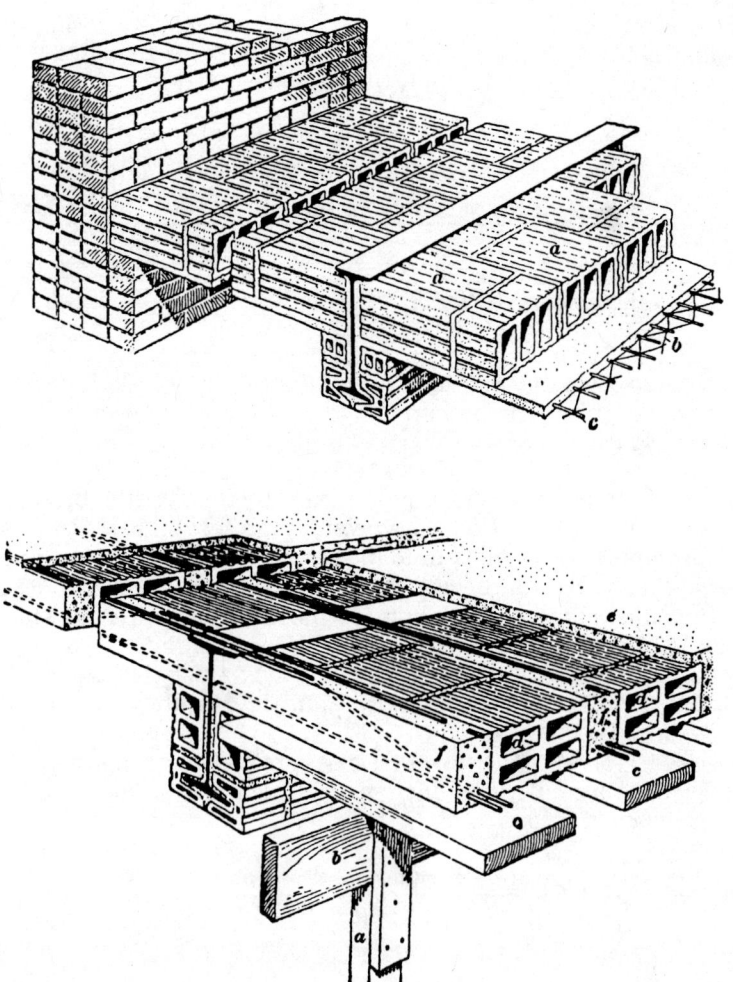

Figure 4-41. Miscellaneous details: fireproof floor systems. (*From White 1932, 14, 15.*)

CAST IRON—WROUGHT IRON BEAMS

Cast iron was originally produced in the United States after the 1620s by a group of colonists sent to start iron smelting. Iron was used in weapons and farm implements until the beginning of the 19th century, when the steam engine and the railroads brought iron into prominence as a commercial enterprise. The first use of cast iron in buildings in England was in the late 1770s and early 1780s with the construction of the Iron Bridge at Coalbrookdale and the malting houses at Shrewsbury.

Cast iron in buildings in the United States began as a nonstructural item, at least in that, in cast iron building facades, it did not carry any building components other than itself. The Arsenal in Watervliet, New York, 1849, by Daniel Badger, is of all-iron construction and is considered the oldest building in America to have cast iron structural elements. Interestingly, the tension members in the trusses of the arsenal are of wrought iron. This indicates that the designers were quite knowledgeable about the tensile vs. compression characteristics of these materials. James Bogardus, the contractor, in 1854, used Cooper's deck beams in the Harper & Brothers Building, New York City (Peters 1987, 14–23).

The advantage of the iron products was in their ability to be cast into almost any shape. Mahan (1885, 89) states that iron is one of the most valuable building materials, owing to its great strength, hardness, and durability. Trautwine (1888) gives formulae to be used in the design of cast iron beams. Trautwine gives design examples of the

Hodgkinson's beams, as shown in Figure 4-42. The formulae were almost empirical, except that the value of the constants must have been derived from experiments. The formulae are difficult to check or verify and therefore they can only be looked at as informative.

Hodgkinson's Formulae:

$$\frac{\text{Center breaking load}}{\text{in tons of 2200 lb}} = \frac{\text{Area of bottom flange, in.}^2 \text{ (depth 00 in.) (constant 2.166)}}{\text{Clear span, ft}}$$

$$\text{Area of bottom flange} = \frac{\text{Load, tons (clear span, ft)}}{2.166 \text{ (Total Depth 00, in.)}}$$

Area of bottom flange by 6 = area of top flange. The top flange is 3–6 times as wide as it is thick. The bottom flange is 6–8 times as wide as it is thick. The stem thickness is usually 1/14–1/24 of the depth of the beam.

Trautwine also presents a series of cast iron shapes with the breaking loads for given clear spans specified for each, as shown on the section (see Figure 4-43). Trautwine does not state which foundry or foundries these members come from. He refers to them as Sterling's cast iron beams. Rivington (1899) gives similar descriptions of the cast iron beams and similar sections to those detailed in Trautwine. However, Rivington does not discuss the design or proportioning of the cast iron beams except to mention the relationship of thickness to width, etc., which seems to be the accepted method of sizing the members, as shown by Trautwine. Figure 4-44 shows the sections and details as given by Rivington.

Kidder (1902, 371), also gives data on cast iron beams and girders and provides another Hodgkinson formula (the difference between the formula given in 1888 and the Kidder formula is the 2000 lb ton).

$$\text{Breaking load, tons} = \frac{\text{Area of bottom flange, in.}^2 \text{ (depth, in.)(2.426)}}{\text{Span, ft}}$$

Kidder also includes members that are combination cast iron and wrought iron, as shown in Figure 4-46. The member is called a cast iron arch girder with wrought iron tension rod. The cast iron and the wrought iron rod are made separately, and the rod is heated to expand and feed into the cast iron girder and then prestress the girder as the rod cools.

The diameter of the tension rod is given by the following formula:

$$d, \text{ in.} = \sqrt{\frac{(\text{load of girder})(\text{span, ft})}{(8)(\text{rise, ft})(7854)}}$$

Cast iron skewbacks were a pair of cast iron thrust blocks that were set into a masonry wall above an opening. The skewbacks would be the bearing point of the ends of a masonry arch. They were restrained by two wrought iron tension rods that held the thrust of the arch. Kidder gives only an elevation of the skewback arch and the formula for the diameter of the tension rods (see Figure 4-47). The skewback arch does not require abutments or buttresses, as the tension rods contain the thrust. The formula for the diameter of the two rods is:

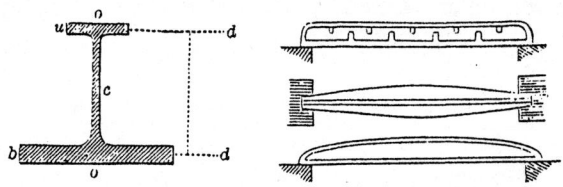

Figure 4-42. Hodgkinson's beams. (*From Trautwine 1888, 518.*)

Table 4-31. Cast Iron Arch Girder with Wrought Iron Tension Rod as Required to Support Walls Shown, Spans 13–26 ft

Height of Wall (ft)	Thickness of Wall (in.)	Top Flange (in.)	Center Web (in.)	Bulb (in.)
40	12	12 × 1	12 × 0.75	3 × 2
50	12	12 × 1.25	12 × 0.875	3 × 2
40	16	12 × 1.25	12 × 0.875	3.5 × 2
50	16	16 × 1.5	12 × 1	4 × 2

$$\text{Diameter each rod, in.} = \sqrt{\frac{(\text{load on girder})(\text{span, ft})}{16(\text{rise, ft})(7854)}}$$

Cast iron beams and girders were very capable of carrying lighter loads at shorter spans. In the flexural action, the tension zone of the stress block would be the detriment to the cast iron beam. That is why cast iron sections were unsymmetrical, with the larger areas being in the bottom or tension zone. Designers and manufacturers knew of the problem and made the areas of the bottom flange larger in an effort to reduce the tensile stress in that region. Cast iron has an average ultimate tensile stress of 17,500 psi, an elastic limit of 6,000 psi, and a modulus of elasticity of 12×10^6 psi. Structural steel for buildings has an average ultimate tensile stress of 60,000 psi, an elastic limit of 30,000 psi, and a modulus of elasticity of 29×10^6 psi. As can be seen, structural steels are far superior to cast iron sections for service as beams. Cast iron has a higher ultimate compressive stress than steel, 80,000 psi as compared to 60,000 psi. Wrought iron sections are closer to steel sections in every area.

Wrought iron has an average ultimate tensile stress of 48,000 psi, an elastic limit of 26,000 psi, and a modulus of elasticity of 28×10^6 psi. The principal components of iron and carbon in cast iron, wrought iron, and steel are shown in Table 4-32.

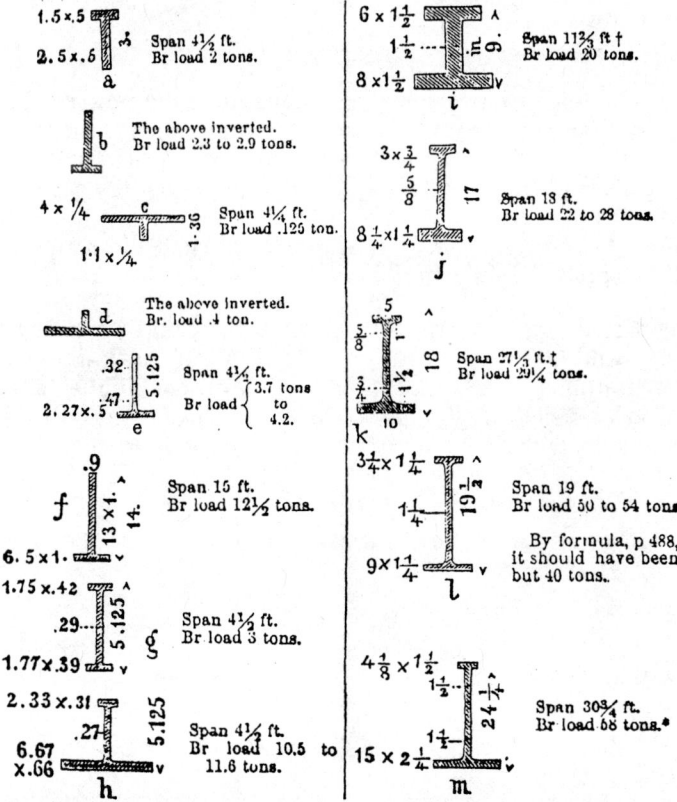

Figure 4-43. Sterling's cast iron beams. (*From Trautwine 1888, 519, 520.*)

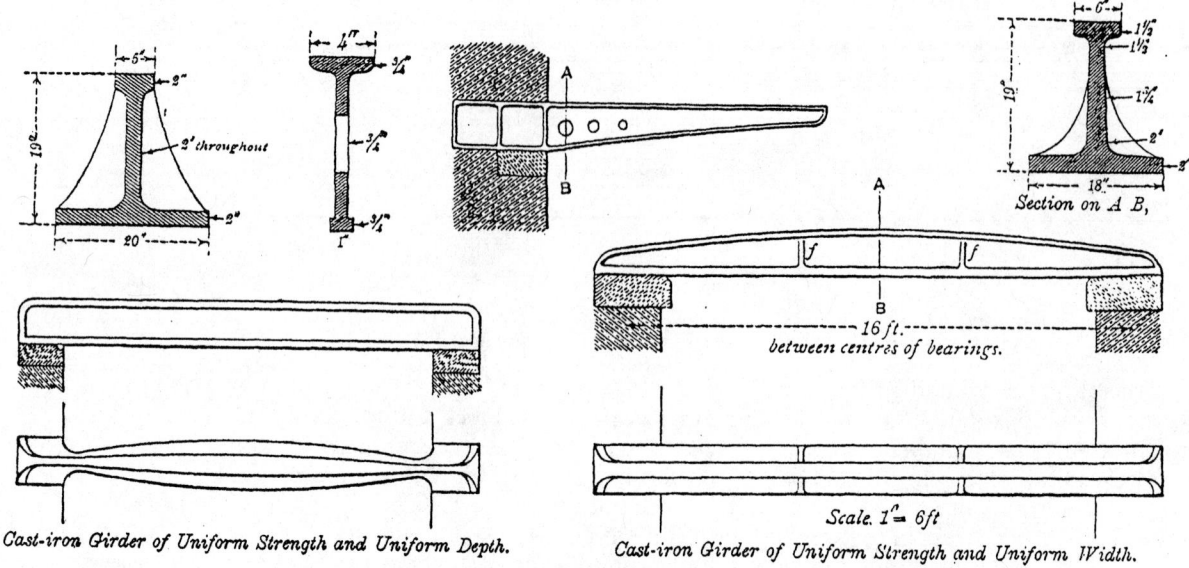

Figure 4-44. Rivington's cast iron beams. (*From Rivington 1899, 105–108.*)

Cast iron beams and girders provided adequate strength for the uses they were required to serve. As stated, designers seemed to work within the limits of the material. Beams and girders of cast iron were easily molded to required shapes, and decorative elements could be set into the molds to form attractive features that could be exposed as a part of the decoration of the building. Steels were rolled into shapes that were uniform in section and could not easily be customized or made into decorative elements. The problems that caused the designers to abandon cast iron for beams and girders were as follows:

1. The members were much heavier than steel and at that time were much harder to handle. (Cranes as we now have them did not exist at this time.)
2. The molding process did not ensure uniform quality. Girders had to be load tested because they were brittle and would fail in places where the casting was defective. There was a larger than desirable quantity of rejected castings.
3. Cast iron beams and girders were subject to a sudden brittle failure if, in the presence of a fire that had significantly preheated the members, they were hit with a spray of cold water.

In the early period of use of cast iron beams, protection from heat was not a part of the technology, and many issues such as chilled water shock were discovered the hard way. Wrought iron and steel were much more restricted in their decorative uses, and designers also learned through experience that the heat of fires in buildings would cause severe problems with those members. The failures were not sudden, nor were

Figure 4-45. Cast iron beam. (*From Kidder 1902, 371.*)

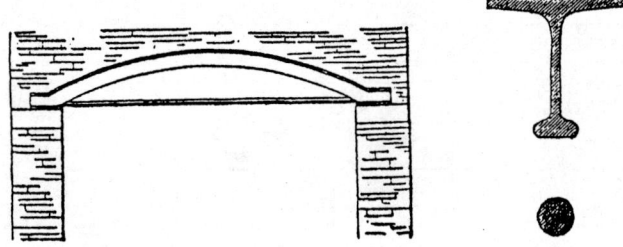

Figure 4-46. Cast iron arch girder and wrought iron tension rod. (*From Kidder 1902, 422.*)

they as complete, in that the significant loss of strength might result only in excessive deflections that would leave the building irreversibly damaged or possibly irreparable. In most cases the building would not suffer a total collapse as would be the case in a sudden failure.

Wrought iron was used in buildings in wood trusses, tension members, etc. for years before beam and girder sections were rolled for structural members. Wrought iron plates and angles were available before I-beams, and they were fabricated to make trusses, columns, and girders through the use of rivets and intricate shape combinations. Trautwine (1888) gives details and drawings of as I-beams, channels, and deck beams, which he states were for use in building construction (see Figure 4-48). Trautwine states that these sections were being made by companies including (ibid., 521):

A. & P. Roberts & Co., Pencoyd Iron Works, Philadelphia
New Jersey Steel & Iron Co., Trenton, N.J.
Carnegie Bros. & Co., Limited, Pittsburgh, Pa.
Phoenix Iron Co., Phoenixville, Pa.
Passaic Rolling Mill Co., Patterson, N.J.
Pottsville Iron & Steel, Pottsville, Pa.

Trautwine also provides tables of the principal sizes (included the largest and smallest) made in the United States. He states that in Belgium, I-beams are rolled as large as 21.5 in. deep and to weights of 334 lb/yard (the early weight specification) and that they were imported into the United States, where the J. H. Jackson Company of New York and Esherick & Co. of Philadelphia were their agents.

Wrought iron is called "malleable iron" in some circles because of its ability to bend in a quite malleable way without failure. Trautwine states that wrought iron sections do not fail by breakage under a steadily increasing load, they just continue to deflect to the point of being useless. The ultimate load is the load that so cripples the beam that it continues to yield indefinitely without increase in load (ibid.). Trau-

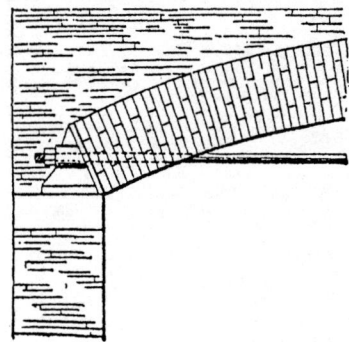

Figure 4-47. Skewback arch and tension rods. (*From Kidder 1902, 423, 424.*)

Table 4-32. Properties of Cast Iron, Wrought Iron, and Steel[a]

Material	Percent Iron	Percent Carbon
Cast iron	91–94	3.5–4.5
Steel	98.1–99.5	0.07–1.30
Wrought iron	99.0–99.8	0.05–0.25

[a] From American Institute of Steel Construction 1930, 57.

twine states that the factor of safety of wrought iron beams should be at least 3, and that when the load is subject to variation, vibration, or likelihood of being suddenly applied, the factor of safety should be 4–6. Trautwine also states that members should be loaded with their webs vertical and that the supports at their ends must be stayed against yielding sideways at the ends and at intermediate points along the span. The distance between the points of lateral support was not to exceed 20 times the width of the flange.

Trautwine provides tables and formulae for the design of Pencoyd channels, I-beams, and deck beams. These members were typical of the beams rolled by most mills, and the I-beams were often used in the early fireproof floors (see Figure 4-49 and Table 4-33).

Pencoyd I-Beams:

$$\text{Center ultimate load} = \frac{\text{coefficient for ultimate center load}}{\text{span, ft}}$$

The above formula is for the ultimate load in lb, including one-half of the member weight. The tabular coefficient given in Table 4-33 is for the ultimate or failure load of a beam of the size in the row of the table that is 1 ft long. The factor of safety for the member should be a minimum of 3 for static loads and 4–6 for loads that are the result of impact.

Uniformly distributed load = 2 × center load
Ultimate center load × 2 = ultimate uniform load
Allowable center load × 2 = allow uniform load
Deflection: for center of member, load ≤ one-half of the ultimate load).

$$\text{Deflection} = \frac{\text{load, lb (span, ft)}^3}{\text{weight of beam (depth, in.)}^2 \text{(constant } K\text{)}}$$

where:

Constant K
For I-beam:
  Loaded at center = 11,200
  Uniform load = 18,000

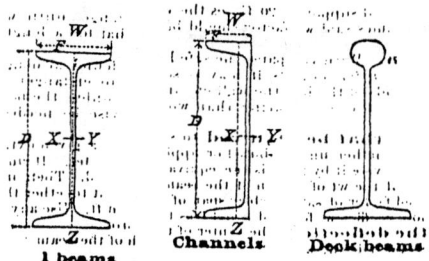

Figure 4-48. Wrought iron shapes (*From Trautwine 1888, 521.*)

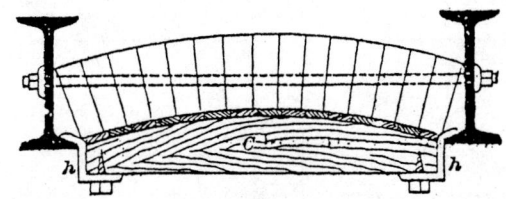

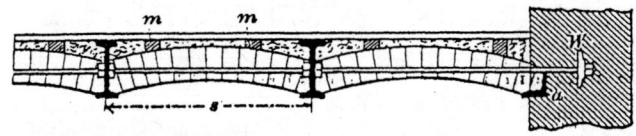

Figure 4-49. Wrought iron Pencoyd I-beams; early fireproof floors. (*From Trautwine 1888, 522.*)

For channel:
    Loaded at center = 10,000
    Uniform load = 16,000

Selecting a wrought iron beam by designing through the Trautwine method and then checking the solution by modern methods of analysis is done as follows:

The Trautwine method of 1888
    Wrought iron floor beam, span = 20 ft, load = 200 psf (uniform)
    spacing = 7 ft, total load = 1400 lb/ft
    Try a 15 in. deep member, Flange = 5.125 in., web thickness = 7/16 in.,
    weight = 145 lb/yard, coefficient = 972,900 lb, I_x = 521.19 in.^{4}

Center ultimate load = $\dfrac{\text{coefficient for ultimate center load}}{\text{span, ft}}$ = $\dfrac{972,900 \text{ lb}}{20 \text{ ft}}$ = 48,645 lb

Ultimate uniform load = 2(48,645 lb) = 97,290 lb

Allowable uniform load = $\dfrac{97,290 \text{ lb} - \dfrac{145 \text{ lb/yard}}{3 \text{ ft/yard}}(10 \text{ ft})}{3(20 \text{ ft})}$ = 1613.5 lb/ft

Deflection of I-beam: $\delta = \dfrac{(\text{load, lbs})(\text{span, ft})^3}{(\text{weight of beam})(d, \text{in.})^2 K} = \dfrac{28,000 \text{ lb }(20 \text{ ft})^3}{145 \text{ lb/yard }(15 \text{ in.})^2\ 18,000}$

    = 0.381 in. (at center of span) for total load

Analysis by Modern Methods
Wrought iron floor beam, span = 20 ft, load = 200 psf (uniform), Spacing = 7 ft,
    total load = 1400 lb/ft

$F_b = \dfrac{\text{modulus of rupture (given in example)}}{\text{factor of safety}} = \dfrac{42,000 \text{ psi}}{3} = 14,000$ psi

Using same I-beam, 15 in. deep member, flange = 5.125 in., web thickness = 7/16 in., weight = 145 lb/yard (48.33 lb/linear ft), I_x = 521.19 in.^{4,} S_x = 68.49 in.^{3}

$$M_{max} = M_{des} = \dfrac{wL^2}{8} = \dfrac{1400 \text{ lb/ft }(20 \text{ ft})^2}{8} = 70,000 \text{ lb/ft}$$

$$S_{required} = \dfrac{M_{des}}{F_b} = \dfrac{70,000 \text{ lb-ft }(12 \text{ in./ft})}{14,000 \text{ lb/in.}^2} = 60 \text{ in.}^3$$

Table 4-33. Wrought Iron Pencoyd I-beams[a]

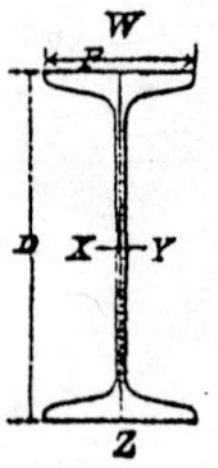

Chart Number	Depth D (in.)	Width of Flange F (in.)	Thickness of Web (in.)	Thickness of Flange (in.)		Weight per Yard (lb)	Coefficient for Ultimate Center Load (lb)	Least Radius of Gyration (in.)	Moments of Inertia	
									About *X Y*	About *W Z*
	*			At root	At edge					
1	15	5 31/32	7/8	1 5/16	3/4	233	1,388,700	1.17	743.60	31.89
	15	5 3/4	31/32			200	1,273,200	1.20	682.08	28.50
2	15	5 1/2	13/16	1 1/8	5/8	201	1,168,800	1.05	626.57	22.16
	15	5 1/8	7/16			145	972,900	1.08	521.19	16.91
3	12	5 23/32	7/8	1 1/4	11/16	194	940,800	1.16	403.48	26.10
	12	5 1/2	21/32			168	867,900	1.17	371.98	23.19
4	12	5 3/32	13/16	1	35/64	163	757,200	.99	324.61	16.02
	12	4 51/64	22/64			120	636,600	1.01	272.86	12.22
5	10 1/2	5 1/2	23/32	1 1/8	21/32	161	708,600	1.17	265.74	22.20
	10 1/2	5 1/4	15/32			134	644,300	1.19	241.63	19.00
5 1/2	10 1/2	5 1/8	21/32	15/16	17/32	134	585,400	1.05	219.63	14.74
	10 1/2	4 7/8	13/32			108	521,100	1.07	195.42	12.45
6	10 1/2	4 11/16	17/32	27/32	7/16	109	480,900	.94	180.34	9.59
	10 1/2	4 1/2	11/32			89	432,700	.97	162.26	8.34
7	10	4 7/8	3/4	1 1/16	1/2	137	544,200	.96	194.41	12.63
	10	4 5/8	1/2			112	486,000	.98	173.58	10.64
8	10	4 17/32	1/2	1	7/16	106	452,400	.93	161.33	9.17
	10	4 3/8	11/32			90	415,200	.95	148.31	8.46
9	9	4 23/32	3/4	23/32	1/2	122	436,800	.94	143.70	10.78
	9	4 3/8	11/32			90	369,600	.96	118.81	8.44
10	9	4 21/64	1/2	25/32	3/8	88	331,500	.87	106.78	6.66
	9	4 1/8	19/64			70	293,700	.89	94.44	5.59
11	8	4 19/32	3/4	27/32	15/32	109	345,900	.93	98.59	9.43
	8	4 1/4	13/32			81	293,700	.94	83.93	7.23
12	8	4 1/8	7/16	3/4	3/8	75	260,700	.86	74.50	5.55
	8	4	5/16			65	242,100	.88	69.17	5.02
14	7	4 1/8	3/4	23/32	11/32	88	232,800	.80	58.60	5.63
	7	3 39/64	15/64			51	172,200	.82	43.06	3.43
16	6	3 3/8	5/8	21/32	5/16	63	144,600	.64	30.77	2.58
	6	3	1/4			40	112,200	.66	24.10	1.80
18	5	2 31/32	7/16	1/2	1/4	40	81,600	.58	14.69	1.35
	5	2 3/4	7/32			30	69,900	.60	12.50	1.09
19	4	3	1/2	1/2	9/32	38	63,000	.62	9.02	1.46
	4	2 3/4	1/4			28	53,100	.63	7.69	1.17
20	4	2 21/64	1/4	3/8	7/32	21.5	39,300	.50	5.55	.54
	4	2 1/4	11/64			18.5	36,000	.51	5.14	.49
21	3	2 11/16	7/16	7/16	1/4	28.6	34,500	.58	3.99	.96
	3	2 1/2	1/4			23	30,600	.59	3.29	.77
22	3	2 13/32	5/16	13/32	13/64	21.7	28,200	.52	3.01	.59
	3	2 1/4	5/32			17	24,600	.53	2.66	.48

[a] From Trautwine 1888, 523a.

Table 4-34. Pencoyd Channels[a]

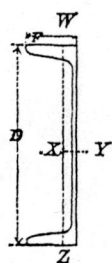

Chart Number	Depth D (in.)	Width of Flange F (in.)	Thickness of Web (in.)	Thickness of Flange (in.) At root	Thickness of Flange (in.) At edge	Weight per Yard (lb)	Coefficient for Ultimate Center Load (lb)	Least Radius of Gyration (in.)	Moments of Inertia About X Y	Moments of Inertia About W Z
30	15 15	4³⁄₈ 4	1 ⁵⁄₈	1	⁵⁄₈	204.5 148.0	1,040,400 842,700	1.10 1.13	557.44 451.51	24.74 19.05
31	12 12	3¹⁷⁄₃₂ 2¹⁵⁄₁₆	1 ¹³⁄₃₂	1	⁹⁄₁₆	160.0 88.5	626,400 426,300	.90 .92	268.51 182.71	12.16 7.42
32	12 12	2⁶¹⁄₆₄ 2²⁹⁄₆₄	⁵⁄₈ ⁹⁄₃₂	¾	¹¹⁄₃₂	101.5 60.0	404,700 274,200	.72 .74	173.51 123.71	5.26 3.22
34	10 10	3¹⁄₁₆ 2¹⁹⁄₃₂	¾ ⁹⁄₃₂	¹⁵⁄₁₆	⁷⁄₁₆	106.0 60.0	366,900 259,500	.82 .84	131.04 92.71	7.13 4.29
35	10 10	2¾ 2⅜	⁵⁄₈ ¼	¾	⅜	86.5 49.0	294,300 206,700	.67 .60	105.16 73.91	3.88 2.33
36	9 9	2⅞ 2⁷⁄₁₆	¾ ⁵⁄₁₆	²⁷⁄₃₂	⅜	93.0 54.0	282,300 200,100	.67 .68	90.66 64.34	4.17 2.47
37	9 9	2¹³⁄₃₂ 2⁹⁄₆₄	½ ¹⁵⁄₆₄	³⁵⁄₆₄	¹⁹⁄₆₄	61.0 37.0	186,300 135,600	.58 .60	59.85 43.65	2.05 1.31
38	8 8	2¾ 2⁹⁄₃₂	¾ ⁹⁄₃₂	¹¹⁄₁₆	⁵⁄₁₆	80.5 43.0	210,000 139,800	.71 .71	60.00 40.00	4.28 2.17
39	8 8	2¹⁹⁄₆₄ 2	½ ¹³⁄₆₄	¹⁵⁄₃₂	⁹⁄₃₂	54.0 30.0	143,400 98,700	.60 .60	41.03 28.23	1.94 1.06
40	7 7	2¾ 2¹⁹⁄₆₄	¾ ¹⁹⁄₆₄	¹¹⁄₁₆	⁹⁄₃₂	73.0 41.0	170,100 117,900	.63 .63	42.57 29.51	3.08 1.71
41	7 7	2⁹⁄₃₂ 1³¹⁄₃₂	½ ³⁄₁₆	⁷⁄₁₆	¼	49.0 26.0	111,300 73,800	.63 .53	27.86 18.46	1.65 .90
42	6 6	2⅝ 2¼	⅝ ¼	⅜	¼	55.4 32.0	126,300 85,500	.63 .67	25.12 18.37	2.56 1.46
43	6 6	2½ 2⅛	⅜ ¼	⅝	³⁄₁₆	54.5 32.0	124,200 72,900	.61 .60	24.35 17.60	2.03 1.15
44	6 6	2¹⁄₃₂ 1¾	½ ⁷⁄₃₂	⅜	¼	39.6 22.7	90,200 54,300	.52 .51	16.73 11.67	1.07 .59
45	5 5	2⅓ 2	⅝ ¼	¹⁹⁄₃₂	¼	46.0 27.3	87,400 57,600	.58 .56	14.20 10.29	1.55 .86
46	5 5	1²⁹⁄₃₂ 1⅝	½ ⁷⁄₃₂	⅜	³⁄₁₆	32.9 18.8	62,500 37,200	.47 .45	12.54 6.67	.73 .37
47	4 4	1³¹⁄₃₂ 1²³⁄₆₄	½ ¼	¹⁹⁄₃₂	¼	31.5 21.5	47,800 36,300	.52 .50	6.49 5.16	.85 .54
48	4 4	1²³⁄₃₂ 1⁹⁄₁₆	⅜ ⁷⁄₃₂	¹³⁄₃₂	¼	23.7 17.5	36,000 28,800	.49 .48	4.97 4.14	.57 .41
49	3 3	1²¹⁄₃₂ 1¹⁷⁄₃₂	¹¹⁄₃₂ ⁷⁄₃₂	¹³⁄₃₂	¼	18.9 15.2	21,600 18,000	.47 .46	2.31 2.03	.42 .32
50	2¼	1⅜	¼	¼	¼	11.3	10,000	.43	.80	.21
51	2 2	1⁵⁄₂₂ 1³⁄₃₂	⁹⁄₃₂ ⁷⁄₃₂	⁵⁄₁₆	³⁄₁₆	10.0 8.75	7,600 6,700	.32 .31	.88 .48	.10 .08
52	1¾	¹⁹⁄₃₂	¹⁹⁄₃₂	¼	³⁄₃₂	8.5	2,300	.17	.38	.01

[a] From Trautwine 1888, 523.

Table 4-35. Standard Beam Separators[a]

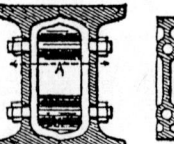 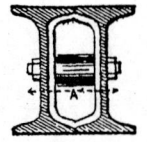 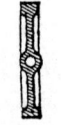

	Beam			Separator		Bolts			
			Minimum Width (in.) Out to Out of Two Flanges	Weight (lb) Off One Sep'r				Weight (lb) of 1 Bolt and Nut	
Chart Number	Depth (in.)	Weight (lb/yard)		For Minimum Width	Additional Each Additional Inch of Width	Number Used	Diameter (in.)	For Minimum Length	Additional for Each Additional Inch of Length
1	15 in. heavy	186 to 233	12	22	3.84	2	¾	1.75	.123
2	15 in. light	145 to 201	11	21	3.13	2	¾	1.62	.123
3	12 in. heavy	168 to 194	11½	16	2.76	2	¾	1.69	.123
4	12 in. light	120 to 163	10⁵⁄₁₆	14½	2.95	2	¾	1.58	.123
5	10½ in. heavy	134 to 161	11	11¼	2.10	1	¾	1.64	.123
5½	10½ in. med.	108 to 135	10¼	11	2.06	1	¾	1.28	.123
6	10½ in. light	89 to 109	9⅜	11	2.03	1	¾	1.53	.123
7	10 in. heavy	112 to 137	9¾	10	1.93	1	¾	1.56	.123
8	10 in. light	90 to 106	9	10	1.93	1	¾	1.52	.123
9	9 in. heavy	90 to 122	9½	9¼	1.63	1	¾	1.52	.123
10	9 in. light	70 to 88	8⅝	9	1.63	1	¾	1.48	.123
11	8 in. heavy	81 to 109	9¼	6¾	1.36	1	¾	1.50	.123
12	8 in. light	65 to 75	8¼	6¼	1.49	1	¾	1.46	.123
13	7 in. heavy	65 to 88	8¼	4	1.26	1	⅝	0.96	.085
14	7 in. light	51 to 88	8¼	4	1.26	1	⅝	0.91	.085
15	6 in. heavy	50 to 63	6¾	3	1.24	1	⅝	0.90	.085
16	6 in. light	40 to 63	6¾	3	1.24	1	⅝	0.87	.085
17	5 in. heavy	34 to 40	6	2¾	1.10	1	½	0.43	.055
18	5 in. light	30 to 40	6	2¾	1.10	1	½	0.42	.055
19	4 in. heavy	28 to 38	6	2	0.85	1	½	0.42	.055
20	4 in. light	18.5 to 21.5	4⅝	2	0.85	1	½	0.39	.055
21	3 in. heavy	23 to 28.6	5⅜	1½	0.69	1	½	0.38	.055
22	3 in. light	17 to 21.7	4¾	1½	0.69	1	½	0.31	.055

[a] From Trautwine 1888, 523b.

The section modulus of the 15 in. I-beam is 68.49 in³, and therefore the member is good for bending.

$$\text{Deflection} = \frac{5wL^4}{384\,EI} = \frac{5\,(1400\text{ lb/ft})(20\text{ ft})^4(1728\text{ in.}^3/\text{ft}^3)}{384\,(28{,}000{,}000\text{ lb/in.}^2)(521.19\text{ in.}^4)} = 0.345\text{ in. (total load)}$$

The deflection shown is again the deflection for the total load.
Today we distinguish between the deflection due to the dead Load and the deflection due to the live load. If the live load for this member was 50 psf, the deflection for the live load = 0.086 in. Shear is not of consequence since the whole section bears on the wall.

See Table 4-34 for the table for Pencoyd channels. Pencoyd channels and Pencoyd I-beams were made of wrought iron in the period 1887 through approximately 1895. The Pencoyd Iron Works may have converted to steel, as the AISC Historical Record indicates that the same sections (approximately) were being rolled in 1896. The actual dates are hard to determine but in this instance should be fairly accurate. In many cases the loads required larger members than were available, and in some instances

Figure 4-50. Compound girders. (*From Rivington 1899, 110.*)

the designers would put two rolled beams in the position where today we would utilize only one member. These double members would be separated and held in alignment by "beam separators" (see Table 4-35).

COMPOUND WROUGHT IRON GIRDERS

Compound girders made up of rolled sections and plates or two rolled sections were used by some designers and could be encountered in buildings of the era 1880–1910 (see Figure 4-50). The computations for the analysis of the compound girders are simple and predictable and would be manually done in most cases. Compound sections are usually of unique design and therefore are not usually set up in tables.

WROUGHT IRON RIVETED GIRDERS

Rivington (1899) gives examples of riveted girders. These members would normally be used in locations where the loads or the spans were too great for the conventional rolled shapes (see Figure 4-51). Riveted girders were usually made up of plates and angles; however, there were cases where other rolled shapes such as channels would be used on these members. Box girders would often be made up of plates and angles to form a hollow box. Box girders would be of particular strength due to the in geometric configuration (see Figure 4-52).

The most popular form of riveted girder was the section built up of plates and angles. This section most resembles the I-beam and has all the characteristics of the cross-section (see Figure 4-53).

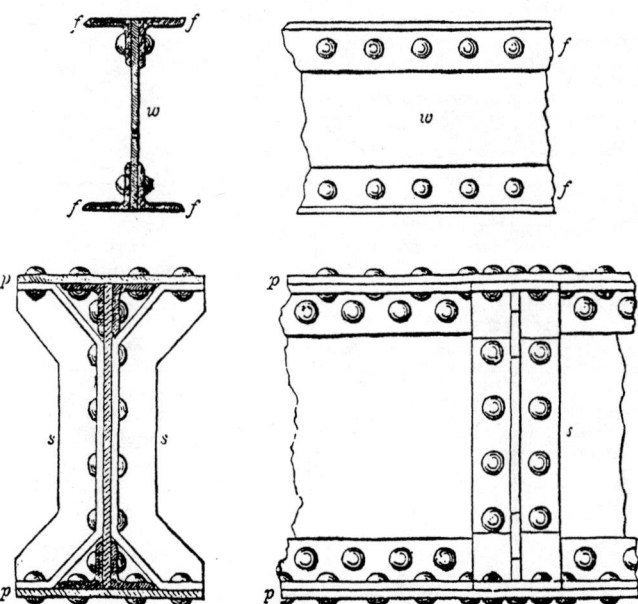

Figure 4-51. Wrought iron riveted girder. Depth d (top to bottom) = one-twelfth of span (simple); f = angles, w = web, p = flange plate, s = stiffener plates. (*From Rivington 1899, III.*)

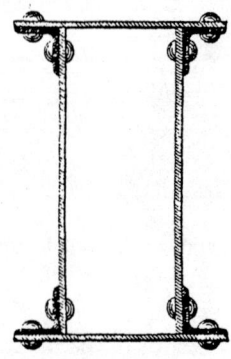

Figure 4-52. Box girder. (*From Rivington 1899, 112.*)

Trautwine gives a set of formulae for the determination of the size of the flanges and web sections on riveted girders (Trautwine 1888, 537):

$$\text{Force, in either flange} = \frac{\text{max. moment, ft-lbs}}{d \text{ (Cg-Cg of flanges, ft)}}$$

$$\text{Area, min., lower flange} = \frac{\text{tensile force, lb}}{\text{allowable tensile stress, lb/in.}^2} \quad \text{(gives effective net area)}$$

$$\text{Stress, psi, of either flange} = \frac{\text{force, lb}}{\text{effective net area of flange section}}$$

The effective cross-section of the tension flange of a riveted girder is taken as the area of the flange components minus the area of the rivet holes. In the case of a single line of rivets or of pairs of rivets that are all in the transverse line, the net section is the gross section minus the rivet holes. In the case of staggered rivets, the Trautwine method is to take the net section as the line through the staggered holes (with stagger lines multiplied by a factor of 0.75) multiplied by the thickness of the individual material.

For wrought iron girders used in buildings the allowable tensile stress in 1888, from Trautwine, was 10,000 psi. The compression flange must be designed as a compression member, with the flange opposing the compression force like a column.

$$r = \frac{\text{(approx.) Square of width of flange, in.}^2}{22.5}$$

$$r = \text{least radii of Gyration} = \sqrt{\frac{I}{A}}$$

$$F_a = \frac{40{,}000 \text{ psi}}{1 + \dfrac{L^2}{r^2(30{,}000)}}$$

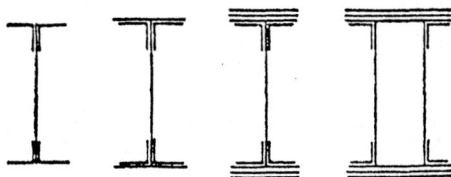

Figure 4-53. Wrought iron riveted girders: combinations of plates and angles. *From Kidder 1902, 410.*)

where:
L = length of beam between lateral supports
F_a = Ultimate compressive stress, psi, factor of safety = 3 min.

Rivets in wrought iron Girders: Trautwine method:
F_p = 16,000 psi, allowable bearing stress on one rivet
Area for bearing = rivet diam. × plate thickness
Number of rivets in the depth of a girder between flange rivet lines or in a length of flange equal to said depth:

$$\text{Number} = \frac{\text{vertical shear}}{\text{allowable load per rivet}}$$

Diagonal Tensional Shear:

$$V_d = \frac{\text{vertical shear (diagonal length across a panel)}}{d, \text{ depth of girder, in.}}$$

Trautwine states that this strain is rarely considered. The vertical shear and stiffeners usually negate the need to check this parameter.

Vertical stiffeners are to be placed on girders at a spacing of 3–5 ft or the depth of girder, whichever is smaller. The spacing should be reduced somewhat (by approximately one-half) at support ends of simple spans.

Use the compression formula and the vertical shear at the point under consideration to determine the area of the stiffener.

Kidder (1902) gives details on the design of wrought iron riveted girders. Kidder gives rules pertaining to the dimensions and the various areas of the girders as follows (ibid., 411):

a. The Depth of a riveted girder may be from 1/10 to 1/16 of the span. The greatest economy was said to be 1/12 of the span.
b. The width of the compression flange shall not be less than 1/20 of the distance between lateral supports. and if there are no lateral supports, then use the distance as the clear span.
c. Arches between girders or floor beams riveted to the sides of girders may be considered as lateral supports.
d. The term "flange" as applied to riveted girders includes all metal at the top or bottom of a girder, exclusive of the web plate material.
e. The term "depth" of a riveted girder means the distance between the centers of gravity of the two flanges.

Kidder also quotes Birkmire as providing 12 points on the details of construction of wrought iron riveted girders (ibid.):

1. All connections and details shall be of such strength that, upon testing, rupture shall occur in the body of the members rather than in any of the details of construction. In members subjected to tensile load full allowance shall be made for the reduction of section by rivet-holes.
2. The webs of plate-girders, when they cannot be had in one length, must be spliced at all joints by a plate on each side of the web. Tees must not be used for splices.
3. Stiffeners shall be used at the ends of all girders and wherever concentrated loads occur, and elsewhere when the shearing strain is greater than the resistance to buckling.
4. The pitch (distance between centers) of rivets shall not exceed 6 in., nor 16 times the thickness of the thinnest outside plate, nor be less than 2.25 inches for 0.75 inch rivets, or 2.625 inches for 0.875 inch rivets, in a straight line.
5. The rivets used shall be 0.75 inch in diameter for plates from 0.375 inch to 0.625 inch thick, and 0.875 inch in diameter for greater thicknesses of plates.
6. The distance between the edge of any piece and the center of a rivet hole must never be less than 1.25 inches.

7. In punching plates or other iron, the diameter of the die shall in no case exceed the diameter of the punch more than 0.0625 inches.
8. All rivet holes must be so accurately punched that when the several parts forming one member are assembled together, a rivet 0.0625 inch less in diameter than the hole can be entered, hot, into any hole without reaming or straining the iron by "drifts".
9. The rivets when driven must completely fill the holes.
10. The rivet heads must be hemispherical, except where flush surfaces are required, and a uniform size for the same sized rivets throughout the work. They must be full and neatly made, and be concentric to the rivet hole.
11. Whenever possible all rivets must be machine driven.
12. The several pieces forming one built member must fit closely together, and, when riveted, shall be free from twists, bends, or open joints.

Splicing of the various components of riveted girders was always a question in designers' minds, it would be today. Kidder provides some rules for the splices or special fabrication methods for making a girder long enough to require splicing of component plates (ibid., 412)

Girders forty feet and less in length will not require any splicing, as the plates and angles can be readily handled in one length.

In splicing the top flange, when of two or more thicknesses, no additional cover-plate will be required over the joint, but the ends should be planed true and butt solidly. The rivets are to be closer near the joint.

The plate covering the bottom flange must be of the same area as the plates joined, and of sufficient length to take a number of rivets equal to the strength of the cover plate.

Kidder also outlines the procedure for designing riveted girders (ibid.):

1. To determine the necessary flange area.
2. To determine the thickness of the web to resist, (a) shearing, and (b) buckling. This step also determines the need for stiffeners.
3. To determine the number and the pitch of the rivets.
4. To determine the length of the outside flange plates. When there is only a single flange plate this step is not required.

Utilization of the rules given above: At this time several theories directed the designers in sizing the flange area of the girder. One theory was that the bending moment was carried entirely by the flanges and that the web area carried the shear. (None of the web area counted for flange area.) Some designers considered one-sixth of the depth of the web as being included in the flange area. Some designers considered the full moment of inertia as working to resist bending moment. The New York Building Law stated that no part of the web should be used for flange area, and no more than half of that area of the angles in contact with the web. The most popular method used in the designs was to include the flange plates and the full section of the angles as the flange area (ibid.). The tension flange should consider the net area, with full consideration for rivet holes, and the compression flange could use the gross area provided the rivets are nested tight in their holes.

Equation: Area of Flange:

$$\text{Max. bending moment} = \text{area of one flange} \times \text{height} \times \text{allowable stress}$$

$$\text{Area of one flange} = \frac{\text{max. bending moment, lb-ft}}{\text{Height of web, in. } (F_t \text{ or } F_c, \text{ psi})}$$

Allowable Stresses:

Compression flange = 9,000 psi for wrought iron, 12,000 psi for steel

Tension flange = 10,000 psi for wrought iron, 13,000 psi for steel

$$\text{Allowable uniform load, lb/ft} = \frac{8(\text{net area of bottom flange})(\text{height, ft})(F_t)}{\text{span, ft}}$$

The actual weight of the girder must be subtracted from the above capacity.
The allowable concentrated load at center would be one-half of the result from the formula minus the total weight of the girder.

Equation: Thickness of Web:

$$\text{Net sectional area of web, in.}^2 = \frac{\text{maximum shear at reaction, lb}}{F_v, \text{ psi}}$$

where:
F_v = 6000 psi for wrought iron, 7,000 psi for steel

The steel value is very conservative. Carnegie Steel Co. and several building laws give an allowable of 10,000 psi for steel.

Equation: Resistance of Web to Buckling

$$\text{Shear requiring no stiffeners, lb} = \frac{10{,}000 \text{ psi (net area web, in.}^2)}{1 + \dfrac{h^2}{2000\,(t^2)}}$$

When the actual shear on the member is larger than the Shear requiring no stiffeners, then stiffeners must be employed.

Minimum stiffener = $3 \times 3 \times 0.375$ in.

Usual maximum = $4 \times 4 \times 0.5$ in. (angles)

Stiffeners must be riveted to the web. Stiffeners may be bent around the flange angles, or the space equal to the thickness of the flange angles may be made up with a filler plate of thickness equal to the thickness of the flange angle.
Stiffener spacing in the area where stiffeners are required shall not be more than 1.25 times the height of the web. The first stiffener is to be at the edge of the support or as near as possible to the connection, or be incorporated into the design of the connection.

where:
h = height of web, in.
t = thickness of web; in.

Bearing of the ends of girders: The general rule was to make the depth of bearing equal to one-half of the web depth and the width equal to the total width of flange.

Spacing of rivets: Rivets attaching flange angles to web of girder:
The force at the point of flange attachment to the web is equal to the moment at that point divided by the depth of the web. The number of rivets between any point and the near end of the girder is as follows:

$$\text{Number of rivets} = \frac{\text{horizontal flange force}}{\text{Shear capacity of one rivet}} \text{ or } \frac{\text{horizontal flange force}}{\text{bearing capacity of one rivet}}$$

Use the larger number of calculated rivets

Total number of rivets in the web angle from end to end:

$$\text{Number of rivets} = \frac{2 \times \text{max bending moment, ft-lb}}{\text{height of web, ft (least resistance of one rivet)}}$$

The rule for rivets was that if the number of rivets required by formulae gave a rivet spacing of more than 6 in., then the number of rivets must be increased to provide a spacing of 6 in. maximum.

Rivets in the flange side of the flange angles:
Single cover plate girders require the same number of rivets as the web side of the angles and at the same spacing, (maximum of 6 in.) and staggered to the center of the spacing of the web side of the angles. The Flange side

required the same number of rivets and spacing as the web side of the angles for the first 3 ft from the end. Beyond that point to the center of the girder, one-half the number of rivets was considered sufficient, provided the spacing does not exceed 6 in.

Girders with two or more cover plates require sufficient numbers of rivets between the end of any plate and the point where its resistance is required to transfer to the angles connecting web to flange. The rivets must develop the safe strength of the plate at points along the plate that correspond to moments and the force of the couple that forms the moment. Toward the center of span, the rivets can be placed according to the rule of greatest pitch.

Example from the Kidder (1902): Method

Wrought iron girder, $F_t = 10,000$ psi, $F_c = 9000$ psi, $F_v = 7000$ psi
Girder span = 50 ft, spacing = 16 ft on centers
Live and dead load = 125 psf
Load on girder = 50 ft × 16 ft. × 125 psf = 100,000 lb

$$\text{Weight of girder} = \frac{W, \text{tons} \times \text{span, ft}}{700} = \frac{50 \text{ tons} \times 50 \text{ ft}}{700} = 3.57 \text{ tons} = 7142.9 \text{ lb}$$

Total uniformly distributed load = 107,142.9 lb = 2142.9 lb/ft

Flange area: assume web depth = 36 in. (about 1/16 of span)

Assume flange width = 12 in. (wood joists brace upper flange)

$$\text{Max. bending moment} = \frac{W, \text{lb} \times L \text{ ft}}{8} = \frac{107,142.9 \text{ lb} \times 50 \text{ ft}}{8} = 669,643.1 \text{ ft-lb}$$

Flange area:

$$\text{Compression flange} = \frac{\text{max. moment, ft-lb}}{\text{height of web, ft} \times F_b} = \frac{669,643.1 \text{ ft-lb}}{3 \text{ ft} (9,000 \text{ psi})} = 24.8 \text{ in.}^2$$

$$\text{Tension flange} = \frac{\text{max. moment, ft-lb}}{\text{height of web, ft} \times F_b} = \frac{669,643.1 \text{ ft-lb}}{3 \text{ ft} (10,000 \text{ psi})} = 22.3 \text{ in.}^2$$

Upper flange:

Compression flange angles = 5 × 3.5 × 0.5 in. (the usual angles for a flange width of up to 12 in. in width, long leg horizontal, area = 4 in.² per angle

Flange plate = 24.8 in.² − 8 in.² = 16.8 in.²

$$\text{Plate thickness} = \frac{16.8 \text{ in}^2}{12 \text{ in.}} = 1.4 \text{ in. (use one } 1 \times 12 \text{ in. and one } 0.5 \times 12 \text{ in. plate)}$$

Lower flange:

Tension flange angles = 5 × 3.5 × 0.5 in.

Use one 1 × 12 in. plate and one 0.5 × 12 in. plate (check for the net area of the tension flange)

Gross area of lower flange = 26.0 in.² (net required = 22.3 in.²)

Area reductions for rivet holes, 0.75 in. rivets:

Reduction = 2.625 in.² + 0.875 in.² = 3.500 in.²

Check of net area = 26.0 in.² − 3.500 in.² = 22.500 in.² . . . o.k.
(12 × 1 & 12 × 0.5 in. plates and 5 × 3.5 × 0.5 in. angles (2) are o.k.)

Length of flange plates:

$$\text{Force at flange attachment} = \frac{668{,}750 \text{ lb-ft}}{3 \text{ ft}} = 222{,}910 \text{ lb}$$

Design of web:

Maximum shear on girder = 53,500 lb (at support ends)

Minimum web thickness = 0.375 in.

Shear resistance of 0.375 × 36 in. plate = 94,500 lb

(Allowable F_v = 7,000 psi and from Table 4-37)

Buckling resistance of a 0.375 × 36 in. web with two 0.75 in. rivets = 31,560 lb
(Table 4-38)

53,500 lb > 31,560 lb, therefore, stiffeners are required

The usual spacing for stiffeners was 3 ft. The stiffener angles were usually 4 × 4 × 0.375 in. angles, one each side of web.
Refer to Figure 4-53 to obtain the value of the shear at the 3-ft spacing of the stiffeners. At the shear value of 27,300 lb, stiffeners are not required; however, Kidder suggests some additional stiffeners throughout the span.

Design of Web rivets and rivet spacing:

Zone 0–3 ft

Force on flange from moment = 50,290 lb

Rivets in double shear, 0.750″ diameter rivet = 6620 lb (one rivet, double shear)

Bearing on 0.375 in. plate = 4220 lb (bearing controls)

Shear and bearing values from Table 4-36.

$$\text{Number of rivets} = \frac{50{,}290 \text{ lb}}{4{,}220 \text{ lb}} = 11.92 = 12 \text{ rivets (from end to 3 ft point)}$$

Spacing of rivets 0–3 ft = 3 in. spacing.

Zone 3–6 ft

Force on flange from moment = 43,870 lb

$$\text{Number of rivets} = \frac{43{,}870 \text{ lb}}{4{,}220 \text{ lb}} = 10.39 = 11 \text{ rivets} \quad \text{(from 3–6 ft)}$$

$$\text{Spacing of rivets 3–6 ft} = \frac{36 \text{ in.}}{11 \text{ rivets}} = 3.27 \text{ in. spacing}$$

Zone 6–9 ft

Force on flange from moment = 37,450 lb

$$\text{Number of rivets} = \frac{37{,}450 \text{ lb}}{4{,}220} = 8.874 = 9 \text{ rivets} \quad \text{(from 6–9 ft point)}$$

$$\text{Spacing of rivets 6–9 ft} = \frac{36 \text{ in.}}{9 \text{ rivets}} = 4.00 \text{ in. spacing}$$

Zone 9–12 ft

Force on flange from moment = 31,030 lb

$$\text{Number of rivets} = \frac{31{,}030 \text{ lb}}{4{,}220 \text{ lb}} = 7.35 = 8 \text{ rivets} \quad \text{(From 9–12 ft point)}$$

$$\text{Spacing of rivets 9–12 ft} = \frac{36''}{8 \text{ rivets}} = 4.50 \text{ in. spacing}$$

Zone 12–25 ft

Force on flange from moment = 60,277 lb

$$\text{Number of rivets} = \frac{60{,}277 \text{ lb}}{4{,}220 \text{ lbs}} = 14.28 = 15 \text{ rivets} \quad \text{(from end to 3 ft point)}$$

$$\text{Spacing of rivets 12–25 ft} = \frac{156 \text{ in.}}{15 \text{ rivets}} = 10.4 \text{ in.} \quad (\text{max.} = 6 \text{ ft spacing})$$

This center zone of 13 ft then requires 26 rivets at 6 in. maximum spacing.

The numbers used for the force at the intersection between the flange and the web are slightly different than those used by Kidder. Kidder scaled his values off the moment diagram,

Table 4-36. Shearing and Bearing Values of Rivets (for Riveted Girders and Wrought Iron)[a]

Diameter of Rivet (in.)		Area of Rivet	Single Shear at 7,500 psi	Bearing Value for Different Thicknesses of Plate at 15,000 pounds per square inch. (= Diameter of Rivet × Thickness of Plate × 15,00 pounds.)										
Fraction	Decimal			1/4 in.	5/16 in.	3/8 in.	7/16 in.	1/2 in.	9/16 in.	5/8 in.	11/16 in.	3/4 in.	13/16 in.	7/8 in.
3/8	0.375	0.1104	828	1,410										
7/16	0.4375	0.1503	1,130	1,640	2,050									
1/2	0.5	0.1963	1,470	1,880	2,340	2,810								
9/16	0.5625	0.2485	1,860	2,110	2,640	3,160	3,690							
5/8	0.625	0.3068	2,300	2,340	2,930	3,520	4,100							
11/16	0.6875	0.3712	2,780	2,580	3,220	3,870	4,510	5,160						
3/4	0.75	0.4418	3,310	2,810	3,520	4,220	4,920	5,630	6,330					
13/16	0.8125	0.5185	3,890	3,050	3,810	4,570	5,330	6,090	6,860	7,620				
7/8	0.875	0.6013	4,510	3,280	4,100	4,920	5,740	6,560	7,380	8,200				
15/16	0.9375	0.6903	5,180	3,520	4,390	5,270	6,150	7,030	7,910	8,790	9,670			
1	1.0	0.7854	5,890	3,750	4,690	5,620	6,560	7,500	8,440	9,380	10,310	11,250		
1 1/16	1.0625	0.8866	6,650	3,980	4,980	5,980	6,970	7,970	8,960	9,960	10,960	11,950	12,950	
1 1/8	1.125	0.9940	7,460	4,220	5,270	6,330	7,380	8,440	9,490	10,550	11,600	12,660	13,710	14,770
1 3/16	1.1875	1.1075	8,310	4,450	5,570	6,680	7,790	8,910	10,020	11,130	12,250	13,360	14,470	15,590

[a] From Kidder 1902, 565.

Table 4-37. Shearing Value of Web Plates, lb[a]

Wrought Steel. Gross Area. Unit Stress 7,000 lb

Depth (in.)	Thickness (in.)						
	3/8	7/16	1/2	9/16	5/8	3/4	7/8
28	73,500	85,750	98,000	110,250	122,500	147,000	171,500
30	78,750	91,875	105,000	118,125	131,250	157,500	183,750
32	84,000	98,000	112,000	126,000	140,000	168,000	196,000
36	94,500	110,250	126,000	141,750	157,500	189,000	220,500
40	105,000	122,500	140,000	157,500	175,000	210,000	245,000
42	110,250	128,625	147,000	165,375	183,750	220,500	257,250
46	120,750	140,875	161,000	181,125	201,250	241,500	281,750
48	126,000	147,000	168,000	189,000	210,000	252,000	294,000
Deduct for One 3/4 in. Rivet							
	2,240	2,660	3,010	3,430	3,850	4,620	5,390
Deduct for One 7/8 in. Rivet							
	2,625	3,080	3,500	3,920	4,375	5,250	6,125

[a] From Kidder 1902, 4212. Kidder provides an example, as follows:
EXAMPLE:—What is the safe shearing value of a 36 × 3/8" web-plate with seven 3/4-inch rivets in stiffeners?

$$\text{Answer. Gross shearing value} = 94{,}500 \text{ lbs.}$$
$$\text{Deduct for 7 rivets, } 7 \times 2{,}240 = \underline{15{,}680} \text{ lbs.}$$
$$\text{Safe resistance} = 78{,}820 \text{ lbs.}$$

Table 4-38. Safe Buckling Value of Web Plates, psi[a]

A. Safe Buckling Value of Web Plates psi

Calculated by formula $p = \dfrac{1000}{1 + \dfrac{d^2}{3000 t^2}}$

where: d = depth, in.
t = thickness, in.

Depth (in.)	Thickness (in.)						
	3/8	7/16	1/2	9/16	5/8	3/4	7/8
28	3,498	4,228	4,890	5,476	5,932		
30	3,192	3,896	4,546	5,133	5,656	6,522	
32	2,889	3,624	4,228	4,787	5,339	6,226	6,920
36	2,456	3,069	3,666	4,229	4,748	5,656	6,392
40	2,087	2,696	3,191	3,724	4,228	5,133	5,882
42	1,930	2,455	2,983	3,498	3,992	4,889	5,649
48	1,548	1,994	2,543	2,918	3,371	4,228	4,992

B. Total Resistance for Plates with Two 3/4 in. Rivets

Depth (in.)	Thickness (in.)						
	3/8	7/16	1/2	9/16	5/8	3/4	7/8
28	34,450	48,580	64,200	80,880	97,340		
30	33,830	48,150	64,230	81,560	99,880	138,200	
36	31,360	46,000	62,800	81,500	101,750	145,300	191,570
42	29,140	43,230	60,040	79,190	100,440	147,600	198,960
48	26,860	40,360	58,820	75,920	97,450	146,670	202,000

[a] From Kidder 1902, 421 m.

and this author calculated the value of the force. The difference did not change the number of rivets in any zone.

Length of upper and lower flange plates:

Net area of outer plate, bottom flange = 6 in.² (12 × 0.5 in. pl.)

Force carried by the bottom outer plate = 6 in.² × 10,000 lb (the maximum ability of the outer plate) = 60,000 lb

One-third of this must be carried by rivets outside of points e and e', which is 20,000 lb force.

$$\text{Number of rivets outside point } e' = \frac{20{,}000 \text{ lb}}{3{,}310 \text{ lb}}$$
$$= 6 \text{ rivets, 3 each side} \quad (3{,}310 \text{ lb is single shear})$$

Point e' is 12.99 ft from the end, say 13 ft, so the plate will extend to 11 ft, 4.5 in. from the end (or 13 ft, 7.5 in. from the center), for development length, 3 rivets plus end distance equals 19.5 in. (also bearing is 18 in. beyond girder end) (see drawing of riveted girder, Figure 4-56.

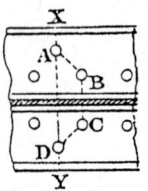

Figure 4-54.

The preceding example from Kidder contains very adequate methods of analysis for the design of wrought iron riveted girders. The allowable stresses on the component materials were very conservative. The modern analysis would use basically the same computational methods, with higher allowable stresses. The modern analysis would tend to favor analysis looking at the moment and shear capacities of the total section, utilizing the moment of inertia of the full section at point of maximum moment and the full section in shear as it is supported by bearing on a wall. The modern analysis looks at the actual conditions of use and framing support method and individually customizes the design. In a modern analysis, to determine the safe loads on

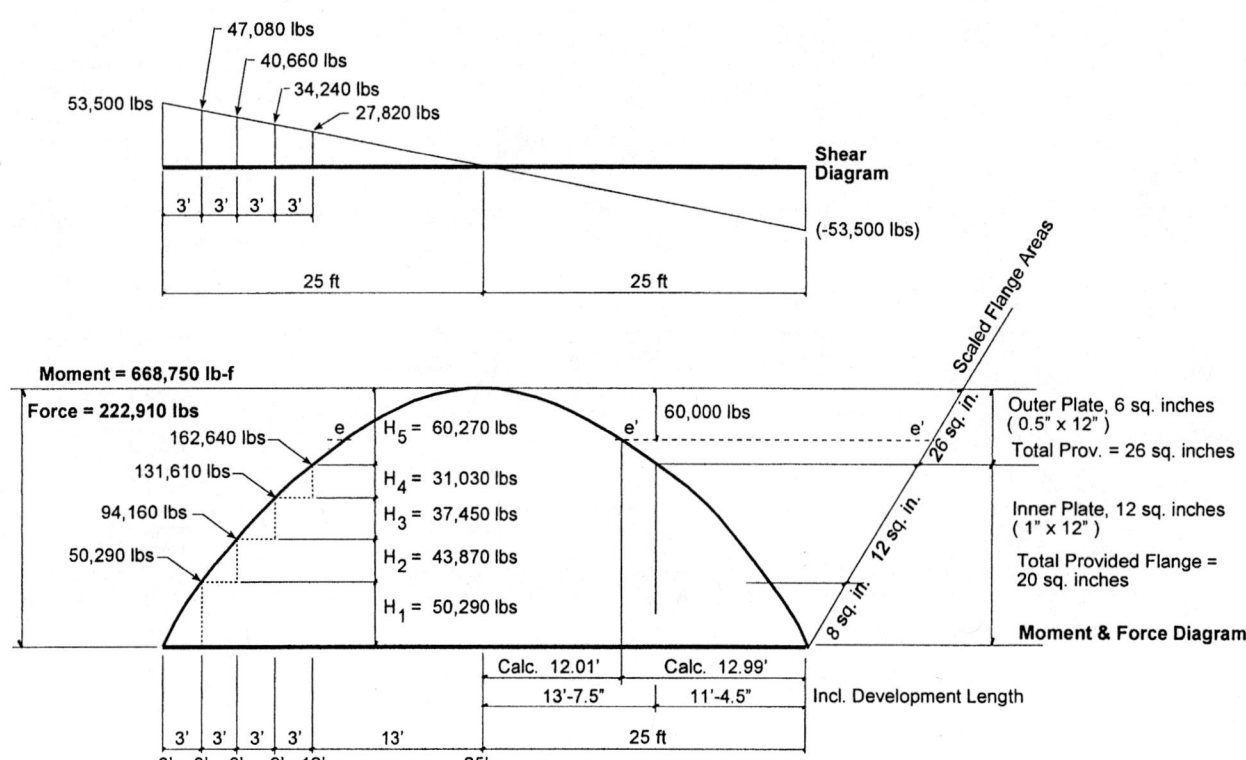

Figure 4-55. Shear and moment diagrams (moment and horizontal flange force).

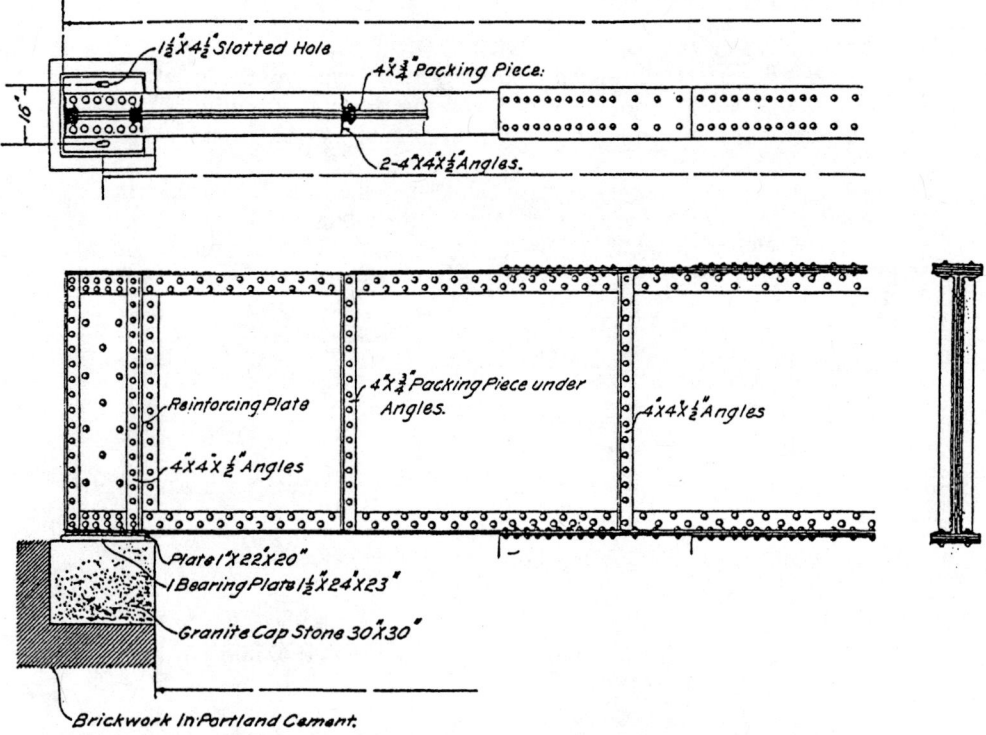

Figure 4-56. Wrought iron riveted girder (*From the example of Kidder method*).

a girder similar to the one shown in the example, it would probably be best to determine the allowable loads based upon the method presented with the allowable stresses shown in the example. If these loads are compatible with the loads from the new use, there will be no need to seek higher allowables. If the need to require heavier loads on the girder exists, then the designer could look into methods of allowing an increase in the allowable stresses. The condition of the member would have a significant impact upon the need to raise the allowable stresses in a relatively modest way, probably between 10 and 20 percent. If these increases in allowable tresses are not sufficient to raise the capacity by the required amount, then other physical changes may be required for the girder.

If a justification for the modest increase in the allowable stresses is sought, the first requirement would be to look at the ultimate stresses for the material as published at approximately the same period.

Wrought iron beams and girders were used in American buildings from the 1870s to the 1890s, a period that saw the end of the use of cast iron girders and the beginning of the use of structural steel. It is difficult to determine the exact dates when one replaced another. We have no idea how much stock was in place after the final day of production at each mill, or how and when each mill stopped production. Old traditions are hard to give up, and while we might think that one product replaced the other at an exact point in time, designers tend to stay with familiar materials and methods. Change, is sometimes a long process. Wrought iron beams and girders probably had one of the shortest periods of use as a principal product of any structural material introduced into American construction—probably 20–25. Wrought iron had certainly been used in, for example, trusses, tension rods, bolts, and plates for a much longer period, even before rolled shapes were introduced. Structural steels were so similar to wrought iron that to many the change may have been only the change in allowable stresses. Kidder (1902) treats wrought iron and steel identically, the only difference being in the allowable stresses, which at that time were not too far apart.

STEEL RIVETED GIRDERS

International Correspondence Schools (1899) has an important section on the design of beams and girders, containing valuable information on the design and fabrication of structural steel riveted girders. The ICS volume presents a method of determining the shear capacity of the web plate of a girder as follows (ibid., para. 6, 63–66):

> Rule,—From the total depth of the web-plate, deduct the sum of the diameters of the rivet holes, which will give the net or efficient depth of the web-plate; multiply the net depth by the safe resistance of the material to shear, and divide the maximum shear in pounds by the product; the quotient will be the required thickness of the material in the web of the girder.

$$t = \frac{R, \text{lb}}{F_v, \text{psi} \, (d_{net}, \text{in.})}$$

where:
R = greatest reaction or max. shear on the member
F_v = allowable shear stress
d_{net} = net depth of the member, = $d_{gross} - n$(hole diameter)

Example

Previous girder, structural steel, allowable $F_v = 0.4(30,000 \text{ psi}) = 12,000$ psi (11,000 psi used by ICS)

$R = 53,000$ lb, $d_{gross} = 36$ in.

$d_{net} = d_{gross} - n(\text{hole diameter}) = 36'' - 11(0.875'') = 26.375$ in.

$$t = \frac{53,000 \text{ lb}}{11,000 \text{ psi} (26.375'')} = 0.183 \text{ in.} \quad \text{(structural steel, min. web } t = 0.3125 \text{ in.)}$$

The result indicates that the web depth may be decreased. However, this depth may be required by bending moment calculations.

Buckling of the Web Plate:
Resistance to buckling is in psi and must be greater than the allowable shear stress before Stiffeners are not required.

$$\text{Resistance to buckling} = \frac{11,000 \text{ psi}}{1 + \frac{d^2}{3,000 t^2}}$$

The usual rule to determine the need for stiffeners was that stiffeners should be provided unless the thickness of the web-plate was at least one-fiftieth of the clear distance between the vertical legs of the flange angles.

where:
d = total depth of girder
t = thickness of girder web

The remainder of the design criteria presented by ICS volume are identical to those in the Kidder method presented above. The ICS does provide a unique method for determining the theoretical length of each of the flange plates for a uniformly loaded girder. This method is based upon a relationship between the area of a given plate and the total area of the flange. The formula to determine the length of individual plates is as follows:

$$L_2 = L_1\sqrt{\frac{a}{A}}$$

where:
- L_2 = length of flange plate in question, ft
- L_1 = span of girder, ft
- a = net area of all plates to and including the plate in question, beginning with the outside plate, sq. in^2.
- A = total net area of the entire flange, in.^{2}

The following example will demonstrate the ICS method and verify many of the features of the Kidder method.

Example

Floor system, steel riveted girder, allowable unit stress = 15,000 psi, span = 60 ft, carry zone = 17 ft, 6 in. (see Figure 4-56).

Total load = 106 psf (live, dead, and girder weight)
Uniform load on girder:

w = 106 psf (17.5 ft.) = 1855 lb/linear ft

Total load on girder:

W = 1855 lb/linear ft (60 ft.) = 111,300 lb

Maximum moment:

M = 0.125 WL = 0.125(111,300)(60) = 834,750 lb-ft

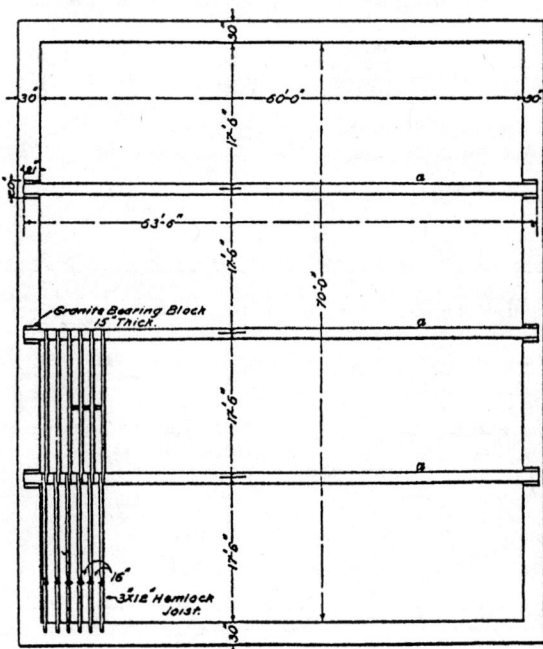

Figure 4-57. Plan drawing.

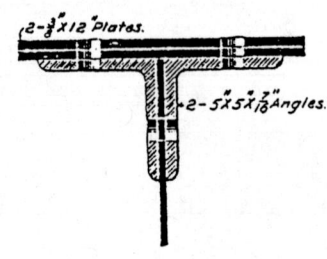

Figure 4-58.

Design of the flanges:

$$A_f = \frac{\text{Moment max. ft-lb}}{d,\text{ ft }(F_b,\text{ psi})} = \frac{834{,}750 \text{ ft-lb}}{4 \text{ ft}(15{,}000 \text{ psi})} = 13.91 \text{ in.}^2$$

2 each 3/8 × 12 in. flange plates = 9.00 in.²

2 each 5 × 5 × 7/16 in. angles = 8.36 in.²

Total = 17.36 in.²

Deduction for rivet holes:

4 each 0.875 in. = 1.312 in.² (2/8 in. plates)

2 each 0.875 in. = 1.531 in.² (7/16 in. angles)

Total = 2.843 in.²

Net area of flanges, tension and compression = 14.517 in.²

Design of Web plate:
Using 11 each 0.875 diameter rivet holes in the vertical line of the web plate Net depth of girder = 48 in. − 11(0.875 in.) = 38.375 in.

$$t = \frac{R \text{ max. shear, lb}}{d \text{ net, in. }(F_v,\text{ psi})} = \frac{55{,}650 \text{ lb}}{38.375 \text{ in. }(11{,}000 \text{ psi})} = 0.131 \text{ in.}$$

$t_{min} = 0.3125$ in. = 5/16 in. (the minimum thickness of web as recommended by all authors of period)

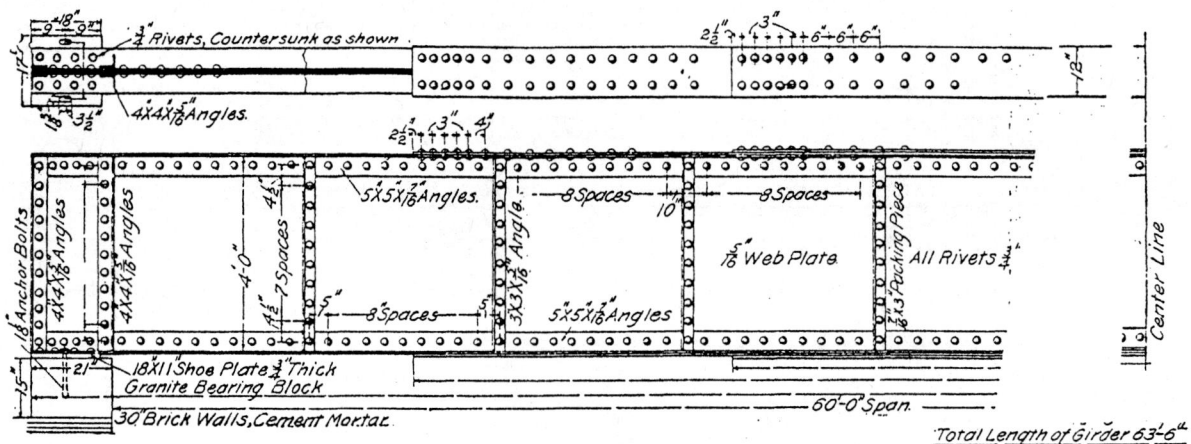

Figure 4-59. Steel Riveted Girder (*from ICS previous example computations*).

Length of flange plates:

$$L_2 = L_1 \sqrt{\frac{a}{A}} = 60 \text{ ft} \sqrt{\frac{3.844}{14.52}} = 30.87'$$

Add 1 ft at each end for rivet development, $L_{outer} = 33'$

$$L_2 = L_1 \sqrt{\frac{a}{A}} = 60 \text{ ft} \sqrt{\frac{7.688}{14.52}} = 43.667'$$

Add 1 ft at each end for rivet development, $L_{inner} = 45$ ft, 8 in.

Most designers of the period believed that it was proper to carry the plate that lies just above the angles all the way to the end of the girder for reasons of geometric composition. Technically it was not necessary to provide this additional plate length.

Stiffeners at end of girder:

Reaction at end of girder = 55,650 lb = max. shear

Allowable compressive stress used in bearing = 13,000 psi

$$\text{Area}_{req'd} \text{ Compression} = \frac{\text{Load, lb}}{\text{allowable } F_c, \text{ psi}} = \frac{55,650 \text{ lb}}{13,000 \text{ psi}} = 4.28 \text{ in.}^2$$

Use 4 each 4 × 4 × 5/16 in. angles holding flanges to web.

The actual design of rivets and spacing of the pitch of the rivets was covered in the last example and will not be covered in detail here. The design of the rivets and the spacing of the web stiffeners were very conservative and should not be a concern in the modern analysis. Splices of web and flange plates appear to have been based upon sound theory, and the need for additional stiffeners at the location of point loads on the girder was shown.

Table 4-39. Ultimate Stress of Wrought Iron, 1893 and 1913

	Ultimate Stress (psi)	Allowable Stress (psi)	Factor of Safety
Carnegie Steel Co. 1893[a]			
Tension	46,000–50,000	10,000	4.6–5
Compression	36,000–40,000	9,000	4–4.5
Shearing	45,000	7,000	6.43
Carnegie Steel Co. 1913[b]			
Tension	48,000	10,000	4.8
Compression	48,000	10,000	4.8
Shearing	40,000	8,000	5.0
Elastic limit	26,000		
Modulus of elasticity	28,000,000		

[a] At 187–189. The yield point for this material would be approximately one-half of the ultimate strength. With the average ultimate tension stress at 48,000 psi and the yield point approximated at 24,000 psi, the allowable stress of 10,000 psi would produce a factor of safety of 2.4 from yield. In the modern allowable stress design method, we utilize a maximum allowable tension stress of 0.50 F_u on net areas, which translates to a factor of safety of 2. For shear, the modern code allows 0.40 F_y, which if applied to wrought iron would be 0.40(24,000 psi) = 9,600 psi.

In a similar manner, compression, while complicated in the modern system, could allow a maximum of 0.60 F_y, which if applied to wrought iron would be 0.60(24,000 psi) = 14,400 psi.

[b] At 335. The material composition is probably exactly the same. As can be seen, the yield point and allowable stresses are now higher than the ones tabulated for 1893, which reflects the development of analytical and testing technologies rather than a change in material properties. A similar rationale would also produce even higher allowable stresses when the modern code ratios of yield to allowable are applied.

As can be seen, the allowables would be significantly higher if used in modern computational analysis. Even when we apply actual situations to further reduce the allowable stresses, i.e., lateral bracing, etc., we would allow greater loads on these members.

If a higher load could be justified by the above methods, a responsible designer would perform load tests to verify the capacity of the girder.

Wrought iron riveted girders and steel riveted girders were massive members designed to carry heavy loads and span over great distances. They were used in circumstances such as over large ballroom or convention spaces in major buildings that had spaces and continued construction over and above the space requiring the large open spans. Hotels, auditoriums, theaters, and municipal buildings would need structural members of this type. Trusses were generally used for large spans in roofs, where the loads and spaces above would not require the massive bending moments and shears that these girders had to take. Integrity of section would make the girder a very unusual member. Many of the designs of the modern period avoid the necessity for heavy girders by not locating the large open span areas in the main body of the architecture, instead making those spaces adjacent to the main building. The moment of inertia of the wrought iron member designed in the first example shown in this section will be informative:

STRUCTURAL STEEL BEAMS

Structural steel beams were apparently available as early as 1888 from Carnegie Bros. & Co., Limited, Pittsburgh, Pennsylvania. Trautwine (1888) publishes a table of the available sizes and design data for Carnegie steel I-beams" (see Table 4-40). Trautwine

Table 4-40. Steel I Beams[a]

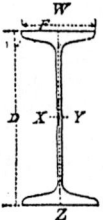

Number of Shape	Depth, D (in.)	Width of Flange (in.)	Thickness of Web (in.)	Thickness of Flange (in.) at Root	Thickness of Flange (in.) at Edge	Weight per yard[b] (lb)	Coefficient for Ultimate Center Load (lb)	Least Radius of Gyration (in.)	Moment of Inertia About XY	Moment of Inertia About WZ
301 b	20	7.00	0.60	1.14	0.66	240	2,415,000	1.39	1,449.0	45.6
301 a	20	6.25	0.50	0.98	0.55	192	1,910,000	1.20	1,146.0	27.3
302 c	15	6.16	0.785	1.25	0.875	240	1,723,800	1.29	775.7	38.8
302 b	15	5.75	0.45	0.95	0.55	150	1,177,000	1.20	529.7	21.0
302 a	15	5.50	0.40	0.78	0.40	123	942,500	1.08	424.1	14.0
303 b	12	5.50	0.39	0.88	0.50	120	781,400	1.20	281.3	16.8
303 a	12	5.25	0.35	0.72	0.35	96	617,500	1.04	222.3	10.3
304 b	10	5.00	0.37	0.82	0.47	99	537,500	1.10	161.3	11.8
304 a	10	4.75	0.32	0.65	0.32	76.5	412,200	0.99	123.7	7.32
305 b	9	4.75	0.31	0.75	0.42	81	409,700	1.07	110.6	9.10
305 a	9	4.50	0.27	0.60	0.28	63	312,300	0.95	84.3	5.56
306 b	8	4.50	0.27	0.67	0.35	66	299,400	1.01	71.9	6.62
306 a	8	4.25	0.25	0.56	0.26	54	240,600	0.91	57.8	4.35
307 b	7	4.25	0.27	0.65	0.35	60	236,500	0.97	49.7	5.52
307 a	7	4.00	0.23	0.53	0.25	46.5	183,700	0.87	38.6	3.47
308 b	6	3.625	0.26	0.59	0.34	48	159,000	0.83	28.6	3.24
308 a	6	3.50	0.23	0.50	0.25	39	130,500	0.77	23.5	2.27
309 b	5	3.13	0.26	0.54	0.33	39	104,700	0.72	15.7	1.99
300 a	5	3.09	0.22	0.44	0.23	30	82,600	0.661	12.4	1.29
310 b	4	2.75	0.24	0.49	0.30	30	64,400	0.645	7.73	1.22
310 a	4	2.625	0.20	0.38	0.20	22.5	49,000	0.584	5.90	0.752

[a] From Trautwine 1888, 523c.
[b] Messrs. Carnegie state the weights of theirr beams in pounds, per *foot*. We give them in pounds per *yard* for the sake of uniformity with our other tables.

also presents formulae for use in the design and detailing of these Carnegie beams (ibid., 523c).

Center Ultimate Load
$$= \frac{\text{Coefficient center ultimate load}}{\text{span, ft}} \quad \text{(in lb, includes half weight of beam)}$$

Coefficients based upon an assumed ultimate fiber stress of 50,000 psi.

extraneous center ultimate load = center ultimate load − half weight of member

chosen   uniformly distributed ultimate load = 2 × center ultimate load

factor of safety = 3 for static loads
= 4 to 6 for vibration or impact-type loads

$$\text{deflection} = \frac{\text{Load lb (span, ft)}^3}{\text{weight lb/yard } (d \text{ in.})(C)}$$

where:
$C = 12,000$ (load at center)
$C = 19,500$ (uniform load)

Steel beams, like wrought iron beams, were often used in pairs. Designers began to use these smaller beams in pairs to increase the strength for carrying masonry wall sections, etc. (Early steel members were relatively small in section, 20 in. deep being the largest for the early period.) Because beams were not uniform between rolling mills (standardization came later), each mill produced its own beam separators, which were cast to fit the members of that mill. Trautwine also tabulates Carnegie beam separators (see Table 4-41). The separators did not add any strength to the two sections; they were used only for the purpose of holding the members in alignment.

Example Beam Selection: Structural Steel, 1888

$F_b = 16,000$ psi

$F_{ult} = 50,000$ psi

Span = 24 ft

Load = 30,000 lb at center

Try: (a Carnegie I-beam):
Carnegie 301b 20 in. I-beam

$$P_{ult} = \frac{2,415,000}{24} = 100,625 \text{ lb}$$

$$P_{all} = P_{ult} = \text{factor of safety} = \frac{100,625 \text{ lbs}}{3} = 33,541 \text{ lb}$$

The beam weighs 240 lb/yard. The actual net load at the center of the member is 31,621 lb.

Check by Modern Analysis:

Carnegie 301b 20 in. I-beam

Moment of inertia = 1449 in.^{4}

$F_y = 25,000$ psi

$F_b = 0.667(25,000 \text{ psi}) = 16,667$ psi

$$M_{max} = \frac{PL}{4} = \frac{31,621 \text{ lb } (24')}{4} = 189,726 \text{ lb-ft}$$

$$M_{wt} = \frac{w \text{ lb/ft}(L \text{ ft})^2}{8} = \frac{80 \text{ lb/ft}(24')^2}{8} = 5,760 \text{ lb-ft}$$

Table 4-41. Carnegie Steel Beam Separators[a]

Beam				Separator		Bolts			
			Minimum Width (in.) Out to Out of the Two Flanges[c]	Weight (lb) of One Separator				Weight (lb) of 1 Bolt and Nut	
Number of Shape	Depth (in.)	Weight (lb) per yard[b]		For Minimum Width[c]	Additional for Each Additional Inch of Width	Number Used	Diameter (in.)	For Minimum Length[c]	Additional for Each Additional Inch of Length
301 b	20	240	14½	18¼	2.43	2	⅞	1.75	0.165
301 a	20	192	13¼	17½	2.46	2	⅞	1.75	0.165
302 c	15	240	12⅝	11	1.72	2	¾	1.75	0.125
302 b	15	150	12¼	12¼	1.81	2	¾	1.625	0.125
302 a	15	123	11½	11½	1.85	2	¾	1.5	0.125
303 b	12	120	11½	9¼	1.43	2	¾	1.5	0.125
303 a	12	96	11¼	9½	1.47	2	¾	1.5	0.125
303 b	12	120	11½	9¼	1.43	1	¾	1.5	0.12
303 a	12	96	11¼	9½	1.47	1	¾	1.5	0.12
304 b	10	99	10½	7	1.18	1	¾	1.5	0.12
304 a	10	76.5	10¼	7¼	1.22	1	¾	1.5	0.12
305 b	9	81	10¼	6½	1.07	1	¾	1.5	0.12
305 a	9	63	9½	6	1.09	1	¾	1.25	0.12
306 b	8	66	9½	5½	0.95	1	¾	1.25	0.12
306 a	8	54	9¼	5½	0.97	1	¾	1.25	0.12
307 b	7	60	9	4½	0.82	1	¾	1.25	0.12
307 a	7	46.5	8¾	4½	0.84	1	¾	1.25	0.12
308 b	6	48	7⅝	2¼	0.52	1	¾	1	0.12
308 a	6	39	7½	2¼	0.54	1	¾	1	0.12
309 b	5	39	6⅝	1½	0.42	1	¾	0.75	0.12
309 a	5	30	6½	1½	0.44	1	¾	0.75	0.12
310 b	4	30	6	1¼	0.33	1	¾	0.75	0.12
310 a	4	22.5	5⅞	1¼	0.35	1	¾	0.75	0.12

Separators for beams 7 inch and larger are made ½ in. thick. Separators for beams smaller than 7 inch are made ⅜ in. thick.

[a] From Trautwine 1888, 523d.
[b] Messrs. Carnegie state the weights of their beams in pounds, per *foot*. We give them in pounds per *yard* for the sake of uniformity with our other tables.
[c] The flanges of the two beams about half an inch apart.

$$M_{tot} = 195{,}486 \text{ lb-ft}$$

$$M = \frac{F_b \text{ psi (I in.}^4)}{c, \text{ in.}}$$

$$= \frac{16{,}667 \text{ psi } (1{,}449 \text{ in.}^4)}{10'' \ (12 \text{ in./ft})} = 201{,}254 \text{ lb-ft} \quad \text{(capacity of beam section)}$$

The moment capacity of the member is 201,254 lb-ft, while the moment on the member from the external load and weight of the beam is 195,486 lb-ft. The beam is satisfactory for bending moment, provided the assumption for the yield and allowable bending stresses is in accordance with sound engineering judgment. The example obviously considers full lateral support of the compression flange.

Shear is not a question, as this member will be assumed to bear on a masonry wall.

Deflection will be checked by the method of 1888 and by the modern formula.

$$\text{1888 deflection} = \frac{\text{Load, lb (span, ft)}^3}{W_b(d \text{ in.})^2(\text{constant})} + \frac{\text{Load, lb (span, ft)}^3}{W_b(d \text{ in.})^2(\text{constant})}$$

$$= \frac{31{,}621 \text{ lb } (24 \text{ ft})^3}{240 \text{ lb/yd}(20 \text{ ft})^2(12{,}000)} + \frac{1{,}920 \text{ lb/ft } (24 \text{ ft})^3}{240 \text{ lbs/yd}(20 \text{ ft})^2(19{,}500)}$$

$$= 0.394 \text{ in. total deflection, live and dead load}$$

$$\text{Deflection by modern analysis} = \frac{PL^3}{48EI} + \frac{5wL^4}{384EI} =$$

$$= \frac{31{,}621 \text{ lb } (24')^3 \, 1{,}728 \text{ in.}^3/\text{ft}^3}{48 \, (29{,}000{,}000 \text{ psi})(1{,}449 \text{ in.}^4)}$$

$$+ \frac{(5)1920 \text{ lb/ft}(24 \text{ ft})^4 \, 1{,}728 \text{ in.}^3/\text{ft}^3}{384(29{,}000{,}000 \text{ psi})(1{,}449 \text{ in.}^4)}$$

$$= 0.375 \text{ in.} + 0.068 \text{ in.}$$

$$= 0.443 \text{ in.} \quad \text{total deflection, } DL + LL$$

The deflections check to be almost exactly the same. The Trautwine method of 1888 checks in each aspect of design, and the members designed under that method should be adequate for today.

Carnegie Steel Co., (1893) presents a detailed design analysis of steel beams and girders and discusses the type of modification required for excessive deflection and the modifications for beams without lateral support or lateral support spacing that exceeds the maximum for full allowable stress. The Carnegie method of 1893 is very sophisticated and even discusses the difference between live load deflection and dead load deflection. Carnegie Steel Co. gives the beam section sizes by diagrams rather than by the tabular method (see Figure 4-60).

The properties of the Carnegie I-beams are tabulated along with their beam coefficients, as shown in Table 4-42.

Additional tables for member selection appear in Carnegie Steel Co. (1893). See Table 4-43 for examples of the load tables for Carnegie I-beams. Note that these safe loads are based upon an allowable bending stress of 16,000 psi for the steel sections available in 1893. The tables utilize the design criteria for members shown in Table 4-42 and present allowable loads for Carnegie I-beams at spans of 12–36 ft and 12–30 ft. Designers were encouraged to specify the Carnegie sections through the availibity of the Pocketbooks, and the tables and design data and examples were presented in a manner that allowed the designer who was not familiar with steel or not at ease with the new material to follow design examples and understand the use of the data. The design methods utilized coefficients determined by the engineers of the Carnegie mills, apparently through conventional flexure formulae. It would be extremely difficult and in some ways unnecessary to rederive these constants. Instead, the approach taken is to verify the reliability of the constants by designing through the modern method to check the results of the member selection through the earlier method.

Example: Steel Beam Design

From Table 4-43, A 24 in. I-beam, 80 lb/linear ft, at a clear span of 24 ft carries safely 38.14 tons or 76,280 lb total (uniformly loaded).

Deflection data, Table 4-43, show a C.S. of 610.2 for the span of 24 ft.

$$\text{Deflection max.} = \frac{\text{deflection coefficient}}{\text{beam depth, in.}} = \text{deflection in 64ths of an inch}$$

$$= \frac{610.2}{24 \text{ in.}} = 25.425 \quad \text{or} \quad \frac{25.425}{64} = 0.397 \text{ in.}$$

From the design properties, Table 4-43, same member, same span $C = 1{,}830{,}500$

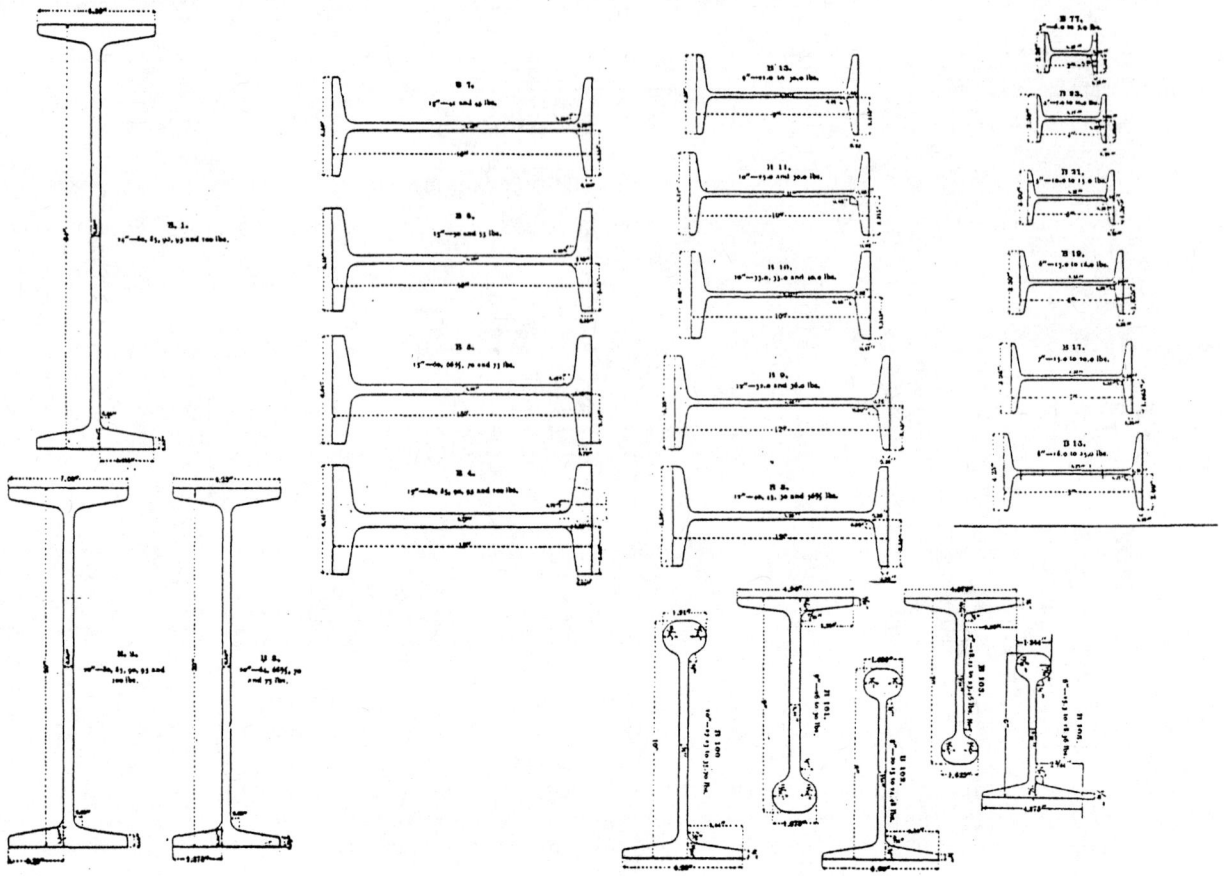

Figure 4-60. Carnegie I-beams, 1893. (*From Carnegie Steel Co. 1893, 1–5.*)

$$\text{Safe load} = \frac{\text{beam coeff.}}{\text{span}} = \frac{1{,}830{,}500}{24 \text{ ft}} = 76{,}270.8 \text{ lb}$$

$$\text{Safe applied uniform load} = \frac{76{,}270.8 \text{ lb}}{24 \text{ ft}} = 3177.95 \text{ lb/ft (minus beam weight of 80 lb/ft)}$$

Table 4-42 contains two sets of constants, C and C'. Constant C is based upon an allowable unit stress of 16,000 psi for buildings, and constant C' is based upon an allowable unit stress of 12,500 for bridges. The state of the design methods required a lower, more conservative approach to allowable stresses.

By Modern Method of Analysis

Uniform load = 3177.95 lb/ft (including beam weight of 80 lb/ft)

Simple span = 24 ft

$$\text{Moment max.} = \frac{wL^2}{8} = \frac{3177.95 \text{ lb/ft } (24 \text{ ft})^2}{8} = 228{,}812.4 \text{ lb-ft}$$

Section modulus = 171.6 in.^{3} for the 24 in. I-beam weight = 80 lb/ft

$$\begin{aligned}
\text{Moment capacity of beam} &= S_x(F_b) \\
&= 171.6 \text{ in}^3 \, (16{,}000 \text{ psi}) \\
&= 228{,}800 \text{ lb-ft (checks of moment capacity)}
\end{aligned}$$

$$\text{Deflection} = \frac{5 \, wL^4}{384 \, EI} = \frac{(5) 3{,}177.95 \text{ lbs/ft} (24 \text{ ft})^4 (1{,}728)}{384 (29{,}000{,}000 \text{ psi}) \, 2059.3 \text{ in.}^4}$$
$$= 0.397 \text{ in. (deflection due to total load)}$$

Table 4-42. Design Properties, Carnegie I-beams, 1893[a]

1	2	3	4	5	6	7	8	9	10	11	12	13	14	15	16
Section Index	Depth of Beam (in.)	Weight per ft (lb)	Area of Section (in.²)	Thickness of Web (in.)	Width of Flange (in.)	Increase of Thickness of Web for Each lb Increase of Weight (in.)	Moment of Inertia, Neutral Axis Perpendicular to Web at Center	Moment of Resistance, Neutral Axis as Before	Radius of Gyration, Neutral Axis as Before	Coefficient of Strength for Fiber Strain of 16,000 psi; Used for Buildings	Add to Coefficient for Every lb Increase in Weight of Beam	Coefficient of Strength for Fiber Strain of 12,500 psi; Used for Bridges	Add to Coefficient for Every lb Increase in Weight of Beam	Mom. of Inertia, Neutral Axis Coincident with Center Line of Web	Radius of Gyration Neutral Axis as Before
B1	24	80	23.5	0.50	6.95	0.0123	2,059.3	171.6	9.42	1,830,500	12,800	1,430,100	10,000	41.6	1.34
B2	20	80	23.5	0.60	7.00	0.015	1,449.2	144.9	7.85	1,545,600	10,450	1,207,500	8,200	45.6	1.39
B3	20	64	18.8	0.50	6.25		1,146.0	114.6	7.80	1,222,400		955,000		27.3	1.20
B4	15	80	23.5	0.77	6.41	0.020	785.9	104.8	5.82	1,117,700	7,800	873,200	6,100	42.2	1.35
B5	15	60	17.6	0.54	6.04		644.0	85.9	6.04	916,300		715,800		30.4	1.32
B6	15	50	14.7	0.45	5.75	0.020	529.7	70.6	6.00	753,300	7,800	588,500	6,100	21.0	1.20
B7	15	41	12.0	0.40	5.50		424.1	56.6	5.94	603,200		471,300		14.0	1.08
B8	12	40	11.7	0.39	5.50	0.025	281.3	46.9	4.90	500,100	6,300	390,700	4,900	16.8	1.20
B9	12	32	9.4	0.35	5.25		222.3	37.0	4.85	395,200		308,800		10.3	1.04
B10	10	33	9.7	0.37	5.00	0.029	161.3	32.3	4.08	344,000	5,200	268,800	4,100	11.8	1.10
B11	10	25	7.5	0.31	4.74		122.5	24.5	4.06	261,200		204,000		7.27	0.99
B13	9	21	6.2	0.27	4.50	0.033	84.3	18.7	3.70	199,900	4,600	156,100	3,600	5.56	0.95
B15	8	18	5.3	0.25	4.25	0.037	57.8	14.4	3.30	154,000	4,200	120,300	3,300	4.35	0.91
B17	7	15	4.4	0.21	3.98	0.042	38.0	10.86	2.92	115,800	3,600	90,500	2,800	3.42	0.87
B19	6	13	3.8	0.23	3.50	0.049	23.5	7.83	2.48	83,500	3,100	65,300	2,400	2.27	0.77
B21	5	10	3.0	0.22	3.00	0.059	12.4	4.96	2.05	52,900	2,600	41,300	2,000	1.29	0.66
B23	4	7	2.1	0.17	2.59	0.074	5.7	2.85	1.66	30,400	2,100	23,800	1,600	0.72	0.59
B77	3	6	1.8	0.20	2.26	0.098	2.6	1.74	1.21	18,560	1,560	14,500	1,220	0.47	0.51

[a] From Carnegie Steel Co. 1893, 99. The following equations are given in the original:

L = Safe Load in lbs. uniformly distributed; 1 = Span in feet,
M = Moment of forces in foot-lbs; C and C' = Coefficients given above,

$$L = \frac{C \text{ or } C'}{1}; \quad M = \frac{C \text{ or } C'}{8}; \quad C \text{ or } C' = L1 = 8M.$$

As can be seen from the above example, the methods of analysis of 1893 results identical to those from modern methods of analysis.

Carnegie Steel Co. (1893) contains an interesting notation about beams without lateral support. The consideration for lateral support of the compression flange of the beam or girder is mentioned in other volumes of about the same age or older, but this is the first mention this author has found concerning any quantitative reduction in allowable stress for members that have only partial lateral support. Table 4-43 gives a tabular summary of the limits of the length of the unbraced compression flange with the corresponding factor to be applied to the allowable stresses. The reduction factor is good for all sections, being dependent upon the width of the compression flange; however, there is no mention of the thickness of the flange. The thickness of the flange for I-beams is a variable that gets thicker towards the web of the member; therefore, the flange stiffness would not be easily calculated. The wide flange sections now utilized much more than I-beams have wider, thinner flanges and must be considered capable of buckling unless they are properly restrained laterally. The Phoenix Iron Company of Philadelphia was one of the early producers of cast iron and then wrought iron members and quickly entered the production of steel members. Phoenix was producing a full range of steel I-beams, deck beams, channels, plates, and angles by 1906. Phoenix also produced design literature, as did all other manufacturers. The Phoenix Iron Co. (1906) contains tables on the properties for designing steel beams and girders and allowable load tables (see Table 4-44).

The values of the coefficients, moments of inertia, widths and thicknesses, etc. for Phoenix steel beam sections are tabulated in Table 4-44. It is interesting to note that the industry is still not standardized, as evidenced by the fact that the values vary slightly between manufacturers. The 20 in. I-beam, 80 lb/ft. from Carnegie, 1893, had a moment of inertia of 1449.2 in.⁴, while the Phoenix section of the same depth and weight had a moment of inertia of 1470.3 in.⁴ The areas of the members would be about the same to provide the same weight, and therefore the geometric properties

Table 4-43. Safe Uniformly Distributed Loads, Design Limits, Deflection Criteria, Carnegie I-beams[a]

A. Safe Loads, Uniformly Distributed, For Carnegie I-beams[b]
(In Tons of 2,000 lb)

Distance Between Supports in feet	24 in. 30 lb	Add for every lb increase in weight	20 in. I-beam 30 in.	20 in. I-beam 34 in.	Add for every lb increase in weight	15 in. I-beam 30 in.	15 in. I-beam 64 in.	15 in. I-beam 50 in.	15 in. I-beam 41 in.	Add for every lb increase in weight	12 in. I-beam 40 lb	12 in. I-beam 32 lb	Add for every lb increase weight
12	76.27	0.53	64.40	50.93	0.44	46.58	38.18	31.39	25.13	0.33	20.84	16.47	0.26
13	70.41	0.49	59.45	47.01	0.40	42.99	35.24	28.97	23.20	0.30	19.24	15.20	0.24
14	65.38	0.46	55.20	43.66	0.37	39.93	32.72	26.90	21.54	0.28	17.86	14.12	0.22
15	61.02	0.43	51.52	40.75	0.35	37.26	30.54	25.11	20.10	0.26	16.67	13.18	0.21
16	57.20	0.40	48.30	38.30	0.33	34.93	28.63	23.54	18.85	0.25	15.63	12.35	0.20
17	53.84	0.38	45.46	35.96	0.31	32.88	26.95	22.16	17.74	0.23	14.71	11.63	0.18
18	50.85	0.36	42.93	33.96	0.29	31.05	26.45	20.93	16.75	0.22	13.90	10.98	0.17
19	48.17	0.34	40.67	32.17	0.28	29.41	24.11	19.82	15.87	0.21	13.17	10.40	0.17
20	45.76	0.32	38.64	30.56	0.26	27.94	22.91	18.83	15.08	0.20	<u>12.51</u>	<u>9.88</u>	0.16
21	43.58	0.30	36.80	29.10	0.25	26.61	21.81	17.93	14.36	0.19	11.91	9.41	0.15
22	41.60	0.29	35.13	27.78	0.24	25.40	20.82	17.12	13.71	0.18	11.37	8.98	0.14
23	39.79	0.28	33.60	26.58	0.23	24.30	19.92	16.37	13.11	0.17	10.87	8.59	0.14
24	38.14	0.27	32.20	25.47	0.22	23.29	19.09	15.69	12.57	0.16	10.42	8.23	0.13
25	36.61	0.26	30.91	24.45	0.21	<u>22.35</u>	<u>18.33</u>	<u>15.06</u>	<u>12.06</u>	0.16	10.01	7.90	0.13
26	35.20	0.25	29.72	23.51	0.20	21.30	17.62	14.48	11.60	0.15	9.62	7.60	0.12
27	33.90	0.24	28.68	22.64	0.19	20.70	16.97	13.96	11.17	0.15	9.26	7.32	0.12
28	32.69	0.23	27.60	21.83	0.19	19.96	16.36	13.45	10.77	0.14	8.93	7.06	0.11
29	31.56	0.22	26.65	21.08	0.18	19.27	15.80	12.98	10.40	0.14	8.62	6.82	0.11
30	30.51	0.21	25.76	20.37	0.17	18.63	15.27	12.55	10.05	0.13	8.34	6.59	0.10
31	29.52	0.21	24.93	19.72	0.17	18.03	14.78	12.15	9.73	0.13	8.07	6.37	0.10
32	28.60	0.20	24.15	19.10	0.16	17.46	14.32	11.77	9.43	0.13	7.81	6.18	0.10
33	27.73	0.19	23.42	18.52	0.16	16.94	13.88	11.41	9.14	0.12	7.58	5.99	0.10
34	26.92	0.19	23.73	17.97	0.15	16.44	13.48	11.08	8.87	0.11	7.36	5.81	0.09
35	26.15	0.18	22.08	17.46	0.15	15.97	13.09	10.76	8.62	0.11	7.14	5.65	0.09
36	25.42	0.18	21.47	16.98	0.15	15.52	13.73	10.46	8.38	0.11	6.95	5.49	0.09

B. Safe Loads, Uniformly Distributed, For Carnegie I-beams[b]
(In Tons of 2,000 lb)

Distance Between Supports in feet	10 in. I-beam 33 lb	10 in. I-beam 25 lb	Add for Every lb Increase in Weight	9 in. I-beam 21 in.	Add for Every lb Increase in Weight	Distance Between Supports in in.	8 in. I-beam 18 in.	Add for Every lb Increase in Weight	7 in. I-beam 15 lb	Add for Every lb Increase in Weight
12	14.33	10.88	0.22	8.33	0.20	5	15.40	0.42	11.58	0.37
13	13.23	10.05	0.20	7.69	0.18	6	12.83	0.35	9.65	0.31
14	12.29	9.33	0.19	7.14	0.17	7	11.00	0.30	8.27	0.26
15	11.47	8.71	0.17	<u>6.66</u>	0.16	8	9.63	0.26	7.24	0.23
16	<u>10.75</u>	<u>8.16</u>	0.16	6.25	0.15	9	8.56	0.23	6.43	0.20
17	10.12	7.68	0.15	5.88	0.14	10	7.70	0.21	5.79	0.18
18	9.56	7.26	0.15	5.55	0.13	11	7.00	0.19	5.27	0.17
19	9.05	6.87	0.14	5.26	0.12	12	6.42	0.17	<u>4.83</u>	0.15
20	8.60	6.54	0.13	5.00	0.12	13	<u>5.92</u>	0.16	4.45	0.14
21	8.19	6.22	0.12	4.76	0.11	14	5.50	0.15	4.14	0.13
22	7.82	5.94	0.12	4.54	0.11	15	5.13	0.14	3.86	0.12
23	7.48	5.69	0.11	4.35	0.10	16	4.81	0.13	3.63	0.11
24	7.17	5.45	0.11	4.17	0.10	17	4.53	0.12	3.41	0.11
25	6.88	5.23	0.10	4.00	0.09	18	4.28	0.12	3.22	0.10
26	6.62	5.02	0.10	3.84	0.09	19	4.05	0.11	3.04	0.10
27	6.37	4.84	0.10	3.70	0.09	20	3.85	0.10	2.90	0.09
28	6.14	4.67	0.09	3.57	0.08	21	3.67	0.10	2.76	0.09
29	5.93	4.51	0.09	3.45	0.08					
30	5.73	4.36	0.09	3.33	0.08					

Table 4-43. (Continued)

	C. Beams without Lateral Support	
	Length of Beam	Proportion of Tabular Load Forming Greatest Safe Load
	20 time flange width	Whole tabular load
	30 time flange width	9/16 tabular load
	40 time flange width	1/16 tabular load
	50 time flange width	7/16 tabular load
	60 time flange width	3/16 tabular load
	70 time flange width	5/16 tabular load

Deflection Coefficients for Carnegie Shapes, Given in 64ths of an Inch[c]

Coefficient Index	Distance Between Supports in Feet								
	6	8	10	12	14	16	18	20	22
C. S.	38.1	67.8	105.9	152.5	207.6	271.2	343.2	423.7	512.7
C.' S.	29.8	53.0	82.8	119.2	162.2	211.8	268.1	381.0	400.5
C. L.	30.7	54.6	85.3	122.9	167.3	218.4	276.5	341.3	413.0
C'. L.	25.6	45.5	71.1	102.4	139.4	182.0	230.4	284.4	344.2

Deflection Coefficients for Carnegie Shapes, Given in 64ths of an Inch[c]

Coefficient Index	Distance Between Supports in Feet								
	24	26	28	30	32	34	36	38	40
C. S.	610.2	716.1	830.5	953.4	1,085.0	1,225.0	1,373.0	1,530.	1,695.
C'. S.	476.6	559.4	648.8	744.8	847.4	956.8	1,073.0	1,195.	1,324.
C. L.	491.5	576.8	669.0	768.0	873.8	986.4	1,106.0	1,232.	1,365.
C'. L.	409.6	480.7	557.5	640.0	728.2	822.0	921.6	1,027.	1,138.

[a] From Carnegie Steel Co. 1893, 69–72.
[b] Safe loads given include weight of beam. Maximum fiber strain, 16,000 lbs. psi.
[c] Figures given opposite C. S. and C'. S. are the deflection coefficients for steel shapes, subject to transverse strain for varying spans, under their maximum uniformly distributed safe loads, derived from a fiber strain of 16,000 and 12,500 respectively; the modulus of elasticity being taken at 29,000,000. Figures given opposite C. I. and C'. I. are for iron beams, under their uniformly distributed safe loads, derived from a fiber strain of 12,000 and 10,000 respectively, the modulus of elasticity being taken at 27,000,000. To find the deflection of any symmetrical shape used as a beam under its corresponding safe load, divide the coefficients given in the above tables by the depth of the beam. This applies to such shapes as I-Beams, channels, Z-bars, etc. For those beams having unsymmetrical axes, such as tees, angles, etc., divide by twice the greatest distance of the neutral axis from the outside fibre. [original note].

would be different in order to produce the apparent differences in design properties. The Phoenix Handbook also restricted the allowable stresses on Phoenix steel beams to 16,000 psi. See Table 4-45 for tables of Phoenix I-Beams, steel box girders, and steel plate girders.

The Cambria Steel Company of Philadelphia was a producer of steel and issued handbooks for designers' use. The 1919 Handbook provides tables of properties and load tables (see Table 4-46).

The Carnegie Steel Co. of (1913) contains the same types of tables. Allowable stresses are given at 16,000 psi. Many new shapes had appeared by this time (see Table 4-47).

The Bethlehem Steel Company of Bethlehem, Pennsylvania, also issued catalogues that included the necessary design data and methodology to assist architects and engineers in the design of buildings. The 1926 edition of Catalogue S-18 contains useful information on all Bethlehem products. See Table 4-48 for examples of the allowable load tables from that catalogue.

Bethlehem Catalogue S-18 states that many city building laws were now (1926) permitting an allowable bending stress of 18,000 psi and that the catalogue contains the beam strength coefficient's C, C', and C'', which represent allowable stresses of 18,000 psi, 16,000 psi, and 12,000 psi. The first two are for buildings and the last is for bridges and buildings with moving loads. The Bethlehem catalogue also states that the coefficient of strength = 0.667(allowable F_b, psi)(section modulus, in.^{3}). See Tables 4-49 and 4-50 for Bethlehem section properties and table examples.

In 1921, the American Institute of Steel Construction was formed to bring about consensus among manufacturers of steel products, standardize shapes where possible,

Table 4-44. Properties of Phoenix I-beams[a]

Section Number	Depth of Beam (in.)	Weight per ft (lb)	Area of Section (in.²)	Thickness of Web (in.)	Width of Flange (in.)	Moment of Inertia, Neutral Axis Perpendicular to Web at Center I	Moment of Inertia, Neutral Axis Coincident with Center Line of Web I'	Radius of Gyration, Neutral Axis Perpendicular to Web at Center r	Radius of Gyration, Neutral Axis Coincident with Center Line of Web r'	Section Modulus, Neutral Axis Perpendicular to Web at Center s	Coefficient of Strength for Fiber Stress of 16,000 psi; Used for Bridges c	Coefficient of Strength for Filter Stress of 12,500 psi, Used for Bridges c'	Distance, Center to Center, required to make Radii of Gyration Equal $\frac{I<n>}{I}$	Depth of Beam (in.)
208	20	100.0	29.41	0.894	7.044	1,667.6	48.93	7.53	1.29	166.8	1,778,800	1,389,700	14.76	20
		95.0	27.94	0.821	6.971	1,618.3	47.28	7.62	1.30	161.8	1,726,200	1,348,600	14.92	
		90.0	26.47	0.747	6.897	1,569.3	45.63	7.71	1.32	156.9	1,673,600	1,307,500	15.10	
		85.0	25.00	0.674	6.824	1,519.6	43.98	7.80	1.33	152.0	1,620,900	1,266,300	15.30	
		80.0	23.53	0.600	6.750	1,470.3	42.33	7.90	1.34	147.0	1,568,300	1,225,300	15.47	
206	20	75.0	22.06	0.649	6.399	1,268.9	30.25	7.58	1.17	126.9	1,353,500	1,057,400	14.98	20
		70.0	20.59	0.575	6.325	1,219.9	29.04	7.70	1.19	122.0	1,301,200	1,016,600	15.21	
		65.0	19.12	0.500	6.250	1,169.6	27.86	7.83	1.21	117.0	1,247,600	974,700	15.47	
207	18	70.0	20.59	0.705	6.245	921.3	24.62	6.69	1.09	102.4	1,091,900	853,000	13.20	18
		65.0	19.12	0.623	6.163	881.5	23.47	6.79	1.11	97.9	1,044,800	816,200	13.40	
		60.0	17.65	0.542	6.082	841.8	22.38	6.91	1.13	93.5	997,700	779,500	13.63	
		55.0	16.18	0.460	6.000	795.6	21.19	7.07	1.15	88.4	943,000	736,700	13.95	
161	15	100.0	29.41	1.192	6.792	900.5	50.98	5.53	1.31	120.1	1,280,700	1,000,600	10.75	15
		95.0	27.94	1.094	6.694	872.9	48.37	5.59	1.32	116.4	1,241,500	969,900	10.86	
		90.0	26.47	0.996	6.596	845.4	45.91	5.65	1.32	112.7	1,202,300	939,300	10.99	
		85.0	25.00	0.898	6.498	817.8	43.57	5.72	1.32	109.0	1,163,000	908,600	11.13	
		80.0	23.53	0.800	6.400	795.5	41.76	5.78	1.32	106.1	1,131,300	883,900	11.25	
162	15	75.0	22.06	0.882	6.292	691.2	30.68	5.60	1.18	92.2	983,000	768,000	10.95	15
		70.0	20.59	0.784	6.194	663.6	29.00	5.68	1.19	88.5	943,800	737,400	11.11	
		65.0	19.12	0.686	6.096	636.0	27.42	5.77	1.20	84.8	904,600	706,700	11.29	
		60.0	17.65	0.590	6.000	609.0	25.96	5.87	1.21	81.2	866,100	676,600	11.49	
164	15	55.0	16.18	0.665	5.755	511.0	17.06	5.62	1.02	68.1	726,800	567,800	11.05	15
		50.0	14.71	0.567	5.657	483.4	16.04	5.73	1.04	64.5	687,500	537,100	11.27	
		45.0	13.24	0.469	5.559	455.8	15.00	5.87	1.07	60.8	648,200	506,400	11.54	
		42.0	12.35	0.410	5.500	441.7	14.62	5.95	1.08	58.9	628,300	490,800	11.70	
165	12	55.0	16.18	0.828	5.618	321.0	17.46	4.45	1.04	53.5	570,600	445,800	8.65	12
		50.0	14.71	0.705	5.495	303.3	16.12	4.54	1.05	50.6	539,200	421,300	8.83	
		45.0	13.24	0.583	5.373	285.7	14.89	4.65	1.06	47.6	507,900	396,800	9.06	
		40.0	11.76	0.460	5.250	268.9	13.81	4.77	1.08	44.8	478,100	373,500	9.29	
166	12	35.0	10.29	0.436	5.086	228.3	10.07	4.71	0.99	38.0	405,800	317,000	9.21	12
		31.5	9.26	0.350	5.000	215.8	9.50	4.83	1.01	36.0	383,700	299,700	9.45	

[a] From Phoenix Iron Co. 1906, 48, 49.

and provide designers with a set of standard specifications for design of structural steel for buildings and bridges. A number of joint committees had proposed several specifications that included load requirements and design methodology for structural steels. (See Chapter 3 for additional discussion of the evolution of these committees.) The Standard Specifications of the American Institute of Steel Construction were first adopted June 1, 1923, and revised November 1, 1923. The first edition of the *Steel Construction Manual* was published in 1930. This manual, which is in its ninth edition today, is the standard for all structural steel design and has all but eliminated the pocketbooks and handbooks issued by individual manufacturers. The A.I.S.C. Manual contains tables on the properties and the allowable loads on steel beams and girders (see Tables 4-51, 4-52, and 4-53). Standardization was not yet in place, and, as can be seen, the A.I.S.C. labels beams by manufacturer in this manual. The specifications contained in the manual reflect the analytical methods of the period. The allowable stresses as determined by the A.I.S.C. specifications are as follows (American Institute of Steel Construction 1930, 211):

Table 4-45. Phoenix I-beams, Steel Box Girders, and Steel Plate Girders[a]

A. Safe Loads, Uniformly Distributed, for I-beams
(in Tons of 2,000 lb):
Maximum Fiber Stress, 16,000 psi

Clear Span (ft)	20 in. I No 206				20 in. I No. 208				15 in. I. No. 162		15 in. I No. 161				18 in. I No. 207			
	65 lb	70 lb	75 lb	80 lb	85 lb	90 lb	95 lb	100 lb	70 lb	75 lb	80 lb	85 lb	90 lb	95 lb	55 lb	60 lb	65 lb	70 lb
10	62.4	65.1	67.7	78.2	80.5	83.1	85.7	88.3	47.2	49.2	56.1	58.0	60.0	61.9	47.1	49.9	52.2	54.6
11	56.7	59.1	61.5	71.1	73.1	75.5	77.9	80.2	42.9	44.7	51.0	52.7	54.5	56.3	42.9	45.4	47.5	49.6
12	52.0	54.2	56.4	65.2	67.0	69.2	71.4	73.6	39.3	41.0	46.8	48.4	50.0	51.6	39.3	41.6	43.5	45.5
13	48.0	50.0	52.1	60.2	61.9	63.9	65.9	67.9	36.3	37.8	43.2	44.6	46.1	47.6	36.3	38.4	40.2	42.0
14	44.6	46.5	48.3	55.9	57.5	59.3	61.2	63.1	33.7	35.1	40.1	41.4	42.8	44.2	33.7	35.6	37.3	39.0
15	41.6	43.4	45.1	52.1	53.6	55.4	57.1	58.9	31.5	32.8	37.4	38.7	40.0	41.3	31.4	33.3	34.8	36.4
16	39.0	40.7	42.3	48.9	50.3	51.9	53.6	55.2	29.5	30.7	35.1	36.3	37.5	38.7	29.5	31.2	32.7	34.1
17	36.7	38.3	39.8	46.0	47.3	48.9	50.4	51.9	27.8	28.9	33.0	34.1	35.3	36.4	27.7	29.3	30.7	32.1
18	34.7	36.1	37.6	43.4	44.7	46.2	47.6	49.1	26.2	27.3	31.2	32.2	33.3	34.4	26.2	27.7	29.0	30.3
19	32.8	34.2	35.6	41.2	42.3	43.7	45.1	46.5	24.8	25.9	29.5	30.5	31.6	32.6	24.8	26.3	27.5	28.7
20	31.2	32.5	33.8	39.1	40.2	41.5	42.8	44.1	23.6	24.6	28.1	29.0	30.0	31.0	23.6	24.9	26.1	27.3
21	29.7	31.0	32.2	37.2	38.3	39.6	40.8	42.0	22.5	23.4	26.7	27.6	28.6	29.5	22.5	23.8	24.9	26.0
22	28.4	29.6	30.8	35.5	36.6	37.8	38.9	40.1	21.5	22.3	25.5	26.4	27.3	28.2	21.4	22.7	23.7	24.8
23	27.1	28.3	29.4	34.0	35.0	36.1	37.2	38.4	20.5	21.4	24.4	25.2	26.1	26.9	20.5	21.7	22.7	23.7
24	26.0	27.1	28.2	32.6	33.5	34.6	35.7	36.8	19.7	20.5	23.4	24.2	25.0	25.8	19.6	20.8	21.8	22.7
25	25.0	26.0	27.1	31.3	32.2	33.2	34.3	35.3	18.9	19.7	22.4	23.2	24.0	24.8	18.9	20.0	20.9	21.8
26	24.0	25.0	26.0	30.1	30.9	31.9	33.0	34.0	18.2	18.9	21.6	22.3	23.1	23.8	18.1	19.2	20.1	21.0
27	23.1	24.1	25.1	29.0	29.8	30.8	31.7	32.7	17.5	18.2	20.8	21.5	22.2	22.9	17.5	18.5	19.3	20.2
28	22.3	23.2	24.2	27.9	28.7	29.7	30.6	31.5	16.9	17.6	20.0	20.7	21.4	22.1	16.8	17.8	18.6	19.5
29	21.5	22.4	23.3	27.0	27.7	28.6	29.5	30.5	16.3	17.0	19.4	20.0	20.7	21.4	16.3	17.2	18.0	18.8
30	20.8	21.7	22.6	26.1	26.8	27.7	28.6	29.4	15.7	16.4	18.7	19.3	20.0	20.6	15.7	16.6	17.4	18.2
31	20.1	21.0	21.8	25.2	26.0	26.8	27.6	28.5	15.2	15.9	18.1	18.7	19.4	20.0	15.2	16.1	16.9	17.6
32	19.5	20.3	21.1	24.4	25.1	26.0	26.8	27.6	14.8	15.4	17.5	18.1	18.7	19.4	14.7	15.6	16.3	17.1
33	18.9	19.7	20.5	23.7	24.4	25.2	26.0	26.8	14.3	14.9	17.0	17.6	18.2	18.8	14.3	15.1	15.8	16.5
34	18.3	19.1	19.9	23.0	23.7	24.4	25.2	26.0	13.9	14.5	16.5	17.1	17.6	18.2	13.9	14.7	15.4	16.1
35	17.8	18.6	19.3	22.3	23.0	23.7	24.5	25.2	13.5	14.0	16.0	16.6	17.1	17.7	13.5	14.3	14.9	15.6
36	17.3	18.1	18.8	21.7	22.4	23.1	23.8	24.5	13.1	13.7	15.6	16.1	16.7	17.2	13.1	13.9	14.5	15.2

Bending stresses:
Total lateral support of compression flange:

$F_b = 18{,}000$ psi

When unsupported length exceeds 15 times b_f:

$$F_b = \frac{20{,}000 \text{ psi}}{1 + \dfrac{L^2}{2{,}000\,(b_f)^2}}$$

where:
L = unsupported lateral length, in.
b_f = width of flange

The laterally unsupported length of beams and girders shall not exceed $40 \times b_f$.

Shearing stresses:
Allowable stress for shear on the gross area of webs:

$F_v = 12{,}000$ psi

Table 4-45. (Continued)

B. Beam Box Girders:
Safe Loads in Tons, Uniformly Distributed[b]

	6¾" c. to c.				7" c. to c.			
	2 Plates 16" × ¾"	18" I-beams 55 lbs per Foot			2 Plates 16" × ¾"	20" I-beams 65 lbs per Foot		
Distance Center to Center of Bearings, in Feet	Safe Load. Uniformly Distributed (including Weight of Girder), in Tons of 2,000 lb	Weight of Girder (including Rivet Heads), in Tons of 2,000 lb	Increase in Safe Load for ¹⁄₁₆" Increase in Thickness of Flange Plates	Increase of Weight in Girder for ¹⁄₁₆ in. Increase in Thickness of Flange Plates	Safe Load, Uniformly Distributed (including Weight of Girder), in Tons of 2,000 lb	Weight of Girder (including Rivet Heads), in Tons of 2,000 lb	Increase in Safe Load for ¹⁄₁₆ in. Increase in Thickness of Flange Plates	Increase of Weight in Girder for ¹⁄₁₆ in. Increase in Thickness of Flange Plates
10	170.23	0.98	7.52		205.88	1.07	8.47	0.03
11	154.76	1.08	6.84		187.15	1.18	7.70	0.04
12	141.86	1.18	6.27		171.57	1.29	7.06	0.04
13	130.95	1.27	5.78		158.37	1.39	6.52	0.04
14	121.59	1.37	5.38		147.06	1.50	6.05	0.05
15	113.49	1.47	5.02		137.25	1.61	5.64	0.05
16	106.39	1.57	4.70		124.68	1.72	5.30	0.05
17	100.14	1.66	4.42		121.11	1.82	4.98	0.06
18	94.57	1.76	4.18		114.31	1.93	4.71	0.06
19	89.60	1.86	3.96		108.36	2.04	4.45	0.06
20	85.12	1.96	3.76		102.93	2.15	4.23	0.07
21	81.06	2.06	3.58		98.05	2.25	4.03	0.07
22	77.88	2.15	3.42		93.58	2.36	3.85	0.07
23	74.01	2.25	3.27		89.52	2.47	3.68	0.08
24	70.93	2.35	3.14		85.78	2.58	3.53	0.08
25	68.09	2.45	3.01		82.36	2.68	3.39	0.08
26	65.47	2.55	2.90		79.19	2.79	3.25	0.09
27	63.05	2.64	2.78		76.25	2.90	3.14	0.09
28	60.80	2.74	2.69		73.52	3.00	3.02	0.09
29	58.70	2.84	2.60		71.00	3.11	2.92	0.10
30	56.74	2.94	2.50		68.63	3.22	2.83	0.10
31	54.91	3.03	2.42		66.42	3.33	2.73	0.10
32	53.20	3.13	2.35		64.34	3.43	2.64	0.11
33	51.59	3.23	2.28		62.39	3.54	2.56	0.11
34	50.07	3.33	2.22		60.55	3.65	2.49	0.11
35	48.64	3.43	2.15		58.82	3.76	2.42	0.12
36	47.29	3.52	2.09		57.18	3.86	2.35	0.12
37	46.01	3.62	2.03		55.64	3.97	2.28	0.12
38	44.80	3.72	1.98		54.17	4.08	2.23	0.13
39	43.65	3.82	1.93		52.80	4.18	2.17	0.13

Where h, the height between flanges (in.) is not more than $60(t)$, the thickness of the web (in.)

If $h \geq$ or greater than $60(t)$, the allowable shear stress is reduced.

$$F_v = \frac{18{,}000 \text{ psi}}{1 + \dfrac{h^2}{7{,}200\, t^2}}$$

American Institute of Steel Construction (1939) provides a detailed section on structural steel specifications in which the ASTM Serial Designation A 9 is the featured

Table 4-45. (Continued)

C. Plate Girders:
Safe Loads in Tons, Uniformly Distributed[c]

30" × ½" Web Plates
12" × ⅜" Flange Plates
5" ts 3½" ts ½" Angles

33" × ½" Web Plates
12" × ⅜" Flange Plates
5" × 3½" × ½" Angles

Distance Center to Center of Bearings, in Feet	Safe Load Including Weight of Girder	Weight of Girder	Increase in Safe Load for 1/16" Increase in Thickness of Flange Plates	Increase in Weight of Girder for 1/16 in. Increase in Thickness of Flange Plates	Safe Load Including Weight of Girder	Weight of Girder	Increase in Safe Load for 1/16 in. Increase in Thickness of Flange Plates	Increase of Weight in Girder for 1/16 in. Increase in Thickness of Flange Plates
20	93.67	1.62	4.62	0.05	105.82	1.70	5.08	0.05
21	89.22	1.69	4.38	0.05	100.78	1.77	4.85	0.05
22	85.15	1.76	4.19	0.06	96.20	1.84	4.62	0.06
23	81.46	1.86	4.00	0.06	92.01	1.95	4.42	0.06
24	78.07	1.93	3.83	0.06	88.18	2.02	4.23	0.06
25	74.94	2.01	3.68	0.06	84.65	2.09	4.06	0.06
26	72.06	2.07	3.54	0.07	81.39	2.17	3.91	0.07
27	69.40	2.14	3.42	0.07	78.38	2.24	3.76	0.07
28	66.91	2.21	3.29	0.07	75.58	2.31	3.63	0.07
29	64.60	2.31	3.17	0.07	72.98	2.42	3.50	0.07
30	62.45	2.38	3.07	0.08	70.55	2.49	3.39	0.08
31	60.44	2.45	2.97	0.08	68.26	2.56	3.29	0.08
32	58.55	2.52	2.88	0.08	66.14	2.64	3.17	0.08
33	56.77	2.59	2.79	0.08	64.13	2.71	3.08	0.08
34	55.11	2.66	2.70	0.09	62.24	2.78	2.99	0.09
35	53.53	2.73	2.63	0.09	60.46	2.85	2.91	0.09
36	52.04	2.83	2.56	0.09	58.79	2.96	2.83	0.09
37	50.63	2.90	2.49	0.09	57.20	3.03	2.75	0.09
38	49.30	2.97	2.42	0.10	55.70	3.11	2.67	0.10
39	48.03	3.04	2.37	0.10	54.27	3.18	2.60	0.10
40	46.83	3.11	2.31	0.10	52.90	3.25	2.55	0.10

[a] From Phoenix Iron Co. 1906, 48, 49.
[b] Values are based on maximum fiber stress of 15,000 psi; 1¾ in. rivet holes in both flanges deducted. Weights of girders correspond to lengths center to center of bearings.
[c] Values are founded on the moments of inertia of the sections using a maximum fiber stress of 15,000 psi; 1¾ in. rivet holes in both flanges deducted. Weights of girders correspond to lengths center to center of bearings and include rivet heads, stiffeners, and fillers.

material as of that date. The allowable unit stresses for the members in flexure for the A.I.S.C. specification and date are as follows (ibid., 264, 265):

Bending:
　Tension on extreme fibers of flexural members:

$F_b = 20,000$ psi

Compression on extreme fibers of flexural members:

For $\dfrac{L}{b_f} \leq 40$, $\quad F_b = \dfrac{22,500 \text{ psi}}{1 + \dfrac{L^2}{1,800(b_f)^2}}$

Table 4-46. Cambria Steel, Properties of I-beams, Allowable Uniform Loads I-beams[a]

Properties of Standard I-beams

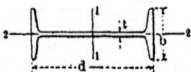

1	2	3	4	5	6	7	8	9	10	11	12	13	14	15	16	1
												Coefficient of Strength		Coefficient of Deflection		
											Increase of Thickness of Web for each Pound Increase in Weight	For Fiber Stress of 16,000 Pounds per Square inch for Buildings	For Fiber Stress of 12,500 Pounds per Square inch for Bridges			
	Depth of Beam	Weight for Foot	Area of Section	Thickness of Web	Width of Flange	Moment of Inertia Axis 1-1	Section Modulus Axis 1-1	Radius of Gyration Axis 1-1	Moment of Inertia Axis 2-2	Radius of Gyration Axis 2-2				Uniform Load	Center Load	
Section Number	d in.	lb	A in.²	t in.	b in.	I in.	S in.	r in.	I' in.	r' in.	t	r	r'	N	N'	Section Number
B65	18	55.0	15.93	0.46	6.00	795.6	88.4	7.07	21.19	1.15	.016	942,880	736,620	.00000098	.00000156	B65
		60.0	17.65	0.56	6.10	841.8	93.5	6.91	22.38	1.13		997,680	779,440	.00000092	.00000148	
		65.0	19.12	0.64	6.18	881.5	97.9	6.79	23.47	1.11		1,044,740	816,200	.00000088	.00000141	
		70.0	20.59	0.72	6.26	921.2	102.4	6.69	24.62	1.09		1,091,800	852,970	.00000084	.00000135	
B73	20	65.0	19.08	0.50	6.25	1,169.5	117.0	7.83	27.86	1.21	.015	1,247,490	974,600	.00000066	.00000106	B73
		70.0	20.59	0.58	6.33	1,219.8	122.0	7.70	29.04	1.19		1,301,110	1,016,490	.00000064	.00000102	
		75.0	22.06	0.65	6.40	1,268.8	126.9	7.58	30.25	1.17		1,353,400	1,057,340	.00000061	.00000098	
B89	24	80.0	23.32	0.50	7.00	2,087.2	173.9	9.46	42.86	1.36	.0123	1,855,310	1,449,460	.00000037	.00000060	B89
		85.0	25.00	0.57	7.07	2,167.8	180.7	9.31	44.35	1.33		1,926,950	1,505,403	.00000036	.00000057	
		90.0	26.47	0.63	7.13	2,238.4	186.5	9.20	45.70	1.31		1,989,700	1,554,450	.00000035	.00000056	
		95.0	27.94	0.69	7.19	2,309.0	192.4	9.09	47.10	1.30		2,052,440	1,603,470	.00000034	.00000054	
		100.0	29.41	0.75	7.25	2,379.6	198.3	8.99	48.55	1.28		2,115,190	1,652,490	.00000033	.00000052	

Properties of Special I-beams

Section Number	d in.	lb	A in.²	t in.	b in.	I in.	S in.	r in.	I' in.	r' in.	t	r	r'	N	N'	Section Number
B105	12	40.0	11.84	0.46	5.25	268.9	44.8	4.77	13.81	1.08	.025	478,130	373,540	.00000288	.00000462	B105
		45.0	13.24	0.58	5.37	285.7	47.6	4.65	14.89	1.06		507,930	396,820	.00000272	.00000435	
		50.0	14.71	0.70	5.49	303.4	50.6	4.54	16.12	1.05		539,300	421,320	.00000256	.00000409	
		55.0	16.18	0.82	5.61	321.0	53.5	4.45	17.46	1.04		570,670	445,830	.00000242	.00000387	
B109	15	60.0	17.67	0.59	6.00	609.0	81.2	5.87	25.96	1.21	0.20	866,130	676,670	.00000127	.00000204	B109
		65.0	19.12	0.69	6.10	636.1	84.8	5.77	27.42	1.20		904,660	706,770	.00000122	.00000195	
		70.0	20.59	0.78	6.19	663.9	88.5	5.68	29.00	1.19		943,870	737,400	.00000117	.00000187	
		75.0	22.06	0.88	6.29	691.2	92.2	5.60	30.68	1.18		983,090	768,040	.00000112	.00000180	
		80.0	23.53	0.98	6.39	718.8	95.8	5.53	32.46	1.17		1,022,300	798,670	.00000108	.00000173	
B113	15	80.0	23.57	0.80	6.40	789.1	105.2	5.79	41.31	1.32	.020	1,122,290	876,790	.00000098	.00000157	B113
		85.0	25.00	0.90	6.50	815.9	108.8	5.71	43.46	1.32		1,160,340	906,520	.00000095	.00000152	
		90.0	26.47	0.99	6.59	843.4	112.5	5.64	45.79	1.32		1,199,550	937,150	.00000092	.00000147	
		95.0	27.94	1.09	6.69	871.0	116.1	5.58	48.25	1.31		1,238,770	967,790	.00000089	.00000143	
		100.0	29.41	1.19	6.79	898.6	119.8	5.58	50.84	1.31		1,277,980	998,420	.00000086	.00000138	
B121	20	80.0	23.73	0.60	7.00	1,466.3	146.6	7.86	45.81	1.39	0.15	1,564,060	1,221,920	.00000053	.00000085	B121
		85.0	25.00	0.66	7.06	1,508.5	150.9	7.77	47.25	1.37		1,609,100	1,257,110	.00000051	.00000082	
		90.0	26.47	0.74	7.14	1,557.5	155.8	7.67	48.98	1.36		1,661,390	1,297,960	.00000050	.00000080	
		95.0	27.94	0.81	7.21	1,606.6	160.7	7.58	50.78	1.35		1,713,670	1,338,810	.00000048	.00000077	
		100.0	29.41	0.88	7.28	1,655.6	165.6	7.50	52.65	1.34		1,765,960	1,379,660	.00000047	.00000075	
B127	24	105.0	30.98	0.68	7.88	2,811.5	234.3	9.53	78.90	1.60	.0123	2,499,090	1,952,420	.00000028	.00000044	B127
		110.0	32.48	0.69	7.94	2,883.5	240.3	9.42	81.04	1.58		2,563,090	2,002,420	.00000027	.00000043	
		115.0	33.98	0.75	8.00	2,955.5	246.3	9.33	83.93	1.56		2,627,090	2,052,420	.00000026	.00000042	

Table 4-46. (Continued)

	Safe Loads in Pounds Uniformly Distributed for Cambria I-beams							
	Safe loads below are figured for fiber stress of 16,000 psi and include weight of beam							
	Standard I-beam					Special I-beam		
Distance Between Supports (ft)	24 in. No. B89					24 in. No. B127		
	80 lb	85 lb	90 lb	95 lb	100 lb	105 lb	110 lb	115 lb
18	103,070	107.050	110,540	114,020	117,510	138,840	142,390	145,950
19	97,650	•101,420	•104,720	108,020	111,330	131,530	134,890	138,270
20	92,770	96,350	99,480	•102,620	•105,760	124,950	128,150	131,350
21	88,350	91,760	94,750	97,740	100,720	119,000	122,050	125,100
22	84,330	87,590	90,440	93,290	96,140	113,590	116,500	119,410
23	80,670	83,780	86,510	89,240	91,960	108,660	111,440	114,220
24	77,300	80,290	82,900	85,520	88,130	•104,130	106,790	109,460
25	74,210	77,080	79,590	82,100	86,410	99,960	•102,530	•105,080
26	71,360	74,110	76,530	78,940	81,350	96,120	98,580	101,040
27	68,720	71,370	73,690	76,020	78,340	92,560	94,930	97,300
28	66,260	68,820	71,060	73,300	75,540	89,250	91,540	93,830
29	63,980	66,450	68,610	70,770	72,940	86,170	88,380	90,590
30	61,840	64,230	66,320	68,410	70,510	83,300	85,440	87,570
31	59,850	62,160	64,180	66,210	68,230	80,620	82,680	84,740
32	57,980	60,220	62,180	64,140	66,100	78,100	80,100	82,100
33	56,220	58,390	60,290	62,200	64,100	75,730	77,670	79,610
34	54,570	56,680	58,520	60,370	62,210	73,500	75,380	77,270
35	53,010	55,060	56,850	58,840	60,430	71,400	73,230	75,060
36	51,540	53,530	55,270	57,010	58,760	69,420	71,200	72,970
37	50,140	52,080	53,780	55,470	57,170	67,540	69,270	71,000
38	48,820	50,710	52,360	54,010	55,660	65,770	67,450	69,130
39	47,570	49,410	51,020	52,630	54,240	64,080	65,720	67,360
40	46,380	48,170	49,740	51,310	52,880	62,480	64,080	65,680
41	45,280	47,000	48,530	50,060	51,590	60,950	62,510	64,080
42	44,170	45,880	47,370	48,870	50,360	59,500	61,030	62,550
43	43,150	44,810	46,270	47,730	49,190	58,120	59,610	61,090
44	42,170	43,790	45,220	46,650	48,070	56,800	58,250	59,710
45	41,230	42,820	44,220	45,610	47,000	55,530	56,960	58,380
46	40,330	41,890	43,250	44,620	45,980	54,330	55,720	57,110
47	39,470	41,000	42,330	43,670	45,000	53,170	54,530	55,890
48	38,650	40,140	41,450	42,760	44,070	52,060	53,400	54,730

[a] From Cambria Steel Co. 1919, 117, 118, 184, 185. Above single dot, safe loads are too great for standard connections.

where:
L = Lateral unbraced length
b_f = Width of flange
Maximum of 20,000 psi

Shear:

Webs of beams, girders, gross section = 13,000 psi

Rivets, single shear = 15,000 psi

American Institute of Steel Construction (1947) provides revisions updating the formulae for the design of beams and girders for building construction. The material at this time was ASTM A 7 Structural Steel. The revised formulae for the allowable stresses are as follows (ibid., 286, 287):

Bending:
Compression on extreme fibers of flanges:

If $\dfrac{L(d)}{b_f t_f} \leq 600 \quad F_c = 20,000$ psi

Table 4-47. Carnegie 1913, Elements of Structural Beams, Allowable Uniform Loads[a]

A. Elements of Structural Beams

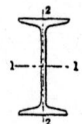

Section Index	Depth of Beam in.	Weight per foot lb	Area of Section in.²	Width of Flange in.	Thickness of Web in.	Axis 1-1			Axis 2-2		
						I in.⁴	r in.	S in²	I in.⁴	r in.	S in.²
B31	27	83.0	24.41	7.500	0.424	2,888.6	10.88	214.0	53.1	1.47	14.1
B24	24	115.0	33.98	8.000	0.750	2,955.5	9.33	246.3	83.2	1.57	20.8
		110.0	32.48	7.938	0.688	2,883.5	9.42	240.3	81.0	1.58	20.4
		105.0	30.98	7.875	0.625	2,811.5	9.53	234.3	78.9	1.60	20.0
B1	24	100.0	29.41	7.254	0.754	2,379.6	9.00	198.3	48.6	1.28	13.4
		95.0	27.94	7.193	0.693	2,309.0	9.09	192.4	47.1	1.30	13.1
		90.0	26.47	7.131	0.631	2,238.4	9.20	186.5	45.7	1.31	12.8
		85.0	25.00	7.070	0.570	2,167.8	9.31	180.7	44.4	1.33	12.6
		80.0	23.32	7.000	0.500	2,087.2	9.46	173.9	42.9	1.36	12.3
B32	24	69.5	20.44	7.000	0.390	1,928.0	9.71	160.7	39.3	1.39	11.2
B33	21	57.5	16.85	6.500	0.357	1,227.5	8.54	116.9	28.4	1.30	8.8
B2	20	100.0	29.41	7.284	0.884	1,655.6	7.50	165.6	52.7	1.34	14.5
		95.0	27.94	7.210	0.810	1,606.6	7.58	160.7	50.8	1.35	14.1
		90.0	26.47	7.137	0.737	1,557.6	7.67	155.8	49.0	1.36	13.7
		85.0	25.00	7.063	0.663	1,508.5	7.77	150.9	47.3	1.37	13.4
		80.0	23.73	7.000	0.600	1,466.3	7.86	146.8	45.8	1.39	13.1
B3	20	75.0	22.06	6.399	0.649	1,268.8	7.58	126.9	30.3	1.17	9.5
		70.0	20.59	6.325	0.575	1,219.8	7.70	122.0	29.0	1.19	9.2
		65.0	19.08	6.250	0.500	1,169.5	7.83	117.0	27.9	1.21	8.9
B81	18	90.0	26.47	7.245	0.807	1,260.4	6.90	140.0	52.0	1.40	14.4
		85.0	25.00	7.163	0.725	1,220.7	6.99	135.6	50.0	1.42	14.0
		80.0	23.53	7.082	0.644	1,181.0	7.09	131.2	48.1	1.43	13.6
		75.0	22.05	7.000	0.562	1,141.3	7.19	126.8	46.2	1.45	13.2
B80	18	70.0	20.59	6.259	0.719	921.2	6.69	102.4	24.6	1.09	7.9
		65.0	19.12	6.177	0.637	881.3	6.79	97.9	23.5	1.11	7.6
		60.0	17.65	6.095	0.555	841.8	6.91	93.5	22.4	1.13	7.3
		55.0	15.93	6.000	0.460	795.6	7.07	88.4	21.2	1.15	7.1
B34	18	46.0	13.53	6.000	0.322	733.2	7.36	81.5	19.9	1.21	6.6
B5	15	75.0	22.06	6.292	0.882	691.2	5.60	92.2	30.7	1.18	9.8
		70.0	20.59	6.194	0.784	663.7	5.68	88.5	29.0	1.19	9.4
		65.0	19.12	6.096	0.686	636.1	5.77	84.8	27.4	1.20	9.0
		60.0	17.67	6.000	0.590	609.0	5.87	81.2	26.0	1.21	8.7
B7	15	55.0	16.18	5.746	0.656	511.0	5.62	68.1	17.1	1.02	5.9
		50.0	14.71	5.648	0.558	483.4	5.73	64.5	16.0	1.04	5.7
		45.0	13.24	5.550	0.460	455.9	5.87	60.8	15.1	1.07	5.4
		42.0	12.48	5.500	0.410	441.8	5.95	58.9	14.6	1.08	5.3
B35	15	36.0	10.63	5.500	0.289	405.1	6.17	54.0	13.5	1.13	4.9

Table 4-47. (Continued)

A. Elements of Structural Beams

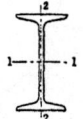

Section Index	Depth of Beam in.	Weight per foot lb	Area of Section in.²	Width of Flange in.	Thickness of Web in.	Axis 1-1			Axis 2-2		
						I in.⁴	r in.	S in.²	I in.⁴	r in.	S in.²
B8	12	55.0	16.18	5.611	0.821	321.0	4.45	53.5	17.5	1.04	6.2
		50.0	14.71	5.489	0.699	303.4	4.54	50.6	16.1	1.05	5.9
		45.0	13.24	5.366	0.576	285.7	4.65	47.6	14.9	1.06	5.6
		40.0	11.84	5.250	0.460	269.0	4.77	44.8	13.8	1.08	5.3
B9	12	35.0	10.29	5.086	0.436	228.3	4.71	38.0	10.1	0.99	4.0
		31.5	9.26	5.000	0.350	215.8	4.83	36.0	9.5	1.01	3.8
B36	12	27.5	8.04	5.000	0.255	199.6	4.98	33.3	8.7	1.04	3.5
B11	10	40.0	11.76	5.099	0.749	158.7	3.67	31.7	9.5	0.90	3.7
		35.0	10.29	4.952	0.602	146.4	3.77	29.3	8.5	0.91	3.4
		30.0	8.82	4.805	0.455	134.2	3.90	26.8	7.7	0.93	3.2
		25.0	7.37	4.660	0.310	122.1	4.07	24.4	6.9	0.97	3.0
B37	10	22.0	6.52	4.670	0.232	113.9	4.18	22.8	6.4	0.99	2.7
B13	9	35.0	10.29	4.772	0.732	111.8	3.29	24.8	7.3	0.84	3.1
		30.0	8.82	4.609	0.569	101.9	3.40	22.6	6.4	0.85	2.8
		25.0	7.35	4.446	0.406	91.9	3.54	20.4	5.7	0.88	2.5
		21.0	6.34	4.330	0.290	84.9	3.67	18.9	5.2	0.90	2.4
B15	8	25.5	7.50	4.271	0.541	68.4	3.02	17.1	4.8	0.80	2.2
		23.0	6.76	4.179	0.449	64.5	3.09	16.1	4.4	0.81	2.1
		20.5	6.03	4.087	0.357	60.6	3.17	15.2	4.1	0.82	2.0
		18.0	5.33	4.000	0.270	56.9	3.27	14.2	3.8	0.84	1.9
B38	8	17.5	5.15	4.330	0.210	58.3	3.37	14.6	4.5	0.93	2.1
B17	7	20.0	5.88	3.868	0.458	42.2	2.68	12.1	3.2	0.74	1.7
		17.5	5.15	3.763	0.363	39.2	2.76	11.2	2.9	0.76	1.6
		15.0	4.42	3.660	0.250	36.2	2.86	10.4	2.7	0.78	1.5
B19	6	17.25	5.07	3.575	0.475	26.2	2.27	8.7	2.4	0.68	1.3
		14.75	4.34	3.452	0.352	24.0	2.35	8.0	2.1	0.69	1.2
		12.25	3.61	3.330	0.230	21.8	2.46	7.3	1.9	0.72	1.1
B21	5	14.75	4.34	3.204	0.504	15.2	1.87	6.1	1.7	0.63	1.0
		12.25	3.60	3.147	0.357	13.6	1.94	5.5	1.5	0.63	0.92
		9.75	2.87	3.000	0.210	12.1	2.05	4.8	1.2	0.65	0.82
B23	4	10.5	3.09	2.880	0.410	7.1	1.52	3.6	1.0	0.57	0.70
		9.5	2.70	2.807	0.337	6.8	1.55	3.4	0.93	0.58	0.66
		8.5	2.50	2.733	0.263	6.4	1.59	3.2	0.85	0.58	0.62
		7.5	2.21	2.660	0.190	6.0	1.64	3.0	0.77	0.59	0.58
B77	3	7.5	2.21	2.521	0.361	2.9	1.15	1.9	0.60	0.52	0.48
		6.5	1.91	2.423	0.263	2.7	1.19	1.8	0.53	0.52	0.44
		5.5	1.63	2.330	0.170	2.5	1.23	1.7	0.46	0.53	0.40

Table 4-47. (Continued)

B. Beams—Allowable Uniform Load in psf

Depth (in.)	Pounds per ft.	Span (ft)																			
		10	11	12	13	14	15	16	17	18	19	20	21	22	23	24	25	26	27	28	30
27	83	22,820	18,860	15,850	13,500	11,640	10,140	8,920	7,900	7,040	6,320	5,710	5,180	4,720	4,310	3,960	3,650	3,380	3,130	2,910	2,540
	115	26,270	21,710	18,240	15,550	13,400	11,680	10,260	9,090	8,110	7,280	6,570	5,960	5,430	4,970	4,560	4,200	3,890	3,600	3,350	2,920
	110	25,630	21,180	17,800	15,170	13,080	11,390	10,010	8,870	7,910	7,100	6,410	5,810	5,300	4,850	4,450	4,100	3,790	3,520	3,270	2,850
	105	24,990	20,650	17,360	14,790	12,750	11,110	9,760	8,650	7,710	6,920	6,250	5,670	5,160	4,720	4,340	4,000	3,700	3,430	3,190	2,780
	100	21,150	17,480	14,690	12,520	10,790	9,400	8,260	7,320	6,530	5,860	5,290	4,800	4,370	4,000	3,670	3,380	3,130	2,900	2,700	2,350
24	95	20,520	16,960	14,250	12,150	10,470	9,120	8,020	7,100	6,340	5,690	5,130	4,650	4,240	3,880	3,560	3,280	3,040	2,820	2,620	2,280
	90	19,900	16,440	13,820	11,770	10,150	8,840	7,770	6,880	6,140	5,510	4,970	4,510	4,110	3,760	3,450	3,180	2,940	2,730	2,540	2,210
	85	19,270	15,930	13,380	11,400	9,830	8,560	7,530	6,670	5,950	5,340	4,820	4,370	3,980	3,640	3,350	3,080	2,850	2,640	2,460	2,140
	80	18,550	15,330	12,880	10,980	9,470	8,250	7,250	6,420	5,730	5,140	4,640	4,210	3,830	3,510	3,220	2,970	2,750	2,550	2,370	2,060
	69.5	17,140	14,160	11,900	10,140	8,740	7,620	6,690	5,330	5,290	4,750	4,280	3,890	3,540	3,240	2,980	2,740	2,540	2,350	2,190	1,900
21	57.5	12,470	10,310	8,680	7,330	6,360	5,540	4,870	4,320	3,850	3,450	3,120	2,830	2,580	2,360	2,170	2,000	1,850	1,710	1,590	1,390
	100	17,660	14,590	12,260	10,450	9,010	7,850	6,900	6,110	5,450	4,890	4,420	4,000	3,650	3,340	3,070	2,830	2,610	2,420	2,250	1,960
	95	17,140	14,160	11,900	10,140	8,740	7,620	6,690	5,930	5,290	4,750	4,280	3,890	3,540	3,240	2,980	2,740	2,540	2,350	2,190	1,900
	90	16,610	13,730	11,540	9,830	8,480	7,380	6,490	5,750	5,130	4,600	4,150	3,770	3,430	3,140	2,880	2,660	2,460	2,280	2,120	1,850
20	85	16,090	13,300	11,170	9,520	8,210	7,150	6,290	5,570	4,970	4,460	4,020	3,650	3,320	3,040	2,790	2,570	2,380	2,210	2,050	1,790
	80	15,640	12,930	10,860	9,260	7,980	6,950	6,110	5,410	4,830	4,330	3,910	3,550	3,230	2,960	2,720	2,500	2,310	2,150	2,000	1,740
	75	13,530	11,180	9,400	8,010	6,910	6,020	5,290	4,680	4,180	3,750	3,380	3,070	2,800	2,560	2,350	2,170	2,000	1,860	1,730	1,500
	70	13,010	10,750	9,040	7,700	6,640	5,780	5,080	4,500	4,020	3,600	3,250	2,950	2,690	2,460	2,260	2,080	1,920	1,790	1,660	1,450
	65	12,480	10,310	8,660	7,380	6,370	5,440	4,870	4,320	3,850	3,460	3,120	2,830	2,580	2,360	2,170	2,000	1,850	1,710	1,590	1,390
	90	14,940	12,350	10,370	8,840	7,620	6,640	5,840	5,170	4,610	4,130	3,730	3,390	3,090	2,820	2,590	2,390	2,210	2,050	1,910	1,660
	85	14,470	11,960	10,050	8,560	7,380	6,430	5,650	5,010	4,470	4,010	3,620	3,280	2,990	2,740	2,510	2,310	2,140	1,980	1,850	1,610
	80	14,000	11,570	9,720	8,280	7,140	6,220	5,470	4,840	4,320	3,880	3,500	3,170	2,890	2,650	2,430	2,240	2,070	1,920	1,790	1,560
	75	13,530	11,180	9,390	8,000	6,900	6,010	5,280	4,680	4,180	3,750	3,380	3,070	2,800	2,560	2,350	2,160	2,000	1,860	1,730	1,500
18	70	10,920	9,020	7,580	6,460	5,570	4,850	4,260	3,780	3,370	3,020	2,730	2,480	2,260	2,060	1,900	1,750	1,620	1,500	1,390	1,210
	65	10,450	8,630	7,260	6,180	5,330	4,640	4,080	3,620	3,220	2,890	2,610	2,370	2,160	1,970	1,810	1,670	1,550	1,430	1,330	1,160
	60	9,980	8,250	6,930	5,900	5,090	4,430	3,900	3,450	3,080	2,760	2,490	2,260	2,060	1,890	1,730	1,600	1,480	1,370	1,270	1,110
	55	9,430	7,790	6,550	5,580	4,810	4,190	3,680	3,260	2,910	2,610	2,330	2,140	1,950	1,780	1,640	1,510	1,400	1,290	1,200	1,050
	46	8,690	7,180	6,040	5,140	4,430	3,860	3,390	3,010	2,680	2,410	2,170	1,970	1,800	1,640	1,510	1,390	1,290	1,190	1,110	970

[a]From Carnegie Steel Co. 1913, 142, 143, 191.

If $\dfrac{L(d)}{b_f t_f} > 600$, $F_c = \dfrac{12{,}000{,}000 \text{ psi}}{\dfrac{L(d)}{b_f t_f}}$ Maximum of 20,000 psi

where:
L = Lateral unbraced length, in.
b_f = width of flange, in.
d = depth of member, in.
t_f = thickness of flange

Tension on extreme fibers of flanges: $F_t = 20{,}000$ psi
Shear:
  Webs of beams, girders, plate girders, gross section: $F_v = 1300$ psi
  Shear on rivets: $F_v = 1500$ psi

In 1953, the A.I.S.C. produced *Historical Record, Dimensions and Properties of Rolled Shapes*, in which the engineers of the A.I.S.C. compiled a summary of dimensions and design data from all of the pocketbooks and manuals that could be obtained. That volume covers the period 1873–1952 and contains wrought iron and steel beams and column shapes rolled in the United States. It also contains a summary of early unit stresses by periods and has been quoted many times in this work. See Table 4-50 for a typical page from the A.I.S.C. *Historical Record*.

Johnson et al. (1894) is one of the only reference books found by this author that discusses the continuous girder. It provides design data summarizing the moments and shears for continuous girders for two-span members through nine-span members (see Figure 4-61).

Table 4-48. Bethlehem Steel Co. Allowable Load Tables[a]

A. Properties of Bethlehem Girder Beams[b]

Section Number	Nominal Depth of Beam (in.)	Weight per ft (lb)	Area of Section (in.²)	Thickness of Web (in.)	Width of Flange (in.)	Axis X-X		
						Moment of Inertia (in.⁴) I	Radius of Gyration (in.) r	Section Modulus (in.²) S
G30	30⅛	200.0	58.52	.76	15.04	9,148.8	12.50	607.5
	30	190.0	55.52	.72	15.00	8,651.1	12.48	576.7
	29⅞	181.0	52.82	.69	14.97	8,181.0	12.45	547.6
G28	28⅛	175.0	51.02	.70	14.29	6,988.7	11.70	497.1
	28	165.0	48.19	.66	14.25	6,577.9	11.68	469.9
G26	26⅛	160.0	46.85	.67	13.79	5,576.6	10.91	427.0
	26	151.0	44.16	.63	13.75	5,237.1	10.89	402.9
	25⅞	144.0	41.99	.61	13.73	4,930.6	10.84	381.0
G24a	24⅛	149.0	43.57	.65	13.29	4,451.1	10.11	369.1
	24	141.0	41.02	.61	13.25	4,174.2	10.09	347.9
	23⅞	133.0	38.71	.58	13.22	3,912.4	10.05	327.7
G24	24⅛	129.0	37.74	.58	12.29	3,844.8	10.09	318.8
	24	121.0	35.30	.54	12.25	3,585.3	10.08	298.8
	23⅞	114.0	33.12	.51	12.22	3,340.6	10.04	279.8
G20a	20⅛	149.0	43.44	.69	12.78	3,106.6	8.46	308.8
	20	142.0	41.31	.66	12.75	2,932.3	8.43	293.2
	19⅞	135.0	39.18	.63	12.72	2,760.6	8.39	277.7
G20	20⅛	120.0	34.95	.59	12.03	2,505.5	8.47	249.1
	20	113.0	32.90	.56	12.00	2,340.2	8.43	234.0
	19⅞	107.0	31.06	.54	11.98	2,184.0	8.39	219.7
G18	18⅛	100.0	29.25	.52	11.54	1,725.7	7.68	190.5
	18	93.0	27.14	.48	11.50	1,593.4	7.66	177.0
	17⅞	87.5	25.40	.46	11.48	1,472.8	7.61	164.7

Concrete and Reinforced Concrete Floor Systems

Concrete floor systems were introduced in approximately 1880 as fill and leveling material over the early arched floor systems, which were supported by wrought iron beams. Concrete was used in foundations somewhat earlier with unreinforced footings under walls and with wrought iron and steel rail grillage foundations. Self-supporting concrete arches existed prior to 1900. They were used in much the same way as brick arches in that the concrete was unreinforced and was used only in compression. The Dennett system shown in Figure 4-62 (1900).

Many engineers and manufacturers produced their own patented floor systems, and these systems were used extensively in buildings during the era 1880–1920 as designers pressed for lightweight, fireproof floor systems. Roebling floor systems were an unreinforced concrete arch floor system that had certain patented elements, specifically the permanent centering (formwork), which was made of wire cloth stiffened with round steel rods of 0.375–0.500 in. diameter. These floors were also supported by wrought iron or steel beams (see Figure 4-63). Roebling floor arches were constructed of a cinder-concrete, generally made of 1 part Portland cement, 2.5 parts sand, and 6 parts anthracite coal cinders (Freitag 1909, 105). The Roebling arch was said to be very strong, but questionable for use as a fireproof floor system because the bottom flanges of the beams were exposed. In many applications of the Roebling system, an expanded wire mesh or flat wire mesh was suspended just under the bottom flanges of the girders and the mesh was plastered to form the finished ceiling

Table 4-48. (Continued)

Section Number	Coefficients of Strength			Maximum Safe Shear on Web (lb)	Axis Y-Y		
	For Fiber Stress of 18,000 psi for Quiescent Loads	For Fiber Stress of 16,000 psi for Quiescent Loads	For Fiber Stress of 12,000 psi for Moving Loads		Moment of Inertia (in.^{4})	Radius of Gyration (in.)	Section Modulus (in.^{2})
	C	C'	C''		I'	r'	S'
G30	7,290,000	6,480,000	4,860,000	217,400	628.5	3.28	83.6
	6,921,000	6,152,000	4,614,000	198,400	589.4	3.26	78.6
	6,571,000	5,841,000	4,381,000	184,200	552.0	3.23	73.7
G28	5,965,000	5,302,000	3,976,000	185,300	496.2	3.12	69.4
	5,638,000	5,012,000	3,759,000	167,700	462.8	3.10	65.0
G-26	5,124,000	4,555,000	3,416,000	168,200	432.8	3.04	62.8
	4,834,000	4,297,000	3,223,000	151,700	402.7	3.02	58.6
	4,572,000	4,064,000	3,048,000	143,200	375.0	2.99	54.6
G24a	4,429,000	3,937,000	2,953,000	155,500	383.3	2.97	57.7
	4,174,000	3,710,000	2,783,000	140,200	356.4	2.95	53.8
	3,932,000	3,495,000	2,621,000	128,700	330.7	2.92	50.0
G24	3,826,000	3,401,000	2,550,000	129,100	278.2	2.72	45.3
	3,585,000	3,187,000	2,390,000	114,100	256.9	2.70	41.9
	3,357,000	2,984,000	2,238,000	102,900	236.7	2.67	38.7
G20a	3,706,000	3,294,000	2,470,000	155,400	384.5	2.97	60.2
	3,519,000	3,128,000	2,346,000	145,400	360.9	2.96	56.6
	3,333,000	2,962,000	2,222,000	135,500	337.6	2.94	53.1
G20	2,989,000	2,657,000	1,992,000	123,500	260.1	2.73	43.2
	2,808,000	2,496,000	1,872,000	113,700	240.8	2.71	40.1
	2,637,000	2,344,000	1,758,000	106,900	222.3	2.68	37.1
G18	2,286,000	2,032,000	1,524,000	97,100	202.6	2.63	35.1
	2,125,000	1,888,000	1,416,000	85,500	185.1	2.61	32.2
	1,977,000	1,757,000	1,318,000	79,500	168.9	2.58	29.4

(see Figure 4-64). The plastered ceiling provided adequate protection for the bottom flanges of the beams for most normal-intensity fires. The advantages of the Roebling system were that the arches were lighter than the brick and terra cotta arches that were in common use at the time, the wire fabric centering that formed the arches was permanent, and the system required no centering or formwork. The actual weight of the arches varied with the span of the arch (the dimension between beams) (see Table 4-51).

The Columbian system was also very popular with designers. It was a patented process that utilized stirrups over the I-beams and bars that run perpendicular to the I-beams at a spacing of approximately 20 in. apart (see Figure 4-65 and Table 4-52). The transverse bars came in three sizes to allow some flexibility in the system. The system required forms (centering) beneath the transverse bars which were hung prior to the concrete pour. The Columbian system was apparently very tough— International Correspondence Schools (1899) describes tests utilizing a 240 lb mass that was dropped repeatedly from a height of 8 ft upon the middle of an 8-ft span, with the result being no damage to the floor. It is assumed that the 8-ft span was the distance between the transverse bars. The Metropolitan Fire Proofing Company of Trenton, New Jersey had a patented system of concrete floor construction that was also very popular with designers (see Figure 4-66).

A number of other floor systems were used on buildings, some more popular than others (see Figure 4-67).

Table 4-48. (Continued)

B. Safe Loads Uniformly Distributed for Bethlehem Girder Beams, in Thousands of Pounds
(Beams Being Secured against Yielding Sideways)[c]

Span in Feet	G24a			G24		
	24⅛"	24"	23⅞"	24⅛"	24"	23⅞"
	140 lb	141 lb	133 lb	129 lb	121 lb	114 lb
	311.1	280.4		258.3		
15	295.3	278.3	257.3	255.1	228.3	
16	276.8	260.9	245.8	239.1	224.1	205.8
17	260.5	245.5	231.3	225.1	210.9	197.5
18	246.1	231.9	218.4	212.6	199.2	186.5
19	233.1	219.7	206.9	201.4	188.7	176.7
20	221.5	208.7	196.6	191.3	179.3	167.9
21	210.9	198.8	187.2	182.2	170.7	159.9
22	201.3	189.7	178.7	173.9	163.0	152.6
23	192.6	181.5	171.0	166.3	155.9	146.0
24	184.5	173.9	163.8	159.4	149.4	139.9
25	177.2	167.0	157.3	153.0	143.4	134.3
26	170.3	160.5	151.2	147.2	137.9	129.1
27	164.0	154.6	145.6	141.7	132.8	124.3
28	158.2	149.1	140.4	136.6	128.0	119.9
29	152.7	143.9	135.6	131.9	123.6	115.8
30	147.6	139.1	131.1	127.5	119.5	111.9
31	142.9	134.6	126.8	123.4	115.6	108.3
32	138.4	130.4	122.9	119.6	112.0	104.9
33	134.2	126.5	119.2	115.9	108.6	101.7
34	130.3	122.8	115.6	112.5	105.4	98.7
35	126.5	119.3	112.3	109.3	102.4	95.9
36	123.0	115.9	109.2	106.3	99.6	93.3
37	119.7	112.8	106.3	103.4	96.9	90.7
38	116.6	109.8	103.5	100.7	94.3	88.3
39	113.6	107.0	100.8	98.1	91.9	86.1
40	110.7	104.4	98.3	95.7	89.6	83.9
41	108.0	101.8	95.9	93.3	87.4	81.9
42	105.5	99.4	93.6	91.1	85.4	79.9
43	103.0	97.1	91.4	89.0	83.4	78.1
44	100.7	94.9	89.4	87.0	81.6	76.3
45	98.4	92.8	87.4	85.0	79.7	74.6

REINFORCED CONCRETE BEAMS AND GIRDERS

Existing reinforced concrete beams and girders are very difficult to analyze because there are no efficient and economical methods of determining the type and size of the reinforcing in the members. The size of the member, width and depth, provides experienced designers some clues as to the probable reinforcing pattern. The concrete itself can be cored and tested, and the strength of the concrete can be determined. Semidestructive methods can be utilized to chip away the concrete to the reinforcing for identification of size and type; however, the locations to expose reinforcing must be chosen selectively to prevent structural damage to members. These locations may not determine needed information about the reinforcing in the critical areas. Load testing is the true method of checking the structural capacity. It is required in situations where there is no information about the design of the building.

Early reinforced concrete was naturally very conservative. The method of determining the ultimate strength was by use of varying-sized cubes, which when tested produced sporadic results. The early concrete mixes would produce significantly higher strengths than the design requirements. The unconfined compression tests that are the standard of the present time produce generally lower results than the cube

Table 4-48. (Continued)

Span in Feet	G20a			G20		
	20⅛"	20"	19⅞"	20⅛"	20"	19⅞"
	149 lb	142 lb	135 lb	120 lb	113 lb	107 lb
12	310.7 308.8	290.8	271.1	247.1	227.3	213.9
13	285.1	270.7	256.4	229.9	216.0	202.8
14	264.7	251.4	238.1	213.5	200.6	188.4
15	247.1	234.6	222.2	199.3	187.2	175.8
16	231.6	219.9	208.3	186.8	175.5	164.8
17	218.0	207.0	196.1	175.8	165.2	155.1
18	205.9	195.5	185.2	166.1	156.0	146.5
19	195.1	185.2	175.4	157.3	147.8	138.8
20	185.3	176.0	166.7	149.5	140.4	131.9
21	176.5	167.6	158.7	142.3	133.7	125.6
22	168.5	160.0	151.5	135.9	127.6	119.9
23	161.1	153.0	144.9	130.0	122.1	114.7
24	154.4	146.6	138.9	124.5	117.0	109.9
25	148.2	140.8	133.3	119.6	112.3	105.5
26	142.5	135.3	128.2	115.0	108.0	101.4
27	137.3	130.3	123.4	110.7	104.0	97.7
28	132.4	125.7	119.0	106.8	100.3	94.2
29	127.8	121.3	114.9	103.1	96.8	90.9
30	123.5	117.3	111.1	99.6	93.6	87.9
31	119.5	113.5	107.5	96.4	90.6	85.1
32	115.8	110.0	104.2	93.4	87.8	82.4
33	112.3	106.6	101.0	90.6	85.1	79.9
34	109.0	103.5	98.0	87.9	82.6	77.6
35	105.9	100.5	95.2	85.4	80.2	75.3
36	102.9	97.8	92.6	83.0	78.0	73.2
37	100.2	95.1	90.1	80.8	75.9	71.3
38	97.5	92.6	87.7	78.7	73.9	69.4
39	95.0	90.2	85.5	76.6	72.0	67.6
40	92.7	88.0	83.3	74.7	70.2	65.9

tests of the early days of structural concrete. Beams and girders were reinforced with a number of types of bars, trussed bars, and welded patterns. Each reinforcing type was patented as introduced by designers and manufacturers and was either embraced by designers or used on specific projects only. Trussed bars were designed to fit beams in an almost custom manner (see Figure 4-68).

Unit bars were another form of reinforcing that was manufactured by bending and assembling the units in a custom manner for specific girders. The width and depth of the girder would determine the dimensions of the units (see Figure 4-69). Steel for concrete reinforcing was produced in three grades as early as 1910. Mild steel reinforcing would have an ultimate strength of 52,000–62,000 psi and an elastic limit of one-half of the ultimate. Medium steel (the most popular) had an ultimate strength of 60,000–70,000 psi and an elastic limit of one-half of ultimate. High-carbon Steel had an ultimate tensile strength of 80,000–100,000 psi and also an elastic limit of one-half of ultimate. High-carbon Steel was considered brittle and unreliable and was not popular with designers.

Table 4-57 the areas and weights of square and round bars. Figure 4-70 gives examples of the types of reinforcing bars used in early concrete construction.

Chapter 2 includes a brief summary of the development of concrete technology in the United States and several tables of mixes, etc. to obtain required strengths. That detail will not be duplicated in this chapter. In the early period of reinforced concrete beam design, there were a number of methods for designing the capacity of a member, just as there were a number of types of reinforcing. Each method of design became known by the name of the inventor or developer of the system, in the same manner as the reinforcing bars. The theory of moments was apparently extremely well

Table 4-48. (Continued)

C. Bethlehem Girder Beams

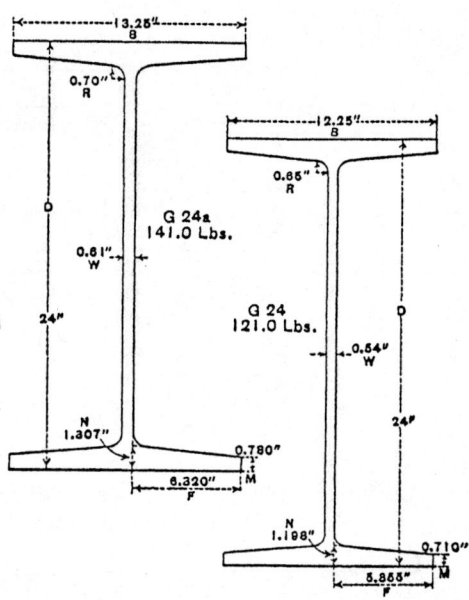

Section Number	Weight per ft (lb)	Nominal Depth of Beam (in.)	Dimensions (in.)						
			Nominal D	B	W	M	N	F	R
	149.0	24⅛	24.12	13.29	.650	.840	1.367	6.320	.70
G24a	141.0	24	24.00	13.25	.610	.780	1.307	6.320	.70
	133.0	23⅞	23.88	13.22	.580	.720	1.247	6.320	.70
	129.0	24⅛	24.12	12.29	.580	.770	1.258	5.855	.65
G24	121.0	24	24.00	12.25	.540	.710	1.198	5.855	.65
	114.0	23⅞	23.88	12.22	.510	.650	1.138	5.855	.65

[a] Reprinted with permission from Bethlehem Steel Co., *Bethlehem Structural Shapes*, Catalogue S-18, pp. 11, 40, 41, 78, 79. © 1926 Bethlehem Steel Co.
[b] W = Safe load, lb, uniformly distributed, including weight of beam.
  L = Span, ft.
  M_f = Bending moment of forces, ft-lb.
  f = Allowable fiber stress, psi.
  S = Section modulus about axis X-X.
C, C', and C" = Coefficients given in the table.
$W = \frac{C}{L}$, or $\frac{C'}{L}$, or $\frac{C''}{L}$; $M_f = \frac{C}{S}$, or $\frac{C'}{S}$, or $\frac{C''}{S}$
C, or C', or C" = $WL = sM_f = \frac{3}{8} f S$
[c] Safe loads given include weight of beam.
Maximum fiber stress, 18,000 pounds per square inch.
Greatest safe loads limited by web shear or buckling are given above the heavy line.
Safe loads below the dotted line produce deflections exceeding 1/360 of the span.

understood almost instantly by these newly begun concrete designers in that the placement of longitudinal reinforcing was in all tension zones. Beam column connections and continuous beams over columns or supports that produce negative moments seem to have been understood from the earliest period. There were many competent designers by 1900, and concrete had been used in foundations for several years, and therefore the properties of concrete had already been recognized and the definition of positive and negative moments was understood. Trussed bars, patented types, were designed such that part of the longitudinal reinforcing moved from the bottom of the beam to the top of the beam in the zone near the columns as required, which was followed in later designs with deformed bars, called "bent bars" (see Figure 4-71).

Table 4-49. Bethlehem Steel Co. Properties of Sections, Allowable Load Tables[a]

A. Properties of Bethlehem I Beams

Section Number	Nominal Depth of Beam (in.)	Weight per ft (lb)	Area of Section (in.²)	Thickness of Web (in.)	Width of Flange (in.)	Axis X-X		
						Moment of Inertia (in.⁴) I	Radius of Gyration (in.) r	Section Modulus (in.²) S
B30	30⅛	129.0	37.52	.580	10.530	5,566.5	12.18	369.6
	30	121.0	35.36	.550	10.500	5,213.6	12.14	347.6
	29⅞	115.0	33.50	.530	10.480	4,886.8	12.08	327.1
B28	28⅛	113.0	32.98	.540	10.030	4,285.5	11.40	304.8
	28	106.0	30.93	.510	10.000	3,993.8	11.36	285.3
	27⅞	100.0	29.18	.490	9.980	3,723.4	11.30	267.1
B26	26⅛	98.0	28.47	.500	9.530	3,200.9	10.60	245.1
	26	91.0	26.55	.470	9.500	2,962.8	10.56	227.9
	25⅞	85.5	24.89	.450	9.480	2,742.2	10.50	211.9
B24b	24³⁄₂₃	104.5	30.63	.550	9.775	2,967.7	9.84	248.4
	24	99.5	29.15	.525	9.750	2,811.7	9.82	234.3
	23²⁹⁄₂₃	95.5	27.79	.505	9.730	2,663.1	9.79	222.8
B24a	24	84.5	24.75	.460	9.500	2,380.1	9.81	198.3
B24	24³⁄₃₂	79.5	23.17	.430	9.035	2,245.3	9.84	186.4
	24	73.5	21.52	.395	9.000	2,087.4	9.85	173.9
B22	22⅛	71.5	20.88	.420	8.535	1,705.2	9.04	154.2
	22¹⁄₁₆	68.5	20.04	.405	8.520	1,629.3	9.02	147.7
	22	65.5	19.08	.385	8.500	1,549.5	9.01	140.9
B20a	20³⁄₃₂	78.0	22.77	.460	8.905	1,568.3	8.30	156.1
	20	73.0	21.37	.430	8.875	1,467.8	8.29	146.8
B20	20⅛	64.5	18.79	.400	8.025	1,283.2	8.26	127.6
	20¹⁄₁₆	62.0	18.11	.390	8.015	1,227.9	8.23	122.4
	20	59.5	17.33	.375	8.000	1,169.7	8.22	117.0

There were a number of reinforcing bars of the patented type in 1910. Many designers and industry engineers were inventing straight bars as well as trussed systems. Figure 4-72 and Table 4-54 give details and member properties for the various types of bars available. The bond between bar and concrete was the cause for all of this experimentation.

International Correspondence Schools 1905A, para. 13, 54 gives a set of rules to be followed by designers in sizing concrete beams.

  6 in. wide concrete girder corresponds to 12 in. I-beams
  8 in. wide concrete girder corresponds to 12–18 in. steel beams
  10 in. wide concrete girder corresponds to 20–24 in. steel beams
  12 in. wide concrete girder corresponding to 30–40 in. riveted girders
  12–24 in. wide concrete girder corresponding to very heavy box girders

Johnson's method was one of the earlier processes for designing reinforced concrete beams. The conservative nature of the early designs is shown by the fact that some designers utilized a standard modulus of elasticity of concrete to be 1,000,000 psi while utilizing 28,000,000 psi for the reinforcing steel. Johnson's method used the transformed section and the above modulus of elasticity (ibid., 55, 56).

Table 4-49. (Continued)

Section Number	Coefficients of Strength			Maximum Safe Shear on Web (lb)	Axis Y-Y		
	For Fiber Stress of 18,000 psi for Quiescent Loads C	For Fiber Stress of 16,000 psi for Quiescent Loads C'	For Fiber Stress of 12,000 psi for Moving Loads C"		Moment of Inertia (in.^{4}) I'	Radius of Gyration (in.) r'	Section Modulus (in.^{2})
B30	4,435,000	3,943,000	2,957,000	134,000	177.5	2.18	33.7
	4,171,000	3,707,000	2,781,000	120,700	164.3	2.16	31.3
	3,925,000	3,489,000	2,617,000	111,900	151.8	2.13	29.0
B28	3,658,000	3,251,000	2,438,000	116,200	142.3	2.08	28.4
	3,423,000	3,043,000	2,282,000	103,800	130.9	2.06	26.2
	3,205,000	2,849,000	2,137,000	95,700	120.2	2.03	24.1
B26	2,941,000	2,614,000	1,961,000	99,600	110.6	1.97	23.2
	2,735,000	2,431,000	1,823,000	88,100	100.9	1.95	21.2
	2,543,000	2,260,000	1,695,000	80,600	91.6	1.92	19.3
B24b	2,957,000	2,628,000	1,971,000	117,300	132.9	2.08	27.2
	2,812,000	2,499,000	1,874,000	107,900	124.8	2.07	25.6
	2,673,000	2,376,000	1,782,000	100,500	117.1	2.05	24.1
B24a	2,380,000	2,116,000	1,587,000	84,200	95.8	1.97	20.2
B24	2,237,000	1,988,000	1,491,000	73,900	81.2	1.87	18.0
	2,087,000	1,855,000	1,392,000	62,200	74.7	1.86	16.6
B22	1,850,000	1,645,000	1,233,000	70,300	65.8	1.78	15.4
	1,773,000	1,576,000	1,182,000	65,500	62.3	1.76	14.6
	1,690,000	1,503,000	1,127,000	59,200	58.8	1.76	13.8
B20a	1,874,000	1,665,000	1,249,000	82,200	84.6	1.93	19.0
	1,761,000	1,566,000	1,174,000	72,900	78.5	1.92	17.7
B20	1,531,000	1,361,000	1,020,000	63,700	54.3	1.70	13.5
	1,469,000	1,306,000	979,400	60,700	51.5	1.69	12.9
	1,404,000	1,248,000	935,700	56,200	48.6	1.68	12.2

Transformed Section:

$$I = bd\left(\frac{d^2}{12} + \frac{e^2}{1+m}\right)$$

$$f_c = \frac{6M}{bd^2}\left(\frac{1 + m + 2\frac{e}{d}}{1 + m + 12\frac{e^2}{d^2}}\right)$$

$$f_t = \frac{6M}{bd^2}\left(\frac{1 + m - 2\frac{e}{d}}{1 + m + 12\frac{e^2}{d^2}}\right)$$

$$f_s = \frac{\frac{12Me}{d^2}}{1 + m + 12\frac{e^2}{d^2}}$$

Table 4-49. (Continued)

B. Safe Loads Uniformly Distributed for Bethlehem I Beams, in TP
(Beams Being Secured against Yielding Sideways)[b]

Span in Feet	B26			B24b		
	26⅛"	26"	25⅞"	24³/₃₂"	24"	23²⁹/₃₂"
	98 lb	91 lb	85.5 lb	104.5 lb	99.5 lb	96.5 lb
13				234.6		
				227.5	215.9	201.0
14	199.1			211.2	200.9	190.9
15	196.1	176.3	161.2	197.1	187.5	178.2
16	183.8	170.9	158.9	184.8	175.8	167.1
17	173.0	160.9	149.6	173.9	165.4	157.2
18	163.4	151.9	141.3	164.3	156.2	148.5
19	154.8	143.9	133.8	155.6	148.0	140.7
20	147.1	136.8	127.2	147.9	140.6	133.7
21	140.0	130.2	121.1	140.8	133.9	127.3
22	133.7	124.3	115.6	134.4	127.8	121.5
23	127.9	118.9	110.6	128.6	122.3	116.2
24	122.5	114.0	106.0	123.2	117.2	111.4
25	117.6	109.4	101.7	118.3	112.5	106.9
26	113.1	105.2	97.8	113.7	108.2	102.8
27	108.9	101.3	94.2	109.5	104.1	99.0
28	105.0	97.7	90.8	105.6	100.4	95.5
29	101.4	94.3	87.7	102.0	97.0	92.2
30	98.0	91.2	84.8	98.6	93.7	89.1
31	94.9	88.2	82.0	95.4	90.7	86.2
32	91.9	85.5	79.5	92.4	87.9	83.5
33	89.1	82.9	77.1	89.6	85.2	81.0
34	86.5	80.4	74.8	87.0	82.7	78.6
35	84.0	78.1	72.7	84.5	80.3	76.4
36	81.7	76.0	70.6	82.1	78.1	74.3
37	79.5	73.9	68.7	79.9	76.0	72.2
38	77.4	72.0	66.9	77.8	74.0	70.3
39	75.4	70.1	65.2	75.8	72.1	68.5
40	73.5	68.4	63.6	73.9	70.3	66.8
41	71.7	66.7	62.0	72.1	68.6	65.2
42	70.0	65.1	60.5	70.4	67.0	63.6
43	68.4	63.6	59.1	68.8	65.4	62.2
44	66.8	62.2	57.8	67.2	63.9	60.8
45	65.4	60.8	56.5	65.7	62.5	59.4
46	63.9	59.5	55.3	64.3	61.1	58.1
47	62.6	58.2	54.1	62.9	59.8	56.9

where:
b = width of concrete beam,
d = depth of concrete beam, overall depth top to bottom of concrete, in.
e = distance from center of beam to center of bars, in.

$$m = \frac{bd}{A}\left(\frac{E_c}{E_s}\right)$$

where:
A = area of steel bars, in.²
E_c = modulus of elasticity of concrete, psi
E_s = modulus of elasticity of steel, psi

Table 4-49. (Continued)

Span in Feet	B24a 24" 34.5 lb	B24 24³⁄₂₂" 79.5 lb	B24 24" 73.5 lb	B22 22⅛" 71.5 lb	B22 22¹⁄₁₆" 65.5 lb	B22 22" 65.5 lb
14	168.4			140.7	131.0	
15	158.7	147.8		132.1	126.6	118.4
16	148.8	139.8	124.4	123.3	118.2	112.7
17	140.0	131.6	122.8	115.6	110.8	105.6
18	132.2	124.3	115.9	108.8	104.3	99.4
19	125.3	117.7	109.8	102.8	98.5	93.9
20	119.0	111.9	104.4	97.4	93.3	88.9
21	113.3	106.5	99.4	92.5	88.7	84.5
22	108.2	101.7	94.9	88.1	84.4	80.5
23	103.5	97.3	90.7	84.1	80.6	76.8
24	99.2	93.2	87.0	80.4	77.1	73.5
25	95.2	89.5	83.5	77.1	73.9	70.4
26	91.5	86.0	80.3	74.0	70.9	67.6
27	88.1	82.9	77.3	71.2	68.2	65.0
28	85.0	79.9	74.5	68.5	65.7	62.6
29	82.1	77.1	72.0	66.1	63.3	60.4
30	79.3	74.6	69.6	63.8	61.1	58.3
31	76.8	72.2	67.3	61.7	59.1	56.3
32	74.4	69.9	65.2	59.7	57.2	54.5
33	72.1	67.8	63.2	57.8	55.4	52.8
34	70.0	65.8	61.4	56.1	53.7	51.2
35	68.0	63.9	59.6	54.4	52.1	49.7
36	66.1	62.1	58.0	52.9	50.7	48.3
37	64.3	60.5	56.4	51.4	49.3	46.9
38	62.6	58.9	54.9	50.0	47.9	45.7
39	61.0	57.4	53.5	48.7	46.7	44.5
40	59.5	55.9	52.2	47.4	45.5	43.3
41	58.0	54.6	50.9	46.3	44.2	42.3
42	56.7	53.3	49.7	45.1	43.3	41.2
43	55.3	52.0	48.5	44.0	42.2	40.2
44	54.1	50.8	47.4	43.0	41.2	39.3
45	52.9	49.7	46.4	42.0	40.3	38.4
46	51.7	48.6	45.4	41.1	39.4	37.6
47	50.6	47.6	44.4			

f_c = compressive stress of concrete, psi
f_t = tensile stress in concrete, psi
f_s = tensile stress in steel, psi
M = maximum moment or design moment, in.-lb

Example Design by Johnson's Method

Simple span 14 ft, total load = 1,000 lb/linear ft.

b = 15 in., d = 24 in., tension bars 0.5 × 1 in. bars (Thacher bar, early form)

f_c = 250 psi, f_t = 100 psi, for concrete

Allowable f_s = 3000 psi, for steel (assumed tensile stress of steel at the tensile stress of 100 psi in the concrete)

Table 4-49. (Continued)

C. Allowable Fiber Stress and Proportion of Safe Loads for Beams Unsupported Laterally

Ratio of Laterally Unsupported Length to Flange Width $\dfrac{L}{B}$	Allowable Fiber Stress (psi) f	Ratio of Allowable Safe Load to Tabular Safe Load	Ratio of Laterally Unsupported Length to Flange Width $\dfrac{L}{B}$	Allowable Fiber Stress (psi) f	Ratio of Allowable Safe Load in Tabular Safe Load
15	18,000	1.000	28	14,368	0.798
16	17,730	0.985	29	14,080	0.782
17	17,475	0.971	30	13,793	0.766
18	17,212	0.956	31	13,509	0.751
19	16,942	0.941	32	13,228	0.735
20	16,667	0.926	33	12,949	0.719
21	16,387	0.910	34	12,674	0.704
22	16,103	0.895	35	12,403	0.689
23	15,817	0.879	36	12,136	0.674
24	15,528	0.863	37	11,873	0.660
25	15,238	0.847	38	11,614	0.645
26	14,948	0.830	39	11,360	0.631
27	14,657	0.814	40	11,111	0.617

When the distance between lateral supports, l, in inches, exceeds 15 times the flange width, B, in inches, the tabular safe loads must be reduced.

The formula for the allowable unit stress, f, in the compression flange of a beam,

$$f = \dfrac{20,000}{1 + \dfrac{l^2}{2,000\,B^2}}$$

is a modification of the column formula given on page 97.

The table below gives the allowable unit stresses, f, and the ratios of the allowable safe load to the tabular safe load, for various values of l/B, based on the above formula.

[a] Reprinted with permission from Bethlehem Steel Co., *Bethlehem Structural Shapes*, Catalogue S-18, pp. 44, 45, 75, 86, 87. © 1926 Bethlehem Steel Co.
[b] Safe loads given include weight of beam.
Maximum fiber stress, 18,000 pounds per square inch.
Greatest safe loads limited by web shear or buckling are given above the heavy line.
Safe loads below the dotted line produce deflections exceeding 1/₁₀₀₀ of the span.

$$M_{max\,pos.} = \dfrac{W(L^2)}{8} = \dfrac{1{,}000\text{ ft }(14\text{ ft})^2}{8} = 24{,}500 \text{ lb-ft} = 294{,}000 \text{ lb-in.}$$

$$f_c = \dfrac{6(294{,}000\text{ lb-in.})}{15\text{ in. }(24'')^2}\left[\dfrac{1 + \dfrac{15\text{ in. }(24\text{ in.})}{5}\left(\dfrac{1}{28}\right) + \dfrac{2(10\text{ in.})}{24\text{ in.}}}{1 + \dfrac{15\text{ in. }(24\text{ in.})}{5}\left(\dfrac{1}{28}\right) + \dfrac{12(10\text{ in.})^2}{(24\text{ in.})^2}}\right] = 159 \text{ lb/in.}^2$$

$$f_t = \dfrac{6(294{,}000\text{ lb-in.})}{15\text{ in. }(24\text{ in.})^2}\left[\dfrac{1 + \dfrac{15\text{ in. }(24\text{ in.})}{5}\left(\dfrac{1}{28}\right) - \dfrac{2(10\text{ in.})}{24\text{ in.}}}{1 + \dfrac{15\text{ in. }(24\text{ in.})}{5}\left(\dfrac{1}{28}\right) + \dfrac{12(10\text{ in.})^2}{(24\text{ in.})^2}}\right] = 98 \text{ lb/in.}^2$$

$$T_s = \dfrac{\dfrac{2(6)(294{,}000\text{ lb-in.})(10\text{ in.})}{(24\text{ in.})^2}}{1 + \dfrac{15''(24'')}{5}\left(\dfrac{1}{28}\right) + \dfrac{12(10'')^2}{(24'')^2}} = 10{,}841 \text{ lb}$$

Table 4-50. A.I.S.C. Historical Record, Table of Beams[a]

24" American Standard Beams
References; See Column (1) and Page 4

1	3	4	5	7	8	9	15	20	12
81907	S13-1922	S19-1926	S19-1926	C1896	C1913	C1921	J&L1900	PE1898-1	CAM 1898
2	S19-1926	S30-1929	S30-1929	C1900	IL1914	C1923	J&L1902	PE1898-2	To 1919
S13-1922	S30-1929	S43-1933	S43-1933	C1903	C1915	C1926	J&L1903	PE1900	incl.
S19-1926	S43-1933	S47-1934	S47-1934	14	C1916	C1930	J&L1905	PE1901	
S30-1929	S47-1934	S51-1938	S51-1938	In 1921	C1917	C1951	J&L1906	22	10–11
S43-1933	S51-1938	S53-1943	S53-1943	16	C1919	IL1932	J&L1908	IN1946	See below
S47-1934	S53-1943	6	S54-1946	J&L1910	C1920	C1934	21	19	
S51-1938	S54-1946	CP 1892	S56-1948	17	18	IL1934	PE1898-2	PE1898-1	
S53-1943	S56-1948	C 1893	13	J&L1916	J&L 1926	CIL 1940	PE1900		
			CAM1921		J&L 1931		PE1901		

Col. (1)	Weight per Foot Lb.	Area Sq. In.	Depth d In.	Flange Width b In.	Web Thick t In.	Dimensions m In.	Dimensions n In.	Dimensions R In.	Dimensions R' In.	Slope Inside Flange %	Axis 1—1 I In.⁴	Axis 1—1 S In.³	Axis 1—1 r In.	Axis 2—2 I In.⁴	Axis 2—2 S In.³	Axis 2—2 r In.
5, 10, 22	120.0	35.13	24.0	8.048	.798	1.404	.800	.60	.30	16⅔	3,010.8	250.9	9.26	84.9	21.1	1.56
8, 12, 14, 17	115.0	33.98	24.0	8.000	.750	1.404	.800	.60	.30	16⅔	2,955.5	246.3	9.33	83.2	20.8	1.57
4, 9, 13, 18	115.0	33.67	24.0	7.987	.737	1.404	.800	.60	.30	16⅔	2,940.5	245.0	9.35	82.8	20.7	1.57
8, 12, 14, 17	110.0	32.48	24.0	7.938	.688	1.404	.800	.60	.30	16⅔	2,883.5	240.3	9.42	81.0	20.4	1.58
4, 9, 13, 18	110.0	32.18	24.0	7.925	.675	1.404	.800	.60	.30	16⅔	2,869.1	239.1	9.44	80.6	20.3	1.58
5, 11, 13, 18, 22	105.9	30.98	24.0	7.875	.625	1.404	.800	.60	.30	16⅔	2,811.5	234.3	9.53	78.9	20.0	1.60
8, 12, 14, 17	105.0	30.98	24.0	7.875	.625	1.404	.800	.60	.30	16⅔	2,811.5	234.3	9.53	78.9	20.0	1.60
244B 21	100.0	29.42	24.0	7.690	.620	1.250	.680	.66	.30	16.1	2,497.3	208.1	9.21	57.53	15.0	1.40
244B 19	100.0	29.42	24.0	7.540	.680	1.210	.640	.66	.30	16.6	2,497.3	208.1	9.21	57.53	15.3	1.40
1, 7, 15	100.0	29.41	24.0	7.254	.754	1.142	.600	.60	.30	16⅔	2,380.3	198.4	9.00	48.56	13.4	1.28
8, 12, 14, 17	100.0	29.41	24.0	7.254	.754	1.142	.600	.60	.30	16⅔	2,379.6	198.3	9.00	48.6	13.4	1.28
B1 6	100.0	29.4	24.0	7.196	.746	1.121	.600	.65	—	16.2	2,342.7	195.2	8.93	46.98	13.1	1.26
3, 11, 13, 16, 18, 22	100.0	29.25	24.0	7.247	.747	1.142	.600	.60	.30	16⅔	2,371.8	197.6	9.05	48.4	13.4	1.29
1, 7, 15	95.0	27.94	24.0	7.192	.692	1.142	.600	.60	.30	16⅔	2,309.6	192.5	9.09	47.1	13.1	1.30
8, 12, 14, 17	95.0	27.94	24.0	7.193	.693	1.142	.600	.60	.30	16⅔	2,309.0	192.4	9.09	47.1	13.1	1.30
243B 19	95.0	27.92	24.0	7.480	.620	1.210	.640	.66	.30	16.6	2,427.0	202.3	9.32	55.93	15.0	1.41
243B 21	95.0	27.92	24.0	7.450	.590	1.250	.680	.66	.30	16.6	2,427.0	202.3	9.32	55.93	15.0	1.41
B1 6	95.0	27.9	24.0	7.135	.685	1.121	.600	.65	—	16.2	2,271.9	189.3	9.02	45.59	12.8	1.28
2, 9, 13, 16, 18	95.0	27.79	24.0	7.186	.686	1.142	.600	.60	.30	16⅔	2,301.5	191.8	9.08	47.0	13.0	1.30

The force on the total area of reinforcing steel:

$$f_s = \frac{10{,}841 \text{ lb}}{5 \text{ in.}^2} = 2168 \text{ lb/in.}^2$$

As can be seen, this method is very conservative. The actual tensile stress on the tension steel is extremely low. This indicates that the area of tension steel is excessive and that the beam is grossly overdesigned.

Check by Modern Analysis:
The check of the design by modern analysis is most suited to an analysis by working stress design.

From ACI 371, 1955 (American Concrete Institute 1941, 17):

$$m = q = \frac{n(A_s)}{bd} = \frac{18(5.00 \text{ in.}^2)}{15 \text{ in.} (22 \text{ in.})} = 0.2728$$

$n = 18$ is assumed due to the fact that the modulus of elasticity of the concrete was taken at a low value, indicating a relatively low compressive stress.

Table 4-50. (Continued)

Col. (1)	Weight per Foot Lb.	Area Sq. In.	Depth d In.	Flange Width b In.	Web Thick t In.	Dimensions m In.	n In.	R In.	R' In.	Slope Inside Flange %	Axis 1—1 I In.⁴	S In.³	r In.	Axis 2—2 I In.⁴	S In.³	r In.
242B 20	90.0	26.47	24.0	7.420	.560	1.210	.640	.66	.30	16.6	2,356.8	196.4	9.44	54.38	14.7	1.44
1, 7, 15	90.0	26.47	24.0	7.131	.631	1.142	.600	.60	.30	16⅔	2,239.1	186.6	9.20	45.7	12.8	1.31
8, 12, 14, 17	90.0	26.47	24.0	7.131	.631	1.142	.600	.60	.30	16⅔	2,238.4	186.5	9.20	45.7	12.8	1.31
B1 6	90.0	26.40	24.0	7.073	.623	1.121	.600	.65	—	16.2	2,201.0	183.4	9.13	44.25	12.5	1.29
3, 11, 13, 16, 18, 22	90.0	26.30	24.0	7.124	.624	1.142	.600	.60	.30	16⅔	2,230.1	185.8	9.21	45.5	12.8	1.32
241B 21	85.0	25.00	24.0	7.220	.540	1.140	.600	.60	.30	16.2	2,181.7	181.8	9.34	44.14	12.2	1.33
1, 7, 15	85.0	25.00	24.0	7.070	.570	1.142	.600	.60	.30	16⅔	2,168.6	180.7	9.31	44.35	12.5	1.33
8, 12, 14, 17	85.0	25.00	24.0	7.070	.570	1.142	.600	.60	.30	16⅔	2,167.8	180.7	9.31	44.40	12.6	1.33
241B 19	85.0	25.00	24.0	7.060	.560	1.140	.600	.60	.30	16⅔	2,181.7	181.8	9.34	44.14	12.5	1.33
2, 9, 13, 16, 18	85.0	24.84	24.0	7.063	.563	1.142	.600	.60	.30	16⅔	2,159.8	180.0	9.33	44.20	12.5	1.33
B1 6	85.0	24.98	24.0	7.012	.562	1.121	.600	.65	—	16.2	2,130.2	177.5	9.23	42.93	12.2	1.31
240B 20	80.0	23.53	24.0	7.000	.500	1.140	.600	.60	.30	16.6	2,111.4	176.0	9.47	42.84	12.2	1.35
B1 6	80.0	23.50	24.0	6.950	.500	1.121	.600	.65	—	16.2	2,059.3	171.6	9.42	41.6	12.0	1.34
1, 7, 15	80.0	23.32	24.0	7.000	.500	1.142	.600	.60	.30	16⅔	2,087.9	174.0	9.46	42.86	12.2	1.36
8, 12, 17	80.0	23.32	24.0	7.000	.500	1.142	.600	.60	.30	16⅔	2,087.2	173.9	9.46	42.90	12.3	1.36
3, 11, 13, 14, 16, 18, 22	79.9	23.33	24.0	7.000	.500	1.142	.600	.60	.30	16⅔	2,087.2	173.9	9.46	42.9	12.2	1.36

References: See Column (1) and Page 4

10			11		
C1923	IL1932	CIL1946	C1921	C1931	CIL1940
C1926	C1934	CIL1948	C1923	IL1932	CIL1946
C1930	IL1934	US1950	C1926	C1934	CIL1948
C1931	CIL1940		C1930	IL1934	US1950

US — United States Steel Company

B	Bethlehem Steel Company 1907		J & L	Jones & Laughlins Limited 1893 to 1902
C	The Carnegie Steel Company, Limited 1893 to 1896		J & L	Jones & Laughlin Steel Company, Beginning 1903
C	Carnegie Steel Company 1900 to 1934		J & L	Jones & Laughlin Steel Corporation, Beginning 1926
C A	Cambria Steel Company		K	Kaiser Steel Corporation
CAM	Cambria Steel Company		L A	Lackawanna Steel Company
C B	Carnegie Brothers & Co., Limited		N J	New Jersey Steel & Iron Co.
CIL	Carnegie–Illinois Steel Corporation		P A	The Passaic Rolling Mill Co.
C X	Carnegie, Kloman & Co., Union Iron Mills		P E	A. & P. Roberts Company (Pencoyd Iron Works)
C P	Carnegie, Phipps & Co., Limited		P H	The Phoenix Iron Company
I L	Illinois Steel Company		P O	Pottsville Iron & Steel Co.
I N	Inland Steel Company		S	Bethlehem Steel Company, Beginning 1909

[a] From H. W. Ferris, ed., *Historical Record, Dimensions and Properties: Rolled Shapes*, pp. 4–103, 4, 12. © 1978 American Institute of Steel Construction, Inc. Reprinted with permission. All rights reserved.

Table 4-51. Roebling Floor Systems[a]

Depth of Concrete at Haunch (in.)	Arch Span	Thickness at liter of Arch (in.)	Weight per ft² of Roebling System (lb)
8	4 ft, 0 in.	3	33
9	4 ft, 6 in.	3	34
10	5 ft, 0 in.	3	36
12	6 ft, 0 in.	3	41
15	7 ft, 6 in.	3	47

[a] From Freiting 1909, 105. Freitag goes on to say that the Roebling arches attain "remarkable strength." The ICS *Treatise on Architecture and Building Construction* states that the Roebling system weight is from 47–50 psf and that they carry from 1,000–2,400 psf on spans of 4.5–5.0 ft (International Correspondence Schools 1899, para. 8, 107).

Table 4-52. Safe Loads and Weights of Floor in Figure 4-65, per ft² Exclusive of I-beams, Plastering, and Flooring[a]

Depth of Bars (in.)	Span (ft)	Thickness of Concrete (in.)	Weight (psf)[a]	Safe Load (psf)
2½	6	4	40	600
2	6	3	30	200
1½	5	2½	24	150

[a] From International Correspondence Schools 1899, para. 8, 108, 109; Reid 1907, 485.

For $m = 0.2728$ and $q = 0.2728$, $k = 0.510$, $j = 0.875$

$$f_s = \frac{12,000\ (M\ \text{lb-ft})}{jd(A_s)} = \frac{12,000\ (24.5\ \text{k-ft})}{0.875\ (22\ \text{in.})(5.00\ \text{in.}^2)} = 3,054.5\ \text{psi}$$

$$f_c = \frac{f_s(k)}{n(1-k)} = \frac{3,054.5\ \text{psi}(0.51)}{18(1-0.51)} = 176.6\ \text{psi}$$

By the working stress method, the stresses prove to be extremely low because the

Table 4-53. Areas and Weights of Bars[a]

	Square ■		Round ●	
Size Inches	Area Inches	Weight per Foot Pounds	Area Inches	Weight per Foot Pounds
1/16	.0039	.013	.0031	.010
1/8	.0156	.053	.0123	.042
3/16	.0352	.120	.0276	.094
1/4	.0625	.213	.0491	.167
5/16	.0977	.332	.0767	.261
3/8	.1406	.478	.1104	.376
7/16	.1914	.651	.1503	.511
1/2	.2500	.850	.1963	.668
9/16	.3164	1.076	.2485	.845
5/8	.3906	1.328	.3068	1.043
11/16	.4727	1.607	.3712	1.262
3/4	.5625	1.913	.4418	1.502
13/16	.6602	2.245	.5185	1.763
7/8	.7656	2.603	.6013	2.044
15/16	.8789	2.989	.6903	2.347
1	1.0000	3.400	.7854	2.670
1 1/16	1.1289	3.838	.8866	3.014
1 1/8	1.2656	4.303	.9940	3.379
1 3/16	1.4102	4.795	1.1075	3.766
1 1/4	1.5625	5.312	1.2272	4.173
1 5/16	1.7227	5.857	1.3530	4.600
1 3/8	1.8906	6.428	1.4849	5.049
1 7/16	2.0664	7.026	1.6230	5.518
1 1/2	2.2500	7.650	1.7671	6.008
1 9/16	2.4414	8.301	1.9175	6.520
1 5/8	2.6406	8.978	2.0739	7.051
1 11/16	2.8477	9.682	2.2365	7.604
1 3/4	3.0625	10.413	2.4053	8.178
1 13/16	3.2852	11.170	2.5802	8.773
1 7/8	3.5156	11.953	2.7612	9.388
1 15/16	3.7539	12.763	2.9483	10.024
2	4.0000	13.600	3.1416	10.681

[a] From International Correspondence Schools 1911, 253.

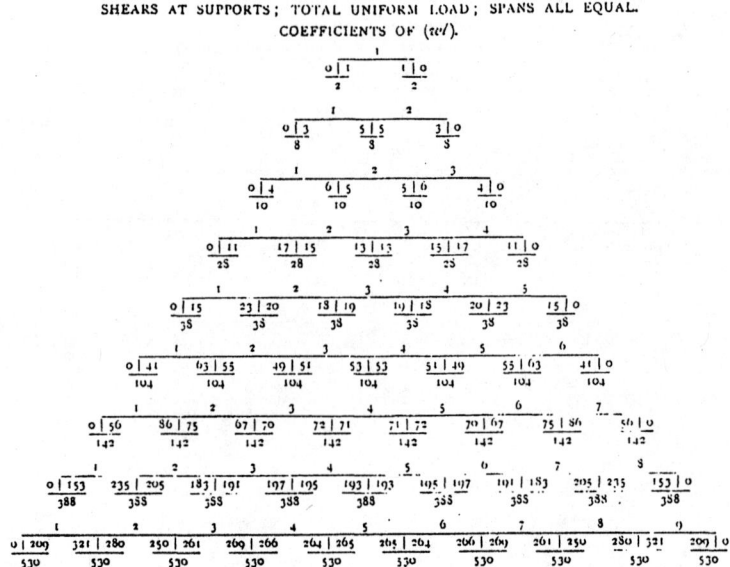

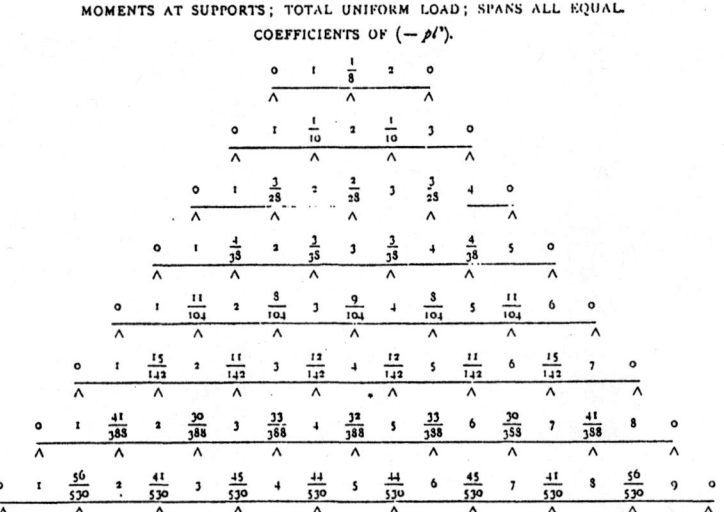

Figure 4-61. Shears and moments, continuous girders. (*From Johnson et al. 1894, 140, 141.*)

member is oversized for the load given in the example. The example did not add the moment from the weight of the beam to the calculations; therefore, to keep the answer comparable the modern solution does not include the weight of the beam. The weight of the beam is 375 lb/linear ft and the additional moment due to this weight is 9,118 lb-ft. The revised stresses due to the weight of the beam are $f_s = 4,200$ psi and $f_c = 243$ psi. The member is still far from overstressed. The original example did not provide the type of concrete or steel. If the concrete were 2,000 psi (ultimate), the allowable stress would be maximum at 450 psi and the reinforcing steel should have had an allowable stress of 15,000 psi.

HISTORIC AMERICAN FLOOR SYSTEMS—BEAMS ◇ 371

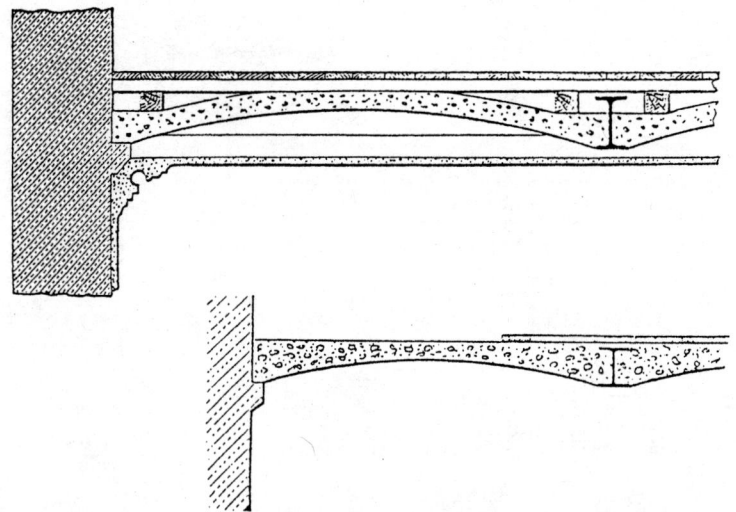

Figure 4-62. Dennett's floor system. (*From Rivington 1900, 140.*)

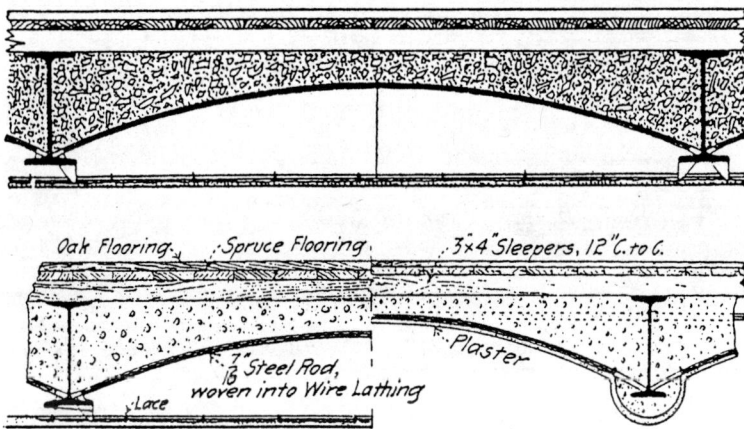

Figure 4-63. Roebling floor system. (*From Freitag 1909, 104; Reid 1909, 105.*)

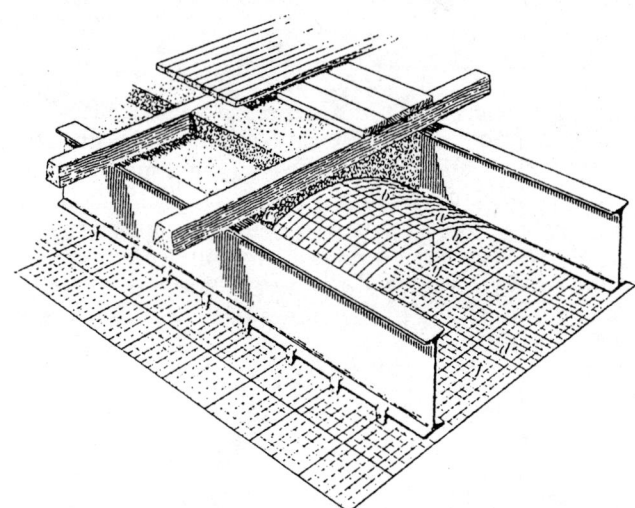

Figure 4-64. Roebling floor system with ceiling. (*From International Correspondence Schools 1899B, para. 8, 107.*)

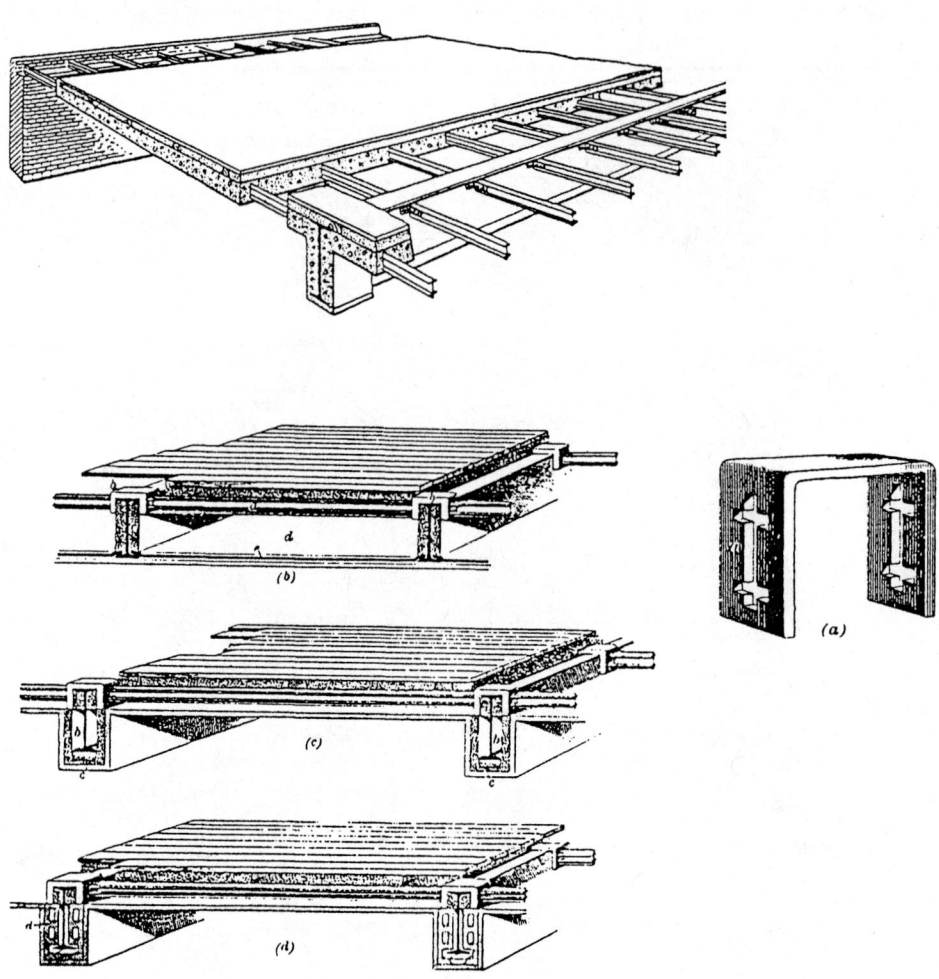

Figure 4-65. Columbian concrete floor system. (*From International Correspondence Schools 1899B, para. 8, 108, 109; Reid 1907, 485.*)

$$M_{design} = A_s f_s (jd) = (5.0 \text{ in.})(4{,}200 \text{ psi})(0.875(22 \text{ in.})) = 404{,}250 \text{ lb-in.}$$
$$= 33{,}687.5 \text{ lb-ft at revised stress}$$

$$M_{actual} = \frac{w(L^2)}{8} = \frac{1{,}375 \text{ lb/ft }(14 \text{ ft})^2}{8} = 33{,}687.5 \text{ lb-ft}$$

The design moment exactly equals the actual moment because the moment used sets the concrete stress block and the *jd* to provide the moment desired.

Mr. E. L. Ransome patented a square, cold twisted reinforcing bar and proposed a system for designing reinforced concrete beams (International Correspondence Schools 1905, para-13, 57, 58). The Ransome system utilized the theory that the compression block of the concrete was at a depth of $d/3$ and that the distance from the center of gravity of the reinforcing to the center of gravity of the compression block *abcd* is $5d/6$. (See Figure 4-73.)

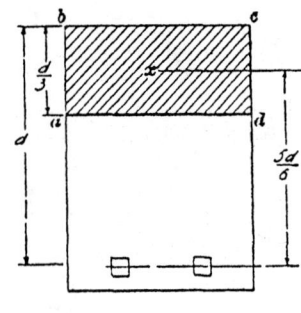

Figure 4-73.

$$M = \frac{f_s(A_s)5d}{6} \quad \text{and} \quad M = \frac{wL^2}{8}$$

From the above, $A_s = \dfrac{w(L^2)}{6.667(d)f_s}$

With cold twisted square reinforcing bars, $A_s = \dfrac{w(L^2)}{7(d)f_s}$

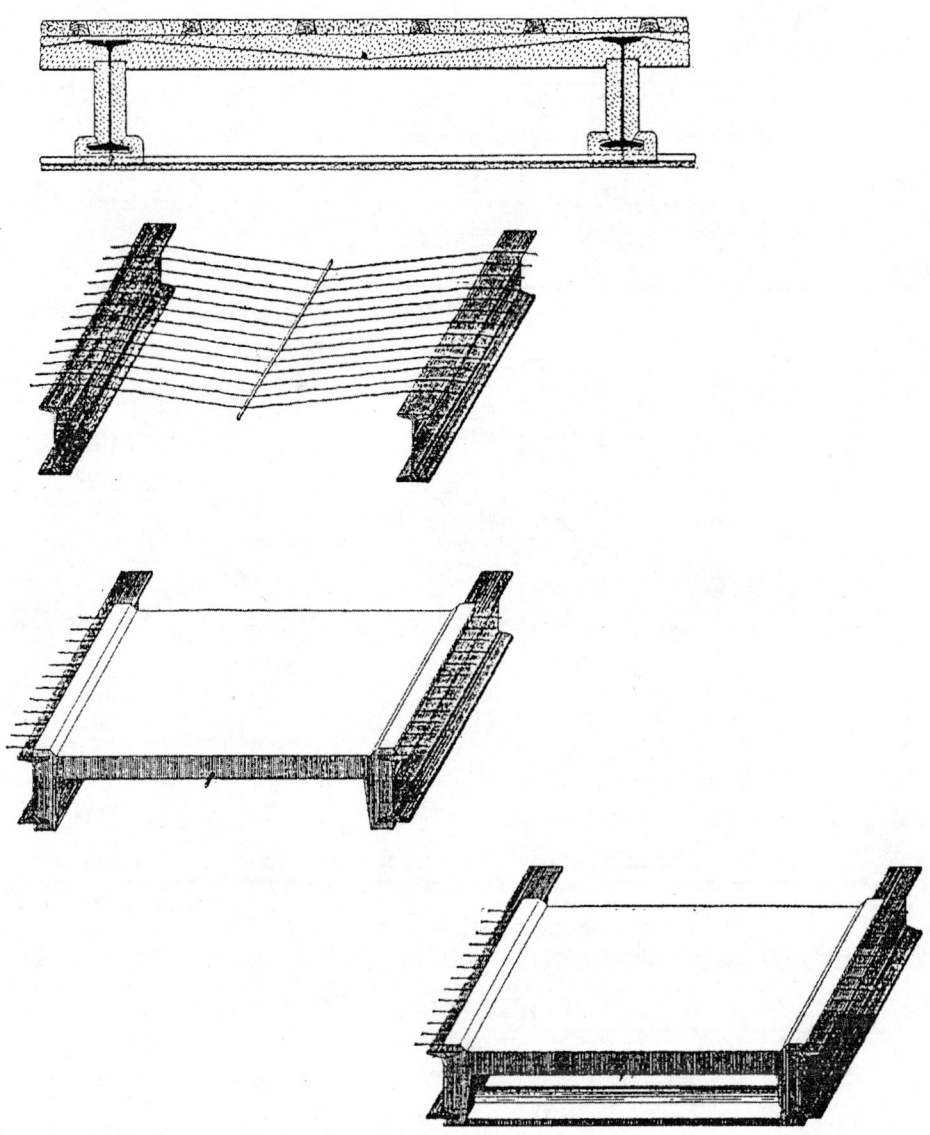

Figure 4-66. Metropolitan Fire Proof Company's floor system. (*From Freitag 1909, 109; Kidder 1902, 452a–452c.*)

where:
- A_s = total area of reinforcing steel, in.^{2}
- w = uniform load on beam, lb/linear ft
- L = span of simple beam, ft
- d = depth of steel, in.
- f_s = stress in reinforcing steel, psi

Hennebique derived a system and formulae based upon the assumption that the compression in the concrete makes an internal force (the internal force couple) that takes half of the moment on the member (ibid., 59, 60). The other half of the force couple is the tensile force in the longitudinal reinforcing. This system assumes that the neutral axis of the beam is at a distance of two-thirds of the depth of the beam. Therefore, the neutral axis is along ad and the center of gravity of the compression block is at one-fourth of the height of the compression block, which is at $d/6$ above the neutral axis (see Figure 4-74).

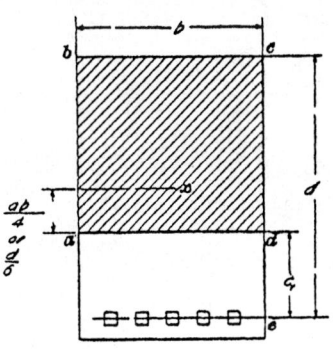

Figure 4-74.

Using the compression block: $f_c = 350$ psi

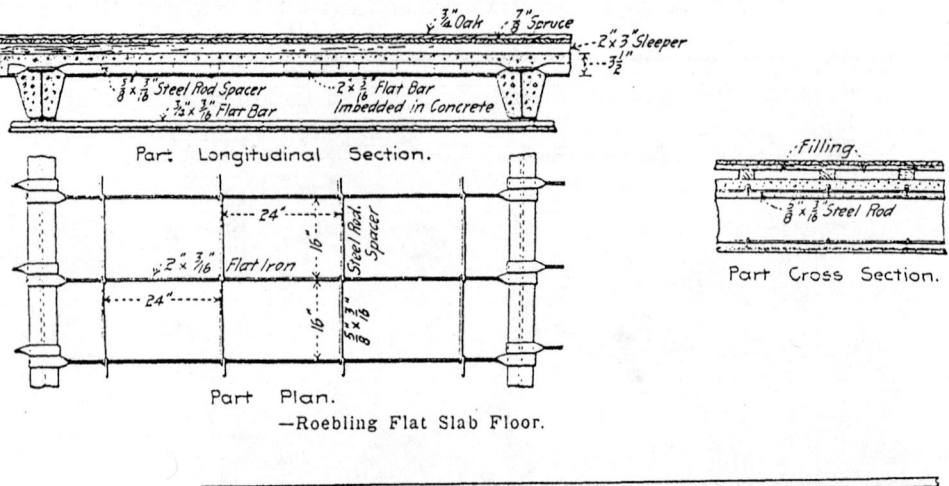
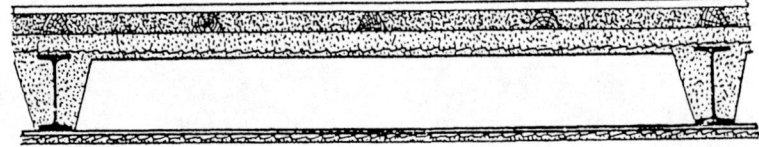
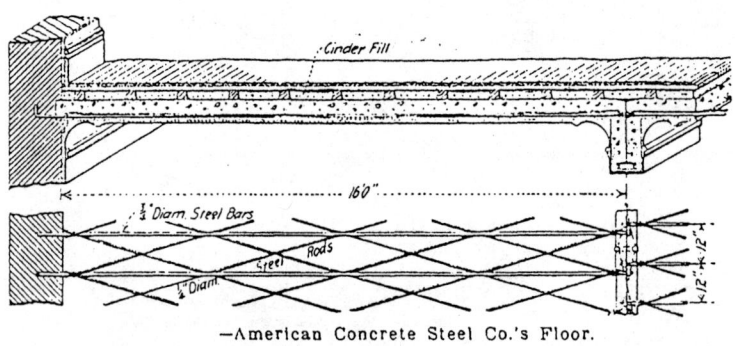
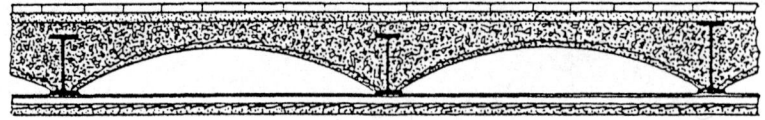

Figure 4-67. Types of concrete floor systems. (*From Freitag 1909, 108.*)

$$\frac{M}{2} = \frac{f_c, \text{psi } (b)(d^2)}{9} = \text{moment, lb-in.}$$

Using the Tension Reinforcing: $f_s = 14{,}000$ psi

$$\frac{M}{2} = (A_s, \text{in.}^2)(f_s, \text{psi})\frac{(d)}{3} = \text{moment, lb-in.}$$

This system is very conservative and would have produced designs that were safe but uneconomical. The Hennebique system was used for many designs, and many

HISTORIC AMERICAN FLOOR SYSTEMS—BEAMS ◇ 375

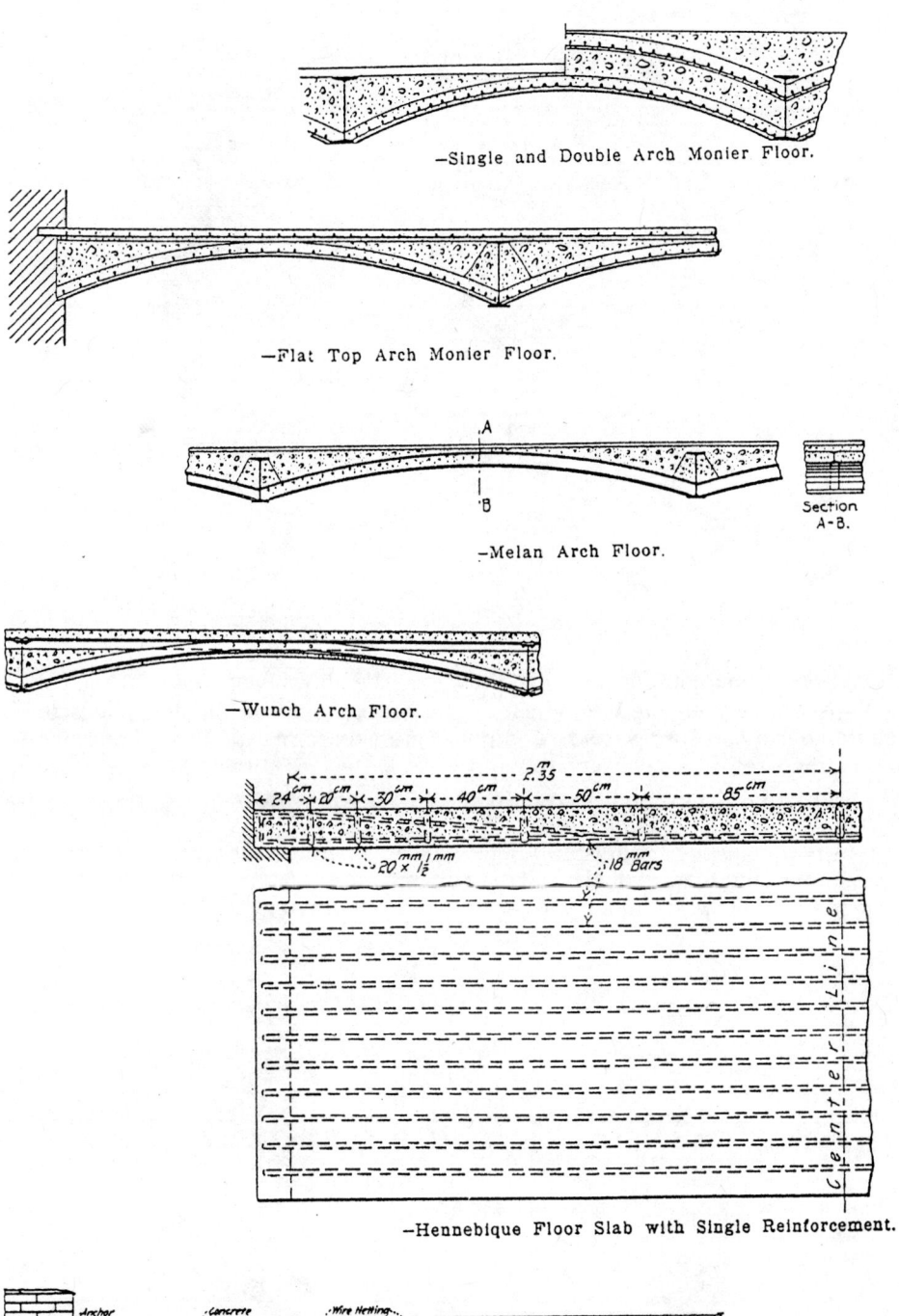

Figure 4-67. (Continued). (From Reid 1907, 490, 491.)

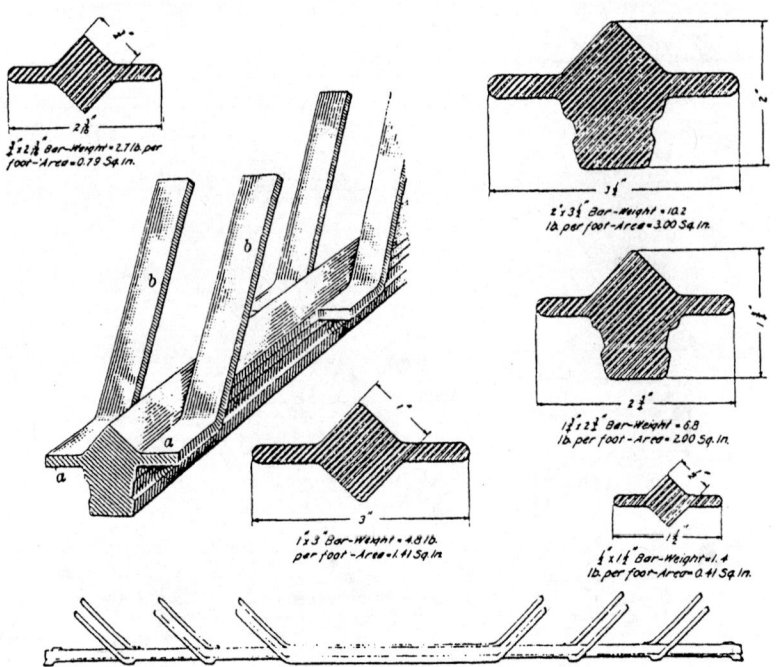

Figure 4-68. Trussed or Kahn bars. (*From International Correspondence Schools 1923B, para. 9, 60; id. 1911, 261.*)

buildings exist that contain this type of design. The allowable stresses of 350 psi and 14,000 psi are very low and demonstrate the cautious use of this new material. The evolution of concrete design had a natural progression from experiment and extremely conservative uses to less conservative approaches. Designs continued to be individual, patented processes until the 1920s, when some standardization of the industry began to occur.

Edwin Thacher, an engineer and designer in concrete, was one of the first to use the relation of the moduli of elasticity of concrete and steel in his formulae. His work provided much of the technology for all of the later work in concrete design. Thacher's

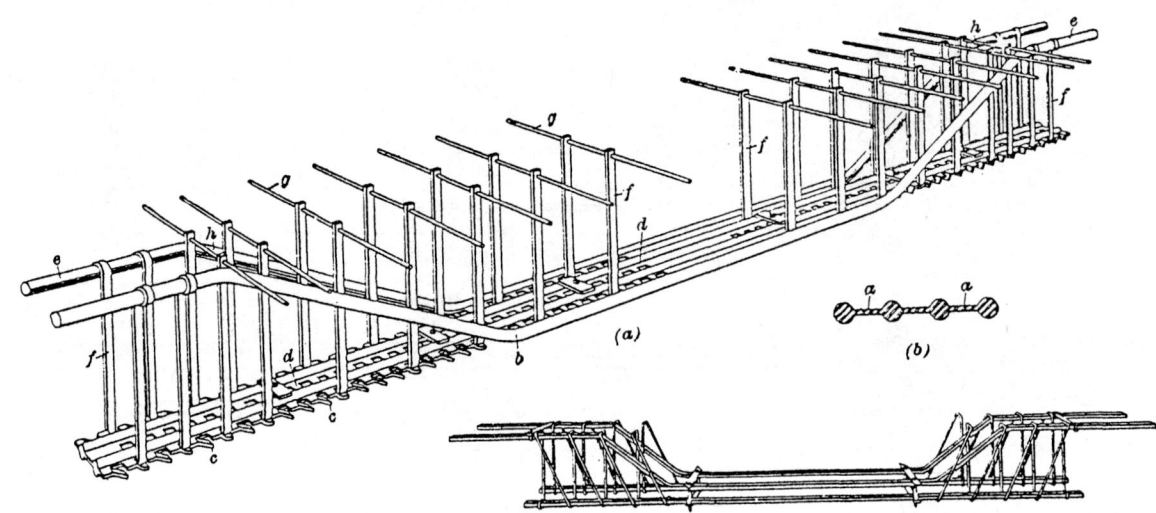

Figure 4-69. Unit bars, manufactured beam trusses. (*From International Correspondence Schools 1908[?], para. 45, 10; id. 1923, para. 9, 60.*

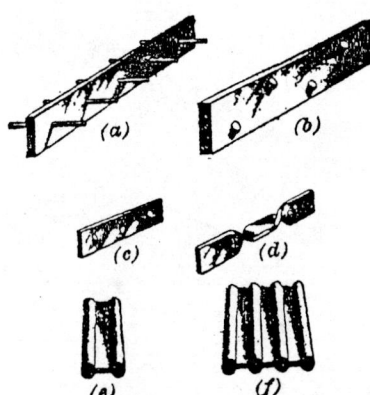

Figure 4-70. Reinforcing bars, flat bars. (a) Hyatt bar; (b) Thacher bar (early form); (c) staff bar; (d) DeMann bar; (e) unit bar (Siamese); (f) unit bar (quad bar). (*From International Correspondence Schools 1911, 254.*)

works were published by the Concrete Steel Engineering Company and were the ones most utilized by designers throughout the country (ibid., 60, 61).

Beam design, Thacher system (see Figure 4-75):
  Required design information: f_s, f_c, E_s, E_c, d, and c

$$y = \frac{d}{\left(\frac{f_c E_s}{f_s E_c} + 1\right)}$$

$$x = d - Y$$

$$A = \frac{d}{2\left(\frac{f_s}{f_c} + \left(\frac{f_s}{f_c}\right)^2 \frac{E_c}{E_s}\right)}$$

$$M_1 = \frac{f_s}{36}\left(\frac{E_c}{E_s}\frac{x^3}{Y} + 3A\right)$$

where:
  f_c = allowable stress in concrete, compressive, psi
  f'_s = allowable stress in steel, compressive, psi
  f_s = allowable stress in steel, tension, psi
  E_c = modulus of elasticity of concrete, psi
  E_s = modulus of elasticity of steel, psi
  A = area of steel, tension, for a 1 in. wide beam, in.^{2}
  nA = area of steel, compression, for a 1 in. wide beam, in.^{2}
  M_1 = moment capacity for a 1 in. wide beam, in.^{2}

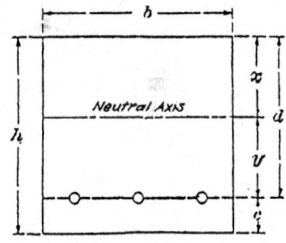

Figure 4-75.

Thacher states that when beams are continuous over two or more supports, make $A = 0.8$ of the above value and reinforce the beams over at the top over the support. Thacher also gives figures for the cover of the steel at the top or bottom of a beam. The cover c is given as a function of the depth of the member, rather than the modern method of determining c from the size of the bar (see Table 4-55).

Thacher also gives values for the modulus of elasticity and ultimate tensile stress of the steel reinforcing bars and a table of the modulus of elasticity and ultimate compressive stress of concrete by mix proportion and at two ages, one month and six months. It is interesting to note that while the ultimate compressive stress of the concrete is given, there is no mention of the type of tests that produced these results.

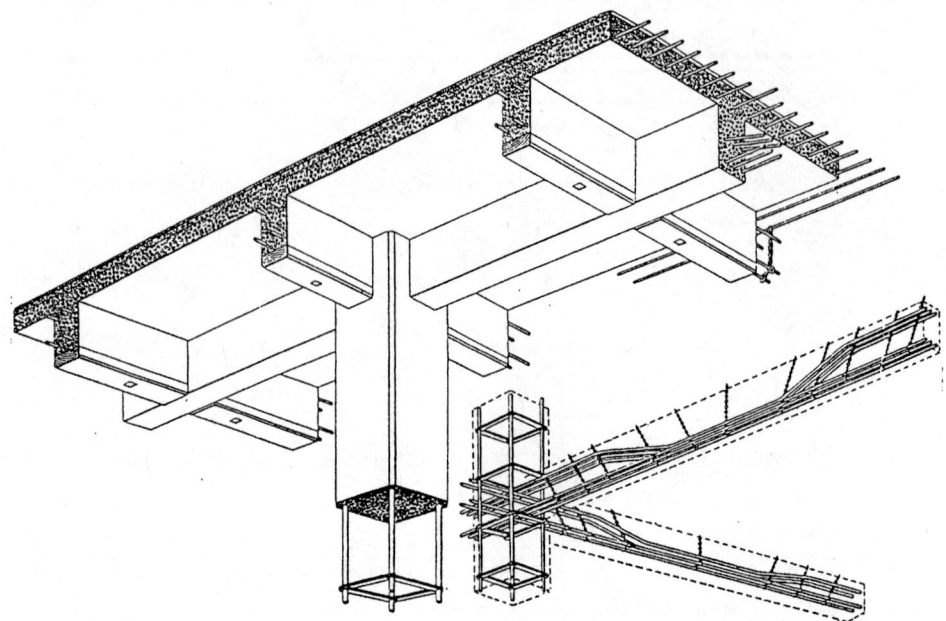

Figure 4-71. Beam reinforcing with trussed bars. (*From International Correspondence Schools 1905A, para. 13, 51.*)

The ILT Volume is dated 1905, and at that time tests were made on concrete cubes of 1, 2, 4, 6, and 12 in. The standard ASTM Tests on 6 in. diameter × 12 in. long cylinders had not yet been adopted by the industry. Thacher's values for the moduli of elasticity of these two companion materials are given in Table 4-56.

Thacher also provides some design tables for reinforced concrete beams and slabs (see Table 4-57).

Factor of safety for one-month old concrete was recommended to be 5.
Factor of safety for 6-month-old concrete was recommended to be 3.5.

Example Problem, Thacher System
Reinforced concrete beam:

$$1\text{-}2\text{-}4 \text{ Mix}, f_c = 2,400 \text{ psi (ult.)}$$

$$E_c = 1,460,000 \text{ psi}$$

$$\text{Span} = 30 \text{ ft}$$

Spacing of beams = 10 ft

Total load = 200 psf

Max. Moment = 225,000 lb-ft (simple span)

Design Moment = 225,500 lb-ft (5) = 1,125,000 lb-ft

Assume $h = 36$ in., $d = 33.75$ in.

From table 4-57 above, $M_1 = 51.25 \, (d^2) = 51.25(33.75 \text{ in.})^2 = 58,380$ ft-lb

$$\text{Width of beam} = \frac{1,125,000 \text{ lb-ft}}{58,380 \text{ lb-ft}} = 20 \text{ in.}$$

$$A_s \text{ (required)} = \frac{d \, (b)}{100} = \frac{33.75 \text{ in.} \, (20 \text{ in.})}{100} = 6.75 \text{ in.}^2$$

Table 4-54 Properties of Deformed Reinforcing Bars[a]

SIZE, WEIGHT, AND ULTIMATE TENSILE STRENGTH OF THACHER BARS

Diameter of Bar Inches	Weight per Foot Pounds	Area of Net Section Inches	Average Ultimate Tensile Strength for Each Bar Pounds
$\frac{1}{4}$	.16	.047	3,000
$\frac{3}{8}$	.34	.10	6,400
$\frac{1}{2}$	.61	.18	11,500
$\frac{5}{8}$	.95	.28	17,900
$\frac{3}{4}$	1.39	.41	26,200
$\frac{7}{8}$	1.87	.55	35,200
1	2.41	.71	45,400
$1\frac{1}{8}$	3.06	.90	57,600
$1\frac{1}{4}$	3.74	1.10	70,400
$1\frac{3}{8}$	4.49	1.32	84,500
$1\frac{1}{2}$	5.30	1.56	99,800
$1\frac{5}{8}$	6.15	1.81	115,800
$1\frac{3}{4}$	7.07	2.08	133,100
$1\frac{7}{8}$	7.99	2.35	150,400
2	9.01	2.65	169,600

SIZE, AREA, AND WEIGHT OF DIAMOND BARS

Nominal Size of Bar Inches	Area of Section Square Inches	Weight per Foot Pounds
$\frac{1}{4}$	062	.22
$\frac{3}{8}$	.14	.48
$\frac{7}{16}$	.19	.65
$\frac{1}{2}$	.25	.85
$\frac{5}{8}$	.39	1.33
$\frac{3}{4}$	.56	1.91
$\frac{7}{8}$	.76	2.60
1	1.00	3.40
$1\frac{1}{4}$	1.56	5.31

PHYSICAL PROPERTIES OF THE RANSOME BAR

Side of Bar Inches	Number of Twists per Foot of Length	Area of Bar Square Inches	Weight per Foot Pounds	Elastic Limit Pounds per Square Inch	Ultimate Tensile Strength Pounds per Square Inch	Elastic Limit of One Bar Pounds	Ultimate Tensile Strength of One Bar Pounds
$\frac{1}{4}$	5	.0625	.213	62,350	86,700	3,897	5,419
$\frac{3}{8}$	3	.1406	.478	61,800	86,600	8,689	12,176
$\frac{1}{2}$	2	.2500	.850	60,120	86,850	15,030	21,713
$\frac{5}{8}$	2	.3906	1.328	57,890	85,820	22,612	33,520
$\frac{3}{4}$	1	.5625	1.913	56,720	85,240	31,905	47,948
$\frac{7}{8}$	1	.7656	2.603	56,150	84,730	42,988	64,869
1	$\frac{3}{4}$	1.0000	3.400	55,760	84,275	55,760	94,275
$1\frac{1}{4}$	$\frac{1}{2}$	1.5625	5.312	55,450	83,150	86,641	129,921

SIZE, NET SECTION, AND WEIGHT OF CORRUGATED, OR JOHNSON, BARS

Size of Bars Inches	Weight per Foot Pounds	Net Section Square Inches
$\frac{1}{4}$	.78	.19
$\frac{3}{4}$	1.56	.38
$\frac{1}{8}$	2.25	.55
1	2.90	.70
$1\frac{1}{4}$	4.56	1.10

Table 4-54 (Continued)

	PROPERTIES OF THE KAHN CUP BAR			
Nominal Size of Bar Inches	Sectional Area Square Inches	Weight per Foot Pounds	Elastic Limit of Each Bar Pounds	Ultimate Tensile Strength of Each Bar Pounds
$\frac{3}{8}$	.1406	.502	5,000	9,800
$\frac{1}{2}$	.2500	.893	9,000	17,500
$\frac{5}{8}$	.3906	1.394	14,000	27,300
$\frac{3}{4}$	.5625	2.008	21,000	39,400
$\frac{7}{8}$	.7656	2.733	28,000	53,600
1	1.0000	3.570	37,000	70,000
$1\frac{1}{8}$	1.2656	4.518	47,000	88,600
$1\frac{1}{4}$	1.5625	5.578	58,000	109,400

	NET SECTIONS AND WEIGHTS OF UNIVERSAL BARS		
No.	Size of Bar Inches	Weight per Foot Pounds	Net Section Square Inch
1	$\frac{1}{4} \times 1$	.73	.19
2	$\frac{3}{8} \times 1\frac{3}{8}$	1.35	.41
3	$\frac{3}{8} \times 1\frac{3}{4}$	1.97	.54
4	$\frac{3}{8} \times 2$	2.27	.65
5	$\frac{3}{8} \times 2\frac{1}{2}$	2.85	.80

	PROPERTIES OF TWISTED LUG BARS		
Size of Bar Inches	Net Sectional Area Square Inches	Weight per Foot Pounds	Safe Working Stress for Each Bar Pounds
$\frac{1}{4}$	.0626	.222	1,250
$\frac{3}{8}$	.1406	.492	2,810
$\frac{1}{2}$	2500	.870	5,000
$\frac{5}{8}$	.3906	1.350	7,810
$\frac{3}{4}$	.5625	1.940	11,250
$\frac{7}{8}$	.7656	2.640	15,310
1	1.0000	3.450	20,000
$1\frac{1}{8}$	1.2656	4.350	25,310
$1\frac{1}{4}$	1.5625	5.370	31,250

[a] From International Correspondence Schools 1911, 255–260.

Final beam dimensions:

$b = 20$ in.

$d = 33.75$ in.

$h = 36$ in.

$A_s = 6.75$ in.^{2}

The Kahn system, another patented type of reinforcing bars (trusses in this case), which was promoted by the manufacturer and many franchised dealers throughout the country. The Kahn system starts with a square bar turned 45° with straight, flat extensions of the cross-section that are separate from the bar except where attached and that turn up to form the image of a Pratt truss (see Figure 4-76).

The Kahn system had its own unique method of calculating the stresses on the concrete and the reinforcing and the allowable moment on the member. Each of these patented systems had a feature that made it special; Kahn bars were designed to be turned up and act as stirrups to resist the shear in the areas of the beams between the midspan and the support. The Kahn system, like all of the others, had specific

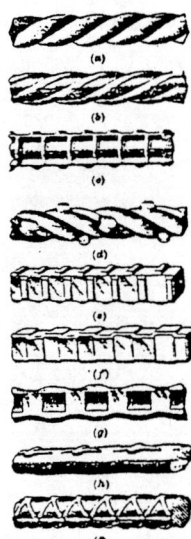

Figure 4-72. Deformed reinforcing bars: (a) Ransome bar; (b) spiral bar; (c) Kahn cup bar; (d) Square-twisted lug bar; (e) Johnson bar (old type); (f) Johnson bar (new type); (g) Universal bar; (h) Thacher bar; (i) Diamond bar. (*From International Correspondence Schools 1911, 260.*)

bar configurations for each area of the concrete structural system. In addition, the bars had to be custom made to length for each position of use. Contrast this to the deformed bar that is standard today, which comes in standard 20 and 40 ft lengths, in 12 sizes, and is either shop cut and tied into specific bundles or field cut to length and tied as required. The Kahn system had design requirements and formulae, as did all of the other systems. Kahn designs used the following values as of 1905 (International Correspondence Schools 1905A, para. 13, 63):

E_s = 30,000,000 psi, modulus of elasticity of steel
E_c = 2,000,000 psi, modulus of elasticity of concrete
f_s = 64,000 psi, ultimate tensile stress of steel
f_t = 200 psi, ultimate tensile stress of concrete
f_c = 3000 psi, ultimate compressive stress of concrete

In each of the above values, the modulus and the ultimate stresses may be changed to the value appropriate for another grade of material.

The Kahn system formulae utilize the ultimate strength of the concrete and the steel in the computations and specifies a factor of safety of 4 or 5.

Table 4-55. Thacher Table of c Cover of Bars[a]

h	d	c	h	d	c	h	d	c
1.5	1.0	0.5	8.0	7.0	1.0	20.0	18.25	1.75
2.0	1.5	0.5	8.5	7.5	1.0	22.0	20.0	2.0
2.5	1.75	0.75	9.0	7.75	1.25	24.0	22.0	2.0
3.0	2.25	0.75	10.0	8.75	1.25	26.0	24.0	2.0
3.5	2.75	0.75	11.0	9.75	1.25	28.0	26.0	2.0
4.0	3.25	0.75	12.0	10.75	1.25	30.0	28.0	2.0
4.5	3.5	1.0	13.0	11.5	1.5	32.0	30.0	2.0
5.0	4.0	1.0	14.0	12.5	1.5	34.0	32.0	2.0
5.5	4.5	1.0	15.0	13.5	1.5	36.0	33.75	2.25
6.0	5.0	1.0	16.0	14.5	1.5	42.0	39.5	2.5
6.5	5.5	1.0	17.0	15.5	1.5	48.0	45.5	2.5
7.0	6.0	1.0	18.0	16.5	1.5	54.0	51.0	3.0
7.5	6.5	1.0	19.0	17.25	1.75	60.0	57.0	3.0

[a] From International Correspondence Schools 1905A, par. 13, 62.

Table 4-56. Thacher's Values for Moduli of Elasticity of Steel and Concrete

Material	Modulus of Elasticity (psi)	Ultimate Stress (psi)
Steel reinforcing	$E_s = 30,000,000$	$f_s = 64,000$
Concrete (1-2-4, 1 month old)	$E_c = 1,460,000$	$f_c = 2,400$
Concrete (1-2-4, 6 months old)	$E_c = 2,580,000$	$f_c = 3,700$
Concrete (1-3-6, 1 month old)	$E_c = 1,220,000$	$f_c = 2,050$
Concrete (1-3-6, 6 months old)	$E_c = 1,860,000$	$f_c = 3,100$

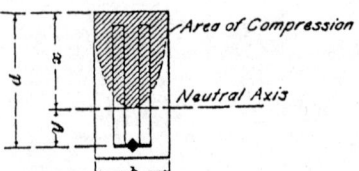

Figure 4-77.

The Kahn system formulae for design and theory of analysis are as follows (see Figures 4-77, 4-78, and Table 4-58):

$$y = \frac{15A + bd^2}{30A + 2bd}$$

where:

b should not be less than $\dfrac{Af_s}{1,800x}$

$$M_1 = (0.625x + y)A(f_s)$$

where:
- A = area of reinforcing bars, in.^{2}
- b = width of beam, in.
- d = depth, top to center of reinforcing bars, in.
- f_s = ultimate tensile stress of steel, psi

There were other patented processes for reinforcing bar trusses, systems, and concrete construction methods in existence during the first quarter of the twentieth century, but the systems shown in this work were the most widely used. Standardization of the concrete industry had its beginnings in 1904, when the Joint Committee on Standard Specifications for Concrete and Reinforced Concrete was formed. The Committee consisted of five representatives from the American Society of Civil Engineers, the American Society for Testing Materials, the American Railway Engineering Association, the American Concrete Institute, and the Portland Cement Association. The Committee spent most of its time in the early periods of organization working on standardizing the production of cement, additives, and mix requirements to produce required strengths and on methods of mixing and casting the concrete. The Committee also standardized testing procedures and methods of determining the ultimate strength of concrete.

Attention was given to establishing standardized design criteria based upon many early features of what was later to be working stress design. The Joint Committee's initial reports were dated 1909 and 1912, and after a final report in 1916 it disbanded. A new Joint Committee was formed in 1919 and presented a final report in 1924. The Second Joint Committee provided detailed specifications for design and construction of buildings with reinforced concrete by working stress methods. Many cities had more restrictive building code requirements, and it literally took years for total standardization to become a reality. One thing that was quick to become standardized was a more uniform use of the straight and/or the deformed reinforcing bar. These

Table 4-57.

Concrete Mix and Age	Area of Steel Required (in.^{2}) for Beam Width of 1 in.	Ultimate Resisting Moment (ft-lb) for a Beam Width of 1 in.
Concrete (1-2-4, 1 month old)	$d/142$	$35.26\,d^2$
Concrete (1-2-4, 6 months old)	$d/100$	$51.25\,d^2$
Concrete (1-3-6, 1 month old)	$d/165$	$30.62\,d^2$
Concrete (1-3-6, 6 months old)	$d/109$	$46.25\,d^2$

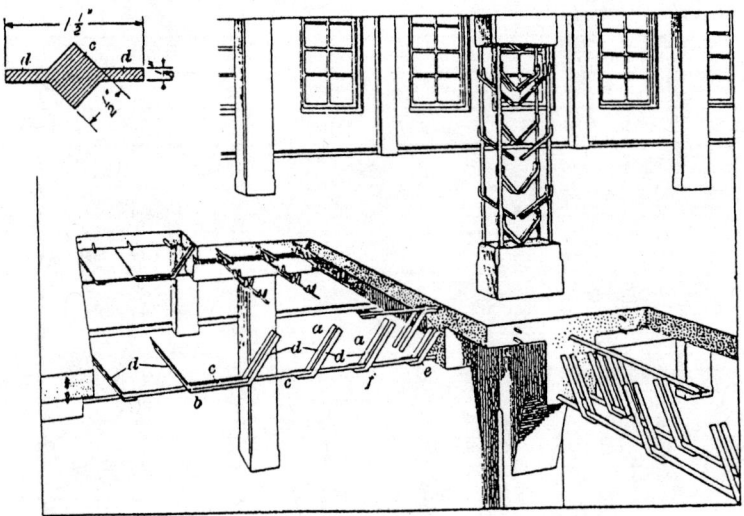

Figure 4-76. Kahn system or reinforced concrete. (*From International Correspondence Schools 1905A, para. 13, 63.*)

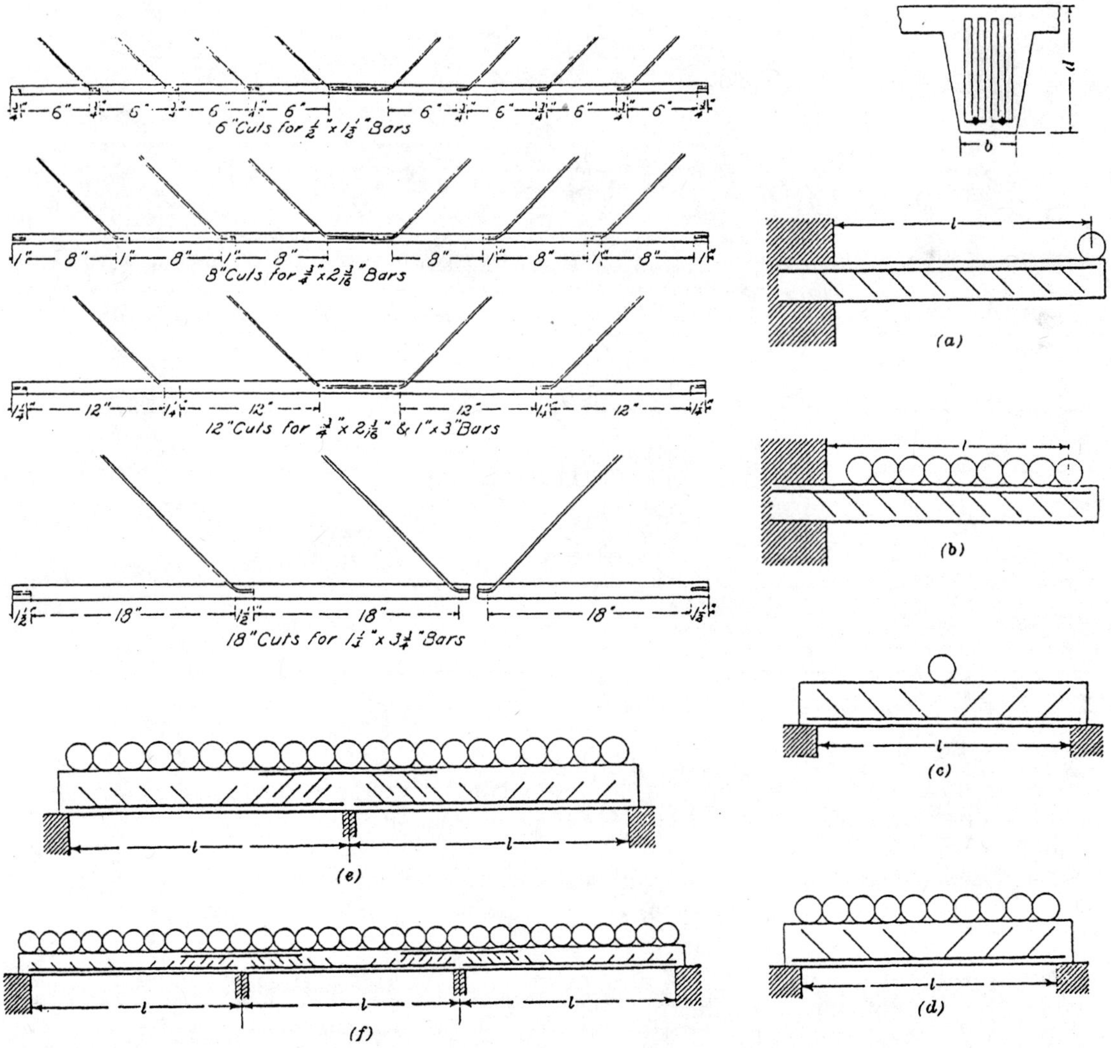

Figure 4-78. Kahn system details, load diagrams. (*From International Correspondence Schools 1905A, para. 13, 67, 70.*)

Table 4-58. Kahn Simple Span Beams, Allowable Uniform Load iin Hundreds of Pounds[a]

	Depth Inches	SAFE UNIFORMLY DISTRIBUTED LOAD, IN HUNDREDS OF POUNDS Distance Between Supports, in Feet																					
		8	9	10	11	12	13	14	15	16	17	18	19	20	21	22	23	24	25	26	27	28	29	30
Case I	6	40	35	32	29	27	25	23	21	20	19	18	17	16										
	8	56	50	45	41	38	35	32	30	28	27	25	24	23	21	20								
	10	74	66	59	53	49	45	42	40	37	35	33	31	30	28	27	26	25	24	23				
	12	93	82	74	67	62	57	53	50	46	44	41	39	37	35	34	32	31	30	28	27			
	14	110	98	88	80	73	68	63	59	55	52	49	46	44	42	40	38	37	35	34	33	31		
	16	123	110	98	89	82	75	70	65	61	58	55	52	49	47	45	43	41	39	38	36	35	34	
	18	138	122	110	100	92	85	79	73	69	65	61	58	55	52	50	48	46	44	42	41	39	38	37
	20	154	137	123	113	102	95	88	82	77	72	68	65	61	58	56	54	51	49	47	46	44	42	41
Case II	8	110	98	88	80	73	68	63	59	55	52	49	46	44	42	40	38	37	35	34	33	31		
	10	147	130	117	106	98	90	84	78	73	69	65	62	59	56	53	51	49	47	45	43	42	40	
	12	190	169	152	138	127	117	108	101	95	90	85	80	76	72	69	66	63	61	58	56	54	52	51
	14	210	188	169	154	141	130	121	113	106	99	94	89	85	81	77	74	71	68	65	63	60	58	56
	16	248	220	198	180	165	152	142	132	124	117	110	104	99	94	90	86	82	79	76	73	71	68	66
	18	280	249	224	203	186	172	160	149	140	132	124	118	112	107	102	97	93	90	86	83	80	77	75
	20	314	279	251	228	209	193	179	168	157	148	139	132	125	120	114	109	105	100	97	93	90	87	84
	22	347	309	278	252	232	214	198	185	174	163	154	146	139	132	126	121	116	111	107	103	99	96	93
Case III	10	244	217	195	177	163	150	140	130	122	115	108	103	98	93	89	85	81	78	75	72	70	68	65
	12	328	290	261	237	218	200	186	174	165	153	145	137	130	124	118	113	107	104	100	97	93	90	87
	14	394	350	315	286	263	242	225	210	197	185	175	166	157	150	143	137	131	126	121	117	112	108	105
	16	448	400	358	326	298	276	256	239	224	211	199	188	179	170	163	155	149	143	138	132	128	123	119
	18	508	452	406	369	338	312	290	270	254	239	226	214	203	193	184	176	169	162	156	150	145	140	135
	20	568	504	453	412	378	348	324	302	283	266	252	238	226	216	206	197	189	181	174	168	162	156	151
	22	634	564	507	461	423	390	362	338	317	298	281	266	254	242	230	220	211	203	195	186	182	175	169
	24	680	605	544	395	455	418	388	362	340	320	302	286	272	259	247	236	226	217	209	201	194	188	181
Case IV	12	452	402	361	328	300	277	258	240	226	212	200	190	182	172	164	157	150	144	139	134	129	125	121
	14	540	480	432	392	360	332	308	288	270	254	240	227	216	206	196	188	180	173	166	160	154	149	144
	16	625	555	499	455	416	382	357	333	312	294	277	263	250	238	227	217	208	200	192	185	178	172	167
	18	707	630	566	515	472	435	405	377	352	333	314	298	283	270	257	246	236	226	218	210	202	195	188
	20	788	700	630	572	525	485	450	420	394	370	350	331	315	300	286	273	262	252	242	233	225	217	210
	22	867	770	694	630	577	535	495	462	434	407	386	365	346	330	315	302	289	278	267	257	248	239	231
	24	960	854	768	700	640	590	550	512	480	452	427	404	382	366	349	334	320	307	295	284	276	265	256
	26	1,040	925	832	757	695	640	595	555	520	490	462	438	415	396	378	361	346	333	320	308	297	287	277
	28	1,125	1,000	900	820	750	692	642	600	563	530	500	474	450	428	410	392	375	360	346	334	321	310	300
	30	1,210	1,075	968	878	805	744	680	645	605	568	536	508	483	460	440	420	404	387	372	358	345	333	321

Case 1, Kahn Bars
$\frac{1}{2}$ inch × $1\frac{1}{2}$ inches, Area = 0.76 in^2
Width b = 8 inches.

Case III, Kahn Bars
1 inch × 3 inches, Area = 2.84 in^2
b = 12 inches.

Taper of beam, not less than 1 inch in 6 inches up to 12 in. deep, vertical for deeper than 12 inches.

Case IV, Kahn Bars
$1\frac{1}{4}$ inch × $3\frac{3}{4}$ inches. Area = 4 in^2
b = 14 inches.

Case II, Kahn Bars
$\frac{3}{4}$ inch × $2\frac{3}{16}$ inchy, Area = 1.56 in^2
b = 10 inches.

[a] From International Correspondence School 1905, par. 13, 68, 69.

bars were more adaptive to unique situations than the premanufactured truss bars. At this time a half-dozen types of straight reinforcing bars were still in use, including twisted square bars.

Urquhart and O'Rourke (1926) provide a summary of working stresses recommended by the Second Joint Committee as follows:

Direct compression on concrete
    Bearing on plain concrete piers and pedestals $0.25 f'_c$

Axial compression, column with longitudinal reinforcement	$0.20 f'_c$
Axial compression, column with longitudinal reinforcement and spirals	$300 \text{ psi} + (0.10 + 4p) f'_c$
Compression in extreme fiber	
Extreme fiber stress in flexure	$0.40 f'_c$
Extreme fiber stress in flexure adjacent to supports of continuous beams	$0.45 f'_c$
Shear	
Beams with no web reinforcement:	
Longitudinal bars anchored	$0.03 f'_c$
Longitudinal bars not anchored	$0.02 f'_c$
Beams with web reinforcement:	
Longitudinal bars anchored	$0.12 f'_c$
Longitudinal bars not anchored	$0.06 f'_c$
Reinforcement	
Tensile or compressive allowable stresses	
Structural steel grade bars	16,000 psi
Intermediate grade bars	18,000 psi
Hard grade bars	18,000 psi
Bond	
Between concrete and plain reinforcing bars	$0.04 f'_c$
Between concrete and approved deformed bars	$0.05 f'_c$
Modulus of elasticity	
Concrete, f_c between 1,500 psi and 2,200 psi	1,933,333 psi
Concrete, f_c between 2,200 psi and 2,900 psi	2,416,667 psi
Concrete, f_c greater than 2,900 psi	2,900,000 psi

As can be seen from the summary above of the recommended working stresses by the Second Joint Committee, the design of concrete building structures has evolved through the initial experimental stages and is now approaching a state of standardized design. Working stress design has now reached some degree of adoption with the publication of Urquhart and O'Rourke in 1926. Rectangular beams and flat slabs, the limits of this investigation, can now be approached through the method of analysis of the cross-section and the force couple, which is the result of assumptions of the stress block in the concrete and the tension on the longitudinal steel. It is now assumed (1926) that the reinforcing bars are embedded so that the union between the steel and concrete is sufficient to make the two materials act as one. Further, it is assumed that the concrete above the neutral axis of the member takes the compression force of the couple and that the concrete below the neutral axis has cracked and that all of the tensile force of the couple is carried by the reinforcing steel. Evidence also exists through review of the reference manuals of the period that designers now are concerned with and understand the concept of the ratio of the area of reinforcing steel to the overall area of the section and the problems that arise from the use of excessive area (cross-sectional area) of tension reinforcing.

Working stress method, 1926:

$$C = \frac{f_c(kd)(b)}{2}$$

$$T = A_s f_s$$

$$jd = d - \frac{kd}{3}$$

$$k = \frac{n}{n + r}$$

$$p = \frac{n}{2r(n + r)}$$

$$p = \frac{A_s}{bd}$$

$$j = 1 - \frac{k}{3}$$

$$f_s = \frac{n(f_c)(1 - k)}{k}$$

$$f_c = \frac{f_s(k)}{n(1 - k)}$$

$$M_s = A_s f_s jd$$

$$M_c = \frac{f_c(k)j(b)d^2}{2}$$

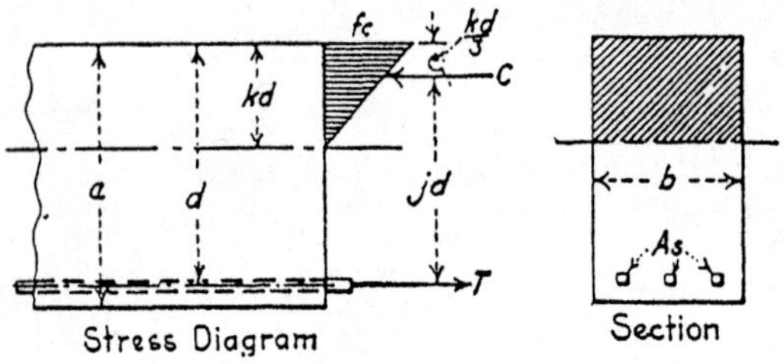

Figure 4-79.

where:
- kd = depth, top to neutral axis, in.
- d = depth of beam, top to c.g. of steel, in.
- f_c = allowable compressive stress of concrete, psi
- A_s = Area of tension reinforcing steel, in.²
- f_s = allowable tensile stress of steel, psi
- r = ratio of f_s over f_c
- M_c = moment from concrete, lb-in.
- M_s = moment from steel, lb-in.
- p = percentage of steel area

The values of M_s and M_c will be equal only when the amount of steel in the beam equals the actual steel ratio.

$$p_{bal} = \frac{A_s}{bd} = \frac{n}{2r(n + r)}$$

If $\frac{A_s}{bd}$ is larger than p_{bal}, the strength of the beam is limited by the concrete

If $\frac{A_s}{bd}$ is smaller than p_{bal}, the strength of the beam is limited by the steel

The recommendations of the Second Joint Committee remained the preferred method of designing and detailing reinforced concrete until American Concrete Institute (ACI) Committee 317 published their *Reinforced Concrete Handbook* in 1941. The Handbook was published cooperatively by the American Concrete Institute, the Portland Cement Association, the Concrete Reinforcing Steel Institute, and the Rail Steel Bar Association. Ultimate strength design was introduced in 1955.

5
Historic American Roof Systems: Lateral Bracing of Buildings

INTRODUCTION

Since the beginning of building, roofs have been one of the most important areas of the systems of enclosure. Early forms of roofs were regional and responded to the materials at hand and the climatic conditions. Stone vaults were prevalent in the Middle East, trusses and sloped roofs in areas of rain and snow, and adobe roofs in arid regions.

The arch or barrel vault depended upon compression in the voussoirs and to some extent the skill of the builder. Flat roofs utilized members that acted as beams and were subject to flexure. Trusses utilized a frame composed of smaller members acting in pure tension and pure compression (theoretically) to span distances longer than a simple beam would span. Opposing rafters tied together by a bottom chord are the simplest form of truss while trussed arches over massive roofs with great open spaces are the most complicated.

American houses and commercial buidings of the period 1840–1940 could include any of the above, but would most likely have either flat roofs, utilizing beams, or trussed roof systems. The traditional materials for flat roof systems were wood, cast iron, wrought iron, steel, and concrete. The materials for trussed roof systems could be any of the above or almost any combination thereof. Utilitarian buildings would normally have a simple, straightforward roof system, while exotic buildings might have specialized or unusual roof systems. A few arched domes over large spaces such as cathedrals and capitol buildings; but they are out numbered by steel skeletal dome framework systems clad to form interior and exterior dome surfaces. The U.S. Capitol is said to have over 8 million pounds of iron in the superstructure of its dome (Kidder 1902, 598).

FLAT ROOF SYSTEMS

Chapter 4 has covered floor systems in historic American buildings in detail. Flat roof systems are extremely similar to floor systems, the main difference being the membrane enclosure and flashing system, which is employed on roofs but does not exist for floors. Structurally, the only difference is the live loads. The trussed girder of wood and steel or cast iron and steel and the flat arch of wrought iron or steel beams supporting transverse arches of masonry are systems common to both floors and roofs. The information in Chapter 4 will not be repeated here. This chapter will concentrate on the truss and the many variations used in trusses for roofs.

Flat trusses for roofs will be included because they do utilize truss action for spanning between supports. The analytical solution of simple trusses is part of an elementary course in statics and will not be dealt with in detail here. Rather, the use of the truss, the many different types of trusses, and the many types of materials and combination of materials in trusses are the topic of this chapter. Truss members are

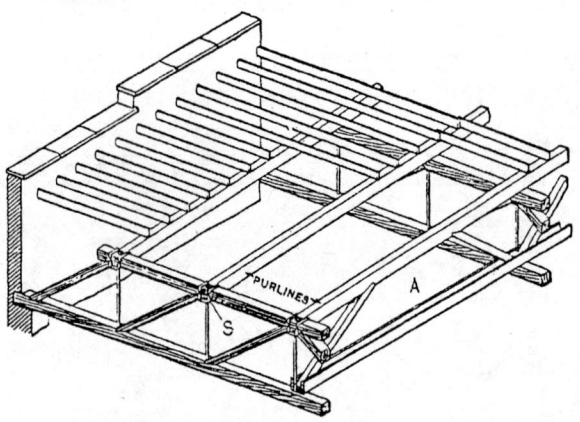

Figure 5-1. Flat truss roof system: Rafters supported on purlins. (*From Kidder 1905, 59.*)

normally pure tension or compression when the loads are applied at the joints of the truss (see Figure 5-1). The purlins span between the trusses and act as beams to support the roof joists; thus the joists span the same direction as the trusses. Figure 5-1 shows an example of the bottom chord bearing type of truss. In this truss, the diagonals are in compression, as is the top chord; therefore, the purlins are required to restrain the top chord of the truss to prevent buckling.

When the joists are spread out to bear on the top chord of the truss, the top chord is in combined compression and bending, as shown in Figure 5-2. This truss is also bottom chord bearing, but the top chord has been extended to bear on the wall because it must carry the rafters, as shown. In this instance, if the rafters are properly attached to the top chord of the truss, the rafters provide the necessary bracing for the top chord of the truss. When trusses span from side wall to side wall of a building that is narrow and deep, as is true of most historic buildings in the urban core of cities, in most cases the "flat" roofs of these long, narrow buildings slope toward the rear. If the bearing course of brick is at a point of change in thickness in the wall, as it usually is, it is impossible to get the slope without changing the truss depth at each bay over the length of the building. The trusses shown in Figure 5-2 are providing this slope. The truss on the right is deeper than the center one, which is deeper than the one on the left, etc. The designer would choose the depth of the shallowest truss as that necessary for the span at the rear, and each truss progressing to the front would usually be 5 in. deeper for every 20 ft bay. The 10-Panel truss shown in Figure 5-3 is the full elevation drawing of a truss similar to the one shown in Figure 5-1. The truss shows both bearing ends and a clear presentation of the chords, diagonals, and verticals along with the purlins and joists. The top and bottom chords are continuous, and the details of the member intersections are clearly visible. The diagonals are let

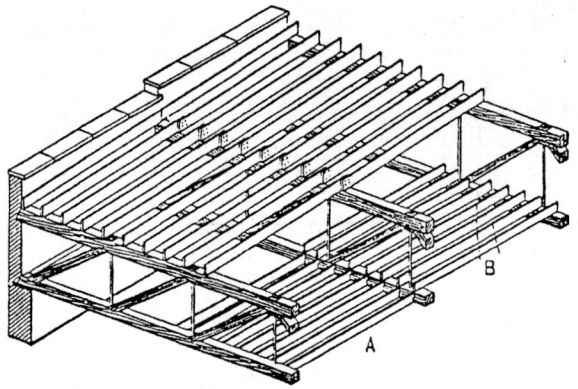

Figure 5-2. Flat truss roof system: rafters rest on top chord of truss. (*From Kidder 1905, 58.*)

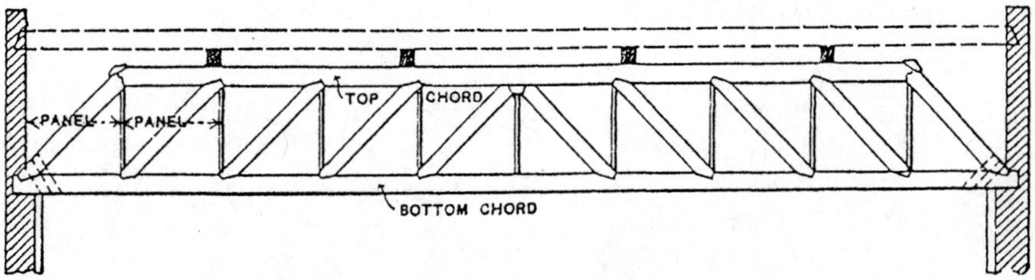

Figure 5-3. A 10-panel flat truss. *(From Kidder 1905, 63.)*

in to the chords, and the wrought iron vertical members are threaded at each end so that the verticals maintain the position of the diagonals (see Figure 5-4). The connection at the bearing point at the ends of the truss is through-bolted, as shown in Figure 5-5. The vertical member acts as the tie rod for connection stability and to hold the shape of the truss.

The trusses shown in the figures are made of medium-sized wood timbers for all of the compression members and wrought iron tension vertical members. The bottom chord is in tension; however, it is a medium-sized timber member. The wrought iron member is substituting for a wood member and, more importantly, is replacing the plates, bolts, and multiple members that would be required if the truss were to be made totally of wood. Wood trusses require multiple members in that the connections must be lapped for bolting and the geometry of the triangular truss panel does not allow layering of the plane of the truss without causing artificial shifts in the center of gravity of the vertical plane. These shifts cause severe stress variations in the truss.

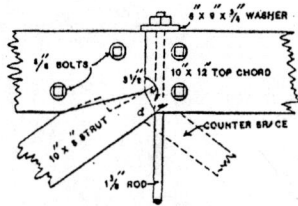

Figure 5-4. Joint in top chord. *(From Kidder 1905, 141.)*

TRUSSES: MULTIPLE-COMPONENT MEMBERS

Designers realized that the only solution to the offset plane problem was to use multiple components for part of the individual members of the truss. The multiple-component members would be collectively sized and would require bolts and cover plates, as shown in Figures 5-6 and 5-7. Wood trusses utilizing medium-heavy timber members of relatively long dimensions required longer sections than were conveniently available, and thus joints in members were required. Unlike steel members, these splices would place additional complications on a truss joint that was already quite complicated. Splices in wood truss members were always placed between truss joints or connections. These splices would be required to carry the full tension or compression of the member. See Figure 5-8 for typical details of such joints.

WOOD LATTICE TRUSSES

Another common all-wood truss used by designers in the period 1885–1940 was the Lattice truss. Kidder states that the lattice truss was designed by Ithiel Towne for use on wooden bridges (Kidder 1925, 29; *id.* 1905, 70). He further states that the lattice

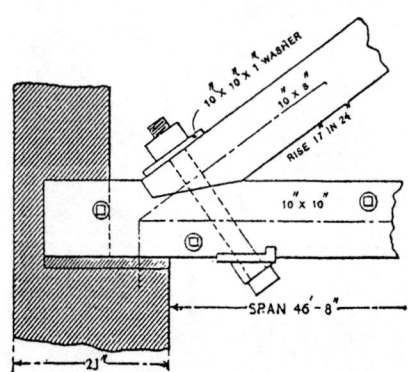

Figure 5-5. Detail connection at bearing. *(From Kidder 1905, 131.)*

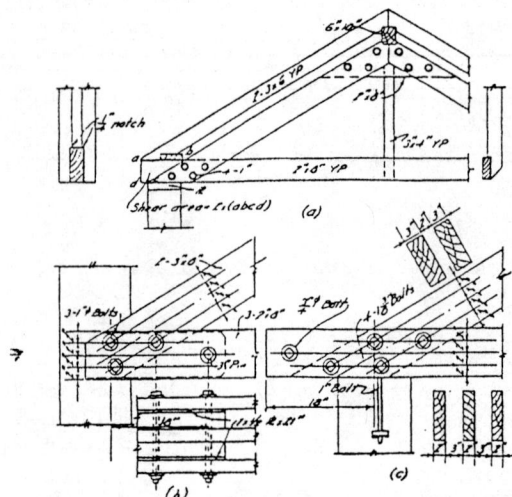

Figure 5-6. Wood truss detail, multiple members. (*From Voss and Varney 1926, 179.*)

truss is very efficient in supporting loads. The lattice truss was considered well suited to flat roofs and an economical substitute for trusses composed of large timbers and wrought iron rods. Kidder states that the one problem with a lattice truss is that it cannot be tightened up as wood and wrought iron trusses can be. See Figure 5-9 for an elevation of a lattice truss. Kidder states that the height of the lattice truss should be from one-eighth to one-sixth of the span and that the braces should be placed at an angle of 45°. The number of spaces equals the span multiplied by two and divided by the height. The lattice truss of Figure 5-9 would therefore contain 16 spaces.

Kidder gives the following rules for computing the stress in the chords and braces (Kidder 1925, 31; *id.* 1905, 73).

1. Under a uniformly distributed load, the maximum stress in the chords may be found by multiplying the total load by the span and dividing by eight times the height, both in feet.
2. The stress in each of the end braces, *a, a, a*, when the angle of inclination approximates 45°, will be one-sixth of the total load, multiplied by 1.4.

As stated, the lattice truss was considered efficient, especially when heavy to medium-sized timber sections were not available. Lattice trusses were generally con-

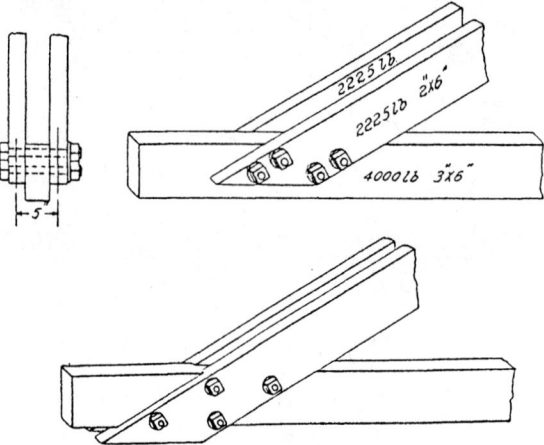

Figure 5-7. Wood truss detail, multiple members. (*From International Correspondence Schools 1905, para. 28, 54.*)

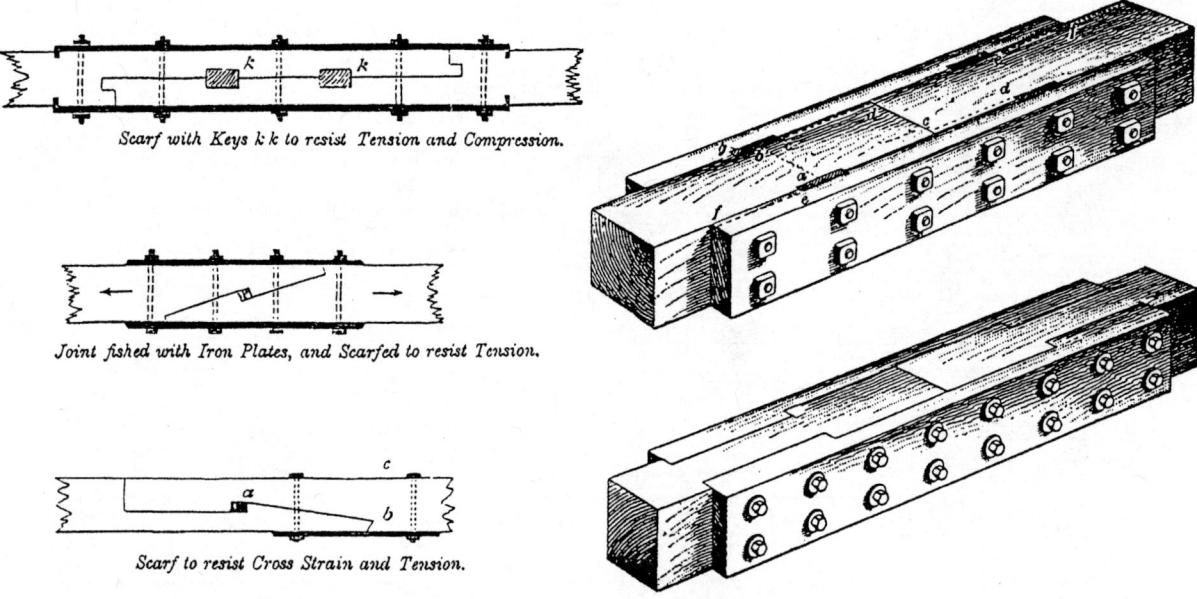

Figure 5-8. Wood member, tension and compression splices. (*From International Correspondence Schools 1905, para. 28, 56, 57; Rivington 1899, 62, 63.*)

structed of multiple 2 in. cross-members, bolted at connections (see Figure 5-10). The lattice truss was designed almost empirically; the rules given above are the result of experience and undetermined analytical methods. The lattice truss is practically indeterminate by today's methods because of the redundant members. Many buildings in America utilized the lattice truss for both floors and roofs. It was more popular for roofs because the designers could not tolerate the depth required for the spans for floors. In addition, floors provide greater loads, which would also need greater depth and member size. Kidder includes a table of member sizes and spans for lattice trusses (see Table 5-1). This table gives the size of members and bolts for several spans of lattice trusses. This table is different from most in that all spans in the table are for the same loads: structural dead loads, gravel roof, plastered ceiling below, and snow load of 20 psf.

GABLE TRUSSES: WOOD AND WOOD/IRON COMPOSITE TRUSSES

Wood trusses were extremely versatile, and designers found them very exciting. Wood was the cheapest building product available at the time and could be custom cut to practically any size and shape. The only real limitation was the complexity of the joints. Decorative trusses were used in churches, and plain trusses were used in manufacturing and commercial applications. Bell (1875) illustrates some early wood trusses (see Figure 5-11). The Bell trusses indicate the state of the art for that period.

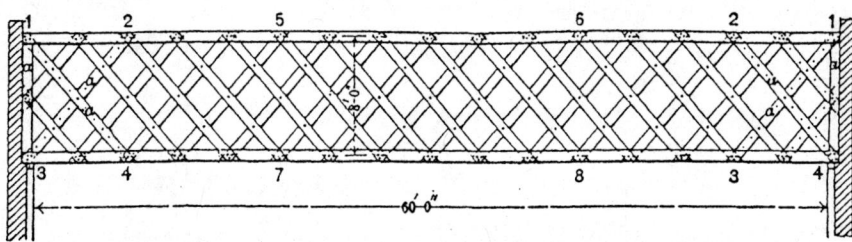

Figure 5-9. Elevation, lattice truss. (*From Kidder 1925, 29; id. 1905, 71.*)

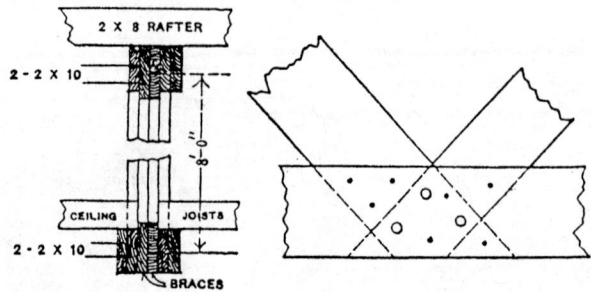

Figure 5-10. Lattice truss details. (*From Kidder 1905, 72.*)

Table 5-1. Lattice Truss Dimensions: First-Quality White Pine or Spruce[a]

To support a gravel roof and plastered ceiling, allowing 20 psf for snow.

Span	Spacing of Trusses	Height Out to Out		Number of Spaces	Size of Bottom Chord	Size of Top Chord	End Braces	Inner Braces	Bolts in Joints 1–5
Ft	Ft	Ft	In.						In.
40	12	5	6	16	4 − 2 × 6	4 − 2 × 6	2 × 6	2 × 6	2 − 1
		7	2	12	4 − 2 × 6	4 − 2 × 6			
	14	5	7	16	4 − 2 × 8	4 − 2 × 8	2 × 6	2 × 6	3 − 7/8
		7	3	12	4 − 2 × 8	4 − 2 × 8			
	16	5	8	16	4 − 2 × 8	4 − 2 × 8	2 × 8	2 × 6	3 − 1
		7	4	12	4 − 2 × 8	4 − 2 × 8			
50	12	6	8	16	4 − 2 × 8	4 − 2 × 8	2 × 10	2 × 8 and 2 × 6	3 − 1
		8	8	12	4 − 2 × 8	4 − 2 × 8			
	14	6	8	16	4 − 2 × 8	4 − 2 × 8	2 × 10	2 × 8 and 2 × 6	3 − 1 1/8
		8	8	12	4 − 2 × 8	4 − 2 × 8			
	16	6	9	16	4 − 2 × 8	4 − 2 × 10	2 × 10	2 × 8 and 2 × 6	3 − 1 1/8
		8	8	12	4 − 2 × 8	4 − 2 × 8			
60	12	8	4	16	4 − 2 × 10	4 − 2 × 10	2 × 10	2 × 8	3 − 1 1/8
		10	10	12	4 − 2 × 10	4 − 2 × 10			
	14	8	4	16	4 − 2 × 10	4 − 2 × 10	2 × 10	2 × 8	3 − 1 1/8
		10	10	12	4 − 2 × 10	4 − 2 × 10			
	16	8	4	16	4 − 2 × 10	4 − 2 × 10	2 × 10	2 × 8	3 − 1 1/4
		10	10	12	4 − 2 × 10	4 − 2 × 10			
	14	9	5	16	4 − 2 × 10	4 − 2 × 12	2 × 10	2 × 8	1 − 2 1/4
		12	4	12	4 − 2 × 10	4 − 2 × 10			
	16	9	5	16	4 − 2 × 10	4 − 2 × 12	2 × 10	2 × 8	1 − 2 3/8
		12	4	12	4 − 2 × 10	4 − 2 × 10			
	18	9	6	16	4 − 2 × 10	4 − 2 × 12	2 × 10	2 × 8	1 − 2 1/2
		12	6	12	4 − 2 × 10	4 − 2 × 12			
80	14	11	0	16	4 − 2 × 12	4 − 2 × 12	2 × 10	2 × 8	1 − 2 3/4
		14	0	12	4 − 2 × 12	4 − 2 × 12			
	16	11	2	16	4 − 2 × 14	4 − 2 × 14	2 × 12	2 × 10 and 2 × 8	1 − 3
		14	0	12	4 − 2 × 12	4 − 2 × 12			
	18	11	2	16	4 − 2 × 14	4 − 2 × 14	2 × 12	2 × 10 and 2 × 8	1 − 3
		14	1	12	4 − 2 × 12	4 − 2 × 14			

Uprights at end same size as end braces.

[a] From Kidder 1925, 32; id. 1905, 74.

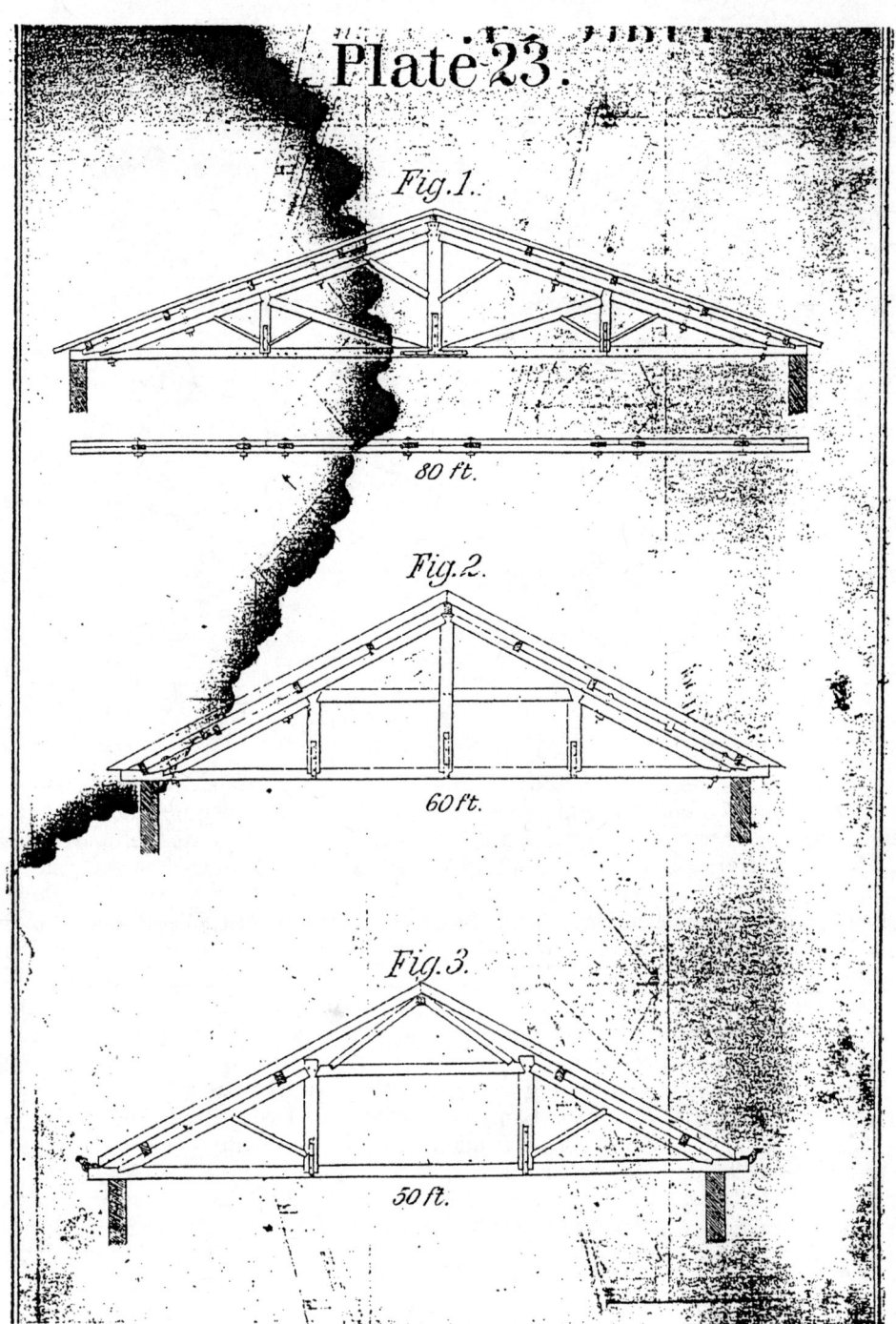

Figure 5-11. Bell's gable-type trusses. (*From Bell (1875), Plate 23.*)

The chords of the trusses are held together by straps and bolts, and at the bearing points the chord intersection is through-bolted, as shown. The Bell trusses and Rivington's trusses have what are called principal rafters and common rafters (see Figures 5-11 and 5-12). Today we would either use more trusses more closely spaced and apply the roof deck or laths directly to the truss rafters or use more purlins to support the deck when the trusses are farther apart.

The connections for wood trusses became more and more of a design specialty, and a number of foundries produced specialty hardware to be used for connections (see Figures 5-13 and 5-14).

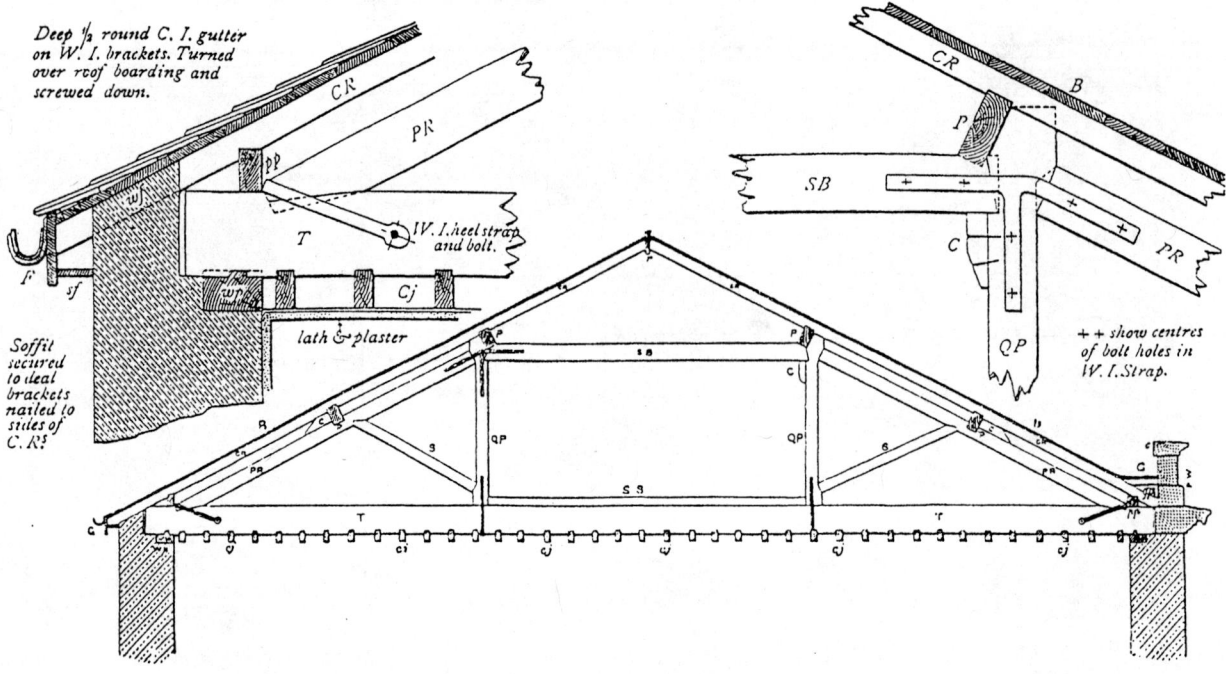

Figure 5-12. Rivington's queen-post truss. (*From Rivington 1899, 170.*)

From the last 10–15 years of the nineteenth century through the 1930s, a large quantity of factory buildings, mill buildings, warehouses, and other medium-sized utility buildings were constructed throughout the United States. Most of these facilities benefited from having large column-free spaces, and truss roofs were quite extensively used. During this period there were many developments in the detailing of trusses, mainly in the detailing and connections of the trusses. Wood and wrought iron trusses were utilized, as well as riveted steel trusses. Wood and wrought iron trusses were developed early, some instances of their use occurring in the 1870s. These trusses were very interesting. Most contained wrought iron for tension members, and they utilized some intricate castings of wrought iron for connection details (see Figure 5-15). In general, buildings became lighter as trusses became common. Roofs became lighter as the details of the roofing system evolved from heavy timber members to wrought iron sections, which were smaller due to their strength and thus lighter in overall weight. As the systems became lighter, cross-bracing and ties soon became a requirement as lateral loads became a more important concern (see Figure 5-16).

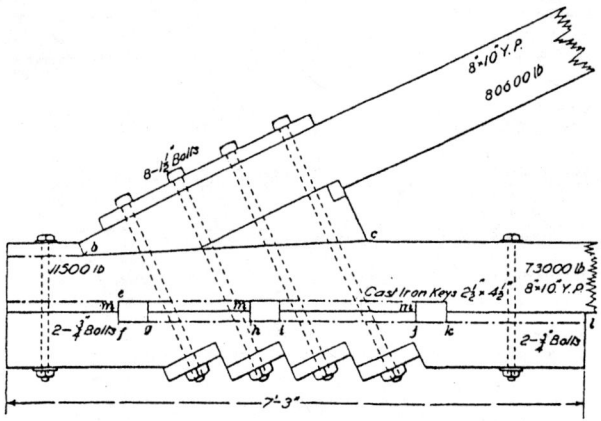

Figure 5-13. Truss connections, splices: specialty hardware. (*From International Correspondence Schools 1905, para. 28, 64, 65.*)

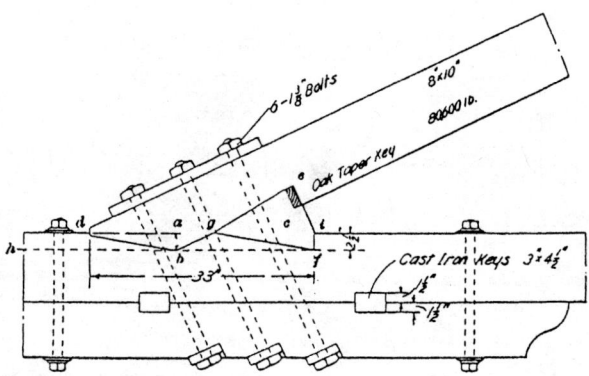

Figure 5-14. Truss connections, castings: specialty hardware. (*From Kidder 1905, 154.*)

The allowable stresses on the wood members determined their sizes. Designers sized compression members and struts in the same way one would design wood columns. Wood tension members were a little more complicated to design because the tension across the bolt shanks would apply a form of tension shear on the sectional area of the material in shear and tear out through splitting at the end of the member away from the connection. In the early periods, these members were empirically sized and were the weakest point in the truss–thus the immediate success of the wrought iron tension members. The members would have been sized on the basis of their allowable stresses for the service intended. Wood compression members would be subject to their allowable stresses by species of wood and length of member (see Chapter 3). Wrought iron members would be checked by utilization of the allowable stresses for the period of original construction (Chapter 4 also gives the allowable stresses and elastic limits for wrought iron by dates.) See Figure 5-16 for a typical truss with wood and wrought iron members and details.

These composite trusses were very popular with designers, and a number of buildings were constructed utilizing this system. The rafter chord is reinforced by the system of tension rods and struts, as is evidenced by the truss in Figure 5-17. These systems are identical to the trussed girders shown in Chapter 6 of this work. The struts act perpendicular to the rafter chord to act as supports, in this case at midspan and

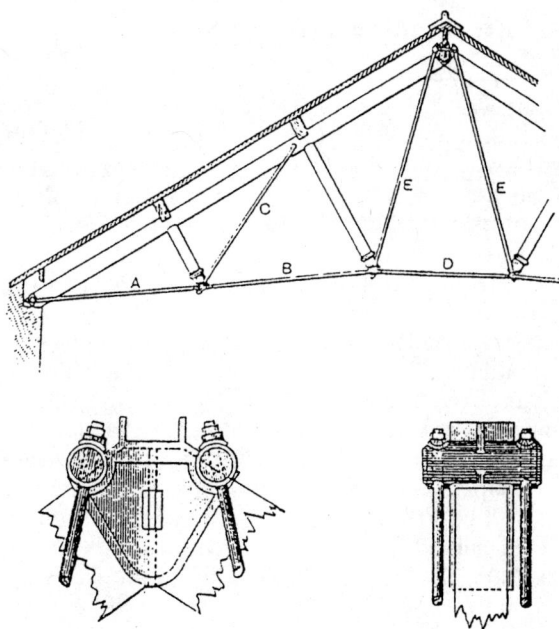

Figure 5-15. Trussed rafter roof (Holt's truss). (*From Rivington 1899, 175, 176.*)

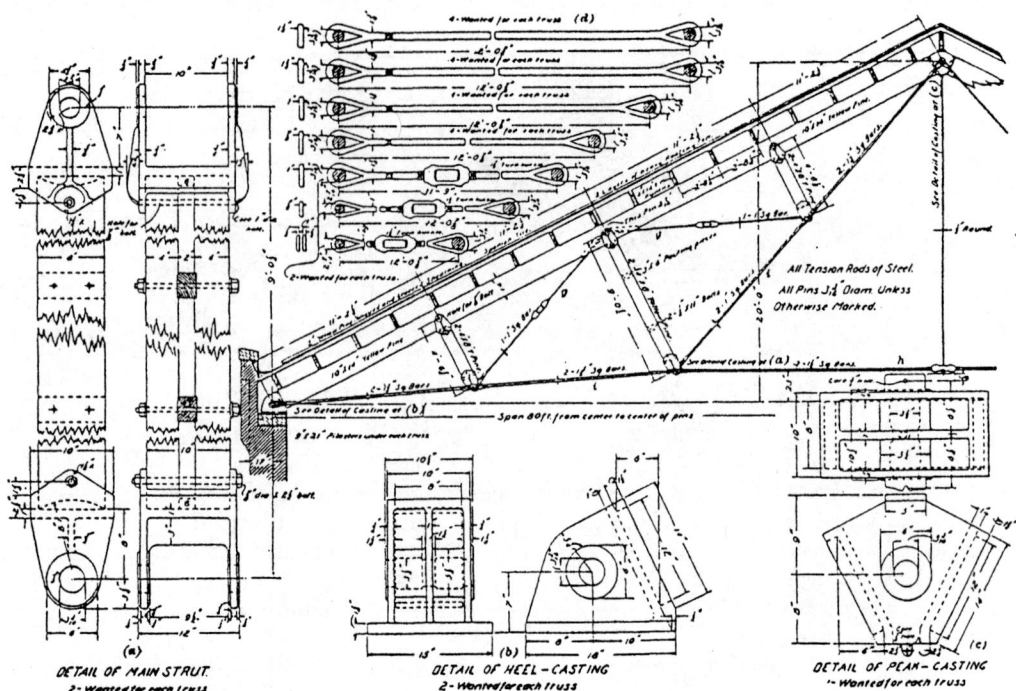

Figure 5-16. Composite truss with wooden struts. (*From International Correspondence Schools 1905, para. 28, 33.*)

at quarterspan. The rafter chords are loaded almost uniformly by the smaller purlins and are therefore subject to flexural loading plus compression (external compression from the tension rods). The tension rods are steel in this drawing; they could also be wrought iron, as earlier examples of this truss would have been. It was common to place turnbuckles in the tension rods to allow for manual adjustment of the tension in the rods. Designers of this period were well aware of the presence of compression in the rafter chord. This is the weakest element in this system and must be checked for two-way or combined stresses.

IRON TRUSSES: WROUGHT IRON AND STEEL TRUSSES

The wrought iron truss was the next phase in the very logical development of roof-spanning devices. At first the all-iron truss was made simply by replacing the wood chords and struts with wrought iron sections and continuing with the wrought iron rods as tension members. The details of the truss needed little modification to be adapted to the wrought iron sections (see Figure 5-17). Connections were made with bolts and wrought iron pins. In many instances, the struts were made of castings, which may have been decorative. See Figure 5-18 for all iron truss elevations and details.

The need for larger, specialized buildings continued to grow through the period. Trusses were very valuable to the designer because their flexibility was almost unlimited. Structural steels and riveting added the then missing product to a market that was ready for it. Mill roofs with monitors to allow air circulation and light into the building were a natural for structural steels. Structural steels were easily combined with plates and other shapes to make built-up members, columns, girders, truss components, and curved surfaces. Domes, which began as the ultimate shape of the arch principle, could be constructed larger and taller utilizing the truss. During the great building era of 1885–1935, the truss was probably the most important component. The steel truss in the latter half of that period was used on almost every building

HISTORIC AMERICAN ROOF SYSTEMS: LATERAL BRACING OF BUILDINGS ◇ 397

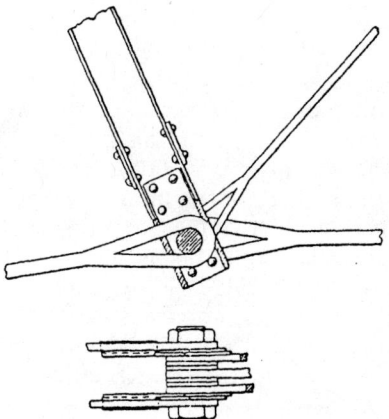

Figure 5-17. Wrought iron truss details. (*From International Correspondence Schools 1905, para. 29, 79.*)

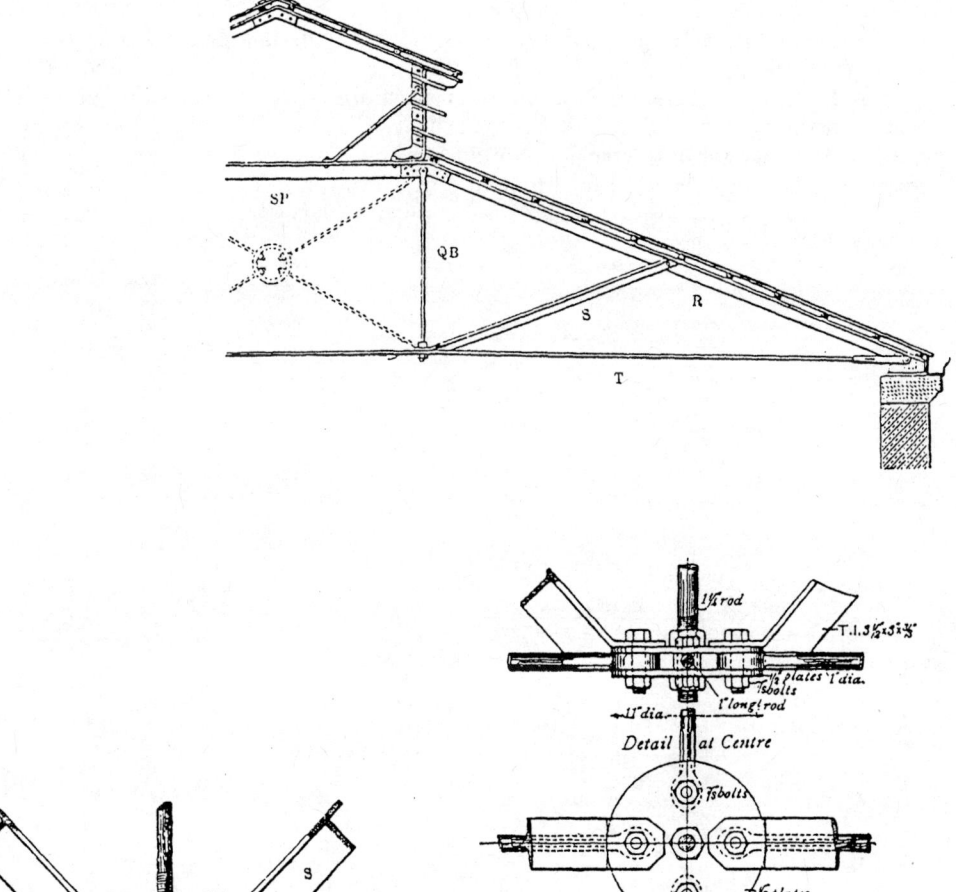

Figure 5-18. Wrought iron truss, iron truss, layout and details. (*From Rivington 1899, 185, 190.*)

possible. The skeletal frame enabled the designer to attach to or hang almost anything from steel members.

The member forces in trusses were most often solved by the use of graphic statics. See Figure 5-19 for the layout of a cantilever and monitor truss for a factory building. This building is divided into two two-story aisles each 50 ft wide and a center aisle 100 ft wide. The solution for member forces shown with the truss layout is by graphic statics. The graphic solutions were, as the name implies, scale drawings of the force vectors at their given directions (slopes), representing the solution of the forces in the members of the truss. The scale would be set in units of force, and the solution for the force in any member would be scaled off the graphic solution. Ricker also provides a table of the solution of the forces in the truss (see Table 5-2).

The graphic solutions were accurate and were the only method of analysis used by most of the designers of the period. Their accuracy depended upon the scale of the diagrams, but they provided very workable force loads for truss members. Today, trusses are solved by mathematical methods, either manually or on the computer. Figure 5-20 and Table 5-3 represent a mathematical solution of a portion of the members of the side roof truss for the P-stress forces only. The tabular result shown can be compared to the result of the graphic solution (Table 5-2).

The graphic solutions for the P-stress loads on the side roof truss indicate a tension force of 1.4 tons in members $X1$ and $X2$ and a compressive force of 1.3 tons in member $Y1$. These loads are slightly different from the ones shown in the mathematical solution, due wholly to the 3.070 kip load on the end panel point at A. Early designers minimized the effect of the panel point load at that end of the truss by passing it directly through to the column section. This has proven to be minimally inaccurate becauseee it affects slightly the members at the joint. The effect is usually minimal, as in this case, and the member sizes are conservative. Member $Y3$ is tabulated in the Ricker graphic solution as 6 tons and the manual solution is 6.01 tons, a difference that cannot be scaled. Similarly, other members check within the scaling tolerance. The truss is sized for the actual maximum and minimum forces produced

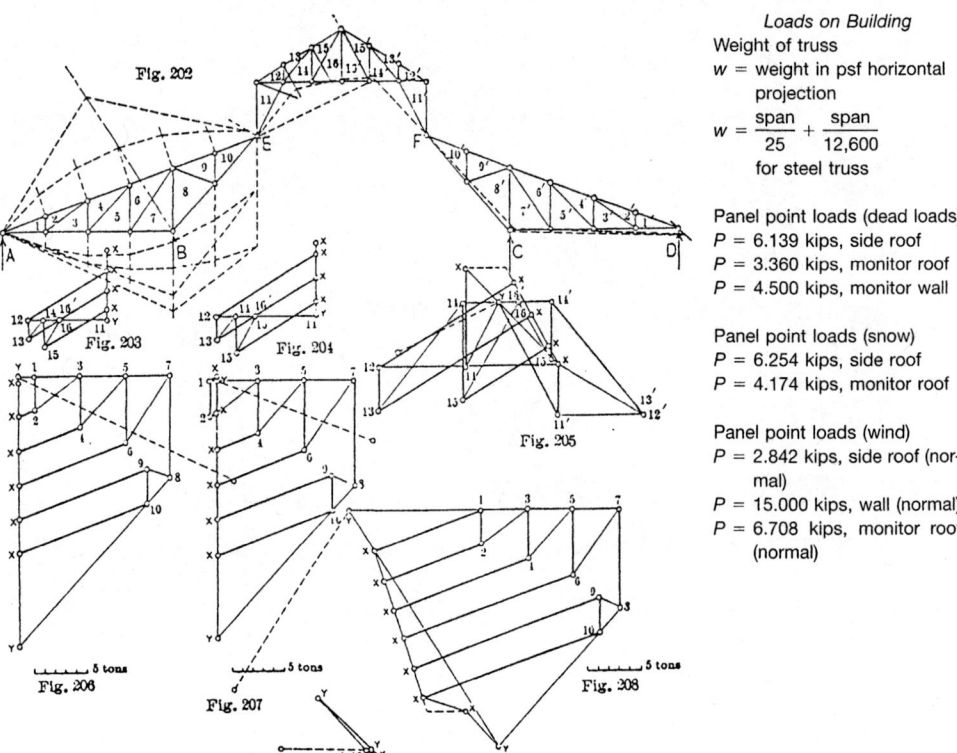

Figure 5-19. Cantilever and monitor truss, graphic statics: columns at points A, B, C, and D. (*From Ricker 1912, 113.*)

Table 5-2. Forces from Graphic Solution of Truss[a]

Member		P-stress[b]	S-stress[c]	W-stress W[d]	W-stress L[e]	Maximum[f]	Minimum[g]
X	1	+1.4	+0.4	+11.5	−0.7	+13.3	+0.7
X	2	+1.4	+0.4	+10.6	−0.7	+12.4	+0.7
X	4	+6.0	+4.2	+14.1	−0.7	+24.3	+5.3
X	6	+10.5	+9.0	+17.6	−0.7	+37.1	+9.8
X	9	+12.6	+10.0	+19.0	−0.7	+41.8	+11.9
X	10	+12.6	+10.2	+18.0	−0.7	+40.8	+11.9
X	11	−5.0	−6.3	−9.0	+4.4	−20.3	−0.6
X	12	−8.0	−10.0	−14.3	+9.2	−32.5	+1.2
X	13	−8.0	−10.0	−16.5	+9.2	−34.3	+1.2
X	15	−6.4	−8.0	−9.0	+0.4	−23.4	−6.0
Y	1	−1.3	−0.4	−14.0	+0.4	−15.7	−0.9
Y	3	−5.6	−4.0	−18.1	+0.4	−27.7	−5.2
Y	5	−10.0	−8.4	−22.5	+0.4	−40.4	−9.6
Y	7	−14.3	−13.0	−26.8	+0.4	−54.1	−13.9
Y	8	−20.6	−18.6	−17.2	−7.5	−56.4	−20.6
Y	10	−17.5	−15.5	−14.5	−7.5	−47.5	−17.5
Y	11	−0.0	−0.0	+6.6	−11.5	−11.5	+6.6
Y	14	+5.5	+7.0	+3.0	−5.3	+15.5	+0.2
Y	16	+4.0	+5.1	+2.4	−2.2	+9.1	+1.8
1	2	−3.0	−3.0	−3.0	−0.0	−9.0	−3.0
3	4	−4.5	−4.7	−4.5	−0.0	−13.7	−4.5
5	6	−6.0	−6.3	−6.0	−0.0	−18.3	−6.0
7	8	−9.3	−9.5	−9.3	−0.0	−28.1	−9.3
9	10	−2.2	−3.1	−3.0	−0.0	−8.3	−2.2
12	13	−1.7	−2.0	−3.9	−0.0	−7.6	−1.7
14	15	−2.5	−3.0	−8.8	+0.5	−14.3	−2.0
16	16'	−0.0	−0.0	−0.0	−0.0	−0.0	−0.0
2	3	+5.2	+5.4	+5.2	+0.0	+15.8	+5.2
4	5	+6.2	+6.5	+6.3	+0.0	+19.0	+6.2
6	7	+7.5	+7.7	+7.4	+0.0	+22.6	+7.5
8	9	+2.4	+2.4	+2.4	+0.0	+7.2	+2.4
11	12	+7.0	+8.5	+7.8	−7.7	+23.3	−0.7
13	14	+2.3	+2.7	+12.7	−13.2	+17.7	−10.9
15	16	+3.0	+3.5	+10.3	+6.4	+16.8	+3.0

[a] From Ricker 1912, 42, 43.
[b] Truss weight plus dead loads.
[c] Snow load, live load, 20 psf.
[d] Windward side wind loads, live load, 17.16 psf (inclined) for side roofs; 27.56 psf (inclined) for monitor roof; 40 psf for vertical monitor wall.
[e] Leeward side wind loads, live load.
[f] Maximum force combination.
[g] Minimum force combination.

Note: The gravity loads on the trusses are accurate and are good for any solution. The wind loads do not consider any negative force on the leeward side of the truss as we would do today. The loads as applied are still conservative. Today we factor down the windward side load and apply a negative leeward side load which does not produce as much force on the windward side members.

by the combined loads. The effect of the wind force on the windward side is to maximize forces on that side of the truss while minimizing or even causing stress reversals on the leeward side of the truss. The difference between the method of applying wind only to the one side (the past method) and applying both reduced windward and a leeward force (vacuum) may be more accurate (today's method); however, the members of the truss are sized for the maximum stress condition. In past and present design methods, wind is applied from all directions. The past methods were very conservative, and the force from the wind pressure was always quite a lot larger than the requirements of a modern analysis. The refinements brought about by tests and actual observations have proven that the actual force from the wind pressure is much lower than the early requirements. Buildings designed utilizing these conservative methods are easily verified and contain a reserve capacity against the lateral forces, provided they are well maintained.

The partial solution by mathematical methods demonstrates the accuracy of the truss solutions in the periods of the early designs. Trusses in older buildings would have been designed and member selection made based upon the formulae of the

400 ◇ Structural Analysis of Historic Buildings

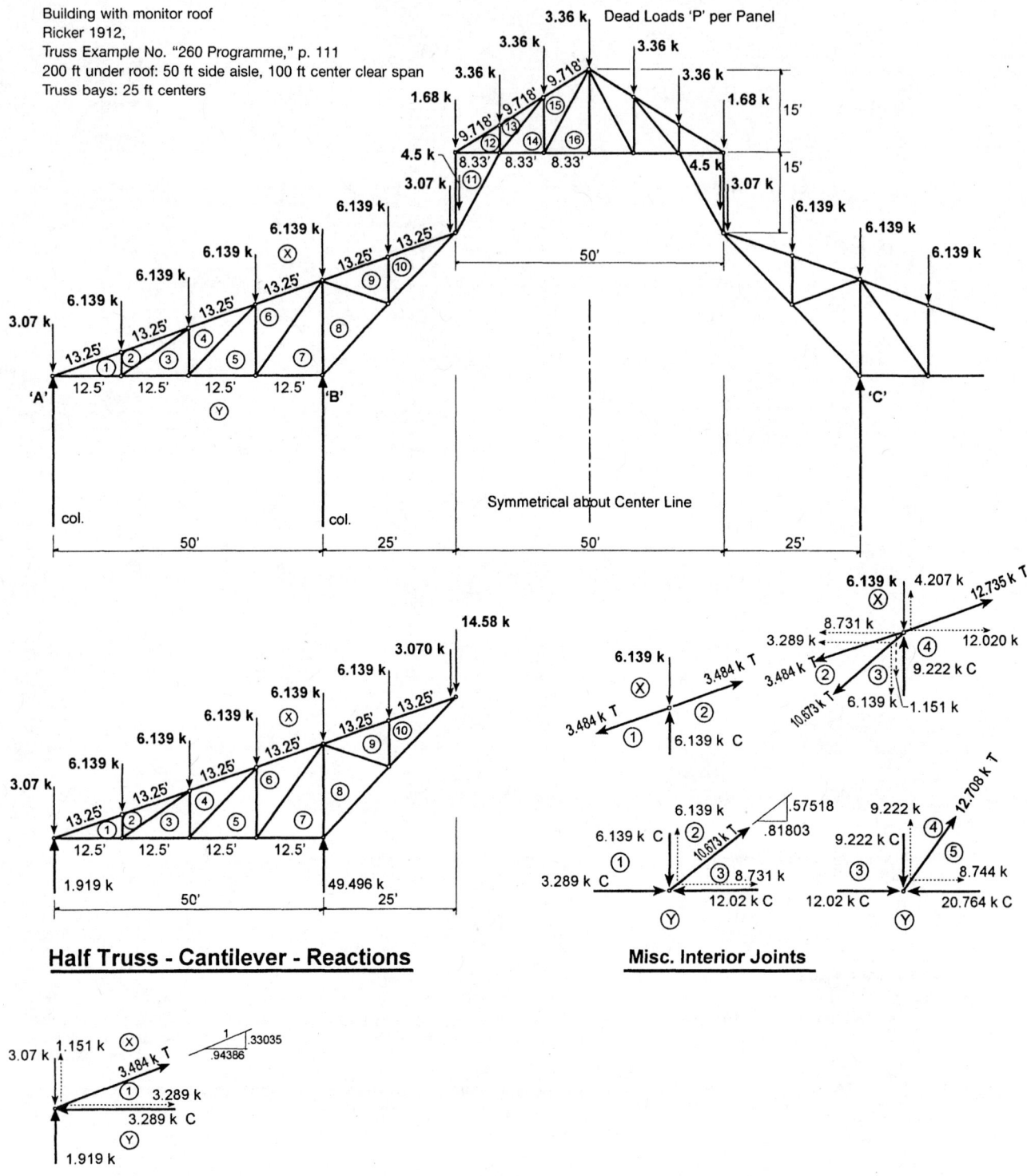

Figure 5-20. Cantilever truss with monitor: partial mathematical solution.

Table 5-3. Member Forces

Member	Force, kips	Force, tons
X-1, X-2	3.484 k T	1.74 tons T
Y-1	3.289 k C	1.65 tons C
Y-3	12.020 k C	6.01 tons C
2-3	10.673 k T	5.54 tons T
3-4	9.222 k C	4.61 tons C
X-4	12.735 k T	6.37 tons T
Y-5	20.764 k C	10.38 tons C

period, with forces in members, tension or compression, and connections based upon forces derived from basic graphic statics. The conservative nature of the member sizing would offset and compensate for any inaccuracies in the graphic statics of the truss solutions. Connections and connection integrity would a be more basic concern for the modern designer wishing to utilize an existing truss for a roof. Care must be taken to identify any change in loading or misuse of the original intended purpose of the truss.

Pin connections are very much a concern for the modern designer; to function, the truss is totally dependent upon the connection. Pin connections in bridge trusses have failed in America, the most notable being the Silver Bridge over the Ohio River at Gallipolis, Ohio, which failed suddenly while loaded with Christmas shopping traffic (automobiles) in December of 1967. The investigation of this incident proved that the pin connection failed, probably from fatigue. As a result, several bridges of the same design were declared unsafe and demolished. Ricker gives a fairly detailed example of a design of a pin connection, which is shown in the following section. The pin connection design is a part of an overall comparison between pin joints and riveted joints for trusses and is given by Ricker in the context of an example of connection detailing for a truss that was an earlier example in his discussion of the graphical analysis of roof trusses. The truss has a 128 ft span and a rise at the center of 25.6 ft of the Fink member configuration (see Figure 5-21 and Table 5-4).

The pins used for joints were machined in the manufacturing process to ensure smooth bearing from the members. They were either threaded at each end or had a hole for a cotter key to retain it in place. Pins and pin nuts were fabricated in standard sizes and dimensions, as were eye bars, upset bars (square and round) with screw ends, loop rods, clevises, turnbuckles, and sleeve nuts. Steel mills and manufacturers such as Carnegie and Cambria provided these items for construction. Most dimensions and sizes were standardized by American Bridge Company standards (see Tables 5-5 and 5-6).

Ricker (1912, 255) gives the following formula for design of pins in connections.

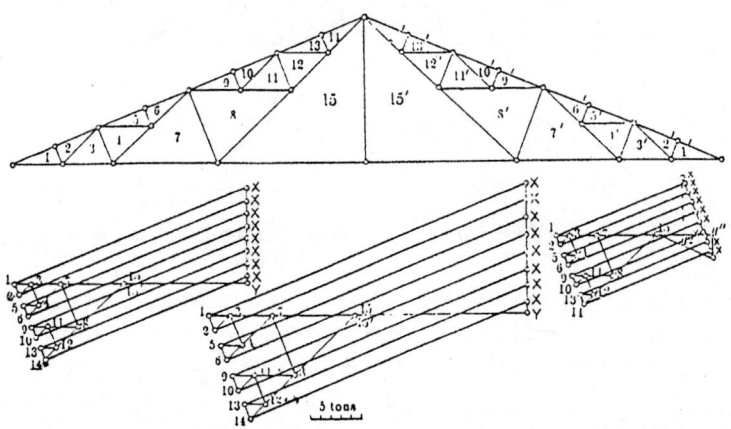

Figure 5-21. Fink-type truss graphic solution. (*From Ricker 1912, 42.*)

Table 5-4 Results for Fink-Type Truss Graphic Solution (Figure 5-21)[a]

Member		P-stress	S-stress	W-stress	Maximum	Minimum
X	1	−23.9	−32.2	−13.2	−69.3	−23.9 T.
X	2	−23.4	−31.7	−13.2	−68.3	−23.4
X	5	−23.0	−31.1	−13.2	−67.3	−23.0
X	6	−22.7	−30.5	−13.2	−66.4	−22.7
X	9	−22.2	−29.9	−13.2	−65.3	−22.2
X	10	−21.7	−29.3	−13.2	−64.2	−21.7
X	13	−21.3	−28.7	−13.2	−63.2	−21.3
X	14	−20.8	−28.1	−13.2	−62.1	−20.8
Y	1	+22.2	+29.9	+15.0	+67.1	+22.2
Y	3	+20.6	+27.9	+13.7	+62.2	+20.6
Y	7	+17.7	+24.0	+12.0	+53.7	+17.7
Y	15	+11.8	+15.9	+5.5	+33.2	+11.8
1	2	−1.1	−1.5	−1.0	−3.6	−1.1
3	4	−2.2	−2.9	−2.0	−7.1	−2.2
5	6	−1.1	−1.5	−1.0	−3.6	−1.1
7	8	−4.3	−6.0	−4.0	−14.3	−4.3
9	10	−1.1	−1.5	−1.0	−3.6	−1.1
11	12	−2.2	−3.0	−2.0	−7.2	−2.2
13	14	−1.1	−1.5	−1.0	−3.6	−1.1
2	3	+1.5	+2.0	+1.4	+4.9	+1.5
4	5	+1.5	+2.0	+1.4	+4.9	+1.5
4	7	+3.0	+3.9	+2.8	+9.7	+3.0
7	6	+5.6	+5.9	+4.2	+15.7	+5.6
8	9	+4.5	+6.0	+4.2	+14.7	+4.5
8	11	+3.0	+3.9	+2.8	+9.7	+3.0
8	15	+5.9	+8.2	+5.5	+19.6	+5.9
10	11	+1.5	+2.0	+1.3	+4.8	+1.5
12	15	+8.8	+12.1	+8.3	+29.2	+8.8
12	13	+1.5	+2.0	+1.4	+4.9	+1.5
14	15	+10.4	+14.0	+9.7	+34.1	+10.4
15	15	+0.0	+0.0	+0.0	+0.0	+0.0

[a]From Ricker 1912, 43.

Shearing Resistance:

$d = 0.4607 \sqrt{Z}$ = diameter of pin for single shear

$d = 0.3257 \sqrt{Z}$ = diameter of pin for double shear

where:
Z = maximum longitudinal force acting in member connected, tons (2,000 lb)

Bearing Resistance:

$d = \dfrac{Z}{12.5(t)} = \dfrac{0.08\,Z}{t}$ = diameter of pin with one bearing at end of member

$d = \dfrac{Z}{15(t)} = \dfrac{0.04\,Z}{t}$ = diameter of pin with two bearings at end of members

where:
Z = maximum longitudinal force acting in member connected tons (2,000 lb)

The formula in Ricker has 0.8 and 0.4, which is a misprint.

Ricker states that t, the thickness of the member, should be $0.377(d)$, and that for flat eye-bars, t should equal 0.167–0.25 of their width.

Moment on the Pin:

$d = 1.0839(M)^{0.333}$

Table 5.5 Truss Hardware with Tables for Sizing[a]

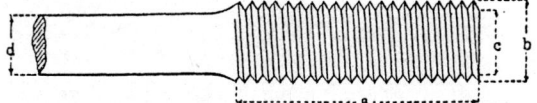

A. Upset Screw Ends for Round Bars
(American Bridge Company Standard)
Pitch and Shape of Thread A. B. Co. Standard

Bar			Upset				Area	
Diameter d (in.)	Area (in.²)	Weight per ft (lb)	Diameter b (in.)	Length a (in.)	Additional Length for Upset +10% (in.)	Diameter at Root of Thread c (in.)	At Root of Thread (in.²)	Excess Over Area of Bar, %
3/4[b]	0.442	1.50	1	4	4	0.838	0.551	24.7
7/8[b]	0.601	2.04	1 1/4	4	5	1.064	0.890	48.0
1	0.785	2.67	1 3/8	4	4	1.158	1.054	34.2
1 1/8	0.994	3.38	1 1/2	4	4	1.283	1.294	30.2
1 1/4	1.227	4.17	1 5/8	4	4	1.389	1.515	23.5
1 3/8	1.485	5.05	1 3/4	4	4	1.490	1.744	17.5
1 1/2	1.767	6.01	2	4 1/2	4 1/2	1.711	2.300	30.2
1 5/8	2.074	7.05	2 1/8	4 1/2	4	1.836	2.649	27.7
1 3/4	2.405	8.18	2 1/4	5	4	1.961	3.021	25.6
1 7/8	2.761	9.39	2 3/8	5	4	2.086	3.419	23.8
2	3.142	10.68	2 1/2	5 1/2	4	2.175	3.716	18.3
2 1/8	3.547	12.06	2 5/8	5 1/2	3 1/2	2.300	4.156	17.2
2 1/4	3.976	13.52	2 7/8	6	4 1/2	2.550	5.108	28.4
2 3/8	4.430	15.06	3	6	4 1/2	2.629	5.428	22.5
2 1/2	4.909	16.69	3 1/4	6 1/2	5 1/2	2.879	6.509	32.6
2 5/8	5.412	18.40	3 1/4	6 1/2	4 1/2	2.879	6.509	20.3
2 3/4	5.940	20.19	3 1/2	7	5 1/2	3.100	7.549	27.1
2 7/8	6.492	22.07	3 3/4	7	6	3.317	8.641	33.1
3	7.069	24.03	3 3/4	7	5	3.317	8.641	22.2
3 1/8	7.670	26.08	4	7 1/2	6	3.567	9.993	30.3
3 1/4	8.296	28.21	4	7 1/2	5	3.567	9.993	20.5
3 3/8	8.946	30.42	4 1/4	8	5 1/2	3.798	11.330	26.6
3 1/2	9.621	32.71	4 1/4	8	5	3.798	11.330	17.8
3 5/8	10.321	35.09	4 1/2	8 1/2	5 1/2	4.028	12.741	23.4
3 3/4	11.045	37.55	4 3/4	8 1/2	6	4.255	14.221	28.8
3 7/8	11.793	40.10	4 3/4	8 1/2	5 1/2	4.255	14.221	20.6

[a] From Carnegie Steel Co. 1923, 85–88.
[b] Special.

where:
  d = diameter of pin required to resist M, the moment on the pin from the spread of the eye bars

The connection of member 7-8 is shown in Figure 5-22 along with details of the composition of the members and the member build-up for the connection. Moment diagrams of the pins are also shown with the details. The detail of the bottom chord intersection with member 7-8 shows member 7-8 and three pairs of eyebars intersecting at the pin. The maximum force (stress, by Ricker) in member 7-8 is 14.3 tons in compression.

The pin size was designed as follows:

Shear of Pin (at top chord):

V = 14.3 T or 28.6 kips

1-13/16 in. diameter = 15.48 tons (from Table 5-7)

Table 5.5 (Continued)

B. Eye Bars
(American Bridge Company Stadards)

Ordinary Eye Bar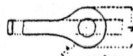

Adjustable Eye Bar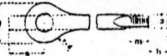

Minimum length of short and from center of pin to end of screw, 6'-8", preferably 7'-0".
Thread on short end to be left hand.
Pitch and Shape of Thread A. B. Co. Standard.

Bar			Head					Screw					
	Thickness			Maximum Pin		Additional Material[a] (ft, in.)		Bar		Excess Upset over Bar, %	Length m (in.)	Additional Material D (ft, in.)	
Width (in.)	Max (in.)	Min (in.)	Diameter d (in.)	Diameter (in.)	Excess Head over Bar, %	For Ordering Bar	For Figuring Weights	Width (in.)	Minimum Thickness (in.)	Diameter u (in.)		For Ordering Bar	For Figuring Weight
2	1	½	4½	1¾	37.5	0–10½	0– 7	2	*½	1¾	39.6	4	1– 0	8
			5½	2¾		1– 2½	0–11		¾	1⅞	36.6	4½	1– 0	7½
			*6½	3¾		1– 7½	1– 4		⅝	2	31.4	4½	0–11	7½
2½	1	⅝	6	2½	40.0	1– ¾	0–10	2½	*¾	2¼	41.2	4½	1– 0	8
			7	3½		1– 5½	1– 2		⅛	2¼	38.1	5	1– 0	8
			*8	4½		1–10¾	1– 7		1	2½	36.7	5	1– 0	7½
3	1½	⅝	7½	3¼	41.7	1– 4½	1– 1	3	*¾	2¼	34.3	5	1– 0	7½
			8½	4¼		1– 9½	1– 3		⅞	2½	41.6	5½	1– 1	9½
			*9½	5½		2– 2½	1–10		1	2½	23.0	5½	1– 1	8½
4	1¾	¾	10	4½	37.5	1– 9	1– 8	4	*¼	2½	23.0	5½	1– 1	8½
		⅞	11	5½		2– 3	1–10		⅞	2¼	32.0	5½	0–11	7½
		1	*12	6½		2– 8	2– 2		1	3	35.7	6	1– 1	8½
									1½	3¼	44.6	6½	1– 2	9½
5	2	¾	12	5¼	35.0	1–10½	1– 8	5	*¾	2½	36.2	6	1– 0	8
		1	13½	6¼		2– 6	2– 2		⅞	3	24.1	8	0–11	7
		1	*15	8¼		3– 3	2– 0		1	3¼	30.2	6½	1– 0	8
									1⅜	3½	34.2	7	1– 1	8½
									1¼	3¼	38.3	7	1– 2	9
6	2	¼	14	5¾	37.5	2– 1	1–10	6	*1	3½	25.8	7	1– 0	7½
		1	14¾	6½		2– 4	2– 1		1¼	3¾	28.0	7	1– 0	8
		1	*16½	8¼		3– 2	2– 8		1¼	4	33.2	7½	1– 1	8½
									1⅜	4½	37.3	8	1– 2	9½
7	2	1	16½	7*	35.7	2– 4½	2– 2	7	*1¼	4	26.9	7½	1– 0	8
		1⅛	17½	8		2–11	2– 6		1¼	4¼	29.6	8	1– 1	8½
		1⅛	*18½	9		3– 4	2–11		1⅜	4½	32.4	8½	1– 2	9
									1½	4¾	35.4	8½	1– 2	9½
8	2	1	18	7	37.5	2– 5½	2– 3	8	*1½	4¼	25.9	8	1– 0	8
		1⅛	10	8		2– 9½	2– 6		1½	4½	27.4	8½	1– 1	8½
		1¼	*20	9		3– 4	2–11		1¾	4¾	20.3	8½	1– 1	8½
									1½	5	31.4	9	1– 2	9
									1⅞	5¼	35.2	9½	1– 3	10
9	2	1⅛	20	7½	38.9	2– 8½	2– 6							
		1¼	22	9½		3– 4½	3– 1							
10	2	1⅛	22½	9	35.0	3– 2½	2–10							
		1¼	24	10½		3– 9	3– 3							
		1⅜	*25	11½		4– 1	3– 7							
12	2	1¼	26½	10	37.5	3– 4	3– 3							
		1⅜	28	11½		4– 2	3– 8							
		1½	*29½	13		4– 3	4– 1							
14	2	1⅜	31	12	35.7	3–11	3– 9							
		1½	33	14		4– 7	4– 4							
		1⅝	34	15		5– 5	4– 8							
16	2	1¾	36	14	37.5	4– 7	4– 6							
		1⅞	37½	16	34.4	4–11	4–10							

Bars marked * should only be used when absolutely unavoidable.
Deduct pin hole when figuring weight.
For 14" Bars, 33" Head, over 1½" thick add 4'-6½".

Table 5-6 Truss Hardware, Tables, Details^a

A. Loop Rods
(American Bridge Company Standard)

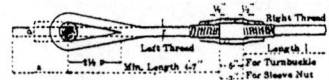

Pitch and Shape of Thread A. B. Co. Standard
Additional Length "A" in Feet and Inches for One Loop
$A = .175 + 5.39r$

Diameter of Pin r	Diameter or Side "r" of Rod in Inches										
	¾	⅞	1	1⅛	1¼	1⅜	1½	1⅝	1¾	1⅞	2
1⅛	0– 9½	0–10	0–11	0–11½							
1¼	0–10	0–10½	0–11½	1– 0	1– 1						
1½	0–11	0–11½	1– 0½	1– 1	1– 2	1– 2½					
1¾	1– 0	1– 0½	1– 1½	1– 2	1– 3	1– 3½	1– 4½	1– 5	1– 6		
2	1– 1	1– 1½	1– 2½	1– 3	1– 4	1– 4½	1– 5½	1– 6	1– 7	1– 7½	1– 8½
2¼	1– 2	1– 3	1– 3½	1– 4½	1– 5	1– 5½	1– 6½	1– 7	1– 8	1– 8½	1– 9½
2½	1– 3	1– 4	1– 4½	1– 5½	1– 6	1– 7	1– 7½	1– 8	1– 9	1– 9½	1–10½
2¾	1– 4	1– 5	1– 5½	1– 6½	1– 7	1– 8	1– 8½	1– 9½	1–10	1–11	1–11½
3	1– 5	1– 6	1– 6½	1– 7½	1– 8	1– 9	1– 9½	1–10½	1–11	2– 0	2– 0½
*3¼	1– 6	1– 7	1– 7½	1– 8½	1– 9	1–10	1–10½	1–11½	2– 0	2– 1	2– 1½
3½	1– 7½	1– 8	1– 8½	1– 9½	1–10	1–11	1–11½	2– 0½	2– 1	2– 2	2– 2½
*3¾	1– 8½	1– 9	1–10	1–10½	1–11	2– 0	2– 0½	2– 1½	2– 2	2– 3	2– 3½
4	1– 9½	1–10	1–11	1–10½	2– 0½	2– 1	2– 2	2– 2½	2– 3	2– 4	2– 4½
*4¼		1–11	2– 0	2– ½	2– 1½	2– 2	2– 3	2– 3½	2– 4½	2– 5	2– 6
4½		2– 0	2– 1	2– 1½	2– 2½	2– 3	2– 4	2– 4½	2– 5½	2– 6	2– 7
*4¾		2– 1	2– 2	2– 2½	2– 3½	2– 4	2– 5	2– 5½	2– 6½	2– 7	2– 8
5		2– 2½	2– 3	2– 3½	2– 4½	2– 5	2– 6	2– 6½	2– 7½	2– 8	2– 9
*5¼			2– 4	2– 5	2– 5½	2– 6	2– 7	2– 7½	2– 8½	2– 9	2–10
5½			2– 5	2– 6	2– 6½	2– 7½	2– 8	2– 9	2– 9½	2–10	2–11
*5¾			2– 6	2– 7	2– 7½	2– 8½	2– 9	2–10	2–10½	2–11½	3– 0
6			2– 7	2– 8	2– 8½	2– 9½	2–10	2–11	2–11½	3– 0½	3– 1
*6¼				2– 9	2– 9½	2–10½	2–11	3– 0	3– 0½	3– 1½	3– 2
6½				2–10	2–10½	2–11½	3– 0	3– 1	3– 1½	3– 2½	3– 3
*6¾				2–11	3– 0	3– 0½	3– 1	3– 2	3– 2½	3– 3½	3– 4
7				3– 0	3– 1	3– 1½	3– 2½	3– 3	3– 3½	3– 4½	3– 5

Pins marked * are special. Maximum shipping length of "1" ??? foot

Bearing on Pin design (single Bearing, Plate Between Two-angle Strut 7-8)

$$d = \frac{0.08\ Z}{t} = \frac{0.08\ (14.3\ \text{tons})}{7.16\ \text{in.}} = 3.051\ \text{in.} = 3\ 1/16\ \text{in.}$$

Bending moment of pin (moment from tension or compression of member):

$$d = 1.0839(M)^{0.333} = 1.0839(12.5125\ \text{in-tons})^{0.333} = 2.5143\ \text{in.}$$

The controlling pin size would be the largest of the three designs, 3-1/16 in. diameter.

The computations for sizing pins for truss joints were as shown above. Ricker also provides a table that summarizes these calculations and gives the capacities of pins from 1–9 in. diameter, which allowed designers to choose steel pins for truss connections directly (see Table 5-7).

Use of the pin connection began to decline in favor of the simpler rivet connection after 1915 as designers saw the opportunities offerby gusset plates and rivets. Riveted connections were easy to assemble, and the center of gravity of the section

Table 5-6 (Continued)

B. Recessed Pin Nuts
American Bridge Company Standard
Dimension in inches

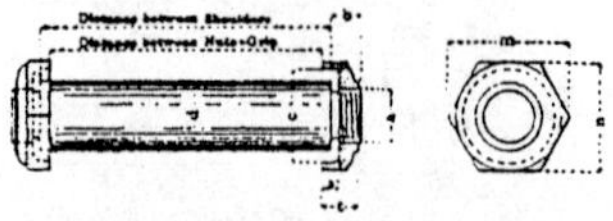

To obtain grip, add 1/16″ for each bar. Nuts threaded 8 threads per inch.
To obtain distance between shoulders, add amount given in table to grip.

Diameter of Pin, d		Pin Thread		Add to Grip	Nut Thickness t	Nut Diameter			Depth d	Diameter rough hole	Weight Pounds	Pattern No.
		a	b			a	m	c				
	2	2¼	1½	1	¼	⅞	2¹⁵⁄₁₆	3⅜	2⅝	¼	1⁵⁄₁₆	1.1	PN 21
	2½	2¼	2	1½	¼	1	3³⁄₁₆	4½	3⅛	¼	1¹¹⁄₁₆	1.7	PN 22
3	*3¼	3½	2½	1¼	¼	1½	4⁵⁄₁₆	5	3⅞	⅛	2⁵⁄₁₆	2.5	PN 23
	*3¾	4	3	1⅛	½	1¼	4⅞	5⅝	4½	⅛	2¹¹⁄₁₆	3.7	PN 24
*4¼	4½	4½	3½	1½	½	1⅜	5¾	6⅝	5¼	½	3⁶⁄₁₆	4.6	PN 25
	5	*5¼	4	1⅝	½	1½	6¼	7⁵⁄₁₆	5¾	½	3¹¹⁄₁₆	6.2	PN 26
5½	5¾	6	4½	1¾	½	1⅝	7	8⅛	6½	½	4⁵⁄₁₆	7.8	PN 27
	*6⅛	*6½	5	1⅞	¾	1¾	7⅝	8⁷⁄₁₆	7	½	4¹¹⁄₁₆	9.9	PN 28
	*6⅜	7	5½	2	¾	1⅞	8⅛	9½	7½	¾	5⁵⁄₁₆	11.8	PN 29
	*7¼	*7½	5½	2	¾	1⅞	8½	10	8	¼	5⁵⁄₁₆	14.3	PN 30
*7¾	8	*8¼	6	2¼	¾	2½	9⅝	10⅞	8½	¼	5¹¹⁄₁₆	18.6	PN 31
*8½		9	6	2¼	¾	2½	10¼	11⅞	9½	½	5¹¹⁄₁₆	23.8	PN 32
*9½		10	6	2½	¾	2¼	11¼	13	10⅜	¼	5¹¹⁄₁₆	31.1	PN 33

Pins marked * are special.

Cotter Pins
American Bridge Company Standard
All Dimensions in Inches

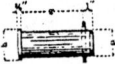

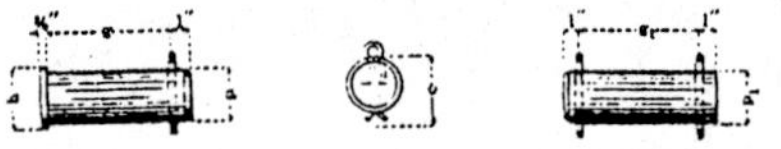

Horizontal or Vertical Pin Finished					Horizontal Pin Rough or Finished			
Pin	Head		Cotter		Pin		Cotter	
p	h	g	c	d	p₁	g₁	c	d
1¼	1½		2	¼	1¼		2	¼
1½	1¾		2½	½	1½		2⅓	½
1¾	2		2¾	½	1¾		2¾	½
2	2⅜		3	½	2		3	½
2¼	2⅝		3¼	½	2¼		3¼	½
2½	2⅞	Net Grip + ⅛″	3¾	½	2½	Net Grip + ¾″	3¾	½
2¾	3½		4	½	2¾		4	½
3	3½		5	½	3		5	½
3¼	3¾		5	½	3¼		5	½
3½	4		6	½	3½		6	½
3¾	4¾		6	½	3¾		6	½

was easily maintained. The double members would attach to either side of the gusset plates, which would be wide and deep enough to serve as a separator for the main chord members and make the connection to the diagonals or struts (see Figure 5-23).

Rivets and riveted connections made the fabrication of structural members relatively easy and also created a new opportunity for field splices of members. Rivets were driven in the shop and in the field with the same result. It may have been more convenient and sometimes easier to drive the rivets in a fabrication shop; however,

Table 5-6 (Continued)

C. Clevises
American Bridge Company Standard
Dimensions in Inches

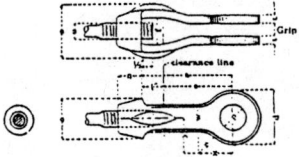

Grip—thickness of plate + ¼" but must not exceed dimension

Clevis Number	Head								Nut					Fork			Weight Pounds
	d	w	t	Max. p	Min. p	T	x	y	a	e	Max. u	Min. u	e	l	a	s	
3	3	1½	½	1½	1	2¼	2¼	3	1½	2¼	1¼	1	3 1/16	1¼	5	4	4
4	4	2	½	2	1¼	3	3	4	1¼	2⅞	1½	1½	3⅜	1¼	6	5	8
5	5	2½	½	2½	1½	3¾	3¾	5	2¼	3¾	2½	1½	4½	2½	7	6	16
6	6	3	¼	3	2	4½	4½	6	2½	4½	2½	2	5½	2½	8	7	26
7	7	3½	½	3½	2½	5½	5½	7	3	5	3	2½	6 3/16	3½	9	8	36

Clevis Numbers for Various Rods and Pins

Rods			Pins										
Round	Square	Upset	1	1¼	1½	1¾	2	2¼	2½	2¾	3	3¼	3½
¼		1	3	3	3								
	¼	1⅛	3	3	3	4	4						
⅞	⅞	1¼		4	4	4	4						
1		1⅜		4	4	4	4						
1½	1	1½		4	4	4	4	5	5				
1¼	1⅛	1⅝		4	4	4	4	5	5				
1⅜		1¾			5	5	5	5	5				
	1¼	1⅞			5	5	5	5	5				
1½	1½	2			5	5	5	5	5	6	6		
1½		2½			5	5	5	5	5	6	6		
1¼	1½	2¼					6	6	6	6	6	7	7
1⅞	1¾	2½				6	6	6	6	6	7	7	7
2	1¾	2½					6	6	6	6	7	7	7
2½		2½					6	6	6	6	7	7	7
	1¾	2¼								7	7	7	7
2¼	2	2½								7	7	7	7
2½	2½	3								7	7	7	7

Clevises above and to right of zigzag line may be used with forks straight, those below and to left of this line should have forks closed so as not to overstrain pin.

field-driven rivets placed during the assembly of members and erection of buildings were driven with equal precision. Rivets were made in several sizes and lengths as required to make the connections. They could have round heads or could be countersunk. Rivet connections were designed for shear of the rivet and the bearing capacity of the material joined. See Table 5-8 for sample rivet values.

USES OF TRUSS SYSTEMS

Trusses were used in many different types of buildings. They became the key element in long-span systems such as armories, churches, coliseums, train sheds, exposition buildings and buildings such as mills and barns. Trusses were efficient at almost any scale or in any type of building. Wooden trusses could be decorative as well as functional. Heavy timber truss members were used in large mill buildings, in many churches, and other types of commercial buildings (see Figures 5-24 and 5-25). Churches made the most unusual use of trusses. There were intricate designs for longitudinal trusses that supported cross-trusses upon which heavy timber purlins

Table 5-6 (Continued)

Turnbuckles and Sleeve Nuts
American Bridge Company Standard
Dimensions in Inches

Turnbuckles

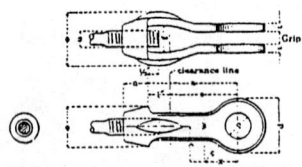

add 6"; add 9" for turnbuckles marked *
Pitch and shape of thread, A. B. Co. Standard

Sleeve Nuts

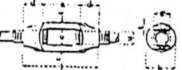

Pitch and shape of thread, A. B. Co. Standard

Diam. of Screw u	Standard Dimensions						Weight Pounds	Diam. of Screw u	Standard Dimensions						Weight Pounds
	d	l	e	f	g	h			d	l	a	b	e	t	
3/8	5/16	7 1/8	3/16	5/16	1/2	1 7/16	1								
7/16	11/16	7 3/16	3/8	1/4	5/8	1 5/8	1								
1/2	3/4	7 1/2	5/8	1/4	5/8	1 3/8	1	*							
9/16	11/16	1 11/16	13/16	3/4	13/16	1 1/2	1 11/12								
5/8	11/16	7 1/8	11/16	5/16	3/4	1 5/16	1 1/2								
3/4	1 1/8	8 1/4	17/16	11/16	7/8	2	2								
7/8	1 5/16	3 3/8	1 1/4	5/8	1	2 1/4	3	7/8	1 1/2	7	1 5/8	1 7/8	1 1/8	1/4	3
1	1 1/2	9	1 5/16	7/16	1 1/4	2 7/16	4	1	1 1/2	7	1 5/8	1 7/8	1 1/8	1/4	3
1 1/8	1 11/16	9 5/8	1 7/16	1/2	1 3/4	2 3/16	5	1 1/8	1 3/4	7 1/2	2	2 3/16	1 5/8	3/16	4
1 1/4	1 7/8	9 1/4	1 3/16	1/2	1 1/2	2 1/4	6	1 1/4	1 3/4	7 1/2	2	2 3/16	1 3/8	1/16	4
1 3/8	2 1/16	10 1/2	1 11/16	1/2	1 5/8	3 1/16	7	1 3/8	2	8	2 5/8	2 3/4	1 5/8	3/8	5
1 1/2	2 1/4	10 1/2	1 1/4	5/8	1 1/4	3 3/16	8	1 1/2	2	8	2 1/2	2 1/4	1 3/8	3/8	6
1 5/8	2 7/16	10 7/8	2	5/8	1 7/8	3 1/8	10	1 5/8	2 1/4	8 1/2	2 3/4	3 7/16	1 7/8	7/16	8
1 3/4	2 5/8	11 1/4	2 1/8	5/8	2	3	11	1 3/4	2 1/4	8 1/2	2 3/4	3 7/16	1 7/8	7/16	9
1 7/8	2 15/16	11 3/8	2 5/16	11/16	2 1/8	3 7/8	12	1 7/8	2 1/2	9	3 1/8	3 5/8	2 1/2	1/2	10
2	3	12	2 5/8	11/16	2 1/4	4 1/4	14	2	2 1/2	9	3 1/4	3 5/8	2 1/2	1/8	11
2 1/8	3 7/16	12 5/8	2 1/2	11/16	2 1/8	4 1/2	17	2 1/8	2 3/4	9 1/2	3 1/2	4 1/16	2 5/8	1/16	14
2 1/4	3 5/8	12 1/4	2 11/16	11/16	2 1/2	4 3/4	20	2 1/4	2 5/8	9 1/2	3 1/2	4 1/16	2 5/8	1/16	15
2 3/8	3 3/16	13 1/8	2 3/4	11/16	2 1/8	4 7/8	22	2 3/8	3	10	3 7/8	4 1/2	2 5/8	5/8	18
2 1/2	3 3/4	13 1/2	3 1/16	11/16	3	5 3/8	25	2 1/2	3	10	3 7/8	4 1/2	2 5/8	5/8	19
2 3/4	4 1/8	14 1/4	3 1/4	15/16	3 1/4	5 3/4	33	2 3/4	3 1/4	10 1/2	4 1/4	4 15/16	2 7/8	11/16	23
2 7/8	4 5/16	14 5/8	3 7/16	1 1/16	3 1/4	6 1/16	36	2 7/8	3 1/2	11	4 5/8	5 5/8	3 1/8	3/4	27
3	4 1/2	15	3 7/8	1 7/16	3 1/2	6 5/8	40	3	3 1/2	11	4 5/8	5 3/8	3 1/2	3/4	28
3 1/4	4 7/8	15 3/8	3 7/8	1 1/16	4	6 3/4	50	3 1/4	6 3/4	11 1/2	5	5 11/16	3 1/2	11/16	35
3 1/2	5 1/4	16 1/2	4 1/4	1 7/16	4	7 1/4	65	3 1/2	4	12	5 5/8	6 1/2	3 5/8	7/8	40
3 3/4	5 5/8	17 1/2	4 7/16	1 3/16	5	8 1/4	95	3 3/4	4 1/4	12 1/2	5 3/4	6 11/16	3 7/8	11/16	47
4	6	18	4 5/8	1 7/16	5	8 3/4	108	4	4 1/2	13	6 1/2	7 1/16	4 1/8	1	55
*4 1/4	6 1/4	21 1/2	4 5/8	1 5/8	5 7/16	9 1/4	140	4 1/4	4 5/8	13 1/8	6 1/2	7 1/2	4 5/8	1 1/2	65
*4 1/2	6 3/4	22 1/8	5 1/2	1 3/4	6 1/2	10 3/4	195	4 1/2	5	14	6 7/8	7 15/16	4 3/4	1 1/16	75
*4 3/4	7 1/4	23 1/2	5 5/8	2	6 1/2	11 1/4	205								
*5	7 1/2	24	6	2 1/4	6 1/2	11 7/8	250								

were to bear, all finally acting together to support the rafters and the roof covering. These trusses utilized heavy timbers as chords, and compression struts and the vertical tension members were wrought iron or steel rods. These trusses were very efficient and could be made to conform to almost any building plan. Heavy timber and steel trusses could be combined with all steel trusses as required by the designs. If truss action occurred over the entire span covered by the truss, there would be no outward thrust on the point of bearing.

Designers of octagon-plan churches with open lantern, such as the one shown in Figure 5-26, would be faced with the problem of outward thrust on the points of bearing if they utilized a series of hip rafters in the form of trusses abutting at a compression ring or octagon in the center as shown in Figure 5-26. The design would require that the outside wall upon which the truss ends bore act as a tension ring in

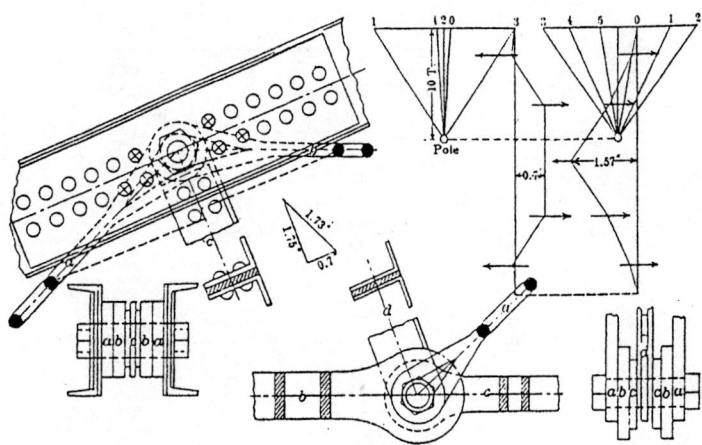

Figure 5-22. Connection design and details, member 7-8. (*From Ricker 1912, 255.*)

order to carry the outward thrust of the trusses. This tension ring bearing wall system was hard to construct. A masonry wall system would require that a steel member, usually a continuous angle, would have to be embedded in the course at the point of bearing. The tension ring embedded in the walls at the point of bearing of the hip rafters would be required to take all of the outward thrust without transferring any horizontal component to the masonry. Designers were well aware of the outward thrust of the hip rafters and the similarity to the forces of the arch—or dome, in this case—and quickly responded by designing clear span trusses to eliminate the rafter and compression ring system. See Figure 5-27 for truss layouts and details.

Steel trusses proved to be much more flexible than their wood or wood and steel predecessors. Not only was steel much stronger than wood, its strength was put to use in many ways with which wood was not compatible. Steel members could be curved or shaped in many ways as required by the designer (see Figures 5-28 and 5-29).

The flexibility of design layout of steel truss and rigid frame systems is evidenced by the plan and elevation of this church structure. The angle change at the intersection of the frame and vertical dome members is the result of the architectural plan of the church. The plan of the building indicates how the frame and truss system was integrated into the plan of the auditorium space. As can be seen, the auditorium space and its windows and doors caused the plan of the structure to shift the column locations. Steel trusses were also used for large churches with vaulted ceilings, such as St. Jerome's Roman Catholic Church in New York City (Kidder 1925, 150). The church structure is a brick and stone veneer and load-bearing masonry combination that is in plan 98 × 150 ft, with vaults rising to 60 ft above the floor (see Figure 5-30). Figure 5-31 shows a plan and half section of the St. Louis Coliseum, ca. 1897.

LATERAL BRACING SYSTEMS: DIAGONAL BRACING

The principle of trusses was extensively utilized for wind bracing or sway bracing for buildings. Every building, regardless of its size, is subject to lateral force from wind. This force does not affect low, heavy masonry buildings with arches or vaulted roofs, such as low level adobe buildings. In some instances these buildings must be checked to be sure, but the weight and wall thickness of these structures normally produce stiff buildings, and the influence of wind is minimal. Taller buildings are an entirely different matter. Buildings of modern lightweight construction and buildings of two stories and up must be braced for the lateral forces of wind. Modern lightweight buildings with curtain wall or veneered construction have several critical areas that must be designed for the positive and negative forces from the lateral pressure of wind. The building frame is not the only portion of the building subject to the wind

Table 5-7. Safe Resistance to Shear, Bearing, Bending Moments[a]

Diameter Rough	Diameter Finished	Shear (Tons)	Bearing Tons per in. Long	Bending Moment in. Tons
1	15/16	4.14	11.72	1.01
1 1/8	1 1/16	5.32	13.28	1.47
1 1/4	1 3/16	6.65	14.84	2.06
1 3/8	1 5/16	8.12	16.41	2.78
1 1/2	1 7/16	9.74	17.97	3.65
1 5/8	1 9/16	11.51	10.53	4.68
1 3/4	1 11/16	13.42	21.09	5.46
1 7/8	1 13/16	15.48	22.66	7.31
2	1 15/16	17.69	24.22	8.93
2 1/8	2 1/16	20.05	25.78	10.77
2 1/4	2 3/16	22.55	27.34	12.85
2 3/8	2 5/16	25.20	28.91	15.18
2 1/2	2 7/16	28.00	30.47	17.77
2 5/8	2 9/16	30.04	32.03	20.65
2 3/4	2 11/16	34.04	33.59	23.82
2 7/8	2 13/16	37.28	35.16	27.30
3	2 15/16	40.66	36.72	31.11
3 1/8	3 1/16	44.20	38.28	35.26
3 1/4	3 3/16	47.88	39.84	39.74
3 3/8	3 5/16	51.71	41.41	44.61
3 1/2	3 7/16	55.68	42.97	49.85
3 5/8	3 9/16	59.81	44.53	55.49
3 3/4	3 11/16	64.08	46.09	61.53
3 7/8	3 13/16	68.50	47.66	68.01
4	3 15/16	73.06	49.22	74.92
4 1/8	4 1/16	77.77	50.78	82.28
4 1/4	4 3/16	82.63	52.34	90.11
4 3/8	4 5/16	87.64	53.91	98.43
4 1/2	4 7/16	92.79	55.47	107.23
4 5/8	4 9/16	98.10	57.03	116.55
4 3/4	4 11/16	103.54	58.59	126.40
4 7/8	4 13/16	109.14	60.16	136.78
5	4 15/16	114.88	61.72	147.72
5 1/8	5 1/16	120.77	63.28	159.22
5 1/4	5 3/16	126.81	64.84	171.31
5 3/8	5 5/16	133.00	66.41	184.00
5 1/2	5 7/16	139.33	67.97	197.29
5 5/8	5 9/16	145.81	69.53	211.22
5 3/4	5 11/16	152.44	71.09	225.78
5 7/8	5 13/16	159.21	72.66	240.99
6	5 15/16	166.13	74.22	256.88
6 1/8	6 1/16	173.20	75.78	273.45
6 1/4	6 3/16	180.41	77.34	290.72
6 3/8	6 5/16	187.78	78.91	308.69
6 1/2	6 7/16	195.29	80.47	327.39
6 5/8	6 9/16	202.94	82.03	346.84
6 3/4	6 11/16	210.75	83.59	367.04
6 7/8	6 13/16	218.71	85.16	388.00
7	6 15/16	226.80	86.72	409.76
7 1/8	7 1/16	235.05	88.28	432.31
7 1/4	7 3/16	243.54	89.84	455.72
7 3/8	7 5/16	251.98	91.41	479.86
7 1/2	7 7/16	260.68	92.97	504.89
7 5/8	7 9/16	269.48	94.53	530.78
7 3/4	7 11/16	278.49	96.09	557.53
7 7/8	7 13/16	287.62	97.66	585.18
8	7 15/16	296.90	99.22	613.71
8 1/8	8 1/16	306.32	100.78	643.16
8 1/4	8 3/16	315.89	102.34	673.54
8 3/8	8 5/16	325.61	103.91	704.86
8 1/2	8 7/16	335.48	105.47	737.15
8 5/8	8 9/16	345.50	107.03	770.40
8 3/4	8 11/16	355.66	108.59	804.63
8 7/8	8 13/16	365.96	110.16	839.86
9	8 15/16	376.42	111.72	876.11

[a] From Ricker 1912, 293.

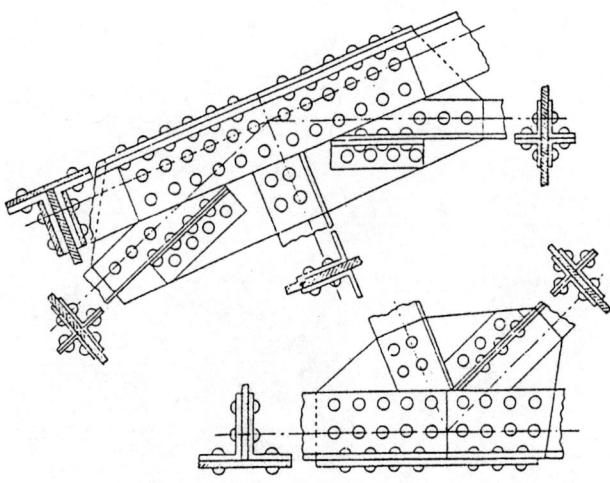

Figure 5-23. Riveted truss connection. (*From Ricker 1912, 257.*)

loads. The cladding system must have sufficient strength to transfer the lateral force of the wind to the structural frame.

Sway bracing, trussed walls, and shear wall systems were utilized for many historical buildings as the need arose from either height of building or increased spans. Larger areas vulnerable to these lateral pressures were thus provided. Kidder (1905) states that buildings of substantial masonry construction (load-bearing walls) with permanent partitions do not require any wind bracing unless the height exceeds one

Table 5-8. Rivets: Shearing and Bearing Values[a]

			½ in. Rivets—Area 0.1963 Square Inch					
Shear	Unit (psi)		7,000	8,000	9,000	10,000	11,000	12,000
	Single Shear per Rivet		1,370	1,570	1,770	1,960	2,160	2,360
	Double Shear per Rivet		2,750	3,140	3,530	3,930	4,320	4,710
Bearing	Unit, Lbs. per Sq. In.		14,000	16,000	18,000	20,000	22,000	24,000
		3/16	1,130	1,500	1,600	1,880	2,060	2,250
		1/4	1,750	2,000	2,250	2,500	2,750	3,000
	Thickness (in.)	5/16	2,190	2,500	2,810	3,130	3,440	3,750
		3/8	2,630	3,000	3,380	3,750	4,130	4,500
		7/16	3,060	3,500	3,940	4,380	4,810	5,250
		1/2	3,500	4,000	4,500	5,000	5,500	6,000
			5/8 in. Rivets—Area 0.3068 in.^{2}					
Shear	Unit (psi)		7,000	8,000	9,000	10,000	11,000	12,000
	Single Shear per Rivet		2,150	2,450	2,760	3,070	3,370	3,680
	Double Shear per Rivet		4,300	4,910	5,520	6,140	6,750	7,360
Bearing	Unit (psi)		14,000	16,000	18,000	20,000	22,000	24,000
		3/16	1,640	1,880	2,110	2,340	2,580	2,810
		1/4	2,190	2,500	2,810	3,130	3,440	3,750
		5/16	2,730	3,130	3,520	3,910	4,300	4,690
	Thickness (in.)	3/8	3,280	3,750	4,220	4,690	5,160	5,630
		7/16	3,830	4,380	4,920	5,470	6,020	6,560
		1/2	4,380	5,000	5,630	6,250	6,880	7,500
		9/16	4,920	5,630	6,330	7,030	7,730	8,440
		5/8	5,470	6,250	7,040	7,810	8,590	9,380

Values below dotted lines are greater than double shear. Values in lb
[a] From Carnegie Steel Co. 1923, 191, 91.

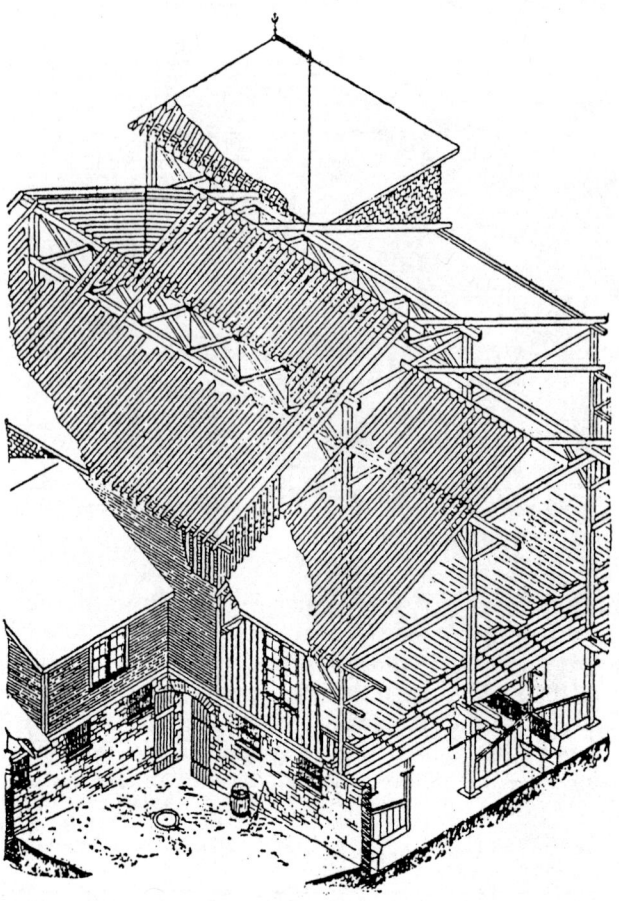

Figure 5-24. Commercial wagon house and stable. (*From Kidder 1925, 95.*)

and one-half times the width of the base (in the direction of the wind). For heights from one and one-half to two times the width of the base, any bracing that is required can generally be provided by a few brick partitions or by trussed wooden partitions. For frame buildings with a height of two or more times the width, some provision should should be made for wind bracing (Kidder 1905, 162). The bays to be braced must be chosen carefully so that there will be no tendency for the building to twist or rotate (when looked at in plan). Masonry load-bearing walls will usually provide partial or full wind resistance for exterior bays, but must be checked for shear. The end walls may be capable of carrying the entire wind pressure on the side walls of the building, provided the length of the building does not build up too large a force for the end walls to dissipate. If the building depth exceeds the maximum length required to dissipate the force from the wind pressure, or if the building is very deep, intermediate bracing bays must be designed. The building width or depth must be checked in any directions, because wind pressure can be applied at any angle to any facade of the building. Long masonry walls must be checked for the effects of one-way or two-way moments from the uniform pressure of the wind (see Chapter 3).

Kidder provides a method for checking a tall building for resistance to overturning: "Multiply the area of the entire side of the building by 30 or 40 psf. (according to the exposure) and then multiply by one-half of the building height and divide by one half the width of the building (in the direction of the wind). If the answer (quotient) is greater than the weight (dead load) of the building, there is a danger of overturning (ibid., 169). Basically the same method is still in use today. We apply a factor of safety of 1.5 or 2 depending upon the code, against a more detailed calculation of wind force (see Chapter 2).

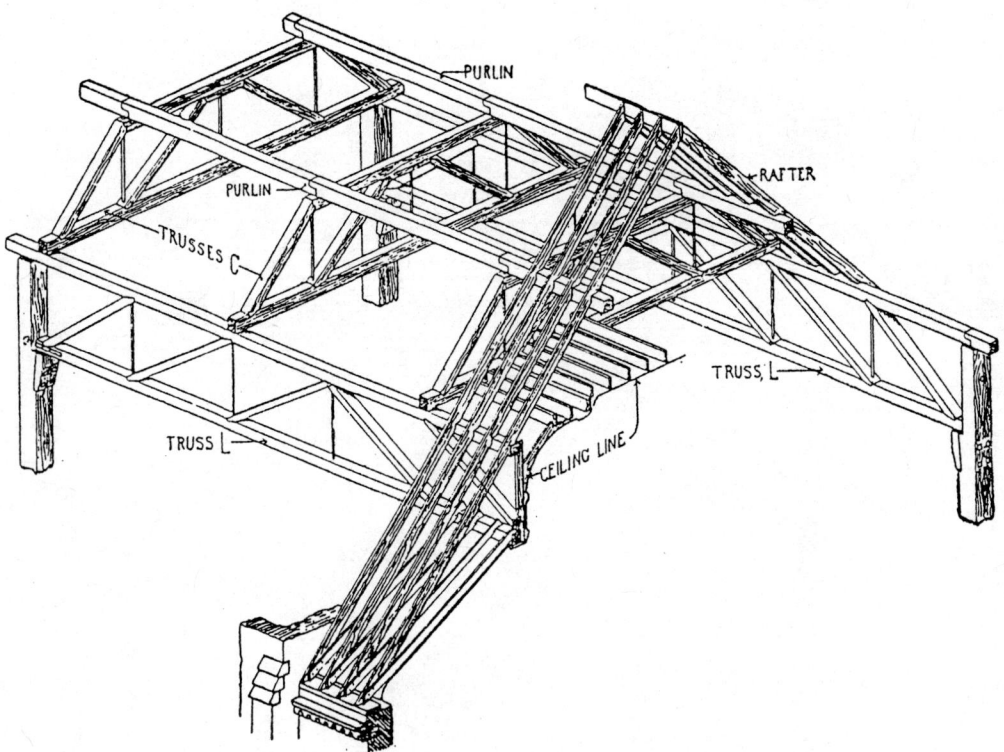

Figure 5-25. Church roof system. (*From Kidder 1925, 133.*)

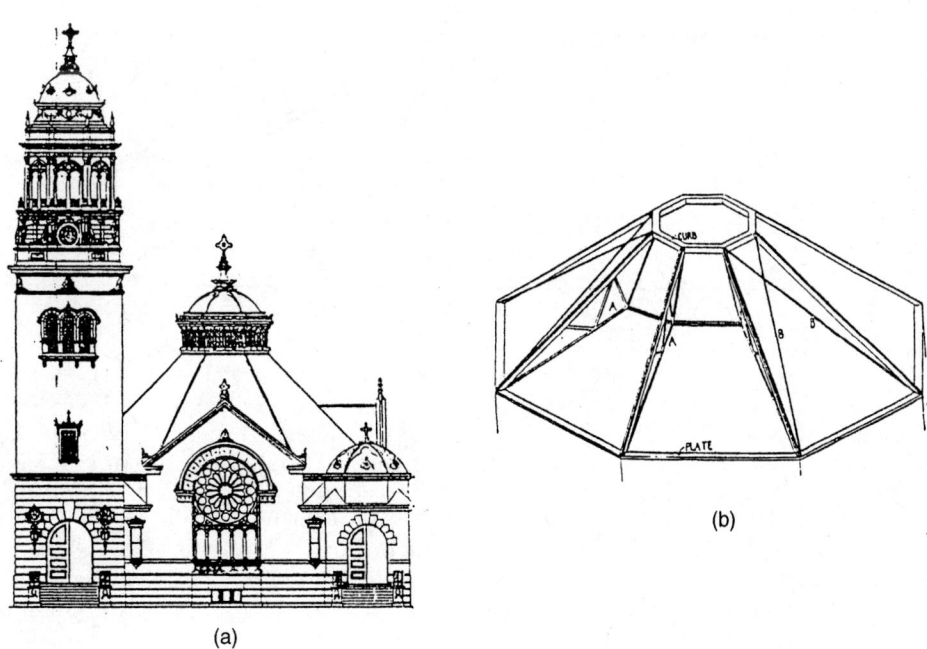

Figure 5-26. (a) Church, octagonal, with tower (from Kidder 1925, 156); (b) octagonal hip-truss roof system (from id., 155).

414 ◇ Structural Analysis of Historic Buildings

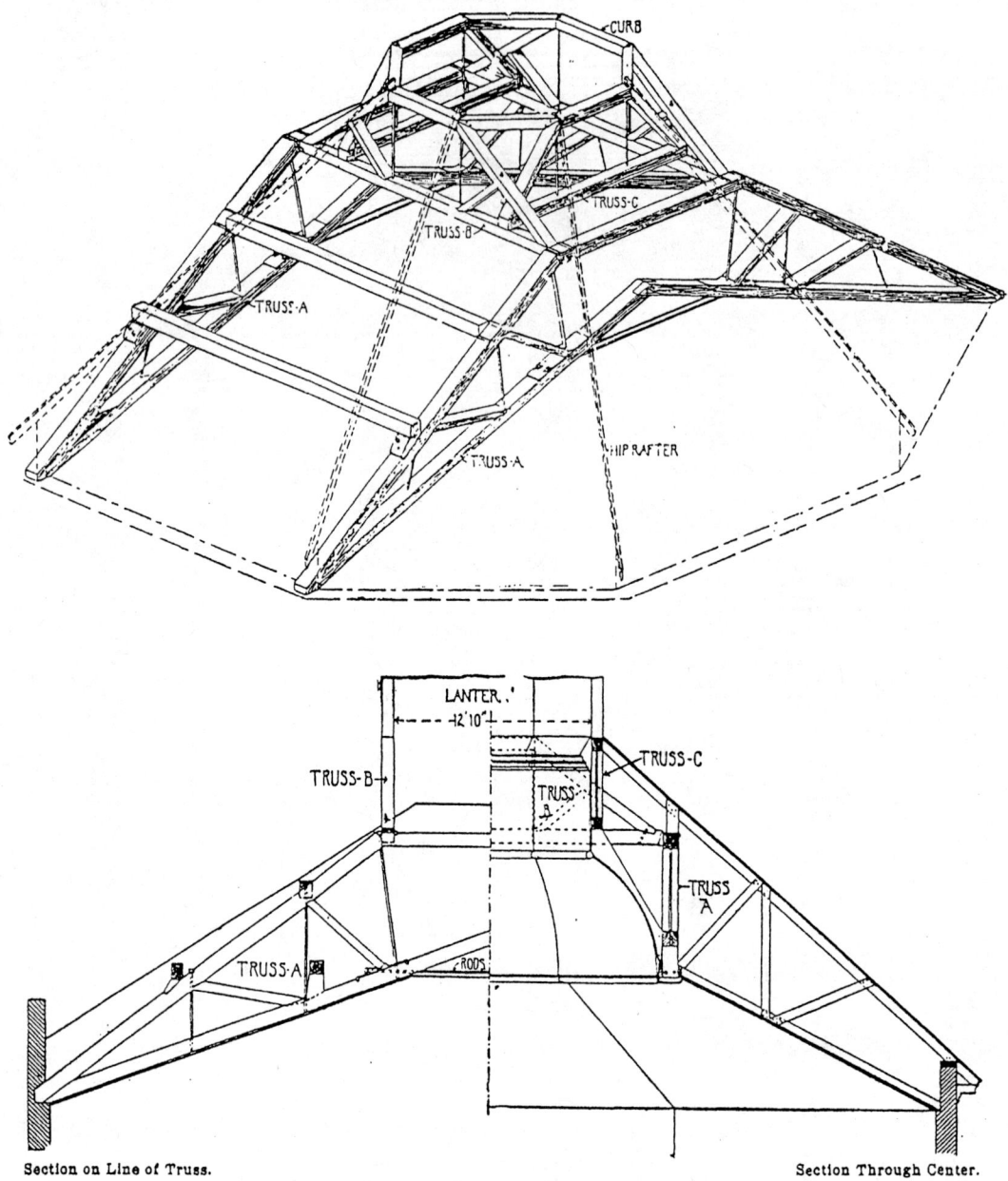

Figure 5-27. Details of octagonal roof with clear span trusses. (*From Kidder 1925, 157, 159.*)

Kidder states that two methods are generally adopted for interior bracing: diagonal systems and knee braces (ibid., 165). With each of these systems some reliance upon the diaphragm of floors and roofs would be required to transfer the lateral loads to the frame bay that is braced. Designers do not want to brace every bay of the building, due to the obstruction of spaces by the bracing. It is not necessary that every bay be braced; however, it is fairly complicated to determine which bays are to be braced or to design the bracing system to avoid conflict with architectural requirements for openings, etc.

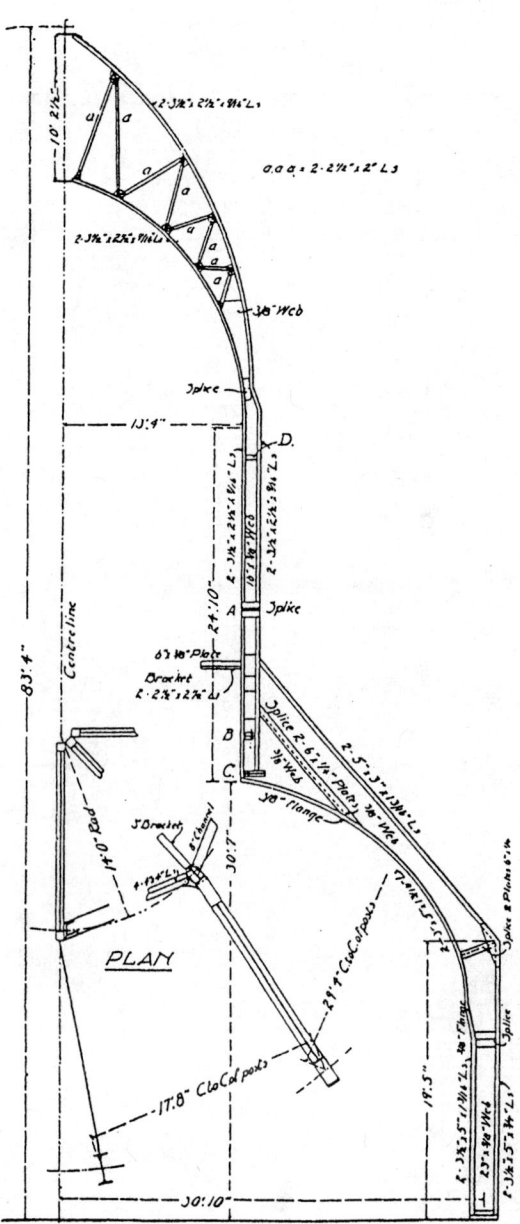

Figure 5-28. Domed church: steel frame. (*From Kidder 1925, 163.*)

Example: Diagonal Bracing System
Building plan: 37 × 114 ft, six stories plus roof, height = 74 ft 6 in.
Kidder assumed the wind pressure to be 30 psf.

Computations (Kidder's Example of Diagonal Bracing Design)
The diagonal bracing in the Kidder example is dimensioned along beam lengths, as shown in the Figure 5-32, from the exterior column line to the column line at the interior corridor at the center of the building. The horizontal dimension of each set of bracing is 13 ft. while the vertical dimension is always from floor to floor which produces a diagonal length that also varies according to floor height.

416 ◇ Structural Analysis of Historic Buildings

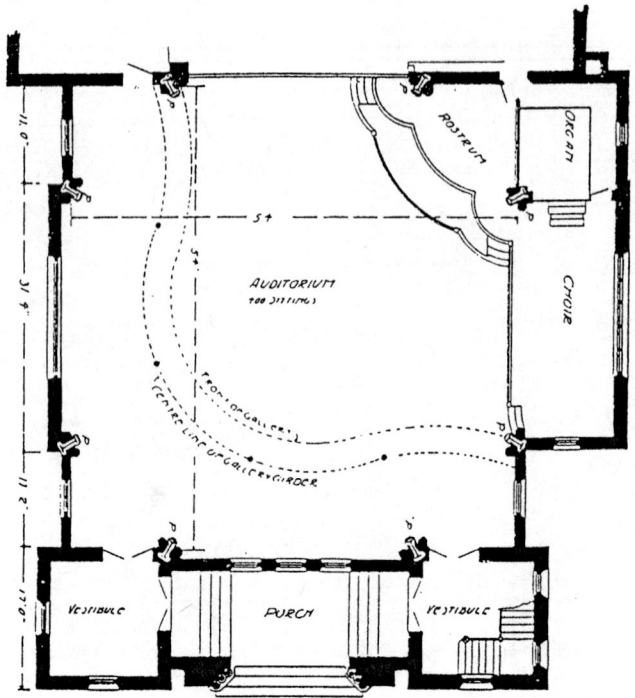

Figure 5-29. Plan of domed church. (*From Kidder 1925, 164.*)

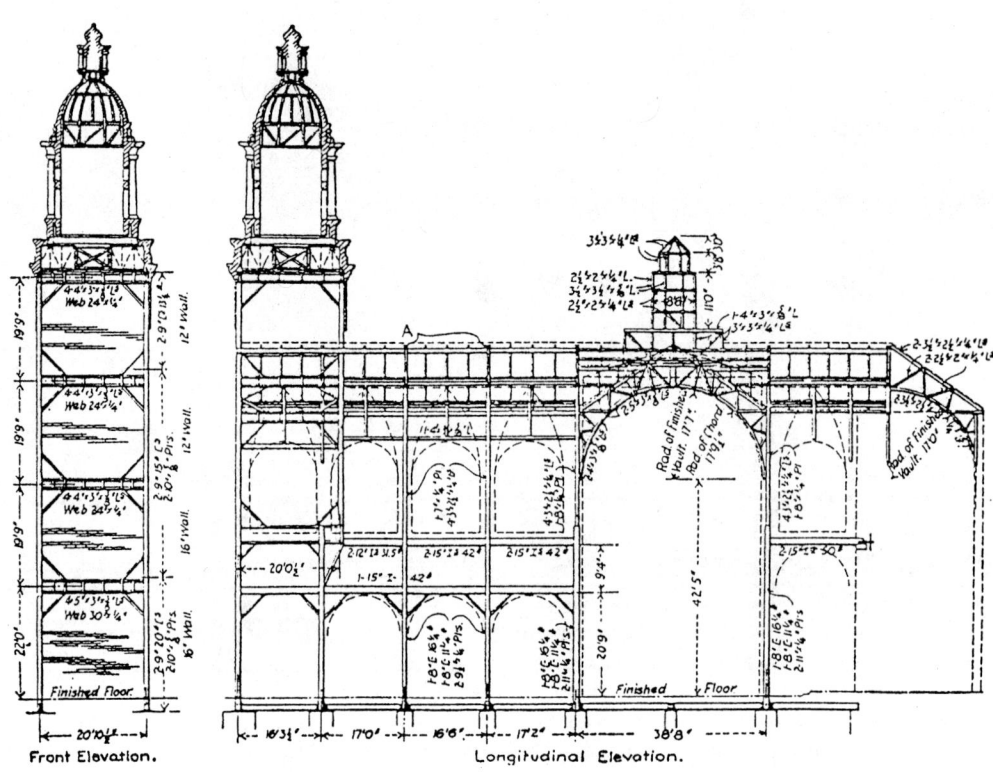

Figure 5-30. St. Jerome's Church, New York City. (*From Kidder 1925, 150.*)

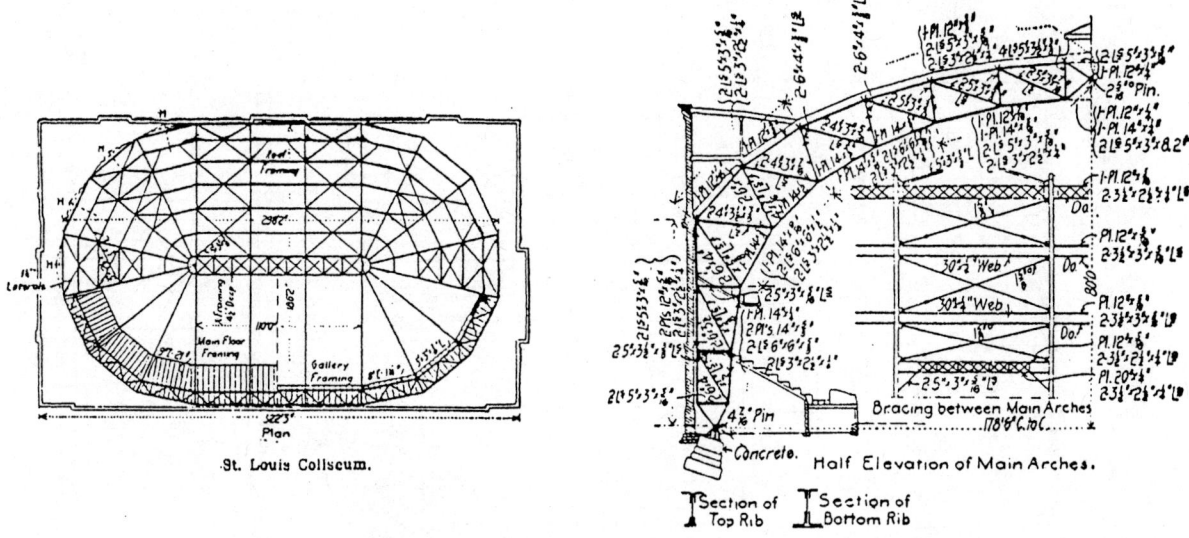

Figure 5-31. St. Louis Coliseum: Plan and Half Section. (*From Kidder 1925, 150.*)

Wind Loads at Line of Floor	Transfer Beam Loads	Diagonals Tension	Leeward Columns Compression (Windward Tension)
P_1 = (9 ft)(84 ft)(30 psf) = 22,680 lb	S_1 = 22,680 lb	D_1 = 22,680 lb (1.4696) D_1 = 33,340 lb	24,400 lb
P_2 = (12 ft)(84 ft)(30 psf) = 30,240 lb	S_2 = 52,920 lb	D_2 = 52,900 lb (1.2615) D_2 = 66,680 lb	65,150 lb
P_3 = (10 ft)(84 ft)(30 psf) = 25,200 lb	S_3 = 78,120 lb	D_3 = 78,120 lb (1.2615) D_3 = 98,430 lb	125,300 lb
P_4 = (10 ft)(84 ft)(30 psf) = 25,200 lb	S_4 = 103,320 lb	D_4 = 103,200 lb (1.2615) D_4 = 130,180 lb	204,850 lb
P_5 = (11 ft)(84 ft)(30 psf) = 27,720 lb	S_5 = 131,040 lb	D_5 = 131,040 lb (1.3615) D_5 = 178,210 lb	325,800 lb
P_6 = (13 ft)(84 ft)(30 psf) = 32,760 lb	S_6 = 163,800 lb	D_6 = 163,800 lb (1.4692) D_6 = 240,790 lb	502,170 lb

The column loads are the product of wind alone and do not have the gravity loads (live and dead loads) included.

The horizontal transfer beam loads are compression (axial) and are wind only. These beams will also have the gravity loads (live and dead loads) from the floors.

Both diagonals are to be sized for the same loads because the tension member will be required to carry all of the wind loads (in either direction).

Kidder states that the allowable stresses are increased in sizing members when wind loads are included in the design loads. It appears that the allowable stresses in the steel were increased by 25 percent.

Note: The above calculations are the total loads from wind pressure on the whole side of the building. If four bays of bracing are provided, the loads shown above must be divided by four; if five, by five, and so on. In addition, the half-bay at each end must be included in the modern design.

The Kidder example shown above does not account for the last one half bay of the building at each end as we do today (see Figure 5-33). Today we would use 99 ft (instead of the 84 ft used by Kidder) as the length of building to be subjected to the lateral pressure of the wind. In addition, there is no provision for, or at least the example does not go as far as discussing, the lateral load from wind pressure on the lower half of the first story. This quantity, which is approximately of the load on P6, must be added to the total lateral load on each column base for the shear on the base components and the anchor bolts at the top of footing to be determined. This method

418 ◇ Structural Analysis of Historic Buildings

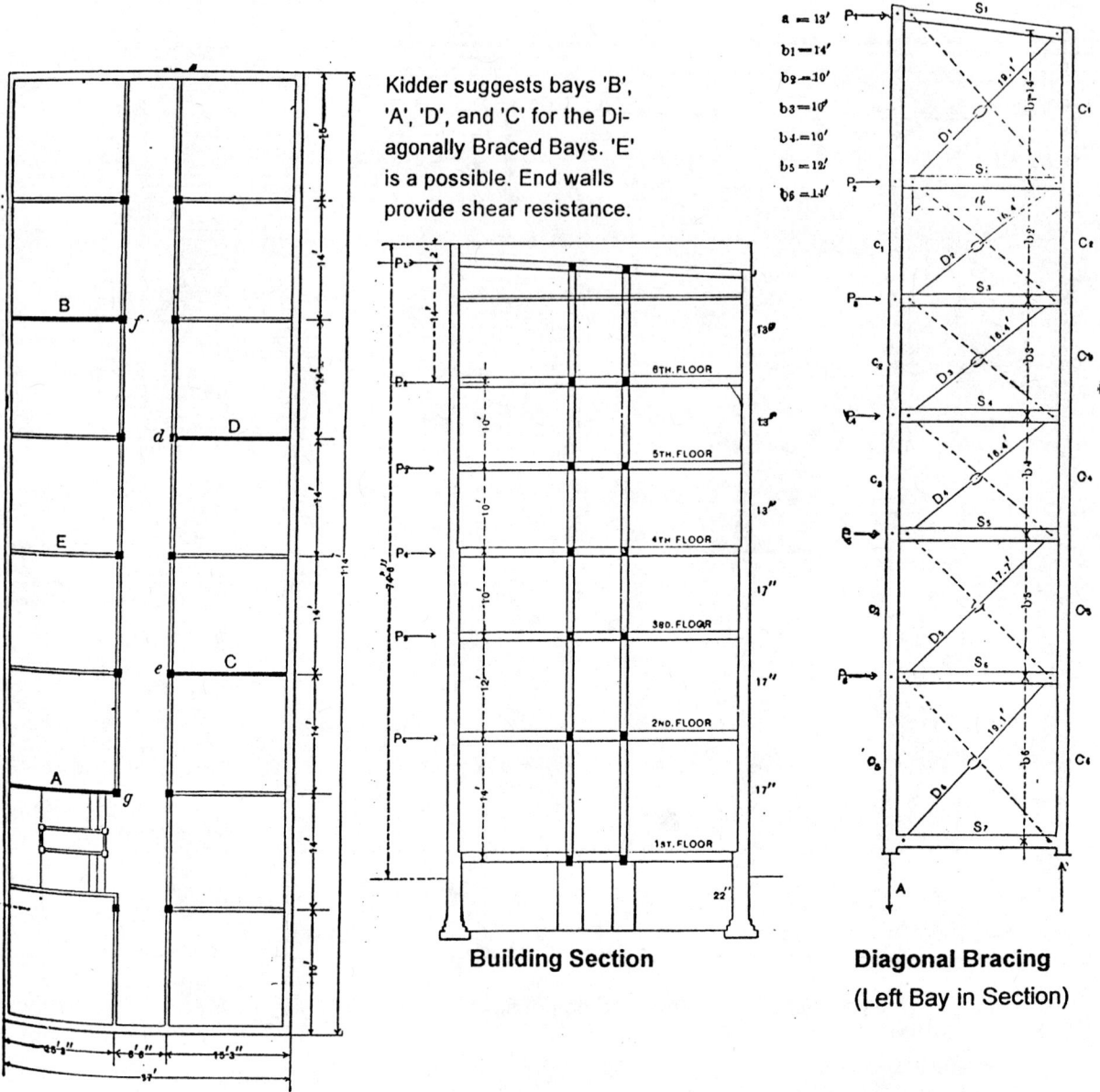

Figure 5-32. Building plan.

of designing the lateral bracing system of the building shown in the example was remarkable for the period and the state of the art at the time. Designers were quite competent in this era; many were self-educated through correspondence courses. They utilized the references and texts studied in this work.

The Kidder example will be checked by modern methods. The same exposure will be checked through modern computations to provide a comparison of the two systems. The location (for wind value) compares to St. Louis, Missouri, where Kidder must have made his design.

St. Louis has a maximum expected 100-year wind speed of 85 mph (BOCA 1989), which translates to the following pressures:

0–30 ft above ground: 14.5 psf
30–50 ft above ground: 20 psf
50–100 ft above ground 24: psf

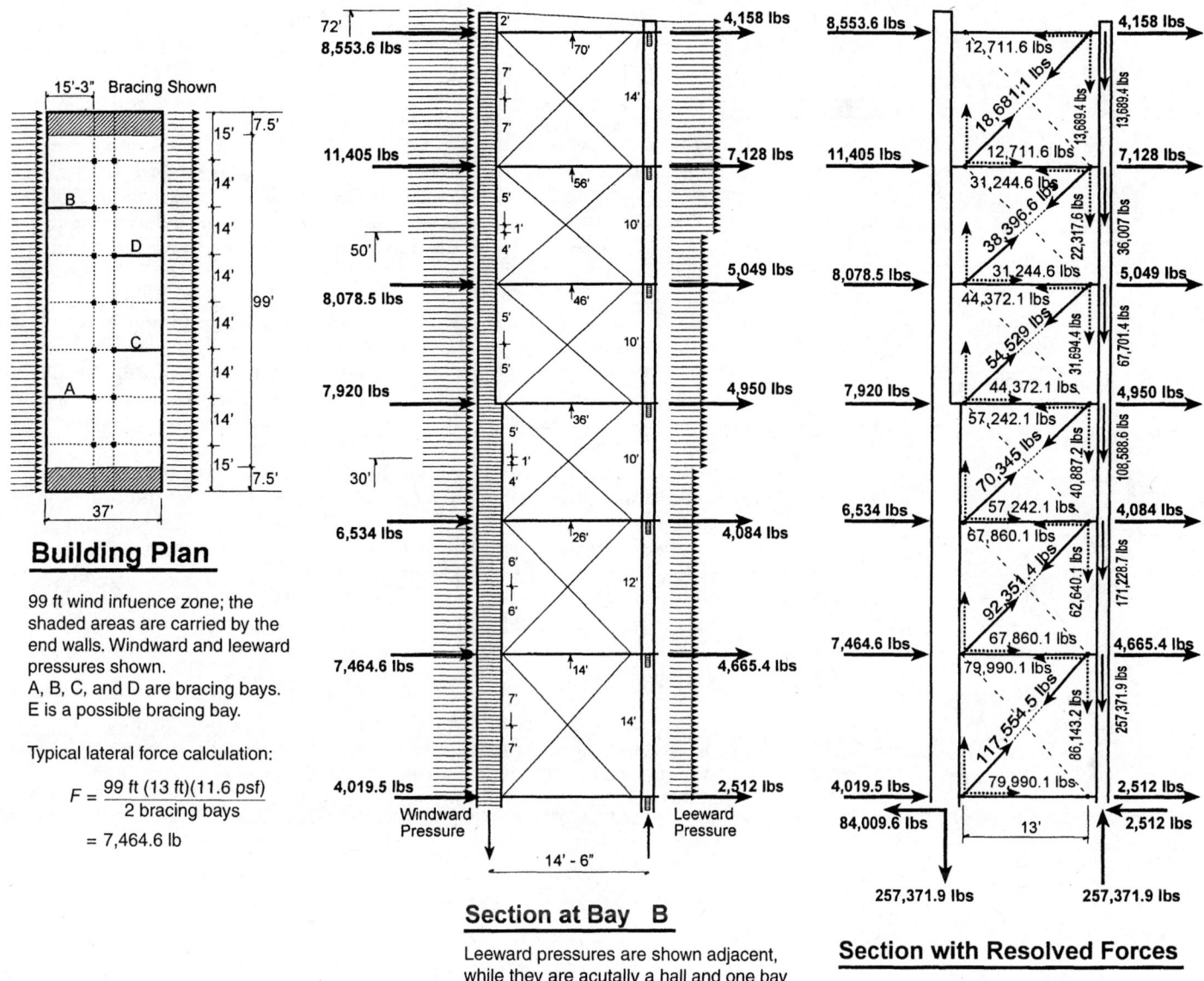

Figure 5-33.

Shape Factors are 1.3 for rectangular prismatic structure (+0.80 windward, −0.50 leeward) to be applied to windward and leeward pressures.

The Kidder example appears to involve load-bearing masonry with wood girders and columns. Diagonal bracing was probably wrought iron rods and turnbuckles. Attachment to the wood members and brick walls would require that tension rods be placed through the building (outside to outside) at each floor near the bracing points to ensure the transfer of loads throughout the system. The Kidder example does not explain the associated details, though they were probably specified. Bracing connections would require details, as would load transfer across beam bearing points at columns.

Steel framing systems are much easier to construct with bracing systems.

Milo S. Ketchum, another noted author of the period, provides details of the wind bracing for the Singer Building, New York City, 1908. The Singer Building consisted of a lower portion 75 × 116 ft in plan and 14 stories tall, with a tower 60 × 60 ft in plan and an additional 27 stories—in all 41 stories including a four-tier lantern. It was the tallest building in the world for a short time (Irace 1988, 44). Figure 5-34 shows the foundation plan of the Singer Building, with a line drawn on the plan to indicate the limits of the 60 × 60 ft tower.

The Singer Building foundations are said to have been "concrete footings sunk by the pneumatic process to a depth of ninety feet" (Ketchum 1924, 125). The drawings of the details in the foundation plan layout of the individual footings appear to show grillage footings, often with two columns bearing on the same lowest level of grillage. The pneumatic process would have been used to clear the excavations of water and mud. There also appears to have been only one basement below ground and therefore the level of foundations would have been well below any basement floor—thus the reason for the pneumatic system. Column spacing was only 12 ft on centers, providing a rather inflexible space for modern use. The Singer Building was demolished in 1967 for a new Skidmore Owings and Merrill Office Building in New York City (Irace 1988, 44), but it is an excellent example for study of the early work.

Figure 5-35 shows the structural floor plan of the 60 × 60 ft tower of the Singer Building; note the column spacing of 12 ft. The diagonal bracing bays are shown in Figure 5-36. The bracing is positioned in the bays as required to ensure stability in the building for a possible wind in any direction. Most of the bracing follows the vertical bays from floor to floor to the basement level (foundation) of the building. The exceptions would transmit the accumulated loads at the bottom of the bracing to other bracing through the diaphragm of the floor and through compression force in the axis of girders. Ketchum indicates the size of the diagonal members but does not give design criteria beyond a uniform wind pressure of 30 psf. Ketchum also provides a small plan of the tower indicating the braced bays in heavy lines and details of the columns and the wind bracing (see Figure 5-37).

LATERAL BRACING METHODS

Ketchum's (1924) states that there are four methods of providing wind bracing for tall buildings: diagonal bracing, knee braces, portal bracing, and brackets (see Figure 5-38). Knee braces and portal braces were popular because they gave more flexibility to designers by allowing them to have door and window openings in bays where

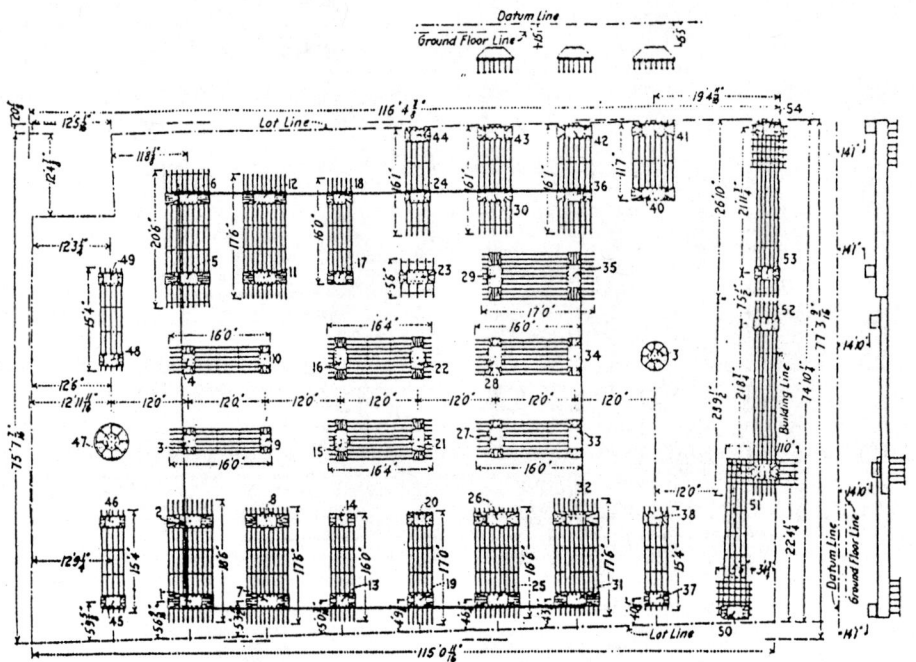

Figure 5-34. Singer Building, New York City: foundation plan, tower limits outlined. (*From Ketchum 1924, 124.*)

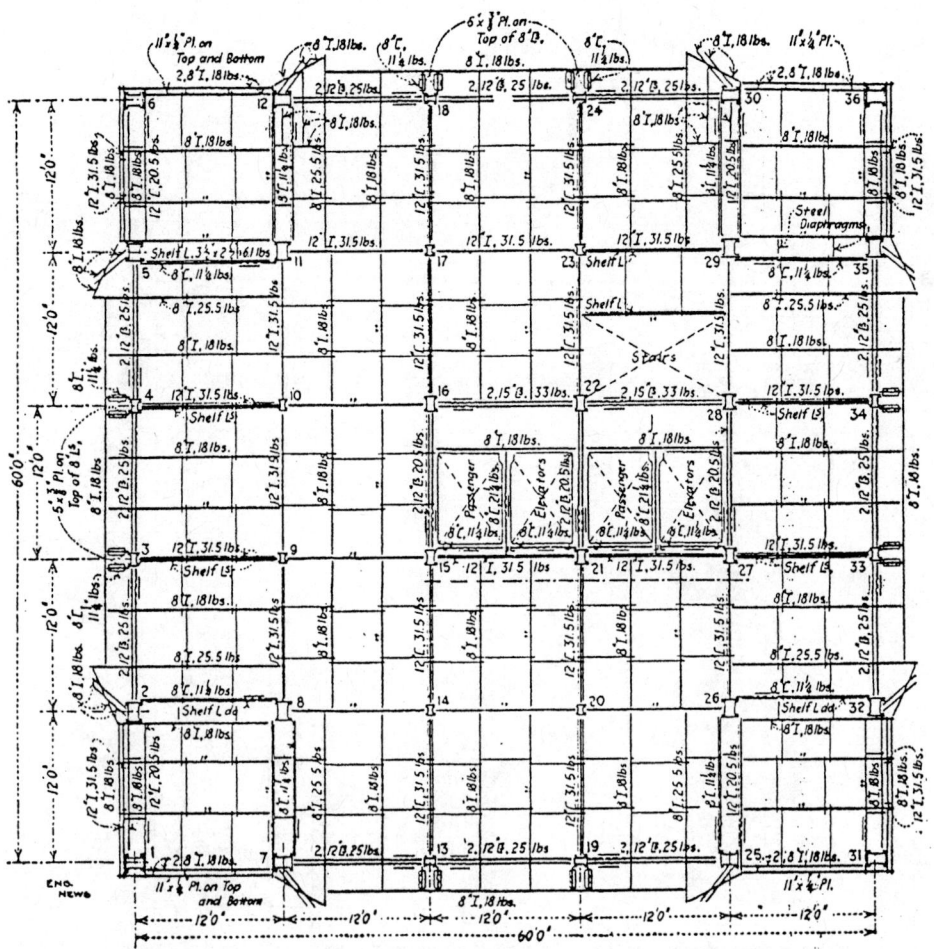

Figure 5-35. Structural floor plan, Singer Building. (*From Ketchum 1924, 123.*)

bracing was designed. Brackets quickly gained popularity for the same reason; however brackets, portals, and knees shared one common problem, the expense of fabrication and construction. The diagonal system was the cheapest and the most efficient in that loads were directly transferred through tension and compression in the diagonals and girders. Each of the other methods required the columns to resist bending moment in addition to the added vertical load from the wind (leeward columns). This was considered a shortcoming common to the three systems that required the moment capacity in the columns. Actually, depending upon the method of transferring the lateral pressure into the frame, the column can be loaded in flexure even in the diagonal system.

The brackets were actually stiffened connections that distributed moments through the connections. See Figure 5-39 for examples of bracket connections in the United Fire Company Building in New York City, constructed ca. 1910. The brackets were designed to make the beam connections to columns have moment-carrying capacity rather than the moment-free shear-only connection. Note that the details of wind bracing are labeled to correspond with detail numbers on the bracing layout (Figure 5-39). The details of the connections show angle shapes for the diagonal members, with beam and diagonals riveted to a gusset plate, with the whole connection attached to the column by the two 6 × 4 × 1/2 in. angles on either side of the gusset. Kidder does not describe the method of analysis utilized by the designers of the United Fire Company Building's bracket connections. He does provide the wind bracing layout for the building (see Figure 5-40).

Freitag (1909) gives five types of wind bracing as shown in Figure 5-41:

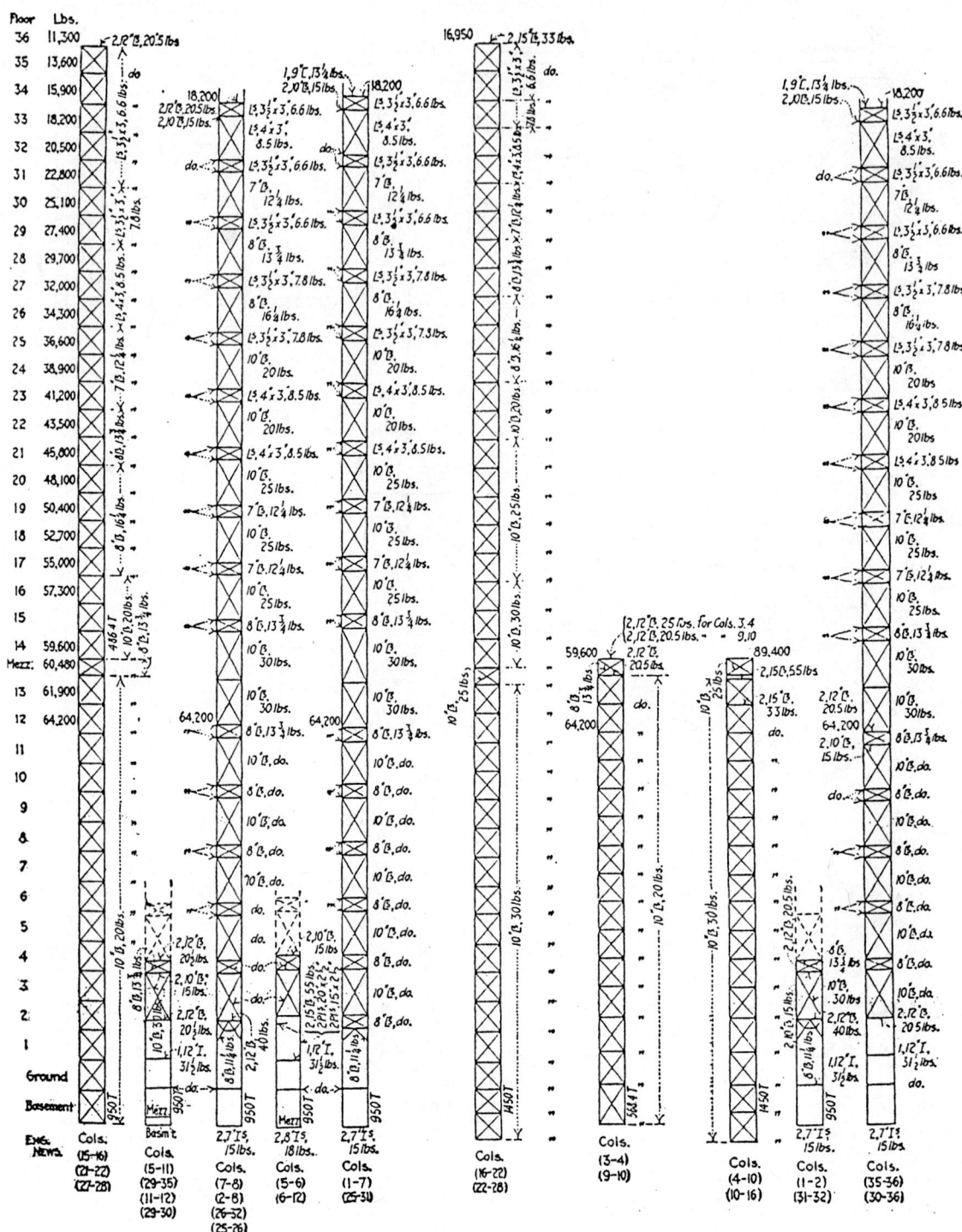

Figure 5-36. Singer Building: wind bracing layout. (*From Ketchum 1924, 125.*)

1. Diagonal bracing (sway rods), single floor level
2. Diagonal bracing (sway rods), two-floor type
3. Portal bracing
4. Knee bracing
5. Lattice girders

The diagonal bracing system, utilizing rods or structural shapes, was the preferred method for Freitag, who relates that the diagonals are more efficient and more eco-

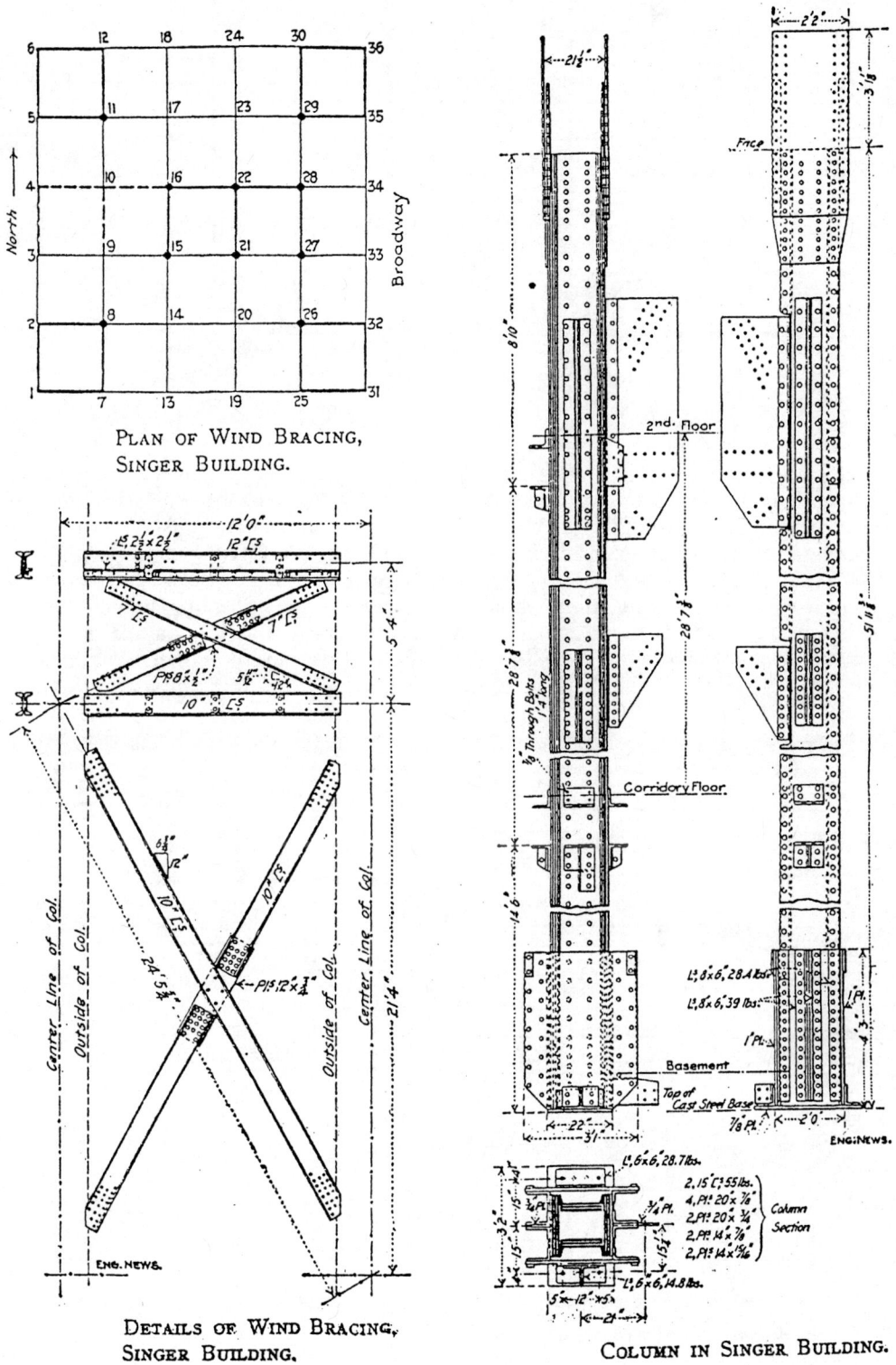

Figure 5-37. Singer Building: bracing plan and column, bracing details. (*From Ketchum 1924, 126.*)

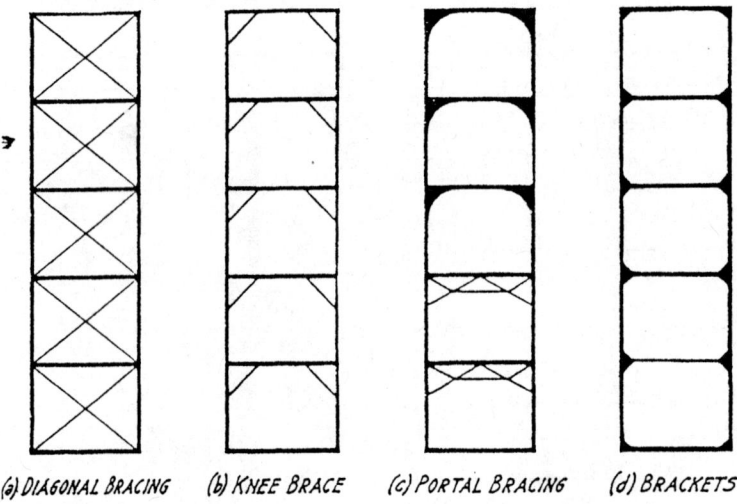

Figure 5-38. Ketchum's types of wind bracing. (*From Ketchum 1924, 121.*)

nomical than the others. Freitag gives examples of each type of wind bracing and relates it to buildings that utilized that system. Most of the buildings he gives as examples are famous buildings by notable master architects who are studied today as pioneers in tall buildings. The choice of type of bracing was dependent upon the designer's preference and the dictates of the architectural scheme. Freitag does not include the bracket connection, which Ketchum and others present. He includes the latticed girder, which requires additional steel but makes a rigid bay in the same manner as a portal. Freitag states that lattice girders were used on many buildings for wind bracing and that the Reliance Building, Chicago, 1894, (one of the most famous

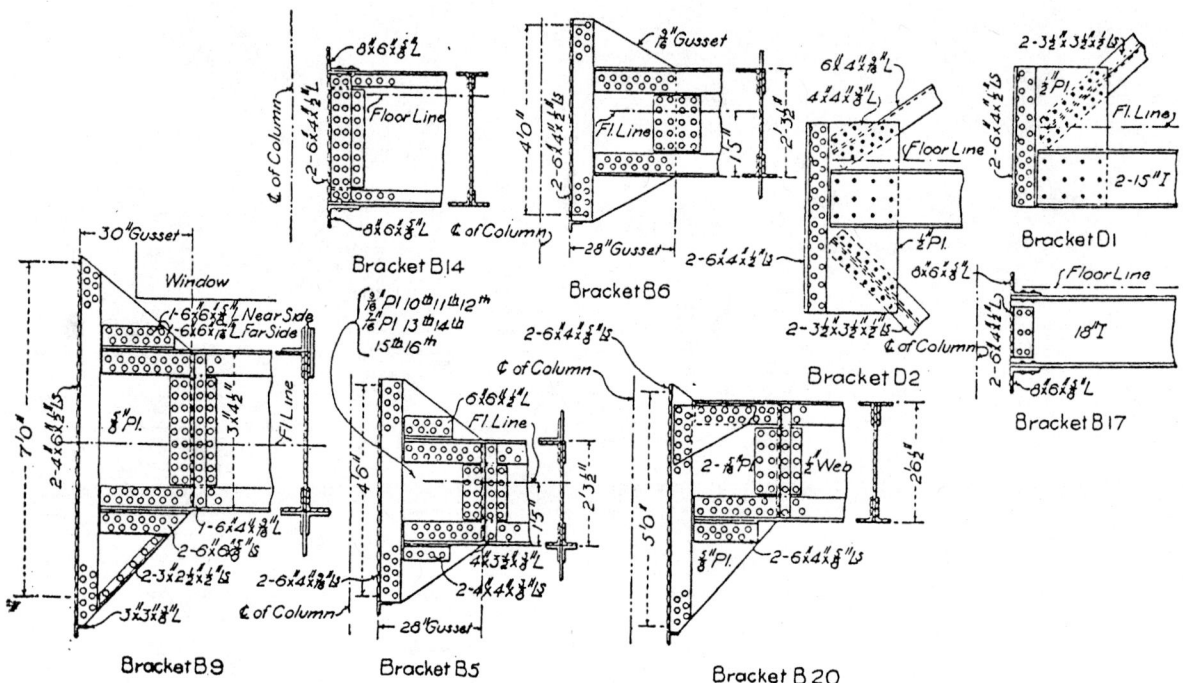

Figure 5-39. Wind bracing details, United Fire Company's Building. (*From Ketchum 1924, 121.*)

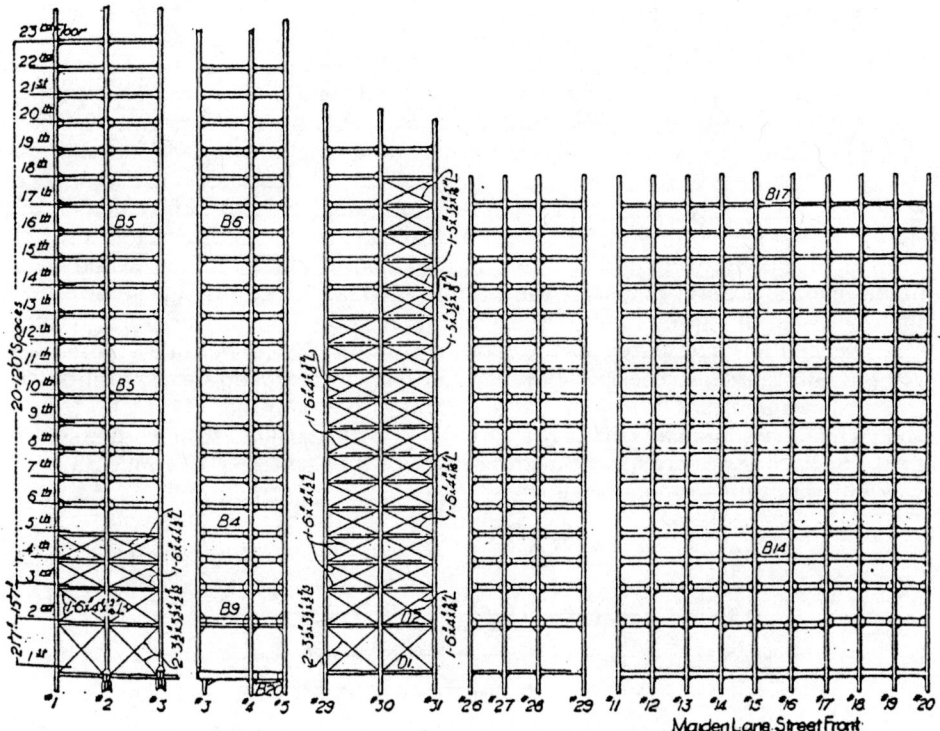

Figure 5-40. Wind bracing layout, United Fire Company's Building. (*From Ketchum 1924, 121.*)

skyscrapers of the 1890s), used a form of the latticed girder. The Reliance's designers utilized full steel plates instead of the lattices, ordinarily they would be called plate girders but these full plate girders were deeper than required to carry floor loads (Freitag 1909, 259). The lattice girder, early portal, and knee form of bracing each transfer the lateral loads from the wind into a frame system through chords and struts being placed in tension and compression to maintain the system rigidity by action of a couple. In these systems, substantial moments exist in the columns and shear is high at column intersection points (with the floor below). The obvious plus in these systems is that there is much more room for flexibility in openings for windows and doors and nonwalled larger spaces exist for offices. Freitag states that "in this type of bracing the wind stresses are transferred to the ground on what is often called the 'table leg principle', that is, each story is made rigid itself, the columns being figured as vertical beams to resist the lateral flexure due to wind stresses" (ibid.)

The shear at the base of each column is magnified by the fact that these shears are the method of transfer from upper floors to lower floors; therefore, as one works the lateral system downward, all shears are cumulative until the ground or foundation level is reached. Every lateral bracing system eventually brings the lateral wind force to the ground.

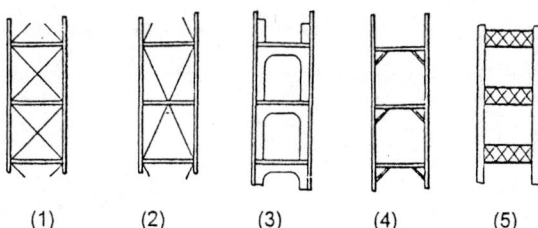

Figure 5-41. Freitag's types of wind bracing. (*From Freitag, 1909, 258.*)

KNEE BRACING SYSTEMS

Knee braces use the principle of triangulation to achieve stiffness in the column beam (strut) connection at the top of the column. The Isabella Building, Chicago, ca. 1892, by the architect W. L. B. Jenney, used knee braces. This typical knee brace is shown in Figure 5-42 in plan and elevation. It is not noted whether the horizontal strut is doubling as a floor beam or not. This would usually be the case, but in a few very early designs these struts were kept separate from floor members. The calculation of the forces in the knee brace and shears, moments, and vertical column loads is given by Freitag as shown in Figure 5-43. For a given floor, the lateral load would be the cumulation of all lateral loads on the building from above, plus the load on that segment. This would apply to any floor with other floors above. The Freitag method will be checked and compared to the type of computation done today. While knee braces are not frequently used now in tall buildings, they are often used in industrial and other low-rise situations where large openings are required. The section shown for the Isabella Building knee brace is a K-truss-type system. Other knee brace systems that do not have the braces meet at the center of the strut (beam) would require a slightly different type of modern analysis.

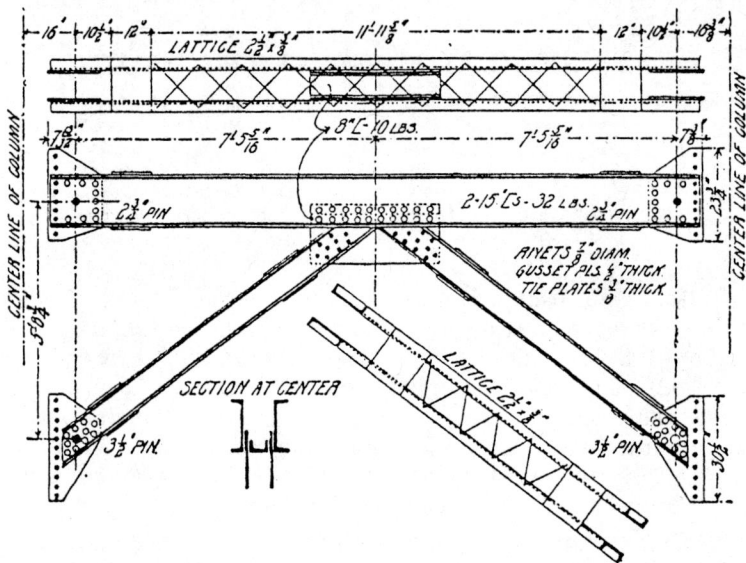

Figure 5-42. Typical knee brace, Isabella Building. (*From Freitag 1909, 274.*)

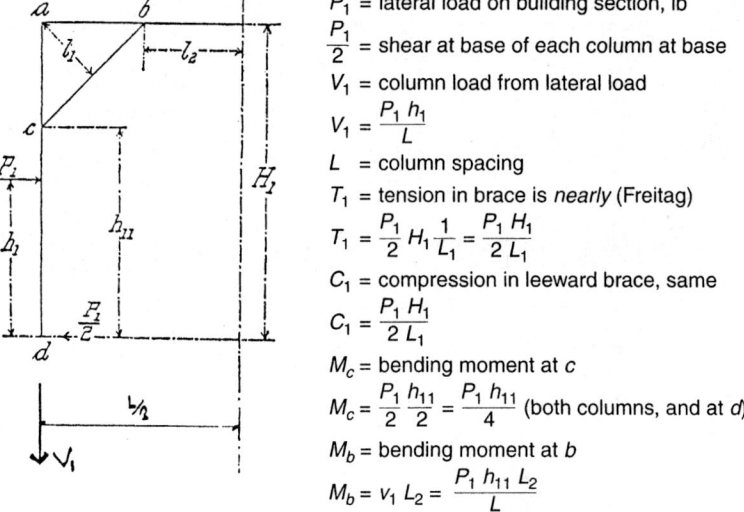

P_1 = lateral load on building section, lb

$\dfrac{P_1}{2}$ = shear at base of each column at base

V_1 = column load from lateral load

$V_1 = \dfrac{P_1 h_1}{L}$

L = column spacing

T_1 = tension in brace is *nearly* (Freitag)

$T_1 = \dfrac{P_1}{2} H_1 \dfrac{1}{L_1} = \dfrac{P_1 H_1}{2 L_1}$

C_1 = compression in leeward brace, same

$C_1 = \dfrac{P_1 H_1}{2 L_1}$

M_c = bending moment at c

$M_c = \dfrac{P_1}{2} \dfrac{h_{11}}{2} = \dfrac{P_1 h_{11}}{4}$ (both columns, and at d)

M_b = bending moment at b

$M_b = v_1 L_2 = \dfrac{P_1 h_{11} L_2}{L}$

Figure 5-43. Analysis of knee braces. (*From Freitag 1909, 273, 274.*)

Check of Knee Brace Statics:

$$V_1 = \frac{P_1(h_1)}{L} = \frac{6.72 \text{ kips }(7 \text{ ft})}{16 \text{ ft}} = 2.94 \text{ kips}$$

$$T_1 = \frac{P_1(H_1)}{2L1} = \frac{6.72 \text{ kips }(14 \text{ ft})}{2(16 \text{ ft})} = 2.94 \text{ kips}$$

$$M = \frac{P_1(h_{11})}{2\,(2)} = \left[\frac{6.72 \text{ kips}}{2}\right]\left[\frac{(10 \text{ ft})}{2}\right] = 16.8 \text{ kip-ft}$$

Validity of the column force (V_1) is in question because of the moment at the bottom of the columns. Equation not written at column inflection points.

Tension in brace ($T1$) is not valid; a new type of computation is now accepted as correct.

The moment at the base of the column (M) is correct.

See Figure 5-44. There are several incorrect assumptions in the Freitag method. The modern analysis would calculate the column loads by taking a free body diagram at the level of the inflection points in the columns because this would be the point of zero moment and the opportunity for summation of moments to determine the column reactions. From that point, the solution of the knee brace and the axial load, shear and moment on the members of the bent can be determined.

Freitag recognizes the moments at the base of the columns but does not utilize the inflection points, resulting in an incorrect answer for the column loads. Freitag also provides incorrect formulae for the load in the knees and does not follow with any formula for moments in the columns or the strut (beam) into which the knees frame.

The Freitag method simplified the calculation for the knee brace and this indicates the inadequacy of this particular computation. Freitag's work is dates to 1909, which is very early in the period of new theories on tall buildings. The 'modern method' of analysis for the knee brace is shown as follows:

The analysis is contingent upon the assumption as to how the curtain wall or enclosure system transfers the lateral pressures from the wind to the internal frame of the buildings.

Freitag shows the load as a concentrated point load at the center height of the column. This may be due to a purlin or spandrel transferring the load to the column at that point, or it may assumption. Today we must check to determine the actual location of the transfer. If an assumption is made, it is usually

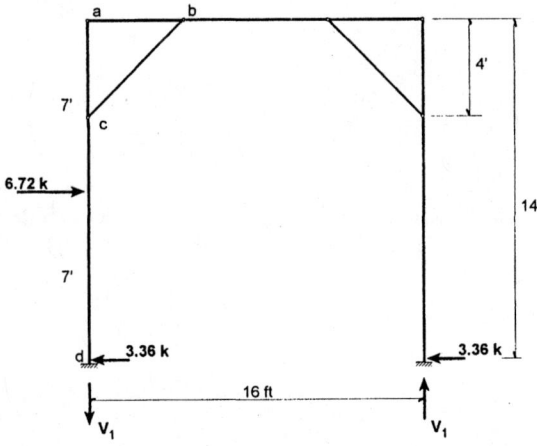

Figure 5-44.

that the load transfers at the line of the floor. The diaphragm or stiffness of the floor transfers uniformly to the frame.

The second parameter that must be considered in the analysis and design is the fixity of the column at its base or intersection with the floor. If the column is continuous from the floor below, it has the fixity required to assume the moment capacity at the base. If the column is changing size at the floor level or just above, that, and the type of connection or splice, must be known. The connection, if nonmoment capacity, would have an effect on the solution and the loads and moments of the frame.

The following example of a knee brace frame bent utilizes a wind load of 30 psf of vertical surface in the manner of the early solution on a frame that is 14 ft tall, center to center of columns at 16 ft and bent spacing at 16 ft. Knees are 4 ft down the column and out the strut, as shown. The building is several stories tall; however the example will look at the top floor and the floor three levels below, which will give adequate treatment to the example.

The upper floors of the building are shown in Figure 5-45. The 6.72 kip load shown by the solid arrow is the load for the level where it bears on the column at midheight (as in the previous example). The bent height is 14 ft and between bents the height is 16 ft, and therefore (14 ft)(16 ft)(30 psf) = 6720 lb at the point of action.

The dotted arrow to the left at the bottom of each column is the reaction of the load from the wind pressure. Freitag assumes that the load is equally divided in the two columns.

Each floor picks up the load from above and also the same 6.72 kips from the wind pressure at its respective level. The point of application of the wind load on the frame is dependent upon the curtain wall system. The assumption that the load is divided equally between the two columns was reasonable and made the analysis statically determinant. Computer analysis does create an imbalance between column shear reactions.

The same example problem, except moving down three floors, provides an example of the modern method when there is moment at the top of the column combined with the accumulation of windward force acting at the level of floor at top of column (for floor being considered). See Figures 5-46 and 5-47.

As stated, the solution will produce different results if the column does not have moment capacity at the floor level of the bent. In addition, if the load is applied at the midheight of the column, the actual horizontal reaction at the base of each column will be different at each column. Equal reactions are most often assumed when the lateral loads are taken to apply at the level of the floor. The following example is a solution for the roof bent, columns pin-connected at their base, with the assumption that the beam or strut has an inflection point at the center of span (see Figure 5-48).

PORTAL FRAMES

Ketchum (1921) shows several examples of portals, including the standard latticed portal, the knee brace, and variations of the trussed portal. Ketchum states that these portals are most often used to transfer wind loads to the foundations of mill buildings and open sheds (see Figure 5-49). The major difference between Ketchum in 1921 and Freitag volume in 1909 is the placement of the lateral load from the wind pressure. Ketchum and others probably recognized by 1921 that the diaphragm of the floor provides the most logical and uniform means of transferring the wind load to the bracing, the columns, and down consecutively to the foundation. The trussed portal provides a logical and useful method of developing the stiffness to brace the building. The following example shows a typical solution for the trussed portal (see Figures 5-50 and 5-51).

According to Freitag (1909), the latticed portal was used in the older portion of the Monadnock Building, Chicago, 1889–91. The original Monadnock contained a portal strut, as shown in Figure 5-52. Freitag states that this was "one of the first attempts

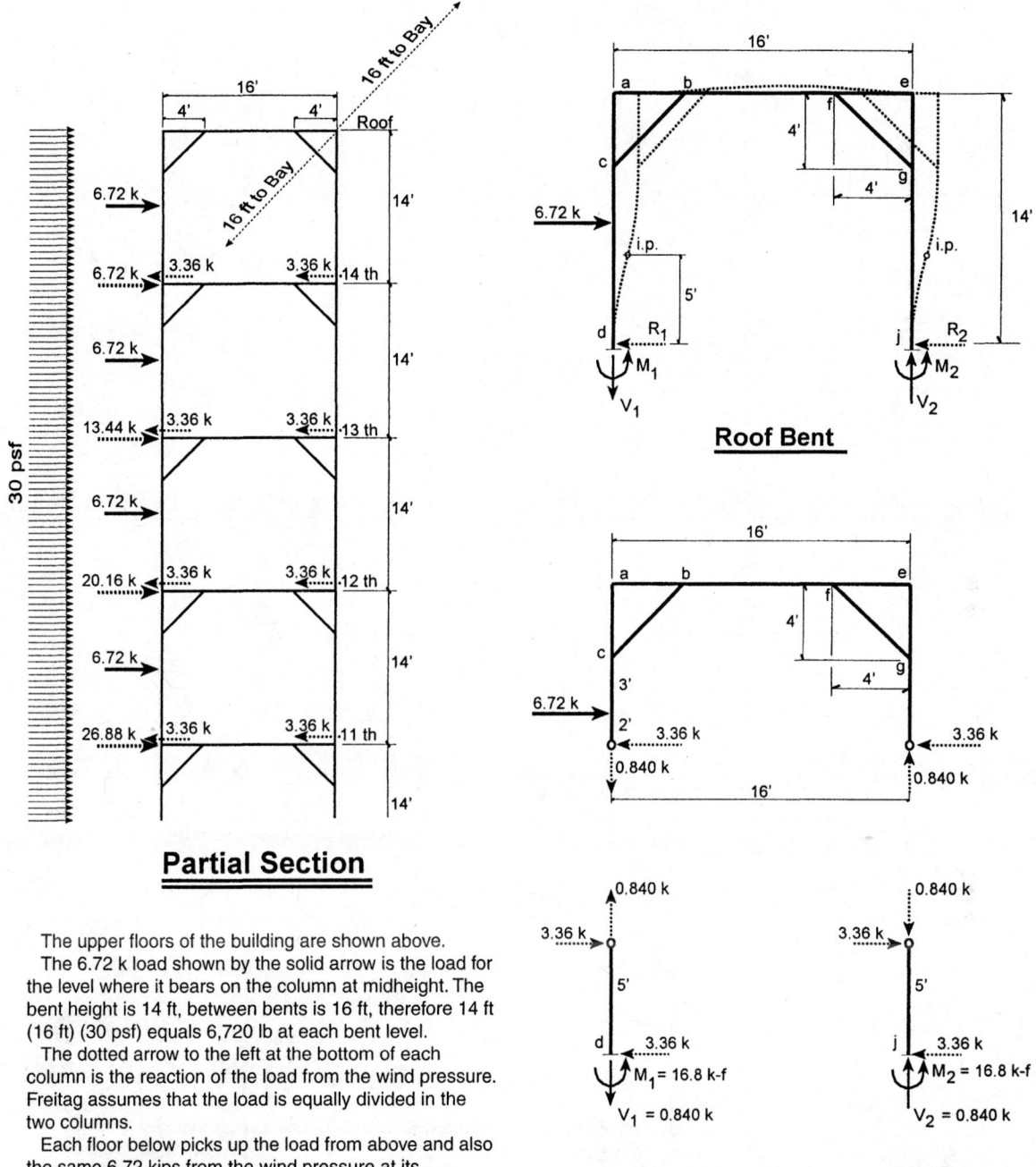

Partial Section

The upper floors of the building are shown above.
The 6.72 k load shown by the solid arrow is the load for the level where it bears on the column at midheight. The bent height is 14 ft, between bents is 16 ft, therefore 14 ft (16 ft) (30 psf) equals 6,720 lb at each bent level.

The dotted arrow to the left at the bottom of each column is the reaction of the load from the wind pressure. Freitag assumes that the load is equally divided in the two columns.

Each floor below picks up the load from above and also the same 6.72 kips from the wind pressure at its respective level. The assumption that the wind load bears at the midheight of the column depends upon the curtain wall details. In addition, there is a question as to the reaction being equal at each column when the load is applied at the midheight.

Figure 5-45.

at a portal system in building construction" (ibid., 271). Condit (1964) discusses the innovative qualities of John Welborn Root, the engineer partner of Daniel Burnham in the firm Burnham and Root, in the design of the original Monadnock Building. The portal strut utilized by Root (Figure 5-52) was attached to the masonry pier at one end and to a column at the other end. The Monadnock was three bays wide, and thus there are two outside bays and one interior bay. Apparently the exterior bays were

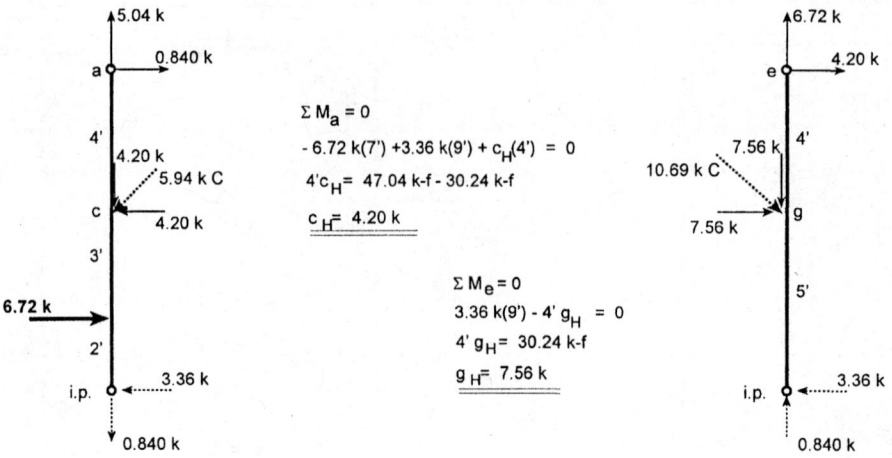

Figure 5-46.

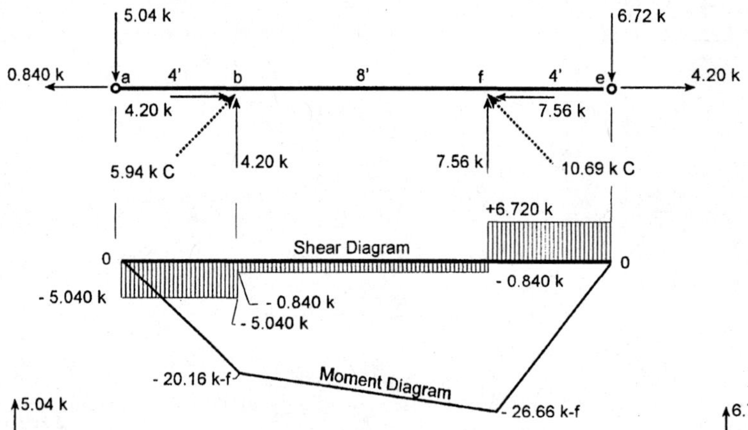

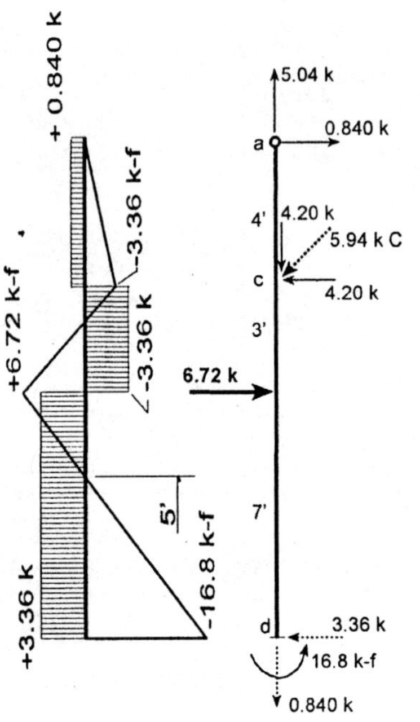

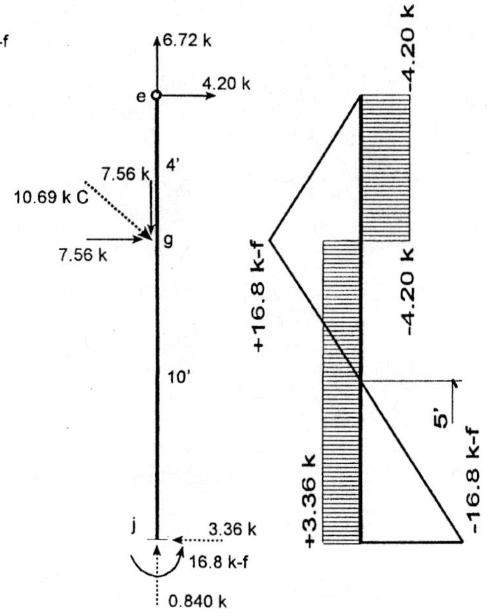

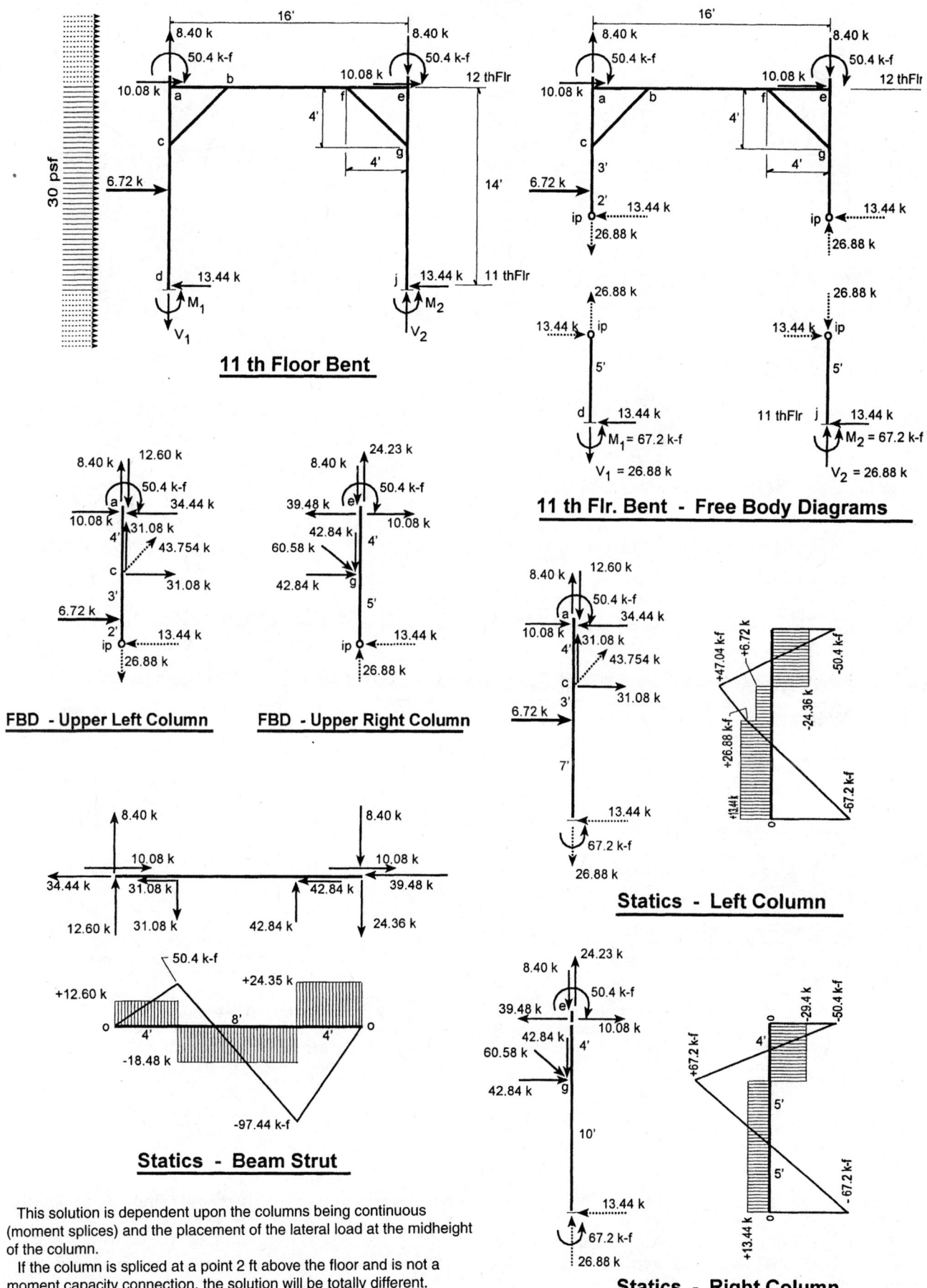

Figure 5-47.

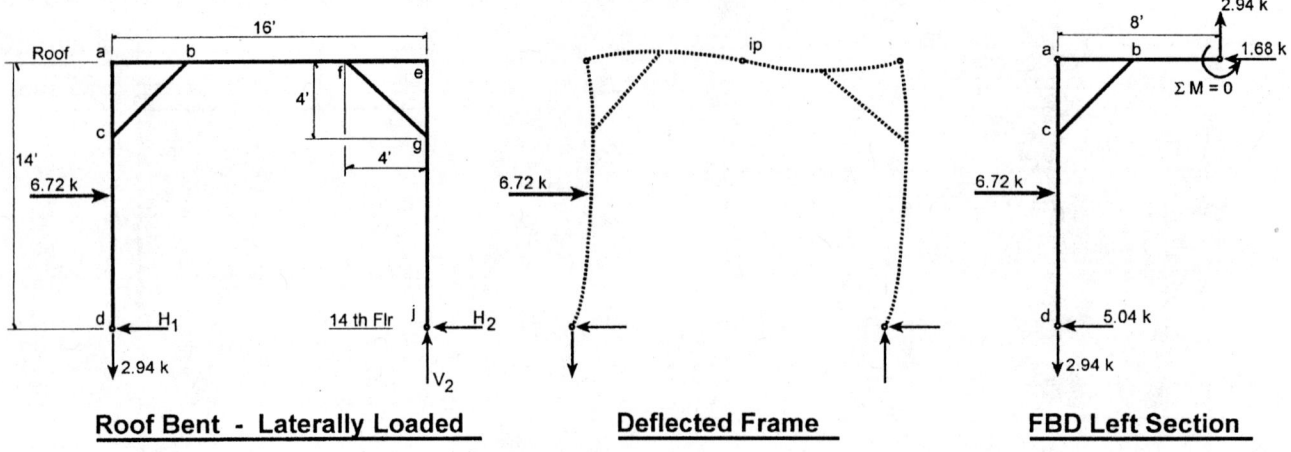

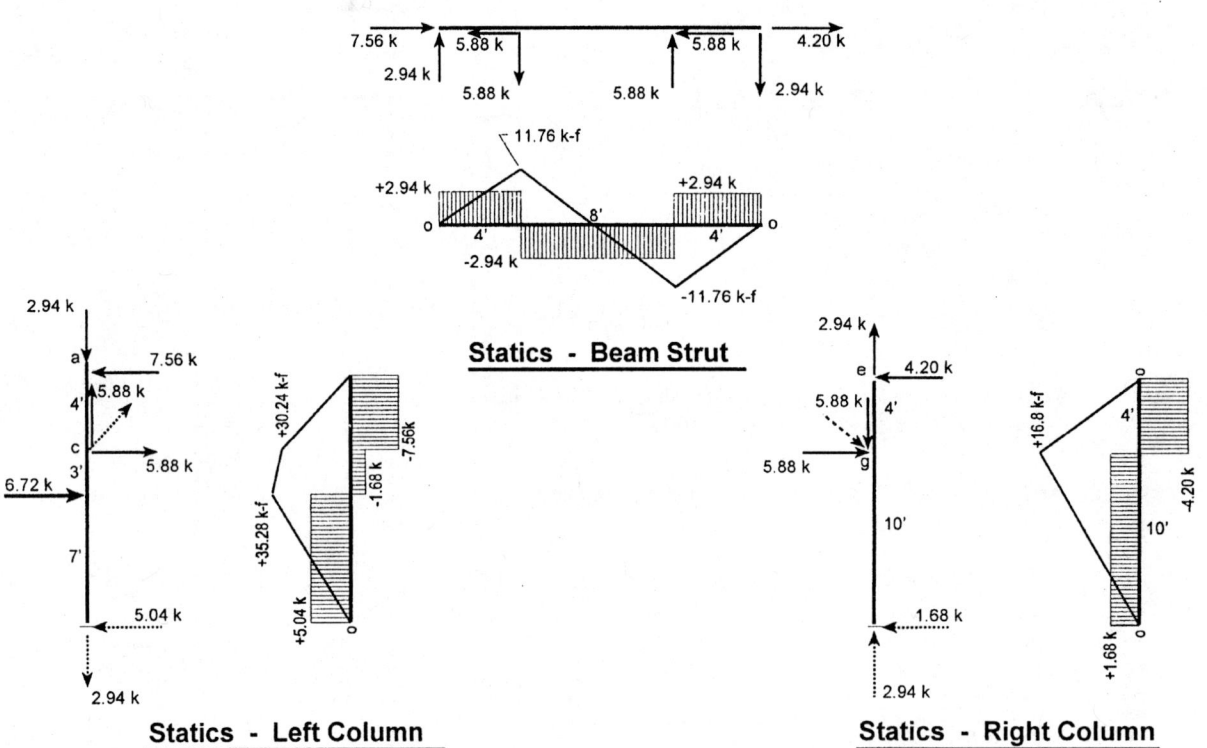

The knee braced bent shown above with no column continuity (moment at column base) shows the difference between this method of bracing and the method utilizing column continuity.

The big difference is in the "knee diagonal" on the windward column. The column with moment at its base has the knee in compression, while the column without the moment at the base has the knee diagonal in tension.

Figure 5-48.

stiffened by the portal strut as shown. It is also assumed that the portal strut was applied only to the upper floors, where masonry walls became thinner. Root felt that the wind force became a problem that required solution. The portal strut by Root for the Monadnock Building appears to be only a strut, not a floor beam, with the portal incorporated into its design. This is evidence of the newness of the technology and of the system not being fully developed. The original Monadnock Building was 16 stories tall and was the last of the great masonry load-bearing buildings. The thickness of the masonry walls at the base of the Monadnock are 72 ft, and thus 6 ft of valuable floor space around the perimeter of the building are lost (Condit 1964, 22). The steel frame and curtain wall system would require only 12–15 ft at the same level. The weight of the building, the thickness of wall-to-height relationship, foundation weight ratios, and the loss of interior rentable space combined to dictate the limit of the masonry load-bearing building. Elevators or the lack therof would also be a part of the equation; however, the single most important factor in changing building construction methods was the introduction of the skeletal steel frame.

SOLID PORTALS

The Old Colony Building, Chicago, 1894, utilized a solid portal made up of web plates and angles to form an arched opening between the columns and below the floor above (see Figures 5-53 and 5-54). This type of portal was to become the item now synonymous with the word "portal". It was very useful in buildings where larger openings in walls were desired, and the shape of the arches could become an architectural element because plaster decorative treatment could be applied to wood frame and furring or wire fabric attached and molded to the shape of the portal. Holabird and Roche utilized the portal as an interior feature or decorative element in the Old Colony Building. Freitag states that the portals were placed at two planes in the building (meaning bays) and that the wind pressure was figured at 27 psf of vertical surface (Freitag 1909, 271). Figure 5-55 shows a section through the entire height of the Old Colony Building, showing its portals. The portal was incorporated into the floor system in the Old Colony Building (served a dual purpose as floor beam and portal), but was not incorporated as a part of the column (see Figure 5-56).

LATTICE GIRDER BRACING

Freitag (1909) gives a method for calculating forces on what he calls the "lattice girder". The lattice girder utilizes a system of steel lattices and a top and bottom strut (horizontal member) to resist the lateral force of wind. It is not totally evident that the system also acts as a girder carrying its portion of the floor, but it is safe to assume that it does—that is, because Freitag talks about the design of columns with a combination of bending and axial load, it is safe to assume that designers of the period would recognize that a beam or girder that had an axial load in addition to the bend-

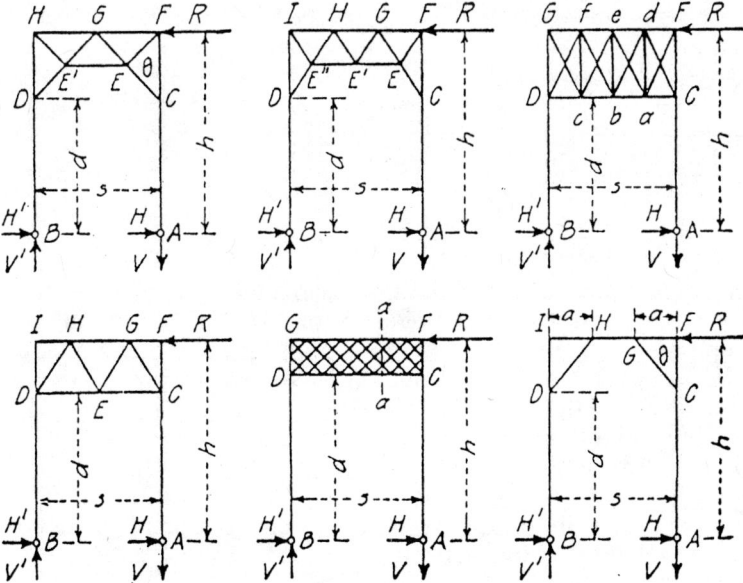

Figure 5-49. Portal frames for buildings. (*From Ketchum 1921, 43.*)

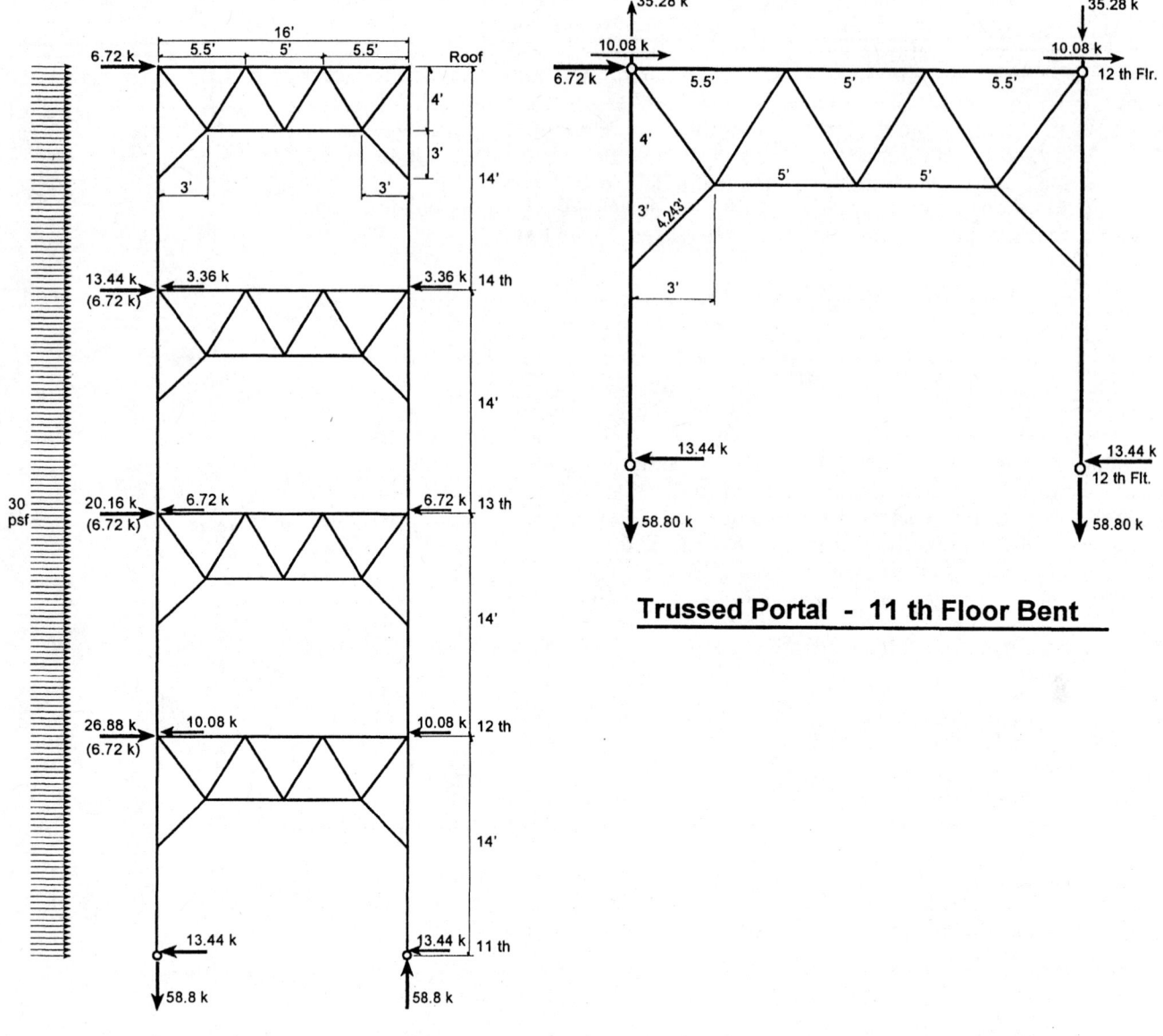

Figure 5-50. Portal frame for building—lateral load.

ing loads would have to be designed in the same manner as a member with combined loading. The lattice girder from the Freitag example also shows a column splice just above the top chord or strut. The lateral load from the wind pressure of the building is also shown as applied in two positions (divided by two) at the top and bottom chords or struts (see Figure 5-57). The shear from all floors above is shown as $Ps/2$ acting in the direction of the wind on each column at the point of splice. The Freitag solution for the lattice girder describes moment at what he calls the columns as "fixed at both ends". Actually, from the description of the formula for the moment at the ends, Freitag has placed an inflection point (zero moment point) at the midheight of the column, or $h_3/2$ (see Figure 5-57).

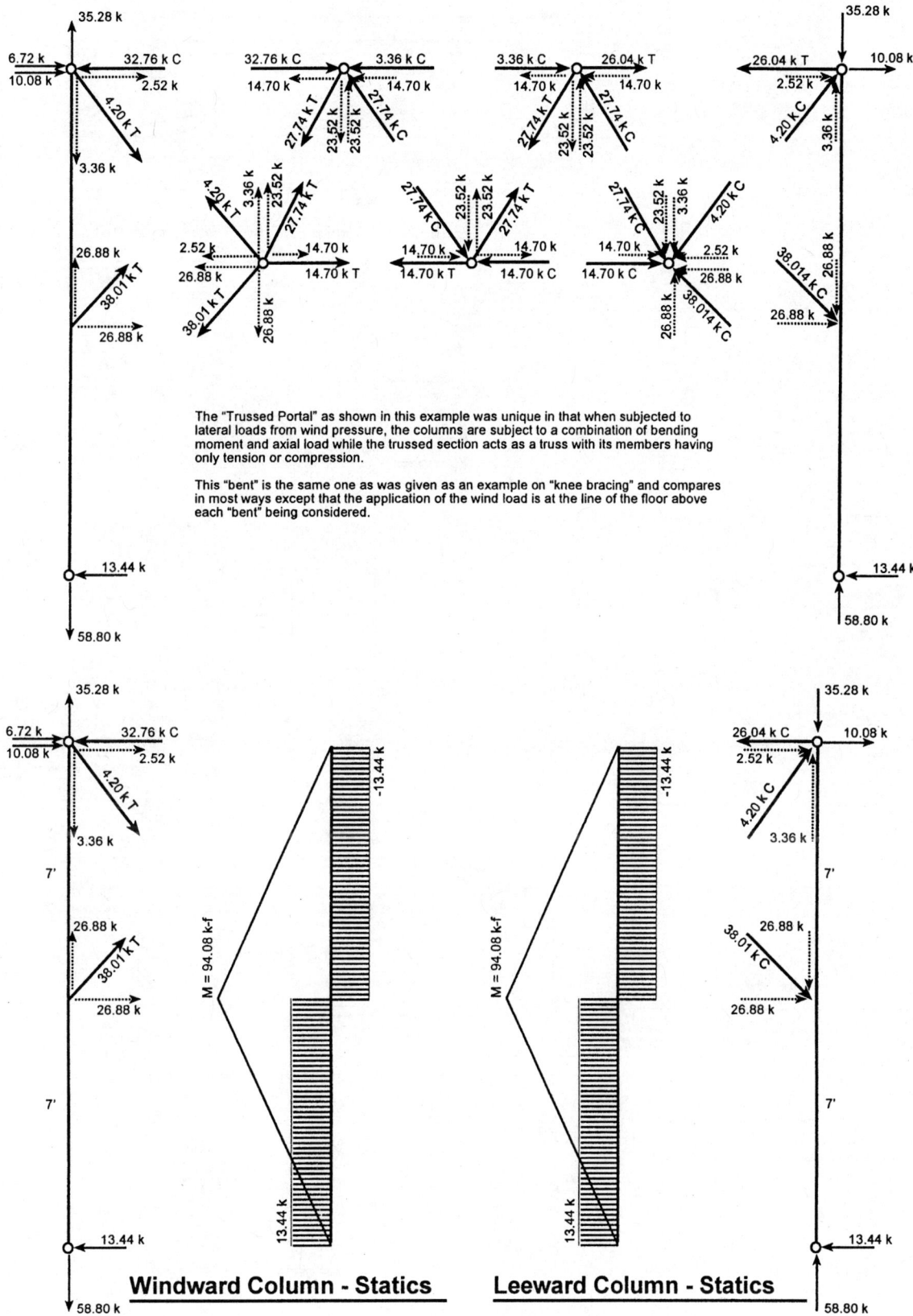

Figure 5-51. Statics: trussed portal.

436 ◇ Structural Analysis of Historic Buildings

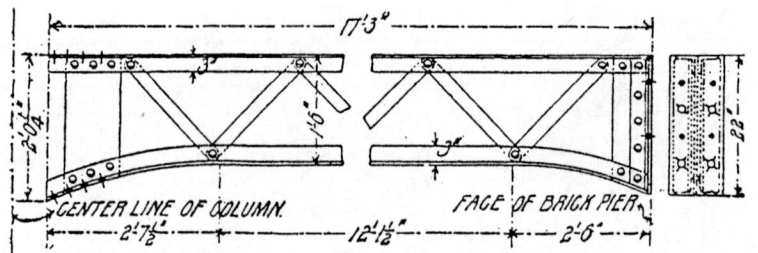

Figure 5-52. Monadnock Building portal. (*From Freitag 1909, 271.*)

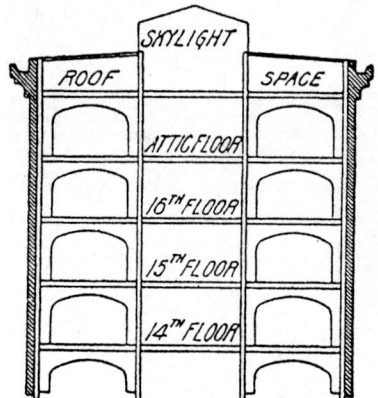

Figure 5-53. Old Colony Building portals. (*From Freitag 1909, 271.*)

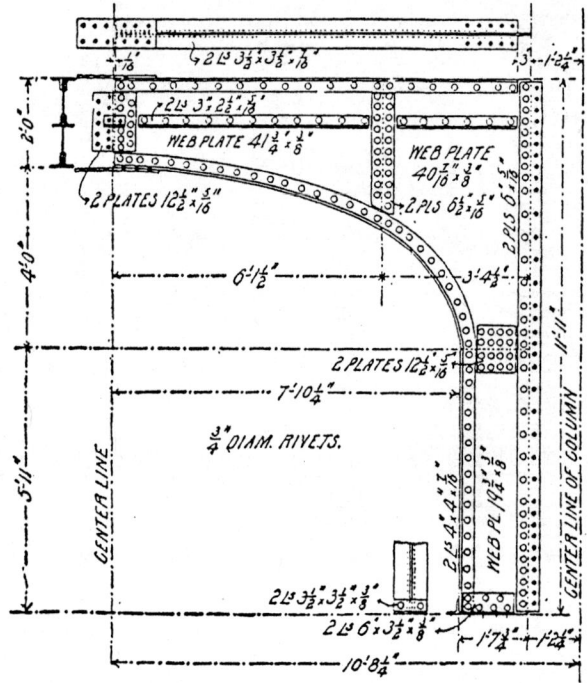

Figure 5-54. Details, portal, Old Colony Building. (*From Freitag 1909, 272.*)

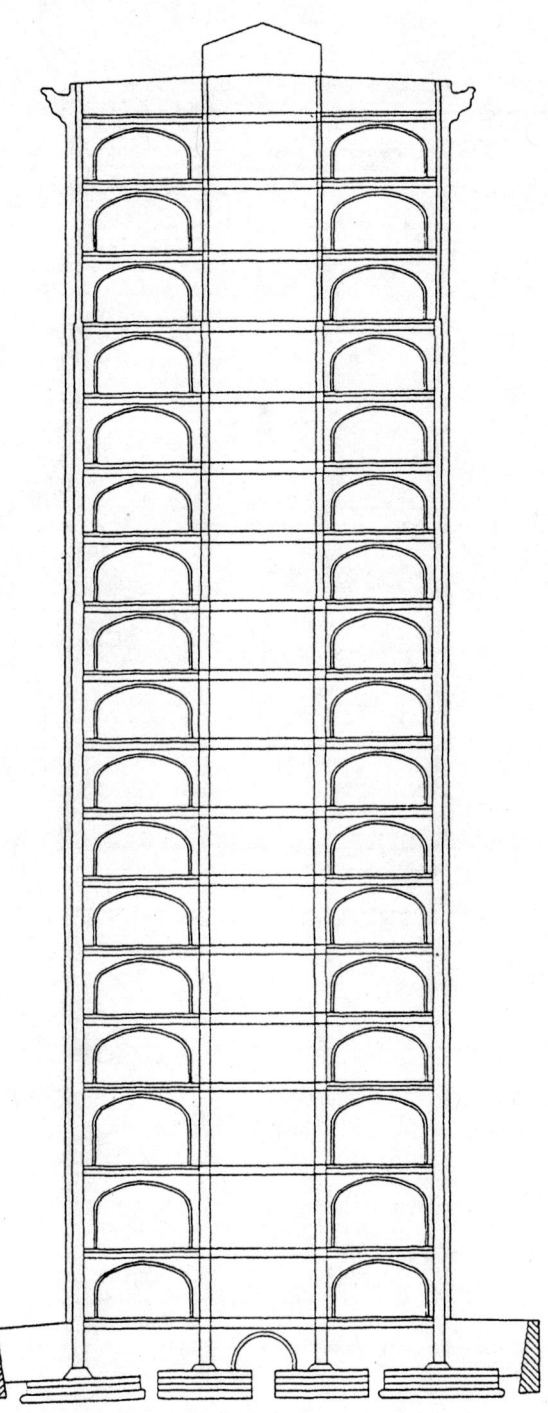

Figure 5-55. Sectional elevation, Old Colony Building. Reprinted with permission from C. W. Condit, *The Chicago School of Architecture: A History of Commercial and Public Building in the Chicago Area, 1875–1925*, Fig. 78, after p. 78. © 1964 The University of Chicago Press.

438 ◇ Structural Analysis of Historic Buildings

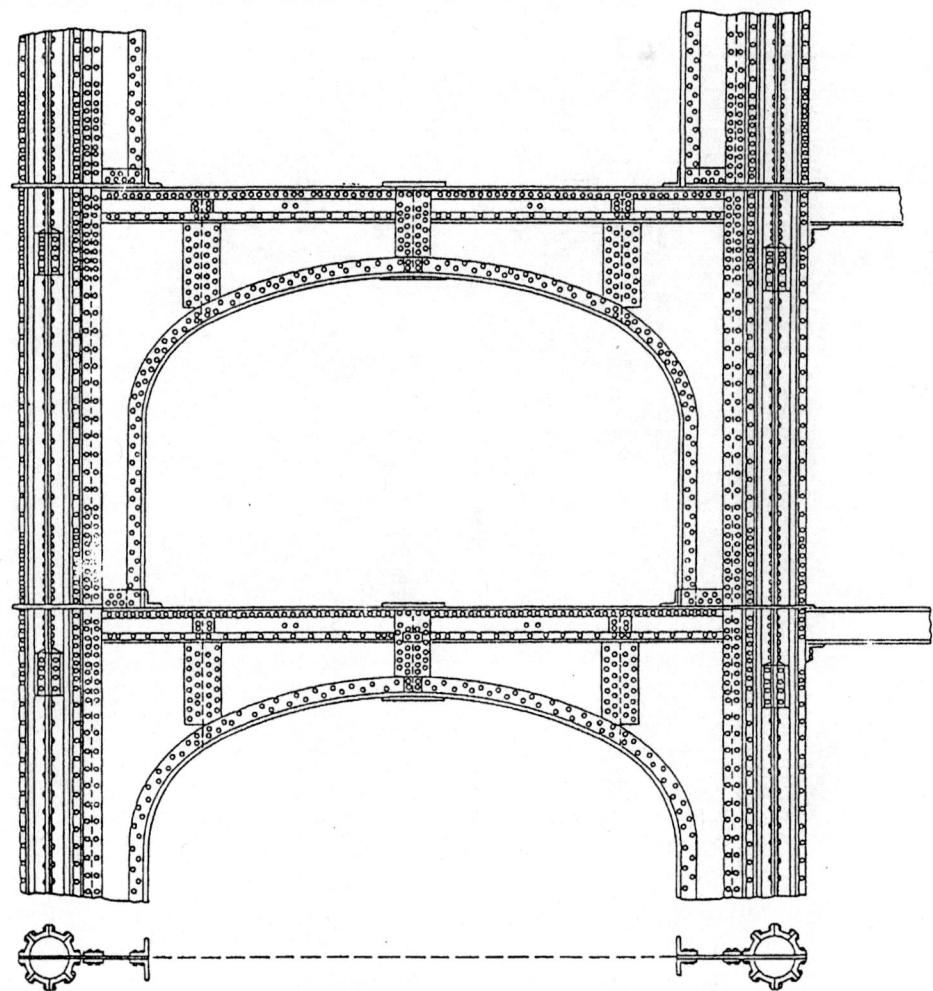

Figure 5-56. Details, portal bracing, Old Colony Buidling. Note: Phoenix columns with web of portal extending through column. Reprinted with permission from C. W. Condit, *The Chicago School of Architecture: A History of Commercial and Public Building in the Chicago Area, 1875–1925*, Fig. 79, after p. 78. © 1964 The University of Chicago Press.

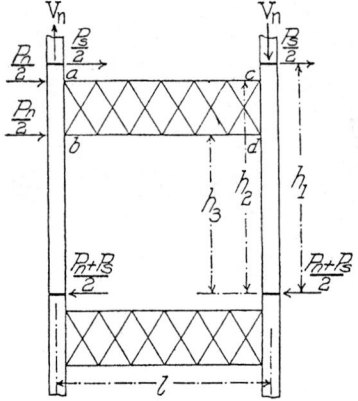

Figure 5-57. Lattice girder. (*From Freitag 1909, 276.*)

Freitag Equations: Lattice Girder:

$$V = V_n + \frac{P_s h_1 + 0.5P_n(h_2-h_3)}{L}$$

$$C_t = \frac{0.5P_s(h_1-h_3) + (P_n + P_s)}{(h_2-h_3)} + \frac{P_n}{2}$$

$$C_b = \frac{0.5P_s(h_1-h_2) + (P_n + P_s)(h_2)}{(h_2-h_3)} + \frac{P_n}{2}$$

$$M = \frac{P_n + P_s}{4}(h_3)$$

where:
 V = column load from wind pressure
 V_n = column load from floors above
 P_s = shear at top of column from wind loads above
 P_n = wind load for this bent level
 C_t = compression force in top flange of lattice girder
 C_b = compression force in bottom flange of lattice girder

LATERAL BRACING: EARLY WIND CONSIDERATIONS

Kidder (1908) has a section on wind bracing of buildings in which he quotes the laws of several cities, giving the designer of the day a valuable reference. Kidder refers to Freitag's *Architectural Engineering*, which he mentions types of bracing. Kidder states that when office and other buildings of 6–10 stories were built with solid masonry walls, no attention was paid to the lateral strains due to wind pressure, except perhaps to make the walls and partitions a little heavier (Kidder 1908[?], 1082). He further states that as these buildings were seldom built of a width less than 50 ft, no other precautions were really necessary. Kidder also states that the modern steel buildings were built to such great heights, especially in proportion to their width, that some efficient means of bracing the steel frame was a necessity. Generally, the use of the skeletal frame demanded the use of wind bracing in one form or another. Cities added this requirement to their building ordinances and also included many requirements about height-to-width ratios. There was also some discussion about buildings in major cities being protected by other buildings. Some evidence exists that designers may have braced buildings above the height of surrounding buildings and expected the influence of wind not to affect lower levels due to adjacent buildings being built in contact with the taller buildings. Chicago had a provision in its building law of 1908 that said, "In the case of all buildings the height of which is more than one and one half times their horizontal dimension, allowance shall be made for wind pressure, which shall not be figured at less than thirty pounds for each square foot of exposed surface. In buildings of skeletal construction, the metal frame must be designed to withstand this wind pressure." New York City building laws required that "all structures exposed to wind shall be designed to resist a horizontal wind pressure of thirty pounds per square foot of surface thus exposed, from the ground to the top of the same, including roof, in any direction." New York City had a further clause, stating that "in buildings under one hundred feet in height, provided the height does not exceed four times the average width of the base, the wind pressure may be disregarded" (ibid, 1083).

Kidder states that the building laws of Boston and Philadelphia contain no reference to wind pressure or the need for bracing. In a very advanced consideration, Kidder states that many designers consider that the connections between the columns and girders, floor systems, and dead weight of the building can resist at least 10 psf of wind pressure. Thus, if the building law requires an additional 30 psf, the overall system would be safe to a total of 40 psf. This is an early reference to the development of the framed structure, which is employed by today's designers in calculating the

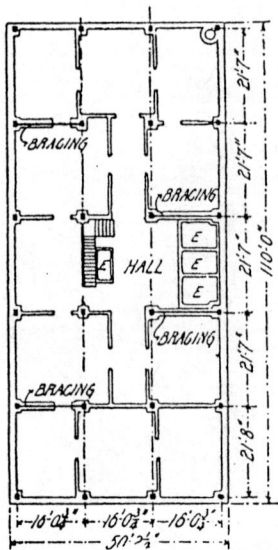

Figure 5-58. Plan, wind bracing, Venetian Building. (*From Kidder 1980, 1086; Freitag 1909, 264.*)

overall resistance from the connections. Kidder gives examples through a discussion of the bracing system for the Venetian Building, Chicago, 1894, which is spaced very adequately in the plan of the building (see Figure 5-58). The bracing in the Venetian Building is of the diagonal type and is continuous from floor to floor, transferring the lateral loads to the ground. However, the applied force from the wind is discontinued at the sixth floor because of the proximity of adjacent buildings that were assumed to protect the Venetian Building (see Figure 5-59).

The wind bracing theory of the period was sound in method and implementation, with the possible exception of the reliance on assistance from the neighboring buildings. We know today that the vulnerability of buildings may result in any building being left without its protective neighbor, and thus we never depend upon any outside influence. It is worthy of note that those buildings that were apparently detailed to depend upon the protective neighbors did carry their diagonal bracing through all floors to the foundation level. The liberal allowance for wind pressure and the conservative allowable stresses on members would probably result in the bracing being adequate for an expected (design) wind. Today's code provisions allow the working stresses in steel members to be increased by an amount of 33 percent (or a factor of 4/3) for wind loading. The short duration of the lateral load is the reason for this provision. Designers and policymakers feel that the short duration loads can be carried through these increased allowable stress with no apparent risk to the building. Kidder also discusses the type of wind bracing used in the Flat-Iron Building, New York City, 1902, designed by Daniel Burnham with the structural engineers Purdy and Henderson. The unique plan of the Flat-Iron Building made it vulnerable to the forces of wind in that the point of the triangle was too narrow to allow the designers to take advantage of the standard bracing systems, such as the portal, knee, and diagonal, and they had to use stiffened connections and the diaphragm of the floors to effectively transfer wind loads back to wider bents. The overall narrowness of the plan made the height-to-width ratio a problem for the designers. The width that is critical to wind load is 86 ft at the widest point to a rounded point (radius about 6 ft at the narrowest). The building is 21 stories tall, rising to a height of 285 ft. Figure 5-60 includes an assortment of details and photographs of the bracing system details and column-girder connections in the Flat-Iron Building from Kidder (1908[?]).

LATERAL BRACING SYSTEMS: EARLY RIGID CONNECTIONS AND STATICAL METHODS

Also notable at this point in time is the absence of any mention of continuity of members or moment distribution (rigid connections) in any of the reference volumes prior to or at the period 1910. The Hardy Cross method of moment distribution was

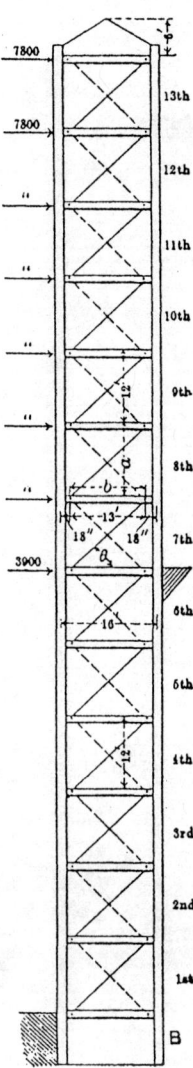

Figure 5-59. Wind bracing, Venetian Building. (*From Kidder 1907, 1087.*)

introduced by Professor Cross in 1932 and was accepted with relative ease and quickness because it provided the solution for continuous members and frames, which was badly needed to facilitate concrete construction. Steel building construction was not as fast to move to this system, and it was after World War II that developments in welding and welded connections made the rigid frame a reality in steel. During the period 1910 through the beginning of World War II, designers of structural steel were using methods described above. In 1930, an important work by Harry E. Schneider, *Practical Wind Bracing*, formalized the use of the cantilever method and the portal method. Schneider describes each method as follows (Schneider 1930, 3):

1. The Cantilever Method: The building acts as a cantilever, fixed at its base, and free to bend in a horizontal direction. The direct stresses in the columns are proportional to their distances from the center of gravity of the bent. The point of inflection of the beams is at mid-span. The vertical shear in the beams, which corresponds to the horizontal shear in an ordinary beam, increases from the outside of the bent to the center of gravity, but the increase is not uniform since the web of the cantilever is cut out between the columns.
2. The Portal Method: The direct stress is taken by the end columns only, the direct stresses in the center columns being zero. At any horizontal plane the wind load is divided equally among all columns. The point of inflection of the beams is not

442 ◇ Structural Analysis of Historic Buildings

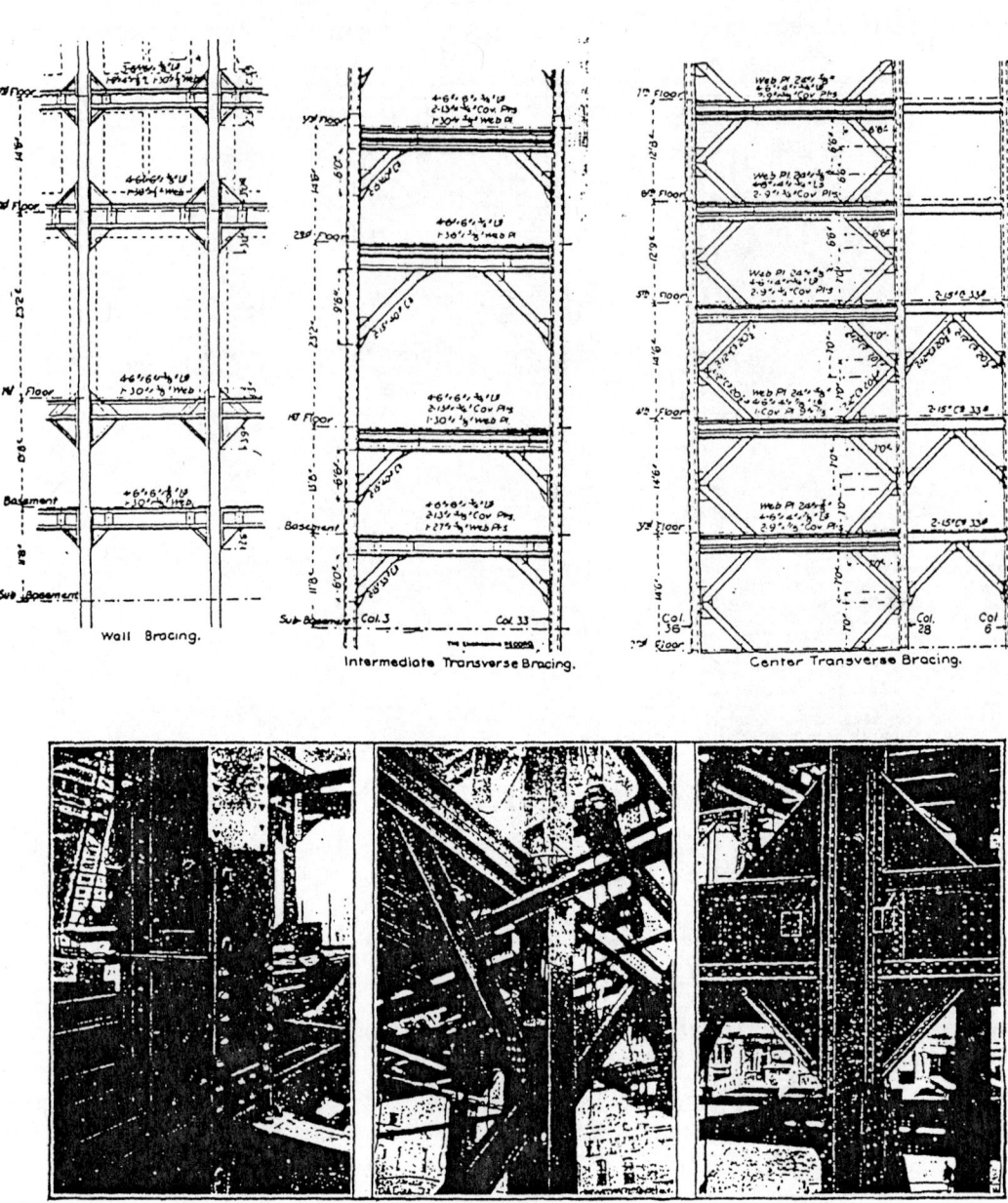

Figure 5-60. Flat-Iron Building, details of wind bracing and connections. (*From Kidder 1908 [?], 1092–1094; Engineering News (March 29, 1902, cited by Kidder).*)

always at mid-span but must be calculated from the beam moment, which is not always the same at both ends of the beam.
3. The Continuous Portal Method: The direct stress in the columns is proportional to their distances from the center of gravity of the bent, the same as in the cantilever method. At any horizontal plane the total wind is divided equally among all the columns, the same as in the portal method. The point of inflection of the beams is not always at mid-span, which is also the same as the portal method.
4. The Prof. Albert Smith Method: The direct stress is taken by the end columns only, the same as in the portal method. At any horizontal plane, the horizontal column shears in the center columns are equal to each other, and twice as much as either end column. The point of inflection of the beams is always at mid-span.

Schneider provides examples of these methods. He states that none of these methods are new, that each has been tried and tested by its individual sponsors and that he is summarizing the existing methods (as of 1930) to allow proper understanding by design professionals. His examples are consistent with methods still in use today, and the examples use consistent loads and frame dimensions that allow for comparison.

The first step in setting up the computations for the solution by any of the methods is to summarize the lateral loads on a building and organize the loads in a tabular system for easy identification. See Figures 5-61, 5-62, 5-63 and 5-64. Figure 5-61 shows a segment of 1 ft in vertical width of building. The loads represent a 1 ft width × vertical floor to floor height for the zone of wall that transfers its load into the frame.

Example: Cantilever Method

    Building: 30 ft bays, transverse
    Four 16 ft bays, longitudinal

1. Center of gravity of the bent

    All columns same stiffness
    Center of gravity = column row

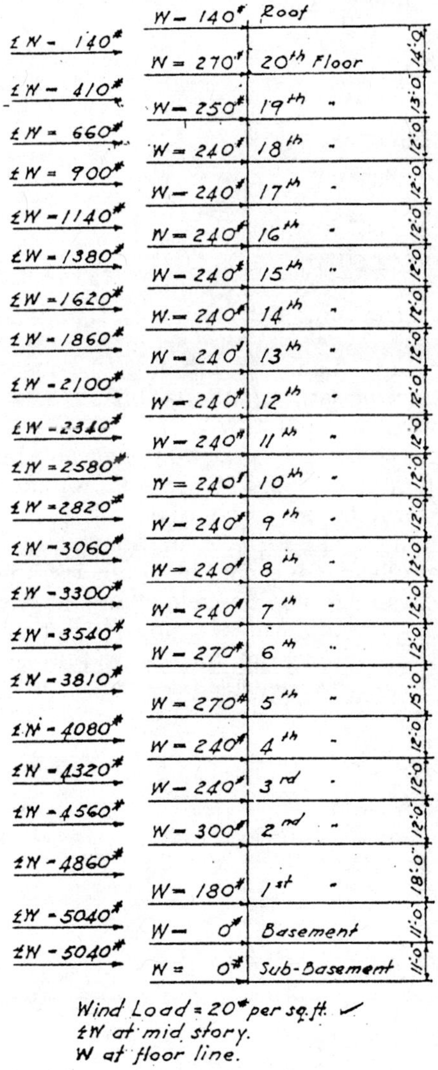

Figure 5-61. Horzontal Load Diagram, 20-story building. Reprinted with permission from H. E. Schneider, *Practical Wind Bracing*, p. 3. © 1930 Lancaster Press, Inc.

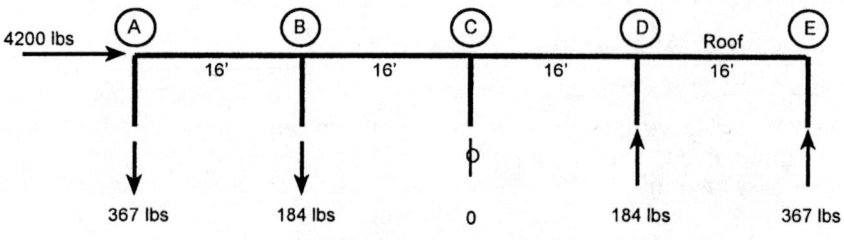

The Roof Bent

Figure 5-62. Cantilever method: bents with equal column spacing.

2. Moment of Inertia of the bent:

   $I = Ad^2$

   $I = (1)(16^2)(2) + (1)(32^2)(2) = 2560 \text{ ft}^4$

3. Column loads: Cantilever method:

   $A \text{ and } E = \dfrac{Mc}{I}$

   $A \text{ and } E = \dfrac{4{,}200 \text{ lb}(7 \text{ ft})(32 \text{ ft})}{2{,}560 \text{ ft}^4} = 367 \text{ lb}$

   $B \text{ and } D = \dfrac{4{,}200 \text{ lb}(7 \text{ ft})(16 \text{ ft})}{2{,}560 \text{ ft}^4} = 184 \text{ lb}$

4. Determine beam shears and moments:

Determine the moment in the upper end of the beams, then the corresponding moment in the columns. In the same manner as before, determine the column loads for the next level downward.

This same procedure is continued throughout the whole building across each level and downward to the foundation. See Figure 5-65 for the solution presented by Schneider for the 17th floor to the roof with axial loads, shears, and moments labeled on the members of the frame. Values shown are from lateral load only. Gravity loads from weight and occupant live loads would be another solution. The wind load and the gravity load solutions would be combined to obtain the design values. The Schneider solution is correct and is still applicable for any design performed today. Structural engineers today have the option of using this method of determining loads, shears, moments, etc. for any building for which they consider it useful. See Figures 5-66, 5-67 and 5-68.

Example: Portal Method: Bents with Equal Column Spacing

Building: Same structure, dimensions, loads, etc. as in the previous example
   Center of gravity, same as previous example
   All columns same stiffness

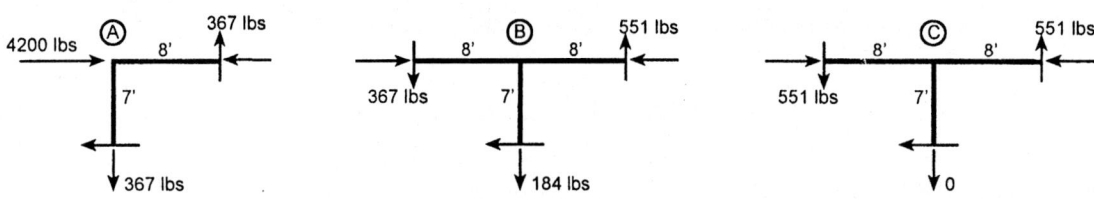

Figure 5-63. Cantilever method: beam shears and moments; initial statics; roof bent.

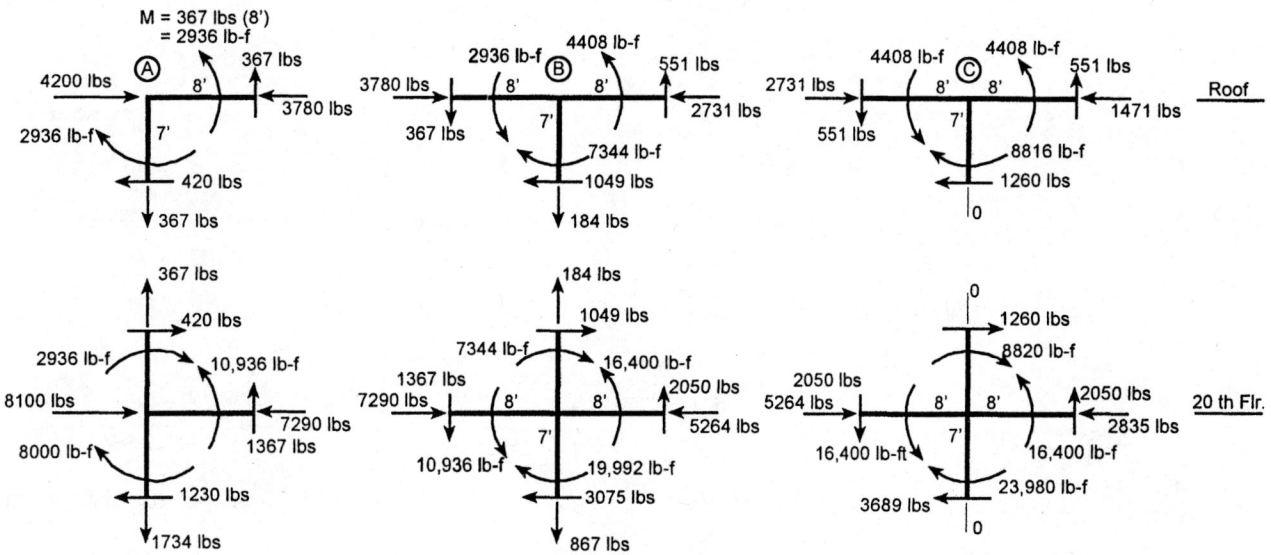

Figure 5-64. Cantilever method: beam shears and moments; further computations.

In this method, only the exterior columns carry axial load produced from the lateral force of the wind.

Interior columns do not carry load resulting from the wind.

Horizontal shear is divided among all columns. Inflection points at column mid-height.

$$E_v \text{ and } A_v = \frac{4{,}200 \text{ lb}(7 \text{ ft})}{64 \text{ ft}} = 459 \text{ lb (column } E \text{ and column } A \text{ axial load)}$$

Determine the column loads for the next level downward:

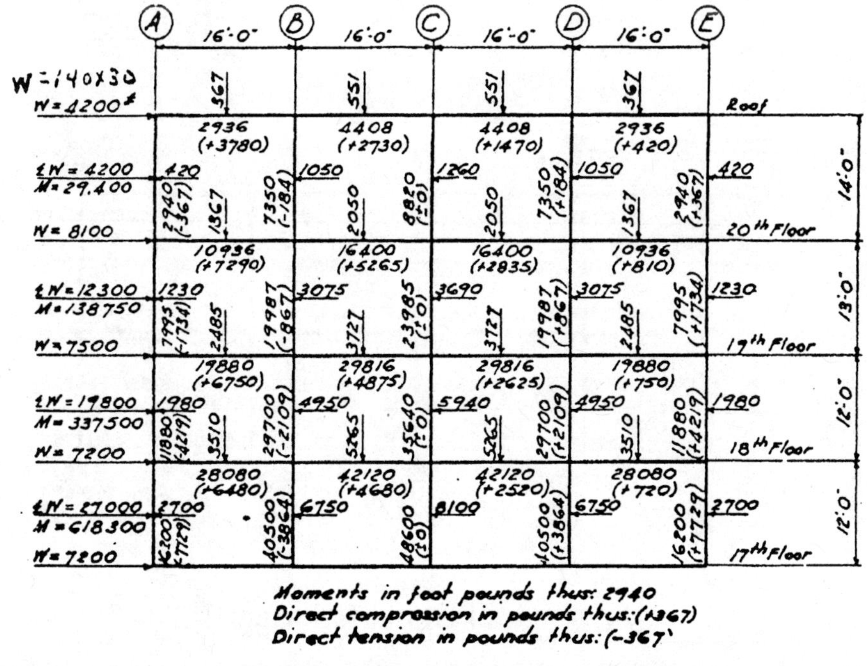

Figure 5-65. Cantilever method: summary upper floors, axial loads, moments, shears. Reprinted with permission from H. E. Schneider, *Practical Wind Bracing*, p. 3. © 1930 Lancaster Press, Inc.

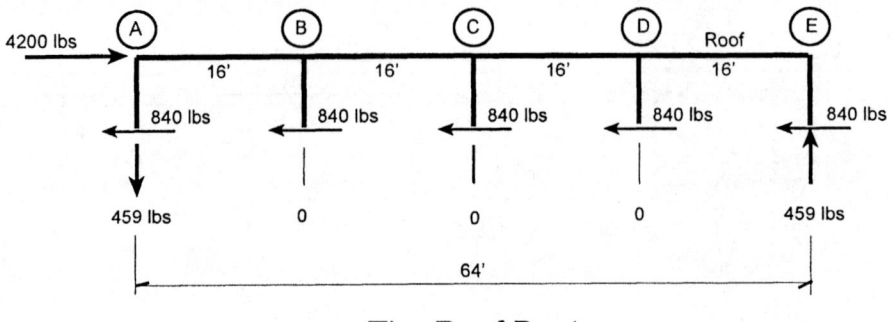

The Roof Bent

Figure 5-66. Portal method: bents with equal column spacing.

In this method, only the exterior columns carry axial load produced from the lateral force of the wind.
Interior columns do not carry load resulting from the wind.
Horizontal shear is divided among all columns. Inflection points at column mid-height.

$$E_v \text{ and } A_v = \frac{4{,}200 \text{ lb}(20.5 \text{ ft}) + 8{,}100 \text{ lb}(6.5 \text{ ft})}{64 \text{ ft}}$$

$$= 2{,}168 \text{ lb (column } E \text{ and column } A \text{ axial load)}$$

Determine beam and column shears and moments, axial loads:

In the same manner as with the preceding cantilever method, Schneider presents a summary of the solution for the upper floors of this 20-story building being used as the example for the wind load and its solution. The values in the Schneider solution agree with the manual computations performed by the same method today. The method is still valid for use today as may be desired
by structural designers. The joints between beams and columns are critical to the performance of these solutions because they must be capable of transferring moments through the joint to opposite side and adjacent members (see Figure 5-69).

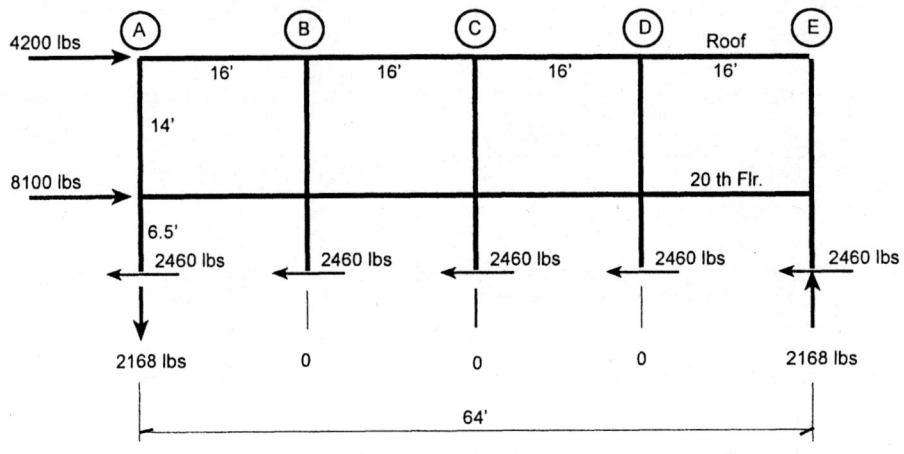

Roof and 20 th Floor Bent

Figure 5-67. Portal method: bents with equal column spacing.

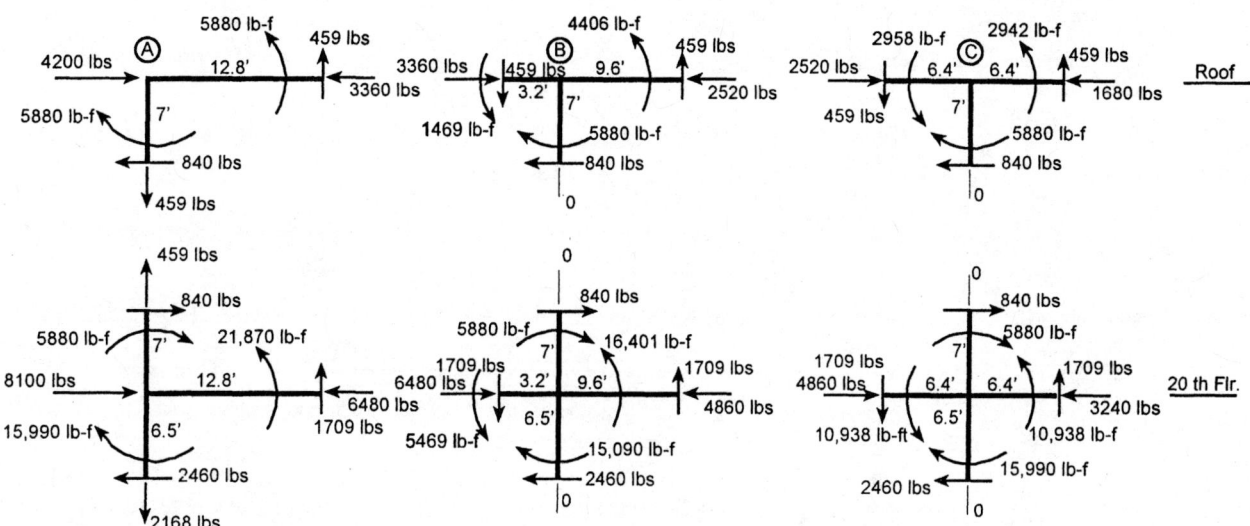

Figure 5-68. Portal method: beam shears and moments; further computations.

Example: Continuous Portal Method

Building: Same structure, dimensions, loads as in the previous examples
   Center of gravity of bent and stiffness of columns, same as in previous examples
The roof bent with partial determination of beam and column, shears, moments, and axial loads. See Figures 5-70 and 5-71.
Axial loads in columns are proportional to their distances from the center of gravity as in the cantilever method.

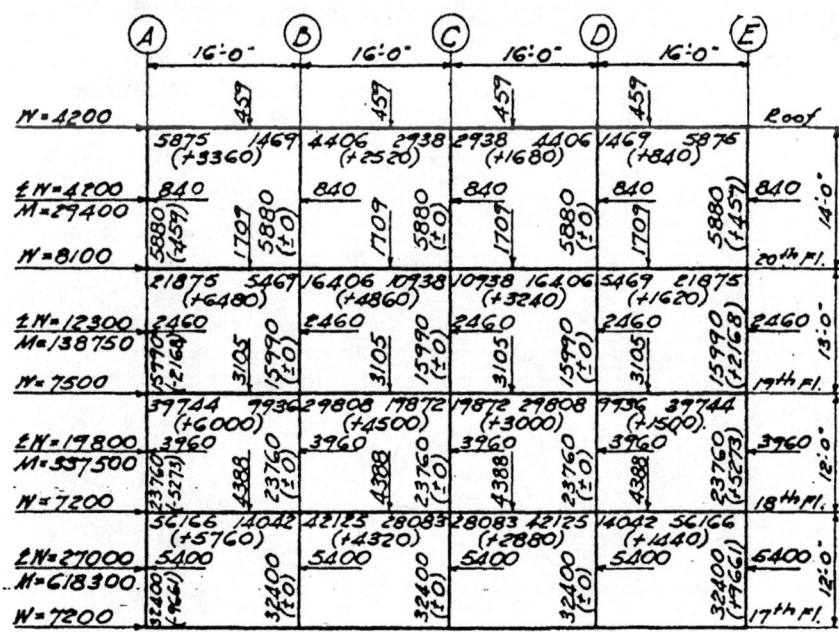

Figure 5-69. Portal method: summary upper floors, axial loads, moments, shears. Reprinted with permission from H. E. Schneider, *Practical Wind Bracing*, p. 70. © 1930 Lancaster Press, Inc.

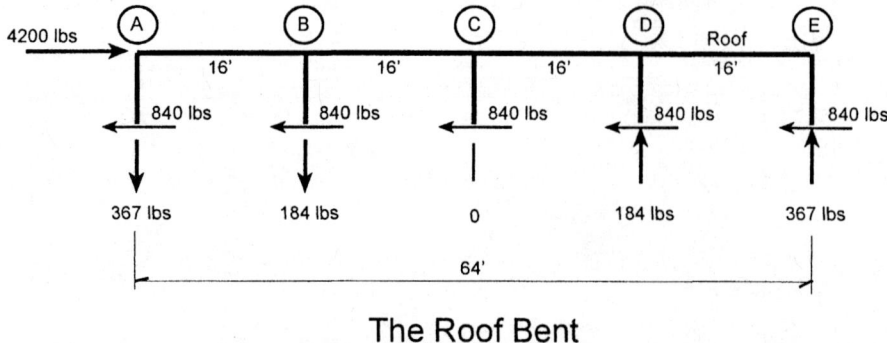

The Roof Bent

Figure 5-70. Continuous portal method: bents with equal column spacing.

Wind shear is divided equally among all columns as in the portal method.

Points of inflection of beams are not always at midspan in the same manner as in the portal method.

The moment in the beam B-A is determined to be 0 (zero) by the computations when the shear on A-B is divided into the moment at the A end and the inflection point is found to be 16 ft away from A. This is the total length of the beam A-B, thus making the connection at B a simple shear connection.

The Schneider solution for the Continuous portal Method is consistent with the other solutions and requires little explanation. A summary of the four methods is provided on the next pages to enable a comparison of the results from each method. See Figure 5-72.

Smith Method

Schneider's summary of the Smith method is clear and requires no lengthy explanation. The loads used in each of the calculations are based upon a 30 ft transverse distance between structural bays, and in the longitudinal direction all columns are of equal stiffness and equally spaced. If the columns are not equally spaced or are actually offset at certain floors, the methods remain the same, but the solutions are different (see Figure 5-73).

The wind bracing in any building of the era in question could be of one of the types discussed or a mixture of any of the types. Designers were very quick to adapt the type that suited the situation on almost a level-by-level or bay-by-bay basis. Any given building could have a combination of systems (see Figure 5-74). Freitag discusses the theoretical limiting height of a building and the potential problem of deflection or vibration. He states that the limiting ratio of height to vulnerable width is approximately 11.5:1. This ratio was based upon the deflection of a cantilever beam, which many of the theoreticians of the day considered the appropriate limit. Freitag states that on a steel skeletal structure 25 ft wide the height would be 285 ft, thus producing a deflection of 8–9 inches. Freitag states that this is not within the practical tolerance, since a deflection of that magnitude would shift the center of gravity of the

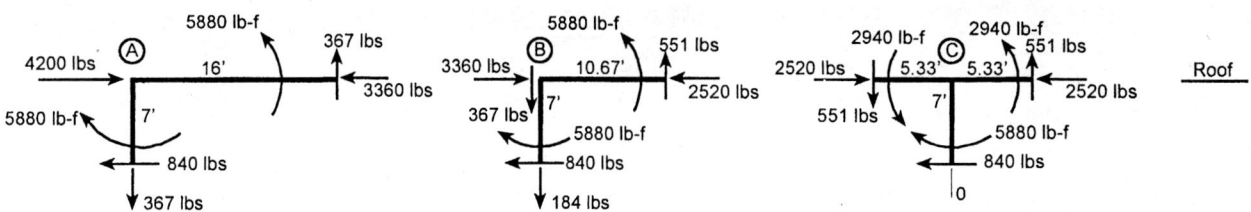

Figure 5-71. Continuous portal method: beam shears and moments; further computations.

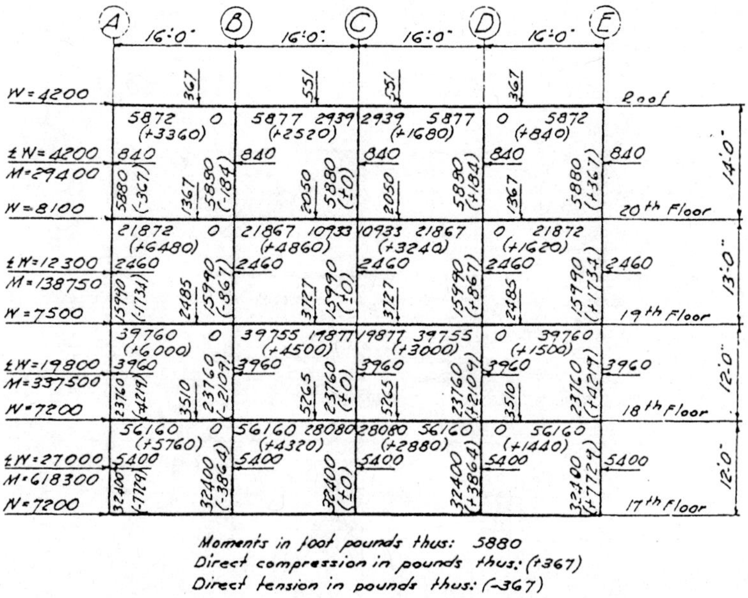

Figure 5-72. Continuous portal method: summary upper floors, axial loads, moments, shears. Reprinted with permission from H. E. Schneider, *Practical Wind Bracing*, p. 119. © 1930 Lancaster Press, Inc.

upper wall section beyond the outer edge of the wall, thus producing failure. Reversing the equation, Freitag states that the maximum allowable deflection would be about 2.5–3 in. which would give a height of 70–95 ft (Freitag 1909, 280).

Freitag states that experiments in Chicago on the deflections in tall skeleton buildings in an 80 mph wind produced only 0.25–0.50 in. of deflection in the Monadnock building. The measurements were taken as independently verified by transit crosshair and by a plumb bob in an interior staircase. These tests were taken in the narrow direction, west to east. He states that the veneer portions of the Monadnock showed

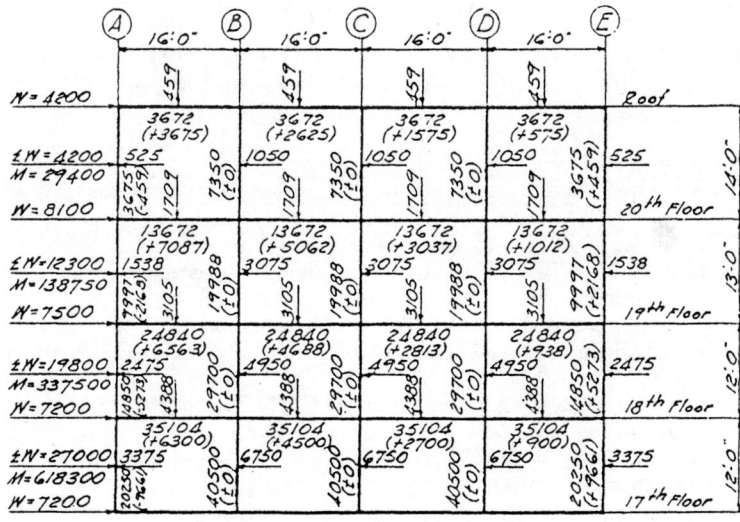

Figure 5-73. Professor Smith's method: summary upper floors, axial loads, moments, shears. Reprinted with permission from H. E. Schneider, *Practical Wind Bracing*, p. 119. © 1930 Lancaster Press, Inc.

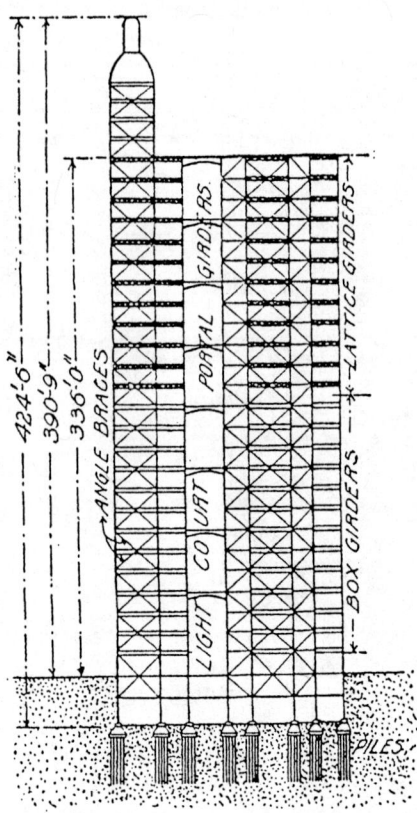

Figure 5-74. Bracing, Park Row Building, New York City. (*From Freitag 1909, 279.*)

more horizontal movement (deflection) than did the solid parts, which he says was to be expected. Also, the longitudinal direction produced a greater deflection, which Freitag says was not to be expected (ibid., 281; *Engineering Record* 1894 (cited by Freitag).

Freitag also states that the New York City Building Law and the Chicago Building Ordinance (approximately 1900+) had the following provisions for wind forces. At the same time the building laws of Boston and Philadelphia had no reference to wind pressures (Freitag 1909, 282, 283).

New York City Code:
All structures exposed to wind shall be designed to resist a horizontal wind pressure of thirty pounds for every square inch foot of surface thus exposed, from the ground to the top of the same, including roof, in any direction.

In no case shall the overturning moment due to wind pressure exceed seventy-five per centum of the moment of stability of the structure.

In all structures exposed to wind, if the resisting moments of the ordinary materials of construction such as masonry, partitions, floors, and connections are not sufficient to resist the moment of distortion due to wind pressure, taken in any direction on any part of the structure, additional bracing shall be introduced sufficient to make up the difference in the moments.

In calculations for wind-bracing, the working stresses set forth in this Code may be increased by fifty per centum.

In buildings under one hundred feet in height, provided the height does not exceed four times the average width of the base, the wind pressure may be disregarded.

Chicago Building Ordinance:
In the case of all buildings, the height of which is more than one and one-half times their least horizontal dimension, allowances shall be made for wind pressure which shall not be figured at less than thirty pounds per square foot of exposed surface. In buildings of skeleton construction the metal frame must be designed to resist this wind pressure.

CONCLUSIONS

As can be seen from the preceding data and examples on roof systems and wind bracing for buildings, the designers of the turn of the century were very innovative. The remarkable thing is that there were no precedents for these designers to follow; for the most part, they were breaking new ground with the designs for their tall buildings, curtain walls, and innovative bracing systems. The designs were very conservative in two ways: the loads were rather liberal—wind pressures were higher than we use today; the allowable stresses on members were lower than those in use today. Grades of steel have changed; however, today we are more liberal with the technical aspects of equations for the allowable stresses and other factors. The methods and equations presented in this chapter and verified by modern analysis are indicative of the level of competence of the designers of the 1890s–1930s. The methods of analysis utilized in the period were sound, and the buildings were equally sound. The design of the roof systems, trusses, and trussed domes of that period was as innovative and forward thinking as that of any building built today, or actually more so. The tall buildings, skeletal structures with curtain wall systems, were constructed in an era of excitement and innovation that may never be duplicated.

6

The Historic Material Assessment

THE CONTEXT OF EVALUATIONS

Many older buildings are revised in good faith today; however, in many cases the actual need for revision is not proved. The system or member capacity is not known, and therefore conservative practice requires that the system or member be reinforced or modified via modern materials that have known characteristics. The lack of data on the materials or components of historic buildings has left modern designers with no other choice but to modify certain systems as insurance against the lack of knowledge. The present state of the U.S. courts, the liability on the part of the designer and the owner, and the expensive consequences dictate that we must be sure of the capacity of systems and members rather than rely too much on intuition. It is not imperative that every building be treated in exactlt the same manner—i.e., the total reconstruction of the original analysis. The designer in charge of the work must exercise professional judgment in deciding how much analysis is to be done. This work presents data and design methodology that have been collected over a period of 20 years and are not available in this quantity in any other format. Comparisons are made of the original design capacity of a member, say a column, the capacity as it would be found using the modified strength of the original member (by the earlier design formulae), and the actual capacity of the column found today by methods of modern analysis utilizing the allowable stresses of the period. The recommended approach is to use modern methods of analysis, the historic section and its geometric properties, and the allowable stresses of the period. This method is still conservative and may require the designer to modify allowable stresses in cases where good engineering judgment permits. For instance, early structural steel allowable stresses were specified at 50 percent of the yield stress of the material. We have been utilizing two-thirds of the yield value since the 1950s with remarkable success. Specific design dimensions, character of end conditions, bearing, and other factors require that certain modifications of allowable stresses be made, and engineering judgment is once again required.

Poul Beckmann, a structural engineer in London, makes some interesting observations in *Structural Aspects of Building Conservation* (1995), in which he discusses the adequacy of existing load-carrying capacity for new use. Beckmann points out the problem of determining the original design loading of a building, or whether there was an actual design loading. The early designers followed empirical traditions, recommendations of the building laws of the city (if such existed), or the recommendations of the textbooks of the period (if they agreed with the loads called for by the author). Beckmann correctly points out that many builders of those periods chose not to follow those recommendations (ibid., 32). It becomes an interesting challenge to attempt to determine the actual loads that were utilized in design. Many times one must work in reverse and attempt to find some consistency in the capacity of existing members and, after subtracting the dead loads, compare the remaining capacity, which would be the live load. Often there is no actual requirement that the original design be known, as it is not critical to the present situation. Therefore, we simply attempt to determine the actual present capacity of the structural components in light of the contemporary or new use.

If the historic use of a building can be determined from records or historical research and the actual loading, the live load, was larger than that proposed with the new use, revisions or reinforcement may not be necessary. The problem becomes one of visual inspection and identification of deterioration. Sustained loading of an amount greater than the new or contemporary use should prove that the capacity of the building is adequate, i.e., the new use has a lower live load requirement than the actual live load that the building has been subjected to in the past. If all members and systems are intact and show no signs of deterioration (and the building has not undergone a long period of nonuse), there is no reason not to expect the building to serve for an extended period of time. Beckmann states, "If structure has carried a certain load for a considerable period in the past, without showing any signs of having been overloaded, and if it has not suffered any significant deterioration (due to rotting of timber, corrosion of metal, etc.) nor been damaged by past alterations, it is clearly capable of carrying the same load for a further term of use" (ibid.). In many instances, the actual calculation of the dead load of the building was very liberal at the time of construction. The contemporary analysis can sometimes accurately calculate the dead load and actually free up additional capacity for live load. Even though the building has only been subjected to the actual self weight and live loads from uses, the original analysis not only may have worked with heavier live loads than the contemporary or new use requires, it may have been overly generous with the dead load of the building fabric.

The period 1820–1940 began with empirical design and ended with analytical design. That statement vastly oversimplifies the sequence of events, materials introduced, systems developed, and conditions prevalent in the expansion and growth of the United States. One effect these many circumstances had on building construction was that the larger cities and population centers built with the latest developments, while nonpopulated areas continued using earlier, traditional systems literally for years beyond. The United States was such a large country in the mid-nineteenth century, with so many degrees of building sophistication acting in parallel, that it is impossible to say that the empirical design methodology ended at a certain time, or that it has ended. Communication was a problem, and new technologies would take years to spread to remote areas. These areas would not necessarily be the farthest point west; they could be any area outside of the population centers. Transportation and communication were slow and tended to be from one population center to another. In many ways, this condition exists even today; not every new construction technique or system developed is utilized immediately throughout the country. In fact, technologies, materials, and systems utilized in significant buildings will usually be employed in construction in the major cities almost simultaneously. The difference between the mid-nineteenth century and today is that instant communication, the mass media, and the education level all inform even the citizens of small towns that new building innovations exist and can be utilized as needed.

The evolution of building laws in the major cities, building codes, and design methodology have been presented in this work, and early design processes have been investigated, with emphasis on the relative adequacy of the results of empirical and early analytical methods. Empirical design was usually the job or responsibility of master builders, and the relative adequacy of the member sizing was a tradition handed down based upon successful experience. Certain rules of thumb existed and in a small way have been considered in this work. Not much is written on this subject, as it was mainly a tradition handed down from journeyman to apprentice. Indigenous building—buildings made by common or untrained builders—would be prevalent in the frontier regions but would be quickly replaced by more substantial construction as populations increased. Early analytical design came to America after the mid-nineteenth century, coinciding with or brought about by the need for larger heavier buildings. At first, spans became longer and interior column and girder systems replaced load-bearing partitions, but building materials remained the same: masonry and wood. The structural engineering profession evolved from 1820–1940, as the structural engineer became the specialist involved in the design of structural systems, members, and foundations. At the beginning of this period, the master builder could be a mason, a craftsman designer, a gentleman architect (self-trained via reading and study), or a traveling crew chief who oversaw the construction process. Over the 120-year period of this study, the designer specialist's role evolved to the point of professional status, and the professional license has become the required credential for the person who specifies the structural system and individual members. The law now requires that all persons in responsible charge of the preparation of plans and specifications be licensed professionals.

The last half of the nineteenth century saw many innovations in building methods and the introduction of many new building materials. The events and significant developments during this period make it one of the most exciting in building history. In the 1850s, there were rarely any buildings over four stories, even in the cities. All buildings were either wood frame or load-bearing masonry walls with wood floor framing. Electrical, plumbing, lighting, and heating systems were nonexistent or very crude. By the end of the nineteenth century, there were elevators, electricity, mechanical systems, plumbing, steel skeletal frame buildings with curtain walls of 20 floors in height, and a high level of general knowledge of building design and construction. More major developments and changes took place in the building industry during that period than during any other before or since. Building laws, material testing, material specifications, and material technologies were pushed to keep up with the advancements and changes in the industry. Standardized specifications and standardized building codes came approximately 20 years later. It is impossible to attempt to develop a broad overview of the developments in the building industry without looking at the true perspective of parallel developments of building legislation, materials, and construction. Enabling legislation did not bring about building innovation; rather, the developments in building methods and systems came first and prompted the changes in building laws.

The ability to design with new systems and perform the analytical computations was also an evolutionary process. Early architects and engineers were technical designers and builders; however, they had to grow with the changes in the industry. As could be expected, many of the changes and innovations were brought about by the users of the technology. The Army Corps of Engineers was heavily involved in design and construction and conducted many experiments and tests on structural components. The Arsenal at Watertown, Massachusetts, was the site of many material developments, column, beam, and truss tests on wood and also in the early periods of cast iron, wrought iron, and steel structural members. Both the Corps of Engineers and notable designers of the time worked at the Arsenal, and many design criteria and formulae came from the Arsenal tests.

This work has investigated the developments in load criteria, building laws, codes, foundations and the aboveground structural components of buildings. Wood, cast iron, wrought iron, steel, and concrete structural members have been followed chronologically from close to the beginnings of analytical computations through roughly 1940. In most instances, where appropriate, examples shown have been verified by developing the original analysis and checked by modern computations through current design procedure. This is consistent with the objectives of this work in that the verification of member sizes and/or system capacities with current methods of analysis has been considered a requirement for certification that the said member or system meets code requirements of today. This certification process would take into account the strength of the early material and the allowable stresses of the period. In an adaptive reuse of a historic building, a new use requires a structural certification, and a rehabilitation/restoration of the building requires that every aspect of the building be brought up to code. An historic building's structural system is usually easily definable, and a thorough investigation would provide the modern designer with the information required both to duplicate the basis of the original design and to verify the capacity of the individual members with regard to the contemporary use. Proper restoration/rehabilitation procedures would dictate that the intended reuse fit the building in a manner compatible with the size, shape, plan, and physical capacity without requiring destructive modifications. Through the data and design procedures presented here, this work investigates the original concept and method of analysis and suggest methods and procedures to verify a member or system.

The methods of early analytical designs discussed here all verify the hypothesis that this work began with. Structural members and systems utilized in historic American buildings were based upon sound design principles and procedures. The design methods employed for individual members of wood, cast iron, wrought iron, tile arches, steel, and concrete throughout the time covered in this work all have one common element: in the early stages of utilization of each type of material in structural systems, the analytical methods started out very conservative. As experience with the various materials was gained through application and use, the conservative nature of the analytical approaches was slightly reduced. An overview of the process by individual materials indicates the same pattern. A time line would start at ultraconservative and gradually decline toward moderately conservative. This does not sound like much movement or change, but in structural language it can be significant. The ultimate strength of members or system units by material was well known from very early on. Each material technology was involved in this evolutionary process, the result being that most design equations used during these early periods were all derived from the same basic ultimate strength equations. Column equations for wood, cast iron, wrought iron, and steel all came from the Euler's work in the 1750s. Each designer who derived column equations for his system or the patented members that his company produced used equations that came from the same Euler equation, with different constants and conversion combinations to arrive at a special formula for use in design. Designers refined their factors of safety downward as they gained experience with the individual materials.

Concrete design was unique. It started with unreinforced concrete columns. Ultimate loads were very low and diameters were very large. Allowable loads were five times lower than ultimate. Reinforcing developed slowly because designers were wary of two very different materials within one conglomerate. Apparently theory was difficult because the two materials were supposedly incompatible. Elastic properties were so different that designers could not derive analytical equations to define the action of columns or beams. It took over 20 years for the theory of the cracked section to be understood.

Each material proved to have an interesting evolution. This writer expected to find reasons to doubt the validity of certain theories or equations. However, the opposite was the case; I have found no theories or equations to be flawed and no designs or design examples that would not check as adequate when modern analysis was used to compare results. In fact, each era of formulae produced adequate member sizes; in most cases they actually produced oversized members when compared to the modern methods of calculations or analytical approaches. Today we use higher allowable stresses or a more reasonable factor of safety, not because we are smarter than the early designers, but probably because we now have over 100 years of experience in the art of building analysis and have constantly refined the methods of determination of the actual loads on buildings. We also have a longer history with these materials and have those same years of experience with the methods of material testing and analysis. The key to the analysis of many historic members or systems is to identify the material properly, the period of its production, and the allowable stresses that can be applied to the member at this time. Most designers now use the computer daily. Present applications in the area of building design and analysis are in load analysis, stress analysis, and very complicated issues such as plastic flow in analysis in zones above the yield point (steel). Member selection is available on some computer programs, and the accuracy of such programs is exact—that is, if the data input to the computer are correct.

Research has not uncovered any reason to distrust the early analytical methods or the product—member sizes, building components—specified by early designers. We must, however, bear in mind that the ratio of commercial buildings built by competent professional designers to those built by nonprofessionals, which is approximately two-to-one today, was probably closer to one-to-one in the period of this study. All building components found today in historic buildings should be checked and verified with two areas of emphasis: (1) identifying the original design concept and verifying through modern analytical methods, such as with columns, and (2) thoroughly checking the structure for external damage from weathering and man-inflicted damage. Neglect is perhaps the most critical concern with older and historic buildings. Years of neglect or of sitting dormant with no occupant, usually happening simultaneously, are the worst problem for any building to endure.

INITIAL CONSIDERATIONS: THE WALK-THROUGH ASSESSMENT

At the start of any project, or preferably during the initial assessment of the building for architectural fit (new use within historic building), the structural engineer and the architect should make a preliminary assessment of the structural condition of the building. The walk-through should be procedurally based, taking into consideration the condition of every aspect of the structure. It should record on tape the spoken thoughts of the engineer, and on still camera and video cassette the visible elements of the structure, interior and exterior. The walk-through should consider architectural decisions, program decisions, and code requirements. It is an assessment of the physical integrity of the building, an identification process to discover gross problems, and an opportunity to estimate or predict structural repairs or alterations concurrent with the extended use or new use of the building. The walk-through must be systematic and should follow a rigid routine to properly identify all aspects of the load paths through the building. Each objective is considered to be important to the cumulative manner of information gathering or focus by degrees of visual analysis.

Objective 1:
  Identify elements that make up the original structural system of the building.
  Identify character-defining features that make the structural system unique.
  Identify or prioritize the system components and how are they performing their function.
  Determine evidence of the date of the original construction.

Objective 2:
  Determine structural integrity, walls, roofs, structural components, etc.—how much is original.
  Visually assess the condition of the structural components.
  Determine the need for further destructive, nondestructive testing or closer visual inspection.

Objective 3:
  Identify problems—are any elements overloaded (excessive deflection, etc.)?
  Identify inherent damage to structural components.
  Identify moisture, faulty use of materials, stone layering, etc.
  Identify rot, decay of wood elements, corrosion, rust, deterioration of metal elements.
  Identfy damages to or cracks in walls, deterioration of walls from moisture problems.
  Identify excessive moisture—rising or ascending dampness in walls.

Objective 4:
  Identify pervasive structural damage.
  Identify differential settling of masonry load-bearing walls.
  Identify fire damage, earthquake damage.

Objective 5:
  Identify additions, intrusions, and changes in the structural system.
  Identify man-inflicted damages or changes that significantly alter the performance of the structural system or structural components.
  Identify replacement or substitute materials or structural components.

Objective 6:
  Estimate probable repairs, etc., degree or quantity of repairs.
  Estimate replacement parts, materials, quantity, and cost.
  Determine the basis of an overall estimate of the costs to rehabilitate the structure.

Objective 7:
  Identify what aspects of the structural system must be retained due to their unique historical character.
  Assess the historic certification process for this historic structure; federal, state, and local requirements.

Objective 8:
: Determine the degree of structural investigation required to complete designs and the eventual structural certification for the building (review architect's first schematic design).
: Assess the work to be done.
: Make initial estimates of cost of completion (± 25 percent estimate).

If the economics indicate an acceptable rate of return and the project proceeds to the next phase, it is at this point that the actual full structural investigation is performed. The data acquired in the walk-through and procedural plan derived from that information give guidance as to the degree of detail required to complete the full structural investigation. Buildings of different periods were built with varying degrees of structural complexity. As expected, most often the older the building the simpler or less sophisticated the structural system. Simple masonry load-bearing buildings with wood joists and floor systems of relatively short span require very little investigation and present the designer with equally simple solutions. The investigation becomes more an assessment of the condition of the walls, the mortar, joist pockets, etc., and of considerations for the building's response to lateral forces. Masonry walls were traditionally constructed three to four wythes thick at ground level regardless of whether they were load bearing. This inherent thickness provided the necessary shear wall capacity, and lateral resistance was high (ibid., 29).

From the 1860s through the 1880s the introduction of many new materials and spatial requirements in buildings made structures more complex, thus requiring a system of analysis and procedures for sizing beams, columns, footings, etc. These buildings require varying degrees of structural investigation. They also have more structural elements—more connections or supports—resulting in a higher degree of determinants to be considered. The period from the 1890s through the 1940s saw a radical change in the concept of building construction. The introduction of the skeletal frame eliminated most load-bearing walls and brought about the construction of tall buildings and the modern era of building utilization. This category of building technology is perhaps the most difficult to analyze because the finishes and curtain wall systems completely conceal the structural system. Semidestructive investigations, exposing the structural fabric (joints, etc.), are necessary but are in direct conflict with the need to preserve historic finishes, etc. Generally, as buildings and structural systems become increasingly complex, structural investigations too are much more complex, as well as time consuming.

The question always exists as to how far to go with the structural investigation. It is not always necessary to reconstruct the entire original analysis. If the live and dead loads are not increased by virtue of the new use, there may be little reason to reverify the entire structure. Modifications for modern conveniences such as fire stairs will require a redesign in such areas. Verification of the heavier, more strategic members may be all that is necessary to certify the structural system for continued use. In some instances, the original design may have utilized heavier live and dead loads than was necessary at the time. Many building laws required 80–100 psf for office buildings in the 1880s and 1890s. If the modern use requires a lighter live load and actual reserve capacity exists in members and systems, there may be no need for redesign or modification. Sound engineering judgment is necessary for determining the degree of analysis to be performed on a building-by-building basis. Beckmann diagrams a procedure to be utilized in this determination (see Figure 6-1).

MATERIAL ASSESSMENT: TESTING

Structural designers always need to know the strength of in situ materials so that the process of assessment (whether through mathematical analysis or through engineering judgment) can proceed. Sometimes all that is needed is material identification, but more often it is material identification and visual inspection to determine the existence of sound fabric. Deterioration of the original strength of members or systems due to external forces, such as rot and corrosion, which may render the element

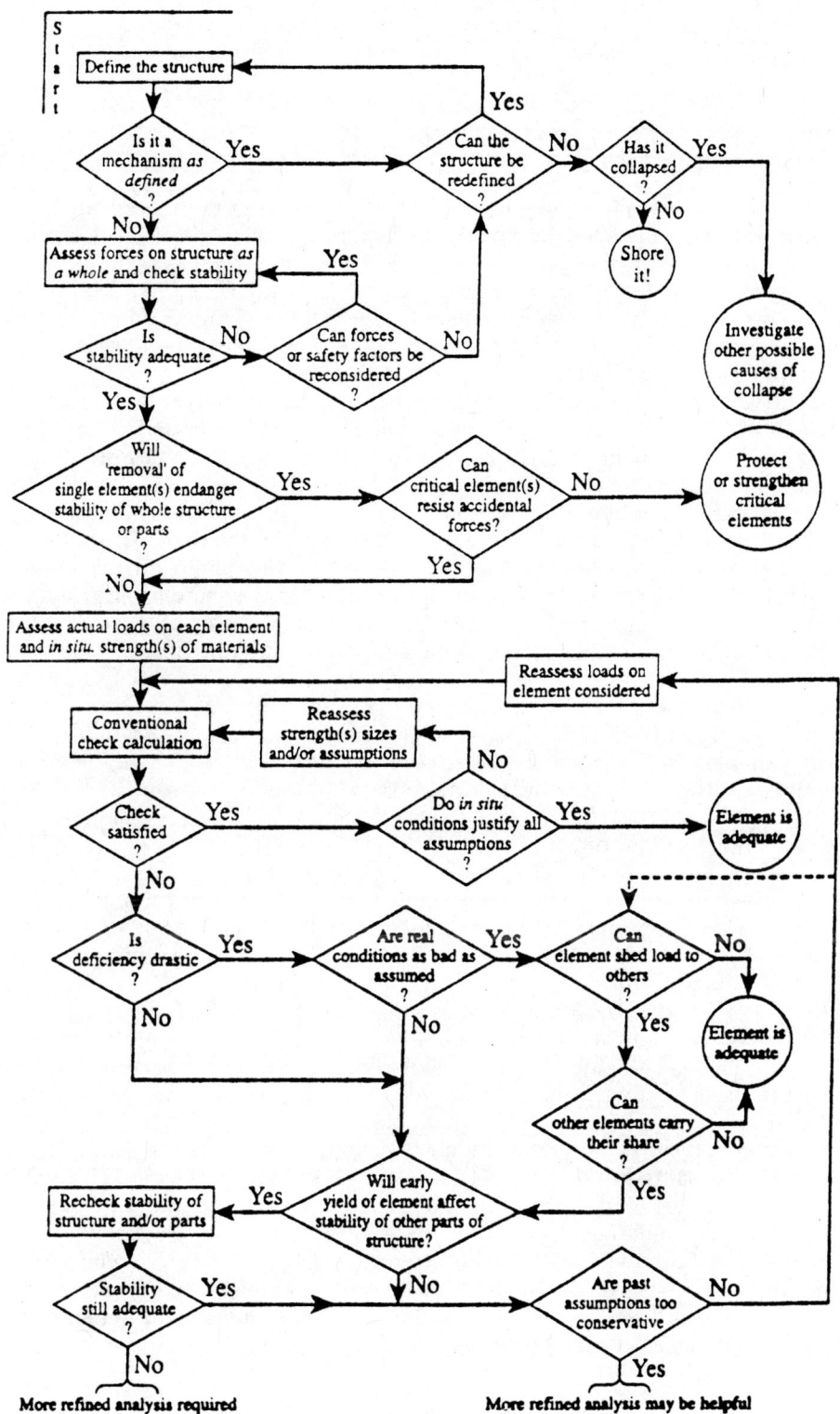

Figure 6-1. Beckmann's chart of the assessment process. (*From P. Beckmann,* Structural Aspects of Building Conservation, *p. 26. (©1994 McGraw-Hill Book Co., with whose kind permission it is reproduced.)*

inadequate, must be located wherever it exists in the structure. Usually signals such as leaks enable the experienced investigator to locate areas of potential damage. Overstresses and structural damage, such as minute cracking in elements, that are the product of excessive loads are very difficult to locate and evaluate.

METHOD OF STRENGTH EVALUATION: MASONRY, BRICK, AND STONE

Historic brick was usually handmade on the site, dried, and fired in kilns that were also constructed on the site. Handmade bricks for individual projects would usually have been made by traveling tradesmen hired by the landowner for the project (to construct his house, for example). In most cases, the landowner would supplement the traveling crew with his own workmen. In the Southern United States, the owner would utilize his slaves as supplement workers. Lime was burned and slaked on the site, and mortars were made with local sands.

As cities grew and fire laws emerged, brick increasingly became the mandated material in the business cores of the larger cities during the 1870s and 1880s. The use of brick in the cities made the remote brick manufacturing yard a necessity because it became impractical to manufacture on the site. The strength of historic brick is a product of several variables, the main one being the heat of the kiln and the position of the brick within the proximity of the fire and the time of the burn. The brick of the period manufactured by these brickyards would still vary in quality, mainly hardness. Until the 1940s, brick for buildings came in two basic grades: hard exterior brick and soft interior brick (salmon brick, named for its color). The soft brick was the brick positioned in the kilns at the greatest distance from the fire (the heat). It was used in exterior walls and in inner wythes of exterior walls because they would not weather well when subjected to the elements.

The strength of historic brick (individual brick) is not of much significance; the crushing strength of individual bricks varies from 2930–5390 psi, while the crushing strength of walls of the same brick in lime-sand mortar varies from 1290–2370 psi (Chapter 3, Table 3-12). The allowable stress for these walls varies from 50–100 psi (in place in lime-sand mortar) (Chapter 3, Table 3-14). The factors of safety for the above walls vary from 25.8 to 23.7, averaging approximately 25.

The tests quoted above were made on a specific group of brick, probably made under controlled conditions. The handmade brick made on the site and manufactured brick of the period 1840–1900 would vary greatly in strength. For this reason, it is fairly unlikely that testing would be of much value to the engineer. On a given building, the strength of the brick may vary as much as ±75–100 percent. Testing would be fairly expensive and would have to be done in several places (as many as 12 on a given wall) to be meaningful. The 75 psi allowable compressive stress would probably still be the most reasonable point of departure for hand-made on-site brick. Later machine-made brick will have higher allowables.

Beckmann (1994, 33) discusses the justification for reducing the safety factor in light of the fact that reduced uncertainties justify reduced safety factors. The problem is reducing the uncertainties; refinement of the accuracy of dead and live loads is the outside approach, while justifying an increase in the allowable stress could be considered the inside approach. Either is difficult to realize; both take calendar time and are expensive due to the professional's time involved. Testing and analysis of masonry walls, brick or stone, will have many common aspects while requiring specialized individual approaches in other areas.

MASONRY WALL TESTING

Nondestructive and destructive methods of evaluation of masonry walls and other building systems are often necessary to assist the designer in the assessment of buildings. The strength of the masonry, in addition to several other factors, needs to be known to enable the determination of the capacity of individual parts or the system comprising a structure. Wilson (1984, 118) provides an excellence reference to assist designers (see Table 6-1).

ASTM Test for Brick

The standard test procedure for brick is ASTM C67-92a, in which brick is tested for compression strength by hydraulic compression testing equipment. The test utilizes a half brick as the sample, and the result is simply the load at failure divided by the area of the half brick. The half-brick test specimens must be capped with gypsum

Table 6-1.

Masonry Assemblages Units and Mortar	Flexural bond strength	Bond strength testing through obtaining a sample from existing wall or making a sample from period brick and mortar of the original mix. Samples from the existing wall are preferred. Tests per ASTM C-67, C-270, and C-518.
	Shear strength	4 × 4 ft sample required; must be cut from existing wall.
	Water absorption	Weighing dry and saturated conditions of test sample. Sample is saturated by submersion in boiling water for 5 hours and cold water for 24 hours.
	Freeze/thaw resistance to damage	Repetitive cycles of wetting, freezing, drying, and weighing for 50 cycles unless specimen fails in testing cycle.
Ceramic glazed structural clay facing tile, facing	Imperviousness	Surface tested by application of permanent ink, allow to dry 5 minutes and wash to determine if surface is stained.
Brick, and solid masonry units	Chemical resistance	End 1.5 in. of test specimen is dipped in 10% solution of HCL for 3 hours. Opposite end is dipped in a 10% solution of KOH for 3 hours. Finishes are then rinsed, dried, and examined visually to observe deterioration, changes in texture or color.
Ceramic glazed structural clay facing tile, facing brick, and solid	Crazing test	Autoclave crazing test. Test specimens are placed in an autoclave with 150 psi steam pressure for 2.5 hours. Specimens are then cooled for at least 3 hours to room temperature. Permanent ink is then applied to glazed surface, then wiped to determine if crazing exists.

Mortars
   Lime-sand mortar
   Portland cement Mortar
   Masonry cement Mortar
	Compressive Strength
	Water retention
	Air content
	
	Efflorescence
		ASTM C-91 Standard tests for compression and water retention are utilized with exceptions per BIA MI-72.
		ASTM C-231 Standard tests for air content.
		
		Efflorescence tendency is determined using the wick test as described in BIA Research Report No. 15, Sec. 4.4, p. 14.

Masonry Face brick Sandlime brick Structural clay tile Concrete block Mortar Including assemblage	Compressive strength	Compression strength of masonry prisms, ASTM E-447, Method B. Materials, masonry to be tested conform to ASTM C-67, ASTM C-140, ASTM C-109 To determine if cells are filled, lightly tap with hammer. Location and uniformity of inner cell grout, wall thickness, use a small masonry bit to drill probe hole. Probe with a stiff wire.
Masonry units and mortars	Continuity compressive strength	Low-frequency ultrasonic testing. Soniscope and transmitter and receiver are used. Strength obtained by comparison of time of transmission through masonry with that of sound test specimen.
	Location of voids and reinforcement	Gamma radiography. Gamma source and x-ray film are used to obtain a photographic image of the interior of a masonry section. Voids show as irregular patches and reinforcing as light area. Hazardous.
	Location of reinforcement	Pachometer, a magnetic detector, is utilized to locate reinforcing. A ferromagnetic component will cause a variation in magnetic field.

capping or sulphur-filler capping to ensure level, plumb test surfaces, and consistent results. The compressive strength, C, equals load at failure divided by the area of the half brick.

Removal of Test Panels

Test panels of masonry wall sections, approximately 9 in. wide by 18 in. tall, are required to provide adequate results in determining the crushing strength of an existing masonry wall. The panels must be sawn with a diamond-tipped blade and transported to the laboratory without damaging the sample. The sawing is difficult; cooling water must be used with extreme caution so as not to deteriorate the lime-sand mortar. Soft mortars are extremely vulnerable to water. Transportation is difficult, and one is rarely assured of getting the sample to the lab without disturbing it. The sample leaves an unsightly void in a wall, which must be repaired. In addition, one sample is hardly enough to provide an accurate determination of the strength of a wall or of a given building. A number of tests are required from different areas of the building to provide a definitive value for the strength of the masonry. Then a factor of safety or some other method of determining the allowable stress on a wall composite must be made on the basis of sound engineering judgment. This method is not recommended by this author, for the following reasons: the results may vary and may not be useful; the building may be significantly weakened by a number of sample panels being cut out, and the overall effect may be unsightly and detrimental to the building, with nothing learned in the process.

ASTM E 447-92b, Method B, is the standard compressive strength test method for existing masonry as well as new construction. This method is considered unreliable because it calls for prisms or test samples to be constructed of the same materials and the same methods as the original construction. The ability of contemporary craftsmen to construct test specimens that are of the same quality and consistency as in situ existing masonry walls will always be in question (ibid., 120).

Using a good physical inspection, an engineer with the experience and judgment to interpret the type and age of the brick and the type and soundness of the mortar can assign an allowable stress that may be as accurate as the result of expensive sample panel tests.

In Situ Testing

In this procedure, a brick column is created in an existing wall by sawing out two sides and the top. The column is load tested by inserting a flat jack at the top of the column and loading the column against the weight of the building overhead. If possible, the load is taken to failure and then the column removed and other brick relaid in the void. This method will produce an accurate portrait of the strength of the wall, but it is destructive and will leave an unsightly blemish in the wall. Even if identical brick can be gathered from a less visible location, such as the parapet on the rear of the building, the patina of age and weathering will still probably not match. The new mortar utilizing pigments for color and sand as close as possible to the original will still be slightly different in appearance. The method of cutting out a brick column provides only one measure representing one area of the wall or the building, and it is difficult to justify a number of these tests to ascertain a proper average. This method is not recommended unless one is willing to rely on a single test or at most a very small number of tests. It is also not a certified ASTM test method at this time.

Test Cores or Cylinders

Horizontal cores extracted from existing masonry walls can be utilized to assess the strength of the unit and can in some cases be correlated with the strength of the wall. This method is relatively new, and fairly extensive development of data per individual building is needed to enable a rational analysis to be precluded. ASTM C-41, C-42, C-496, and ASTM C-283 provide data on the taking and testing of horizontal and vertical cores in existing concrete. These test cores can be as small as 2 in. in diameter. However, there are currently no criteria for the use of the same specifications on masonry walls of brick or stone. This method appears to have a lot of promise as it minimizes the amount of material extracted from the wall. The test method is to subject the cylinder to a compressive test along the side of the cylinder with loading plates perpendicular to a vertical diameter of the cylinder. The compressive load provides a uniform tensile stress in the cylinder perpendicular to the plane of loading. The tests must be correlated with material of a known strength to allow the calibration to be set. The basis for testing of the core sample requires that a sample be taken with the center of the core being the center of a horizontal joint in the masonry wall. Equal sizes of masonry must be above and below the joint. A number of tests will give sufficient information to enable assessment of the strength of the masonry wall. The test was developed in Brazil for concrete and appears to give very accurate measures of the tensile stress capacity of a test section, which can then be correlated with the compression stress capacity. Figure 6-2 gives a diagrammatic representation of the stresses acting on the cylinder.

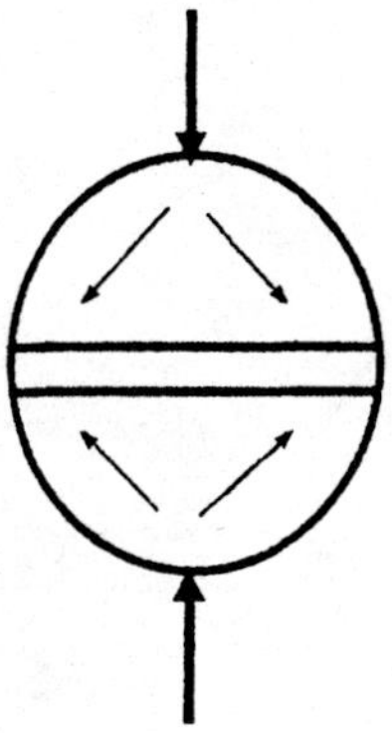

Figure 6-2. Stresses on test cylinder.

Ultrasonic Testing

The strength of a masonry wall can be assessed with relatively good accuracy provided the ultrasonic equipment can be properly calibrated and the operator has adequate experience. The ultrasonic test determines whether the density or relative uniformity of the wall is consistent. Ultrasonic testing normally looks for cracks or cavities in the mass being tested. The ultrasonic pulse is measured in time of travel (usually milliseconds) through the masonry wall. The time is correlated to a set calibration that can be done on core samples from a section or duplicate test specimen of the same thickness and of known compressive stress capacity. Experienced operators should be able to produce satisfactory results; however, the density of the mortar and of the brick or stone are different and will generate different times. Therefore, care must be taken to avoid allowing the pulse to go along a mortar joint and thus give false readings. The method utilized is through-transmission of the pulse from the transmitter to the receiver via the test material. Computer imaging methods such as acoustical tomography can depict internal profiles of stress in a solid three-dimensional medium. Figure 6-3 gives a simplified flow diagram of ultrasonic testing.

Any form of testing is expensive. The more sophisticated the tests, the more complex and expensive is the equipment and the higher the degree or operator expertise

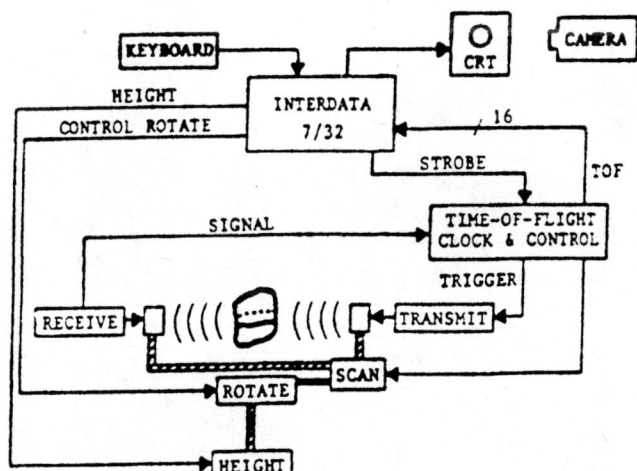

Transmitter and Receiver on Opposite Sides of Item Tested

Time-of-flight signal is measured with a 16-bit, 100MHz clock read by the counter at up to 400 points along a scan profile with each point separated by 0.15°. There are 120 views, each containing 400 samples, and each sample containing up to 8 channels of 16 bit time-of-flight data.

Figure 6-3. Ultrasound computed tomography system.

required. Testing may be a solution for very important, high-profile projects; however, most owners of common buildings barely justify expenditures to maintain or upgrade their structures. Many contractors and substitute material manufacturers will work against reasonable measures while promoting their replacement products. At times that replacement product is not a coating or new cladding; it is in fact a replacement building.

A more rational and less expensive approach would be to attempt to determine the capacity of a wall in light of its actual condition. The size, mass, exposure, and use of the new building will all play an important role in determining the sustainability of the masonry building. A brick wall with an allowable compressive stress of 75 psi would have to be 90 ft tall to develop that actual compressive stress at the base, and that is including an average factor of safety of 25. Could the wall actually be 2250 ft tall before crushing the base brick? The above statements consider only the dead load of the brick and mortar and not any imposed floor or roof loads or compression from lateral loads. Methods demonstrated in this work show an analysis of a masonry load-bearing wall of typical two- and three-story buildings. Lateral loads present the real problems for exposed masonry structures. The presence of shear walls, trussed partitions, and floor diaphragms (in good working order) unifies the overall composition of a structure and makes it easier to justify higher stress levels through sound engineering judgment. Eccentric loading on a brick masonry wall can come from several conditions, as demonstrated in this work. The Masonry Society and the Brick Institute of America have produced extensive literature presenting reliable data and design criteria for load bearing masonry construction. The Specification ACI 530-88/ASCE 5-88 provides building code requirements for masonry structures for designers' use that have been adopted by most code jurisdictions in the United States. The Masonry Society gives formulae and diagrams for the strength of walls with no tensile strength as shown in Figure 6-4 (Abrams 1992, 41, crediting the derivation to Sohlin 1971).

The prevailing feeling in the United States is that the cracked section method is in error because it results in values of the compressive stress at wall edge that are far too high. Most designers prefer to allow a very small allowable tensile stress (about 4–5 percent of the allowable compressive stress) on the opposite edge of the wall, that solution being a much more rational approach. Others prefer to limit the tensile stress on the opposite edge of the wall at zero stress, which automatically produces a maximum compressive stress. The eccentricity is thus limited by either method to an amount that produces either the minimal stress or zero stress. (See Figures 6-5 and 6-6.)

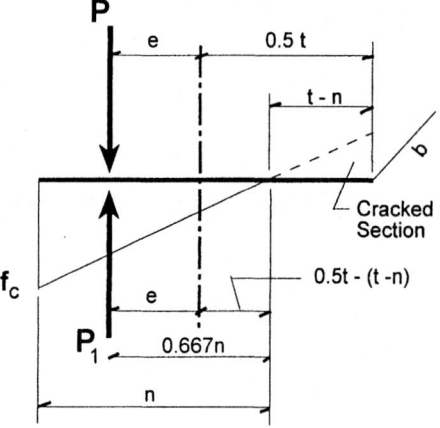

$$f_c = \frac{2P}{bn} = \frac{4P}{3bt\left(1-\frac{2e}{t}\right)}$$

$$n = 3\left(\frac{t}{2} - e\right)$$

The Masonry Society also gives a disclaimer for the method, stating that the approach gives "higer values for P than that which would result in some limiting allowable stress." It also states that one should exercise great care in applying this approach.

Many designers are conservative and do not allow any tension stress in a historic masonry wall. Others limit the tension stress to a very small amount, approximately 5 percent of the compressive allowable. A thorough inspection of the wall must precede any decision on

Figure 6-4. Cracked section analysis.

Example: Minimum Tensile Stress

24 in. wall section, load = 1440 lb
Eccentricity set for 10 psi tension

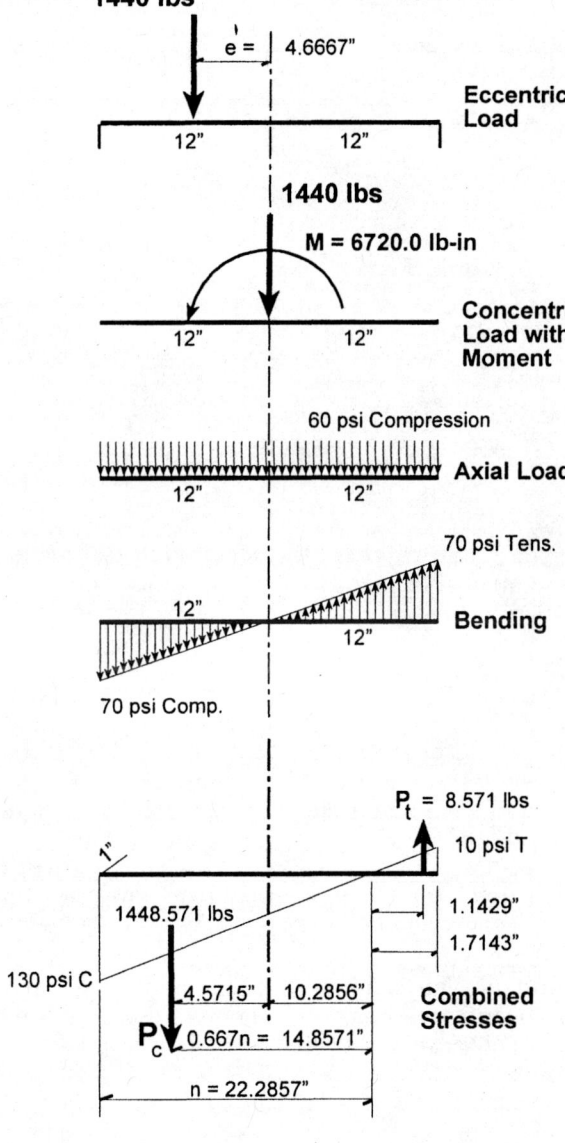

ΣF_V = 1448.571 lb − 8.571 lb
= 1440 lb (the original load)

ΣM = 1448.571 lb(4.5714 in.) + 8.571 lb(10.2856 in. + 1.1429 in.)
= 6720.0 lb-in.

Figure 6-5a.

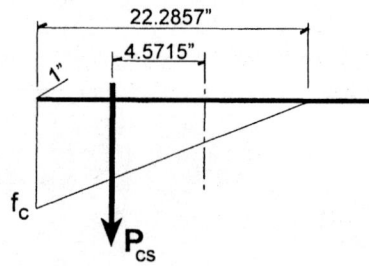

Constants for the 24 in. wall, 1 in. long:

$I = 1152$ in.^{4} $S = 96$ in.^{3}

The portion of the stress block that is in compression has for its base the length n, which goes from the point of zero stress to the outer edge of the section as determined by the conventional combined stress methods.

$$f = \frac{P}{A} \pm \frac{M}{S} \quad \text{tension or compression}$$

The conventional method of stress analysis assumes that the material is capable of developing tensile stresses as well as compressive stresses, and that the material deformations and/or modulus of elasticity are uniform for either tension or compression.

Check of the section by cracked section analysis:

$0.5\, f_c(22.2857\text{ in.})(1\text{ in.})(4.5715\text{ in.}) = 6720$ lb-in.

$f_c = 131.921$ psi

The modified stress block, which indicates a higher compressive stress at the outside edge of the section, produces a force and line of action that equates to the same moment (resultant moment) as the moment caused by the eccentricity of the load. The same load and eccentricity, when subjected to the Masonry Society's cracked section analysis, as shown above, produce a compressive stress of 131.921 psi.

The increase is slight because the tensile stress in the conventional analysis is very low.

Figure 6-5*b*.

Example: Zero Tensile Stress

24 in. wall section, load = 1440 lb
Eccentricity set for 0 psi tension

$P = 1440$ lb, $e = 4$ in., $M = 5760$ lb-in.

Tables 6-2–6-4 and Figures 6-7, 6-9, and 6-10 present tabular and graphical results of the conventional methods of analysis and show the locations of the eccentricities that produce zero tension in the walls and the point of the maximum tensile stress of 10 psi. The conventional method is shown in the graphs, along with the Masonry Society Method for cracked section, which produces much higher compressive stresses in the wall after the section has cracked. The graphs are of 12 in., 16 in., and 24 in. wall thickness, 1 in. long, with base compressive stresses of 40 psi, 50 psi, and 60 psi plotted against a set of eccentricities for a constant load that produced the base stresses. (See Figure 6-8.)

Brick masonry structures are significant in number throughout the United States. Brick utilized in load-bearing masonry from the period 1820 through approximately 1890 varies greatly in quality (allowable compressive stress). Better quality and more consistent strength and dimension came during the 1870s as commercially manufactured bricks became available. The question still remains as to how to evaluate historic brick load-bearing walls and how to assign or determine the allowable compressive stresses for a given wall. Visual inspection of the brick and the mortar joints and a determination of the age of the building, combined with sound engineering judgment are the basic requirements. A determination of the loads is required, and any condition of eccentric loading should be identified. An analysis of the combined axial and eccentric loading utilizing standard combined stress principles and the effects of cracked section theory will, in the opinion of this author, be adequate to determine the existence of any problem areas. Lateral loads and unbraced wall heights and

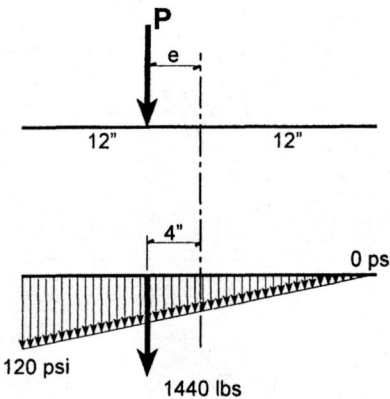

$$f = \frac{P}{A} \pm \frac{M}{S}$$

$$= \frac{1440 \text{ lb}}{24 \text{ in.}^2} \pm \frac{5760 \text{ lb-in.}}{96 \text{ in.}^3}$$

$$= 60 \text{ psi} \pm 60 \text{ psi}$$

$$= 120 \text{ psi } C \text{ and } 0 \text{ psi}$$

Figure 6-6.

Table 6-2. Loads and Eccentricities, Compressive Stresses: 12 in. Thick Wall

Conventional Analysis Zero Tension and 25 psi Tension Maximums									Masonry Society Method Cracked Section Analysis		

Wall Thickness = 12 in. Moment of Inertia = 144 in.^{4}

$$f_b = \frac{\text{Moment lb-in.}}{24} \text{ for 12 in. wall}$$

$$n = 3\left(\frac{t}{2} - e\right)$$

$$f'_c = \frac{2P}{bn}$$

t (in.)	f_{co} (psi)	P (lb)	e (in.)	M (lb-in.)	Z (in.)	f_b (psi)	f_c (psi)	f_t (psi)	e (in.)	n (in.)	f'_c (psi)
12	40	480	0	0	0	0	40	0	0	0	40
12	40	480	1	480	18	20	60	−20	1	15	60
12	40	480	2	960	12	40	80	0	2	12	80
12	40	480	3	1,440	10	60	100	20	3	9	107
12	40	480	3.3	1,560	9.7	65	105	25	3.25	8.25	116
12	40	480	4	1,920	9	80	120	40	4	6	160
12	40	480	5	2,400	8.4	100	140	60	5	3	320
12	50	600	0	0	0	0	50	0	0	0	50
12	50	600	1	600	18	25	75	−25	1	15	75
12	50	600	2	1,200	12	50	100	20	2	12	100
12	50	600	3	1,800	10	75	125	25	3	9	133
12	50	600	4	2,400	9	100	150	50	4	6	200
12	50	600	5	3,000	8.4	125	175	75	5	3	400
12	60	720	0	0	0	0	60	0	0	0	60
12	60	720	1	720	18	30	90	−30	1	15	90
12	60	720	2	1,440	12	60	120	0	2	12	120
12	60	720	2.8	2,040	10	85	145	25	2.83	9.5	152
12	60	720	3	2,160	10	90	150	30	3	9	160
12	60	720	4	2,880	9	120	180	60	4	6	240
12	60	720	5	3,600	8.4	150	210	90	5	3	480

468 ◇ Structural Analysis of Historic Buildings

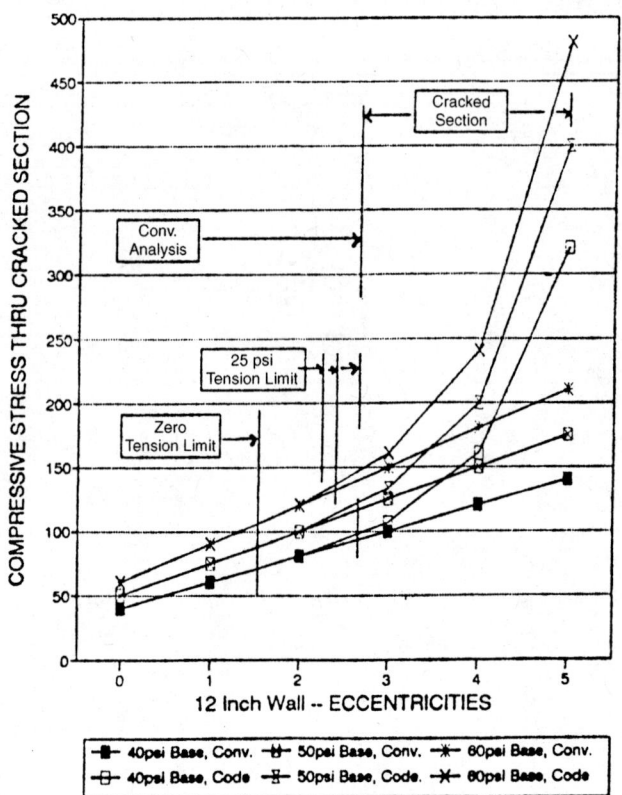

Figure 6-7. Twelve in. wall eccentricities: conventional analysis vs. Masonry Soceity method.

Table 6-3. Loads and Eccentricities, Compressive Stresses: 16 in. Thick Wall

Conventional Analysis Zero Tension and 25 psi Tension Maximums									Masonry Society Method Cracked Section Analysis		
Wall Thickness = 16 in. Moment of Inertia = 341 in.⁴									$n = 3(t/2 \cdot e)$		
f_b = Moment lb · in./42.6667 (for 16 in. wall)									$f'_c = (2P)/(bn)$		
t (in.)	f_{co} (psi)	P (lb)	e (in.)	M (lb-in.)	Z (in.)	f_b (psi)	f_c (psi)	f_t (psi)	e (in.)	n (in.)	f'_c (psi)
16	40	640	0	0	0	0	40	0	0	16	40
16	40	640	1	640	29.3	15	55	−25	1	16	55
16	40	640	2	1,280	18.7	30	70	−10	2	16	70
16	40	640	3	1,920	15.1	45	85	5	3	15	85.33
16	40	640	4	2,560	13.3	60	100	20	4	12	108.7
16	40	640	4.33	2,773	12.9	65	105	25	4.33	11	116.4
16	40	640	5	3,200	12.3	75	115	35	5	9	142.2
16	40	640	6	3,840	11.6	90	130	50	6	6	213.3
16	40	640	7	4,480	11	105	145	65	7	3	426.7
16	50	800	0	0	0	0	50	0	0	0	50
16	50	800	1	800	29.3	18.7	68.7	−31	1	0	68.7
16	50	800	2	1,600	10.7	37.5	87.5	−13	2	0	87.5
16	50	800	3	2,400	15.1	56.2	106	6.25	3	15	106.7
16	50	800	4	3,200	13.3	75	125	25	4	12	133.3
16	50	800	5	4,000	12.3	93.7	144	43.7	5	9	177.8
16	50	800	6	4,800	11.6	112	162	62.5	6	6	266.7
16	50	800	7	5,600	11	131	181	81.2	7	3	533.3
16	60	960	0	0	0	0	60	0	0	0	60
16	60	960	1	962	29.3	22.5	62.5	−38	1	21	82.5
16	60	960	2	1,920	16.7	45	105	−15	2	18	105
16	60	960	3	2,660	15.1	67.5	127	7.5	3	15	128
16	60	960	2.79	3,627	13.7	85	145	25	3.78	12.66	151.7
16	60	960	4	3,840	13.3	90	150	30	4	12	160
16	60	960	5	4,800	12.3	112	172	52.5	5	9	213.3
16	60	960	6	5,760	11.6	135	195	75	6	6	320
16	60	960	7	6,720	11	157	217	97.5	7	3	640

Masony Society Method—Cracked Section

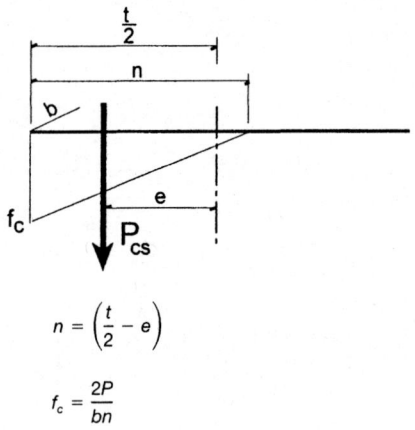

$$n = \left(\frac{t}{2} - e\right)$$

$$f_c = \frac{2P}{bn}$$

Figure 6-8.

lengths that produce conditions of one-way or two-way bending and methods of determination of shear resistance in cross-walls are identifiable and need to be checked in accordance with the methods shown in Chapter 5. Brick masonry construction that can be identified or closely associated with a specific date can be safely assumed to have the following allowable stresses:

Building, 1820–1860	75 psi allowable compressive stress.
Building, 1860–1890	100 psi allowable compressive stress.
Building, 1890–1940	150 to 300 psi allowable compressive stress (over period)

The following mortar mixes compare in strength and expansion characteristics to the period brick and should be used in pointing and relaying sections of historic brick

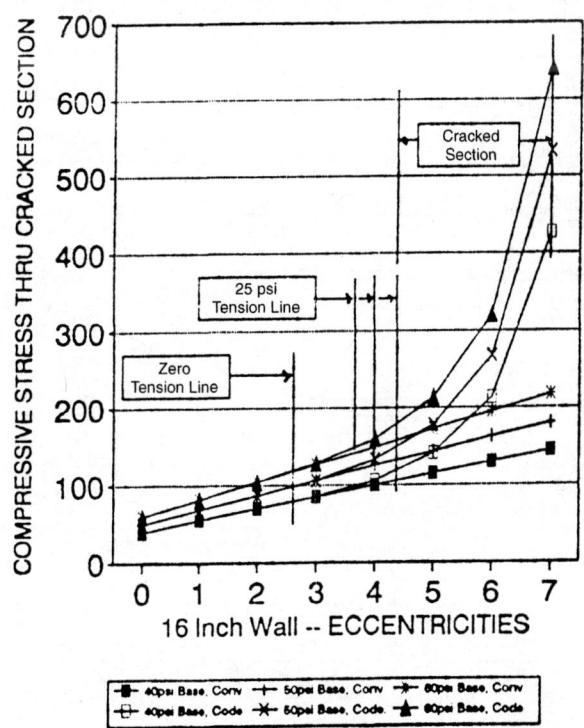

Figure 6-9. Sixteen in. wall eccentricities: Conventional analysis vs. Masonry Society method.

470 ◇ Structural Analysis of Historic Buildings

Table 6-4. Loads and Eccentricities, Compressive Stresses: 24 in. Thick Wall

Conventional Analysis									Masonry Society Method		
Zero Tension and 25 psi Tension Maximums									Cracked Section Analysis		

Wall Thickness = 24 in. Moment of Inertia = 1,152 in.^{4} $n = 3(t/2 \cdot e)$

f_b = Moment lb · in./96 (for 24 in. wall) $f'_c = (2P)/(bn)$

t (in.)	f_{co} (psi)	P (lb)	e (in.)	M (lb-in.)	Z (in.)	f_b (psi)	f_c (psi)	f_t (psi)	e (in.)	n (in.)	f'_c (psi)
24	40	960	0	0	0	0	40	0	0	0	40
24	40	960	2	1,920	24	20	60	−20	2	30	60
24	40	960	4	3,840	24	40	80	0	4	24	80
24	40	960	6	5,760	20	60	100	20	6	18	106.7
24	40	960	6.5	6,240	19.4	65	105	25	6.5	16.5	116.4
24	40	960	8	7,680	18	80	120	40	8	12	160
24	40	960	10	9,600	16.8	100	140	60	10	6	320
24	50	1,200	0	0	0	0	50	0	0	0	50
24	50	1,200	2	2,400		25	75	−25	2	24	75
24	50	1,200	4	4,800	24	50	100	0	4	24	100
24	50	1,200	6	7,200	20	75	125	25	6	18	133.3
24	50	1,200	8	9,600	18	100	150	50	8	12	200
24	50	1,200	10	12,000	16.8	125	175	75	10	6	400
24	60	1,440	0	0	0	0	60	0	0	0	60
24	60	1,440	2	2,880	24	30	90	−30	2	30	90
24	60	1,440	4	5,760	24	60	120	0	4	24	120
24	60	1,440	5.67	8,160	20.5	85	145	25	5.67	19	151.7
24	60	1,440	6	8,640	20	90	150	30	6	18	160
24	60	1,440	8	11,520	18	120	180	60	8	12	240
24	60	1,440	10	14,400	16.8	150	210	90	10	6	480

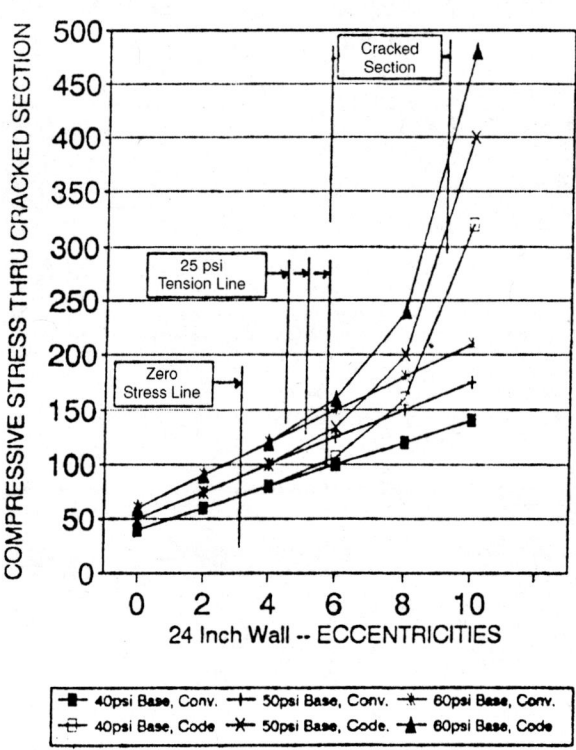

Figure 6-10. Twenty-four in. wall eccentricities: conventional analysis vs. Masonry Society method.

as required. Both the brick and mortar allowables are for in composition, in wall, allowables. Compatible mortars are as follows:

Historic mix: 1 part lime, 3 parts sand, 75 psi allowable compressive stress
Buildings, pre-1860: 1 part Portland, 4 parts lime, 15 parts sand, 75 psi allowable compressive stress
Buildings, 1860–1890: 1 part Portland, 2 parts lime, 15 parts sand, 100 psi allowable compressive stress
Buildings, 1890–1940: 1 part Portland, 2 parts lime, 12 parts sand, 150 psi allowable compressive stress
Type O mortar: 1 part Portland, 2 parts lime, 9 parts sand, 350 psi allowable compressive stress
Type N mortar: 1 part Portland, 1 parts lime, 6 parts sand, 750 psi. allowable compressive stress

This author proposes to evaluate brick or stone masonry walls utilizing the above allowables, gravity dead and live loads, eccentric loads (causing moment due to the location of joist loads, etc.), the cracked section, and lateral live load from wind in the method described in the conclusion to Chapter 2 and at this point allow the designer to see visually that the combined stresses are within the maximum stress envelope. Figure 6-12 shows this method of visual interpretation for the same wall as in the conclusion to Chapter 2, with the addition of the lateral live load. The diagrams in Figure 6-12 show, from left to right, the wall with its dead and live loads and displacement diagram as caused by the wind loads, the stress diagram produced by the wind loads, the stress diagram from the dead and live gravity loads, and finally the stress diagram, which is the product of the combination of all of the stresses on the wall.

As can be determined by inspection, the stresses on the wall are very low when looked at individually; however, the combined stresses are somewhat higher, with the compressive stress at the interior face of the wall reaching as high as 81.854 psi at the lower level. In some cases the combined stresses are complementary in that they actually lower the overall effect, and in some cases they are additive. In one instance the exterior, windward, wall reaches a level of tensile stress that is questionable. At that location, 5 ft below the top of the wall at the bearing point, the eccentricity of

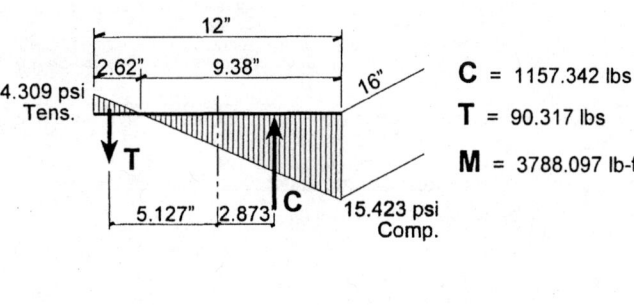

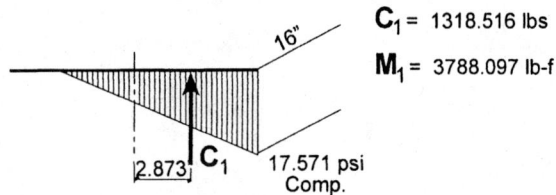

Figure 6-11. Stress diagram of wall at roof joist: standard section and cracked section.

472 ◇ Structural Analysis of Historic Buildings

	Wind Loading		Gravity: D. + L.		Gravity: D. Only		Combined: W + D + L		
	Outside Face psi	Inside Face psi	Outside Face psi	Inside Face psi	Outside Face psi	Inside Face psi	Outside Face psi	Inside Face psi	Adjusted Cracked Sec. Ins. F. psi
Roof	3.656 T	3.656 C	2.500 C	2.500 C	2.500 C	2.500 C	1.156 T	6.156 C	6.598 C*
Joist Brg.	3.600 T	3.600 C	2.667 T	9.333 C	1.333 C	5.333 C	6.267 T	12.933 C	17.659 C*
								13.435 C*	
Mid. Ht.	5.760 C	5.760 T	2.333 C	14.333 C	6.333 C	10.333 C	8.093 C	8.573 C	13.299 C*
3rd Flr	3.600 T	3.600 C	7.333 C	19.333 C	11.333 C	15.333 C	3.733 C	22.933 C	27.659 C*
Joist Brg.	2.880 C	2.880 C	3.367 C	31.967 C	8.967 C	19.633 C	6.247 C	34.847 C	41.687 C*
								39.573 C*	
Mid. Ht.	4.608 C	4.608 T	1.633 C	36.967 C	13.967 C	24.633 C	6.241 C	32.355 C	39.195 C*
2nd Flr	2.880 T	2.880 C	6.633 C	41.967 C	18.967 C	29.633 C	3.753 C	44.847 C	51.687 C*
Joist Brg.	5.120 T	5.120 C	4.074 C	54.594 C	16.600 C	33.932 C	9.194 T	59.714 C	69.372 C*
								66.554 C*	
Mid. Ht.	8.192 C	8.194 C	2.599 C	61.267 C	23.267 C	40.599 C	10.791 C	53.075 C	62.733 C*
1st Flr	5.120 C	5.120 C	9.265 C	67.932 C	29.933 C	44.265 C	4.145 C	73.052 C	82.710 C*

Gravity Load of D.L. + L.L. provides the highest stresses in the wall due to the floor loads on the joists which bear on the wall with an eccentricity of 4".

The Lateral Load on the Windward and Leeward Wall causes vertical reactions on the foundations which would decrease the Windward Wall Stresses and Increase the Leeward Wall Stresses by about 5 psi.

* **Note:** The adjusted Compressive Stress on the Inside Face of Wall is from the "Cracked Section" method.

Masonry Wall
Gravity and Wind Loading

Joists spaced at 16" on centers, analyze as 1.3333' long)
Joist Span = 24' (Net, Wall to Wall)
Roof: D.L. = 12 psf, L.L. = 24 psf
    Joist Reaction = 576 lbs
Floors: D.L. = 20 psf, L.L. = 50 psf
    Joist Reaction = 1120 lbs

Wall Weight = 120 lbs per cubic foot.

Windward Face of Wall: 0' to 30' --- 12psf(0.8) = 9.6 psf
                30' to 50' -- 15psf(0.8) = 12 psf
Leeward Face of Parapet: 30' to 50' -- 15psf(0.5) = 7.5 psf

Constants for 12" thick x 16" long wall

$I = \dfrac{16'' (12'')^3}{12} = 2304 \text{ in}^4$

$S = \dfrac{16'' (12'')^2}{6} = 384 \text{ in}^3$

$A = 12'' (16'') = 192 \text{ in}^2$

Maximum Stresses in the Brick Masonry are to be compared to the Allowable Stresses for the type of Brick and Mortar and the date of construction of the building. The building configuration, transverse walls, trussed partitions, presence of tension rods, and unbraced length of walls must be considered.

The ability of the floors to act as a diaphragm is an important issue when the designer allows the loads to be transferred to transverse shear walls.

Cracked Section Example:

Original Section - Joist Brg, 2nd Flr

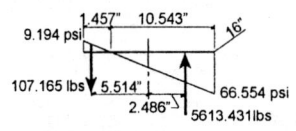

Mom. = 14,545.897 lb-in

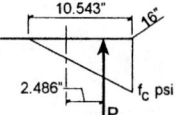

Setting the Cracked Section equal to the Moment of 14,545.897 lb-in

$\dfrac{f_c}{2} (10.543'')(16'')(2.486'') = 14{,}545.897$ lb-in

$f_c = 69.372$ psi

$P = 69.372$ psi $(0.5)(10.543'')(16'')$
  $= 5851.112$ lbs

$M = 5851.112$ lbs $(2.486$ in$)$
  $= 14{,}545.864$ lb-in (Confirmed)

Figure 6-12. Masonry wall stress analysis: wall with loads, lateral load and gravity load stresses and the combined stresses shown with the limiting stress envelopes.

the joist load and wind loading has brought about an excessive tensile stress. Reanalyzing the wall section to include the Rabun method of crack section analysis at that point produces a revised compressive stress of 17.593 psi at the inside face of the wall at that point. This additional stress is no problem for the inside face of the wall, and the section that has cracked is only 2.62 in. into the wall, as shown in Figure 6-11.

Stone

Stone load-bearing masonry buildings represent only a small fraction of the existing building inventory in the United States. This is unfortunate, as America has some areas with outstanding stone quality. In some areas indigenous construction of homesteads of stone occurred because it was the material at hand. These buildings were not refined, nor did they evolve in refinement. Most of the stone load-bearing structures in America are government and religious buildings. Stone may be the most permanent natural building material utilized by civilizations throughout the world, but it is labor intensive and in the capitalistic system it has all but been eliminated. Stone is still utilized in America as a veneer, and even that type of construction requires more labor than is felt desirable. Remarkably, most polished stones utilized in America in curtain wall construction are imported from Italy due to the cost of quarrying and polishing.

The same tests that apply to brick masonry can be used effectively on stone masonry. The ultrasonic method has specific uses in checking for delamination of sedimentary stones, particularly stones that have been laid with their bedding planes vertical. Deterioration in stone should also be detected through inconsistent readings in the ultrasonic method. Core sampling can be utilized to determine compression capabilities, and the walls can be analyzed in the same manner as brick masonry. Cracked section analysis would also be good for stone load-bearing construction.

The effects of the environment upon the stone are not to be ignored. The twentieth century and all of the environmental problems we have generated have aided in the deterioration of stone that is susceptible to acids and other deteriorating factors. The natural binders in many stones begin to break down with simple exposure to the quality of air that exists today. Conditions in the eighteenth and nineteenth centuries were also detrimental to the stone due to the use of high-sulfur-content coal to heat almost every building in the United States during this period; this practice continued well into the twentieth century. The natural porosity of stones, even polished stones, allows moisture to carry pollutants into the area behind the surface of the stone. These pollutants are deposited in the pores of the stone, usually only a few millimeters below the surface. As the moisture evaporates, the pollutants are left behind, and as these deposits grow through repeated cycles of deposit and evaporation, the dynamic balance is disturbed. The presence of the deposits retards the free and natural movement of moisture by not allowing moisture to move to the surface. As evaporation occurs, a dry area just behind the surface is created. The expansion of the deposits as they crystallize creates spalling and destruction of the surface of the stone. As one surface is destroyed and a new surface from behind is exposed, the cycle repeats. Silicone treatment to seal the porosity of the stone is conceptually a good idea, but the natural breathing process of the stone is interrupted by the sealants and the moisture content in the stone cannot maintain its balance via free water and water vapor. The presence of moisture in stone at depths beyond natural surface breathing can be determined by the neutron moisture probe, which measures electrical conductivity in the stone without any detrimental effect. An infrared scanner can also be utilized to identify moisture-laden areas through greater heat conduction.

Water is always seen as the trigger that starts the cycle of decay in stone, the breakdown of chemical binders. Even changes in electrical and magnetic properties are activated by moisture. Moisture attacks in three basic forms: vapor, liquid, and ice. The chemical breakdown of rock is very slow in some cases and relatively fast in others. Porosity and the size of pores, chemical composition, availability of pollutants, temperature, and humidity all contribute to the inevitable weathering process.

Good building stone as selected by master masons and quarried by trained technicians should last indefinitely when utilized properly in building construction. The subsequent exposure to air and moisture begins a process of decay that may be relatively short when accelerated by pollutants and the formation of weak acids when mixed with moisture. The ultimate strength of stone is determined by unconfined compressive tests on samples sawn from quarry blocks (ASTM C-170).

Finally, as in brick masonry construction, visual inspection to determine the actual condition of the stone in a wall may be the most valuable determinant. Stone that is weathering badly will be obvious, as will stone that has held up well during its exposure to the elements. If the building has been uniformly and adequately maintained and the stone appears to be sound, the designer can assign a high compressive stress allowable for walls constructed with lime-sand mortar and dimension stone. If the joints appear sound and the mortar is consistent and shows no evidence of weathering or breaking down, the wall may be capable of supporting loads at or above the original design values. Rusticated exterior surfaces are able to protect joints in dimension laid stonework, and the irregular surfaces that are exposed to the atmosphere can also weather at a much slower pace than smooth face stones. Irregular stones laid in a coursed rubble pattern will have more irregular joints, and the thickness of the mortar will vary greatly. Joints in these walls are more succeptible to weathering due to the exposure of the wider, less protected areas of the mortar.

Engineering judgment can be relied upon when the stone and joints are sound and there is no reason to question the condition of the stone. Stone that is in an advanced state of decay due to weathering and acid attack via pollutants and moisture is obvious to the experienced designer. The need for testing should be determined by an experienced designer. Chemical cleaning of the stone is almost always wanted by clients to change the image of the building. Care must be taken and the consultation of experts is needed to prevent damaging the stone by harsh chemicals. Both testing and cleaning are expensive and should be required only if there is sufficient justification.

Brick masonry walls require the same careful consideration in regard to chemical cleaning. Many patented processes are harsh but may be safe when used in diluted form. Water used in cleaning and rinsing does not damage the wall when workmen use normal wide-angle spray tips and low pressures. Too many cleaning contractors use too strong a solution and run their water pressures at very high levels. A safe, nondestructive water pressure is usually 80–100 psi with a wide-angle tip. When the water pressure is 1000–1500 psi, damage is imminent; lime-sand mortar can be washed completely out of joints. Sand blasting can destroy layers of stone and the hard or vitrified surface of brick can be lost, exposing a softer interior portion of brick. At the least, the surface of stone that has been sandblasted will undergo a change of surface texture, and sometimes the character of a building will be radically changed. The first rule of cleaning may well be not to clean at all. The change in a building that was supposed to create a new image, attract new business development, or bring about higher rents may not achieve any of those desired results. The cleaning may damage the building to a degree that remedial work and even more expense may be required. Unless there is real justification for cleaning, it is best to follow the first rule.

ASSESSMENT OF METALS AND THEIR CONDITION

The use of metals in building construction is almost as old as building. The Greeks and Romans poured molten lead, copper, and brass at various times into slots and holes in tops of stones to tie two stones together in a type of early bonding between wythes of walls or to secure stones that were butted end to end. These tie bars would act in tension and may have done the job of keeping stones in alignment. Because neither material would be able to withstand much tensile force, they would hold alignment under static conditions while not offering any significant strength to prevent failure of the system. If a colonade with a layer of horizontal stonework above the arches began to fail through loss of a column or foundation support, the failure could not be prevented by any type of tie bar.

Metal in many different compositions was utilized for armament from the Dark Ages forward. The types and relative hardness of metals are not fully known, as examples of each type and era are not all available. Obtaining the heat of higher temperatures than could be formerly attained by blacksmithing was the main contribution of the early blast furnaces of the fifteenth century. Size restrictions limited the production of these furnaces, but they began the process of production of iron in the liquid state, which could be made into many shapes with sand molds. Cast iron utilization for structural purposes began in the eighteenth century with the work of Darby and others in England. Coke made from coal made the furnaces much more efficient, thus allowing production to increase tremendously. Cast iron contains 2–5 percent carbon and has as its properties a relatively high compressive stress, a natural resistance to rust (rusts only on surface and becomes a protective coating), and a very low tensile capacity. The low tensile capacity made the material difficult to use as beams because the brittleness of the material made failures sudden and without warning. Some beams that were utilized in cast iron were very large and spanned very short distances to prevent excessive tension in the bending section. The ability to be cast with relative material uniformity and in complex shapes made cast iron a natural material for the types and sizes of columns that were needed in the nineteenth century. Thus many thousands of buildings were constructed of masonry exterior bearing walls with cast iron columns along interior bays. At the climax of the cast iron period, total building fronts to six stories in height were of cast iron.

Wrought iron is a product in which the carbon content is extremely low, usually between 0.1 and 0.5 percent. Wrought iron is very malleable and was sometimes called malleable iron. It was made in a puddling process where the carbon was burned out of the iron by a process of heating with an oxygen-rich flame. The additional oxygen burned away the carbon while the men worked the iron; thus the term *puddling*. The manufacture of wrought iron generally produced plates, angles, and a few channels and I-beams. The channel and I-beam shapes were used individually and with plate and angle shapes to form some of the built-up members, such as columns and girders. These members would be held together by rivets as would field connections. In addition, during the process the wrought iron was steam hammered and layered and rehammered to remove encrustations of slag and other impurities. The final sheets of wrought iron would actually be laminations, which may be visible under a magnifying glass, especially when deterioration has begun to occur. This can be utilized as a method of identifying wrought iron.

The Bessemer furnace, invented in the mid-1850s, produced early structural steels sometimes called mild steel. Steel has a carbon content of approximately 1.2 percent, depending upon the grade. The Bessemer process used a conical designed furnace and compressed air injection into the molten iron to reduce the carbon to a certain level. The process had the advantage of being almost continuous. Material that was ready would be poured out of the funnel by tipping the conical furnace, and a new batch of iron ore would be added to the molten steel remaining in the furnace. The amount of material poured out of the furnace would be approximately the top one-third with each pour. The operator made the pour when the heat had reached a level where the molten material was of a specific hue. The molten steel was poured into molds or ingots, and then, while still above a certain temperature, it was shaped by being forced through rolling mills. William Siemens began to obtain results in a more efficient system in 1861 with his gas-burning open-hearth process, which enabled operators to produce a steel with more precise control of chemical and physical properties. See Table 6-5 for a chart of the chemical properties of cast iron, wrought iron, and steel.

By the early 1890s, steel shapes had practically eliminated wrought iron shapes in America. The ease of hot-rolling structural steels compared to the rerolling process of welding smaller wrought iron bars together to make shapes speeded the change. Steel manufacture has continued to evolve through improvements to the furnace processes and the modern types of rolling mills. Steel shapes made by the hot-rolling process are almost like an extrusion. Properties are consistent throughout, and there are no actual joints in the sections that could delaminate as a product of deterioration or oxidation. The compressive force of the rolling mill can be varied, and certain alloy steels obtain high strength characteristics from the application of great pressures against the steel during the rolling operation.

Table 6-5. Chemical Properties of Iron and Steel[a]

Name	Iron	PERCENT. OF				
		Carbon	Manganese	Sulphur	Phosphorus	Silicon
Pig Iron	91 —94	3.50—4.50	.50—2.50	.018—.100	.030—1.00	.25—3.50
Plain Steel	98.1—99.5	.07—1.30	.30—1.00 (.03—.10 as cast)	.020—.060 (.120)	.002—.100	.005— .50
Wrought Iron	99.0—99.8	.05— .25	.01—.10	.010—.100	.050—.20	.02— .20

[a] From American Institute of Steel Construction 1930, 55, 57.

Steel shapes are joined or attached by the use of rivets, bolts, or welding. Historic structures do not contain welded connections, as welding is a modern method (1930s to present). It is interesting to note that welding has now replaced riveting and even bolting in the fabrication process. Beams and columns or bar joists rarely leave a fabrication shop without containing extensive welding in critical areas, such as connections. The modern welded frame will contain nearly 100 percent welded connections, both shop and field. The recent earthquakes in southern California caused many thousands of weld failures, which has brought questions about the entire process to the forefront. All new construction in California that includes welded frames has been put on indefinite hold. Developers interested in moving their projects along are forced to return to the braced frame. Welding methods will be reevaluated, especially procedures for certification of field welders. We now seem to be on the threshold of a new significant development in the structural design process.

TEST METHODS FOR METAL

The identification of cast iron as the material in question is usually rather easy. The etched surface from the sand molds and the radii of fillets are initial indicators. Beams of cast iron will usually have unequal flanges because the area of the tension flange will need to be larger to increase the capacity of the beam by lowering the stresses in the tension flange. If the product is in situ in a building, the spans will be no more than 10–12 ft.

Wrought iron and steel can be identified by chemical analysis of small samples taken from drilling one-quarter to one-half in. cores through the material. Even the metal shavings from the drilling may be enough for determination. Drilling is preferred to any type of flame cutting because the heat of the material being cut would alter the properties in the zone of the cutting. Beckman states that for early steel a full chemical analysis should be done because chemical control was more difficult with earlier processes. Some chemical impurities, such as sulphur and phosphorus, may cause loss of strength with time if they are present in too large a percentage (Beckmann 1994, 181).

Several methods of nondestructive evaluation are available to determine, among other things, decay, cracks, and flaws and to some extent strengths of metals in existing buildings. A summary of the method, the application, and the results and limitations as outlined by Wilson is given in Table 6-6.

Accessibility of the iron or steel sections and connections is very difficult in many instances, especially in structures from the late 1880s forward where masonry or terra cotta curtain wall systems were employed. Very often these buildings would utilize clay tile inner walls with the three-coat plaster system, the result being masonry on both sides of the iron or steel framing. In addition, columns, bracing in interior bays, and floor framing may also be covered with masonry for fireproofing and may be inaccessible. The chosen test method may be nondestructive, but the required destruction to get to the framing may be sizable. How many tests and where are always a concern. Every member is essential under certain circumstances or else it would not be in the structure. Testing every member or connection in a structure is not necessary. Some preliminary assessment of the system, loads and member sizes, end

Table 6-6. Methods of Nondestructive Evaluation[a]

Visual-optical	Borescopes Fiber optics Panoramic camera	Defects at surface, cracks, corrusion, pits, etc. Limited but can be utilized in places where only the instrument can get. Not too expensive.
Liquid penetrant	Liquid penetrant containing dye	Penetrant drawn into defects by capillary action. Used on nonmagnetic metals only. Inexpensive but requires cleaning of test specimen, emulsion, drying, temperature control, residence time for developing powder. Inexpensive but time consuming, surface examination and cracks that exist at surface.
Ultrasonic	Vibrations above 20,000 Hz	Will detect surface and subsurface defects. Accurately measures thickness. Detects voids, cracks, porosity, segregated inclusions, welding flaws, etc. Transmitter and receiver must be on opposite side of test specimen. Pulse-echo method works by reflection of pulse on the same side as transmitted but not as accurate, can be used if only one side is accessible.
Magnetic particles	Magnetic particles are attracted to magnetic lines of leakage force	Detects cracks, seams, laps, voids, porosity, and inclusions. Senses flaws down to one-fourth in. below surface.
X-ray, gamma ray	X-rays	Detects both internal and external flaws. Voids or low-density areas show up on film. Can detect flaws or stress fatigue in welds as well as metal. Costly. Dangerous.
Eddy current	Electrical conductivity	Detects discontinuities, cracks, seams, and variations in allow. Qualitative comparisons only.
Test specimen	Removal of sample	Stress-strain data, yield point, ultimate strength, tension or compression. Sample must be large enough to provide usable data. Too large to take out of existing member without leaving the member in distress.

[a] From Wilson 1984, 193–195.

reactions, and connection capacities will enable the experienced designer to pinpoint certain typical locations in the structure to expose and test. Inspection of the building's enclosure system, with specific attention to evidence of water infiltration through walls and leaks in the roof, may indicate specific areas in need of visual inspection, evaluation, and possibly testing, if suspect.

Masonry and terra cotta curtain wall systems of the 1890s–1930s are very unlike the modern curtain wall system. These systems are much more permanent in that they cannot be removed and reattached. Many of the hooks and bars involved in the terra cotta hanging system have seriously deteriorated on many buildings and are located in a position deeply embedded in the masonry curtain wall system. Terra cotta sections are hollow or have open ends, and cracks can allow water to infiltrate. Solid masonry systems have ties and hooks between wythes but are not as subject to oxidation as terra cotta. Material evaluation and testing may be required for certain structures for a number of reasons. A change in use or an adaptive reuse that changes the live load requirement of the structure may necessitate an extensive examination and testing of structural framing and hanging or curtain wall support components as a part of the certification process. The serious implications and legal liabilities associated with such assessments necessitate a systematic and cautious approach to the evaluation process.

Masonry load-bearing structures built prior to the 1880s may have cast iron columns (in fact, this can almost be expected if the columns are not wood), and in most cases the columns will not be protected or covered by masonry. Identification of the material in the columns should be relatively easy. Cast iron is usually identifiable by surface texture and architectural fenestration. If the cast iron column is visibly in good condition, with no cracks or flaws, it is probably capable of carrying its rated loads. Consistent thickness around the perimeter should be verified. The column is cast lying horizontal, and there may be a casting problem of nonconcentric interior and exterior surfaces (see Figure 6-13). If the wall thickness is irregular or there is a thin edge or section of the wall the column will have eccentric characteristics and will not take on

Figure 6-13. Casting problems: cast iron columns.

load evenly across its cross-section. This is an inherent weakness that will seriously weaken the column. A slight inconsistency in wall thickness along a portion of the column would be acceptable provided the variation in thickness is not more than 10 percent. Some H-shapes and I-shapes, channels, and a few solid round shapes, if unprotected, are easy to inspect visually. In the early 1880s, the average allowable compressive stress for cast iron columns varied from 5,600–8,000 psi (each would be reduced for longer column lengths), depending upon the source (see Chapter 3).

Wrought iron structural shapes were manufactured in the United States as early as the 1870s; however, only a couple of producers were making limited numbers and types of sections. The 1880s to the very early 1900s were the years of peak production and use of wrought iron. Chapter 3 records data from wrought iron sources in a summary of the unit stresses recommended by manufacturers. The average of these unit stresses was 10,000 psi as the base with stress reductions for columns of scheduled lengths and end restraint conditions. Beams of wrought iron shapes would have the same unit stresses, but formulae that took into account the lateral bracing system as a function of the span would reduce the unit stresses to a lower allowable stress for flexural members.

Structural steels began to be produced in the late 1880s with basic unit stresses averaging just under 16,000 psi. A vast array of structural steels were produced by a number of manufacturers over the next 50 years, as outlined in Chapter 5. Each producer provided its own design criteria and formulae until the industry began to standardize in the 1930s. ASTM criteria and standardization of methods of testing began at the turn of the century; however, manufacturers were still attempting to take shares of the market by featuring distinct steel qualities, alloys, and individual shapes in their attempts to supersede wrought iron.

Wilson (1984, 192) issues guidelines for use in structural assemblies of certain dates. (See Table 6-7.) Wilson states that these values may be used when the characteristics of the steel are unknown. Manufacturers of the early periods represented by the chart recommended using working stresses (allowable stresses) of 50 percent for steel after 1963, allowable stresses are 24,000 psi. This is reasonable and relatively correct, provided the material in situ has not been damaged due to deterioration and or oxidation.

Chapters 3 and 4 provide a more detailed tabulation of wrought iron and structural steel stresses. The data present allowable, yield, and ultimate stresses for wrought iron and most grades of structural steel by date and in many examples by manufacturer. The material that follows is an expansion of written documentation on the

Table 6-7.

	Yield Point (psi)
Wrought iron members	25,000
Structural steel, constructed prior to 1905	25,000
Structural steel, constructed 1905–1932	30,000
Structural steel, constructed 1933–1963	33,000
Structural steel, constructed after 1963	36,000

capabilities of the testing methods that are the most appropriate for the types of metals used in American construction.

> *Ultrasonic pulse testing:* Ultrasonic testing for flaws in metals, and more specifically cracks in welds, has been done for years on pressure vessels, atomic-related equipment, and welds of structural elements. Ultrasonics has been utilized on stress analysis or, more precisely, comparative data between sound material and material with flaws. If a wave from an ultrasonic transmitter encounters a discontinuity in the test specimen, a crack or void, it will be diffracted around the flaw and back on course. This additional time of travel is measurable, and the machine can locate and plot an image of the flaw. Variations in thickness, laminations that are separating, and deterioration would all show up as irregularities on the cathode ray tube. The method is accurate and as sensitive as the operator desires. Operator experience is very important to these tests.
>
> *X-ray, Gamma ray testing:* X-ray evaluation methods are used in metals to make comparative observation of densities, chemical composition, and thickness of a test specimen. The ultrasonic method and the X-ray method produce similar results in metals of the type used in building structural systems. X-ray testing is much more expensive than ultrasonic testing, and the equipment is bulky, and dangerous unless operated properly. The camera is required to be on one side of the specimen and the film on the other side. The one advantage is a film record of the results of the test. Another disadvantage is the number of tests to be made to yield proper data. The quantity of tests to be made by ultrasonic and X ray are not that different; neither method covers more than a few square inches of material per test.
>
> *Eddy current testing:* The eddy current system of testing metals is utilized to locate the same types of discontinuities as the ultrasonic and the X-ray methods. The tests measure the changes in the impedance of a coil due to changes in the flow of eddy currents in a test specimen. The tests can be used to measure thickness, flaw detection, hardness, grain size, and several other factors. The system is usually effective only to a depth of approximately one-eighth to one-fourth in. below the surface.
>
> *Magnetic particle testing:* Magnetic particle testing is easy, inexpensive, and fast and works well on in situ testing of surface flaws in metals. The method uses three different currents for different results. Alternating current (ac) can be utilized to detect surface discontinuities. It is good for locating fatigue cracks in structural members. Direct current (dc) can detect both surface and subsurface discontinuities. It is best for subsurface investigations. Half-wave rectification works best for subsurface investigations.

The methods described above provide little more than locations of discontinuities or flaws in metal components. Very little quantitative data is gained from these tests. They merely identify the type, size, and length of cracks or flaws in the metal. In many instances that is all one requires of a test, such as in reactor vessels, boilers, or pressure piping. A single flaw in an important metal forging can make the entire piece unusable. However, structural components in buildings are critical in certain other ways, or the information is sought for a specific reason. We seek to investigate the soundness of a member as well as verify its ability to carry load. A crack at the surface or even below would be important to know about; however, we usually look for deterioration, oxidation, cracks in welds, and even fatigue cracks in certain circumstances. As in masonry, the costs versus the results in the nondestructive methods may prohibit the use of extensive testing on an existing building. The importance of the building and the adaptive use requirements may be the determinant in the decision to test structural components in a given building. Wrought irons and structural steels are supposed to be uniform in material composition by nature, and the process of heating the iron ore to the liquid state, maintaining a temperature for a period of time, and pouring off the compound in such a manner as to not allow surface impurities or slag to enter the bloom or the ingot insure a quality shape. If the building can be dated and there is evidence that it has not gone through any periods of sustained water intrusion or other type of physical abuse, the building may not require testing—especially if the period of the original design and construction would have been one of conservative design. The conservativeness could have resulted from the assumed live loads, the calculation of the dead loads, or the ratio of the ultimate stress of structural components to the allowable stress. Sound engineering judgment coupled with experience may be the best determinant in deciding the need for testing.

ASSESSMENT OF CONCRETE AND ITS CONDITION

Modern structural concrete is almost a twentieth century product. Concrete is a proportionate mix of cement, sand, and gravel (or fine and coarse aggregates). When it is mixed thoroughly and water is added, a process of setting called *dehydration* begins. In the early days, concrete was used on foundations, it being a logical experimental replacement for beton, which had been in use in Europe and America for a number of years. Beton was basically the same type of mixture, with lime instead of cement, and was a extension of lime-sand mortar. Beton (a French name for cement) was in use in France for most of the nineteenth century. These cements were of inconsistent quality and were unreinforced. They depended upon the bulk of massive size (lime, sand, small aggregates, and bricks and rock) to develop sufficient compressive strength to make adequate foundations for buildings, bridges, and other structures. Modern reinforced concrete construction and the development of a properly fired limestone cement (called portland cement for its color) started in the United States in about 1904 with the construction of the first reinforced concrete framed structure, the Ingals Building in Cincinnati, Ohio.

At first, even reinforced concrete in the United States seemed to be accepted only for foundations and foundation systems. The grillage footing, the predecessor of the spread footing, was first conceived by designers who recognized that wood grillages did something beyond spreading the load from a wall to a larger area of soil. The grillage foundation, which combined railroad rails and beton or concrete, carried its load by a combination of soil pressure and flexural bending in the footing.

Many concrete pioneers, such as Ransome, Ward, Hyatt, Hennebique, and Thacher, worked on concrete and reinforced concrete systems for 30 to 40 years before the turn of the century. There were concrete arched bridges in the United States, both reinforced and unreinforced, by the 1890s, and a one-story frame warehouse at Port Chester, New York, in 1875. All of the early reinforced concrete structures, beams, and columns were constructed with patented reinforcing systems developed experimentally by those pioneers. Many of the proprietary systems made their way into early reinforced concrete buildings before individual tied bars were introduced and accepted. A prominent designer would be engaged to construct a building or structure, and his shapes (often experimental shapes) would be utilized in the structure. There were more than a dozen types of individual bars, square, round, square twisted, lug bars, and several types of deformed shapes, and trussed bars of almost as many types and grades of steel as there were inventors by the year 1900. The "billet" bar, or deformed bar as we know it today, was adopted as a standard specification by 1911. There were at least three concrete specifying agencies by 1915 setting specifications and standards for construction.

The problem we face in the analysis of historic concrete structures is that the concrete, by its own nature and the nature of the technology, completely encases the steel reinforcement. One cannot look at a reinforced concrete beam and determine the type or placement of its reinforcing. We can date the structure and make intelligent guesses as to the location of the reinforcing, but we will not know for sure without chipping away concrete to expose reinforcement. This is destructive, and the degree of the destruction cannot help but significantly affect the structural ability of the beam or column in question. In order to perform an analytical evaluation on a reinforced concrete structure, we must obtain the size, dimensions, and information about the strength of the reinforcing. To date there are only a couple of methods to evaluate or identify reinforcing that is embedded within the volume of concrete members and slabs.

The strength of the concrete is also a variable that is dependent upon not only the components—cement, sand, and stone—but the proportions of the mix. The quality and amount of water added to the mix is of critical significance. Early users of this new technology were experimenting with mixes and reinforcing and developing their proprietary systems. Not until 1928 did concrete construction begin to be standardized and codes address concrete in a uniform, organized manner.

TEST METHODS FOR CONCRETE

Wilson (1894) presents a very good table of concrete test methods and their expected results (see Figure 6-14). A number of evaluation methods for concrete and reinforced concrete have been developed for the nuclear power industry, pavements and bridge construction, and other nonbuilding applications. We in the building industry are fortunate to have a number of tests whose primary function is other than for building evaluation that can be utilized to provide data for our use. This work will look at nondestructive and semidestructive evaluation methods that can be of benefit in evaluating existing structural concrete in buildings. Engineers in this business can benefit from the results and potentially save owners significant sums of money. Wilson outlines a number of methods of nondestructive evaluation and semidestructive evaluation of structural concrete, as shown in Table 6-8.

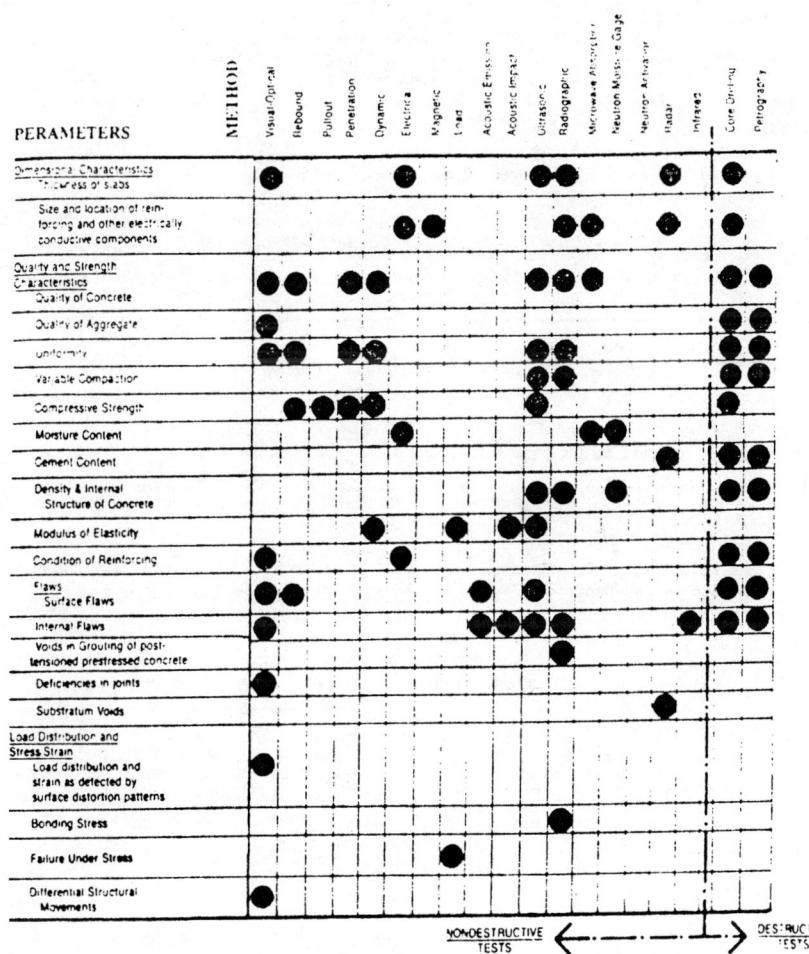

Figure 6-14. Concrete test methods. (*From Wilson 1984, 231.*)

Table 6-8. Methods of Nondestructive and Semidestructive Evaluation of Structural Concrete[a]

Visual/optical		
Visual	Magnifying glass Monocular lens	Detection of cracks 2–3 microns wide. In situ concrete surfaces. Surface information only. Can yield valuable information.
Surveying vertical and horizontal movement	"Tell-tale" measuring instruments	Long-term observations to observe critical movements in concrete structures. Cyclical observations.
Joint survey	Manual measuring calipers, etc.	Check for expansion-contraction. Joint fillers and chemical attack.
Fiber optics	Working length 30–1,330 mm	Detects internal cracks, voids, or flaws if path to surface exists. Clear, high-resolution images of remote inspection subjects. Use at cracks or areas where cores have been taken. Can make boreholes to extend investigation.
Rebound/penetration		
Schmidt rebound hammer	Surface hardness Strength of concrete	Concrete strength tested by rebound distance correlation to calibration curve. Results affected by surface condition of concrete. Determine scale for hammer readings.
Windsor probe	Depth penetration Strength correlation	Difficult to correlate between penetration measurement and actual strength of concrete. Requires multiple tests to determine scale for penetration depth.
Electrical/magnetic		
Dielectric	Tests for moisture content only	Can determine the moisture content to ±0.25% accuracy. Equipment is expensive and limited to one use.
Electrical resistivity	Tests thickness Rebar location	Limited to testing of pavements and on-grade slabs. Affected by air-entrainment, moisture, density, salts.
Magnetic Cover Meters Pachometers	Detects rebar location and depth of bars	Can detect rebar only to a depth of 7 to 8 inches within Conc. Confused by multiple layers of rebar. Does not work with welded wire fabric.
Acoustical		
Acoustic emission	Monitoring of high-frequency acoustic signals	Detection of growing internal cracks. Must be used when structure is being loaded. Cracks must be forming. Expensive and not too useful for structural purposes.
Acoustic impact	Impact energy to locate cracks and voids	Used mainly on slabs on grads. Equipment is expensive.
High-energy ultrasonics		
Ultrasonics	Evaluates thickness, quality, uniformity	Accurate measurements of concrete thickness is the only feature of this method to date. Both surfaces of concrete must be accessible.
Radiographs		
X-ray	Density and internal structure. Rebar location.	Hazardous testing method. Expensive. Good only for concrete slabs. Record of results on film. Both sides of concrete must be accessible.
Gamma ray	Location and condition of rebars, density voids, and thickness of concrete	Hazardous testing method. Expensive. Good for any in situ concrete structural components. Both sides of concrete must be accessible.
Microwave absorption		
Microwave	Measurement of moisture content	Can be used for moisture content only. Not very accurate. Both sides of concrete must be accessible.
Dynamic or vibration		
Ultrasonic pulse	Pulse velocity and resonate frequency	Measures travel time of pulses through concrete. Measures concrete uniformity. Requires field calibration. Measures modulus of elasticity, rigidity, and durability of concrete. Does not measure strength of concrete.
Nuclear		
Neutron scattering	Moisture content	Measurement of moisture content only through a measurement of decreased neutron energy. Expensive, calibration not yet conclusive.
Neutron activation analysis	Cement content in concrete	Expensive. Calibration not yet conclusive. Presently of little value to building assessment.

Table 6-8. *(Continued)*

Infrared			
	Infrared test	Passive heat patterns	Various heat patterns are identified with defects, voids, cracks, etc. Potentially accurate method of detecting concrete defects. Inexpensive, still being developed.
Load testing			
	Load tests	Application of design load	Test structure by virtue of loading to or above design load. Overloading can cause cracks or even failure; care must be taken not to load too high. High degree of reliability, moderately expensive, requires importing and removal of load system.
Radar			
	Radar	Detection of voids	Used on slabs; reinforcing confuses the instrument. Not useful on reinforced concrete structures. Can scan large areas. Potential, but not well enough worked out at present time.

[a] From Wilson 1984, 232–235.

Only a few of the methods of concrete testing outlined above are of value to the engineer in the assessment of the concrete and the reinforcing. The strength of the concrete needs to be known as do the type, quantity, and strength of the reinforcing. Concrete cracks (micro cracks through visible flexure cracks) are natural in the tension zone and can actually help in the analysis of the components of the structural system. Through the crack locations and the projection of the cracks (from the edge to approximately the point of zero flexural stress), we are able to identify tension and compression zones in the structural system. We can also ascertain other factors, such as the structure's reaction to lateral loading and differential settlement when one of the visual/optical methods is used. In many instances, such as with slabs and beams above grade, it is possible to obtain a physical measurement of the thickness, but without the original construction drawings it is not possible to determine the location and size of the reinforcing. If the building has performed through its life without visible damage from the loading, there may be no need to know all of the details concerning the reinforcing. A small number of load tests of certain critical beams, made by individually stacking bags of cement directly on top of the beam to achieve at first half the assumed design live load, then progressively adding load uniformly to the three-quarter point while observing the flexural action, cracking, strain gage movement, deflection, etc., will provide the experienced designer with an extensive amount of usable data. The cement bags are suggested because they weigh exactly 94 lb each. In most beams the location of the rebar is not altogether adequate for complete analysis of the beam. The area of the steel and the tensile strength need to be known for a complete analysis of an existing member, and to date the most accurate method of obtaining rebar size and location is through X-ray and gamma ray testing. It is expensive, the safety precautions are extensive and costly, and one or more pictures are required of several if not all of the members, columns, and beams for the capacity to be verified and thus the structure certified. A simple elevation or side view is not all that is required; the rebars are in alignment and some will be hidden behind the one closest to the surface. Therefore, photographs from the top and at least two from angles would be required in addition at each member not only to indicate the size of an individual bar but to enable the bars to be counted. This would also enable the designer to obtain required data on earlier proprietary systems, such as trussed bars.

Obtaining the strength of the concrete is a relatively simple matter, as the concrete can be bored and tested for compression and/or a form of tensile/compression failure. Drilled cores for determination of strength of concrete can be as small as 2 in. in diameter. These cores should be obtained by professional operators and taken with diamond bit core drilling saws as per ASTM C-41 requirements. If there is a question as to the strength, hardness, or general condition of the concrete, it may be necessary to have a laboratory run two additional tests: ASTM C-295 Petrographic Analysis of Aggregate for Concrete and ASTM C-856 Petrographic Analysis of Hardened Concrete. These tests will evaluate the original mix, cement, and aggregates and assist in determining the condition of the concrete. The number of tests to make and the location of the of places to sample should be decided in conjunction with a professional who has experience in this area.

Concrete or reinforced concrete is one of the youngest construction materials that we have to deal with. In many ways it is also one of the most difficult. In early days it was often job mixed (mixed in the field), and cement, sand, and aggregates are each a variable, with not only their mix proportions but their quality varying. The chemical composition of the cement, sand (fine aggregate), and stone (coarse aggregate) vary from region to region, and the activator in the process of hydration, the water, has highly varying chemical substances. Couple all of these variables with the fact that the concrete is cast into forms containing reinforcing that is supposed to end up at a specific place in the final product, and there is room for even the most optimistic assessor to look at the quality of structural concrete. The buildings use, its history of maintenance, and the condition of the surface of the concrete all have a bearing on the overall judgment of the condition of the concrete. Concrete is alkaline and has a natural tendency to protect its reinforcement. Tension and temperature cracks can allow moisture, and carbon dioxide produces carbonates that lower the alkalinity. Salts and sulfates attack the reinforcing, and the corrosion process begins. If the reinforcing is very near the surface, the expansion of the steel as it corrodes can crack and spall the concrete.

Reinforced concrete buildings are for obvious reasons the hardest of the older buildings to work with. While they may not be as old as many of the masonry, stone, or wooden structures in our inventory, they are the ones that are the most difficult to assess. The tests that are most usable for assessment of concrete structures are explained in more detail as follows.

Visual/Optical Inspection

In most instances, hairline cracks or cracks so small that they cannot be seen without a magnifying optical instrument are not detrimental in an ordinary building structure. Normally even cracks that are visible to the naked eye may not be letting moisture into the section and thus corroding reinforcing bars. This is true especially in the case of interior or protected structural elements. The opportunity to use optical instruments (5–10 = power magnification) may allow the designer to learn a significant amount about a structures reaction to load. Through the location of tension zones in the concrete, the designer may utilize optical instrumentation to predict tension zones and make informed choices as to where to perform other, more costly tests.

Surveying: Horizontal and Vertical Movement

Important buildings that show indications of movement through cracking patterns may need to be monitored over a sufficient time period to allow several cycles of movement to occur. This may require that observations be taken over a period of two or more years. It is not best for the structure or the ideal design environment to rush into repairs on a building of any type. A building in the monument classification should be handled with extreme care. Immediate decisions and remedial work that is not carefully considered may do more harm than good and also be a waste of money. Tell-tale instruments mounted on opposite sides of cracks or suspected movement locations can give valuable information as to movements of a structure. They will enable designers to make history plots of the movements and determine if the hysterisis is from normal movement or if an actual structural deficiency exists. Expansion joint surveys may also reveal abnormal movements in a structure.

Rebound Test

Utilization of the Schmidt rebound hammer can determine the compressive strength of existing concrete in situ. The hammer is placed against the surface of the concrete member and a plunger strikes the concrete surface. The amount of rebound is measured by a scale on the instrument. The scale is calibrated to give an instant reading of the compressive strength of the existing concrete in the given member. The user is cautioned to respect the possible inaccuracies of this method. Data from the hammer test should be correlated with the results of core sampling. The quantity of core samples required to analyze the structure may can be reduced when core sampling and the rebound method are used together. ASTM C-805 Test for Rebound Number of Hardened Concrete provides method for the rebound test.

Windsor Probe Device

This test method provides indications of the compressive strength of existing concrete. The test consists of a type of gun that drives a high-strength steel probe into the existing concrete via a charge of powder. The depth of the penetration, determined by measuring the length of the probe remaining outside of the concrete, is correlated with charts for the compression strength of the concrete. ASTM C-803 Test for Penetration Resistance of Hardened Concrete provides data for this type of test. One caution exists: the probe test is said to give higher strength readings than actually exist on concrete 40–50 years old. This method should also be correlated with core sampling methods.

Electrical Resistivity Test

To date, this method has been used successfully in determining the thickness of slabs on grade (slabs on ground) and the location of reinforcing bars. The electrical resistivity of a slab of concrete is different than that of the underlying stratum, as is the resistivity of the reinforcing bars. The method is new and seems to be affected by moisture, salt, and/or temperature gradients.

Magnetic-Electrical Field Evaluation

This test has the possibility of being a valuable asset to the assessment field. It is based upon the fact that ferromagnetic and electrically conductive materials affect the field of an electromagnet. The limitation to date is that the test cannot provide readings at a depth beyond 7 in. in the concrete. There are instances where this test will not provide data, and its contribution may be in the fact that it can be used with other methods. It is inexpensive, and the data provided by the tests when conducted by an experienced operator are considered to be highly accurate. This method could be utilized on building slabs and along the faces of beams and columns while a more expensive test, such as gamma ray, could be used to probe deeper as needed in beams and columns.

High-Energy Ultrasonics

Ultrasonics clearly has one of the best potentials for evaluation of existing concrete. It is currently being developed further and will soon be capable of providing even more data on the concrete material and the reinforcing. At this point the thickness of concrete members can be determined to an accuracy of 5 percent. The equipment needs a transmitter on one side of the sample and a receiver on the other side to produce accurate results.

X-rays and Gamma Rays

X-rays can provide either a film reproduction or a live image of the internal structure of the concrete, including the size and location of the reinforcement. Gamma ray equipment is much lighter and is easily portable and is therefore becoming extremely popular for construction testing. Certain of the gamma ray sources (cesium 137, cobalt 60, and iridium 192) do not require electrical power at the site. The tests are expensive and dangerous to operators and the public, and operators must be licensed by the Nuclear Regulatory Commission.

Microwave Absorption

Microwave emissions are electromagnetic and can be reflected, diffracted, and absorbed by different materials at differing rates. Microwave tests can be utilized to determine reinforcing size and location. The transmitter produces a 3000 MHz (100 mm wavelength), and the receiver is rated at 3 kHz. The transmitter and receiver must be placed on opposite sides of the test specimen. The results can be seen on a computer screen with data manually read or printed images for permanent record.

Dynamic Testing: Ultrasonic Pulse Velocity and Resonant Frequency:

The ultrasonic pulse velocity assumes that vibrational wave propagation is affected by the quality of the consistency of the concrete. The path of travel through a concrete member from the transmitter to the receiver can be converted to an average pulse velocity for different strengths of concrete. On the scale of pulse velocity, the less impedance (velocity slowing) of the pulse, the stronger the concrete:

Pulse velocity	Strength evaluation
Above 4575 m/sec	Excellent
3660–4574 m/sec	Good
3050–3659 m/sec	Questionable
2135–3046 m/sec	Poor
Below 2135 m/sec	Very Poor

Further work on this system may allow a direct reading of the strength of the concrete when the equipment can be gauged by given calibration samples prior to reading on the actual member.

ASSESSMENT OF WOODS AND THEIR CONDITION

Wood has been utilized for structural purposes since man moved out of the cave. More importantly, and of more concern, to this subject, wood has been a major structural element in America since the colonization of the continent began. It could also be argued that the Native Americans were using wood for a long time before the fifteenth century. In the United States there are wood structures in the areas of the original colonies dating to the mid-seventeenth century. When the colonists arrived in America, there were vast virgin forests with a variety of types and species of woods in various regions all over the continent. Wood was used for building, fuel, and any number of agrarian tools. Forests had to be cleared to allow space for farming. The fertile fields left behind as a forest was cleared represented literally hundreds of years of forestation and the leavings of the many rich cycles of decay of a forest renewing itself naturally. There are no virgin forests left in the United States except those of the Rocky Mountains and the Pacific Northwest.

It is impossible to describe the properties of wood and to detail the highly involved evolution of the building codes for wood in the United States in a format such as this. A synopsis of the analytical design methods for wood in construction is given in the body of this volume but within the methodology there is little mention of the condition of wood or individual wood members in an existing or historic building. The condition of a wood beam or column must be a portion of any quantitative abstract concerning wood and its ability to carry load. We cannot ignore the condition of the wood, but that condition may be very difficult to ascertain. The basic strength characteristic of wood is in its uniformity of grain, density, and moisture content. The grain and density are inherent to the member that is, when a member is cut or sawn from a log, the patterns of growth of the tree show up in the member. The master craftsmen of the eighteenth and nineteenth centuries would saw the logs in an optimum way to realize the best framing timber. There were no grading schemes until the twentieth century, and inferior wood was simply not placed in service. Basically all of the inherited methods changed after the building moratorium of World War II. Trees were cut in the manner that produced the most wood (quality not withstanding), the straight-sawn or mill-run method contrasting to the earlier quarter-sawn method. Many aspects of construction changed at the end of World War II with the post-war construction boom, especially in housing. The nation was faced with the largest demand for housing in its history.

The work of the conservation architect and structural engineer is complicated by the rather long history of wood construction in the United States. There were several subperiods in this span of time. The last 50 years have contained several subperiods, divisions within which have been marked by material size changes. All wood members were empirically sized until the last quarter of the nineteenth century. Standardization of sizes began in about the 1880s but has never remained consistent even since. A 2 in. member in 1880 would be about 2⅛ in. thick (to allow for shrinkage) or thicker when sawn. As time elapsed, the 2 in. member went to 1⅞, 1¾, 1⅝, and is now at 1½. This is unique to this country and normally causes a great deal of consternation among designers. It does, however, provide us with some fairly conclusive dating techniques.

Since about 1930, lumber species have been graded under a series of grading agencies that are still not totally standardized. The main body of code specifications is the National Design Specification for Wood Construction, in which all standard construction grade timber is graded by species and stress capacity by material thickness and depth. The grade classifications designate the allowable stresses for compression parallel to the grain, bending, compression perpendicular to the grain, horizontal shear, etc. The grade classifications are given to the wood at the mill by inspectors at a point where the timber can be separated to various lines by grade as directed by the inspector on a visual basis. In the United States a very small percentage of lumber in certain sizes is machine stress rated. The expense of this premium material has made its use extremely rare. One main difference between wood for construction today and the wood of the empirical methods of the seventeenth and eighteenth centuries is the process of kiln drying. Introduced in the twentieth century (or in the 1880s), kiln drying served to accelerate the process of reducing the moisture content in the wood. The moisture content in structural wood at the point of equilibrium is approximately 14 percent. Kiln-dried material was available only in areas where the lumber demand justified the expense of building and operating a kiln. The larger metropolitan areas would support a kiln and finishing mill in much the same way that brick manufacture became a commercial trade in the cities first. The shipping of finished lumber as we know it today did not really become a commercial reality until the 1930s. The less populated rural areas depended upon the small sawmill and the air-drying processes of the previous era, or builders or landowners dried lumber in barns or sheds for their own purposes. Air-drying took two to five years and required that the lumber be stacked in a special way to prevent board-on-board contact and allow for free flow of air around the lumber. The overall quality of air-dried lumber is just as good as that of kiln-dried. It takes experience to determine when the moisture content has reached the 14–18 percent range. Final milling and surfacing of the material took place after the drying process.

Water and temperature are the two greatest enemies of any species of wood that has been cut, sawn, and processed for construction. Both are activators for rot and fungi of many varieties. Wood that is kept dry (14–15 percent moisture content) will not deteriorate or be subject to fungal attack. Wood does not lose strength over a period of time. At the constant moisture content of 14–15 percent, wood that has been in service for 100–200 years has not lost more than 1–2 percent of its strength. The direction and pattern of the grain may warp a member slightly during a long period of load, such as through the effect of long-term deflection, but the wood will be as sound as it originally was. There are no studies to disallow continuing to use the same allowable stress used in the original design. Earlier in this work indications were provided of the period allowable stresses and the use of original design formulae. Early wood structural members must be thoroughly inspected, and if necessary the assessment may include analysis based upon today's design formulae and methods of evaluation of lateral bracing, etc. while using the period allowable stresses. Engineering judgment may indicate a reduced allowable stress if there is any reason for concern about an individual member. If a load test should show deflection in excess of that calculated for a given live load, the allowable bending stress or modulus of elasticity may have been liberal for the member or there may be reason to suspect the condition of the member. If there are signs of earlier moisture or other deterioration of strength, such as insect attack, then further investigation must be included.

Wet rot and dry rot are both the product of moisture intrusion and continuous dampness or periods of wet or dry on the surface of the timber. Moisture alone is insufficient to cause rapid decay if it is relatively short-term. Moisture and temperatures above 72°F are both required before fungi can develop. Wet rot and dry rot are easily detected through a number of surface indications. The presence of wet or dry rot, moisture, fungi, swelling, and the lot usually means that a member is not capable of performing anymore. Once the deterioration has started, the only solution is to cut away all of the decayed material, even the entire end section of a member up to several feet in length. Any portion of a decayed section that remains will continue to attack the unaffected area of the member unless it is removed.

The evaluation of wood structural members is much more dependent on visual inspection than are the methods for steel, masonry, and concrete. There are only two methods that measure strength qualities in wood, while other scientific methods deal mostly with moisture content (which in a way relates to strength or loss of strength), and none of the above can be utilized without significant reference to visual inspection and evaluation. Wilson suggests the methods outlined in Table 6-9.

CONCLUSIONS AND RECOMMENDATIONS

The historic building is unique because of its place in history, its architectural style, its craftsmanship, or special circumstances such as structural system, material innovation, or one-of-a-kind features. The evolution of structural systems has also been unique. The medieval braced frame systems that came to America often with the indentured woodworker from England followed a tradition that had evolved from structural connections, mortise and tenon. The braced frame had some influence upon the architecture of the period because beams and columns and frames of the walls of the floor above could not all intersect at the same point as they do today. The mortises would be in all directions at the same point and thus weaken all of the connections. The need was to offset the connections, and the gunstock post and second-floor cantilever (only a foot or two) were developed to separate connections; thus the medieval form came about. Many houses and early buildings in New England followed medieval traditions. The abundance of large, hardwoods in the New England colony provided the material for the heavy timbers utilized in the system. These structures were inherently strong, and the mortise and tenon system gave them great resistance to lateral loads and overall stiffness. The braced frame continued to be used well into the twentieth century in barns and sheds, mainly on farms. In many instances, the braced frame evolved to a modified frame system that utilized smaller members and corner posts (columns), and in residential and light commercial structures the braced frame system evolved into the balloon frame as mechanical sawing of timber began to produce uniformity in sizes.

Masonry load-bearing wall construction began as the need for larger, more substantial buildings developed. The resistance of masonry buildings to the transfer of fires from one building to another made masonry structures popular and later mandated in the urban areas. As much as the protection from the transfer of fires from one building to another enhanced safety in the urban areas, masonry load-bearing buildings with wood floors and floor framing and interior partitioning were far from fireproof. These buildings were just as susceptible to fires originating in their interior as were wood frame building. The true fireproof building was not yet designed. It took iron and then steel joists and the masonry jack arches to in-fill the space between the joists to form floors and hollow clay tile partitions with plaster for walls to create a building that contained no wood structural components. Although the "fireproof" building had reached the level of non combustible structural components, fires still occurred due to the need for heat in the winter, which left fireplaces and open flame devices in buildings. The fireproof building did eliminate 80–90 percent of the fires in buildings and the resultant loss of life.

Masonry buildings had limitations that precluded their attaining the heights and volumes of the type of building that was necessary in the 1880s as urban areas needed to increase building volumes in centralized areas. The caged building with masonry

Table 6-9. Methods of Evaluation of Wood Structural Members[a]

Visual/optical		
Visual inspection	Determine species, decay, Rot, Fungi, Insect	Visual inspection to determine the location or possible location of dampness, rot, or fungal decay. Inspect potential locations for water infiltration, roof, gutters, flashings. Determine the need for further investigation.
Visual stress grading	Determine strength grading, wood quality	Visual inspection to determine the stress grading of the material, (if material has grading marks). Visually inspect the material to determine if any knots, checks, shakes etc., and direction of grain to determine if the wood can be fully stressed to the maximum allowable.
Manual probing	Determine extent of decay	Manually probe the wood member with a straight wire prick to pull away decayed wood to determine the depth of the decay.
Penetration tests		
Pilodyn penetrometer	Determine degree of degradation	Can estimate in situ degree of decay by measuring the range of penetration of a steel pin into the surface of a wood member. Compare to the penetration of a sound member of the same species. Relatively accurate, must compare to other tests.
Electrical tests		
Dielectric moisture meters	Measures moisture content	Two types: capacitance meter—a change in oscillation frequency; power-loss meter—a loss of amplitude of electric wave. Either type gives fairly accurate measurement of the moisture content in wood (at the surface). Use to make preliminary determination to determine if further tests are necessary.
Resistance-type moisture meter	Measures moisture content	Measures electrical resistance between two probes. Probes can be on opposite sides of wood three-quarter in. wide. Probes can be left in place with recorder or tested at intervals. Relatively accurate, preservatives, paint, and decay effect result.
Electrical resistance probe	Measures moisture content	Wooden probe inserted into member. Two electrode faces measure the moisture content of the material. Can be built into the structure for long-term monitoring. Probes can show long-term drift as strain gages and a hysteresis curve can be developed.
Pulse velocity		
Ultrasonic longitudinal wave propagation	Measures modulus of elasticity, strength	Discontinuities measured by variation of velocity of longitudinal wave propagation. Transmitter and receiver on opposite sides of member. Degree of decay is measured by establishing a modulus of elasticity, translates to strength of wood member.
Stress-wave propagation equipment	Measures modulus of elasticity, strength	Transmitter and receiver on opposite sides of member. Wave velocity (transverse) is influenced by inconsistencies in the wood member. Density and wave velocity are translated to modulus of elasticity. Strength is correlated to the modulus of elasticity. Relatively low cost, accuracy medium to good.
Weight test		
Over drying	Measures moisture content	Requires samples to be removed from member and weighed prior to and after oven drying. Can determine the moisture content as related to fully dried sample. Very accurate. Laboratory equipment required.
Radiographic evaluation		
X-ray	Irregularities, decay, splits, knots, insect damage, hidden members	Requires access to both sides of member. Permanent record on film. Can detect internal density variations. Special equipment for wood. Harmful. High cost.

[a] From Wilson 1984, 104–106.

exterior walls and wood or iron framed interior bays was also limited in size, especially height. The availability of iron sections, then Wrought Iron, to increase spans, and finally structural steel and the steel frame and curtain wall system (the skeletal-frame building) made heights practical and economical. The skeletal frame and the greater heights it allows would not have been practical without the elevator and of course wrought iron or steel. These buildings were a marvel to the layman; the erection of one floor per day made the new building system a phenomenon to observe. The notoriety of the tall building did not go without comment in the newspapers and on government debate floors. Many people and officials were concerned about the safety of the buildings and the social implications of having all so many people concentrated in these building in a confined space. Questions were raised about the "people jams" that would occur at the end of the day as all of those people poured on to the streets and began their homeward journeys. Critics said that a fire in one of these buildings with so many people confined to one enclosure would constitute a major problem. The newspapers were full of complaints about the safety of these high-rise (20-story) buildings. Editorials warned that they would topple over in a moderate wind and there were all kinds of rumors about the air not being safe at these elevations. Birkmire (1898) discusses these fears. One is of particular interest, as it rather sums up the whole era. A tenant in the American Tract Society Building in New York City moved a grandfather clock to his offices near the top floor (21 stories) and reported that the clock would not operate. The possibilities considered were that the pendulum would not swing because of the movement (sidesway) of the building, simply that the clock would not run at that elevation. In either case, the clock was sufficient proof that man did not belong at that elevation and that we should limit the height of these modern wonders to less than 10 stories.

The 1890s were a period of a great economic boom, and new high-rise buildings were being constructed throughout the larger cities. Architects were puzzled as to how to adapt styles to the high-rise building. There were many attempts to construct these buildings of the same style as their lower (three- to five-story) counterparts, but the scale of the skyscraper was not compatible with any of the existing styles. Many attempts ended up being very presentable buildings, such as the American Surety Building in New York City, 1895, by the architect Bruce Price (see Figure 6-15). The debate continues and may actually never be resolved.

The modern skyscraper is still struggling for an identity of style, to no avail. While many of the technical problems of the early skyscrapers have been resolved or dealt with, the social problems predicted by the critics of the skyscraper have not been totally confirmed or denied. The danger of falling or sustaining structural damage from natural causes has all but been proven wrong. However, as one looks back at the mood of the times of the 1890s, one catastrophic incident could have postponed the tall building era for many years. The designers of those times and the new concepts and new material, structural steel, have proven to be the true innovators of that dynamic period the end of a century. We are at that same point today; however, the emphasis now is upon social problems rather than new building technologies. Today we face two problems, or one problem and one threat. The threat is in the fact that there are so many people concentrated in the modern skyscraper. Terrorists have already attempted to blow up several tall buildings, the most notable being the World Trade Center in New York; and they have succeeded with others, such as the Federal Building in Oklahoma City. We do not have the technology at this point to build a building that a terrorist cannot be prevented from destroying, provided he has the will and the means. The second problem is the environmental contamination that attacks the fabric of buildings. Many forms of erosion are almost not preventable, as we have now evolved into a society of unaccountability for many of these contaminants.

The whole business of assessment of the structural fabric of a building, the masonry, the steel (wrought iron and cast iron), structural concrete, and wood, is going to take on a new importance in the years to come. We in this coutnry are removing buildings for little reason except that bankers and society in general distrust older structures or desire to create jobs or spend capitol on new structures rather than rehabilitate the older structure. In every case, we replace a building with a new build-

Figure 6-15. American Surety Building. (*From Birkmire 1898, 21.*)

ing of inferior quality with an expected life cycle of 25 years before radical remodeling or destruction is required. We have entered an era of disposable buildings without even knowing it. Businesses come and go, spatial requirements change, and it is cheaper and easier to finance a new building than to rehabilitate an existing building. The social aspect of "newness is goodness" is more a part of the society than ever before, as is the dislike and distrust of something old.

Conservationists and people who admire the architecture and human scale of the historic building must work to contain this trend by working on public opinion and creating an atmosphere of knowledge and trust of the materials and methods of the past forms of architecture. It is hoped that this work will assist in this effort.

BIBLIOGRAPHY

A58 Committee. 1945. *The American Standard Building Code Requirements for Minimum Design Loads in Buildings and Other Structures.* A58.1-1945.

Abrams, D. P. 1992. *Masonry Structures.* Boulder, Colo: Masonry Society.

American Concrete Institute. 1988. ACI Publication No. 530-88, Section 6.3. Detroit: American Concrete Institute.

———. 1989. *Building Code Requirements for Reinforced Concrete.* ACI Publication No. 318–89. Detroit: American Concrete Institute.

———. 1941. *Reinforced Concrete Design Handbook.* Detroit: American Concrete Institute.

American Institute of Steel Construction. 1930. *Manual of Steel Construction.* 1st ed. New York: American Institute of Steel Construction.

———. 1939. *Manual of Steel Construction.* 3d ed. New York: American Institute of Steel Construction.

Baker, I. O. 1892. *A Treatise on Masonry Construction.* 7th ed. New York: John Wiley & Sons.

———. 1910. *A Treatise on Masonry Construction.* 10th ed. New York: John Wiley & Sons.

Beckmann, P. 1994. *Structural Aspects of Building Conservation.* London and New York: McGraw-Hill.

Beedle, L. S. 1964. *Structural Steel Design.* New York: Roland Press.

Bell, W. E. 1857. *Carpentry Made Easy, or, The Science and Art of Framing.* Philadelphia: Howard Challen.

Bethlehem Steel Co. 1926. *Bethlehem Structural Shapes.* Catalogue S-18. Bethlehem, Pa.: Bethlehem Steel Co.

Birkmire, W. H. 1894. *Skeletal Construction in Buildings.* 2d ed. New York: John Wiley & Sons.

———. 1898. *The Planning and Construction of High Office-Buildings.* 1st ed. New York: John Wiley & Sons; London: Chapman & Hall.

Bray, D. E., and D. McBride. 1992. *Nondestructive Testing Techniques.* New York: John Wiley & Sons.

Building Officials and Code Administrators International, Inc. 1987. *BOCA National Building Code, 1987.*

Cambria Steel Co. 1919. *A Handbook of Information Relating to Structural Steel.* 12th ed. Philadelphia, Pa.: Cambria Steel Co.

Carnegie Steel Co. 1893. *Pocket Companion.* Pittsburgh, Pa.: Carnegie Steel Co.

———. 1913. *Pocket Companion.* 16th ed. Pittsburgh, Pa.: Carnegie Steel Co.

———. 1920. *Pocket Companion.* 21st ed. Pittsburgh, Pa: Carnegie Steel Co.

———. 1923. *Pocket Companion for Engineers, Architects and Builders.* 23d ed. Pittsburgh, Pa.: Carnegie Steel Co.

Comer, J. P. 1942. *New York City Building Control.* New York: Columbia University Press.

Concrete Reinforcing Steel Institute. 1978. *CRSI Handbook.* 3d ed. Chicago: Concrete Reinforcing Steel Institute.

Condit, C. W. 1964. *The Chicago School of Architecture: A History of Commercial and Public Building in the Chicago Area, 1875–1925.* Chicago and London: University of Chicago Press.

Dunham, C. W. 1939. *The Theory and Practice of Reinforced Concrete.* 1st ed. New York and London: McGraw-Hill.

Engineering News. 1902. March 29.

Engineering Record. 1894. March 3.

Ferguson, P. M. 1973. *Reinforced Concrete Fundamentals.* 3d ed. New York: John Wiley & Sons.

Ferris, H. W., ed. 1978. *Historical Record, Dimensions and Properties: Rolled Shapes.* New York: American Institute of Steel Construction.

Fitzgerald, R. W. 1967. *Strength of Materials.* Reading Mass., and London: Addison-Wesley.

Freitag, J. K. 1895. *Architectural Engineering.* 1st ed. New York: John Wiley & Sons.

———. 1909. *Architectural Engineering*. 2d ed. New York: John Wiley & Sons; London: Chapman & Hall.

Hamlin, Talbot. 1953. *Architecture Through the Ages*. Rev. ed. New York: G. P. Putnam's Sons.

Harper, R. H. 1985. *The Evolution of the English Building Regulations*. London and New York: Mansell.

Hool, G. A., and N. C. Johnson. 1929. *Handbook of Building Construction*. New York and London: McGraw-Hill.

Hool, G. A., N. C. Johnson, and S. C. Hollister. *The Concrete Engineer's Handbook*. 1st ed. New York and London: McGraw-Hill.

Hool, G. A., and W. S. Kinne. 1923. *Foundations, Abutments and Footings*. 1st ed. New York and London: McGraw-Hill.

———. 1924. *Steel and Timber Structures*. New York and London: McGraw-Hill.

Huntington, W. C. 1929. *Building Construction*. 1st ed. New York: John Wiley & Sons; London: Chapman & Hall.

International Correspondence Schools. 1899A. *The Building Trades Handbook*, 2d ed. Scranton, Pa.: International Textbook Co.

———. 1899B. *A Treatise on Architecture and Building Construction*. Vol. 2, *Masonry, Carpentry, Joinery*. Scranton, Pa.: Colliery Engineer Co.

———. 1905A. *Loads in Structures; Properties of Sections; Materials of Structural Engineering; Beams and Girders; Columns and Struts; Details of Construction; Graphical Analysis of Stresses*. International Library of Technology, vol. 51. Scranton, Pa.: International Textbook Co.

———. 1905B. *Statics of Masonry, Heavy Foundations, Retaining Walls, Wireproofing, Roof-Truss Design, Wind Bracing, Specifications*. International Library of Technology, vol. 52. Scranton, Pa.: International Textbook Co.

———. 1908. *Operations Preliminary to Building, Limes, Cements, and Mortars, Excavation, Shoring and Piling, Foundations, and Others*. International Library of Technology, vol. 34C. Scranton, Pa.: International Textbook Co.

———. 1911. *The Concrete Engineer's Handbook*. 1st ed. Scranton, Pa.: International Textbook Co.

———. 1912. *The Building Trades Handbook*. 2d ed. Scranton, Pa.: International Textbook Co.

———. 1923A *Fireproofing of Buildings, Stair Building, Ornamental Metal Work, Builders' Hardware, Roofing, Sheet-Metal Work, Mill Design*. International Library of Technology, vol. 33C. Scranton, Pa.: International Textbook Co.

———. 1923B. *Operations Preliminary to Building; Limes, Cements, and Mortar; Excavating, Shoring, and Piling; Foundations; Stone Masonry; Concrete Construction; Areas, Vaults, and Retaining Walls; Carpentry; Mechanics of Carpentry; Joinery; The Steel Square*. Volume 30C of International Library of Technology. Scranton, Pa.: International Textbook Co.

———. 1924. *Design of Beams*. International Library of Technology, vol. 11E. Scranton, Pa.: International Textbook Co.

———. 1960. *The Building Trades Handbook*. 8th ed. Scranton, Pa.: International Textbook Co.

Irace, F. 1988. *Emerging Skylines: The New American Skyscrapers*. New York: Whitney Library of Design.

Jandl, H. W., ed. 1983. *The Technology of Historic American Buildings*. Washington, D.C.: Foundation for Preservation Technology.

Johnson, J. B., C. W. Bryan, and F. E. Turneaure. 1894. *The Theory and Practice of Modern Framed Structures*. 3d rev. ed. New York: John Wiley & Sons.

Johnston, L. J. 1904. In *Engineering News*, April 14.

Ketchum, M. S. 1918. *The Structural Engineer's Handbook*. 2d ed. New York and London: McGraw-Hill.

———. 1924. *The Structural Engineer's Handbook*. 3d ed. New York and London: McGraw-Hill.

———. 1921. *The Design of Steel Mill Buildings*. 4th ed. New York and London: McGraw-Hill.

Kidder, F. E. 1900. *Carpenter's Work*. Vol. 2 of *Building Construction and Superintendence*., 3d ed. New York: William T. Comstock.

———. 1902. *The Architect's and Builder's Pocket-book*. 13th ed. New York: John Wiley & Sons; London: Chapman & Hall.

———. 1905A. *Mason's Work*. Volume 1 of *Building Construction and Superintendence*, 7th ed. New York: William T. Comstock.

———. 1905B. *Strength of Beams, Floors and Roofs*. New York: David Williams.

———. 1908[?]. *The Architect's and Builder's Pocket-book*. Ed. unknown. New York: John Wiley & Sons.———. 1925. *Trussed Roofs and Roof Trusses*. Volume 3 of *Building Construction and Superintendence*, 4th ed. New York: William T. Comstock.

Kidder, F. E., and H. Parker. 1949. *Kidder-Parker Architects' and Builders' Handbook: Data for Architects, Structural Engineers, Contractors, and Draughtsmen*, 18th ed. New York: John Wiley & Sons. [This is a continuation of Kidder's *Architect and Builder's Handbook*.]

Lord, A. R. 1928. *A Handbook of Reinforced Concrete Building Design, in Accordance with the 1928 Joint Standard Building Code*, 1st ed. Detroit: American Concrete Institute.

Mahan, D. H. 1885. *A Treatise on Civil Engineering*. Rev. ed., ed. D. Wood. New York: John Wiley & Sons.

Merriman, T., and T. H. Wiggin. 1947. *American Civil Engineer's Handbook*. 5th ed. New York: John Wiley & Sons.

Moxon, J. 1703. *Mechanick Exercises, or, the Doctrine of Handy-Works*. 3d ed.; reprint, Scarsdale, N.Y.: Early American Industries Assn., 1975.

Mulligan, J. A. 1942. *Handbook of Brick Masonry Construction*. 1st ed. New York and London: McGraw-Hill.

National Bureau of Standards. 1955. *Minimum Design Loads in Buildings and Other Structures*. ASA A58.1-1955.

National Forest Products Assn. 1988. *National Design Specification for Wood Construction.* Washington, D.C.: National Forest Products Assn.

———. 1991. *National Design Specification for Wood Construction.* Washington, D. C.: National Forest Products Assn.

O'Rourke, C. E. 1940. *General Engineering Handbook.* 2d ed. New York and London: McGraw-Hill.

Patton, W. M. *A Practical Treatise on Foundations,* 1st ed. New York: John Wiley & Sons.

Paul, C. E. 1916. *Heavy Timber Mill Construction Buildings.* 3d ed. Chicago: National Lumber Manufacturers' Assn.; reprint, Southern Pine Assn.: 1918.

Peters, T. F. "The Rise of the Skyscraper from the Ashes of Chicago." In *American Heritage of Invention and Technology,* Fall 1987.

Peterson, C., ed. 1976. *Building Early America: Contributions Toward the History of a Great Industry.* Radnor, Pa.: Chilton Book Co.

Phoenix Iron Co. 1906. *Handbook of Useful Information, Tables, Rules, Data, and Formulae Appertaining to the Use of Steel.* Philadelphia, Pa.: Phoenix Iron Co.

Ramsey, C. G., and H. R. Sleeper. 1936. *Architectural Graphic Standards.* 2d ed. New York: John Wiley & Sons.

Reid, H. A. 1907. *Concrete and Reinforced Concrete Construction.* 1st ed. New York: Myron C. Clark.

Ricker, N. C. 1912. *A Treatise on the Design and Construction of Roofs.* 1st ed. New York: John Wiley & Sons; London: Chapman & Hall.

Rivington. 1899. *Elementary Stage.* Part 1 of *Notes on Building Construction.* London: Longmans, Green & Co.

———. 1900. *Advanced Stage.* Part 2 of *Notes on Building Construction.* London: Longmans, Green & Co.

Sahlin, S. 1971. *Structural Masonry.* Englewood Cliffs, N.J.: Prentice-Hall.

Schneider, H. E. 1930. *Practical Wind Bracing.* Buffalo, N.Y.: Lancaster Press.

Shedd, T. C. 1934. *Structural Design in Steel.* 1st ed. New York: John Wiley & Sons.

Sprague, P. E. 1983. "Chicago Balloon Frame." In H. W. Jandl, ed., *The Technology of Historic American Buildings.* Washington, D.C.: Foundation for Preservation Technology.

Thurston, R. H. 1892. *Materials of Construction.* 5th ed. New York: John Wiley & Sons.

Timoshenko, S. P. 1930. *History of Strength of Materials.* New York: McGraw-Hill; reprint, New York: Dover, 1983.

Trautwine, J. C. 1888. *The Civil Engineer's Pocketbook.* 13th ed. New York: John Wiley & Sons.

Turneaure, F. E., ed. 1908. *The Cyclopedia of Civil Engineering.* Chicago: American School of Correspondence.

Turneaure, F. E., and E. R. Maurer. 1914. *Principles of Reinforced Concrete Construction.* 2d ed. New York: John Wiley & Sons; London: Chapman & Hall.

Urquhart, L. C., and C. E. O'Rourke. 1941. *Elementary Structural Engineering.* 1st ed. New York and London: McGraw-Hill.

Voss, W. C., and E. A. Varney. 1926. *Architectural Construction.* Vol. 2, bk 1, "Wood Construction."

White, c. E. 1932 *Hollow Tile.* Part 1 of *Hollow Tile and Fireproofing,* ed. W. S. Lowndes. Scranton, Pa.: International Textbook Co.

Wilson, F. 1984. *Building Materials Evaluation Handbook.* New York: Van Nostrand Reinhold.

Index

Allowable Stress Design (ASD), Phase Two Assessment, 9
American Institute of Steel Construction (AISC), 193, 206, 215, 218
American Society for Testing of Materials (ASTM), 193, 206, 461–462, 483
American Standards Association (ASA), 36–37
Arches, fireproofing, floor systems, 301–315
Architecture:
   education, 19–20
   licensing, 40
Asbestos, Phase One Assessment, 4
Assessment methodology, 1–51. *See also* Material assessment
   building law:
      American, 16–19
      early, 11–14
      strength requirements and, 14–16
   certification process, 2–3
   education and manuals, 19–20
   foundation systems, problems/solutions, 106–115
   live loads:
      occupant loads and, 20–40
      reductions in, 47–51
   material chronology, 9–11
   overview, 1–2
   Phase One Assessment, 3–6
   Phase Two Assessment, 6–9
   wind loads:
      modern building codes, 45–47
      snow loads and, 40–45
Auger borings, soil testing, foundation systems, 58–59

Balloon frame construction, walls and columns, 118–119
Beam footings, concrete/rail and concrete/beam footings, 84–98
Beams. *See* Floor systems
Bearing capacity, substrata, foundation systems, 61–65
Beton bed foundations, foundation systems, 74–75
Braced frame construction, walls and columns, 117–118
Brick:
   fireproofing, floor systems, 300–301
   masonry load-bearing walls, walls and columns, 138, 141–148
   strength evaluation, material assessment, 460
   test procedures, 461–462
Brick footings, foundation systems, 68–74
Brick piers, foundation systems, degrees of failure, 112–113
Building Code Committee (U.S. Department of Commerce), 34–35, 36, 50

Building Code of the National Board of Fire Underwriters, 65
Building Law of 1882 (New York), 16, 17–19
Building Law of 1913 (New York), 12
Building laws:
   curtain walls, 29
   early, assessment, 11–14
   modern, wind loads, 45–47
   strength requirements and, assessment, 14–16
Building life cycle, live loads/occupant loads, assessment, 25
Building Officials and Code Administrators International, Inc. (BOCA), 37–38, 45, 47, 50

Caissons/cylinders, pneumatic, foundation systems, 78–84
Cast iron beams, floor systems, 315–325
Cast iron columns, walls and columns, 162–183
Certification process, assessment methodology, 2–3
Chicago Building Ordinance of 1894, 29
Chicago Building Ordinance of 1911, 26–27
Columns, *see* Walls and columns
Compound beams:
   trussed girders, floor systems, 289–299
   wood and steel beams, floor systems, 282–289
Compound wrought iron girders, floor systems, 325
Concrete assessment, 480–486
   generally, 480–481
   methods, 481–486
Concrete beams and girders, reinforced, floor systems, 359–386
Concrete columns, reinforced, walls and columns, 227–255
Concrete floor systems, 357–359
Concrete/rail and concrete/beam footings, foundation systems, 84–98
Connections, statical methods and, lateral bracing systems, 440–450
Construction industry, Great Depression, 35–36
Core tests, masonry walls, 462–463
Cornices, parapet walls and, walls and columns, 148–151
Correspondence schools, architecture and engineering, 19–20
Curtain walls:
   architectural styles, 30–32
   building laws, 29
Cylinders/caissons, pneumatic, foundation systems, 78–84
Cylinder tests, masonry walls, 462–463

Dead loads, assessment, 20–40
Degrees of failure, foundation systems, 110–115
Diagonal bracing, lateral bracing systems, roof systems, 409–420

Diamond drill, soil testing, foundation systems, 59

Early building laws, assessment methodology, 11–14
Eddy current testing, metals, 479
Education, architecture and engineering, 19–20
Electrical resistivity test, concrete assessment, 485
Engineering:
   education, 19–20
   licensing, 40
England:
   building traditions, 13, 16
   construction technology, 55

Ferrous systems, fireproofing, walls and columns, 219–227
Fireproofing:
   ferrous systems, walls and columns, 219–227
   floor systems, 299–315
Flat roof systems, described, 387–389
Floor systems, 257–386
   concrete and reinforced concrete, 357–359
   fireproofing, 299–315
   reinforced concrete beams and girders, 359–386
   steel riveted girders, 336–340
   structural steel beams, 340–357
   trussed girders, 289–299
   wood and steel beams, 282–289
   wood joists and beams, 257–282
   wrought iron beams, 315–325
   wrought iron girders, compound, 325
   wrought iron riveted girders, 325–335
Foundation systems, 53–115
   Beton bed foundations, 74–75
   brick and stone footings, 68–74
   concrete/rail and concrete/beam footings, 84–98
   early (to 1860), 53–56
   pile foundations, 75–78
   pneumatic cylinders and caissons, 78–84
   problems/solutions, 106–115
     analysis methods, 106–108
     degrees of failure, 110–115
     openings in masonry walls, 108–110
   reinforced concrete spread footings, 98–106
   soil testing (early), 56–61
   substrata bearing capacity, 61–65
   timber grillage, platform foundations, and wood spread footings, 66–68

Gable truss (wood/iron composite), roof systems, 391–396
Gamma-ray testing:
   concrete assessment, 485
   metals, 479
Girders:
   compound wrought iron, floor systems, 325
   reinforced concrete, floor systems, 359–386
   steel riveted, floor systems, 336–340
   trussed, floor systems, 289–299
   wrought iron riveted, floor systems, 325–335
Great Depression, construction industry, 35–36

Hennebique column, reinforced concrete columns, 230–232
High-energy ultrasonics, concrete assessment, 485
Historic materials. *See also* Material assessment
   assessment methodology, 2
   material chronology, assessment methods, 9–11

International Correspondence Schools, 19–20, 42, 48, 66
Inverted arch foundation, brick and stone footings, 72–74
Investment Tax Act, 2, 3
Iron truss, roof systems, 396–407
Iron/wood composite gable truss, roof systems, 391–396

Joists, *see* Floor systems

Knee bracing systems, roof systems, 426–428

Lateral bracing systems:
   connections and statical methods, roof systems, 440–450
   diagonal bracing, roof systems, 409–420
   methods, roof systems, 420–425
   wind considerations, roof systems, 439–440
Lattice girder bracing, roof systems, 433–439
Lattice truss, wood, roof systems, 389–391
Licensing, architects and engineers, 40
Live loads:
   occupant loads and, assessment, 20–40
   reductions in, assessment, 47–51
Load-bearing walls, masonry, walls and columns, 138, 141–148
Load Resistance Factored Design, Phase Two Assessment, 9
Loads:
   live loads/occupant loads, assessment, 20–40
   wind loads/snow loads, assessment, 40–45
Load tests, soil testing, foundation systems, 59–61
London Building Law of 1877, 14–16

Magnetic-electrical field evaluation, concrete assessment, 485
Magnetic particle testing, metals, 479
Manuals, architecture and engineering, 19–20, 33, 50
Masonry, strength evaluation, material assessment, 460
Masonry building walls, walls and columns, 151–161
   one-way vs. two-way moments, 160–161
Masonry load-bearing walls, walls and columns, 138, 141–148
Masonry piers and columns, walls and columns, 161–162
Masonry walls:
   live loads/occupant loads, assessment, 28
   openings in, foundation systems, 108–110
Masonry wall testing, 460–474
   brick (ASTM) test, 461–462
   core or cylinder tests, 462–463

in situ testing, 462
stone, 473–474
test panel removal, 462
ultrasonic testing, 463–473
Material assessment, 453–491. *See also*
Assessment methodology
concrete, 480–486
generally, 480–481
methods, 481–486
evaluation context, 453–456
masonry wall testing, 460–474
metals, 474–480
generally, 474–476
methods, 476–480
recommendations, 488, 490–491
strength evaluation, 460
testing, 458–460
walk-through assessment, 457–458
wood, 486–488, 489
Material chronology, assessment methodology, 9–11
Metal assessment, 474–480
generally, 474–476
methods, 476–480
Microwave absorption test, concrete assessment, 485–486
Mill construction, walls and columns, 137–138, 139–141
Modern building laws, wind loads, 45–47
Multiple-component trusses, roof systems, 389

National Building Code, 11
Needle beam system, foundation systems, degrees of failure, 111
New York Building Law of 1892, 29

Occupant loads, live loads and, assessment, 20–40
One-way vs. two-way moments, masonry building walls, 160–161
Openings, in masonry walls, foundation systems, 108–110
Optical inspection, concrete assessment, 484

Parapet walls, cornices and, walls and columns, 148–151
Phase One Assessment, methodology, 3–6
Phase Two Assessment, methodology, 6–9
Pigeon droppings, Phase One Assessment, 4
Pile foundations, foundation systems, 75–78
Platform foundations, foundation systems, 66–68
Platform (western) frame construction, walls and columns, 119–123
Pneumatic cylinders/caissons, foundation systems, 78–84
Portals:
frame, roof systems, 428–433
solid, roof systems, 433

Rail footings, concrete/rail and concrete/beam footings, 84–98
Rebound test, concrete assessment, 484–485
Reinforced concrete beams and girders, floor systems, 359–386

Reinforced concrete columns, walls and columns, 227–255
Reinforced concrete floor systems, 357–359
Reinforced concrete spread footings, foundation systems, 98–106
Rigid connections, statical methods and, lateral bracing systems, 440–450
Rod test (sounding rod), soil testing, foundation systems, 57–58
Roof systems, 387–451
flat systems, 387–389
gable truss (wood/iron composite), 391–396
iron truss, 396–407
knee bracing systems, 426–428
lateral bracing systems:
connections and statical methods, 440–450
diagonal bracing, 409–420
methods, 420–425
wind considerations, 439–440
lattice girder bracing, 433–439
multiple-component members, 389
overview, 387
portals:
frame, 428–433
solid, 433
truss system uses, 407–409
wood lattice trusses, 389–391

Section 106 review, 2, 3
Skeletal construction, live loads/occupant loads, assessment, 27–29
Skyscraper, term of, 27–28
Snow loads, wind loads and, assessment, 40–45
Soils, substrata bearing capacity, foundation systems, 61–65
Soil testing, early, foundation systems, 56–61
Sounding rod (rod test), soil testing, foundation systems, 57–58
Spread footings, reinforced concrete, foundation systems, 98–106
Standard Building Code, masonry load-bearing walls, 147–148
Standards for Rehabilitation, 2
Statical methods, connections and, lateral bracing systems, 440–450
Steel beams, structural, floor systems, 340–357
Steel riveted girders, floor systems, 336–340
Steel/wrought iron truss, roof systems, 396–407
Stone:
masonry wall testing, 473–474
strength evaluation, material assessment, 460
Stone footings, foundation systems, 68–74
Strength evaluation, material assessment, 460
Strength requirements, building laws and, assessment, 14–16
Structural steel beams, floor systems, 340–357
Structural steel columns, walls and columns, 192–219
Substrata bearing capacity, foundation systems, 61–65
Surveying, concrete assessment, 484

Terra cotta:
floor systems, fireproofing, 301–315
walls and columns, fireproofing, 221–223, 225

Testing. *See also* Masonry wall testing; Material assessment
   masonry walls, 460–474
   material assessment, 458–460
Test pits, soil testing, foundation systems, 57
Timber column analysis, walls and columns, 133–137
Timber grillage, foundation systems, 66–68
Trussed girders, floor systems, 289–299
Trusses. *See* Roof systems
Two-way vs. one-way moments, masonry building walls, 160–161

Ultrasonic testing:
   concrete assessment, 485, 486
   masonry walls, 463–473
   metals, 479
Underpinning, foundation systems, degrees of failure, 113–115
U.S. Department of Commerce, 34–35, 36, 50

Visual inspection, concrete assessment, 484

Walk-through assessment, materials, 457–458
Walls and columns, 117–255
   balloon frame construction, 118–119
   braced frame construction, 117–118
   cast iron columns, 162–183
   fireproofing, ferrous systems, 219–227
   masonry building walls, 151–161
      one-way vs. two-way moments, 160–161
   masonry load-bearing walls, 138, 141–148
   masonry piers and columns, 161–162
   mill construction, 137–138, 139–141
   parapet walls and cornices, 148–151
   platform (western) frame construction, 119–123
   reinforced concrete columns, 227–255
   structural steel columns, 192–219
   timber column analysis, 133–137
   wood column sizing, 123–133
   wrought iron columns, 183–192
Warehouses, live loads/occupant loads, assessment, 23
Wash borings, soil testing, foundation systems, 59
Western (platform) frame construction, walls and columns, 119–123
Wind loads:
   lateral bracing systems, roof systems, 439–440
   modern building codes, assessment, 45–47
   SBC, masonry load-bearing walls, 147–148
   snow loads and, assessment, 40–45
Windsor probe, concrete assessment, 485
Wood:
   material assessment, 486–488, 489
   timber column analysis, walls and columns, 133–137
Wood column sizing, walls and columns, 123–133
Wood/iron composite gable truss, roof systems, 391–396
Wood joists and beams, floor systems, 257–282
Wood lattice trusses, roof systems, 389–391
Wood spread footings, foundation systems, 66–68
Wood/steel beams, floor systems, 282–289
Wrought iron beams, floor systems, 315–325
Wrought iron columns, walls and columns, 183–192
Wrought iron girders, compound, floor systems, 325
Wrought iron riveted girders, floor systems, 325–335
Wrought iron/steel truss, roof systems, 396–407

X-ray testing:
   concrete assessment, 485
   metals, 479